# Statistical Methods in Food and Consumer Research

## Second Edition

# Food Science and Technology
# International Series

# Statistical Methods in Food and Consumer Research

## Second Edition

**Maximo C. Gacula, Jr.**
Gacula Associates Consulting, Scottsdale, Arizona

**Jagbir Singh**
Department of Statistics, Temple University, Philadelphia, Pennsylvania

**Jian Bi**
Sensometrics Research and Service, Richmond, Virginia

**Stan Altan**
J & J Pharmaceutical Research and Development, Raritan, New Jersey

AMSTERDAM • BOSTON • HEIDELBERG • LONDON
NEW YORK • OXFORD • PARIS • SAN DIEGO
SAN FRANCISCO • SINGAPORE • SYDNEY • TOKYO
Academic Press is an imprint of Elsevier

Academic Press is an imprint of Elsevier
30 Corporate Drive, Suite 400, Burlington, MA 01803, USA
525 B Street, Suite 1900, San Diego, California 92101-4495, USA
84 Theobald's Road, London WC1X 8RR, UK

**Library of Congress Cataloging-in-Publication Data**
APPLICATION SUBMITTED

**British Library Cataloguing-in-Publication Data**
A catalogue record for this book is available from the British Library.

ISBN: 978-0-12-373716-8

For information on all Academic Press publications
visit our Web site at www.elsevierdirect.com

Printed in the United States of America
08  09  10  9  8  7  6  5  4  3  2  1

# Table of Contents

# Preface to the Second Edition

*Statistical Methods in Food and Consumer Research,* published in 1984, is the first book that dealt with the application of statistics to food science and sensory evaluation. Since then, statistical software has become indispensable in the statistical analysis of research data. The use of multivariate analysis in sensory evaluation has emerged in part due to the availability of statistical software that sensory and research analysts can effectively use. The use of nonparametric statistics in sensory evaluation resulted in several scientific research publications which are included in the second edition.

For updating applications of statistics to advancement of sensory science we believe there is a need for a second edition of the book. We are pleased to have Jian Bi with years of experience in applications of statistics, in particular sensometrics, and Dr. Stan Altan with years of expertise in experimental designs as coauthors. Chapters 9–11 of the book were completely re-written and expanded from the basic principles of several statistical procedures to sensory applications. Two chapters were added—Chapter 12, Perceptual Mapping and Descriptive Analysis, and Chapter 13, Sensory Evaluation in Cosmetic Studies. For data analysis we have provided computer programs or codes in S-Plus® and SAS® (Statistical Analysis System) for Chapters 9–13. For Chapter 7 which deals with Response Surface Designs, Design-Expert® is highly recommended.

The revised book can be used as text for a first course in applied statistics for students majoring in Food Science and Technology, and for other students in agricultural and biological sciences. As a first course, Chapters 1–8 will be used. For graduate students majoring in sensory science, a second course would cover Chapters 9–13.

We thank the academic and industrial professionals who found the book useful and encouraged us to update it. We are grateful to Academic Press/Elsevier, Inc. for their interest in publishing the second edition, in particular to Nancy Maragioglio, Carrie Bolger, and Christie Jozwiak for their continuous and readily available support.

Maximo C. Gacula, Jr.
Jagbir Singh
Jian Bi
Stan Altan

# Introduction

This introductory chapter is written to familiarize the reader with common symbols, notation, concepts, and some probability tools that are used in the subsequent chapters. This chapter, therefore, is a minimal prerequisite for the remainder of this book. The details of the material presented here can be found in many beginning statistical methods textbooks, such as those by Dixon and Massey (1969) and Huntsberger and Billingsley (1981).

## 1.1 A BRIEF REVIEW OF TOOLS FOR STATISTICAL INFERENCE

The purpose of this section is to review briefly the tools used in *statistical inference*. Statistical inference is the process of making inductive statements based on *random* samples from target populations. The inference process includes the estimation of population parameters and the formulation and testing of hypotheses about the parameters: a *parameter* is a numerical quantity that describes some characteristics of the population. For instance, the mean $\mu$ and the variance $\sigma^2$ are examples of population parameters. In the hypothesized simple linear regression model, $Y = \alpha + \beta X + \varepsilon$, the intercept $\alpha$ and the slope $\beta$ are the parameters of the model.

Because random sampling is a prerequisite for valid statistical inference, it is important that the samples be collected following an accepted random sampling procedure. If this is not done, the inferences about the population are not reliable. A precise and reliable inference can be reached only through the use of an appropriate *experimental design* for random sampling and data collection. The general principles of experimental designs are summarized in Section 1.2. The sample size plays a major role in ensuring that the selected random sample will lead to statistical inferences with a specified amount of confidence.

## Notation and Symbolism

The science of statistics is concerned with the collection, summarization, and interpretation of experimental data. The data may be summarized in the form of graphs, charts, or tables. Because of the many statistical operations to which the data are subjected, a general system of notation is used to facilitate the operations. A working knowledge of this system is essential in understanding the computational procedures involved in statistical analyses. A brief description of the notation that will be encountered in succeeding chapters is given here.

A commonly encountered symbol is sigma, $\Sigma$, used to denote a sum or total. For example, the sum of $n$ numbers $X_1, X_2, \ldots, X_n$ is denoted by

$$\sum_{i=1}^{n} X_i = X_1 + X_2 + \cdots + X_n. \tag{1.1-1}$$

The lower and upper limits on $\Sigma$ indicate that the $X_i$ is added from $i = 1$ to $i = n$; for simplicity, the limits on $\Sigma$ may be omitted if no confusion is likely. Similarly, we have the following:

$$\sum X_i^2 = X_1^2 + X_2^2 + \cdots + X_n^2,$$

$$\sum X_i Y_i = X_1 Y_1 + X_2 Y_2 + \cdots + X_n Y_n, \tag{1.1-2}$$

$$\sum cX = cX_1 + cX_2 + \cdots + cX_n = c\sum_{i=1}^{n} X_i,$$

where $c$ is a constant. Note that $(\Sigma X_i)^2$, which denotes the square of the sum defined by (1.1-1), is not the same as that given by (1.1-2).

Whereas the population mean is symbolized by $\mu$, the mean of a sample $X_1, X_2, \ldots, X_n$ of $n$ items is denoted by $\overline{X}$, where

$$\overline{X} = \sum X_i / n. \tag{1.1-3}$$

Similarly, the population variance is denoted by $\sigma^2$, where the sample variance is denoted by $S^2$, where

$$S^2 = \sum (X_i - \overline{X})^2 / n - 1. \tag{1.1-4}$$

Whenever the population parameters are unknown, as is usually the case, they are estimated by appropriate sample quantities called *statistics*. Thus, if population mean $\mu$ and variance $\sigma^2$ are unknown, they may be estimated by statistics $\overline{X}$ and $S^2$, respectively. At times an estimate of a parameter, say $\theta$, is denoted by $\hat{\theta}$. Thus, an estimate of the population $\mu$ can be denoted by $\hat{\mu}$, which may or may not be $\overline{X}$.

Depending on the populations under investigation, appropriate symbols can be made more explicit. For example, if more than one population is under investigation in a study, their means can be denoted by $\mu_1$, $\mu_2$, and so on.

In a multiclassified data structure, the dot notation is used to indicate totals. For example, let $X_{ijk}$ denote the $k$th observation in the $(i, j)$th cell. Then $X_{ij}$ can be used to denote the sum of $n_{ij}$ observations in the $(i, j)$th cell. That is $X_{ij.} = \sum_{k=1}^{nij} X_{ijk}$. If there is only one observation per cell, that is, $n_{ij} = 1$, then there is no use in retaining the subscript $k$, and one may express the data in a row and column structure. For example, Table 1.1 consists of $r$ rows and $c$ columns; the observation in the $i$th row and the $j$th column is denoted by $X_{ij}$. As noted earlier, when $i$ or $j$ or both are replaced by a dot $(\cdot)$, the dot denotes the sum over the replaced subscript. Therefore,

$$X.. = \sum_{j=1}^{c} \sum_{i=1}^{r} X_{ij} = G$$

is the grand total, and the grand mean is given by

$$\overline{X} = X../rc = \widehat{\mu}.$$

It follows that (see Table 1.1) $X_{i.}$ and $X_{.j}$ are the sums of the $i$th row and $j$th column, respectively. Thus, $\overline{X}_{i.} = X_{i.}/c$, and $\overline{X}_{.j} = X_{.j}/r$. This system is easily extended to a more complex data structure such as $X_{ijk...p}$, where the number of subscripts corresponds to the number of classifications. If $n_{ijk}$ denotes the number of observations in the $(i, j)$th cell, the dot notation representation is similarly used. For

**Table 1.1** An $r \times c$ Table with One Observation per Cell

| | Columns | | | | | |
| Rows | 1 | 2 | $\cdots$ | $c$ | Row sums | Row means |
| --- | --- | --- | --- | --- | --- | --- |
| 1 | $X_{11}$ | $X_{12}$ | $\cdots$ | $X_{1j}$ | $X_{1.}$ | $\overline{X}_{1.}$ |
| 2 | $X_{21}$ | $X_{22}$ | $\cdots$ | $X_{2j}$ | $X_{2.}$ | $\overline{X}_{2.}$ |
| $\vdots$ | $\vdots$ | $\vdots$ | $\vdots$ | $\vdots$ | $\vdots$ | $\vdots$ |
| $r$ | $X_{i1}$ | $X_{i2}$ | $\cdots$ | $X_{ij}$ | $X_{i.}$ | $\overline{X}_{i.}$ |
| Column sums | $X_{.1}$ | $X_{.2}$ | $\cdots$ | $X_{.c}$ | $X..$ (grand total) $= G$ | |
| Column means | $\overline{X}_{.1}$ | $\overline{X}_{.2}$ | $\cdots$ | $\overline{X}_{.c}$ | | $\overline{X}$ (grand mean) |

example, $n\ldots$ denotes the total number of observations in the experiment; $n_{i..}$ denotes the number of observations summed over the $j$ and $k$ classifications.

## The Normal Distribution

Researchers design and conduct experiments to observe one or more quantities whose numerical values cannot be predicted with certainty in advance. The quantities of interest may be the shelf life or the preference score of one item, or some sensory characteristics of, say, food products. The shelf life, the preference score, etc., of an item is not known in advance but could be any one of a number of possible values. A quantity that can assume any one numerical value from a range of possible values, depending on the experimental outcomes, is called a *random variable* (r.v.) in statistical terminology. Since the numerical values of an r.v. depend on the experimental outcomes, the values of an r.v. are governed by a chance model or a probability distribution, as it is called in statistical language. In short, associated with each r.v. is a probability distribution, and statisticians use this structure to study various aspects of the quantities (r.v.'s) of interest.

Some probability distributions can be described by close mathematical functions called the *probability density functions* (pdf's). Once the pdf of an r.v. $X$ is given, it can be used to calculate probabilities of events of interest in decision making. In what follows now we introduce one important probability distribution through its pdf and a number of others without writing down their pdf's.

The most important probability distribution in applied statistics is the *normal* or *Gaussian probability distribution*. An r.v. $X$ is said to have a normal probability distribution if its pdf is of the mathematical form

$$f(X) = \frac{1}{\sqrt{2\pi}\sigma} \exp\left[-\frac{(X-\mu)^2}{2\sigma^2}\right], \qquad (1.1\text{-}5)$$

where $-\infty < X < \infty$, $-\infty < \mu < \infty$, and $\sigma^2 > 0$. The graph of $f(X)$ is sketched in Figure 1.1. The graph is a bell-shaped curve that is symmetric around $\mu$. Hence, the parameter $\mu$ is the center of the distribution of the values of $X$. The shape of the curve depends on $\sigma^2$. The curve tends to be flat in shape as the shape parameter $\sigma^2$ increases. The normal probability distribution is characterized completely by

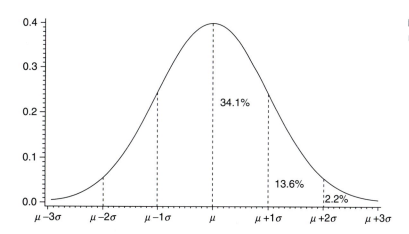

its two parameters, $\mu$ and $\sigma^2$. Indeed, $\mu$ and $\sigma^2$ are, respectively, the mean and the variance of the distribution. The positive square root of $\sigma^2$, denoted by $\sigma$, is the standard deviation.

A normal curve, as stated before, is symmetric around $\mu$. That is, the heights of the curve at $\mu - c$ and $\mu + c$ are the same for any value of $c$. The curve extends to infinity approaching the horizontal line quite closely in each direction. The total area under the curve is equal to 1.0, of which about 95% is concentrated above the interval $\mu - 2\sigma$ to $\mu + 2\sigma$, and about 99% above the interval $\mu - 3\sigma$ to $\mu + 3\sigma$.

When an r.v. $X$ has a normal distribution with mean $\mu$ and standard deviation $\sigma$, we write it as $X$ is $N(\mu, \sigma)$. A useful result in computing probabilities associated with the normal distribution is the following:

Let $X$ be $N(\mu, \sigma)$ and define

$$Z = (X - \mu)/\sigma. \tag{1.1-6}$$

The quantity $Z$ is called the *standardized score* of $X$. The distribution of $Z$ is also normal with mean 0 and variance 1. That is, $Z$ is $N(0, 1)$, which is called the *standard normal distribution*.

Thus, any normally distributed r.v. can be transformed into the standardized score $Z$. Hence, if $X$ is $N(\mu, \sigma)$ and we want to find the probability, say $P(a \leq X \leq b)$ for any two constants $a$ and $b$, we can find it by standardizing $X$ to $Z$. That is,

$$P(a \leq X \leq b) = P\left(\frac{a - \mu}{\sigma} \leq Z \leq \frac{b - \mu}{\sigma}\right). \tag{1.1-7}$$

The probabilities associated with the standard normal distribution are tabulated in Table A.1.[1]

If $Z$ is $N(0, 1)$, then the following are true as seen by the symmetry of the distribution around the mean value 0:

1. $P(Z \geq 0) = P(Z \leq 0) = 0.500$;
2. $P(Z \geq c) = P(Z \leq -c)$ for any constant $c$;
3. $P(Z \geq c) = 1 - P(Z \leq c)$ for any constant $c$;
4. $P(Z \leq c) = 1 - P(Z \leq -c)$ for any constant $c$.

Using the preceding results and Figure 1.2, one can easily find the probabilities of various events from Table A.1. For example,

■ **FIGURE 1.2** Standard normal distribution curve. Ordinate, $f(X)$.

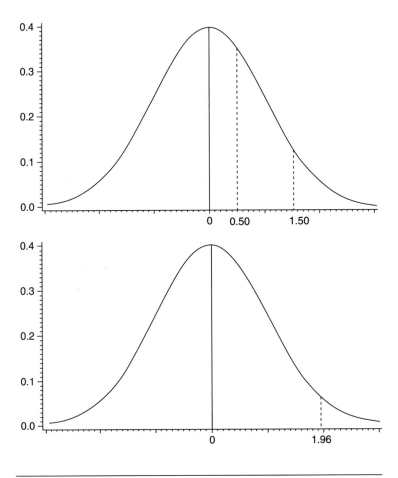

---

[1]Tables with numbers prefixed with A are to be found in the Appendix.

**5.** $P(Z \geq 1.960) = 1 - P(Z \leq 1.960)$
$$= 1 - 0.975 = 0.025;$$
**6.** $P(Z \leq -1.960) = P(Z \geq 1.960)$
$$= 0.025;$$
**7.** $P(-1.960 \leq Z \leq 1.960) = P(Z \leq 1.960) - P(Z \leq -1.960)$
$$= 0.975 - 0.025$$
$$= 0.950;$$
**8.** $P(0.500 \leq Z \leq 1.500) = P(Z < 1.500) - P(Z < 0.500)$
$$= 0.933 - 0.692$$
$$= 0.242.$$

The pdf of the standard normal distribution, denoted by $\phi(X)$, is readily seen from (1.1-5) when $\mu = 0$ and $\sigma^2 = 1$. That is,

$$\phi(X) = (1/\sqrt{2\pi})\exp\left(-\frac{1}{2}X^2\right), \qquad (1.1\text{-}8)$$

where $-\infty < X < \infty$. If $Z$ is $N(0, 1)$, the cumulative probability $P(Z \leq X)$ is denoted by $\Phi(X)$ and is called the *cumulative distribution function* (cdf) of the standard normal r.v. $Z$.

In statistical inference we also come across three other probability distributions. They are called the $t$ distribution, the chi-square $(\chi^2)$ distribution, and the $F$ distribution. Both the $t$ and $\chi^2$ distributions depend on only one parameter, whereas the $F$ distribution depends on two. In statistical terminology, the parameters of these distributions are called the degrees of freedom (DF) parameters. For the $F$ distribution the two parameters are identified as the "numerator" and the "denominator" DF. The percentiles of the $t$ distribution are given in Table A.2. As the DF increases, the percentiles of the $t$ distribution approach those of the standard normal distribution. That is, for large DF ($>30$), the $t$ distribution can be approximated by the standard normal distribution. Some selected percentiles of the $\chi^2$ distribution for various values of the DF parameters are given in Table A.3. Frequently used percentiles of the $F$ distribution for various combinations of the numerator and the denominator DF are given in Table A.4.

Throughout this book we shall use $Z_\alpha$, $t_{\alpha,\upsilon}$, $\chi^2_{\alpha,\upsilon}$, and $F_{\alpha,\upsilon_1,\upsilon_2}$ to denote, respectively, the $\alpha$th percentile of the standard normal distribution, the $t$ distribution with $\upsilon$ DF, the $\chi^2$ distribution with $\upsilon$ DF, and the $F$ distribution with the numerator DF $\upsilon_1$ and the denominator DF $\upsilon_2$. It is useful to know that

$$1/F_{\alpha,v_1,v_2} = F_{1-\alpha,v_2,v_1}.$$

There are other probability distributions, such as the lognormal, the Weibull, the binomial, and the exponential, which will be introduced whenever needed. The probability background discussed here should suffice as a beginning.

## Estimation

As mentioned, the population parameters are usually unknown and are to be estimated by appropriate sample quantities called *statistics*. For example, statistic $\overline{X}$, the sample mean, may be used to estimate population mean $\mu$. Also $S^2$, the sample variance, can be used to estimate population variance $\sigma^2$. A statistic, when it is used to estimate a parameter, is called an *estimator*. Since an estimator is a sample quantity, it is subject to sampling variation and sampling errors. That is, the values assumed by an estimator vary from one sample to another. In this sense, the possible range of values of an estimator is governed by a chance or probability model, which quantifies the extent of sampling variation and sampling error. If an estimator $\widehat{\theta}$ for estimating $\theta$ is such that the mean of the distribution of $\widehat{\theta}$ is $\theta$, then $\widehat{\theta}$ is called an *unbiased* estimator. In statistical language the mean of the distribution of $\widehat{\theta}$ is also called the *expected value* of $\widehat{\theta}$ and is denoted by $E(\widehat{\theta})$. In this notation, an estimator $\widehat{\theta}$ of $\theta$ is unbiased if $E(\widehat{\theta}) = \theta$. Both the sample mean $\overline{X}$ and variance $S^2$ are unbiased estimators of the population mean $\mu$ and variance $\sigma^2$, respectively. That is, $E(\overline{X}) = \mu$, and $E(S^2) = \sigma^2$. However, sample standard deviation $S$ is not an unbiased estimator of the population standard deviation $\sigma$. In our notation, $E(S) \neq \sigma$. But other criteria in addition to the unbiasedness, such as consistency or efficiency, may be used in deciding appropriate estimators. We need not go into the theory and details because that is not the aim of this book, but we shall use only the statistically established "best" estimators of the parameters of concern to us.

An estimator when used to estimate a parameter by just one number is called a *point estimator*, and the resulting estimate is called a *point estimate*. Similarly, if a parameter is estimated by a range of values, then we have an *interval estimate*.

For estimating population mean $\mu$ by an interval estimate, we can specify an interval, for example, $(\overline{X} - S, \overline{X} + S)$, for the likely values of $\mu$, where $S$ is the sample standard deviation. Obviously, the range of an interval estimate may or may not contain the true parameter value. But we can ask: How confident are we in using interval $(\overline{X} - S, \overline{X} + S)$ to contain the true value of $\mu$, whatever that is? To answer this sort of question, statisticians use the so-called confidence intervals for estimation. For example, the degree of confidence is about $(1 - \alpha)$ 100% that the interval

$$\overline{X} - t_{\alpha/2,n-1}(S_{\overline{x}}), \quad \overline{X} + t_{\alpha/2,n-1}(S_{\overline{x}}),$$

contains population mean $\mu$, where $S_{\overline{x}} = S/\sqrt{n}$ is an estimate of the standard deviation or the standard error of $\overline{X}$. We may emphasize here the $\overline{X}$ is a sample quantity and, therefore, is subject to sampling variations and sampling errors. It is the quantification of the sampling variation and errors in the distribution of $\overline{X}$ that lets statisticians declare the degree of confidence associated with an interval estimate based on $\overline{X}$. A very useful result pertaining to the probability distribution of $\overline{X}$ is the following.

Consider a population with mean $\mu$ and variance $\sigma^2$. Suppose that we can list all possible random samples of size $n$ from this population and compute $\overline{X}$ for each sample, thus generating the distribution of $\overline{X}$. Theoretical statisticians have shown (the central limit theorem) that the distribution of $\overline{X}$, for all practical purposes, is normal with mean $\mu$ and variance $\sigma^2/n$. That is, $\overline{X}$ is $N(\mu, \sigma/\sqrt{n})$.

## Testing of Hypotheses

Another area of statistical inference is that of testing statistical hypotheses. The testing of hypotheses consists of

1. Formulation of hypotheses;
2. Collection, analysis of data, and choosing a test statistic;
3. Specification of a decision rule by means of a test statistic for accepting or rejecting hypotheses.

The formulation of the hypotheses deals with the objectives of the proposed experiment. For instance, the researcher may want to find out whether a process modification has any effect on product texture. The researcher proceeds by producing the product using the old process and the modified process. Let $\mu$ denote the mean texture using the modified process, while $\mu_0$, the mean texture of the old process, is known. Then the *null hypothesis* is written as

$$H_0 : \mu = \mu_0, \tag{1.1-9}$$

which states that on the average there is no texture improvement using the modified process. The *alternative hypothesis* may be

$$H_a : \mu \neq \mu_0, \tag{1.1-10}$$

which states that there is a change in the mean texture by using the modified process. The alternative $H_a$ in (1.1-10) is a *two-sided* hypothesis because in it $\mu$ may be less than $\mu_0(\mu < \mu_0)$ or greater than $\mu_0(\mu > \mu_0)$. A *one-sided alternative* hypothesis is either

$$H_a : \mu < \mu_0, \quad \text{or} \quad H_a : \mu > \mu_0. \tag{1.1-11}$$

The formulation of a one-sided hypothesis is based on the purpose and beliefs of the experimenter.

If, instead of in the mean, the interest is in other parameters, one can similarly formulate null and alternative hypotheses about them. After null and alternative hypotheses have been formulated, appropriate sample statistics are used to develop statistical decision rules for accepting or rejecting hypotheses.

Since a statistical decision rule is based on sample statistics, it must account for sampling variations and errors. Hence, the acceptance or rejection of the null hypothesis on the basis of a *statistical test* is subject to error. The rejection of the null hypothesis when in fact it is true results in a Type I error, the probability of which is denoted by $\alpha$. For example, a probability of $\alpha = 0.05$ indicates that on the average the test would wrongly reject the null hypothesis 5 times in 100 cases. The value of $\alpha$ is called the *significance level* of the statistical test; values of 0.05 and 0.01 are standard levels of significance used in experimentations. Another type of error is known as the Type II error, which results when the test accepts the null hypothesis when in fact it is false. The probability of a Type II error is denoted by $\beta$.

When a researcher is setting up a statistical test, both types of errors should be controlled. If a Type I error is more serious than a Type II error, we want a test statistic with a small specified value of $\alpha$ and to control $\beta$ as much as possible. If both types of errors are equally serious, then we want the size of $\alpha$ and $\beta$ to be close to each other. Usually, the size of $\alpha$ is specified by the researcher during the planning of the experiment. Then it is possible to find a statistical test that meets the specified $\alpha$ level and at the same time keeping the probability of a Type II error to a minimum for all the points in the alternative hypothesis.

Table 1.2 summarizes how Type I and II errors are committed, and their probabilities are shown within parentheses. If $H_0$ is true, the probability of the correct choice is given by the level of

| **Table 1.2** Two Types of Errors in Testing Hypotheses | | |
|---|---|---|
| | **$H_0$ is** | |
| **Decision** | **True** | **False** |
| Accept $H_0$ | No error $(1 - \alpha)$ | Type II error $(\beta)$ |
| Reject $H_0$ | Type I error $(\alpha)$ | No error $(1 - \beta)$ |

confidence $(1 - \alpha)$. Similarly, if $H_0$ is false, the probability of a correct choice is given by the *power of the test* $(1 - \beta)$, which is the ability of the test to reject $H_0$ when it is false. Power values range from 0 to 1 with higher values indicating a higher probability for rejecting a false null hypothesis.

Let us consider the hypotheses about population mean formulated in (1.1-9), (1.1-10), and (1.1-11). Suppose for the time being that $\sigma^2$ is known. Since the hypotheses pertain to population mean $\mu$, it seems reasonable to develop a statistical decision rule based on the sample mean $\overline{X}$ of a random sample $X_1, X_2, \cdots X_n$ of $n$ observations. If the new process is indeed as good as the old one, then we do not expect $(\overline{X} - \mu_0)$ to be significantly different from zero; otherwise, the difference would be significantly different from zero. Thus, the central quantity on which the test is to depend is $(\overline{X} - \mu_0)$ or its standardized version, which is

$$Z = \frac{(\overline{X} - \mu_0)}{\sigma/\sqrt{n}}, \qquad (1.1\text{-}12)$$

where $\sigma/\sqrt{n}$ is the standard error (deviation) of the distribution of $\overline{X}$. In nearly all practical situations, as stated before, $\overline{X}$ is $N(\mu, \sigma/\sqrt{n})$. Therefore, when indeed $H_0$ in (1.1-9) is true, $Z$ is $N(0, 1)$. Hence, for the two-sided alternative $H_a$: $\mu \neq \mu_0$ and for the $\alpha$ level of significance, the decision rule is: reject $H_0$ if either $Z < Z_{\alpha/2}$ or $Z > Z_{1-\alpha/2}$ (Figure 1.3); for the alternative $H_a$: $\mu < \mu_0$, which is one-sided, reject $H_0$ if $Z < Z_\alpha$ (Figure 1.4); finally, for the alternative $H_a$: $\mu > \mu_0$, reject $H_0$ if $Z > Z_{1-\alpha}$ (Figure 1.5). For instance, if $\alpha = 0.05$, $Z_{\alpha/2} = Z_{0.025} = -1.960$, $Z_{1-\alpha/2} = Z_{0.975} = 1.960$, $Z_{0.05} = -1.645$, and $Z_{0.95} = 1.645$.

If $\sigma$ is unknown, as is the case in practical situations, modify $Z$ to

$$t = \frac{\overline{X} - \mu_0}{S/\sqrt{n}}, \qquad (1.1\text{-}13)$$

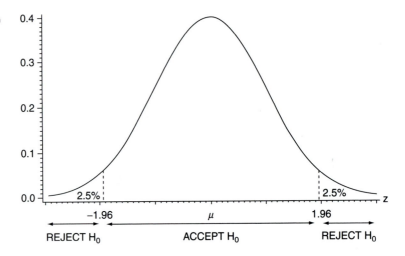

■ **FIGURE 1.3** Two-sided test for $H_0: \mu = \mu_0$ against $H_a: \mu \neq \mu_0$; $\alpha = 0.05$. Ordinate, $f(X)$.

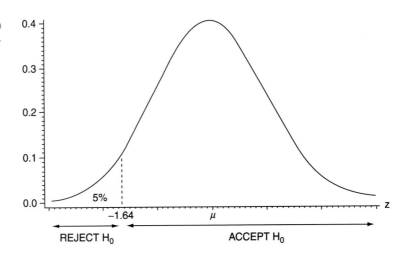

■ **FIGURE 1.4** One-sided test for $H_0: \mu = \mu_0$ against $H_a: \mu < \mu_0$; $\alpha = 0.05$. Ordinate, $f(X)$.

where $\sigma$ in $(1.1–12)$ is replaced by $S$, the sample standard deviation. Now reject $H_0$ in favor of $H_a: \mu \neq \mu_0$ if either $t < t_{\alpha/2,n-1}$ or $t > t_{1-\alpha/2,n-1}$. When the alternatives are one-sided, reject $H_0$ in favor of $H_a: \mu < \mu_0$ if $t < t_{\alpha,n-1}$, and in favor of $H_a: \mu > \mu_0$ if $t > t_{1-\alpha,n-1}$. For $n > 30$, one may use $Z_\alpha$ to approximate $t_{\alpha,n-1}$ for any $\alpha$ values. Two examples that follow illustrate the statistical inference concepts.

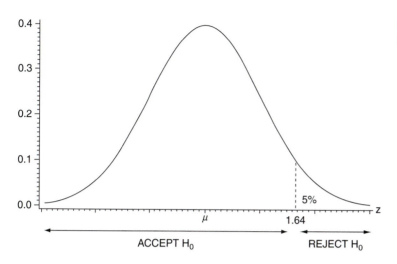

**Example 1.1-1**

The mean shelf life of a particular food product is claimed to be 60 days from the date of product manufacture. A store's audit based on a random sample of 10 items yielded $\overline{X} = 56$ days with a standard deviation $S$ of 6 days. On the basis of this information, the null hypothesis $H_0$: $\mu = 60$ was rejected in favor of $H_a$: $\mu < 60$. Let us compute $\alpha$, the probability of Type I error. Note that sample standard deviation $S$ was found to be 6 days, but we pretend for ease of computation that indeed $\sigma = 6$. Since we decide to reject $H_0$: $\mu = 60$ in favor of $H_a$: $\mu < 60$ after having observed $\overline{X} = 56$ days, we would also certainly reject $H_0$ for any value of $\overline{X}$ less than 56. That is, the statistical decision rule used is

Reject $H_0$ whenever $\overline{X} \leq 56$. Therefore,

$$\alpha = P(\overline{X} \leq 56 \text{ when } \mu = 60)$$

$$= P\left(Z \leq \frac{56 - 60}{\sigma/\sqrt{n}}\right) \tag{1.1-14}$$

$$= P(Z \leq -2.11) = 0.017 \qquad \text{(from Table A.1).}$$

Had we not pretended that $\sigma = 6$ and used $S = 6$ as an estimate of $\sigma$, we would have required the use of the $t$ distribution instead of the standard normal for computing the probability in (1.1-14). Unfortunately, our Table A.2 has the $t$ distribution tabulated only for selected percentiles, and we would not have been able to compute the desired probability.

**Example 1.1-2**

A cattle population is expected to read 17 lb on the Armour Tenderometer with a standard deviation of 3 lb. We want to specify a decision rule for testing $H_0$: $\mu = 17$ versus $H_a$: $\mu < 17$ at the $\alpha = 0.05$ level of significance on the basis of a sample of size $n = 10$ and also assess the probabilities of a Type II error for various points in the alternative. The decision rule is to be based on the $Z$ score. Since the alternative is one-sided, the decision rule is: Reject $H_0$ if $Z < Z_{0.05} = -1.645$. We may use the relationship in (1.1–6) to restate the decision rule in terms of $\overline{X}$. That is, $\overline{X} < \mu_0 - [\sigma(1.645)]/\sqrt{n}$, where $\mu_0 = 17$, $\sigma = 3$, and $n = 10$. Hence, reject $H_0$ if $\overline{X} < 15.44$ lb and accept it otherwise (see Figure 1.6).

■ **FIGURE 1.6** Areas of $\beta$ for two values of the alternative hypothesis.

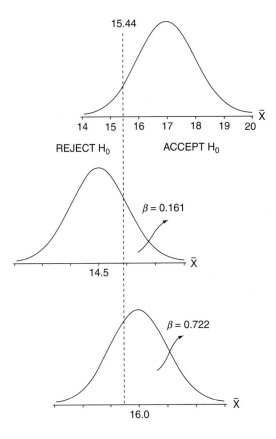

Now suppose that indeed $\mu = 14.5$. For this point in the alternative hypothesis $H_a$, what is the probability of Type II error? That is,

$$\beta = P(\text{accept } H_0 \text{ when } \mu = 14.50)$$

$$= P(\overline{X} > 15.44 \text{ when } \mu = 14.50)$$

$$= P\left(Z > \frac{15.44 - 14.50}{3/\sqrt{10}}\right)$$

$$= P(Z > 0.991) = 0.161.$$

This means that our decision rule will wrongly declare that $\mu = 17$ when indeed $\mu = 14.5$ with probability 0.161. In other words the power $(1 - \beta)$ of our rule, when $\mu = 14.5$ is $1 - 0.161$ equals 0.839. Equivalently, using our rule, the probability of rejecting $H_0$ when $H_a$ is true with $\mu = 14.5$ is nearly 84%.

Table 1.3 shows the probabilities of Type II error and the power of the test for several values in the alternative hypothesis. If $\mu = 16.0$, the probability of Type II error is 0.722. This result indicates that the further away the value of $H_a$ is from $H_0$, the smaller will be the probability of committing the Type II error. In general, it is difficult to control $\beta$ uniformly for all points in $H_a$. However, we can reduce $\beta$ by increasing the sample size of the experiment and/or by lowering the size of the significance level $\alpha$.

**Table 1.3** Probability of Type II Error as a Function of Values in $H_a$

| $H_a: \mu$ | $Z = \dfrac{15.44 - \mu}{3/\sqrt{10}}$ | $\beta =$ Probability of Type II Error | Power $= (1 - \beta)$ |
|---|---|---|---|
| 14.0 | 1.517 | 0.064 | 0.936 |
| 14.5 | 0.991 | 0.161 | 0.839 |
| 15.0 | 0.464 | 0.322 | 0.678 |
| 15.5 | −0.063 | 0.536 | 0.464 |
| 16.0 | −0.590 | 0.722 | 0.278 |
| 16.5 | −1.116 | 0.846 | 0.154 |
| 17.0 | −1.644 | 0.959 | 0.050 |

### Experimental Significance

A statistically significant difference between the true and the hypothesized parameter values is declared when the null hypothesis is rejected. That is, a real difference probably exists, and such a difference cannot be explained only by sampling errors. The acceptance of this result should be translated into terms of experimental or practical significance. For example, the investigator should look at the magnitude of the difference between treatment groups considered useful in a study.

When the standard levels of significance used in statistical significance testing are raised to, say, $\alpha = 0.10$ (10%) from $\alpha = 0.05$, the effect is to reduce the $\beta$ error. If a small $\beta$ error is crucial in the experiment, then it would be justified to raise the $\alpha$ level. However, if larger values of $\alpha$ are of serious experimental consequences, the $\alpha$ level should not be raised. The change in the values of $\alpha$ and $\beta$ can be seen in Figure 1.6 by moving the rejection region either to the left or right on the space of the sampling distribution.

### 1.2 PRINCIPLES OF EXPERIMENTAL DESIGN

Much is known about the foundation of experimental design since the book by R. A. Fisher, *The Design of Experiments*, was first published in 1925. Although experimental designs were first used in agricultural experimentation, Fisher's conception of the design of experiments has found important applications in practically all fields of research and development. Experimental design is defined in several ways. For example, Ostle and Mensing (1975) defined experimental design as a complete sequence of steps taken ahead of time to ensure that the appropriate data will be obtained in a way that will permit an objective analysis leading to valid inferences with respect to the stated problem. The definition implies that the person formulating the design clearly understands the objectives of the proposed investigation. It also implies that a statistical design must come before the experiment or that the time to design an experiment is before the experiment is begun. In another context, Finney (1960) defined the design of experiment to mean (1) the set of treatments selected for comparison, (2) the specifications of the plots to which the treatments are to be applied, (3) the rules by which the treatments are to be allocated to plots, and (4) the specifications of the measurements or other records to be made on each plot.

*Randomization, replication,* and *local control* are the three basic principles of experimental design. The development of a design is generally founded on these principles. We briefly discuss these principles in general terms. The reader may refer to the interesting papers by Greenberg (1951), Harville (1975), and White (1975), and to the 1960 edition of R. A. Fisher's *The Design of Experiments*.

## Randomization

Randomization is defined as a procedure for allocating the experimental units to specified groups, with each unit having an equal chance of being selected as a member of that group. In the context of experimental design, the experimental unit refers to that part of the material to which the treatment is applied for taking observations. The members of the specified group are said to be a random sample of the target population, about which inferences are to be made. Random number tables are generally used to accomplish randomization in an experiment. Generally speaking, random assignment of experimental units to treatments is needed to fulfill the statistical assumptions in the analysis of variance and to minimize the effects of systematic and personal biases.

## Replication

Replication generally denotes independent repetitions of an experiment under identical experimental conditions. Each repetition constitutes a replicate of that experiment. Suppose that, in an accelerated life testing study, a researcher wishes to determine the shelf life of hot dogs using three storage temperatures ($t = 3$) and it is determined that 15 packages should be used in each storage temperature. Here, a package of hot dogs is the experimental unit and $r = 15$ constitutes the number of replications in the study. For this example, the estimated standard error of the mean shelf life for each temperature setting is

$$S_{\bar{x}} = S/\sqrt{r}, \qquad (1.2\text{-}1)$$

where $S^2$ is an estimate of the error variance $\sigma^2$. As can be seen, $S_{\bar{x}}$ can be reduced by increasing the number of replications $r$, thus providing increased precision and sensitivity in measuring treatment differences.

## Local Control

Replication and randomization, when properly carried out, make statistical tests of significance valid. Local control, on the other hand, makes statistical tests more sensitive by reducing the experimental error. Reduction of error by local control is accomplished by blocking, grouping, homogeneous pairing, and proper choice of the size and shape of the experimental units or blocks. The use of local control is seen in Chapters 4–7 and 11. The choice of a design to accomplish the reduction of error has been thoroughly discussed by Kempthorne (1952), Federer (1955), Cochran and Cox (1957), and in many other books.

## Blinding

We have briefly discussed randomization, replication, and local control as the statistical elements of a sound experimental design. These elements, when prudently carried out, result in sound data for testing various hypotheses in the experiment. A nonstatistical element, so important in consumer testing and other studies where the data are subjectively obtained, is *blinding*. Blinding refers to the concealment of the identification of treatments from the experimenter and from the panelists or judges. Blinding is accomplished by *coding* the treatments using two- or three-digit numbers. The purpose of blinding is to control bias due to the experimenter's knowledge of the study and the panelists' knowledge of the treatments. For instance, comparisons of products that are brand identified are biased owing to brand effects; package design and other appearance characteristics, when not concealed, generally contribute to biased evaluation by panelists. Such bias is evident in the data shown in Table 1.4. In the first comparison of treatments A and B, the container design used in the experiment was the same for both treatments, except for the contents of the container. In the second comparison, the container design was different for each treatment. As shown in Table 1.4, the hypothesis of A and B being equal was rejected in the first comparison, but accepted in the second; the latter comparison is obviously biased due to the effect of container design.

| Table 1.4 Experimental Data to Show the Effect of Blinding[a] | | | | |
|---|---|---|---|---|
| | **First Comparison (Blinded)** | | **Second Comparison (Not Blinded)** | |
| **Attributes** | **A** | **B** | **A** | **B** |
| Wet skinfeel | 7.2** | 6.6 | 7.5 | 7.2 |
| Dry skinfeel | 7.1** | 6.4 | 7.2 | 6.9 |
| Overall liking | 7.1** | 6.3 | 7.3 | 7.0 |

*\*\*p < .01 means that A is significantly greater than B on the average on a 9-point scale.*
*[a]Source: Armour Research Center File.*

## Planning the Test Program

An appreciation of the logic of scientific experimentation may be had by studying the flowchart shown in Figure 1.7. A close examination of the flowchart shows that the sequence of events in the planning of an experiment is consistent with the definition of experimental design.

An experimental problem originates with the researcher. In collaboration with the statistician, the null and alternative hypotheses as well as the desired treatment comparisons and level of significance, are specified. On the basis of this collaboration, an experimental design that specifies a statistical model, the number of replications, and the randomization procedure is developed. The cost and the physical limitations of the study are also considered in the choice of an experimental plan.

There is no statistical analysis that can compensate for a haphazardly executed experiment. There should be pride, personal involvement, and common sense in the conduct of the experiment. There should be accountability among the members of the research team in every step of the study. The experimenter should be cognizant of techniques for achieving precision and accuracy and how to minimize measurement errors.

A source of error in the statistical analyses of data is the incorrect use of computer programs. The computer program used should be the correct one for the assumed statistical model. In statistical analyses the data should be reduced, reporting only the statistics relevant to the study. Finally, the interpretation of the result requires a careful scrutiny of the data. The result of an experiment often leads to new experimental problems, and a new cycle of experimentation may start from the beginning of the cycle shown in the flowchart (Figure 1.7). The important details involved in the planning of experimental investigation are outlined as follows (Bicking, 1954):

A. Obtain a clear statement of the problem.
   1. Identify the new and important problem area.
   2. Outline the specific problem within current limitations.
   3. Define the exact scope of the test program.
   4. Determine the relationship of the particular problem to the whole research or development program.

B. Collect available background information.
   1. Investigate all available sources of information.
   2. Tabulate the data pertinent to planning a new program.

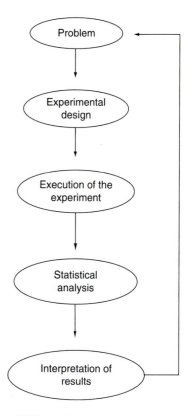

■ **FIGURE** **1.7** Flowchart for planning scientific investigations.

C. Design the test program.
   1. Hold a conference of all parties concerned.
      a. State the proposition to be studied.
      b. Agree on the magnitude of differences considered worthwhile.
      c. Outline the possible alternative outcomes.
      d. Choose the factors to be studied.
      e. Determine the practical range of these factors and the specific levels at which tests will be made.
      f. Choose the end measurements that are to be made.
      g. Consider the effects of sampling variability and of the precision of the test methods.
      h. Consider possible interrelationships or interactions of the factors.
      i. Determine the limitations of time, cost, materials, instrumentation, and other facilities and of extraneous conditions, such as weather.
      j. Consider the human relation aspects of the program.
   2. Design the program in preliminary form.
      a. Prepare a systematic and inclusive schedule.
      b. Provide for stepwise performance or adaptation of the schedule if necessary.
      c. Eliminate the effects of variables not under study by controlling, balancing, or randomizing them.
      d. Minimize the number of experimental runs.
      e. Choose the method of statistical analysis.
      f. Arrange for orderly accumulation of data.
   3. Review the design with all concerned.
      a. Adjust the program in accordance with comments.
      b. Spell out the steps to be followed in unmistakable terms.

D. Plan and carry out the experimental work.
   1. Develop methods, materials, and equipment.
   2. Apply the methods or techniques.
   3. Attend to and check details; modify methods if necessary.
   4. Record any modification of program design.
   5. Take precautions in the collection of data.
   6. Record the progress of the program.

E. Analyze the data.
   1. Reduce recorded data, if necessary, to numerical form.
   2. Apply proper mathematical statistical techniques.

F. Interpret the results.
   1. Consider all the observed data.
   2. Confine conclusions to strict deductions from the evidence at hand.

3. Test questions suggested by the data by independent experiments.
4. Arrive at conclusions with respect to the technical meaning of results, as well as to their statistical significance.
5. Point out the implications of the findings for application and further work.
6. Account for any limitations imposed by the methods used.
7. State the results in terms of verifiable probabilities.

G. Prepare the report.
   1. Describe the work clearly, giving the background, pertinence of the problems, and meaning of the results.
   2. Use tabular and graphic methods of presenting data in good form for future use.
   3. Supply sufficient information to permit the readers to verify the results and draw their own conclusions.
   4. Limit the conclusions to an objective summary of the evidence so that the work recommends itself for prompt consideration and decisive action.

## 1.3  THE ROLE OF THE STATISTICIAN IN RESEARCH

Let us look into the relationship between the statistician and researcher. Two relationships are distinguished: partnership and client–consultant relationships.

### Partnership

The statistician and researcher are partners in research from the inception of the experiment to the termination of the study. Ideally, a symbiotic relationship between them should exist to be fruitful and rewarding. In the process of association, the statistician learns about the subject of the experiment, and, in turn, the researcher acquires a view of the statistical principles as related to the experimental problem. It is through this constant association that statisticians develop new statistical techniques and improve the existing methodologies.

### Client–Consultant Relationship

Statistical consulting is like a child–parent or doctor–patient relationship. Such a relationship is often characterized by communication difficulties due to seemingly different expectations and behaviors of both consultants and clients (Hyams, 1971). The question that often leads to difficulties is, Who is the child and who is the parent? This question has been considered by Boen (1972). The client (researcher) who insists on a parent-to-child relationship is one who makes it clear that he is the boss and will decide what will be done and when. This type of relationship should be discouraged and should not exist in a scientific laboratory. The reverse is also true. Both the researcher and statistician should be good listeners in a professional manner, and both should have the ability to work as a team.

The existence of a good client–consultant relationship requires three conditions (Daniel, 1969): (1) a good experimental problem, (2) a ready client, and (3) a favorable organizational situation. A good problem is one that is clearly formulated, important to the client, and within the consultant's area of competence. A ready client is responsive, amenable to instruction, and a willing and able researcher. The third condition defines the closeness and awareness of the management to the actual problem, the work required, and the role of statistics as a support tool in scientific work.

Finally, the statistician in a scientific laboratory plays the role of a teacher. Cameron (1969) emphasized the importance of the consultant's educational training in providing lectures, courses, manuals, etc., that are tailor-made for the needs of his organization.

## EXERCISES

**1.1.** Define Type I and Type II errors. On the basis of a random sample of size $n$, state the decision rule for each of the following hypotheses:

    **1.** $H_0$: $\mu = \mu_0$,
        $H_a$: $\mu \neq \mu_0$;
    **2.** $H_0$: $\mu \leq \mu_0$,
        $H_a$: $\mu > \mu_0$;
    **3.** $H_0$: $\mu \geq \mu_0$,
        $H_a$: $\mu < \mu_0$;

Assuming a normal distribution, specify the decision rules when $\alpha = .05$, $\sigma^2 = 16$, and $n = 36$. How would you proceed to find $\beta$ in each of the preceding setups?

**1.2.** Using your decision rule for testing the following hypotheses at $\alpha = .05$, what is the probability of rejecting a false null hypothesis if $\mu = 1.5$?

$$H_0: \quad \mu_0 = 2.0, \quad \sigma = 0.4;$$
$$H_a: \quad \mu_0 \neq 2.0, \quad n = 6.$$

**1.3.** Discuss the measures that should be taken to have a scientifically sound experiment. Hints: aims of the experiment, role of randomization, etc.

**1.4.** Let $X$ be the cost of treating urinary calculus in patients without an operation. The cost is assumed to have a normal probability distribution with a mean of $280 and a standard deviation of $76. Find the percentage of patients having incurred a cost of

1. Less than $150
2. Between $250 and $450
3. Exactly $300

**1.5.** The length of stay, in days, for patients of a certain disease is assumed to be normal with a mean of 7.54 days and a standard deviation of 4.50 days. Out of 300 patients with the disease, how many will stay for

1. One day or less
2. No more than 21 days

**1.6.** Define and explain the following terms:

1. Random variable
2. Point estimate
3. Interval estimate
4. Power of a statistical decision rule
5. Sampling variation
6. Statistics
7. Estimator and parameter
8. Distribution of $\overline{X}$

# Statistical Sensory Testing

Sensory impressions are difficult to quantify, but quantification is needed to aid in decision making. Decisions are often suspended when panel members cannot agree because of undefined criteria in the decision process. Once the criteria for quantifying subjective judgments have been agreed on, statistical sensory testing experiments can be carried out. The experimental data obtained from such sensory experiments are measures of human judgments and are called *subjective* or *sensory data*. The sensory data can be analyzed using appropriate statistical models dictated by the sensory experimental designs. The unique features of sensory data are that they are unstable and skewed; thus, special methods of design and analysis are used. In the chapters to follow, we will discuss a number of experimental designs and statistical analyses of sensory data. In particular, Chapters 5 and 11 deal with balanced incomplete block designs and paired comparison designs, respectively, which are important tools in many sensory testing experiments. In this chapter we consider, among others, the process of quantifying subjective judgments and the construction of scales.

## 2.1 PSYCHOPHYSICAL ASPECTS OF SENSORY DATA

The quantification of subjective judgments belongs to the realm of what psychologists called *psychophysics*, generally defined as the science that deals with the quantitative relationships between stimulus and response (Guilford, 1954). In this relationship two continua are assumed: the psychological and the physical continua. A continuum is visualized as a linear representation of continuous events. Once we have assigned numerical values to the psychological and physical continua, we have produced scales of measurements, referred to as the *psychological* and the *physical scale*, respectively. The

■ **FIGURE** 2.1 Some relationships between the psychological and the physical continua.

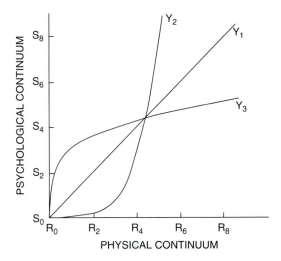

psychological scale, which represents the perceived magnitude of a stimulus, is to be determined so that its values correspond to each known numerical value assigned to the physical continuum.

Figure 2.1 illustrates some relationships between the psychological and physical continua. To have a meaningful scale for measuring sensations, we must assume that the physical continuum has a one-to-one relationship with the psychological continuum as given by curves $Y_1$, $Y_2$, and $Y_3$ in Figure 2.1. The $Y_2$ curve illustrates a relationship in which the length of the psychological continuum is longer than the physical continuum, whereas that of $Y_3$ represents the contrary. Once we know such a relationship, it will then become possible to assign physical units to sensory experiences. In many practical considerations, transformations are used on the created scales so that the relationship between the psychological and physical scales becomes linear. The advantage of a linear relationship is that the fixed distances between two points on the physical scale are mapped into fixed distances on the psychological scale.

In this section we briefly cover four psychophysical laws or models that relate the stimulus magnitude on a physical scale to the sensory response on the psychological scale. For detailed coverage of this subject, readers are referred to Guilford (1954), Torgerson (1958), Thurstone (1959), and Stevens (1961a).

## Weber's Law

Two weights differ by a just noticeable amount when separated by a given increment. If the weights are increased, the increment must also be proportionally increased for the difference to remain just noticeable. If $R$ is the stimulus magnitude and $\Delta R$ an increment of the stimulus, then Weber's law states that $\Delta R/R$ is constant for the just noticeable difference. Hence, according to Weber's law,

$$\Delta R/R = k, \qquad (2.1\text{-}1)$$

which states that the just noticeable difference (JND) $\Delta R$ is a constant proportion $k$ of the stimulus magnitude $R$.

A classical example of Weber's law is weight lifting. We can discriminate between weights of 3.0 and 3.5 lb and between weights of 7.0 and 8.2 lb 50% of the time. Both pairs of weights have $k = 0.17$, but the $\Delta R$ of the first pair is 0.5 lb, whereas that of the second pair is 1.20 lb. This example illustrates that the noticeable difference is perceived when we have attained a certain ratio between the stimulus magnitude and its increment. Within a range of stimulus magnitudes, Weber's law is shown to operate with our senses of touch, smell, hearing, and taste. A gustatory example of Weber's law was reported by Schutz and Pilgrim (1957) for $R$ defined as the point of subjective equality (PSE):

|        | PSE($R$) | $\Delta R$ | $\Delta R/R$ |
|--------|----------|------------|--------------|
| Salty  | 0.993    | 0.152      | 0.153        |
| Sweet  | 1.008    | 0.173      | 0.172        |
| Sour   | 1.006    | 0.226      | 0.224        |
| Bitter | 1.003    | 0.303      | 0.302        |

Note that the ratio of $\Delta R$ to $R$, although constant for a given stimulus, varies from stimulus to stimulus.

## Fechner's Law

Fechner (1860) chanced on Weber's law and undertook to use it for measuring the intensity of a sensation $S$. If $\Delta S$ represents an increment in the intensity of sensation $S$, then Fechner went ahead and assumed that all equal increments of sensation must be proportional to $\Delta R/R$. That is, according to Fechner,

$$\Delta S = C(\Delta R/R), \qquad (2.1\text{-}2)$$

where $C$ is the constant of proportionality. Let this equation be integrated, with the increments treated mathematically as differentials in the limit to provide

$$\int dS = C \int \frac{dR}{R}.$$ (2.1-3)

If $R$ is measured in terms of threshold stimulus, that is, the value of the stimulus at which sensation $S$ is zero and just ready to appear, and if constants are changed to common logarithms, we get

$$S = \beta \log R + \alpha,$$ (2.1-4)

where $\beta$ is the slope coefficient and $\alpha$ the intercept equal to $-\beta \log R_0$ for threshold stimulus value $R_0$. Expression (2.1-4) is known as the Weber–Fechner formula, which states that sensation intensity $S$ is a linear function of $\log R$. Using the standard least squares procedures, we can estimate parameters $\alpha$ and $\beta$ when regressing $S$ on $\log R$. The plot of $S$ and $R$ on semilog graph paper should be approximately a straight line if indeed the Weber–Fechner principle applies.

## Stevens' Power Law

In addition to Fechner's equation, other expressions that relate stimulus to sensation have been found. Stevens (1961b) argued that a power function, not Fechner's logarithmic function, describes the operating characteristics of a sensory system. Advocates of ratio scales, as exemplified by Stevens' method of magnitude estimation, subscribe to a power relationship. Supported by volumes of experimental evidence, Stevens (1953, 1961a,b, 1970) showed that sensation intensity $S$ grows in proportion to the physical stimulus $R$ raised to a power $\beta$. Mathematically written, the power relationship is

$$S = \alpha R^{\beta},$$ (2.1-5)

where $\beta$ is the exponent of the power function and constant $\alpha > 0$. By taking the logarithms of both sides of (2.1-5), we have

$$\log S = \beta \log R + \log \alpha,$$ (2.1-6)

which is a straight line on log–log graph paper. The two parameters $\beta$ and $\alpha$ can be determined by regression methods when regressing $\log S$ on $\log R$. A more general form of the power law that accounts for the threshold is

$$S = \alpha (R - R_0)^{\beta},$$ (2.1-7)

**Table 2.1** Estimates of $\beta$ for the Stevens Power Law

| Variable | Physical Stimulus $R$ | Exponent $\beta$ | Reference |
|---|---|---|---|
| Sourness | Citric acid molarity | 0.72 | Moskowitz (1971) |
| | Lactic acid molarity | 0.84 | |
| Odor | 1-Butanol | 0.48 | Piggott and Harper (1975) |
| | Coffee odor | 0.55 | Reese and Stevens (1960) |
| Sweetness | Sucrose | 1.60 | Moskowitz (1970) |
| | Cyclamate salts (calcium and sodium) | 1.00 | |
| | Sucrose | 1.30 | Stevens (1969) |

where $R_0$ is a threshold constant. On this scale stimuli are measured in terms of their subjective distances from the threshold. For ranges of stimuli well above the minimum detectable level, the threshold value $R_0$ is relatively negligible, but it is not so otherwise (Stevens, 1961b).

Table 2.1 shows power relationships between sensation and stimulus. The exponent $\beta$ determined by the so-called method of magnitude estimation is a constant value for a particular stimulus, as has been experimentally verified. The method of magnitude estimation has been described by Stevens (1956, 1958) and in several papers by Moskowitz and associates (1970, 1971).

Stevens' relationship can be obtained by a simple modification of Fechner's argument. By writing $dS = CR^{\beta-1} \, dR$, integrating, and choosing the measurement scales appropriately, we obtain Fechner's law when $\beta = 0$, and Stevens' relationship otherwise.

## Thurstone's Law of Comparative Judgment

Thurstone's law of comparative judgment was published in 1927 and later reprinted in 1959 in Thurstone's book *The Measurement of Values*. The law is considered again in Chapter 11. The law of comparative judgment equates the unknown psychological scale to the observed frequency of a stimulus. Let $P_{12}$ be the observed proportion by which stimulus $R_2$ is judged greater than stimulus $R_1$. Also let $Z_{12}$ denote the values of the standard normal variate $Z$ such that $P(Z > Z_{12}) = P_{12}$. Denote by $S_1$ and $S_2$ the mean sensations evoked by stimuli $R_1$ and $R_2$, respectively. If $\rho_{12}$ denotes the correlation between the two sensation distributions generated by stimuli $R_1$ and $R_2$, then Thurstone's law of comparative judgment is

$$S_1 - S_2 = Z_{12}\sqrt{\sigma_1^2 + \sigma_2^2 - 2\rho_{12}\sigma_1\sigma_2}. \qquad (2.1\text{-}8)$$

The distance $(S_1 - S_2)$ on the psychological scale corresponds to the distance between $R_1$ and $R_2$. The experimental design for this law is the paired comparison method discussed and illustrated by examples in Chapter 11.

## 2.2 SCALES OF MEASUREMENT

The subjective measurements of attributes and other physical properties of objects can be classified on the basis of one of four scales: the nominal, ordinal, interval, and ratio scales (Stevens, 1946, 1951). Of the four, the ordinal, interval, and ratio scales are the most useful in practice. Each scale provides a measure for judging objects. Statistical analyses depend, apart from other considerations, on the type of measurement scale used to collect the data. Table 2.2 lists some of

**Table 2.2** Statistical Methods Permissible for the Four Types of Scales[a]

| Scale | Measurement Scale Value X Invariant under Mathematical Transformation, $Y = f(X)$ | Permissible Statistics | Example |
|---|---|---|---|
| Nominal | Permutation groups $Y = f(X)$, $f(X)$ means any one-to-one substitution | Number of cases, mode, contingency table correlation | Numbers assigned to players for identification |
| Ordinal | Isotonic group $Y = f(X)$, $f(X)$ means monotonic function | In addition to foregoing statistics, median, percentiles | Hardness of minerals; grades of leather, wood, etc. |
| Interval | General linear group, $Y = \alpha + \beta X$, $\beta > 0$ | In addition to all statistics for nominal and ordinal scales, mean, standard deviation, correlation, analysis of variance | Temperature (°F and °C); calendar dates |
| Ratio | Similarity group, $Y = \beta X$, $\beta > 0$ | In addition to all statistics appropriate for the other three scales, geometric mean, harmonic mean, coefficient of variation | Length, weight, loudness scale |

[a]Source: Stevens (1946, 1951, 1961a).

the permissible statistics for each type of scale. Note that the listing is hierarchal in the sense that if a statistic is permissible for the nominal scale, it is also permissible for the ordinal, interval, and ratio scales. The second column gives mathematical transformations that leave the scale form invariant.

Briefly, the nominal scale is the most unrestricted and the weakest form of numerical assignment. This form includes operations used in identification and classification of items or objects. It involves an arbitrary assignment of numbers, such as the numbering of football players for identification or the coding of items in general. An important use of the nominal scale is the categorization of items or stimuli into two classes—perceived and not perceived, as used in threshold determination.

The ordinal scale involves the assignment of rank so that the rank order corresponds to the order of magnitude of the property being measured. A classic example of the ordinal scale is the Table of Hardness of Minerals. An example in biology is the pecking order or social dominance of fowl and cattle. In sensory testing, products are ranked on the basis of a desired attribute, where number 1 may be assigned to the highest ranking product, 2 to the second highest, and so on to the last ranking. The ordinal scale is monotonic since the ranks imply order of magnitude. The ordinal scale does not measure differences between objects being ranked. For instance, the difference between objects having ranks 1 and 2 is not necessarily the same as that between objects having ranks 3 and 4. Thus, the ordinal scale is limited to rank ordering of items.

The interval scale possesses a unit of measurement whereby the differences between the assigned numerals on the scale have meaning and correspond to the differences in magnitude of the property being measured. Most statistical data encountered in practice are in the form of an interval scale. Examples of interval scales are the hedonic scales and intensity rating scales discussed later in this section.

A ratio scale has the property of an interval scale, except that a true origin or zero point exists on the ratio scale. Therefore, for $x$ and $y$ measured on this scale, not only can we say that $x$ is *many units* greater than $y$, but also that $x$ is *many times* greater than $y$. A technique known as magnitude estimation, popularized by Moskowitz and his associates (1970, 1971) in the food and consumer sciences, is based on the ratio scale. All types of statistical measures and analyses are permissible for data in the form of ratio scales. We may summarize the important properties of ordinal, interval, and ratio scales in a form given by Torgerson (1958):

|  | **No Natural Origin** | **Natural Origin** |
| --- | --- | --- |
| No Distance | Ordinal scale | Ordinal scale |
| Distance | Interval scale | Ratio scale |

The absolute threshold, a point on the scale at which a transition occurs from no sensation to sensation, usually serves as the natural origin of the scale. Since the interval and ratio scales provide a meaningful difference that has metric information, it is possible to measure the magnitude of the distance between attributes on these scales.

## Hedonic Scale

Several types of scales used in sensory evaluation and consumer testing are effective in characterizing various food products. The most popular for preference evaluation is the hedonic scale, developed in 1947 at the Quarter-master Food and Container Institute for the Armed Forces (Peryam and Girardot, 1952). This scale defines the psychological states of "like" and "dislike" on a linear scale with like on the upper end and dislike on the lower end of the scale. A sample questionnaire is shown in Figure 2.2. For each hedonic description along the 9-point scale, a numerical value ranging from 1 (dislike extremely) to 9 (like extremely) is assigned successively on the assumption that a continuum of psychological scale defines directly the physical categories of response. The assignment of numerals is a psychophysical problem, which is briefly discussed in Section 2.4. The descriptive words or phrases that are anchors on the scale reflect the individual's sensory experiences about the stimuli under a given set of conditions.

The hedonic scale is bipolar; that is, both ends of the scale have descriptive adjectives that may not necessarily be opposite in sensory meaning. The neutral or zero point of the scale is the point that divides the scale into the like and dislike category sets, or the point of "neither like nor dislike" category in Figure 2.2. It is not necessary to have a neutral point on the scale. There are pros and cons regarding the neutral point issue, and the choice is left to the investigator. Some guidelines, however, are pointed out in Section 2.4.

<u>QUESTIONNAIRE</u>

Instructions

You will be given three servings of food to eat and you are asked to say about each how <u>you like it</u> or <u>dislike it</u>. Eat the entire portion which is served you before you make up your mind unless you decide immediately that it is definitely unpleasant. Rinse you<u>r</u> mouth with the water provided after you have finished with each sample and then wait for the next. There will be approximately two minutes between samples.

Use the scale below to indicate your attitude. Write the code number of the sample in the space above and check at the point on the scale which best describes your feeling about the food. Also <u>your comments</u> are invited. They are generally meaningful.

Keep in mind that you are the judge. You are the only one who can tell what you like. Nobody knows whether this food should be considered good, bad or indifferent. An honest expression of your personal feeling will help us to decide.

<u>SHOW YOUR REACTION BY CHECKING ON THE SCALE</u>

Code:_____

| Like extremely | Like very much | Like moderately | Like slightly | Neither like nor dislike | Dislike slightly | Dislike moderately | Dislike very much | Dislike extremely |
|---|---|---|---|---|---|---|---|---|
| | | | | | | | | |

Comments:

Code:_____

| Like extremely | Like very much | Like moderately | Like slightly | Neither like nor dislike | Dislike slightly | Dislike moderately | Dislike very much | Dislike extremely |
|---|---|---|---|---|---|---|---|---|
| | | | | | | | | |

Comments:

Code:_____

| Like extremely | Like very much | Like moderately | Like slightly | Neither like nor dislike | Dislike slightly | Dislike moderately | Dislike very much | Dislike extremely |
|---|---|---|---|---|---|---|---|---|
| | | | | | | | | |

Comments:

■ **FIGURE 2.2** Sample questionnaire for the hedonic scale. *Source:* Peryam and Girardot (1952).

## Intensity Rating Scale

Another type of scale is used for intensity rating. Like the hedonic scale, the intensity rating scale possesses the property of an interval scale. The intensity rating scale is unipolar with descriptions such as "none" and "very weak" at one end of the scale and "large," "extreme," and "very strong" at the other end. This scale is illustrated for off-flavor consisting of 7 categories:

| Category | None | Very Slight | Slight | Moderate | Moderately Strong | Strong | Very Strong |
|---|---|---|---|---|---|---|---|
| Rating | 1 | 2 | 3 | 4 | 5 | 6 | 7 |

Instead of 1, 2, 3, 4, 5, 6, and 7 to anchor categories in the intensity rating scale, we may use 0, 1, 2, 3, 4, 5, and 6. Category 7 (very strong) is not the limit of the scale. The limit is set by our inability to find an expression for the underlying psychological continuum. The psychological scale limits of the scale just given are $1 \leq X < \infty$. The lower limit is fixed by the verbal description assigned to category 1. It is when we associate a physical scale with the psychological continuum that the limits are explicitly defined. As in the off-flavor scale, the lower and upper limits are 1 and 7, respectively. A sample questionnaire is given in Figure 2.3.

■ **FIGURE 2.3** Sample questionnaire for the intensity rating scale.

QUESTIONNAIRE

Please taste and compare the samples from left to right and indicate your judgment on Smoke Level by a check mark in the space provided.

| | Sample: | Sample: | Sample: |
|---|---|---|---|
| None | | | |
| Very light | | | |
| Light | | | |
| Medium | | | |
| Heavy | | | |
| Very heavy | | | |

Comments:

QUESTIONNAIRE

You are given <u>4</u> samples at the same time.  Please rank the samples on the basis of

<u>Preference</u>

| <u>Rank</u> | <u>Sample Number</u> |
|------|---------------|
| First | _____ |
| Second | _____ |
| Third | _____ |
| Fourth | _____ |

Reasons for samples ranked first and fourth:

_____

_____

_____

■ **FIGURE 2.4** Sample questionnaire for ranking four samples.

## Rankings

The ordinal scale provides measurements based on rank order. Ranking is a simple and popular method for gathering data. The main drawback is that the differences between ranks have no meaningful distance or metric information relevant to the magnitude of the property being measured. The usefulness of rank lies in situations in which a meaningful interval or ratio scale is not feasible. A sample questionnaire for rankings is shown in Figure 2.4. The data obtained from using Figure 2.4 are already in the form of ranks. To analyze the ranks, one must transform them into normal scores by the aid of Table A.14. For example, if we have four treatments to be ranked, the observed ranks 1, 2, 3, and 4 for each treatment are replaced by the scores 1.029, 0.297, –0.297, and –1.029, respectively. The normal scores are used in the statistical analyses. In Chapter 9, we discuss various nonparametric statistical methods to analyze quantitative data that have been assigned ranks according to their magnitude.

## 2.3  SCALE STRUCTURES

The scale forms shown in Figures 2.2 and 2.3 are *structured scales*. Each point along the continuum is anchored. The panelists check the anchored categories that represent their judgments about the items in the sample. The numerical value assigned to each category is tabulated, generating discrete values. In contrast, the *unstructured*

Name_____ Sample No. _____ Date_____

Please evaluate the sample by placing a short vertical line across the scale where in your opinion the sample should be classified for that characteristic.

Appearance

    Exterior color

           Light                        Dark

    Surface moisture

           Dry                        Wet

Flavor

    Smoke

           None                     Strong

    Sweetness

           None                     Strong

■ **FIGURE 2.5** Sample questionnaire for an unstructured scale.

*scale*, also known as the graphic scale, is not anchored. Two descriptive words or phrases are indicated on both ends of the scale to show direction. A sample questionnaire is given in Figure 2.5. The panelists are asked to check along the line their sensory feelings about the items. The check marks are later converted into quantitative scores based on a predetermined length of the scale. The scale length usually varies. In Section 2.4 we discuss the psychophysical aspect of the appropriate length of the scale. The center of the scale may be anchored; however, anchoring the scale is discouraged because it often leads to bimodal distribution.

The data generated by the unstructured scale tend to be continuous and normally distributed. The reason is that the bounds of the categories on the scale are no longer fixed but are allowed to vary according to the taste acuity and training of the individual panelist. In contrast to the structured scale, the unstructured scale assumes nothing about the equality of width between intervals on the psychological continuum. Both the structured and unstructured scales are consistent in measuring the extent and direction of many sensory attributes. Baten (1946) reported a better comparison between items when using the unstructured scale. In another report (Raffensperger et al., 1956) it was shown that the structured scale had a significantly greater amount of systematic error than the unstructured scale. The large amount of systematic error in using the structured scale may be attributed to the fact that sensations are forced to be equally spaced on the psychological scale. The equally spaced aspects of sensation are considered in Section 2.4. Many studies recommend the unstructured scale and support the point of view that this scale provides a finer sensory discrimination. A drawback of the unstructured scale is that, in the absence of electronic devices such as a digitizer, the scale requires additional work in transcribing the data into quantitative values; also, a certain amount of panelist training is necessary in using the scale.

## 2.4 **CONSTRUCTION OF A SCALE**

A subjective response should be properly scaled so that a meaningful number corresponds closely to the underlying psychological continuum. Our present knowledge of psychological scaling is due to the work of experimental psychologists. References on this subject include Torgerson (1958), Thurstone (1959), and Gulliksen and Messick (1960). One important feature of a scale is its length. For the structured scale, the length pertains to the number of categories or intervals. For the unstructured scale, it is simply the length of the scale. Another feature of a scale is the assignment of appropriate adjectives to anchor or describe each category on the scale. Another feature is symmetry, pertaining to the number of positive and negative scale values around the origin. Other important features are the width between scale values and the linearity of the scale.

The effectiveness of a scale for measuring subjective responses from physical stimuli depends on how these features are met and defined. The choice of the number of categories on the scale is the initial step in the development of a scale. Jones et al. (1955), studying the hedonic scale, concluded as follows: (1) Longer scales, up to 9 intervals, tend to be more sensitive to differences among foods; anywhere from 5 to 9 intervals are adequate for measuring sensory differences; (2) the elimination of a neutral category is beneficial; and (3) symmetry, such as an equal number of positive and negative intervals, is not an essential feature of a rating scale. In earlier studies (Garner and Hake, 1951; Pollack, 1952) it was found that the amount of information increases as the number of categories increases. Bendig and Hughes (1953) recommended 9 categories on the basis of their comparative study of rating scales with 3, 5, 7, 9, and 11 categories. Later, Raffensperger et al. (1956) suggested an 8-point balanced scale for measuring tenderness.

In general, rating scales with a small number of categories do not provide enough range for discrimination, especially for individuals with high sensory sensitivity. On the other hand, scales with a large number of categories may exaggerate physical differences and destroy scale continuum. The results of various studies reviewed here provide a guide to optimum scale development.

A first step in the development of a scale consists of conducting a "word meaning experiment" to arrive at the appropriate words or phrases that will serve as anchors for the scale categories. We will describe a study dealing with finding word anchors for the hedonic scale.

## Edwards' Method of Successive Intervals

The method of successive intervals is a psychophysical technique used to scale stimuli on a psychological scale when the relative positions of stimuli on a physical scale are unknown. The detailed description of this method was given by Edwards (1952). As Edwards outlined, the method of successive intervals for scaling possesses three important properties: (1) the method requires only one judgment from each subject for each stimulus; (2) the method yields scale values that are linearly related to those obtained by the method of paired comparisons; and (3) the method provides its own internal consistency check on the validity of the various assumptions made.

The basic data of the Edwards method are ratings of stimuli corresponding to successive categories representing increasing amounts of defined attributes. Such ratings are used to construct psychological scales. On this psychological scale, sensations from each stimulus are assumed to be normally distributed. The psychological scale value of each stimulus is considered to be the mean of each sensation distribution.

The steps in the determination of psychological scale values are summarized:

1. Prepare a relative frequency distribution and the corresponding cumulative frequency distribution of the ratings of each stimulus.
2. For each relative cumulative frequency $\alpha$, find the standard normal deviate $Z_\alpha$ ($\alpha$th percentile of the standard normal distribution)

from Table A.11. Then determine the category widths by successive differences between $Z_\alpha$ values.

3. Locate the category containing the median of the frequency distribution and let $\alpha_1$ and $\alpha_2$ be the cumulative frequencies corresponding to the lower and upper limits of the median category. Let $X_\alpha$ denote the cumulative width of all categories up to and including the category corresponding to cumulative frequency $\alpha$.

4. The median sensation scale value $S$ is estimated by $\widehat{S}$:

$$\widehat{S} = X_{\alpha_1} + (X_{\alpha_2} - X_{\alpha_1})\left(\frac{0.50 - \alpha_1}{\alpha_2 - \alpha_1}\right). \qquad (2.4\text{-}1)$$

Alternatively, one may also estimate the mean sensation scale value graphically by plotting the cumulative frequency $\alpha$ (Step 1) against the cumulative mean width (Step 3). Determine the scale value of a stimulus as the 50th percentile median of the cumulative frequency plot. The standard deviation of the sensation distribution can be estimated by the difference between the 84th percentile $Z_{0.84}$ and the 50th percentile $Z_{0.50}$ of the cumulative frequency plot.

These steps are illustrated in Example 2.4-2.

---

**Example 2.4-1**

The first step in the construction of a scale consists of conducting a "word meaning experiment" to select words and phrases as anchors for the scale categories. The present example is taken from a food preference study reported by Jones and Thurstone (1955) and Jones et al. (1955). A total of 51 words and phrases that could be used as anchors for the hedonic scale were examined. Figure 2.6 displays the questionnaire used in the word meaning study.

Using Edwards' method of successive intervals, the psychological scale values and standard deviations for each word or phrase were derived graphically. The result is shown in Table 2.3. A scale value represents the "average meaning" of the word or phrase, whereas the standard deviation reflects the ambiguity of the meaning of the word as used by subjects. In the selection of verbal anchors, an anchor with an estimated scale value in close agreement with the assigned physical scale value and with low ambiguity as measured by the standard deviation is to be preferred. For example, on the basis of the estimated scale value 4.05, it can be seen

Instructions to Respondents

WORD MEANING TEST

In this test are words and phrases that people use to show like or dislike for food. For each word or phrase make an "x" mark to show what the word or phrase means to you. Look at the examples.

*Example I*

Suppose you heard a man in the mess hall say that he "barely liked" creamed corn. You would probably decide that he likes it only a little. To show the meaning of the phrase "barely like," you probably mark under +1 on the scale below.

| | Greatest Dislike | | | | Neither Like Nor Dislike | | | | Greatest Like |
|---|---|---|---|---|---|---|---|---|---|
| | −4 | −3 | −2 | −1 | 0 | +1 | +2 | +3 | +4 |
| Barely like | | | | | | x | | | |

*Example II*

If you heard someone say he had the "greatest possible dislike" for a certain food, you would probably mark under −4, as shown on the scale below.

| | Greatest Dislike | | | | Neither Like Nor Dislike | | | | Greatest Like |
|---|---|---|---|---|---|---|---|---|---|
| | −4 | −3 | −2 | −1 | 0 | +1 | +2 | +3 | +4 |
| Greatest possible dislike | x | | | | | | | | |

For each phrase on the following pages, mark along the scale to show how much like or dislike the phrase means.

■ **FIGURE 2.6** Sample questionnaire for a word meaning test. *Source:* Jones and Thurstone (1955).

from Table 2.3 that the word phrase "like intensely" is to be preferred to all other such phrases in each class with assigned physical value 4.0. Table 2.3 serves as a valuable guide in preference scale development. The magnitude of the standard deviation (Table 2.3) indicates that phrases at either extreme of the meaning scales show greater ambiguity than word phrases falling near the middle.

A simple graphical test for normality on normal probability paper is shown in Figure 2.7. Departures from linearity illustrate failures of the normality assumption. The phrase "dislike moderately" depicts a positively skewed distribution, whereas "average" exhibits a bimodal or two-peak distribution. A nonnormal distribution indicates confusion regarding the meaning of the phrase. Of the 51 stimuli studied, 46 did not depart appreciably from normality.

**Table 2.3** Scale Values and Standard Deviations for 51 Descriptive Phrases Included in the Word Meaning Study[a]

| Phrase | Scale Value | Standard Deviation |
|---|---|---|
| Best of all | 6.15 | 2.48 |
| Favorite | 4.68 | 2.18 |
| Like extremely | 4.16 | 1.62 |
| Like intensely | 4.05 | 1.59 |
| Excellent | 3.71 | 1.01 |
| Wonderful | 3.51 | 0.97 |
| Strongly like | 2.96 | 0.69 |
| Like very much | 2.91 | 0.60 |
| Mighty fine | 2.88 | 0.67 |
| Especially good | 2.86 | 0.82 |
| Highly favorable | 2.81 | 0.66 |
| Like very well | 2.60 | 0.78 |
| Very good | 2.56 | 0.87 |
| Like quite a bit | 2.32 | 0.52 |
| Enjoy | 2.21 | 0.86 |
| Preferred | 1.98 | 1.17 |
| Good | 1.91 | 0.76 |
| Welcome | 1.77 | 1.18 |
| Tasty | 1.76 | 0.92 |
| Pleasing | 1.58 | 0.65 |
| Like fairly well | 1.51 | 0.59 |
| Like | 1.35 | 0.77 |
| Like moderately | 1.12 | 0.61 |
| OK | 0.87 | 1.24 |
| Average | 0.86 | 1.08 |
| Mildly like | 0.85 | 0.47 |

*(continued)*

**Table 2.3** Scale Values and Standard Deviations for 51 Descriptive Phrases Included in the Word Meaning Study[a]—*cont...*

| Phrase | Scale Value | Standard Deviation |
|---|---|---|
| Fair | 0.78 | 0.85 |
| Acceptable | 0.73 | 0.66 |
| Only fair | 0.71 | 0.64 |
| Like slightly | 0.69 | 0.32 |
| Neutral | 0.02 | 0.18 |
| Like not so well | −0.30 | 1.07 |
| Like not so much | −0.41 | 0.94 |
| Dislike slightly | −0.59 | 0.27 |
| Mildly dislike | −0.74 | 0.35 |
| Not pleasing | −0.83 | 0.67 |
| Don't care for it | −1.10 | 0.84 |
| Dislike moderately | −1.20 | 0.41 |
| Poor | −1.55 | 0.87 |
| Dislike | −1.58 | 0.94 |
| Don't like | −1.81 | 0.97 |
| Bad | −2.02 | 0.80 |
| Highly unfavorable | −2.16 | 1.37 |
| Strongly dislike | −2.37 | 0.53 |
| Dislike very much | −2.49 | 0.64 |
| Very bad | −2.53 | 0.64 |
| Terrible | −3.09 | 0.98 |
| Dislike intensely | −3.33 | 1.39 |
| Loathe | −3.76 | 3.54 |
| Dislike extremely | −4.32 | 1.86 |
| Despise | −6.44 | 3.62 |

[a]Source: Jones et al. (1955).

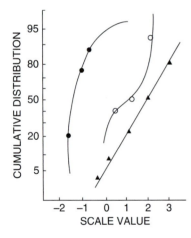

■ **FIGURE** 2.7 Cumulative plots for three descriptive phrases displaying different types of distribution: ○, dislike moderately; ■, average; ▲, preferred. *Source:* Jones et al. (1955)

**Example 2.4-2**

The "likeability" data from 96 judges shown in Table 2.4 pertain to the appearance and the flavor of barbecue sauce and are already summarized in the relative frequency distribution form. Here, appearance and flavor are the stimuli to be scaled. None of the judges rated the sauce in category 9 (like extremely), so the cumulative frequency distribution is over categories 1–8. In this example we examine the width of the scale as projected from the psychological continuum and the scale values of the two stimuli as they relate to the physical scale using Edwards' method of successive intervals.

From the relative frequencies $f$ (Table 2.4), we obtain the relative cumulative frequencies $\alpha$. The normal deviate $Z_\alpha$, corresponding to cumulative frequency $\alpha$, is obtained from Table A.11. For $\alpha = 0.010$, $Z_\alpha$ is $-2.326$, and so on. The normal deviates are estimates of the upper limits of the categories. For instance, for stimulus "appearance," deviates $-2.326$, $-1.739$, $-1.103$, $-0.613$, $0.261$, $1.146$, and $2.014$ are estimates of the upper limits for categories 1, 2, 3, 4, 5, 6, and 7, respectively. Hence, successive

**Table 2.4** Estimates of Category Widths for Two Stimuli Using Edwards' Method of Successive Intervals

| | Dislike Extremely 1 | 2 | 3 | 4 | 5 | 6 | 7 | 8 | Like Extremely 9 |
|---|---|---|---|---|---|---|---|---|---|
| **Frequency Distribution $f$** | | | | | | | | | |
| Appearance | 0.010 | 0.031 | 0.094 | 0.135 | 0.333 | 0.271 | 0.104 | 0.021 | 0.0 |
| Flavor | 0.031 | 0.042 | 0.135 | 0.115 | 0.271 | 0.302 | 0.094 | 0.010 | 0.0 |
| **Cumulative Distribution $\alpha$** | | | | | | | | | |
| Appearance | 0.010 | 0.041 | 0.135 | 0.270 | 0.603 | 0.874 | 0.978 | 1.0 | — |
| Flavor | 0.031 | 0.073 | 0.208 | 0.323 | 0.594 | 0.896 | 0.990 | 1.0 | — |
| **Normal Deviate $Z_\alpha$** | | | | | | | | | |
| Appearance | −2.326 | −1.739 | −1.103 | −0.613 | 0.261 | 1.146 | 2.014 | — | — |
| Flavor | −1.866 | −1.454 | −0.813 | −0.459 | 0.238 | 1.259 | 2.326 | — | — |
| **Category Width $W$** | | | | | | | | | |
| Appearance | — | 0.587 | 0.636 | 0.490 | 0.874 | 0.885 | 0.868 | — | — |
| Flavor | — | 0.412 | 0.641 | 0.354 | 0.697 | 1.021 | 1.067 | — | — |
| Mean Width $\bar{W}$ | — | 0.500 | 0.639 | 0.422 | 0.786 | 0.953 | 0.968 | — | — |

differences between the normal deviates provide estimates of the widths of successive categories. These quantities are shown as category width $W$ in Table 2.4. For example, for categories 2 and 3, we estimate their widths by $-1.739 + 2.326 = 0.587$, and $-1.103 + 1.739 = 0.636$, respectively.

If the sensation distribution evoked by the two stimuli (appearance and flavor) are assumed to be the same, then an improved estimate of the category widths may be obtained by averaging the widths of the two stimuli. These average widths $\overline{W}$ are also given in Table 2.4.

The calculations for estimating the mean psychological scale values are given in Table 2.5. The first row in the table is simply the physical scale values assigned to the categories. The next two rows are the psychological scale values $X_\alpha$ (cumulative widths) corresponding to the two stimuli. The fourth row is the average scale values $\overline{W}_\alpha$ of the entries in the preceding two rows. The next two rows are the cumulative frequencies $\alpha$ corresponding to the two stimuli. The mean or median sensation scale values $\widehat{S}$ are calculated using formula (2.4-1). The estimates are 2.32 and 1.86 for "appearance" and "flavor," respectively.

**Table 2.5** Estimation of the Median Sensation

| | Assigned Physical Scale | | | | | |
|---|---|---|---|---|---|---|
| | **2** | **3** | **4** | **5** | **6** | **7** |
| Psychological Scale (Cumulative Mean Width) | | | | | | |
| Appearance $X_\alpha$ | 0.587 | 1.223 | 1.713 | 2.587 | 3.472 | 4.340 |
| Flavor $X_\alpha$ | 0.412 | 1.053 | 1.407 | 2.104 | 3.125 | 4.192 |
| Average $\overline{X}_\alpha$ | 0.500 | 1.138 | 1.560 | 2.346 | 3.299 | 4.266 |
| Cumulative Distribution | | | | | | |
| Appearance $\alpha$ | 0.041 | 0.135 | 0.270 | 0.603 | 0.874 | 0.978 |
| Flavor | 0.073 | 0.208 | 0.323 | 0.594 | 0.896 | 0.990 |

Calculation of Median Sensation Using (2.4-1)

$$\text{Appearance } \widehat{S} = 1.713 + (2.587 - 1.713)\left[\frac{0.500 - 0.270}{0.603 - 0.270}\right] = 2.317$$

$$\text{Flavor } \widehat{S} = 1.407 + (2.104 - 1.407)\left[\frac{0.500 - 0.323}{0.594 - 0.323}\right] = 1.862$$

$$\text{Average } \overline{S} = 1.560 + (2.346 - 1.560)\left[\frac{0.500 - 0.297}{0.599 - 0.297}\right] = 2.088$$

We note that the median sensation value, as well as the standard deviation, can be estimated from the normal probability plot in Figure 2.8. The median sensation corresponds to the 50th percentile. For example, from the normal probability plot corresponding to stimulus "appearance," the median is estimated by 2.30. Furthermore, since $Z_{0.84}$ is 3.30 and $Z_{0.50}$ is 2.30, the standard deviation of the sensation distribution evoked by stimulus "appearance" is estimated by $Z_{0.84} - Z_{0.50}$, which is 1.00. Similarly, for stimulus flavor, the standard deviation is estimated to be $2.82 - 1.90 = 0.92$.

The scaling of stimuli by the method of successive intervals assumes that the sensation distribution is normal on the psychological scale. It is possible to check the validity of this assumption. The assumption holds if the graph of the observed cumulative frequency distribution on normal probability paper is a straight line (see Figure 2.8). Or we may compute the expected cumulative frequencies assuming normality and compare them with the observed cumulative frequencies $\alpha$. To do this, we convert $X_\alpha$ values in the sensation scale to $U$ values as follows:

$$U_\alpha = (X_\alpha - \widehat{S}),$$

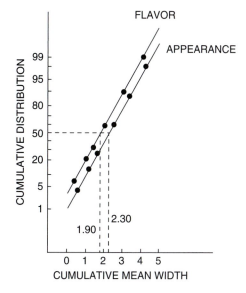

■ **FIGURE 2.8** Normal probability plot for appearance and flavor.

**Table 2.6** Calculations of the Expected Frequencies under the Normality Assumption

| | Assigned Physical Scale | | | | | |
|---|---|---|---|---|---|---|
| | **2** | **3** | **4** | **5** | **6** | **7** |
| Psychological Scale | | | | | | |
| Appearance $X_\alpha$ | 0.587 | 1.223 | 1.713 | 2.587 | 3.472 | 4.340 |
| $U_\alpha = X_\alpha - \widehat{S}$ | −1.730 | −1.094 | −0.604 | 0.270 | 1.155 | 2.023 |
| Expected Cumulative Frequencies $\alpha'$ | 0.042 | 0.138 | 0.274 | 0.606 | 0.877 | 0.97 |
| Observed Cumulative Frequencies $\alpha$ | 0.041 | 0.135 | 0.270 | 0.603 | 0.874 | 0.978 |
| Discrepancy $\alpha - \alpha'$ | −0.001 | −0.003 | −0.004 | −0.003 | −0.003 | 0.00 |

where $\widehat{S}$ is the estimated median value of the sensation scale. Then using Table A.12, we find $\alpha'$ so that

$$\alpha' = \Phi(U_\alpha),$$

where $\Phi$ denotes the cumulative distribution function of the standard normal distribution. If the normality assumption is valid, then observed frequencies $\alpha$ must be in "agreement" with the expected frequency $\alpha'$. A crude measure of agreement is the $\Sigma(\alpha - \alpha')$ for all categories. The sum should be nearly zero if the assumption holds. A better measure of agreement is a so-called $\chi^2$ goodness-of-fit measure $\Sigma(\alpha - \alpha')^2/\alpha'$, where the sum extends over all categories. Such a measure is considered in Chapter 9. Table 2.6 shows calculations for $\alpha'$ following the psychological scale in Table 2.5 for stimulus appearance.

## Scaling of Categories

Having developed a psychological scale, we may proceed to the scaling of physical categories. Note from Table 2.4 that the $Z$ values −2.326, −1.739, −1.103, −0.613, 0.261, 1.146, and 2.014 are estimates of the upper limits for categories 1, 2, 3, 4, 5, 6, and 7, respectively. To scale the categories, we need to estimate their midpoint. The midpoint of category 2, for example, is estimated by the average of the upper limits of categories 1 and 2. That is, $(-2.326 - 1.739)/2 = -2.032$ is an estimate of the midpoint of category 2. Estimates of all the midpoints of other categories are shown in Table 2.7. The

**Table 2.7** Scaling of Categories by the Edwards Method

| | Assigned Physical Scale | | | | | | | | |
|---|---|---|---|---|---|---|---|---|---|
| | **1** | **2** | **3** | **4** | **5** | **6** | **7** | **8** | **9** |
| $Z_\alpha$ for Appearance (Category Upper Limit) | −2.326 | −1.739 | −1.103 | −0.613 | 0.261 | 1.146 | 2.014 | ∞ | — |
| Estimates of Category Midpoints | — | −2.032 | −1.421 | −0.858 | −0.176 | 0.704 | 1.580 | — | — |
| Category Midpoints −(−0.176) | — | −1.856 | −1.245 | −0.682 | 0.000 | 0.880 | 1.756 | — | — |

reader should note that it is not possible to scale category 1 since for the Edwards procedure the lower limit of category 1 is not estimated. For a similar reason we cannot scale the last category either.

We may relocate the center of the psychological scale to zero, which corresponds to the neutral category or natural origin on the physical scale. Note that in our Example 2.4-2 numeral 5 on the physical scale corresponds to the neutral category or natural origin on the physical scale. The midpoint of this category is estimated by −0.176, which is to be relocated at zero. That is, a new scale for categories with the center at zero is given by the last row of Table 2.7. The plot of the psychological scale versus the physical scale is shown in Figure 2.9. The plot shows unequal interval widths between categories.

As indicated, Edwards' method does not scale the extreme categories. But if the scaling of all categories is necessary, we may use Guilford's (1954) scaling method, which we now illustrate.

According to Guilford, the midpoint $m_i$ of the $i$th category is estimated by

$$m_i = \frac{Y_i - Y_{i+1}}{\alpha_{i+1} - \alpha_i}, \qquad i = 1,\dots,k, \tag{2.4-2}$$

where $Y_i$ is the ordinate of the normal curve at the lower limit of the $i$th category (by definition we let $Y_1 = 0$ for the lower limit of category 1), $Y_{i+1}$ is the ordinate of the normal curve at the upper limit of the $i$th category [by definition we let $Y_{k+1} = 0$ at the upper limit of the $k$th (highest) category], $\alpha_i$ is the cumulative proportion of judgments below the lower limit of the $i$th category, and $\alpha_{i+1}$ is the cumulative proportion of judgments below the upper limit of the $i$th category. The following example illustrates Guilford's method of scaling categories.

PHYSICAL SCALE

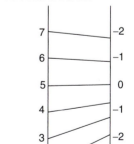

PSYCHOLOGICAL SCALE

■ **FIGURE 2.9** Plot of the physical and psychological scales obtained by the Edwards' method.

**Example 2.4-3**  This example uses the data in Example 2.4-2. Table 2.8 shows the steps required for scaling using Guilford's method. The ordinates of the normal deviates are obtained from Table A.13.

Differences $Y_i - Y_{i+1}$ between adjacent ordinates for stimulus appearance are

$$Y_1 - Y_2 = 0.000 - 0.027 = -0.027,$$

$$Y_2 - Y_3 = 0.027 - 0.088 = -0.061,$$

and so on. All these differences are shown for both stimuli in Table 2.8. The last two rows of the table are the differences between the adjacent cumulative proportions for the two stimuli. For instance, for stimulus appearance, these successive differences are

$$\alpha_2 - \alpha_1 = 0.010 - 0.000 = 0.010,$$

$$\alpha_3 - \alpha_2 = 0.041 - 0.010 = 0.031,$$

and so on. Now using (2.4-2), we can estimate midpoints of all categories. For stimulus appearance, estimated midpoints are

**Table 2.8** Steps for Scaling Categories Using the Guilford Method

| | **Assigned Physical Scale** | | | | | | | | |
|---|---|---|---|---|---|---|---|---|---|
| | **1** | **2** | **3** | **4** | **5** | **6** | **7** | **8** | **9** |
| Cumulative Proportions $\alpha$ | | | | | | | | | |
| Appearance | 0.010 | 0.041 | 0.135 | 0.270 | 0.603 | 0.874 | 0.978 | 1.000 | — |
| Flavor | 0.031 | 0.073 | 0.208 | 0.323 | 0.594 | 0.896 | 0.990 | 1.000 | — |
| Normal Deviates $Z_\alpha$ | | | | | | | | | |
| Appearance | −2.326 | −1.739 | −1.103 | −0.613 | −0.261 | 1.146 | 2.014 | — | — |
| Flavor | −1.866 | −1.454 | −0.813 | −0.459 | 0.238 | 1.259 | 2.326 | — | — |
| Normal Ordinates at $Z_\alpha$ | | | | | | | | | |
| Appearance $Y_i$ | 0.027 | 0.088 | 0.217 | 0.331 | 0.386 | 0.207 | 0.053 | — | — |
| Flavor $Y_i$ | 0.070 | 0.139 | 0.287 | 0.359 | 0.388 | 0.181 | 0.027 | — | — |
| Differences between Ordinates $Y_{i+1} - Y_i$ | | | | | | | | | |
| Appearance | −0.027 | −0.061 | −0.129 | −0.114 | −0.055 | 0.179 | 0.154 | 0.053 | — |
| Flavor | −0.070 | −0.069 | −0.148 | −0.072 | −0.029 | 0.207 | 0.154 | 0.027 | — |
| Differences between Proportion of Judgment in Each Category $\alpha_{i+1} - \alpha_i$ | | | | | | | | | |
| Appearance | 0.010 | 0.031 | 0.094 | 0.135 | 0.333 | 0.271 | 0.104 | 0.022 | — |
| Flavor | 0.031 | 0.042 | 0.135 | 0.115 | 0.271 | 0.302 | 0.094 | 0.010 | — |

$$m_1 = -0.027/0.010 = -2.700,$$

$$m_2 = -0.061/0.031 = -1.968,$$

and so on. These values for the two stimuli are shown in Table 2.9. As for Edwards' method for scaling categories, one can use the neutral category 5 as the natural origin to relocate the center of the psychological scale at zero. In Table 2.9. this has been done for both stimuli in rows 3 and 4. The psychological scale of stimulus appearance is plotted against the physical scale in Figure 2.10. The plot is similar to that obtained by the Edwards' method; that is, the category widths in the lower portion of the scales are narrower than those in the upper portion.

The unequal widths of the category intervals indicate that panelists were not uniform in their sensitivities to differences over the scale continuum. An implication of unequal category widths is that an interval of one category separating two highly acceptable foods may represent greater psychological difference than an interval of the same length for foods that are disliked or liked only moderately (Moskowitz and Sidel, 1971). A method for adjusting unequal category width is described by Guilford (1954) and is illustrated by Example 2.4-4.

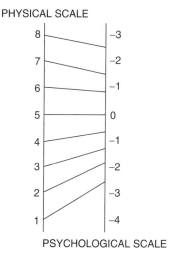

PHYSICAL SCALE

PSYCHOLOGICAL SCALE

■ **FIGURE 2.10** Plot of the physical and psychological scales for appearance obtained by the Guilford method.

**Table 2.9** Scaling of Categories by the Guilford Method

| | Assigned Physical Scale | | | | | | | | |
|---|---|---|---|---|---|---|---|---|---|
| | **1** | **2** | **3** | **4** | **5** | **6** | **7** | **8** | **9** |
| Midpoints $m_i$ | | | | | | | | | |
| Appearance | −2.700 | −1.968 | −1.372 | −0.844 | −0.165 | 0.661 | 1.481 | 2.409 | — |
| Flavor | −2.258 | −1.643 | −1.096 | −0.626 | −0.107 | 0.685 | 1.638 | 2.700 | — |
| $m_i - (-0.165)$ | | | | | | | | | |
| Appearance | −2.535 | −1.803 | −1.207 | −0.679 | 0.000 | 0.826 | 1.646 | 2.574 | — |
| $m_i - (-0.107)$ | | | | | | | | | |
| Flavor | −2.151 | −1.536 | −0.989 | −0.519 | 0.000 | 0.792 | 1.745 | 2.807 | — |

**Example 2.4-4**

In Table 2.10, we copied from Table 2.9 the estimates of the midpoints of categories for both stimulus appearance and flavor. To scale categories, we need to find successive differences of the midpoints. To find better possible estimates of category scale values, we use the successive

**Table 2.10** Calculations of the Adjusted Category Scale Using the Guilford Method (Example 2.4-4)

| | Category (Physical) Scale $X$ | | | | | | | | |
|---|---|---|---|---|---|---|---|---|---|
| | **1** | **2** | **3** | **4** | **5** | **6** | **7** | **8** | **9** |
| Midpoints $m_i$ | | | | | | | | | |
|   Appearance | −2.700 | −1.968 | −1.372 | −0.844 | −0.165 | 0.661 | 1.481 | 2.409 | — |
|   Flavor | −2.258 | −1.643 | −1.096 | −0.626 | −0.107 | 0.685 | 1.638 | 2.700 | — |
| Sum | −4.958 | −3.611 | −2.468 | −1.470 | −0.272 | 1.346 | 3.119 | 5.224 | — |
| Successive Differences of Sums | | 1.347 | 1.143 | 0.998 | 1.198 | 1.618 | 1.773 | 2.105 | — |
| Average | | 0.674 | 0.572 | 0.599 | 0.499 | 0.809 | 0.887 | 1.053 | — |
| Cumulative Average | 0.0 | 0.674 | 1.246 | 1.745 | 2.344 | 3.153 | 4.040 | 5.093 | — |
| Guilford's Adjusted Category Scale Values | 1.0 | 1.674 | 2.246 | 2.745 | 3.344 | 4.153 | 5.040 | 6.093 | — |

average difference of the midpoints of the two stimuli. These calculations are shown in Table 2.10. The Guilford adjusted category values are obtained by adding 1.0 to all estimated category scale values. In adjusting raw scores resulting from sensory evaluation experiments, we must replace the raw scores by their corresponding adjusted category scale values. For instance, the raw score 2 is replaced by the adjusted score 1.674, 3 is replaced by 2.246, and so on. The adjusted scores are used in the subsequent statistical analyses of the data. An application of the use of this adjustment procedure in sensory tests is reported by Cloninger et al. (1976).

## 2.5 DISTRIBUTION OF SENSORY DATA

Some knowledge of the distributional properties of sensory data is necessary for using statistical methods. For instance, if sensory data are not normally distributed, it would be of interest to know the factors causing the nonnormality and the extent of nonnormality.

### Measures of Shape Distribution

Distribution of data can be characterized to a great extent by its variance $S^2$, skewness $g_1$, and kurtosis $g_2$. The variance describes the spread of the sample observations from the arithmetic mean. Skewness measures the symmetry of distributions having extended tails. For a distribution with negative skewness, the tail is extended to the left (Figure 2.11), whereas for positive skewness the tail is extended to the right (Figure 2.12). Thus, $g_1$ is essentially a measure of the symmetry of the data. The degree of skewness is measured from sample values $X_1, X_2, \ldots, X_n$ by

$$g_1 = \sum (X_i - \overline{X})^3 / nS^3, \qquad (2.5\text{-}1)$$

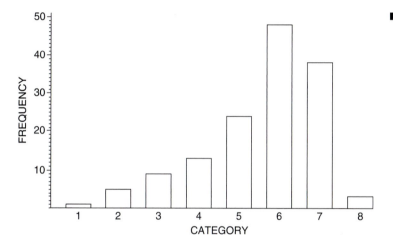

■ **FIGURE 2.11** Illustration of negative skewness.

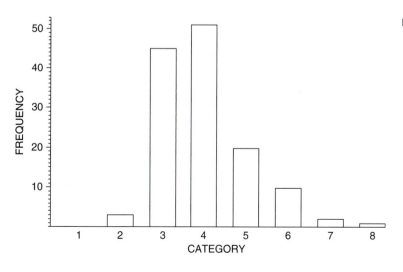

■ **FIGURE 2.12** Illustration of positive skewness.

where $S$ is the sample standard deviation. For a perfectly symmetrical distribution, $g_1$ is zero.

Kurtosis is a measure of abnormality characterized by peakedness or flatness of the distribution. More specifically, it is a measure of unimodality or bimodality of the distribution (Darlington, 1970). Figures 2.13 and 2.14 illustrate this type of abnormality. An estimate of $g_2$ from a sample $X_1, \ldots, X_n$ is obtained by

$$g_2 = \left[ \sum (X_i - \overline{X})^4 / ns^4 \right] - 3. \tag{2.5-2}$$

A negative $g_2$ indicates bimodality (two peaks), whereas a positive $g_2$ indicates unimodality (peakedness). In sensory testing, unimodality often suggests good agreement among panel members as to the choice of categories for expressing their sensory experiences.

## Statistical Properties of Subjective Data

The discussion that follows pertains to data generated by categorical judgments. Such data are the most often used in practice. In this section we consider some of their properties and point out guidelines for their proper use.

■ FIGURE 2.13 Illustration of kurtosis (unimodal).

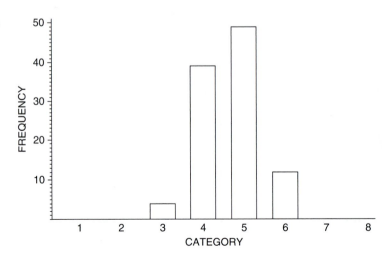

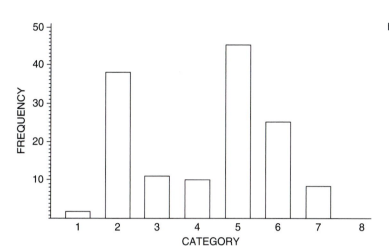

Three main factors govern the distribution of categorical data:

1. The nature of attributes being measured such as acceptability of products, off-flavor development in stored products, and sweetness level;
2. The type of rating scale used such as bipolar, unipolar, and the number of categories;
3. The position of the mean or median on the scale.

Let us examine Figures 2.15 through 2.17, which are distributions of off-flavor scores based on an intensity scale of 1 for "none" and 7 for "extremely strong."

The intensity rating scale for off-flavor is unipolar; that is, category 1 of the scale is fixed by the verbal description used, and the remaining categories toward the other end are elastic. In stability studies of food products, barring differences in the panelists' thresholds, there is no detectable off-flavor at the initial evaluation (Figure 2.15). Consequently, the distribution of the scores clusters around categories 1 and 2. At the initial period of evaluation, the exponential distribution seems to describe the behavior of the sensory scores. When the off-flavor begins to develop, the median score is located around the center of the rating scale and the distribution tends to

■ **FIGURE** 2.15 Distribution of sensory scores at the initial evaluation.

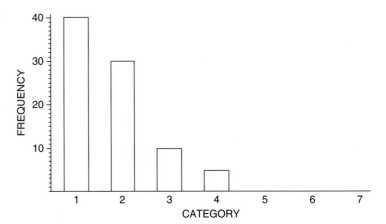

■ **FIGURE** 2.16 Distribution of sensory scores at the middle evaluation.

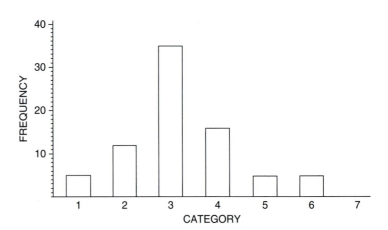

■ **FIGURE** 2.17 Distribution of sensory scores at the final evaluation.

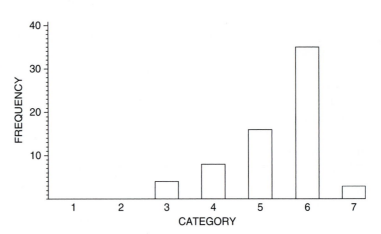

be normal (Figure 2.16). At the final period of evaluation, the distribution is approximately normal, being somewhat negatively skewed (Figure 2.17).

The centering of acceptable anchors is a characteristic of bipolar rating scales such as the hedonic scale. The mean score of average quality products is generally near the center of the hedonic scale. Peryam and Girardot (1952), using a 9-point hedonic scale, observed that most foods fall in the range of 5.5 to 7.5 units on the scale and the distribution of scores tends to be normal unless the product is poor or well liked, whereby a skewed distribution would often arise. Another factor that contributes to the normality of the bipolar scales is the judgmental tendency of the panelists to avoid the extremes and thus score around the middle of the rating scale.

Hopkins (1950) observed that the variance and mean of the distribution of sensory data are not independent if the data are observed over the full range of the scale, but they are independent when most of the data fall near the center of the rating scale. Read (1964) noted that the variance tends to be smaller for the data observed at either end of the scale and remains relatively constant for scores near the center of the scale. Another property of sensory data often overlooked is that of relativity. The magnitudes of sensory scores in a given situation are relative. For example, in comparative studies, the score an item receives is dependent on the quality of the items to which it is being compared. Therefore, it is not recommended that these scores be associated with data obtained from other experiments. However, if sensory data are derived from judgments of highly trained panelists, they may be related to the results of other studies.

## Transformation

It is often useful to transform experimental data to a scale that provides a simple and meaningful interpretation of the relationship among variables. An equally important consideration is statistical. For instance, the assumptions of normality, constancy of variance, and independence may not be valid for the data on the original scale. Often such data can be transformed to a new scale where these assumptions are approximately satisfied. Transformation should be avoided wherever possible, however, because of the complexity in interpreting the transformed data. We are not discussing transformations for achieving linearity and variance stability; these are extensively discussed elsewhere (e.g., Aitchison and Brown, 1957; Snedecor and Cochran, 1967; Thoni, 1967). Our interest here is in transformations of categorical data.

One of the many causes of the nonnormality of sensory data is the unequal widths of category intervals. The unequal interval width results in nonadditivity of effects and heterogeneous variance along the scale continuum. The width of the category intervals varies with the type and length of the scale and with the nature of the sensory attributes being measured.

An attractive transformation by Wright (1968) and Aitchison and Brown (1957), which treats a negatively skewed distribution, is

$$Y = \log(U - X), \qquad U > X, \qquad (2.5\text{-}3)$$

where $X$ on the original scale is transformed to $Y$, and $U$ is the upper limit of the scale. The log transformed scale is to the base 10. On a rating scale of 1 to 9, one choice of $U$ is 10.5 to avoid a possibility of $(U - X)$ being negative.

Another transformation that may be used to adjust for effects of the upper and the lower limits of the scale is

$$Y = \log\left(\frac{X - L}{U - X}\right), \qquad L < X < U, \qquad (2.5\text{-}4)$$

where $L$ is the lower limit of the scale. The transformed data $Y$ are used in statistical analyses. Once a statistical analysis has been completed, results of the analysis may be interpreted on the original scale by means of inverse transformation.

---

**Example 2.5-1**

Table 2.11 displays the like/dislike data for stimulus appearance of a barbecue sauce. Notice that the relative frequency distribution is negatively skewed. Let us apply the transformation $Y = \log(10.5 - X)$, where $Y$ is the transformed value corresponding to the category scale $X$. The effectiveness of a transformation to normalize the data can be determined by plotting the cumulative frequencies versus $Y$ values from the transformed scale on normal probability paper. We have done this for the data in Table 2.11, and the result is plotted in Figure 2.18. The plot is linear, indicating that the distribution of the transformed $Y$ values approaches a normal distribution.

**Table 2.11** Distribution of Barbecue Sauce Data

| Category Scale (X) | Relative Frequencies (N = 100) | Cumulative Frequencies | Y = log (10.5 − X) |
|---|---|---|---|
| Dislike extremely | | | |
| 1 | 0 | 0 | 0.978 |
| 2 | 1 | 1 | 0.929 |
| 3 | 1 | 2 | 0.875 |
| 4 | 3 | 5 | 0.813 |
| 5 | 10 | 15 | 0.740 |
| 6 | 30 | 45 | 0.653 |
| 7 | 50 | 95 | 0.544 |
| 8 | 5 | 100 | 0.398 |
| 9 | 0 | 100 | 0.176 |
| Like extremely | | | |

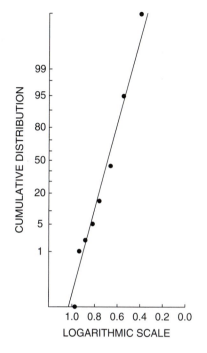

■ **FIGURE 2.18** Normal probability plot of $Y$ scores from the transformation $Y = \log(10.5 - X)$.

## 2.6  SELECTION OF PANEL MEMBERS

### Criteria Based on Binomial Response

To obtain reliable results from sensory evaluation, we should select the members of the panel of judges for taste acuity and internal consistency. The training can be compared to the calibration of an instrument, except that the calibration of human beings as measuring instruments is a more complicated and difficult task. In this section we discuss a statistical method for selecting panelists or judges capable of quantifying and making consistent judgments.

One method of quantifying the selection of panel members is by the application of sequential analysis (Wald, 1947). In sequential analysis the sample size is not fixed in advance of the experiment. Experimental observations are taken sequentially. After each experimental outcome, a decision is made as to whether to terminate the experiment on the basis of the observations obtained up to that point or to continue further experimentation. The application of sequential analysis for the selection of judges was first reported by Lombardi (1951). Since then, papers have been published for its application in sensory evaluation of food products (Steiner, 1966; Gacula et al., 1974; Cross et al., 1978). In this section we illustrate the method of sequential analysis when the underlying statistical model is binomial. When the statistical model is for the normal random variable, in particular degree of difference, it will be discussed in the next section.

For a binomial model, responses are classified as correct or incorrect, defective or nondefective, acceptable or not acceptable. In sensory evaluation, too many incorrect responses from judges may indicate that they are guessing or making random judgments. Let $p$ be the proportion of incorrect responses in $n$ testings by a judge; thus, $1 - p$ is the proportion of correct responses.

For practical considerations, we can specify $p_0$ and $p_1$ such that $p_0 < p_1$ with the following criteria: If a prospective member is judged to have made incorrect selections more than $p_1$ proportion of the times, that member will not be included in the panel of judges. On the other hand, if the prospective member is judged to have made incorrect selections fewer than $p_0$ of the times, the member will be included in the panel. The task of including or excluding a prospective member on the basis of the stated criteria can be approached through formulation of the following null and alternative hypotheses:

$$H_0 : p \leq p_0; \quad H_a : p > p_1. \tag{2.6-1}$$

The acceptance of $H_0$ means that the member is to be included in the panel, whereas the acceptance of $H_a$ means that the member is not to be included. Of course, in testing hypotheses, one runs the risk of making Type I and Type II errors (see Chapter 1). In our formulation here, a Type I error amounts to not including a member in the panel when indeed the member is capable and would not make incorrect selections more than $p_0$ of the times. Similarly, a Type II error is to include a member in the panel when indeed the member will make incorrect selections more than $p_1$ of the times. Usually, the probabilities of Type I and Type II errors of a decision rule are denoted by $\alpha$ and $\beta$, respectively.

For an optimum rule for testing the hypotheses in (2.6-1), we may use the sequential decision rule. The decision rule follows from three regions known as the *acceptance region, rejection region,* and *region of indecision.* These regions are specified in terms of $L_0$ and $L_1$.

$$L_0 = \frac{\log[\beta/(1-\alpha)]}{\log(p_1/p_0) - \log[(1-p_1)/(1-p_0)]}$$
$$+ n \frac{\log[(1-p_0)/(1-p_1)]}{\log(p_1/p_0) - \log[(1-p_1)/(1-p_0)].}$$

(2.6-2)

and

$$L_1 = \frac{\log[(1-\beta)/\alpha]}{\log(p_1/p_0) - \log[(1-p_1)/(1-p_0)]}$$
$$+ n \frac{\log[(1-p_0)/(1-p_1)]}{\log(p_1/p_0) - \log[(1-p_1)/(1-p_0)].}$$

(2.6-3)

For a member, if the number of incorrect selections out of $n$ (at the $n$th stage) is found to exceed $L_1$, $H_0$ is rejected and the member is not included in the panel. If the number of incorrect selections at the $n$th stage is less than $L_0$, $H_0$ is accepted and the member is included in the panel. If at the $n$th stage the member is neither accepted nor rejected, the experiment proceeds to the $(n + 1)$th stage. The experiment at each stage amounts to having a member make an additional test. The three regions are shown in Figure 2.19 where the number of tests $n$ is the horizontal axis and the accumulated number of incorrect decisions is the vertical axis.

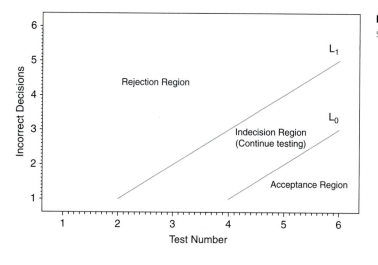

■ **FIGURE 2.19** The three decision regions in a sequential test.

We may briefly note the four steps used in the sequential analysis for selecting judges:

1. After undergoing training in sensory terminologies and definitions, a prospective judge is given a sensory test such as a triangle, duo–trio, or pair test. The result is recorded in suitable form as part of the individual candidate's permanent record.
2. If the accumulated number of incorrect responses lies in the indecision region, the testing of the candidate continues.
3. If the accumulated number of incorrect responses lies in the rejection region, the candidate is disqualified. As shown in Figure 2.19, this area is above $L_1$.
4. If the accumulated number of incorrect responses lies in the acceptance region, the candidate is qualified as a member of the panel. This area is below $L_0$ in Figure 2.19.

The amount of risk involved in making incorrect decisions depends on the specified values of parameters $p_0$, $p_1$, $\alpha$, and $\beta$. These parameters should be carefully specified to avoid setting them so high that they result in the rejection of many reasonably good judges. A small difference between $p_0$ and $p_1$, coupled with low specified values of $\alpha$ and $\beta$, results in a reduced number of tests and a greater risk of excluding fairly good judges. Another important consideration in using sequential analysis is knowing how many tests to require. Because this knowledge is valuable in planning experiments, Wald (1947) provided a formula for the average number $n$ of tests required to reach a decision. For values $p = p_0$ and $p = p_1$, the formulas are

$$n_{p0} = \frac{(1-\alpha)\log[\beta/(1-\alpha)] + \alpha\log[(1-\beta)/\alpha]}{p_0\log(p_1/p_0) + (1-p_0)\log[(1-p_1)/(1-p_0)]}, \qquad (2.6\text{-}4)$$

$$n_{p1} = \frac{\beta\log[\beta/(1-\alpha)] + (1-\beta)\log[(1-\beta)/\alpha]}{p_1\log(p_1/p_0) + (1-p_1)\log[(1-p_1)/(1-p_0)]}, \qquad (2.6\text{-}5)$$

respectively. Taking into consideration various experimental limitations, we can reach an appropriate estimate of the number of tests required.

---

**Example 2.6-1**

A triangle test, using a 2% sucrose solution, was conducted to determine the taste acuity of prospective judges. The following are the hypotheses and error rates:

$$H_0: \quad p = p_0 = 0.33, \qquad \alpha = 0.10$$
$$H_a: \quad p = p_1 = 0.40, \qquad \beta = 0.10$$

We have $1 - p_0 = 0.67$ and $1 - p_1 = 0.60$. The outcomes of the triangle test are given in Table 2.12. Using (2.6-2) and (2.6-3), we obtain

**Table 2.12** Performance of Two Judges Using the Triangle Test (Example 2.6-1)

| Judge Number | Test Number | Decisions[a] | Accumulated Number of Incorrect Decisions |
|---|---|---|---|
| 28 | 1 | C | 0 |
| | 2 | C | 0 |
| | 3 | I | 1 |
| | 4 | C | 1 |
| | 5 | I | 2 |
| | 6 | I | 3 |
| | 7 | I | 4 |
| | 8 | C | 4 |
| | 9 | I | 5 |
| | 10 | I | 6 |
| | 11 | I | 7 |
| 30 | 1 | C | 0 |
| | 2 | I | 1 |
| | 3 | C | 1 |
| | 4 | C | 1 |
| | 5 | C | 1 |
| | 6 | C | 1 |
| | 7 | C | 1 |
| | 8 | C | 1 |
| | 9 | C | 1 |
| | 10 | C | 1 |
| | 11 | C | 1 |
| | 12 | C | 1 |
| | 13 | C | 1 |
| | 14 | I | 2 |
| | 15 | C | 2 |

*(continued)*

**Table 2.12** Performance of Two Judges Using the Triangle Test (Example 2.6-1)—*cont...*

| Judge Number | Test Number | Decisions[a] | Accumulated Number of Incorrect Decisions |
|---|---|---|---|
| | 17 | C | 2 |
| | 18 | C | 2 |
| | 19 | C | 2 |
| | 20 | C | 2 |
| | 21 | C | 2 |
| | 22 | C | 2 |
| | 23 | C | 2 |
| | 24 | C | 2 |
| | 25 | C | 2 |
| | 26 | C | 2 |
| | 27 | C | 2 |

[a]C, correct; I, incorrect.

$$L_0 = \frac{\log(0.10/0.90)}{\log(0.40/0.33) - \log(0.60/0.67)} + n\frac{\log(0.67/0.60)}{\log(0.40/0.33) - \log(0.60/0.67)}$$
$$= -7.2618 + 0.3645n,$$

and

$$L_1 = \frac{\log(0.90/0.10)}{\log(0.40/0.33) - \log(0.60/0.67)} + n\frac{\log(0.67/0.60)}{\log(0.40/0.33) - \log(0.60/0.67)}$$
$$= 7.2618 + 0.3645n.$$

Both $L_0$ and $L_1$ are linearly related to $n$ with the same slope 0.3645. Therefore, any two values of $n$ can be used to construct the two lines $L_0$ and $L_1$. The graphs of these lines are shown in Figure 2.20 along with the performance of the two judges recorded in Table 2.12. Note that judge 30 was accepted after 26 tests, whereas judge 28 remained in the region of indecision up to 11 tests; hence, further taste testing was continued.

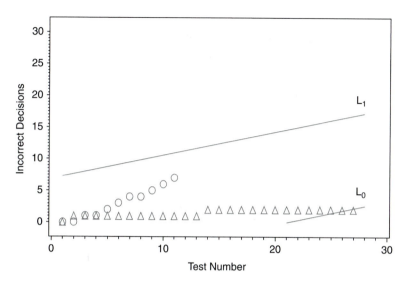

■**FIGURE 2.20** Plot of performance of two judges (O, judge 28; △, judge 30) in a sequential test.

---

For parameter values, $p_0 = 0.20$, $p_1 = 0.30$, $\alpha = 0.10$, and $\beta = 0.10$, let us compute the average number of tests required for reaching a decision using (2.6-4) and (2.6-5). We find

$$n_{p0} = 68.4, \quad n_{p1} = 62.4.$$

On the basis of these average values, one may design up to 69 tests—a liberal estimate.

**Example 2.6-2**

---

## Criteria Based on Rating Scale

The use of rating scale is widely used in obtaining intensity and hedonic data. Of course, for panelist selection and training, the intensity scale is used. One problem with the sensory scale is the variance across the scale category is unequal, and it is well known that skewness and kurtosis affect the application of the standard statistical procedures, resulting in some biased estimates of the mean and variance derived from the statistical model. To minimize this problem, researchers have studied the use of degree of difference approach (Aust et al., 1987; Castura et al., 2006; Pecore et al., 2006). The use of a statistical model in measuring sensory panel performance was reported by Latreille et al. (2006), with a review of several methods of modeling the response, and by Bi (2003). In this section, we discuss the sequential analysis by Wald (1947) and the traditional quality control procedure.

## Sequential Analysis

The theory of sequential analysis is discussed in books by Wald (1947) and Wetherill (1966). Based on degree of difference as the data, the sequential probability ratio can be written as

$$\frac{\sigma_d^2}{\mu_1 - \mu_0}\log\frac{\beta}{1 - \alpha} + n\frac{\mu_1 + \mu_0}{2} < \sum_i^n d_i < \frac{\sigma_d^2}{\mu_1 - \mu_0}\log\frac{1 - \beta}{\alpha} + n\frac{\mu_1 + \mu_0}{2}$$

(2.6-6)

where

$\sigma_d^2 = $ the variance of the differences between identical samples;

$\alpha = $ the Type I error (see Chapter 1);

$\beta = $ the Type II error (see Chapter 1);

$\mu_1 - \mu_0 = $ the average difference between identical samples to be denoted by $d$;

$\sum_i^n d_i = $ the sum of the differences between samples for $n$ test evaluations.

It can be shown that the cumulative value of $\sum_i^n d_i$ when plotted against $n$ can be used as a basis for a sequential decision since the boundary lines of $\sum_i^n d_i$ are given by the inequalities in (2.6-6). In $n$ repeated evaluations, a truly good judge would have

$$\sum_i^n d_i = 0$$

and as a consequence of this, the inequalities

$$\sum_i^n d_i > 0$$

and

$$\sum_i^n d_i < 0$$

represent rejection regions of the opposite direction. To accommodate both inequalities, we need a double-sided sequential test. This test is obtained from the relation $L_0 = -L_0$ and $L_1 = -L_1$. From this relation, (2.6-6) can be written in a form of regression line where the slope $S = \mu_1 + \mu_0/2$:

$$\sum_{i}^{n} d_i > 0 : L_0 = \frac{\sigma_d^2}{d} \log \frac{\beta}{1 - \alpha/2} + n(S) \qquad (2.6\text{-}7)$$

$$L_1 = \frac{\sigma_d^2}{d} \log \frac{1 - \beta}{\alpha/2} + n(S)$$

$$\sum_{i}^{n} d_i < 0 : L_0' = -\frac{\sigma_d^2}{d} \log \frac{\beta}{1 - \alpha/2} - n(S) \qquad (2.6\text{-}8)$$

$$L_1' = \frac{\sigma_d^2}{d} \log \frac{1 - \beta}{\alpha/2} - n(S)$$

The steps in sequential panel selection are as follows:

1. A prospective judge, after undergoing instructions on sensory terminologies and definitions, is given $n$ paired difference tests using identical samples.
2. Decide on the values of $\alpha$, $\beta$, $S$, and $\sigma_d^2$. Based on these preassigned values, compute the regression equations using (2.6-7) and (2.6-8). Construct the boundary lines with $\sum_{i}^{n} d_i$ as the ordinate and $n$ as the abscissa.
3. If the cumulative value of the differences, $\sum_{i}^{n} d_i$, lies between $L_0$ and $L_1$ (region of indecision) or between $L_0'$ and $L_1'$ (region of rejection), the candidate judge is rejected.
4. If $\sum_{i}^{n} d_i$ is above $L_1$ or below $L_1'$ (region of rejection), the candidate judge is rejected.
5. If $\sum_{i}^{n} d_i$ is between $L_0$ and below $L_0'$ (region of acceptance), the candidate judge is accepted as a member of the panel. The plot pertaining to Steps 2–5 is illustrated in Example 2.6-3.

---

**Example 2.6-3**

An anti-aging product was tested for the purpose of screening panel members. The same product was applied on the right (R) and left (L) side of the area below the eye. It is expected that $d = R - L = 0$. A fine line under the eye was the sensory attribute rated using the following scale: 1 = none, 2 = very slight, 3 = slight, 4 = moderate, 5 = moderately severe, 6 = severe, 7 = very severe. The results for two judge candidates are shown in Table 2.13. The construction of the plot used the following preassigned values:

$$\sigma_d^2 = 2.25 \qquad \alpha = 0.10 \qquad \beta = 0.20 \qquad d = 1.0 \qquad S = d/2 = 0.5$$

The cumulative value, $\sum_{i=1}^{n} d_i$, for both candidates also is plotted in this graph. Using candidate 1 as an example, the cumulative value is obtained as follows. For $n = 2$,

**Table 2.13** Calculation of Plotting Coordinates $\left( n, \sum_{i=1}^{n} d_i \right)$ in a Variable Sequential Panel Selection

| Candidate | Subject | R | L | $d_i = R - L$ | Sum of $d_i$ |
|-----------|---------|---|---|---------------|--------------|
| 1 | 1 | 6 | 7 | −1 | −1 |
| | 2 | 7 | 6 | 1 | 0 |
| | 3 | 5 | 4 | 1 | 1 |
| | 4 | 5 | 5 | 0 | 1 |
| | 5 | 7 | 6 | 1 | 2 |
| | 6 | 6 | 5 | 1 | 3 |
| | 7 | 6 | 5 | 1 | 4 |
| | 8 | 7 | 7 | 0 | 4 |
| | 9 | 6 | 6 | 0 | 4 |
| | 10 | 7 | 6 | 1 | 5 |
| 2 | 1 | 4 | 3 | 1 | 1 |
| | 2 | 4 | 4 | 0 | 1 |
| | 3 | 3 | 2 | 1 | 2 |
| | 4 | 4 | 5 | −1 | 1 |
| | 5 | 5 | 5 | 0 | 1 |
| | 6 | 4 | 3 | 1 | 2 |
| | 7 | 4 | 4 | 0 | 2 |
| | 8 | 5 | 5 | 0 | 2 |
| | 9 | 3 | 3 | 0 | 2 |
| | 10 | 4 | 5 | −1 | 1 |

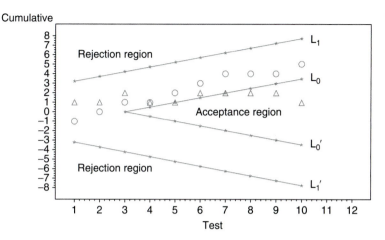

$$\sum_{i=1}^{2} d_i = d_1 + d_2 = -1 + 1 = 0$$

and the plotting coordinate is (0,2). For $n = 4$,

$$\sum_{i=1}^{4} d_i = d_1 + d_2 + d_3 + d_4 = -1 + 1 + 1 + 0 = 1$$

and the coordinate is (1,4). These coordinates can be verified in Figure 2.21.

After 7 tests, candidate 2 is accepted as a panel member since the plot of $\sum_{i=1}^{7} d_i$ falls in the region of acceptance. The plot for candidate 1 falls in the region of indecision; therefore, more sensory testing is required. In practice, occasional testing is encouraged to update an individual's record. In so doing, the maximum records of an individual judge are fully utilized.

Table 2.14 contains the SAS program to produce Figure 2.21. In the INPUT statement, *cumulative* is the data for candidate 1 and *cumulative 1* is the data for candidate 2, which can be found as the last column in Table 2.13. The regression equations derived from (2.6-7) and (2.6-8) are shown after the INPUT statement.

**Table 2.14** SAS Program for Sequential Analysis of Rating Scale Data

```
*prog sequential analysis plot ex 2.6-3.sas;
options nodate;
data a;
input Test Cumulative Cumulative1 n;

y1=-1.5226 + 0.5*n;
y2=2.7093 + 0.5*test;
y3=1.5226 - 0.5*n;
y4=-2.7093 - 0.5*test;

cards;
1  -1  1  .
2   0  1  .
3   1  2  3
4   1  1  4
5   2  1  5
6   3  2  6
7   4  2  7
8   4  2  8
9   4  2  9
10  5  1  10
;
goptions reset=global gunit=pct
     ftext=swissb htitle=4 htext=4;

     symbol1 color=red value=circle        height=5;
     symbol2 color=red value=triangle      height=5;
     symbol3 color=red value=star i=join   height=1;
     symbol4 color=red value=star i=join   height=1;
     symbol5 color=red value=star i=join   height=1;
     symbol6 color=red value=star i=join   height=1;

     axis1 order=(-8 to 8 by 1)
     offset=(5)
     width=3;

     axis2 order=(1 to 12 by 1)
     offset=(5)
     width=3;

proc gplot data=a;
     plot Cumulative*Test Cumulative1*Test y1*Test y2*Test
       y3*Test y4*Test/overlay
     vaxis=axis1
     haxis=axis2;
     run;
```

## Quality Control Chart for Degree of Difference

In this section, we discuss the degree of difference from the target sample using the intensity scale and analyzed using the quality control chart. The key information in this method of evaluating panelist or judge performance is the target sample. The target sample is the basis for determining how far away each judge rated the test sample from the target. The target sample is a reference standard with known intensity for a particular attribute. Attributes for aroma, flavor, texture, and appearance are known in descriptive analysis protocol and thoroughly discussed in books by Meilgaard et al. (1999) and Lawless and Heymann (1998).

The hypotheses for degree of difference are as follows:

| | |
|---|---|
| Null hypothesis: | $H_0$: $\mu_1 - \mu_2 = 0$ |
| Alternative hypothesis: | $H_a$: $\mu_1 - \mu_2 \neq 0$ |

where $\mu_1$ is the mean of the sample to be evaluated and $\mu_2$ is the mean of the target reference sample. In theory, if the judges are well trained, the null hypothesis will be accepted. But because of sampling error, upper and lower confidence levels of the difference are estimated to provide judge acceptance ranges. The sampling error of the difference denoted by $S^2$ is obtained by pooling the variance of each judge; that is,

$$S_p^2 = \frac{S_1^2(n_1 - 1) + S_2^2(n_2 - 1) + .. + S_p^2(n_p - 1)}{(n_1 + n_2 + .. + n_p) - n_p} \qquad (2.6\text{-}9)$$

where $S_p^2$ is the pooled variance of judges for all products and $n_p$ is the number of products. If the number of judges is equal for each product evaluation, (2.6-9) reduces to

$$S_p^2 = (S_1^2 + S_2^2 + \cdots + S_p^2)/np \qquad (2.6\text{-}10)$$

The pooled standard deviation is

$$S_p = \sqrt{S_p^2}. \qquad (2.6\text{-}11)$$

Using (2.6-11), the 95% confidence limits (CL) that represent all judges is

$$CL = \pm 1.96(S_p) \qquad (2.6\text{-}12)$$

where 1.96 is the $t$-value in Table A.2 at infinity degrees of freedom.

Equation (2.6-12) is a formula the sensory analyst can easily implement by computing the mean and standard deviation of the data. This will be illustrated in Example 2.6-4.

| **Example 2.6-4** | In this example 9 products plus the reference standard were evaluated by 5 judges using a 15-point descriptive analysis rating scale. The sweetness of the samples was evaluated such that 1 = not sweet enough, 9 = too sweet. The sensory data are shown in Table 2.15 where p1 to p9 are products and Ref is the reference standard; a sample calculation is shown at the bottom part of this table. The SAS code for the analysis is given in Table 2.16. |

The next step in the analysis is to plot the differences from reference standards (Table 2.17) for each subject with the 95% upper and lower limits indicated in the plot. Figure 2.22 shows the plot for products p1–p3. Ratings given by judge 5 for p1 and p2 were outside the lower limit of the confidence interval. For products p4–p6, judges 2 and 5 rated product 6 above the upper limit (Figure 2.23). For products p6–p9 all the judges were in full agreement, since all product degree of difference ratings were inside the 95% confidence limits. The interaction between judges and differences from target should be examined by connecting the same symbols in Figures 2.22 to 2.24 to form a line. The crossing of the lines indicates the presence of interaction, which suggests

**Table 2.15** Sensory Data for 9 Products (p1–p9) with a Reference (Ref) Standard for 5 Judges

| Judge | p1 | p2 | p3 | p4 | p5 | p6 | p7 | p8 | p9 | Ref |
|-------|----|----|----|----|----|----|----|----|----|-----|
| 1 | 4 | 4 | 5 | 4 | 5 | 6 | 4 | 5 | 4 | 5 |
| 2 | 4 | 5 | 4 | 4 | 4 | 7 | 4 | 5 | 5 | 5 |
| 3 | 4 | 4 | 5 | 5 | 5 | 6 | 5 | 4 | 6 | 5 |
| 4 | 6 | 5 | 5 | 4 | 5 | 5 | 4 | 5 | 4 | 5 |
| 5 | 3 | 3 | 5 | 4 | 4 | 7 | 4 | 5 | 5 | 5 |

Calculation of differences from reference for 5 judges denoted by d1 – d5 for product 1. The details are shown in Table 2.17.
Differences, $d1 = 4{-}5 = -1$ $d2 = 4{-}5 = -1$ $d3 = 4{-}5 = -1$ $d4 = 6{-}5 = 1$ $d5 = 3{-}5 = -2$

Mean, $M1 = \sum_{1}^{5}(d1 + d2 + d3 + d4 + d5)/5 = -0.80$

Product p1(DiffA) variance, $S_1^2 = \sum_{i=1}^{5}(d_i - (-0.80))^2/5 - 1 = 1.20$

Pooled variance, $S_p^2 = (1.20 + 0.70 + 0.20 + 0.20 + 0.30 + 0.70 + 0.20 + 0.20 + 0.70\ )\ /\ 9 = 0.4889$

$S_p = \sqrt{0.4889} = 0.6992$

95% CL $= 1.96(0.6992) = \pm1.40$, where 1.96 is the t-value at infinity degrees of freedom (see Table A.2).

**Table 2.16** SAS Program for Example 2.6-4

```
*prog example 2.6-4 panel selection.sas;
options nodate;
data a;
input Judge x1-x10;
DiffA=x1-x10;
d2=x2-x10;
d3=x3-x10;
DiffB=x4-x10;
d5=x5-x10;
d6=x6-x10;
DiffC=x7-x10;
d8=x8-x10;
d9=x9-x10;

cards;
1 4 4 5 4 5 6 4 5 4 5
2 4 5 4 4 4 7 4 5 5 5
3 4 4 5 5 5 6 5 4 6 5
4 6 5 5 4 5 5 4 5 4 5
5 3 3 5 4 4 7 4 5 5 5
;
proc print data=a;
var judge diffA d2-d3 diffB d5-d6 diffC d8-d9;
title 'Degree of difference Target=x10';
run;

proc means mean n std var maxdec=2;
var diffA d2-d3 diffB d5-d6 diffC d8-d9;
title 'Degree of difference Target=x10';
run;

goptions reset=global gunit=pct
    ftext=swissb htitle=5 htext=5;
    symbol1 color=red value=dot    height=5;
    symbol2 color=red value=x      height=5;
    symbol3 color=red value=square height=5;

    axis1 order=(-3 to 3 by 1)
    offset=(5)
    width=3;

    axis2 order=(1 to 5 by 1)
    offset=(5)
    width=3;
proc gplot data=a;
    plot diffA*judge d2*judge d3*judge /overlay
    vaxis=axis1
    haxis=axis2
    vref=-1.4 1.4
    lvref=2;
    run;
```

*(continued)*

**Table 2.16** SAS Program for Example 2.6-4—*cont...*

```
proc gplot data=a;
    plot diffB*judge d5*judge d6*judge /overlay
    vaxis=axis1
    haxis=axis2
    vref=-1.4 1.4
    lvref=2;
    run;

proc gplot data=a;
    plot diffC*judge d8*judge d9*judge /overlay
    vaxis=axis1
    haxis=axis2
    vref=-1.4 1.4
    lvref=2;
    run;
```

**Table 2.17** SAS Output for Example 6.2-4

| | | Degree of Difference from Target Ref | | | | | | | | |
|---|---|---|---|---|---|---|---|---|---|---|
| | | **Diff** | | | **Diff** | | | **Diff** | | |
| Obs | Judge | A | d2 | d3 | B | d5 | d6 | C | d8 | d9 |
| 1 | 1 | −1 | −1 | 0 | −1 | 0 | 1 | −1 | 0 | −1 |
| 2 | 2 | −1 | 0 | −1 | −1 | −1 | 2 | −1 | 0 | 0 |
| 3 | 3 | −1 | −1 | 0 | 0 | 0 | 1 | 0 | −1 | 1 |
| 4 | 4 | 1 | 0 | 0 | −1 | 0 | 0 | −1 | 0 | −1 |
| 5 | 5 | −2 | −2 | 0 | −1 | −1 | 2 | −1 | 0 | 0 |

| The MEANS Procedure | | | | |
|---|---|---|---|---|
| Variable | Mean | N | Std Dev | Variance |
| DiffA | −0.80 | 5 | 1.10 | 1.20 |
| d2 | −0.80 | 5 | 0.84 | 0.70 |
| d3 | −0.20 | 5 | 0.45 | 0.20 |
| DiffB | −0.80 | 5 | 0.45 | 0.20 |
| d5 | −0.40 | 5 | 0.55 | 0.30 |
| d6 | 1.20 | 5 | 0.84 | 0.70 |
| DiffC | −0.80 | 5 | 0.45 | 0.20 |
| d8 | −0.20 | 5 | 0.45 | 0.20 |
| d9 | −0.20 | 5 | 0.84 | 0.70 |

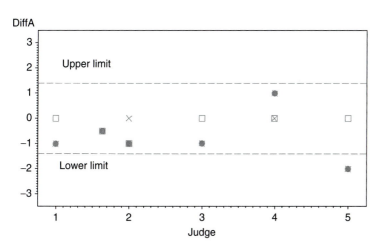

■ **FIGURE 2.22** Plot of differences from target sample by judges and the 95% lower and upper limits (●, product 1; x, product 2; □, product 3).

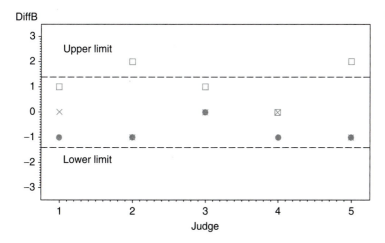

■ **FIGURE 2.23** Plot of differences from target sample by judges and the 95% lower and upper limits (●, product 4; x, product 5; □, product 6).

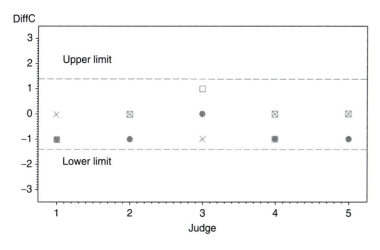

■ **FIGURE 2.24** Plot of differences from target sample by judges and the 95% lower and upper limits (●, product 7; x, product 8; □, product 9).

disagreement between judges. In practice, it is recommended that the judges should perform the task of forming the lines in the plot. Then the sensory analyst can discuss the plots with the judges to determine the next step in the training process.

## EXERCISES

**2.1.** Discuss the basic differences and relationships, if any, between the two psychophysical laws advanced by Fechner and Stevens.

**2.2.** How is an interval scale different from a ratio scale?

**2.3.** The frequency distribution of scale categories for a 7-point off-flavor scale follows:

| Category | Frequency |
|---|---|
| 1 (none) | 75 |
| 2 | 86 |
| 3 | 93 |
| 4 | 74 |
| 5 | 60 |
| 6 | 36 |
| 7 (very strong) | 21 |

1. What is the rational origin of the foregoing 7-point scale?
2. Scale the categories using Guilford's method and interpret the results. Also find Guilford's adjusted scores.
3. Determine the mean and the standard deviation of the estimated category scale.

**2.4.** The difference between the 84th percentile and 50th percentile is an estimate of the standard deviation. Justify this using the fact that 68.26% of the normally distributed population is within $\pm 1$ standard deviation from the mean.

**2.5.** The following data are taken from a psychophysical experiment. The response for each concentration is the average of the responses of 34 panelists.

| Concentration | Response |
|---|---|
| 0.014 | 2.11 |
| 0.058 | 2.75 |
| 0.220 | 4.46 |
| 0.890 | 8.51 |
| 3.570 | 10.74 |

1. Graphically plot the data using Fechner and Stevens' psychophysical laws. Interpret your plots.
2. Use Eq. (2.1-6) to estimate $\beta$, the slope of the relationship, by plotting log $S$ versus log $R$.

**2.6.** The relative frequency distribution of an 8-point hedonic scale is as follows:

| Category | Frequency |
|---|---|
| 1 | 1 |
| 2 | 3 |
| 3 | 5 |
| 4 | 13 |
| 5 | 32 |
| 6 | 25 |
| 7 | 12 |
| 8 | 0 |

1. Transform the data to a suitable scale and plot results on normal probability graph paper.
2. Scale the categories following Guilford's scaling method. Then find Guilford's adjusted scale values.

**2.7.** Using Example 2.6-3, do a complete sequential analysis using $\alpha = 0.05$ and $\beta = 0.30$. Then discuss the results.

**2.8.** Compute the variances for judges 2–5 on the data given in Table 2.15.

**2.9.** Discuss the interaction between judge and difference from target for Example 2.6-4.

# The Analysis of Variance and Multiple Comparison Tests

Chapter 1 dealt with a number of basic statistical concepts and methods, including rules for testing hypotheses about the mean of a single population. In many practical situations, interest would be not only in testing the equality of means of two or more populations, but also in other hypotheses. The technique of the analysis of variance is an important statistical tool for testing various interesting hypotheses. Details of the analysis of variance technique rest with the type of experimental design used. We will encounter various experimental designs and their analyses in Chapters 4–13. Here, we discuss the analysis of variance technique to understand basic concepts.

## 3.1 ANALYSIS OF VARIANCE

A procedure for partitioning the total variation in observed data into various components and assigning them to respective causes is called the *analysis of variance*. The source of each component variation is identified and tested for its significance as a source of variation in data. A statistical model incorporating the nature of the study is developed to identify sources of variation. For instance, let us consider $a$ treatments, which are to be compared on the basis of responses $X_{ij}$, where $X_{ij}$ is the $j$th response for the $i$th treatment. A statistical model for this type of study may be written as

$$X_{ij} = \mu + T_i + \varepsilon_{ij}, \qquad \begin{array}{l} i = 1, \ldots, a, \\ j = 1, \ldots, n_i, \end{array} \qquad (3.1\text{-}1)$$

where $\mu$ is the grand mean response estimated by $\overline{X} = \Sigma_i \Sigma_j X_{ij}/N$, $N = \Sigma_i n_i$; $T_i$ is the effect of the $i$th treatment, which is estimated by $t_i = \overline{X}_i - \overline{X}$ with $\Sigma_i t_i = 0$; and $\varepsilon_{ij}$ are the random errors, which are assumed to be independently and normally distributed with mean 0 and variance $\sigma^2$.

Clearly, according to Model (3.1-1) each observation $X_{ij}$ is made up of the grand mean effect $\mu$, a treatment effect $T_i$, and a random error effect $\varepsilon_{ij}$. Therefore, the total variation among $X_{ij}$'s is partitioned into three components, one for each component effect. The total variation and its components are called the *sum of squares* (SS) in the analysis of variance terminology. Thus, we have the total sum of squares partitioned into the sum of squares due to treatments (SStr), the sum of squares due to error (SSe), and the sum of squares due to the grand mean. Usually, the total sum of squares is adjusted for the sum of squares due to the grand mean. This is why analysis of variance tables do not list the sum of squares due to the grand mean separately. The total sum of squares adjusted for the grand mean is denoted by SSto and is defined as

$$\text{SSto} = \sum\sum(X_{ij} - \overline{X})^2 = \sum\sum X_{ij}^2 - \text{CF}, \qquad (3.1\text{-}2)$$

where CF, known as the correction factor or the sum of squares due to the grand mean, is $X^2../N$.

The treatment sum of squares is the weighted sum of squared deviations of sample treatment means from the grand mean. That is,

$$\text{SStr} = \sum_i n_i(\overline{X}_{i\cdot} - \overline{X})^2 = \sum_i n_i \overline{X}_{i\cdot}^2 - \text{CF}. \qquad (3.1\text{-}3)$$

The sum of squares due to error or the error sum of squares is defined as

$$\text{SSe} = \sum\sum(X_{ij} - \overline{X}_{i\cdot})^2. \qquad (3.1\text{-}4)$$

It is true that SSto = SStr + SSe. Definitions of the sum of squares for other statistical models with many sources of variation are appropriately extended.

The analysis of variance table for this model is as shown in Table 3.1. The DF column in this table is known as the *degrees of freedom* for the respective SS; the MS column has the *mean squares*. An SS divided by its DF is called the mean square. The MS is an estimate of the variance contributed by its source to the total. A test statistic for testing the equality of treatment effects is the $F$ ratio MStr/MSe. The observed $F$ ratio is

**Table 3.1** One-Way Analysis of Variance of Model (3.1-1)

| Source of Variance | DF | SS | MS | F Ratio |
|---|---|---|---|---|
| Total | $N - 1$ | SSto [see (3.1-2)] | | |
| Between Treatments | $a - 1$ | SStr [see (3.1-3)] | MStr | MStr/MSe |
| Within Treatments | $N - a$ | SSe [see (3.1-4)] | MSe | |

compared with percentiles of the $F$ distribution in Table A.4. The null hypothesis of no treatment differences is rejected if the observed $F$ ratio is greater than the tabled $F$ at the desired significance level. A rationale for using the $F$ ratio statistic is given later in this section.

**Example 3.1-1**

The effects of aging and the injection of a tenderizing enzyme into beef muscle on meat tenderness as measured by the Warner–Bratzler shearing device were investigated. A portion of the data of this experiment is shown in Table 3.2. The measurements are in pounds of pressure per square inch. The SSto, SStr, and SSe were obtained using (3.1-2), (3.1-3), and (3.1-4), respectively. The analysis of variance in Table 3.3 shows that the $F$ ratio is $48.577/7.459 = 6.51$ with 3 and 16 DF. This $F$ ratio is higher than the 99th percentile of the $F$ distribution from Table A.4 at the 1% level of significance. Therefore, we reject the null hypothesis that the four treated muscles have equal tenderness values.

A two-way analysis of variance is an extension of the one-way analysis of variance and is given in Table 3.4. The total variation adjusted for the

**Table 3.2** Tenderness Readings of Loin Longissimus Dorsi Muscle Corresponding to Four Treatments (Example 3.1-1)

|  | Treatments | | | |  |
| --- | --- | --- | --- | --- | --- |
|  | **1** | **2** | **3** | **4** |  |
|  | 14.2 | 10.4 | 10.1 | 23.4 |  |
|  | 11.3 | 7.8 | 11.4 | 11.2 |  |
|  | 11.8 | 12.5 | 11.4 | 17.6 |  |
|  | 16.9 | 10.2 | 9.6 | 15.6 |  |
|  | 12.2 | 11.0 | 7.8 | 16.0 |  |
| Treatment Totals | 66.4 | 51.9 | 50.3 | 83.8 | $X.. = 252.4$ |
| Treatment Means | 13.3 | 10.4 | 10.1 | 16.8 | $\bar{X} = 12.62$ |

*Note:*

$SSto = [(14.2)^2 + (11.3)^2 + \cdots + (15.6)^2 + (16.0)^2] - \dfrac{(252.4)^2}{20} = 265.072.$

$SStr = \dfrac{[(66.4)^2 + (51.9)^2 + (50.3)^2 + (83.8)^2]}{5} - \dfrac{(252.4)^2}{20} = 145.732.$

$SSe = 265.072 - 145.732 = 119.340.$

**Table 3.3** Analysis of Variance of Beef Tenderness Data in Table 3.2

| Source of Variance | DF | SS | MS | F Ratio |
|---|---|---|---|---|
| Total | 19 | 265.072 | | |
| Treatments | 3 | 145.732 | 48.577 | 6.51* |
| Error | 16 | 119.340 | 7.459 | |
| *p < .01. | | | | |

**Table 3.4** Two-Way Analysis of Variance[a]

| Source of Variance | DF | SS |
|---|---|---|
| Total | $abr - 1$ | $\text{SSto} = \sum_i \sum_j \sum_k (X_{ijk} - \bar{X})^2 = \sum_i \sum_j \sum_k X_{ijk}^2 - \text{CF}$ |
| A | $a - 1$ | $\text{SSa} = br \sum_i (\bar{X}_{i..} - \bar{X})^2 = \dfrac{\sum_i (X_{i..})^2}{br} - \text{CF}$ |
| B | $b - 1$ | $\text{SSb} = ar \sum_j (\bar{X}_{.j.} - \bar{X})^2 = \dfrac{\sum_i (X_{.j.})^2}{ar} - \text{CF}$ |
| AB | $(a - 1)(b - 1)$ | $\text{SSab} = r \sum_i \sum_j (\bar{X}_{ij.} - \bar{X}_{i..} - \bar{X}_{.j.} + \bar{X})^2$ $= \dfrac{\sum_i \sum_j (X_{ij.})^2}{r} - \text{CF} - \text{SSa} - \text{SSb}$ |
| Error | $ab(r - 1)$ | $\text{SSe} = \sum_i \sum_j \sum_k (X_{ijk} - \bar{X}_{ij.})^2$ $= \text{SSto} - \text{SSa} - \text{SSb} - \text{SSab}$ |

[a]Model: $\mu + A_i + B_j + (AB)_{ij} + \varepsilon_{ijk}$, $i = 1, \ldots, a$; $j = 1, \ldots, b$; $k = 1, \ldots, r$.

mean is partitioned into SS due to sources of variation, which are factor $A$ at $a$ levels, factor $B$ at $b$ levels, their interaction factor $AB$ at $ab$ levels, and the error factor. The effects of sources $A$ and $B$ are known as *main effects*. Likewise, an $n$-way analysis of variance would consist of $n$ main effects and several interaction effects.

## Assumptions in the Analysis of Variance

The analysis of variance as a technique for statistical inference requires certain assumptions to be theoretically valid. We summarize these assumptions briefly; see Cochran (1947) and Eisenhart (1947) for details. For theoretical validity of the analysis of variance in a parametric setup, one assumes that observations follow the normal distribution and that the error terms are independently distributed with mean 0 and common variance $\sigma_e^2$.

Let us consider the following model:

$$X_{ij} = \mu + A_i + B_j + \varepsilon_{ij}, \qquad \begin{aligned} i &= 1, \ldots, a, \\ j &= 1, \ldots, b, \end{aligned} \qquad (3.1\text{-}5)$$

where $\mu$ is the grand mean estimated by $\overline{X}$, $A_i$ is the effect of the $i$th level of factor $A$ estimated by $a_i = \overline{X}_{i\cdot} - \overline{X}$ with $\Sigma a_i = 0$, $B_j$ is the effect of the $j$th level of factor $B$ estimated by $b_j = \overline{X}_{\cdot j} - \overline{X}$ with $\Sigma b_j = 0$, and $\varepsilon_{ij}$ are the random errors assumed to be independently and normally distributed with mean 0 and variance $\sigma_e^2$. Hence, observations $X_{ij}$ are mutually independent and normally distributed with mean $(\mu + A_i + B_j)$ and variance $\sigma_e^2$. A departure from normality affects statistical tests through reduced power of the $F$ tests. However, the nonnormality should be extreme before the robustness of the $F$ tests is severely affected. The additivity assumption is that the interaction of factors $A$ and $B$ is negligible. If the additivity holds, the difference between any two levels of $A$ is independent of the levels of $B$, and vice versa. For example, the expected difference between the first and second level of $A$ under the additivity assumption is $E(X_{1j}) - E(X_{2j}) = A_1 - A_2$, which is independent of the levels of $B$. Should the nonadditivity exist and the model be rewritten to account for it, the difference $E(X_{1j}) - E(X_{2j})$ will depend on the $j$th level of $B$. Nonadditivity when it exists and is not accounted for will bias the estimation of $\sigma_e^2$. Sometimes additivity can be achieved by a suitable transformation of the data. For instance, microbial counts are usually transformed into logarithmic scale to achieve additivity. If either the assumption of independence or the common variance, or both, is not valid, there will be a loss of sensitivity in tests of significance. With correlated observations the variance of $\overline{X}_{i\cdot}$ will tend to be biased. For example, if $X_{ij}$ in Model (3.1-5) is correlated with correlation $\rho$, then the variance of $\overline{X}_{i\cdot}$ is the quantity $\sigma_e^2 \{1 + (b - 1)\rho\}$ divided by $b$ instead of merely $\sigma_e^2/b$.

## Fixed- and Random-Effects Models

An analysis of variance model is classified into one of the three types: the fixed-effects model (Model I), the random-effects model (Model II), and the mixed-effects model (Model III). In a fixed-effects model, the interest is only in those treatments that are being used in the experiment and their effects. Hence, all conclusions from a fixed-effects model are limited to only those treatments that are being observed in the experiment. If, on the other hand, treatments in an experiment are considered to be randomly selected from a population of treatments, then the analysis of variance would follow a random-effects model. The effects are considered random variables. The variability among all treatments of the population is of interest rather than only the randomly selected set of treatments. In a mixed-effects model, as the name suggests, some effects are treated as random variables, whereas others are considered fixed.

The main differences and similarities between Models I and II are summarized in Table 3.5 for a one-way analysis of variance model. The partitioning of the total SS into components due to various effects for both Models I and II is identical. But interpretations of statistical analyses are different for the two models. Consider a one-way model given in Table 3.5, where $A_i$ may be the effect due to the $i$th instrument used to measure tenderness. If there are 5 instruments in the laboratory and our interest is in only these 5 instruments, then the instrument effects are fixed effects. The instrument effects become random effects if the 5 instruments were selected at random from a population of, say, 50 instruments in a laboratory and an estimate of the variability among instruments is sought on the basis of a sample of 5.

The following null and alternative hypotheses are of interest for a fixed-effects model:

$$H_0: \quad A_1 = A_2 = \cdots = A_5,$$

**Table 3.5** Fixed- Versus Random-Effects Model for One-Way Classification

| Basis for Comparison | Model I (Fixed Effects) | Model II (Random Effects) |
|---|---|---|
| Statistical model | $X_{ij} = \mu + A_i + \varepsilon_{ij}$, $\Sigma A_i = 0$ | $X_{ij} = \mu + A_i + \varepsilon_{ij}$ |
| Assumptions | Effects $A_i$ are fixed; $\mu$ is the grand mean considered fixed; $\varepsilon_{ij}$ are independent and normally distributed | Effects $A_i$ are random with variance $\sigma_a^2$; $\mu$ is the grand mean considered fixed; $A_i$ and $\varepsilon_{ij}$ are independent and normally distributed |
| Estimation of interests | Fixed effects $A_i$ and linear contrasts among $A_i$ | Variance components $\sigma_a^2$ |
| Hypotheses | $H_0: A_i = A_j$ for all pairs $(i, j)$ | $H_0: \sigma_a^2 = 0$ |
| | $H_a: A_i \neq A_j$ for at least one pair $(i, j)$ | $H_a: \sigma_a^2 > 0$ |
| Analysis and interpretation | The inferences are limited to the fixed effects of treatments used in the experiment | The inferences are extended to the population of treatments |

against the alternative hypothesis

$H_a$ : At least one pair of effects differs significantly.

The null hypothesis for the random-effects model is

$$H_0 : \quad \sigma_a^2 = 0,$$

where $\sigma_a^2$ is the variance of the population of all instrument effects against the alternative

$$H_a : \quad \sigma_a^2 > 0.$$

Tests of significance of the preceding hypotheses can be rationalized following expressions of the expected mean squares. There are many papers outlining rules of thumb for writing down expressions for the expected mean squares in the analysis of variance. We may, for example, refer to papers by Schultz (1955), Henderson (1959), Lentner (1965), and Millman (1967). With appropriate simplifying modifications for corresponding designs, the rules may be summarized as follows:

1. Regardless of the model being fixed or random effects, the expected mean square $E(MSe)$ associated with the error component is always $\sigma_e^2$.

2. For the fixed-effects models, the expected mean square $E(MS)$ of any source is $\sigma_e^2$ plus a fixed-effects variance component of the source. The fixed-effects variance component of a source is obtained if observations in its mean square expression are replaced by fixed effects parameters of the source.

3. For random-effects models, the expected mean square $E(MS)$ of a source is $\sigma_e^2$ plus
   a. Variance component of the random effect of the source multiplied by the number of observations for all levels of the source;
   b. Sum of the variance components of interaction effects involving the source; the variance component of each interaction effect in the sum is multiplied by the number of observations corresponding to the interaction effect.

For illustration of these rules, we consider a balanced design with two sources $A$ and $B$. The design is called balanced when, for each combination of levels of $A$ and $B$, an equal number of observations is taken. For unbalanced designs, we may consult Kempthorne (1952),

King and Henderson (1954), Harvey (1960), Goldsmith and Gaylor (1970), and Searle (1971). The model for our illustration is

$$X_{ijk} = \mu + A_i + B_j + (AB)_{ij} + \varepsilon_{ijk}, \qquad \begin{matrix} i = 1, \ldots, a, \\ j = 1, \ldots, b, \\ k = 1, \ldots, r, \end{matrix} \qquad (3.1\text{-}6)$$

where $A_i$ and $B_j$ denote the effects of the $i$th level of source $A$ and the $j$th level of source $B$, respectively. The interaction effect of the $i$th level of source $A$ and the $j$th level of source $B$ is $(AB)_{ij}$. These effects are fixed parameters if the model is a fixed-effects model with $\Sigma A_i = 0$, $\Sigma B_j = 0$, and $\Sigma (AB)_{ij} = 0$. The effects $A_i$, $B_j$, and $(AB)_{ij}$ are random effects if the model is a random-effects model with variance components $\sigma_a^2$, $\sigma_b^2$, and $\sigma_{ab}^2$, respectively. Regardless of the model type, $\mu$ is the grand mean, which is always treated as a fixed effect, whereas the $\varepsilon_{ijk}$ are random effects with variance $\sigma_e^2$. The algebraic expressions for various SSs are shown in an analysis of variance Table 3.4.

Following Rule 1, $E(\text{MSe}) = \sigma_e^2$. If it is a fixed-effects model, then the expected mean square $E(\text{MSa})$ for source $A$ is obtained following Rule 2 and is displayed in Table 3.6. Note from Table 3.4 that

$$\text{MSa} = \frac{br}{a-1} \sum_i (\bar{X}_{i..} - \bar{X})^2$$

$$= \frac{br}{a-1} \sum_i \left( \sum_j \sum_k \frac{X_{ijk}}{br} - \sum_i \sum_j \sum_k \frac{X_{ijk}}{abr} \right)^2.$$

**Table 3.6** A Two-Way Analysis of Variance Table with $E(\text{MS})$ for the Fixed-Effects Model (3.1-6)

| Source of Variance | DF | SS | MS | E(MS) |
|---|---|---|---|---|
| Total | $abr - 1$ | SSto | | |
| Source A | $a - 1$ | SSa | MSa | $\sigma_e^2 + \dfrac{rb}{a-1} \sum_i A_i^2$ |
| Source B | $b - 1$ | SSb | MSb | $\sigma_e^2 + \dfrac{ra}{b-1} \sum_j B_j^2$ |
| AB Interaction | $(a-1)(b-1)$ | SSab | MSab | $\sigma_e^2 + \dfrac{r}{(a-1)(b-1)} \sum_j \sum_j (AB)_{ij}^2$ |
| Error | $ab(r-1)$ | SSe | MSe | $\sigma_e^2$ |

Now using Rule 2, we have

$$E(MSa) = \sigma_e^2 + \frac{br}{a-1} \sum_i \left( \sum_j \sum_k \frac{A_i}{br} - \sum_i \sum_j \sum_k \frac{A_i}{abr} \right)^2.$$

Since $\sum_j \sum_k A_i = brA_i$ and $\sum_i \sum_j \sum_k A_i = 0$, then

$$E(MSa) = \sigma_e^2 + \frac{br}{a-1} \sum_i A_i^2.$$

Similarly,

$$E(MSb) = \sigma_e^2 + \frac{ar}{b-1} \sum_j B_j^2$$

and

$$E(MSab) = \sigma_a^2 + \frac{r}{(a-1)(b-1)} \sum_i \sum_j (AB)_{ij}^2.$$

For our fixed-effects model, hypotheses of interest are as follows. For testing the significance among levels of source $A$, the null hypothesis is

$$H_0: \quad A_1 = A_2 = \cdots = A_a$$

versus the alternative

$$H_a: \quad A_i \neq A_j \text{ for at least one pair } (i, j).$$

We note that the null hypothesis is equivalent to stating that $\sum A_i^2 = 0$. Hence, if indeed the null hypothesis is true, then $E(MSa) = E(MSe) = \sigma_e^2$. On the other hand, if the alternative is true, then $E(MSa)$ exceeds $E(MSe)$ by $[br/(a-1)] \sum A_i^2$. In view of these considerations, an appropriate test statistic for testing the significance of source $A$ is the $F$ ratio

$$F = MSa/MSe,$$

which is compared with the desired percentile of the $F$ distribution with $(a-1)$ numerator DF and $ab(r-1)$ denominator DF.

Similarly, for testing the significance of source $B$, we can formulate

$$H_0: \quad B_1 = B_2 = \cdots = B_b$$

against the alternative

$$H_a: \quad B_i \neq B_j \text{ for at least one pair } (i, j).$$

To test these hypotheses, one can use this appropriate test statistic

$$F = \text{MS}b/\text{MS}e,$$

which has the $F$ distribution with $(b - 1)$ numerator DF and $ab(r - 1)$ denominator DF. For testing the significance of interaction effects, one may formulate

$$H_0 : \quad (AB)_{ij} = 0 \text{ for all pairs } (i, j)$$

against the alternative

$$H_a : \quad (AB)_{ij} \neq 0 \text{ for at least one pair } (i, j).$$

The following $F$ ratio statistic is used:

$$F = \text{MS}ab/\text{MS}e,$$

with $(a - 1)(b - 1)$ numerator DF and $ab(r - 1)$ denominator DF.

If Model (3.1-6) is a random-effects model, then the expected mean square expressions can be obtained following Rule 3 and are given in Table 3.7. Suppose we want to test the significance of the interaction variance component $\sigma_{ab}^2$. That is,

$$H_0 : \quad \sigma_{ab}^2 = 0$$

**Table 3.7** A Two-Way Analysis of Variance Table with $E(\text{MS})$ If (3.1-6) Is a Random-Effects Model

| Source of Variance | DF | SS | MS | E(MS) |
|---|---|---|---|---|
| Total | $abr - 1$ | SSto | | |
| Source $A$ | $a - 1$ | SS$a$ | MS$a$ | $\sigma_e^2 + r\sigma_{ab}^2 + rb\sigma_a^2$ |
| Source $B$ | $b - 1$ | SS$b$ | MS$b$ | $\sigma_e^2 + r\sigma_{ab}^2 + ra\sigma_a^2$ |
| AB Interaction | $(a - 1)(b - 1)$ | SS$ab$ | MS$ab$ | $\sigma_e^2 + r\sigma_{ab}^2$ |
| Error | $ab(r - 1)$ | SS$e$ | MS$e$ | $\sigma_e^2$ |

against the alternative

$$H_a: \quad \sigma_{ab}^2 > 0.$$

If the null hypothesis is true, then $E(MSab)$ reduces to $E(MSe) = \sigma_e^2$. Hence, an appropriate test statistic is

$$F = MSab/MSe,$$

having the $F$ distribution with $(a - 1)(b - 1)$ numerator DF and $ab$ $(r - 1)$ denominator DF.

Suppose we wish to test the significance of the variance component of source $A$; we then formulate

$$H_0: \quad \sigma_a^2 = 0$$

against the alternative

$$H_a: \quad \sigma_a^2 > 0.$$

Under this null hypothesis, note from Table 3.7 that $E(MSa) = E(MSab)$. Hence, an appropriate test statistic is

$$F = MSa/MSab$$

having the $F$ distribution with $(a - 1)$ numerator DF and $(a - 1)(b - 1)$ denominator DF. Similarly, for testing

$$H_0: \quad \sigma_b^2 = 0$$

against the alternative

$$H_a: \quad \sigma_b^2 > 0,$$

we use

$$F = MSb/MSab,$$

which has the $F$ distribution with $(b - 1)$ numerator DF and $(a - 1)$ $(b - 1)$ denominator DF. The $F$ ratio statistics are judged to be significant for the rejection of their null hypotheses if the corresponding observed $F$ ratio is larger than the specified percentile of the $F$ distribution for the given level of significance.

To estimate variance components of a random-effects model, one may equate the expected mean square expressions with their respective observed mean square values. For example, if (3.1-6) is a random-effects model, then its variance components can be estimated by equating the MS column with the $E(MS)$ column of Table 3.7. The resulting set of equations for solving $\sigma_a^2$, $\sigma_b^2$, $\sigma_{ab}^2$, and $\sigma_e^2$ is given by (3.1-7). For our Model (3.1-6) the system of equations to be solved for $\sigma_a^2$, $\sigma_b^2$, $\sigma_{ab}^2$, and $\sigma_e^2$ is

$$
\begin{bmatrix}
\sigma_e^2 + r\sigma_{ab}^2 + rb\sigma_a^2 \\
\sigma_e^2 + r\sigma_{ab}^2 + ra\sigma_b^2 \\
\sigma_e^2 + r\sigma_{ab}^2 \\
\sigma_e^2
\end{bmatrix}
=
\begin{bmatrix}
MSa \\
MSb \\
MSab \\
MSe
\end{bmatrix}.
\tag{3.1-7}
$$

**Example 3.1-2**

Bar life loss for four brands of soap was studied using a mechanical bar life tester. Two technicians made three replications for each brand. The data are shown in Table 3.8, along with marginal totals and other information. Using the formulas given in Table 3.4, we obtain the SS for total, soap, technician, soap × technician interaction, and error as follows:

**Table 3.8** Percentage Bar Life Loss Data of Four Soap Brands (Example 3.1-2)

| Technicians | Soap Brands | | | | Totals |
|---|---|---|---|---|---|
| | **A** | **B** | **C** | **D** | **Totals** |
| 1 | 60.6 | 73.9 | 66.1 | 76.7 | 822.5 |
| | 59.0 | 74.2 | 64.4 | 75.6 | |
| | 59.2 | 73.8 | 65.0 | 74.0 | |
| 2 | 59.9 | 73.9 | 63.5 | 73.8 | 817.9 |
| | 61.6 | 74.0 | 64.9 | 73.0 | |
| | 62.0 | 72.3 | 63.0 | 76.0 | |
| Totals | 362.3 | 442.1 | 386.9 | 449.1 | 1640.4 |
| Means | 60.4 | 73.7 | 64.5 | 74.9 | |
| Brand by Technician Subtotals | | | | | |
| 1 | 178.8 | 221.9 | 195.5 | 226.3 | |
| 2 | 183.5 | 220.2 | 191.4 | 222.8 | |

$$CF = (1,640.4)^2/24 = 112,121.34,$$

$$\begin{aligned}
SSto &= [(60.6)^2 + (59.0)^2 + \cdots + (76.0)^2] - CF \\
&= 921.54,
\end{aligned}$$

$$\begin{aligned}
SSsoap &= [(362.3)^2 + (442.1)^2 + \cdots + (449.1)^2]/6 - CF \\
&= 894.68,
\end{aligned}$$

$$\begin{aligned}
SStech &= [(822.5)^2 + (817.9)^2]/12 - CF \\
&= 0.88,
\end{aligned}$$

$$\begin{aligned}
SSsoap \times tech &= [(178.8)^2 + (183.5)^2 + \cdots + (222.8)^2]/3 \\
&\quad -894.68 - 0.88 - CF \\
&= 8.12,
\end{aligned}$$

$$\begin{aligned}
SSe &= 921.54 - 894.68 - 0.88 - 8.12 \\
&= 17.85.
\end{aligned}$$

The analysis of variance is shown in Table 3.9. When the fixed-effects model is used, the $F$ ratio $298.23/1.86 = 160.33$ tests the null hypothesis that soap brands have equal bar life loss. Since 160.33 is far greater than the table value with 3 numerator and 16 denominator DF, we reject the null hypothesis. Table 3.10 corresponds to the analysis of variance for a random-effects model.

Let us also consider a three-factor cross-balanced design with a layout such as that in Figure 3.1. In the layout, levels $B_1$ and $B_2$ of factor $B$ appear with each level of factor $A$. Likewise, levels $C_1$ and $C_2$ of factor $C$ appear with each level of $B$. If the effect of a level of factor $B$ varies with the levels of factor $A$, the interaction effect between the two factors enters the statistical model. For three-factor cross-balanced designs, we may use the following model:

$$X_{ijkm} = \mu + A_i + B_j + C_k + (AB)_{ij} + (AC)_{ik} \\
+ (BC)_{jk} + (ABC)_{ijk} + \varepsilon_{ijkm}, \tag{3.1-8}$$

**Table 3.9** Analysis of Variance of Bar Life Loss Assuming Fixed-Effects Model

| Source | DF | SS | MS | F Ratio |
|---|---|---|---|---|
| Total | 23 | 921.54 | | |
| Soaps | 3 | 894.68 | 298.23 | 160.33 |
| Technicians | 1 | 0.88 | 0.88 | 0.47 |
| Soaps × Technicians | 3 | 8.12 | 2.71 | 1.46 |
| Error | 16 | 17.86 | 1.86 | |

**Table 3.10** Analysis of Variance of Bar Life Loss Assuming Random-Effects Model

| Source of Variance | DF | MS | F Ratio |
|---|---|---|---|
| Total | 23 | | |
| Soaps | 3 | 298.23 | 298.23/2.71 is the ratio for testing the soap variance component |
| Technicians | 1 | 0.88 | 0.88/2.71 is the ratio for testing the technician variance component |
| Soaps × Technicians | 3 | 2.71 | 2.71/1.86 is the ratio for testing the interaction variance component |
| Error | 16 | 1.86 | |

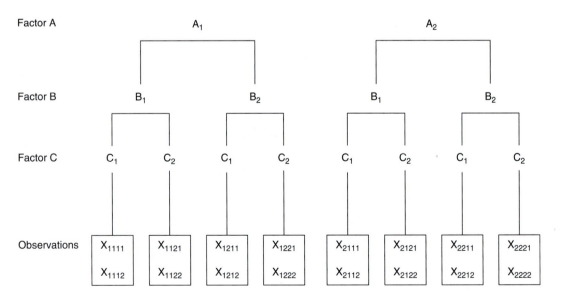

■ **FIGURE 3.1** Layout of a cross–balanced design.

where

$$i = 1, \ldots, a; \quad j = 1, \ldots, b; \quad k = 1, \ldots, c; \quad m = 1, \ldots, r;$$

and $X_{ijkm}$ is the observation in the $(ijkm)$th cell; $\mu$ is the grand mean estimated by $\bar{X}$; $A_i$ is the effect of the $i$th level of factor $A$ estimated by $a_i = \bar{X}_{i\ldots} - \bar{X}$ with $\Sigma a_i = 0$; $B_j$ is the effect of the $j$th level of factor $B$ estimated by $b_j = \bar{X}_{\cdot j \cdot \cdot} - \bar{X}$ with $\Sigma b_j = 0$; $C_k$ is the effect of the $k$th level of factor $C$ estimated by $c_k = \bar{X}_{\cdot \cdot k \cdot} - \bar{X}$ with $\Sigma c_k = 0$; $(AB)_{ij}$ is the interaction effect between $A$ and $B$ factors estimated by $(ab)_{ij} = \bar{X}_{ij\cdot\cdot} - \bar{X}_{i\ldots} - \bar{X}_{\cdot j\cdot\cdot} + \bar{X}$ with $\Sigma(ab)_{ij} = 0$; $(AC)_{ik}$ is the interaction effect between factors $A$ and $C$ estimated by $(ac)_{ik} = \bar{X}_{i\cdot k\cdot} - \bar{X}_{i\ldots} - \bar{X}_{\cdot\cdot k\cdot} + \bar{X}$ with $\Sigma(ac)_{ik} = 0$; $(BC)_{jk}$ is the interaction effect between factors $B$ and $C$ estimated by $(bc)_{jk} = \bar{X}_{\cdot jk\cdot} - \bar{X}_{\cdot j\cdot\cdot} - \bar{X}_{\cdot\cdot k\cdot} + \bar{X}$ with $\Sigma(bc)_{jk} = 0$; $(ABC)_{ijk}$ is the interaction effect between factors $A$, $B$, and $C$ estimated by $(abc)_{ijk} = \bar{X}_{ijk\cdot} + \bar{X}_{i\ldots} + \bar{X}_{\cdot j\cdot\cdot} + \bar{X}_{\cdot\cdot k\cdot} - \bar{X}_{ij\cdot\cdot} - \bar{X}_{i\cdot k\cdot} - \bar{X}_{\cdot jk\cdot} - \bar{X}$ with $\Sigma(abc)_{ijk} = 0$; and $\varepsilon_{ijkm}$ are random errors assumed to be independently and normally distributed with mean 0 and variance $\sigma_e^2$.

The analysis of variance of the three-factor cross-balanced design model is summarized in Table 3.11 and the $E(MS)$ expressions in Table 3.12. We note that $E(MS)$ of source $A$ includes all interaction variance components involving factor $A$:

$$E(MSa) = \sigma_e^2 + r\sigma_{abc}^2 + rc\sigma_{ab}^2 + rb\sigma_{ac}^2 + rbc\sigma_a^2.$$

**Table 3.11** Analysis of Variance of Three-Factor Cross-Balanced Design Model (3.1-8)

| Source of Variance | DF | SS |
|---|---|---|
| A | $a - 1$ | $SSa = (\Sigma_i X_i^2 \ldots /bcr) - CF$ |
| B | $b - 1$ | $SSb = (\Sigma_j X_{\cdot j}^2 \ldots /acr) - CF$ |
| C | $c - 1$ | $SSc = (\Sigma_k X_{\cdot\cdot k}^2 \cdot /abr) - CF$ |
| AB | $(a - 1)(b - 1)$ | $SSab = (\Sigma_i \Sigma_j X_{ij}^2 \cdot\cdot /cr) - SSa - SSb - CF$ |
| AC | $(a - 1)(c - 1)$ | $SSac = (\Sigma_i \Sigma_k X_{i\cdot k}^2 /br) - SSa - SSc - CF$ |
| BC | $(b - 1)(c - 1)$ | $SSbc = (\Sigma_j \Sigma_k X_{\cdot jk}^2 /ar) - SSb - SSc - CF$ |
| ABC | $(a - 1)(b - 1)(c - 1)$ | $SSabc = (\Sigma_i \Sigma_j \Sigma_k X_{ijk}^2 \cdot /r) - SSa - SSb - SSc - SSab - SSac - SSbc - CF$ |
| Error | $abc(r - 1)$ | $SSe = SSto - SSa - SSb - SSc - SSab - SSac - SSbc - SSabc - CF$ |
| Total | $abcr - 1$ | $SSto = \Sigma_i \Sigma_j \Sigma_k \Sigma \Sigma_m X_{ijkm}^2 - CF, \ CF = (X \ldots)^2 /abcr$ |

**Table 3.12** $E(MS)$ of Three-Way Cross-Balanced Design Model (3.1-8) for Random-Effects and Fixed-Effects Models

| Source of Variance | DF | MS | Random Effects $E(MS)$ | Fixed Effects $E(MS)$ |
|---|---|---|---|---|
| A | $a - 1$ | MSa | $\sigma_e^2 + r\sigma_{abc}^2 + rc\sigma_{ab}^2 + rb\sigma_{ac}^2 + rbc\sigma_a^2$ | $\sigma_e^2 + \dfrac{rbc}{a-1}\sum_i A_i^2$ |
| B | $b - 1$ | MSb | $\sigma_e^2 + r\sigma_{abc}^2 + rc\sigma_{ab}^2 + ra\sigma_{bc}^2 + rac\sigma_b^2$ | $\sigma_e^2 + \dfrac{rac}{b-1}\sum_j B_j^2$ |
| C | $c - 1$ | MSc | $\sigma_e^2 + r\sigma_{abc}^2 + rb\sigma_{ac}^2 + ra\sigma_{bc}^2 + rab\sigma_c^2$ | $\sigma_e^2 + \dfrac{rab}{c-1}\sum_k C_k^2$ |
| AB | $(a-1)(b-1)$ | MSab | $\sigma_e^2 + r\sigma_{abc}^2 + rc\sigma_{ab}^2$ | $\sigma_e^2 + \dfrac{rc}{(a-1)(b-1)}\sum_i\sum_j (AB)_{ij}^2$ |
| AC | $(a-1)(c-1)$ | MSac | $\sigma_e^2 + r\sigma_{abc}^2 + rc\sigma_{ac}^2$ | $\sigma_e^2 + \dfrac{rb}{(a-1)(c-1)}\sum_j\sum_k (AC)_{ik}^2$ |
| BC | $(b-1)(c-1)$ | MSbc | $\sigma_e^2 + r\sigma_{abc}^2 + rc\sigma_{bc}^2$ | $\sigma_e^2 + \dfrac{ra}{(b-1)(c-1)}\sum_j\sum_k (BC)_{jk}^2$ |
| ABC | $(a-1)(b-1)(c-1)$ | MSabc | $\sigma_e^2 + r\sigma_{abc}^2$ | $\sigma_e^2 + \dfrac{r}{(a-1)(b-1)(c-1)} \times \sum_i\sum_j\sum_k (ABC)_{ijk}^2$ |
| Error | $abc(r-1)$ | MSe | $\sigma_e^2$ | $\sigma_e^2$ |
| Total | $abcr - 1$ | | | |

Likewise, $E(MS)$ of source $AB$ is

$$E(MSab) = \sigma_e^2 + r\sigma_{abc}^2 + rc\sigma_{ab}^2,$$

which includes all interaction variance components involving factors $A$ and $B$ together.

If effects are random, then from the $E(MS)$ expressions we note that, to test the significance of source $A$, one may use either $F = MSa/MSab$ or $F = MSa/MSac$ with appropriate numerator and denominator DF parameters. The first $F$ ratio statistic is biased upward by $rb\sigma_{ac}^2$ and the second $F$ statistic is biased by $rc\sigma_{ab}^2$. However, if $\sigma_{ac}^2$ and $\sigma_{ab}^2$ are not significantly different from zero, either $F$ ratio is appropriate. When two-way-interaction variance components are large, one can use methods due to Satterthwaite (1946), Cochran (1951), and Naik (1974). The Satterthwaite method is illustrated in

Kirk (1968) and Ostle and Mensing (1975). The $F$ ratios for testing the two-way-interaction variance components are

$$F = \text{MS}ab/\text{MS}abc, \quad \text{for testing the significance of} \quad \sigma_{ab}^2,$$

$$F = \text{MS}ac/\text{MS}abc, \quad \text{for testing the significance of} \quad \sigma_{ac}^2,$$

$$F = \text{MS}bc/\text{MS}abc, \quad \text{for testing the significance of} \quad \sigma_{bc}^2,$$

$$F = \text{MS}abc/\text{MS}e, \quad \text{for testing the significance of} \quad \sigma_{abc}^2,$$

with appropriate DF parameters. If effects are assumed fixed, all $F$ ratios for testing fixed effects will have MSe in their denominators.

There are designs with a layout such as that in Figure 3.2 where levels of factor $B$ are nested within those of factor $A$, and levels of factor $C$ are nested within those of factor $B$. The reason is that, for example, level $C_1$ of factor $C$ appears only with level $A_1$ of factor $A$, and so on. For each combination of levels in the layout, there are two observations $X_{ijkm}$.

A hierarchic notation ":" is used to symbolize effects and to write models for nested designs. Let $B_{j:i}$ denote the effect of the $j$th level of factor $B$ nested within the $i$th level of factor $A$, and $C_{k:ji}$ denote the effect of the $k$th level of factor $C$ nested within the $j$th level of $B$ and $i$th level of $A$. In this notation an observation $X_{ijkm}$ can be represented in the following way:

$$X_{ijkm} = \mu + A_i + B_{j:i} + C_{k:ji} + \varepsilon_{ijkm}, \tag{3.1-9}$$

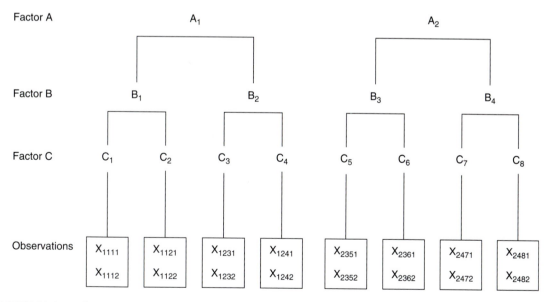

**FIGURE 3.2** Layout of a nested–balanced design.

where

$$i = 1, \ldots, a,$$
$$j = 1, \ldots, b \quad \text{nested within } a \text{ levels},$$
$$j = 1, \ldots, c \quad \text{nested within } b \text{ levels},$$
$$m = 1, \ldots, r \quad \text{nested within } c \text{ levels},$$

$\mu$ is the grand mean, $A_i$ is the effect of the $i$th level of $A$, and $\varepsilon_{ijkm}$ are random-error components of the model. Suppose (3.1-9) is a random-effects model. The analysis of variance of this model is shown in Table 3.13. The notation in the table follows the hierarchical notation introduced earlier. For example, $B : A$ means that the source of variation is factor $B$ nested within $A$, and $\sigma_{b:a}^2$ is the variance component of the source $B : A$. All other symbols are to be similarly understood. Because of nesting, the variance $\sigma_{b:a}^2$ associated with the source $B : A$ appears in the $E(MS)$ of source $A$. Likewise, the variance component $\sigma_{c:ba}^2$ appears in the $E(MS)$ of sources $A$, as well as in that of $B : A$. Thus, for the random-effects model, the variance component of a source appears in the $E(MS)$ of all sources in the hierarchical sequence of nested designs. For example, both variance components $\sigma_{b:a}^2$ and $\sigma_{c:ba}^2$ would appear in the $E(MS)$ expression of source $A$. Notice that the error variance component $\sigma_e^2$ appears in all $E(MS)$ expressions. In Table 3.13, notice also that the DF follow a particular pattern of nesting. The breakdown of the total sum of squares into various components is as follows:

**Table 3.13** Analysis of Variance of a Three-Factor Nested Balanced Design Assuming Random-Effects Model (3.1-9)

| Source of Variance | DF | SS | MS | E(MS) |
|---|---|---|---|---|
| A | $a - 1$ | $SSa = \dfrac{\Sigma_i X_{i\ldots}^2}{bcr} - \dfrac{X_{\ldots}^2}{abcr}$ | MSa | $\sigma_e^2 + r\sigma_{c:ba}^2 + cr\sigma_{b:a}^2 + bcr\sigma_a^2$ |
| B : A | $a(b - 1)$ | $SSb : a = \dfrac{\Sigma_i \Sigma_j X_{ij\ldots}^2}{cr} - \dfrac{\Sigma_i X_{i\ldots}^2}{bcr}$ | MSb : a | $\sigma_e^2 + r\sigma_{c:ba}^2 + cr\sigma_{b:a}^2$ |
| C : BA | $ab(c - 1)$ | $SSc : ba = \dfrac{\Sigma_i \Sigma_j \Sigma_k X_{ijk\cdot}^2}{r} - \dfrac{\Sigma_i \Sigma_j X_{ij\ldots}^2}{cr}$ | MSc : ba | $\sigma_e^2 + r\sigma_{c:ba}^2$ |
| Error | $abc(r - 1)$ | $SSe = \displaystyle\sum_i \sum_j \sum_k \sum_m X_{ijkm}^2 - \dfrac{\Sigma_i \Sigma_j \Sigma_k X_{ijk\cdot}^2}{r}$ | MSe | $\sigma_e^2$ |
| Total | $abcr - 1$ | $SSto = \displaystyle\sum_i \sum_j \sum_k \sum_m X_{ijkm}^2 - \dfrac{X_{\ldots}^2}{abcr}$ | | |

$$\sum_i \sum_j \sum_k \sum_m (X_{ijkm} - \bar{X})^2 = bcr \sum_i (\bar{X}_i \ldots - \bar{X})^2$$

$$+ cr \sum_i \sum_j (\bar{X}_{ij}.. - \bar{X}_i \ldots)^2$$

$$+ r \sum_i \sum_j \sum_k (\bar{X}_{ijk.} - \bar{X}_{ij..})^2 \quad\quad (3.1\text{-}10)$$

$$+ \sum_i \sum_j \sum_k \sum_m (X_{ijkm} - \bar{X}_{ijk.})^2.$$

Equivalently,

$$SSto = SSa + SSb : a + SSc : ba + SSe.$$

Usually, for computing purposes, alternative formulas in Table 3.13 are used. The $F$ ratios are obtained following the $E(MS)$ expressions.

To test for the significance of source $A$, we form

$$F = MSa/MSb : a \quad\quad (3.1\text{-}11)$$

with $(a-1)$ numerator DF and $a(b-1)$ denominator DF. Similarly for testing the significance of source $B$, we use

$$F = MSb : a/MSc : ba \qu\quad (3.1\text{-}12)$$

with $a(b-1)$ numerator DF and $ab(c-1)$ denominator DF. Finally, for testing the significance of source $C$, we use

$$F = MSc : ba/MSe \qu\quad (3.1\text{-}13)$$

with $ab(c-1)$ numerator DF and $abc(r-1)$ denominator DF. One may estimate the variance components by solving a system of equations such as that in (3.1-7).

---

**Example 3.1-3**

An experiment was conducted to determine the effects of distributional routes on the total bacterial counts of hot dogs from points of origin (plants) to display cases in the stores. The routes or factors randomly selected were plants, cities, and stores. The data are shown in Table 3.14. The information given at the bottom part of this table facilitates calculations of SS (as outlined in Table 3.13):

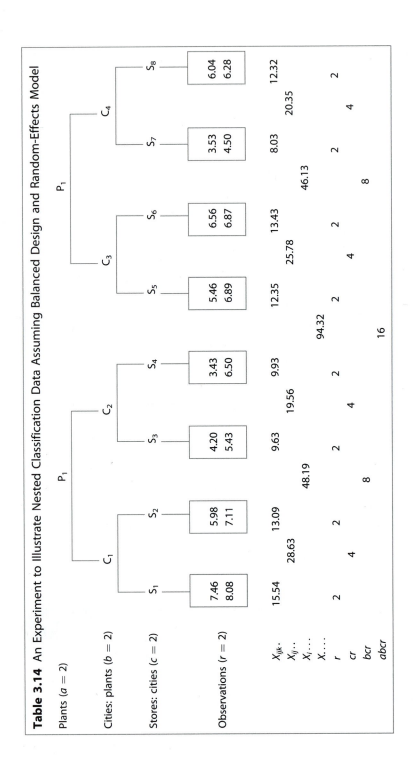

**Table 3.14** An Experiment to Illustrate Nested Classification Data Assuming Balanced Design and Random-Effects Model

$$SSp = \frac{[(48.19)^2 + (46.13)^2]}{8} - \frac{(94.32)^2}{16}$$
$$= 556.2816 - 556.0164$$
$$= 0.2652,$$

$$SSc:p = \frac{[(28.63)^2 + (19.56)^2 + (25.78)^2 + (20.35)^2]}{4} - 556.2816$$
$$= 570.2503 - 556.2816$$
$$= 13.9687,$$

$$SSs:cp = \frac{[(15.54)^2 + (13.09)^2 + \cdots + (12.32)^2]}{2} - 570.2503$$
$$= 576.6664 - 570.2503$$
$$= 6.4161,$$

$$SSe = [(7.46)^2 + (8.08)^2 + \cdots + (6.28)^2] - 576.6664$$
$$= 584.5354 - 576.6664$$
$$= 7.8690,$$

$$SSto = 584.5354 - 556.0164$$
$$= 28.5190.$$

The analysis of variance is shown in Table 3.15. The $F$ ratios following (3.1-11) through (3.1-13) showed that only city was found significant at the 1% level of significance.

The variance components are estimated using the equations in (3.1-7). Equating the $E$(MS) in Table 3.13 to the MS in Table 3.15, we have

$$\begin{bmatrix} \sigma_e^2 + 2\sigma_{s:cp}^2 + 4\sigma_{c:p}^2 + 8\sigma_p^2 \\ \sigma_e^2 + 2\sigma_{s:cp}^2 + 4\sigma_{c:p}^2 \\ \sigma_e^2 + 2\sigma_{s:cp}^2 \\ \sigma_e^2 \end{bmatrix} = \begin{bmatrix} 0.2652 \\ 6.9844 \\ 1.6040 \\ 0.9836 \end{bmatrix}$$

**Table 3.15** Analysis of Variance of a Nested Design of Total Bacteria Counts in Hot Dogs (Example 3.1-3)

| Source of Variance | DF | SS | MS | F Ratio |
|---|---|---|---|---|
| Total | 15 | 28.5190 | | |
| Plants | 1 | 0.2652 | 0.2652 | 0.04 |
| Cities: plants | 2 | 13.9687 | 6.9844 | 4.35 |
| Stores: cities: plants | 4 | 6.4161 | 1.6040 | 1.63 |
| Error | 8 | 7.8690 | 0.9836 | |

Note that $\hat{\sigma}_e^2 = 0.9836$; thus

$$\hat{\sigma}_{s:cp}^2 = (1.6040 - 0.9836)/2 = 0.3102,$$

$$\hat{\sigma}_{c:p}^2 = (6.9844 - 1.6040)/4 = 1.3451,$$

$$\hat{\sigma}_p^2 = (0.2652 - 6.9844)/8 = -0.8399.$$

The estimate of variance component for $\sigma_p^2$ comes out to be negative and therefore should be considered zero. To appreciate the practical significance of variance components, we may express the components in percentages of their sum. We find $\hat{\sigma}_p^2 = 0\%$, $\hat{\sigma}_{c:p}^2 = 50.97\%$, $\hat{\sigma}_{s:cp}^2 = 11.75\%$, and $\hat{\sigma}_e^2 = 37.27\%$. Thus, city as a source of variation accounts for 50.97% of the total variance components of bacterial counts in hot dogs. With this information, we can reduce the variance component due to city by increasing the number of cities in future studies.

## 3.2 MULTIPLE-COMPARISON TESTS

For the fixed-effects model the analysis of variance $F$ ratio statistic, if found significant, implies only that the treatments or factor levels are different. In this case, the researcher naturally proceeds to identify treatments that are alike and to create statistically homogeneous groups. The grouping of treatments into homogeneous groups is accomplished by means of multiple-comparison procedures. Some well-known procedures are the least significant difference (LSD) test, Dunnett's test, Tukey's Studentized range test, Duncan's multiple-range test, the Student–Newman–Keuls test, and the Scheffé test, among others. These tests are described in several books (Federer, 1955; Steel and Torrie, 1980; Kirk, 1968). We consider here only the first four tests mentioned.

It may happen that not each multiple-comparison test used on the same data yields the same homogeneous treatment groups. The disagreement may be partly due to the manner in which the significance level $\alpha$ is viewed from one multiple-comparison test to another. In the literature on multiple-comparison tests, the significance level has been viewed in essentially two ways following the *comparisonwise error rate* and the *experimentwise error rate*. For discussion of these two types of error rates, the reader may refer to Tukey (1953), Harter (1957), Chen (1960), and Steel (1961). A criticism of the comparisonwise error rate is that when large numbers of comparisons are made, the probability of declaring a false significant difference is high. The LSD and Duncan's multiple-range tests use the comparison-wise error rate. A major criticism of the experimentwise error rate is that the probability of finding a true treatment difference decreases with an increasing number of treatments. Dunnett's test and Tukey's Studentized range test use the experimentwise error rate.

To discuss the four multiple-comparison tests, we consider a one-way fixed-effects model for $k$ populations with possibly different means $\mu_1, \mu_2, \ldots, \mu_k$. The hypotheses of interest are

$$H_0 : \quad \mu_i = \mu_j, \quad i \neq j, \quad i, j = 1, \ldots, k;$$
$$H_a : \quad \mu_i \neq \mu_j, \quad \text{for at least one pair} \quad (i, j).$$

We denote the sample means of $k$ populations by $\overline{X}_1, \overline{X}_2, \ldots, \overline{X}_k$ and let $d_{ij} = |\overline{X}_i - \overline{X}_j|$ for all pairs $(i, j)$. It is assumed that all the populations have the same variance $\sigma^2$ and are normally distributed.

## Least Significant Difference Test

The LSD test was described by R. A. Fisher (1960) in the well-known book, *The Design of Experiments*. The test procedure uses the $t$ distribution successively to judge the significance of pairwise comparisons. An observed paired difference $d_{ij}$ is compared with its LSD, defined as

$$\text{LSD} = t_{(1-\alpha/2),v}(S_{d_{ij}}), \tag{3.2-1}$$

where

$$S_{d_{ij}} = \sqrt{\text{MSe}\left(\frac{1}{n_i} + \frac{1}{n_j}\right)}, \tag{3.2-2}$$

and $t_{(1-\alpha/2),v}$ is the $(1 - \alpha/2)100$th percentile of the $t$ distribution with $v$ DF associated with the MSe in the analysis of variance. If, say, $n_i = n_j = n$, then

$$S_{d_{ij}} = \sqrt{2\text{MSe}/n}.$$

If $d_{ij}$ is greater than its LSD, then $\mu_i$ is significantly different from $\mu_j$.

A disadvantage of the LSD method is that none of the individual paired comparisons may be judged significant even if the overall $F$ test rejected the hypothesis of the equality of means. Fisher (1960) and LeClerg (1957) suggested some guidelines for using the LSD test. When the $F$ test is significant, the LSD test should be applied to test for the significance of only those comparisons that are planned in advance.

For illustration, let us consider again the data in Example 3.1-1. As shown in Table 3.3, the $F$ ratio 6.51 exceeds at the 1% level the tabled $F$ with 3 numerators and 16 denominators DF (Table A.4). Therefore, there is evidence that at least two means are significantly different from each other. The treatment means in this example are as follows:

$$\overline{X}_1 = 13.3, \quad \overline{X}_2 = 10.4, \quad \overline{X}_3 = 10.1, \quad \overline{X}_4 = 16.8.$$

The first step in applying the LSD test is to compute the standard error of pairwise differences between means using (3.2-2). Since $n_i = n_j = 5$ for each pair $(i, j)$, we have

$$S_{d_{ij}} = \sqrt{2(7.459)/5} = 1.727.$$

The second step is to obtain the $t_{(1-\alpha/2),v}$ for $\alpha = 0.01$ and $v = 16$ from Table A.2. This value is $t_{0.995,16} = 2.921$. Hence, the LSD for each pair from (3.2-1) is

$$\text{LSD} = (2.921)(1.727) = 5.05.$$

Each pairwise difference $d_{ij}$ is to be compared with this LSD value. Let us suppose that the following paired differences were planned for comparisons:

$$d_{14} = |13.3 - 16.8| = 3.5,$$
$$d_{24} = |10.4 - 16.8| = 6.4,$$
$$d_{34} = |10.1 - 16.8| = 6.7.$$

Since $d_{24}$ and $d_{34}$ exceed the LSD value of 5.05, these are significant differences at the 1% level of significance.

## Dunnett's Test for Multiple Comparisons with a Control

The test was given by Dunnett (1955, 1964) for comparing means $\mu_1$, $\mu_2, \ldots, \mu_k$ of $k$ populations with the mean $\mu_0$ of a control population. The null hypothesis is

$$H_0: \quad \mu_i = \mu_0 \quad \text{for all} \quad i$$

versus the alternative that at least one $\mu_i$ is different from $\mu_0$.

We estimate $\mu_i$ by $\overline{X}_i$, $i = 0, 1, \ldots, k$, and the common variance $\sigma^2$ of the $(k + 1)$ populations by the MSe of the appropriate analysis of variance.

Differences

$$d_i = |\overline{X}_i - \overline{X}_0|, \quad i = 1, \ldots, k,$$

are compared with the critical difference $d$

$$d = D_{\alpha,k,v}(\sqrt{2\text{MSe}/n}), \tag{3.2-3}$$

where $D_{\alpha,k,v}$ is given by Table A.5; $v = (k+1)(n-1)$ is the DF associated with MSe, and $n$ is the number of observations in each of the $(k+1)$ samples. A difference $d_i$ is significant if $d_i > d$.

For illustration let us pretend that in our previous example $\overline{X}_4$ corresponds to the sample mean $\overline{X}_0$ from a control population and that there are three populations with sample means $\overline{X}_1$, $\overline{X}_2$, and $\overline{X}_3$ to be compared with the control. We find differences

$$d_1 = |\overline{X}_1 - \overline{X}_0| = 13.3 - 16.8 = 3.5,$$

$$d_2 = |\overline{X}_2 - \overline{X}_0| = 10.4 - 16.8 = 6.4,$$

$$d_2 = |\overline{X}_3 - \overline{X}_0| = 10.1 - 16.8 = 6.7.$$

From Table A.5 corresponding to $\alpha = 0.01$ and for a two-tailed test, $D_{\alpha,k,v} = 3.39$. Hence, from (3.2-3),

$$d = (3.39)\sqrt{2(7.459)/5} = 5.85.$$

Comparisons of $d_i$ with $d$ show that means $\mu_2$ and $\mu_3$ are significantly different from $\mu_0$.

Robson (1961) applied Dunnett's test to the balanced incomplete block design (Chapter 5). Because of the balanced property of the design, all adjusted treatment means have the same variance and all pairs have the same covariance. Now instead of (3.2-3) the critical difference $d$ is defined as

$$d = D_{\alpha,k,v}(\sqrt{2\text{MSe}/nE}) \tag{3.2-4}$$

where $E$ is an efficiency factor defined for each incomplete block design.

## Tukey's Studentized Range Test

Tukey (1953) proposed a multiple-comparison method for pairwise comparisons of $k$ means and for the simultaneous estimation of differences between means by confidence intervals with a specified confidence coefficient $(1 - \alpha)$. Suppose $n$ observations are taken in each of $k$ samples and the analysis of variance $F$ test is significant for testing the equality of $k$ means. The critical difference to be exceeded for a pair of means to be significantly different is the so-called honestly significant difference (HSD), given by

$$\text{HSD} = Q_{\alpha,k,v}(\sqrt{\text{MSe}/n}), \tag{3.2-5}$$

where $v$ is the DF of MSe, and the critical value $Q_{\alpha,k,v}$ of the Studentized range test is obtained from Table A.6. For our previous example we find $Q_{0.01,4,16} = 5.19$. The standard error of $\bar{X}_i$ is estimated by

$$(\text{MSe}/n)^{1/2} = (7.459/5)^{1/2} = 1.221.$$

Therefore, the critical difference is

$$\text{HSD} = 5.19(1.221) = 6.34.$$

Any two means with the estimated difference $d_{ij} > 6.34$ would be declared significantly different from zero. For our example with four means, there are six pairwise differences:

$$d_{12} = |13.3 - 10.4| = 2.9,$$
$$d_{13} = |13.3 - 10.1| = 3.2,$$
$$d_{14} = |13.3 - 16.8| = 3.5,$$
$$d_{23} = |10.4 - 10.1| = 0.3,$$
$$d_{24} = |10.4 - 16.8| = 6.4^{*,}$$
$$d_{34} = |10.1 - 16.8| = 6.7^{*.}$$

The $d_{ij}$ with asterisks are significant differences.

Simultaneous confidence intervals for differences $(\mu_i - \mu_j)$ are given by $d_{ij} \pm \text{HSD}$. For our example, the six simultaneous intervals with 99% confidence coefficient are given by $d_{ij} \pm 6.34$. They are

$$\mu_1 - \mu_2 : \quad (-3.44, 9.24),$$
$$\mu_1 - \mu_3 : \quad (-3.14, 9.54),$$
$$\mu_1 - \mu_4 : \quad (-9.84, 2.84),$$
$$\mu_2 - \mu_3 : \quad (-6.04, 6.64),$$
$$\mu_2 - \mu_4 : \quad (-12.74, -0.06),$$
$$\mu_3 - \mu_4 : \quad (-13.04, -0.36).$$

We may note that if $H_0 : \mu_1 = \mu_2 = \mu_3 = \mu_4$ is true, then all the intervals must include zero. The range of intervals for $(\mu_2 - \mu_4)$ and $(\mu_3 - \mu_4)$ does not include zero.

## Duncan's Multiple-Range Test

Duncan (1955) proposed a multiple-comparison test that makes pairwise comparisons within each subset of means. The level of significance for each subset is fixed and is determined by the number of elements in it. If we let $p$ denote the number of means in a subset of $k$ means, then for each pairwise comparison of means in this subset, the level of significance is given by

$$\alpha_p = 1 - (1 - \alpha)^{p-1}, \tag{3.2-6}$$

where $\alpha$ is the desired significance level of any pairwise comparison. The critical difference $R_p$ of Duncan's test for $p$ means, known as the shortest significant range, is given by

$$R_p = r_{\alpha,p,v}(\sqrt{MSe/n}), \tag{3.2-7}$$

where $r_{\alpha,p,v}$ are tabulated in Table A.7 for $\alpha = 0.05$ and $\alpha = 0.01$ and various values of $p$ and error DF $v$. The test criterion for Duncan's test is to compare $d_{ij}$ with $R_p$. The steps in performing the test are as follows:

1. Arrange all the means in increasing order of magnitude.
2. Compute $R_p$ for each size of subset of means using (3.2-7).
3. Compute $d_{ij}$ for all the $k(k - 1)/2$ pairs of means.
4. Declare $d_{ij}$ significant when $d_{ij} > R_p$.
5. Any two or more means not significantly different from each other are underlined.

For our continuing example the total number of pairwise comparisons is $4(4 - 1)/2 = 6$. The means, arranged in increasing order, are $\bar{X}_3 < \bar{X}_2 < \bar{X}_1 < \bar{X}_4$. The six pairwise differences and the subset size $p$ are as follows:

| $d_{ij}$ | Subset size $p$ |
|---|---|
| $\bar{X}_3 - \bar{X}_4$ | 4 |
| $\bar{X}_3 - \bar{X}_1$ | 3 |
| $\bar{X}_3 - \bar{X}_2$ | 2 |
| $\bar{X}_2 - \bar{X}_4$ | 3 |
| $\bar{X}_2 - \bar{X}_1$ | 2 |
| $\bar{X}_1 - \bar{X}_4$ | 2 |

The first step is to compute $(MSe/n)^{1/2} = (7.459/5)^{1/2} = 1.221$. Since in Example 3.1-1 the $F$ ratio was found significant at the 1% level, it is customary to use the 1% level in Duncan's table to compute $R_p$. For this example from Table A.7, we obtain values of $r_{\alpha,p,v}$ corresponding to $v = 16$, $\alpha = 0.01$, and $p = 2,3,4$. Then from (3.2-7), we find $R_p$ values as shown here:

$$
\begin{array}{llll}
p: & 2 & 3 & 4 \\
r_{0.01,p,16}: & 4.12 & 4.34 & 4.45 \\
R_p: & 5.04 & 5.30 & 5.43
\end{array}
$$

Any difference between means of subset size $p = 4 > R_p = 5.43$ is declared significant; a mean difference $d_{ij}$ for subset size $p = 3$ is significant if it is greater than 5.30 and so on. The differences between means and their significance are summarized here:

| $d_{ij}$ | $R_p$ | $p$ |
|---|---|---|
| $16.8 - 10.1 = 6.7^*$ | 5.43 | 4 |
| $16.8 - 10.4 = 6.4^*$ | 5.30 | 3 |
| $16.8 - 13.3 = 3.5$ | 5.04 | 2 |
| $13.3 - 10.1 = 3.2$ | 5.30 | 3 |
| $13.3 - 10.4 = 2.9$ | 5.04 | 2 |
| $10.4 - 10.1 = 0.3$ | 5.04 | 2 |

The first two differences are such that $d_{ij} > R_p$, and, therefore, they are declared significant at the 1% level of significance. We may summarize and present the results of Duncan's test by underlining the means that are not significantly different from each other. Thus for our example we summarize as follows:

$$
\underline{16.8 \quad\quad \underline{13.3 \quad\quad 10.4 \quad\quad 10.1}}
$$

All means underscored by the same line are not significantly different.

Duncan's multiple-range test can be used for an unbalanced design when an unequal number of observations is taken for treatments (Kramer, 1956). A difference $d_{ij}$ is significant if

$$
d_{ij} > R_p \sqrt{(n_i + n_j)/2n_i n_j},
$$

where now $R_p = r_{\alpha,p,v}(\text{MSe})^{1/2}$, and $n_i$ and $n_j$ refer to the number of observations for the $i$th and $j$th treatments, respectively. Now one can follow the steps outlined earlier to carry out Duncan's test.

## Choice of Multiple-Comparison Tests

The choice of a test to use subsequent to the significance of the analysis of variance $F$ test is not obvious. If comparisons to be tested for significance are known in advance, the use of the LSD or Dunnett's test is appropriate. Frequently, experiments in research and development are designed to compare all treatments, and each paired comparison is independently interpreted with respect to the other comparisons. If a treatment is found "superior" over others, the superior treatment is then put to more laboratory testing. In such situations the comparison-wise error rate is an appropriate measure of risk, and Duncan's multiple-range test should be used. If a composite inference is desired on all sets of treatments, the use of the experimentwise error rate measure is appropriate, and Tukey's Studentized range test may be used.

Figures 3.3 and 3.4 show the comparisonwise and the experimentwise error rates, respectively, for three multiple-comparison tests. As to be expected, when the number of treatments $k = 2$, all three multiple-comparison tests have the same protection level for both error rates. Tukey's HSD test is conservative regardless of the type of error rate. The error rates are less than 5% with an increasing number of treatments. Contrariwise, the LSD is liberal. The comparisonwise error rate is a little over the 5% level, particularly when $k > 4$ (Figure 3.3). For the experimentwise error rate when $k = 9$, the LSD

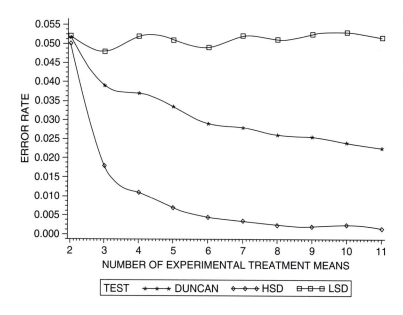

■ **FIGURE 3.3** Per comparison error rate of multiple-comparison tests with $n = 15$ observations per treatment mean. *Source:* Boardman and Moffitt (1971). With permission of The Biometric Society.

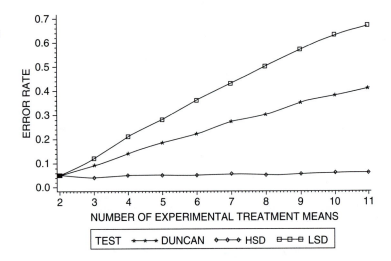

■ **FIGURE 3.4** Experimentwise error rate of multiple-comparison tests with $n = 15$ observations per treatment mean. *Source:* Boardman and Moffitt (1971). With permission of The Biometric Society.

indicates at least one false significant difference ~57% of the time (Figure 3.4). The error rates for Duncan's test are in between the LSD and Tukey's HSD test, as shown in Figure 3.3.

Although the LSD is liberal with respect to the comparisonwise error rate, it is powerful for detecting a true difference (Figures 3.5 and 3.6). On the other hand, Tukey's HSD test commits Type II errors more often than the Duncan and LSD tests. The power curve for Duncan's test is in between the two tests. For more details on the comparison of the various multiple-comparison tests, see Balaam (1963), Petrinovich and Hardych (1969), and Carmer and Swanson (1973).

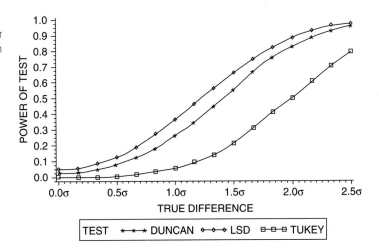

■ **FIGURE 3.5** Empirical power curves of multiple-comparison tests at the 5% significance level for $k = 10$ treatments. *Source:* Chen (1960). With permission of the author.

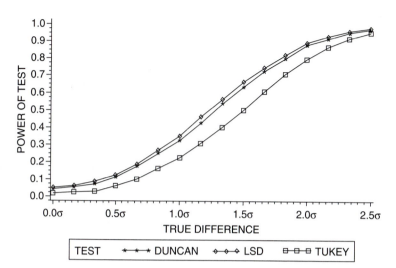

## 3.3 **SAMPLE SIZE ESTIMATION**

A consideration in statistical planning of scientific experiments is the estimation of sample size for reliable decision making. There are statistical formulas that help experimenters reach adequate estimates of sample size. We consider only a few formulas in this section to familiarize the reader with the basics of the statistical nature of sample size determination. The reader must also keep in mind that, although statistical formulas serve their purpose, many important nonstatistical considerations play significant roles in sample size estimation. For example, there are budget and time limitations; also, aims and the significance of resulting statistical decisions are important considerations.

Let us consider a simple formulation of hypothesis testing where under the null hypothesis the mean $\mu$ of a population is hypothesized to be $\mu_0$ against an appropriate alternative. Suppose the statistical test to be used is such that it should not run the risks of rejecting the null hypothesis incorrectly (Type I error) and accepting the false null hypothesis (Type II error) more than $100\,\alpha\%$ and $100\,\beta\%$ times, respectively. Suppose that the experimenter can provide an initial estimate of the population variance $\sigma^2$ from previous studies or from a preliminary small sample or simply by guessing. In this situation an approximate statistical formula for providing an estimate of sample size $n$ to meet Type I and Type II error constraints is given by

$$n = \sigma^2 (Z_\alpha + Z_\beta)^2 / (\mu - \mu_0)^2. \qquad (3.3\text{-}1)$$

| Example 3.3-1 | An investigator wishes to test the mean preference score of a product against the mean preference score of a regularly produced product. The investigator wishes to detect the difference $(\mu - \mu_0)$ of at least 0.5 units on a preference scale. The two types of error rates $\alpha$ and $\beta$ are to be no more than 5% and 10%, respectively. If, from such preference studies, an estimate of $\sigma^2$ is 1.0, then (3.3-1) gives |
|---|---|

$$n = (1.0)^2 (Z_{0.05} + Z_{0.10})^2 / (0.5)^2,$$

where from Table A.1, $Z_{0.05} = -1.645$ and $Z_{0.10} = -1.290$. Hence,

$$n = 4(-1.645 - 1.290)^2 = 34.46 \approx 35.$$

Tang (1938) has discussed the determination of sample size when the analysis of variance technique is to be used for comparing $k \geq 2$ means. Tang's method, being iterative, is rather difficult to use. Kastenbaum et al. (1970) simplified Tang's method. They used the standardized maximum difference $\tau$ between pairs of means to estimate $n$. That is, they found

$$\tau = (\mu_{max} - \mu_{min}) / \sigma, \tag{3.3-2}$$

where $\mu_{max}$ and $\mu_{min}$ are some appropriately arrived at estimates of the largest and smallest means based on prior information. Remember that an assumption in the analysis of variance is that all populations have equal standard deviation denoted by $\sigma$. Of course $\sigma$ is also not known. One must again use some appropriate estimate of it in (3.3-2). Using an estimated $\tau$ value from (3.3-2), one can then use Tables A.8 and A.9 for estimating $n$ for completely randomized and randomized block designs, respectively. Both tables are given for values of $k = 2 - 6$ and for different combinations of $\alpha$ and $\beta$ error rates.

| Example 3.3-2 | This example was taken from Kastenbaum and Hoel (1970). The mean life span of a particular strain of mice is estimated to be 125 weeks. Radiation of these animals at sublethal doses of X-rays above 100 R (roentgens) is known to reduce their survival time. In fact, every increase of 50 R of X radiation in the range 100–400 R appears to reduce longevity by ~3 weeks. A biologist wishes to know whether a corresponding reduction of longevity can be observed in the range 0–50 R. He proposes to experiment with two groups of animals, one of which will be unirradiated (control) and the other radiated with 50 R of X-rays. The question is, How |
|---|---|

many animals should be placed in each of the two groups? In this example, the number of populations is $k = 2$, $\mu_{max} = 125$ weeks, and $\mu_{min} = 122$ weeks. Suppose the risk levels chosen by the experimenter are $\alpha = 0.10$ and $\beta = 0.05$ and the standard deviation, based on previous experience, is $\sigma = 20$. Thus, $\tau = (\mu_{max} - \mu_{min})/\sigma = 3/20 = 0.15$ and from Table A.8 the number of mice required for each treatment is $n = {\sim}1000$.

---

In the same situation as in Example 3.3-2, the experimenter might ask, How large a difference could I detect with 500 observations in each group? Referring to Table A.8 for $\alpha = 0.10$, $\beta = 0.05$, and $n = 500$, we find $\tau = 0.208$. Thus, $(\mu_{max} - \mu_{min}) = \sigma(0.208) = 20(0.208) = 4.16$ weeks.

**Example 3.3-3**

## Sample Size for Confidence Interval Estimation

Harris et al. (1948) used (3.3-3) for estimating the sample size to obtain a $(1 - \alpha)$ 100% confidence interval with a specified rate of $\beta$ error (equivalently referred to as $1 - \beta$ assurance level) and specified width of the confidence interval. Suppose we desire to estimate the mean of a population with $(1 - \alpha)$ 100% confidence interval of width $2d$. Then the sample size $n$ may be found using

$$d/S = \sqrt{(t_\alpha^2 F_{1-\beta})/n + 1)}, \qquad (3.3\text{-}3)$$

where $S$ stands for an estimate of standard deviation from previous samples. Note that the percentiles $t_\alpha$ and $F_{1-\beta}$ depend on $n$ and, therefore, Eq. (3.3-3) is not explicit in $n$. Thus, the equation would be solved for $n$ through trial and error. But Johnson (1963) constructed tables that may be used to find $n$ once we know $d/S$. Table A.10 in the Appendix contains $d/S$ values corresponding to the 95% confidence interval and the 95% assurance level.

---

A researcher wishes to conduct a chemical audit for percent moisture of a product at the store level. He wants to estimate the mean moisture content with a 95% degree of confidence interval of width no more than 2%. A previous audit based on 10 observations had shown that an estimate of the standard deviation of the moisture content is 1.5%. Since half of the interval width is $d = 1$ and $S = 0.015$, $d/S = 0.667$. The DF associated

**Example 3.3-4**

with $S$ is always one less than the number of observations used in calculating it. Hence, $DF = 9$. From Table A.10 we see that the entry for $d/S$ closest to 0.667 is 0.703. Corresponding to this entry, the sample size required is $n = 25$. Hence, to have a 95% confidence interval with 95% assurance, we require at least 25 observations.

## Sample Size for Dichotomous Responses

The problem of estimating the proportion of items in a population having several characteristics of interest or comparing two proportions is quite frequent. The formulas for sample size estimation when estimating a single proportion or testing a hypothesis about it are given in many texts for a first basic course in statistics. We give a formula for sample size estimation when comparing two proportions. Let $p_1$ and $p_2$ be the proportions of items in two populations having several characteristics of interest. Suppose we want to detect the difference $(p_1 - p_2)$ if it is more than a specified amount $d$, and in doing this we do not want to exceed the risk of Type I and II errors beyond specified levels $\alpha$ and $\beta$, respectively. An approximate but simple formula for estimating $n$ is given by

$$n = [(Z_\alpha + Z_\beta)/d]^2 (p_1 q_1 + p_2 q_2), \tag{3.3-4}$$

where $q_1 = 1 - p_1$, $q_2 = 1 - p_2$, and $d$ is the specified amount of difference to be detected if it exists between $p_1$ and $p_2$. Since $p_1$ and $p_2$ are not known, some initial estimates are substituted for them in (3.3-4). An estimate of $n$ biased upward may be obtained by (3.3-5) instead of (3.3-4):

$$n = \frac{1}{2}[(Z_\alpha + Z_\beta)/d]^2. \tag{3.3-5}$$

---

**Example 3.3-5**

A food technologist wishes to run a preference trial on two foods. He wishes to find if preference proportions of two foods differ by 20% or more. How many panelists should be used to remain within the risk levels $\alpha = 0.05$ and $\beta = 0.10$ in his decision making?

We note that $d = 0.20$, $Z_{0.05} = -1.645$, and $Z_{0.10} = -1.290$. Since no initial estimates are available for $p_1$ and $p_2$, we use (3.3-5) to estimate $n$. Hence

$$n = \frac{1}{2}[(-1.645 - 1.290)/0.20]^2 \approx 108.$$

---

## EXERCISES

**3.1.** An experimenter wishes to study the effects of three levels of tenderizing enzyme. Four animal carcasses were available for this study. With three observations per enzyme level and animal combinations, discuss the following:

**1.** Layout of the data;
**2.** Statistical model;
**3.** Analysis of variance table, including the expected mean square;
**4.** *F* ratio for comparing the three levels of enzyme.

**3.2.** Consider the system of equations in (3.1-7) and solve it for $\sigma_e^2$, $\sigma_a^2$, $\sigma_b^2$, and $\sigma_{ab}^2$ in terms of MSe, MSa, MSb, and MSab.

**3.3.** Consider the following model:

$$Y_{ijk} = \mu + H_i + S_{j:i} + e_{ijk}, \quad \begin{aligned} i &= 1, \ldots, 8, \\ j &= 1, \ldots, 3, \\ k &= 1, \ldots, 6, \end{aligned}$$

where $\mu$ is the grand mean, $H_i$ is the effect of the *i*th herd, $S_{j:i}$ is the effect of the *j*th sire nested within the herd, and $e_{ijk}$ are the random errors. The mean square values were calculated as follows:

| Source | MS |
|---|---|
| Herd | 0.580 |
| Sire:herd | 0.081 |
| Error | 0.062 |

**1.** Complete the column for DF.
**2.** Find the expected mean square expressions and use them to estimate all variance components.
**3.** Let $\sigma_h^2$ and $\sigma_{s:h}^2$ be the variance components of random effects $H_i$ and $S_{j:i}$, respectively. Test the null hypotheses $H_0$: $\sigma_h^2 = 0$ and $H_0$: $\sigma_{s:h}^2 = 0$.

**3.4.** The data given in the following are off-flavor scores of luncheon meats from a shelf-life study. Two packages from each of the five stores were evaluated following a predetermined experimental design.

| Stores : | A | | B | | C | | D | | E | |
|---|---|---|---|---|---|---|---|---|---|---|
| Packages : | 1 | 2 | 3 | 4 | 5 | 6 | 7 | 8 | 9 | 10 |
| | 1 | 2 | 7 | 7 | 7 | 7 | 3 | 4 | 5 | 2 |
| | 1 | 5 | 7 | 7 | 7 | 7 | 4 | 6 | 6 | 5 |
| | 4 | 4 | 5 | 5 | 4 | 5 | 2 | 5 | 3 | 2 |
| | 1 | 6 | 4 | 7 | 1 | 7 | 7 | 6 | 7 | 3 |
| | 7 | 6 | 2 | 6 | 7 | 6 | 6 | 5 | 6 | 4 |

1. Suggest a statistical model and its analysis of variance table giving the expected mean square expressions.
2. Compute the mean squares. Assuming a random-effects model, estimate the variance components.
3. Does your model allow for testing the significance of packages? Explain.

3.5. Using the analysis of variance in Table 3.10, compute the variance components for soap, technician, soap × technician, and error. Interpret your results.

# Experimental Design

In this chapter we deal with a number of experimental designs. Some well-known designs are covered for the sake of completeness and effectiveness in presenting basic design concepts. The others are of a specialized nature and not usually found in textbooks. The balanced incomplete block design, important for its wide application in the area of sensory evaluation and consumer testing of foods and materials, is covered separately in Chapter 5.

## 4.1  SIMPLE COMPARATIVE EXPERIMENTS

The simplest design pertains to the comparison of two treatments. The two most commonly used designs are group-comparison and paired comparison designs.

### Group-Comparison Designs

Let there be $N$ homogeneous or like experimental units available for experimentation. In a group-comparison design, one treatment is applied to $n_1$ randomly selected units out of $N$ and the other to the remaining $n_2 = N - n_1$ units. So the two groups differ only with respect to the treatments being applied. Let $X_{1i}$ and $X_{2i}$ denote the $i$th observations on treatments 1 and 2, respectively. Thus, for the group-comparison design, we have the following layout of two random samples:

| Treatment 1 | Treatment 2 |
|:---:|:---:|
| $X_{11}$ | $X_{21}$ |
| $X_{12}$ | $X_{22}$ |
| $X_{13}$ | $X_{23}$ |
| — | — |
| — | — |
| $X_{1n_1}$ | $X_{2n_2}$ |

In group-comparison designs, the two random samples are independent of each other. If $\mu_1$ and $\mu_2$ denote, respectively, the means of the two groups, then interest would be in testing one of the following hypotheses:

$$H_0: \quad \mu_1 - \mu_2 = \Delta \quad \text{versus} \quad H_a: \quad \mu_1 - \mu_2 \neq \Delta, \qquad (4.1\text{-}1)$$

$$H_0: \quad \mu_1 - \mu_2 < \Delta \quad \text{versus} \quad H_a: \quad \mu_1 - \mu_2 \geq \Delta, \qquad (4.1\text{-}2)$$

$$H_0: \quad \mu_1 - \mu_2 > \Delta \quad \text{versus} \quad H_a: \quad \mu_1 - \mu_2 \leq \Delta, \qquad (4.1\text{-}3)$$

where $\Delta$ is some specified amount. Often hypotheses with $\Delta = 0$ are of interest. An important statistic for testing the preceding hypotheses is

$$t = \frac{(\bar{X}_1 - \bar{X}_2) - \Delta}{S_p \sqrt{(1/n_1) + (1/n_2)}}, \qquad (4.1\text{-}4)$$

where $\bar{X}_1$ and $\bar{X}_2$ are the means of the two random samples from groups 1 and 2, respectively, and

$$S_p = \sqrt{\frac{S_1^2(n_1 - 1) + (S_2^2(n_2 - 1)}{n_1 + n_2 - 2}} \qquad (4.1\text{-}5)$$

is the pooled standard deviation obtained from two sample variances $S_1^2$ and $S_2^2$, which are calculated as follows:

$$S_1^2 = \left( \sum_i X_{1i}^2 - n_1 \bar{X}_1^2 \right) \Big/ (n_1 - 1), \qquad (4.1\text{-}6)$$

$$S_2^2 = \left( \sum_i X_{2i}^2 - n_2 \bar{X}_2^2 \right) \Big/ (n_2 - 1). \qquad (4.1\text{-}7)$$

The underlying assumptions for using the $t$ statistic are that both groups are normally distributed populations with a common variance $\sigma^2$ but with different means. Under these assumptions, the $t$ statistic follows the $t$ distribution with $(n_1 + n_2 - 2)$ DF. If a setup of the hypotheses is to be tested at the $\alpha$ level of significance, then the computed value of $t$ is compared with an appropriate percentile of the $t$ distribution with $(n_1 + n_2 - 2)$ DF. That is, reject $H_0$ if

$$|t| \geq t_{(1-\alpha/2),(n_1+n_2-2)},$$

provided that the alternative hypothesis is two-sided as in (4.1-1). For the setup of hypotheses (4.1-2), reject $H_0$ if

$$t > t_{(1-\alpha),(n_1+n_2-2)}.$$

For the setup of the hypotheses in (4.1-3), reject $H_0$ if

$$t < t_{\alpha,(n_1+n_2-2)}.$$

In particular, if $\Delta = 0$, the one-way analysis of variance $F$ ratio can also be used to test the equality of the two means. The reason is that when $\Delta = 0$, $F = t^2$.

A $(1 - \alpha)$ 100% confidence interval using the $t$ statistic in (4.1-4) can be set up as

$$\overline{X}_1 - \overline{X}_2 \pm [t_{(1-\alpha/2),(n_1+n_2-2)}]S_{\overline{X}_1-\overline{X}_2}, \qquad (4.1\text{-}8)$$

where $S_{\overline{X}_1-\overline{X}_2}$ is the standard deviation, also known as the standard error of the difference $X_1 - X_2$. This standard error is

$$S_{\overline{X}_1-\overline{X}_2} = S_p\sqrt{(n_1 + n_2)/n_1n_2}, \qquad (4.1\text{-}9)$$

where $S_p$ is given by (4.1-5). The calculation of (4.1-8) will yield the upper and lower bounds of the confidence interval using the positive and negative signs, respectively. If $n_1 = n_2 = n$, then expression (4.1-4) reduces to

$$t = [\overline{X}_1 - \overline{X}_2 - \Delta]/S_p\sqrt{2/n}. \qquad (4.1\text{-}10)$$

An example illustrating the statistical analysis of the group-comparison design is given in Table 4.1. Using expressions (4.1-6) and (4.1-7), we find

$$S_1^2 = \{[(5.5)^2 + (4.6)^2 + \cdots + (4.9)^2] - 10(4.96)^2\}/9$$
$$= 0.738,$$

$$S_2^2 = \{[(5.3)^2 + (4.6)^2 + \cdots + (4.5)^2] - 10(4.31)^2\}/9$$
$$= 1.043.$$

Substituting these values into (4.1-5), we obtain $S_p^2 = 0.891$. For testing hypotheses as in (4.1-1) with $\Delta = 0$, the calculated value of the $t$ statistic is 1.54. The tabled value $t_{0.975,18} = 2.101$ for $\alpha = 0.05$ and $n_1 + n_2 - 2 = 18$ DF. Since the calculated $t$ is less than tabled $t$, we conclude that treatments 1 and 2 do not differ significantly. For a 95% confidence interval, $t_{0.95,18} = 2.101$. Substitution in (4.1-8) gives a confidence interval for the difference between treatment means, which is $(-0.24, 1.54)$. Table 4.2 illustrates the computations for unequal number of observations from two groups. We point out that $p < .01$ indicates significant difference at the 1% level.

## Paired Comparison Designs

Homogeneity of experimental units is a prerequisite for a sound group-comparison design, and randomization is used to eliminate any inherent uncontrolled biases in the experimental units. A paired comparison

**Table 4.1** Calculations of the $t$ Statistic for Group-Comparison Design

| Observation Number | Treatment 1 $X_{1i}$ | Treatment 2 $X_{2i}$ |
|---|---|---|
| 1 | 5.5 | 5.3 |
| 2 | 4.6 | 4.6 |
| 3 | 3.7 | 2.5 |
| 4 | 3.9 | 3.0 |
| 5 | 5.0 | 4.0 |
| 6 | 4.7 | 4.4 |
| 7 | 5.0 | 4.0 |
| 8 | 6.7 | 6.0 |
| 9 | 5.6 | 4.8 |
| 10 | 4.9 | 4.5 |
| | $n_1 = 10$ | $n_2 = 10$ |
| | $\overline{X}_1 = 4.96$ | $\overline{X}_2 = 4.31$ |
| | $S_1^2 = 0.738$ | $S_2^2 = 1.043$ |
| | $S_1 = 0.86$ | $S_2 = 1.02$ |

Note:
$S_p = \sqrt{[0.738(9) + 1.043(9)]/18} = 0.944,$
$S_{\overline{x}_1 - \overline{x}_2} = 0.944\sqrt{(10 + 10)/10(10)} = 0.422.$
Test statistic is
$t = \overline{X}_1 - \overline{X}_2 / S_{\overline{x}_1 - \overline{x}_2} = 0.65/0.422 = 1.54.$
For a 95% confidence interval for $\mu_1 - \mu_2$, calculate
$(\overline{X}_1 - \overline{X}_2) + 2.101(0.422) = 1.54,$
$(\overline{X}_1 - \overline{X}_2) - 2.101(0.422) = -0.24.$
Hence, the 95% confidence interval is
$-0.24 \leq \mu_1 - \mu_2 \leq 1.54.$

design consists of $n$ pairs, each pair with two like units. Both treatments are applied randomly to the two like units of each pair. Since units of a pair are alike, differences that exist for a pair are essentially due to treatment effects and uncontrolled random errors. Some examples where the pairing of units comes about naturally are

1. The right and the left sides of biological materials, such as the use of the right and left carcass in meat science experiments;

**Table 4.2** Calculations of the $t$ Statistic for Group-Comparison Design with Unequal Number of Observations

| Group 1 | Group 2 |
|---|---|
| 10.1 | 18.0 |
| 15.2 | 18.0 |
| 18.0 | 20.9 |
| 16.0 | 29.8 |
| 19.7 | 30.1 |
| 21.0 | 25.7 |
| 16.1 | 25.1 |
| 18.6 | 28.5 |
| 11.9 | |
| 12.1 | |
| 10.2 | |
| 13.8 | |
| $n_1 = 12$ | $n_2 = 8$ |
| $\overline{X}_1 = 15.23$ | $\overline{X}_2 = 24.51$ |
| $S_1^2 = 13.473$ | $S_2^2 = 24.959$ |
| $S_1 = 3.67$ | $S_2 = 5.00$ |

Note:
$$S_p = \sqrt{[13.473(11) + 24.959(7)]/18} = 4.235,$$
$$S_{\overline{X}_1 - \overline{X}_2} = 4.235\sqrt{(12 + 8)/12(8)} = 1.933.$$
Test statistic is
$$t = \overline{X}_1 - \overline{X}_2/S_{\overline{X}_1 - \overline{X}_2} = 9.28/1.933 = 4.80, p < .01.$$
For a 95% confidence interval for $\mu_1 - \mu_2$, calculate
$$(\overline{X}_1 - \overline{X}_2) + 2.101(1.933) = 13.34,$$
$$(\overline{X}_1 - \overline{X}_2) - 2.101(1.933) = 5.22.$$
Hence, the 95% confidence interval is
$$5.22 \leq \mu_1 - \mu_2 \leq 13.34.$$

2. Biological units of related genetic constitution, such as the use of identical twins and littermates in genetics and nutrition studies;
3. Self-pairing in which a single individual or unit is measured on two occasions, such as "before and after" treatment measurements.

The general idea of the paired comparison design is to form homogeneous pairs of like units so that comparisons between the units of a pair measure differences due to treatments rather than of units. This arrangement leads to dependency between observations on units of the same pair.

Obviously, in the paired comparison designs, observations come in pairs $(X_{1i}, X_{2i})$, where $X_{1i}$ and $X_{2i}$ denote, respectively, observations for the first and second treatment corresponding to the $i$th pair. Since units of a pair are alike, differences $d_i = X_{1i} - X_{2i}$ measure only the differences between treatments. For the paired comparison design also, we are interested in hypotheses (4.1-1) through (4.1-3). An appropriate test statistic now is

$$t = \bar{d} - \Delta/(S_d/\sqrt{n}), \qquad (4.1\text{-}11)$$

where $\bar{d} = \bar{X}_1 - \bar{X}_2$, and

$$S_d = \sqrt{\sum_i (d_i - \bar{d})^2/(n-1)}. \qquad (4.1\text{-}12)$$

Assuming that the differences between pairs are normally distributed, the $t$ statistic in (4.1-11) follows the $t$ distribution with $(n-1)$ DF. Using this statistic, reject $H_0$ at the $\alpha$ level of significance if

$$|t| > t_{(1-\alpha/2),(n-1)}$$

for the setup of the hypotheses in (4.1-1). Reject $H_0$ in (4.1-2) if

$$t > t_{(1-\alpha),(n-1)},$$

and reject $H_0$ in (4.1-3) if

$$t < t_{\alpha,(n-1)},$$

A $(1-\alpha)$ 100% confidence interval estimator for $\Delta = \mu_1 - \mu_2$ is

$$\bar{d} \pm [t_{(1-\alpha/2),(n-1)}]S_d/\sqrt{n}. \qquad (4.1\text{-}13)$$

Table 4.3 displays a numerical example of paired comparison data and its analysis.

One application of the paired comparison designs is in sensory evaluation of products. Each member of a panel compares two products and assigns scores on a sensory scale. Thus, the scores $X_{1i}$ and $X_{2i}$ being made by the $i$th individual are not independent and constitute a pair $(X_{1i}, X_{2i})$. Observations $X_{1i}$ and $X_{2i}$ are correlated because panelists who are high raters tend to give high scores to both experimental units being compared. Likewise, low raters tend to give low scores to both experimental units. Another application of the paired comparison is the case in which a single measurement $X$ is obtained with device 1, and two measurements, $Y$ and $Z$, are obtained with device

**Table 4.3** An Example of Paired Comparison Data Analysis

| Pair | $X_1$ | $X_2$ | $d_i = X_{1i} - X_{2i}$ | $d_i - \bar{d}$ | $(d_i - \bar{d})^2$ |
|------|-------|-------|-------------------------|-----------------|---------------------|
| 1 | 63 | 55 | 8 | 2.7 | 7.29 |
| 2 | 50 | 42 | 8 | 2.7 | 7.29 |
| 3 | 60 | 60 | 0 | −5.3 | 28.09 |
| 4 | 49 | 50 | −1 | −6.3 | 39.69 |
| 5 | 60 | 55 | 5 | −0.3 | 0.09 |
| 6 | 70 | 66 | 4 | −1.3 | 1.69 |
| 7 | 56 | 51 | 5 | −0.3 | 0.09 |
| 8 | 71 | 65 | 6 | 0.7 | 0.49 |
| 9 | 75 | 66 | 9 | 3.7 | 13.69 |
| 10 | 59 | 50 | 9 | 3.7 | 13.69 |
| | 613 | 560 | 53 | 0 | 112.10 |
| Mean | 61.3 | 56.0 | $\bar{d} = 5.3$ | | |

*Note:*
$S_d = \sqrt{112.10/9} = 3.529$
*Test statistic is*
$t = 5.3/(3.529/\sqrt{10}) = 4.75, \quad p < .01.$
*For a 95% confidence interval for $\mu_1 - \mu_2 = \Delta$, calculate*
$5.3 + 2.262(3.529/\sqrt{10}) = 7.82,$
$5.3 - 2.262(3.529/\sqrt{10}) = 2.78.$
*Hence, the 95% confidence interval for $\Delta$ is*
$2.78 \leq \mu_1 - \mu_2 \leq 7.82.$

2. The analysis of such data was considered by Hahn and Nelson (1970). Their procedure is summarized next.

The Hahn–Nelson computational procedure is basically an extension of the paired comparison data analysis. In the Hahn–Nelson procedure, the differences between devices are obtained as follows:

$$d_i = X_i = \tfrac{1}{2}(Y_i + Z_i), \quad i = 1, \ldots, n. \qquad (4.1\text{-}14)$$

It is these differences that are used for calculating the $t$ statistic in (4.1-11), and so on.

Table 4.4 is an example taken from a paper by Hahn and Nelson (1970) involving the depths of craters on a metal surface after the firing of a steel ball of known velocity. Using (4.1-14), we obtain $d_i$ and

**Table 4.4** A Special Example of a Paired Comparison Design and Its Analysis

| Test Units | Readings on Device 1, $X_i$ | First Reading on Device 2, $Y_i$ | Second Reading on Device 2, $Z_i$ | $d_i = X_i - \left(\frac{Y_i + Z_i}{2}\right)$ | $d_i - \bar{d}$ | $(d_i - \bar{d})^2$ |
|---|---|---|---|---|---|---|
| 1 | 71 | 77 | 80 | −7.5 | −0.9 | 0.81 |
| 2 | 108 | 105 | 96 | 7.5 | 14.1 | 198.81 |
| 3 | 72 | 71 | 74 | −0.5 | 6.1 | 37.21 |
| 4 | 140 | 152 | 146 | −9.0 | −2.4 | 5.76 |
| 5 | 61 | 88 | 83 | −24.5 | −17.9 | 320.41 |
| 6 | 97 | 117 | 120 | −21.5 | −14.9 | 222.01 |
| 7 | 90 | 93 | 103 | −8.0 | −1.4 | 1.96 |
| 8 | 127 | 130 | 119 | 2.5 | 9.1 | 82.81 |
| 9 | 101 | 112 | 108 | −9.0 | −2.4 | 5.76 |
| 10 | 114 | 105 | 115 | 4.0 | 10.6 | 112.36 |
| | | | | $\sum_i d_i = -66.0$ | 0.0 | 987.90 |

Note:

Test statistic is

$t = -6.6/(10.48)/\sqrt{10} = -1.99.$

For a 95% confidence interval $\mu_d$ calculate

$-6.6 + 2.262(10.48/\sqrt{10}) = 0.89,$

$-6.6 - 2.262(10.48/\sqrt{10}) = -14.09.$

Hence, the 95% confidence interval is

$-14.09 \leq \mu_d \leq 0.89.$

then $\bar{d} = -66.0/10 = -6.6$. The standard deviation of $\bar{d}$ using (4.1-12) is $S_{\bar{d}} = \sqrt{987.90/9} = 10.48$. The test of significance and the calculation of the confidence interval are shown in Table 4.4. At $\alpha = 0.05$, $t_{0.95,9} = 2.262$. Since $|-1.99| < 2.262$, we conclude that the readings between devices 1 and 2 do not differ significantly at the 5% level of significance.

## 4.2 COMPLETELY RANDOMIZED DESIGNS

The completely randomized (CR) design is an extension of the group-comparison designs of Section 4.1 when the number of treatments exceeds two. The experimental units in the CR design are randomly assigned to $t$ treatments so that each unit has an equal chance of being assigned to one of the treatments. A CR design is used if the experimental units are alike. Its layout appears in Table 4.5. In this layout, the variation from column to column is due to differences between treatments. The variation among rows is random and beyond control if the experimental units are alike. Therefore, the row variation in a CR design is called *experimental error* or *error variance* and is denoted by $\sigma_2^2$. A statistical model to describe an observation $X_{ij}$ in a CR design is given by

$$X_{ij} = \mu + T_i + \varepsilon_{ij}, \qquad \begin{array}{l} i = 1, \ldots, t, \\ j = 1, \ldots, n_i, \end{array} \qquad (4.2\text{-}1)$$

where $\mu$ is the grand mean, $T_i$ is the effect due to the $i$th treatment so defined that $\Sigma T_i = 0$, and $\varepsilon_{ij}$ are the random experimental errors that

**Table 4.5** Layout of a Completely Randomized Design

|  | Treatments | | | |
|---|---|---|---|---|
|  | **1** | **2** | $\cdots$ | **t** |
|  | $X_{11}$ | $X_{21}$ | $\cdots$ | $X_{t1}$ |
|  | $X_{12}$ | $X_{22}$ | $\cdots$ | $X_{t2}$ |
|  | $\vdots$ | $\vdots$ | $\cdots$ | $\vdots$ |
|  | $X_{1n}$ | $X_{2n}$ | $\cdots$ | $X_{tn}$ |
| Totals | $X_1.$ | $X_2.$ | $\cdots$ | $X_t.$ $\qquad X..$ |

take into account all sources of variation not controlled by the design. Usually, the $\varepsilon_{ij}$ are assumed to be independently and normally distributed with mean 0 and constant variance $\sigma_e^2$. For the model in (4.2-1), the breakdown of total variation in the data is summarized by a one-way analysis of variance table, as discussed in Chapter 3, Section 3.1. The calculations of the various SS associated with the model are also given in Chapter 3. The main advantages of the CR design are its simplicity and its easy execution. The main disadvantage is the requirement of alike (homogeneous) experimental units, which may be difficult to fulfill, particularly when the number of treatments is large.

| **Example 4.2-1** | Consider four chemical mixtures of a compound at varying concentrations denoted by $T_1$, $T_2$, $T_3$, and $T_4$. The effect of the compound is known to minimize moisture losses in ground meats. Forty tubes of ground lean meat were prepared and later randomly and equally distributed among the four treatments. To minimize experimental error, investigators took extreme care in the application of the four compounds to the four groups of meat. |

Table 4.6 displays the percent amount of water loss after heating the tubes in a hot water bath. Each observation is assumed to follow Model (4.2-1). The hypotheses of interest are

$H_0:$   $T_1 = T_2 = T_3 = T_4,$
$H_a:$   At least for one pair, treatment means are significantly different.

The standard deviation $S_i$ for each treatment and the coefficient of variation $CV = (S_i/\overline{X}_i)100$ are given in Table 4.6. The statistic CV is a measure of relative variation independent of the unit of measurement. In this example, the variation in $T_1$ is roughly twice that in $T_4$. The calculations for the SS are shown in the footnote to Table 4.6.

The analysis of variance table corresponding to Model (4.2-1) is shown in Table 4.7. The $F$ ratio at 3 numerator and 36 denominator DF is significant at the 1% level. Therefore, we reject the null hypothesis of equal treatment means or effects. Which of the four means are significantly different? The answer is not given by the $F$ test. For an answer, one must run a multiple-comparison test, as discussed in Chapter 3.

**Table 4.6** Example of the Statistical Analysis of a Completely Randomized Design with Four Treatments

|  | $T_1$ | $T_2$ | $T_3$ | $T_4$ |  |
|---|---|---|---|---|---|
|  | 24.54 | 22.73 | 22.02 | 20.83 |  |
|  | 23.97 | 22.28 | 21.97 | 24.40 |  |
|  | 22.85 | 23.99 | 21.34 | 23.01 |  |
|  | 24.34 | 23.23 | 22.83 | 23.54 |  |
|  | 24.47 | 23.94 | 22.90 | 21.60 |  |
|  | 24.36 | 23.52 | 22.28 | 21.86 |  |
|  | 23.37 | 24.41 | 21.14 | 24.57 |  |
|  | 24.46 | 22.80 | 21.85 | 22.28 |  |
|  | 23.89 | 21.52 | 22.16 | 22.41 |  |
|  | 24.00 | 23.73 | 22.56 | 22.81 |  |
| $X_i.$ | 240.25 | 232.15 | 221.05 | 227.31 | $X.. = 920.76$ |
| $\overline{X}_i.$ | 24.03 | 23.22 | 22.11 | 22.73 |  |
| $S_i$ | 0.55 | 0.89 | 0.58 | 1.20 |  |
| CV (%) | 2.28 | 3.82 | 2.61 | 5.26 |  |

Note:
$CF = (920.76)^2/40 = 21,194.97.$
$SSto = [(24.54)^2 + (23.97)^2 + \cdots + (22.81)^2] - CF = 45.32.$
$SStr = \{[(240.25)^2 + (232.15)^2 + (221.05)^2 + (227.31)^2]/10\} - CF = 19.69.$
$SSe = 45.32 - 19.69 = 25.63.$

**Table 4.7** Analysis of Variance for the Data of a Completely Randomized Design (Example 4.2-1)

| Source of Variance | DF | SS | MS | F Ratio |
|---|---|---|---|---|
| Total | 39 | 45.32 |  |  |
| Between Treatments | 3 | 19.69 | 6.56 | 9.22* |
| Within Treatments | 36 | 25.63 | 0.71 |  |

*$p < .01.$

## 4.3 **RANDOMIZED COMPLETE BLOCK DESIGNS**

A drawback of the CR design is that it requires all the experimental units to be alike. If this assumption is not satisfied, it is not feasible to use the CR design. To account for the variation among the units, blocks of homogeneous units are formed. Each block consists of as many units as the number of treatments. The treatments are applied randomly to units within a block. Such a design is called the randomized complete block (RCB) design. Thus, in an RCB design, each block consists of homogeneous experimental units in order to increase the sensitivity of treatment comparisons within blocks.

An RCB design experiment proceeds as follows:

1. Define the experimental unit. A can or package of ham, a stick of sausage, a package of frankfurters or luncheon meats, and a fabricated meat cut are some examples of experimental units in food research.
2. On the basis of several criteria, form blocks consisting of like experimental units.
3. Randomly assign treatments to the experimental units within each block. For instance, consider a layout of five treatments ($t = 5$) arranged in three blocks ($b = 3$). The random allocation of treatments to experimental units may yield the following layout:

Treatments

|        |   |   |   |   |   |   |
|--------|---|---|---|---|---|---|
|        | 1 | 1 | 2 | 4 | 3 | 5 |
| Blocks | 2 | 5 | 3 | 4 | 2 | 1 |
|        | 3 | 1 | 3 | 2 | 5 | 4 |

The experimental outcomes of an RCB design can be arranged in a two-way layout according to blocks and treatments. A layout is given in Table 4.8. The $(i, j)$th observation $X_{ij}$ for the $i$th treatment and the $j$th block is assumed to follow the statistical model

$$X_{ij} = \mu + T_i + B_j + \varepsilon_{ij}, \qquad \begin{matrix} i = 1, \ldots, t, \\ j = 1, \ldots, b, \end{matrix} \qquad (4.3\text{-}1)$$

where $\mu$ is the grand mean, estimated by $\overline{X} = X../N$, where $N = tb$; $T_i$ is the effect of the $i$th treatment, estimated by $t_i = \overline{X}_{i.} - \overline{X}$, with $\Sigma_i t_i = 0$; $B_j$ is the effect of the $j$th block estimated by $b_j = \overline{X}_{.j} - \overline{X}$, with $\Sigma_j b_j = 0$; and $\varepsilon_{ij}$ are random errors assumed to be independently and normally distributed, with mean 0 and variance $\sigma_e^2$.

**Table 4.8** Layout of Experimental Outcomes of a Randomized Complete Block Design

| Blocks | Treatments | | | | Block Sums | Block Means |
|---|---|---|---|---|---|---|
| | **1** | **2** | $\cdots$ | **t** | | |
| 1 | $X_{11}$ | $X_{21}$ | $\cdots$ | $X_{t1}$ | $X_{\cdot 1}$ | $\overline{X}_{\cdot 1}$ |
| 2 | $X_{12}$ | $X_{22}$ | $\cdots$ | $X_{t2}$ | $X_{\cdot 2}$ | $\overline{X}_{\cdot 2}$ |
| $\vdots$ | $\vdots$ | $\vdots$ | $\vdots$ | $\vdots$ | $\vdots$ | $\vdots$ |
| b | $X_{1b}$ | $X_{2b}$ | $\cdots$ | $X_{tb}$ | $X_{\cdot b}$ | $\overline{X}_{\cdot b}$ |
| Treatment Sums | $X_{1\cdot}$ | $X_{2\cdot}$ | $\cdots$ | $X_{t\cdot}$ | $X_{\cdot\cdot}$ | |
| Treatment Means | $\overline{X}_{1\cdot}$ | $\overline{X}_{2\cdot}$ | $\cdots$ | $\overline{X}_{t\cdot}$ | | $\overline{X}$ |

In this model it is assumed that the treatment by block interactions, symbolized by $(TB)_{ij}$, is negligible. If this is true, then we say that there is complete additivity between treatments and blocks, in the sense that differences between treatments in different blocks are constant regardless of the blocks. If indeed interaction effects are not negligible, then in the preceding model they are confounded with $\varepsilon_{ij}$ and hence inseparable.

Table 4.9 displays the analysis of variance for Model (4.3-1). The expected mean square, $E(MS)$, expressions for treatments, blocks, and errors are also shown in the table.

A hypothesis of interest in an RCB is the equality of treatment effects. That is,

$$H_0: \quad T_i = T_j \quad \text{for each pair} \quad (i, j);$$
$$H_a: \quad T_i \neq T_j \quad \text{for at least a pair} \quad (i, j).$$

**Table 4.9** Analysis of Variance for the Randomized Complete Block Design Corresponding to Model (4.3-1)

| Source of Variance | DF | SS | MS | E(MS) |
|---|---|---|---|---|
| Total | $N - 1$ | SSto | | |
| Treatments | $t - 1$ | SStr | MStr | $\sigma_e^2 + b\sum_i T_i^2 / t - 1$ |
| Blocks | $b - 1$ | SSbl | MSbl | $\sigma_e^2 + t\sum_j B_j^2 / b - 1$ |
| Experimental Units | $(t - 1)(b - 1)$ | SSe | MSe | $\sigma_e^2$ |

The hypothesis $H_0$ is tested by the $F$ ratio statistic,

$$F = MStr/MSe,$$

with $(t - 1)$ and $(t - 1)(b - 1)$ DF.

By definitions of the various sums of squares, we have

$$SSto = \sum_i \sum_j (X_{ij} - \overline{X})^2 = \sum_i \sum_j X_{ij}^2 - CF, \qquad (4.3\text{-}2)$$

$$SStr = b \sum_i (\overline{X}_{i\cdot} - \overline{X})^2 = \frac{\sum_i X_{i\cdot}^2}{b} - CF, \qquad (4.3\text{-}3)$$

$$SSbl = t \sum_j (\overline{X}_{\cdot j} - \overline{X})^2 = \frac{\sum_j X_{\cdot j}^2}{t} - CF, \qquad (4.3\text{-}4)$$

where $CF = X_{\cdot\cdot}^2/N$. The error sum of squares is defined by

$$SSe = \sum_i \sum_j (X_{ij} - \overline{X}_{i\cdot} - \overline{X}_{\cdot j} + \overline{X})^2,$$

which can also be found by the following relationship:

$$SSe = SSto - SStr - SSbl. \qquad (4.3\text{-}5)$$

---

**Example 4.3-1**

Four samples of beef patties with varying amounts of flavor additives were evaluated for preference by a 32-member panel. Half of the preference data is reproduced in Table 4.10. The beef patties are the experimental units, and the panelists are the blocks. The preference scores within a block are expected to be homogenous because they come from the same individual. The treatments denoted by $T_1$, $T_2$, $T_3$, and $T_4$ were served simultaneously to the panelists in random fashion to minimize the order effect. A hedonic scale ranging from "1 = dislike extremely" to "8 = like extremely" was used. The calculations for the sums of squares following formulas (4.3-2) through (4.3-5) are given in the footnote to Table 4.10, and the analysis of variance is shown in Table 4.11. For testing a null hypothesis of the equality of the four treatment effects, the appropriate $F$ ratio $= 7.58/1.85 = 4.10$. At the 5% level of significance, the tabulated $F$ value with 3 and 45 DF is 2.84, which is exceeded by 4.10. This leads to the rejection of the null hypothesis and the decision that the amount of flavor additives does matter. The estimated standard error of any treatment effect is

**Table 4.10** Preference Scores for Beef Patties (Example 4.3-1)

| Blocks | Treatments | | | | $X_{\cdot j}$ | $\bar{X}_{\cdot j}$ |
|---|---|---|---|---|---|---|
| | $T_1$ | $T_2$ | $T_3$ | $T_4$ | | |
| 1 | 6 | 7 | 7 | 8 | 28 | 7.00 |
| 2 | 4 | 6 | 3 | 5 | 18 | 4.50 |
| 3 | 5 | 6 | 4 | 6 | 21 | 5.25 |
| 4 | 4 | 4 | 4 | 4 | 16 | 4.00 |
| 5 | 8 | 4 | 3 | 4 | 19 | 4.75 |
| 6 | 3 | 5 | 3 | 6 | 17 | 4.25 |
| 7 | 3 | 5 | 4 | 4 | 16 | 4.00 |
| 8 | 4 | 5 | 2 | 6 | 17 | 4.25 |
| 9 | 3 | 4 | 6 | 5 | 18 | 4.50 |
| 10 | 3 | 6 | 4 | 4 | 17 | 4.25 |
| 11 | 1 | 5 | 2 | 6 | 14 | 3.50 |
| 12 | 3 | 5 | 6 | 7 | 21 | 5.25 |
| 13 | 2 | 5 | 6 | 7 | 20 | 5.00 |
| 14 | 7 | 5 | 7 | 7 | 26 | 6.50 |
| 15 | 7 | 4 | 5 | 5 | 21 | 5.25 |
| 16 | 3 | 6 | 4 | 6 | 19 | 4.75 |
| $X_{i\cdot}$ | 66 | 82 | 70 | 90 | $X.. = 308$ | |
| $\bar{X}_i$ | 4.13 | 5.13 | 4.38 | 5.62 | $\bar{X} = 4.81$ | |

Note:
$CF = (308)^2/64 = 1482.25.$
$SSto = (6^2 + 4^2 + \cdots + 5^2 + 6^2) - CF = 155.75.$
$SStr = \{[(66)^2 + (82)^2 + (70)^2 + (90)^2]/16\} - CF = 22.75.$
$SSbl = \{[(28)^2 + (18)^2 + \cdots + 19^2]/4\} - CF = 49.75.$
$SSe = 155.75 - 22.75 - 49.75 = 83.25.$

$$SE(t_i) = \sqrt{MSe/b} = \sqrt{1.85/16} = 0.34,$$

and the estimated standard error of the difference between two treatment effects $t_i$ and $t_j$, $i \neq j$, is

$$SE(t_i - t_j) = \sqrt{2MSe/b} = \sqrt{2(1.85)/16} = 0.48.$$

**Table 4.11** Analysis of Variance of Preference Scores for a Randomized Complete Block Design (Example 4.3-1)

| Source of Variance | DF | SS | MS | F Ratio |
|---|---|---|---|---|
| Total | 63 | 155.75 | | |
| Treatments | 3 | 22.75 | 7.58 | 4.10* |
| Blocks (Panelists) | 15 | 49.75 | 3.32 | |
| Error (Experimental Units) | 45 | 83.25 | 1.85 | |
| *$p < .05$. | | | | |

## Randomized Complete Block Designs with More Than One Observation per Experimental Unit

We may point out that, if only one observation is planned for each treatment in each block, it will not be possible to estimate treatment by block interactions. If the RCB design in Example 4.3-1 is repeated to take more than one observation for each treatment in each block, then it would be possible to estimate treatment by block interactions. The statistical model for an RCB design model with interaction is given by

$$X_{ijk} = \mu + T_i + B_j + (TB)_{ij} + \varepsilon_{ijk}, \quad \begin{aligned} i &= 1,\ldots,t, \\ j &= 1,\ldots,b, \\ k &= 1,\ldots,r, \end{aligned} \quad (4.3\text{-}6)$$

where $(TB)_{ij}$ are the interaction effects, and all other effects of the model are as defined in (4.3-1). The effects $T_i$, $B_j$, and $(TB)_{ij}$ are estimated by $t_i$, $b_j$, and $(tb)_{ij}$, respectively, as follows:

$$t_i = \overline{X}_{i..} - \overline{X}, \qquad \sum_i t_i = 0,$$

$$b_j = \overline{X}_{.j.} - \overline{X}, \qquad \sum_i b_j = 0,$$

$$(tb)_{ij} = \overline{X}_{ij.} - \overline{X}_{i..} - \overline{X}_{.j.} + \overline{X}, \qquad \sum_i (tb)_{ij} = \sum_j (tb)_{ij} = 0.$$

The SS due to treatments, blocks, interactions, and errors are as follows:

**Table 4.12** Analysis of Variance for a Randomized Complete Block Design with More Than One Observation per Experimental Unit

| Source of Variance | DF | MS | E(MS) | F Ratio |
|---|---|---|---|---|
| Total | $N - 1$ | | | |
| Treatments ($T$) | $t - 1$ | MStr | $\sigma_e^3 + rb\sum_i T_i^2/t - 1$ | MStr/MSe |
| Blocks ($B$) | $b - 1$ | MSbl | $\sigma_e^2 + rt\sum_j B_j^2/b - 1$ | |
| TB Interactions | $(t - 1)(b - 1)$ | MStb | $\sigma_e^2 + r\sum_i \sum_j (TB)_{ij}^2/(t - 1)(b - 1)$ | |
| Error | $tb(r - 1)$ | MSe | $\sigma_e^2$ | |

$$\text{CF} = X^2.../N, \qquad N = tbr; \tag{4.3-7}$$

$$\text{SSto} = \sum_i \sum_j \sum_k (X_{ijk} - \bar{X})^2 = \sum_i \sum_j \sum_k X_{ijk}^2 - \text{CF}; \tag{4.3-8}$$

$$\text{SStr} = br \sum_i (\bar{X}_{i..} - \bar{X})^2 = \left(\sum_i X_{i..}^2/br\right) - \text{CF}; \tag{4.3-9}$$

$$\text{SSbl} = tr \sum_i (\bar{X}_{.j.} - \bar{X})^2 = \left(\sum_j X_{.j.}^2/tr\right) - \text{CF}; \tag{4.3-10}$$

$$\text{SStb} = r\sum_i \sum_j \left(\bar{X}_{ij.} - \bar{X}_{i..} - \bar{X}_{.j.} + \bar{X}\right)^2$$
$$= \left(\sum_i \sum_j X_{ij.}^2/r\right) - \text{CF} - \text{SStr} - \text{SSbl}; \tag{4.3-11}$$

$$\text{SSe} = \text{SSto} - \text{SStr} - \text{SSbl} - \text{SStb}. \tag{4.3-12}$$

Table 4.12 shows how to test the null hypothesis of equality of treatment effects when interactions $(TB)_{ij}$ are accounted for in the statistical model. The $F$ ratio under Model I (fixed effects) for testing $H_0: T_i = T_j$, for all pairs $(i, j)$ is $F = \text{MStr/MSe}$ with $(t - 1)$ and $tb(r - 1)$ DF.

**Example 4.3-2**

The data in Table 4.13 are taken from a large set of data pertaining to sensory off-flavor development in luncheon meats. In this study, a package, defined as the experimental unit, contains six slices representing the observational units; four treatments were involved. Products were served simultaneously to the panelists (blocks) in a random fashion to

**Table 4.13** Off-Flavor Scores of Luncheon Meats (Example 4.3-2)

| Blocks | $T_1$ | $T_2$ | $T_3$ | $T_4$ | Block Totals |
|---|---|---|---|---|---|
| | | Treatments | | | |
| 1 | 4 | 2 | 2 | 3 | |
| | $\frac{2}{6}$ | $\frac{2}{4}$ | $\frac{3}{5}$ | $\frac{3}{6}$ | 21 |
| 2 | 3 | 2 | 4 | 1 | |
| | $\frac{4}{7}$ | $\frac{3}{5}$ | $\frac{3}{7}$ | $\frac{1}{2}$ | 21 |
| 3 | 3 | 3 | 3 | 2 | |
| | $\frac{5}{8}$ | $\frac{2}{5}$ | $\frac{3}{6}$ | $\frac{2}{4}$ | 23 |
| 4 | 4 | 4 | 2 | 1 | |
| | $\frac{3}{7}$ | $\frac{1}{5}$ | $\frac{4}{6}$ | $\frac{1}{2}$ | 20 |
| 5 | 3 | 3 | 1 | 2 | |
| | $\frac{3}{6}$ | $\frac{2}{5}$ | $\frac{2}{3}$ | $\frac{1}{3}$ | 17 |
| Treatment Totals | 34 | 24 | 27 | 17 | 102 |
| Treatment Means | 3.4 | 2.4 | 2.7 | 1.7 | $\overline{X} = 2.6$ |

Note:

$CF = (102)^2/40 = 260.1.$

$SSto = (4^2 + 2^2 + \cdots + 2^2 + 1^2) - CF = 41.9.$

$SStr = \{[(34)^2 + (24)^2 + (27)^2 + (17)^2]/10\} - CF = 14.9.$

$SSbl = \{[(21)^2 + (21)^2 + (23)^2 + (20)^2 + (17)^2]/8\} - CF = 2.4.$

$SStb = [(6^2 + 4^2 + \cdots + 3^2 + 3^3)/2] - CF - SStr - SSbl$

$\qquad = 287.0 - 260.1 - 14.9 - 2.4 = 9.6.$

$SSe = 41.9 - 14.9 - 2.4 - 9.6 = 15.0.$

minimize order effects. An intensity rating scale ranging from "1 = none" through "7 = very strong off-flavor" was used. Each panelist made two sensory evaluations at various times for each package.

The calculation of the various sums of squares uses formulas (4.3-7) through (4.3-12). The calculations are shown at the bottom of Table 4.13. Assuming a fixed-effects model, the $F$ ratio for testing the significance of treatment effects is $F = 4.97/0.75 = 6.63$, which is significant at the 1% level (Table 4.14). A multiple-comparison test may be carried out if one wishes to compare pairs of treatments.

**Table 4.14** Analysis of Variance of Off-Flavor Scores (Example 4.3-2)

| Source of Variance | DF | SS | MS | F Ratio |
|---|---|---|---|---|
| Total | 39 | 41.9 | | |
| Treatments (T) | 3 | 14.9 | 4.97 | 6.63* |
| Blocks (B) | 4 | 2.4 | 0.60 | |
| TB Interactions | 12 | 9.6 | 0.80 | |
| Error | 20 | 15.0 | 0.75 | |
| *p < .01. | | | | |

## 4.4 LATIN SQUARE DESIGNS

In the RCB design the sensitivity of treatment comparisons is achieved by having blocks (rows) with homogeneous units. The Latin square (LS) designs consist of blocking of experimental units into rows and columns so that each treatment appears once in a row and once in a column. The blocking by rows and columns is effective for controlling the variability among the experimental units. A main drawback of this design is that the number of treatments in the experiment must equal the number of rows, which in turn must equal the number of columns. A listing of the LS designs was given by Fisher and Yates (1963).

A LS design is given in Table 4.15, where *A*, *B*, *C*, and *D* denote four treatments. Note that the number of treatments is equal to the number of rows and the number of columns. Let $X_{ij(k)}$ denote the

**Table 4.15** Layout for a 4 × 4 Latin Square

| Rows | Columns | | | |
|---|---|---|---|---|
| | 1 | 2 | 3 | 4 |
| 1 | A | B | C | D |
| 2 | B | C | D | A |
| 3 | C | D | A | B |
| 4 | D | A | B | C |

observation on the $k$th treatment in the $i$th row and the $j$th column position. In general, if there are $m$ treatments in an LS design, its statistical model can be written as

$$X_{ij(k)} = \mu + C_i + R_j + T_k + \varepsilon_{ij(k)}, \quad \begin{array}{l} i = 1, \ldots, m, \\ j = 1, \ldots, m, \\ k = 1, \ldots, m, \end{array} \quad (4.4\text{-}1)$$

where $\mu$ is the grand mean estimated by $X.../N$, where $N = m^2$; $C_i$ is the effect of the $i$th column estimated by $c_i = \overline{X}_{i..} - \overline{X}$, with $\Sigma_i c_i = 0$; $R_j$ is the effect of the $j$th row estimated by $r_j = \overline{X}_{.j.} - \overline{X}$, with $\Sigma_j r_j = 0$; $T_k$ is the effect of the $k$th treatment estimated by $t_k = \overline{X}_{..(k)} - \overline{X}$, with $\Sigma_k t_k = 0$; and $\varepsilon_{ij(k)}$ are random errors assumed to be independently and normally distributed with mean 0 and variance $\sigma_e^2$.

We emphasize that an LS design yields $m^2$ observations rather than $m^3$, as the subscript may erroneously seem to indicate. To emphasize this, we put the subscript $k$ in parentheses. The model in (4.4-1) does not account for the interaction effects. The analysis of variance of this model is given in Table 4.16. The sums of squares associated with each source of variance are given by

$$CF = X^2.../N,$$

$$SSto = \sum_{i,j} (X_{ij(k)} - \overline{X})^2 = \sum_{i,j} X_{ij(k)}^2 - CF, \quad (4.4\text{-}2)$$

$$SSc = m \sum_i (\overline{X}_{i..} - \overline{X})^2 = \frac{\sum_i X_{.i.}^2}{m} - CF, \quad (4.4\text{-}3)$$

$$SSr = m \sum_j (\overline{X}_{.j.} - \overline{X})^2 = \frac{\sum_j X_{.j.}^2}{m} - CF, \quad (4.4\text{-}4)$$

**Table 4.16** Analysis of Variance for Latin Square Designs Assuming Fixed-Effects Model (4.4-1)

| Source of Variance | DF | MS | E(MS) |
|---|---|---|---|
| Total | $N - 1$ | | |
| Columns | $m - 1$ | MSc | $\sigma_e^2 + m \sum_i C_i^2 / m - 1$ |
| Rows | $m - 1$ | MSr | $\sigma_e^2 + m \sum_j R_j^2 / m - 1$ |
| Treatments | $m - 1$ | MStr | $\sigma_e^2 + m \sum_k T_k^2 / m - 1$ |
| Error | $(m - 1)(m - 2)$ | MSe | $\sigma_e^2$ |

$$SStr = m \sum_k (\overline{X}_{..(k)} - \overline{X})^2 = \frac{\sum_k X_{..(k)}^2}{m} - CF, \qquad (4.4\text{-}5)$$

$$SSe = SSto - SSc - SSr - SStr. \qquad (4.4\text{-}6)$$

Note that the $\Sigma$ notation used in (4.4-2) means that the sum is over all the $m^2$ units in the LS design.

---

Four different doses of insulin denoted by A, B, C, and D were tested on rabbits and compared in terms of the subsequent sugar content in the rabbits' blood. The data and relevant calculations are given in Table 4.17, and the analysis of variance is shown in Table 4.18. The reader may verify how the treatment sums were obtained in Table 4.17.

**Example 4.4-1**

Assuming a fixed-effects model, the $F$ ratio $521.1/19.8 = 26.32$ is significant at the 1% level. Therefore, we conclude that the four doses of insulin differed in their effects on the level of blood sugar in the rabbit.

**Table 4.17** Latin Square Experiment to Study the Effect of Insulin on Blood Sugar in Rabbits[a]

| Days | Rabbits I | II | III | IV | Totals |
|------|------|------|------|------|--------|
| 1 | B: 47 | A: 90 | C: 79 | D: 50 | 266 |
| 2 | D: 46 | C: 74 | B: 63 | A: 69 | 252 |
| 3 | A: 62 | B: 61 | D: 58 | C: 66 | 247 |
| 4 | C: 76 | D: 63 | A: 87 | B: 59 | 285 |
|  | 231 | 288 | 287 | 244 | 1050 |

[a]*Source: De Lury (1946).*
*Note:*
*Treatment sums:*
  *A = 62 + 90 + 87 + 69 = 308,*
  *B = 47 + 61 + 63 + 59 = 230,*
  *C = 76 + 74 + 79 + 66 = 295,*
  *D = 46 + 63 + 58 + 50 = 217.*
  *CF = (1050)²/16 = 68,906.25.*
  *SSto = [(47)² + (46)² + ⋯ + (66)² + (59)²] − CF = 2,545.75.*
  *SSc = {[(231)² + (288)² + (287)² + (244)²]/4} − CF = 646.25.*
  *SSr = {[(266)² + (252)² + (247)² + (285)²]/4} − CF = 217.25.*
  *SStr = {[(308)² + (230)² + (295)² + (217)²]/4} − CF = 1,563.25.*
  *SSe = 2545.75 − 646.25 − 217.25 − 1563.25 = 119.00.*

**Table 4.18** Analysis of Variance of Insulin Data on Rabbits (Example 4.4-1)

| Source of Variance | DF | SS | MS | F Ratio |
|---|---|---|---|---|
| Total | 15 | 2545.75 | | |
| Rabbits | 3 | 646.25 | 215.4 | |
| Days | 3 | 217.25 | 72.4 | |
| Treatments | 3 | 1563.25 | 521.1 | 26.32* |
| Error | 6 | 119.00 | 19.8 | |

*$p < .01$.

## Replicating LS Designs

When there are only four treatments to be studied using the LS design, the DF for error will be small. In such cases we cannot have a reliable estimate of the experimental error $\sigma_e^2$ from the LS design. Hence, the LS design is often replicated to have more observations. For replicating LS designs, we may begin with a standard Latin square and generate other Latin squares through randomization. A standard square is one in which the first row and the first column are in alphabetic or in numeric order. Four such standard LS designs are shown:

$$
\begin{array}{cccccc}
\text{I} & A & B & C & D & \quad \text{II} & A & B & C & D \\
& B & A & D & C & & B & C & D & A \\
& C & D & B & A & & C & D & A & B \\
& D & C & A & B & & D & A & B & C \\
\end{array}
$$

$$
\begin{array}{cccccc}
\text{III} & A & B & C & C & \quad \text{IV} & A & B & C & D \\
& B & D & A & C & & B & A & D & C \\
& C & A & D & B & & C & D & A & B \\
& D & C & B & A & & D & C & B & A \\
\end{array}
$$

For Latin squares of order $5 \times 5$ or less, all rows except the first and all columns are randomized. For the higher order squares, the rows and columns are independently randomized and the treatments assigned randomly to the letters $A$, $B$, $C$, ..., etc. Suppose we have selected standard square I and label each row and each column from 1 to 4:

Columns

|   | 1 | 2 | 3 | 4 |
|---|---|---|---|---|
| 1 | A | B | C | D |
| 2 | B | A | D | C |
| 3 | C | D | B | A |
| 4 | D | C | A | B |

Rows

We can generate other Latin squares through the following steps in a randomizing process:

1. Randomize rows 2, 3, and 4. If a random arrangement yields 3, 4, 2, then we have

Columns

|   | 1 | 2 | 3 | 4 |
|---|---|---|---|---|
| 1 | A | B | C | D |
| 3 | C | D | B | A |
| 4 | D | C | A | B |
| 2 | B | A | D | C |

Rows

Thus, row 3 moves up into the second position, and so on.

2. Randomize columns 1, 2, 3, and 4. If a random arrangement yields 1, 4, 3, 2, we have the following Latin square:

$$
\begin{array}{cccc}
A & D & C & B \\
C & A & B & D \\
D & B & A & C \\
B & C & D & A
\end{array}
$$

3. Last, randomly assign four treatments to the four letters. Suppose the random assignment results in the following:

$$
\begin{array}{cccc}
T_3 & T_1 & T_4 & T_2 \\
A & B & C & D
\end{array}
$$

Thus, the above randomization leads to the following $4 \times 4$ LS design:

$$
\begin{array}{cccc}
T_3 & T_2 & T_4 & T_1 \\
T_4 & T_3 & T_1 & T_2 \\
T_2 & T_1 & T_3 & T_4 \\
T_1 & T_4 & T_2 & T_3
\end{array}
$$

If we need another replication, the preceding randomizing process can be repeated all over again.

A statistical model for the LS designs with replications can be written as

$$
X_{i(j)kp} = \mu + S_i + T_j + (ST)_{ij} + C_{k:i} + R_{p:i} + \varepsilon_{i(j)kp}, \tag{4.4-7}
$$

where

$$i = 1, \ldots, s; \quad j = 1, \ldots, m; \quad k = 1, \ldots, m; \quad p = 1, \ldots, m;$$

$X_{i(j)kp}$ is the observed response for the $i$th square, the $j$th treatment in the $k$th column, and the $p$th row; $\mu$ is the grand mean obtained by $X \ldots / N$, where $N = sm^2$; $S_i$ is the effect of the $i$th replication or square estimated by $s_i = \overline{X}_{i \cdots} - \overline{X}$, with $\Sigma_i \, s_i = 0$; $T_j$ is the effect of the $j$th treatment estimated by $t_j = \overline{X}_{\cdot(j) \cdots} - \overline{X}$, with $\Sigma_j \, t_j = 0$; $(ST)_{ij}$ is the interaction effect between the $i$th square and $j$th treatment estimated by $(st)_{ij} = \overline{X}_{i(j) \cdots} - \overline{X}_{i \cdots} - \overline{X}_{\cdot(j) \cdots} + \overline{X}$, with $\Sigma_i \, \Sigma_j^* \, (st)_{ij} = 0$; $C_{k:i}$ is the effect of the $k$th column within a square estimated by $c_{k:i} = \overline{X}_{i \cdot k \cdot} - \overline{X}_{i \cdots}$, with $\Sigma_k \, c_{k: \, i} = 0$; $R_{p:i}$ is the effect of the $p$th row within a square estimated by $r_{p:i} = \overline{X}_{i \cdots p} - \overline{X}_{i \cdots}$, with $\sigma_p \, r_{p:i} = 0$; and $\varepsilon_{i \cdot (j)kp}$ are random errors assumed to be independently and normally distributed with mean 0 and variance $\sigma_e^2$.

Notice that the column and row effects are nested within squares because they were independently randomized from one square to another. The analysis of variance for the model in (4.4-7) is shown in Table 4.19.

**Table 4.19** Analysis of Variance for Latin Square Design with Replications for Model (4.4-7)

| Source of Variance | DF | SS |
|---|---|---|
| Total | $sm^2 - 1$ | $SSto = \sum\limits_{i,(j),k,p} X_{i(j)kp}^2 - CF$ |
| Squares (S) | $s - 1$ | $SSs = \left( \sum\limits_i X_i^2 \ldots / m^2 \right) - CF$ |
| Treatments (T) | $m - 1$ | $SStr = \left( \sum\limits_j X_{\cdot j}^2 \ldots / sm \right) - CF$ |
| ST Interactions | $(s - 1)(m - 1)$ | $SSst = \left( \sum\limits_{i,(j)} X_{i(j) \cdots}^2 / m \right) - CF - SSs - SStr$ |
| Columns:squares | $s(m - 1)$ | $SSc : s = \left( \sum\limits_{i,(j)} X_{i \cdot k \cdot}^2 / m \right) - \left( \sum\limits_i X_i^2 \ldots / m^2 \right)$ |
| Rows:squares | $s(m - 1)$ | $SSr : s = \left( \sum\limits_{i,p} X_{i \cdots p}^2 / m \right) - \left( \sum\limits_i X_i^2 \ldots / m^2 \right)$ |
| Error | $s(m - 1)(m - 2)$ | $SSe = SSto - SSs - SStr - SSst - SSc{:}s - SSr{:}s$ |

**Example 4.4-2**

The data for this example are taken from a study that pertains to tenderization of heavy pork loins. Four treatments were studied with respect to their effectiveness in improving tenderness using a 4 × 4 Latin square. Four animals make up the columns in a square, and the two lefts and two rights make up the rows.

Two replicate squares from this experiment are reproduced in Table 4.20. The marginal totals for columns, rows, and squares are listed in this table. The treatment totals are found by summing each treatment score separately over the squares. For example, the total for treatment A is $6 + 7 + 5 + 5 + 6 + 4 + 3 + 6 = 42$. The calculations for the various sums of squares are given in the footnote to Table 4.20. The analysis of variance is given in Table 4.21. The F ratio with numerator DF $= 3$ and denominator DF $= 12$ for testing the significance of treatment effects is $F = 3.79/0.96 = 3.95$, which is found significant at the 5% level. The hypotheses about squares, rows, and columns may be similarly tested.

**Table 4.20** Tenderness Data for the Heavy Pork Loins Experiment (Example 4.4-2)

| | Square I | | | | | Square II | | | | | |
|---|---|---|---|---|---|---|---|---|---|---|---|
| **Columns** | **1** | **2** | **3** | **4** | **Totals** | **5** | **6** | **7** | **8** | **Totals** | |
| Rows | | | | | | | | | | | |
| 1 | D: 7 | A: 7 | C: 7 | B: 7 | 28 | A: 6 | D: 7 | C: 8 | B: 5 | 26 | |
| 2 | B: 5 | C: 6 | A: 5 | D: 7 | 23 | B: 3 | A: 4 | D: 5 | C: 4 | 16 | |
| 3 | A: 6 | B: 7 | D: 7 | C: 6 | 26 | D: 7 | C: 6 | B: 5 | A: 6 | 24 | |
| 4 | C: 8 | D: 8 | B: 6 | A: 5 | 27 | C: 5 | B: 6 | A: 3 | D: 6 | 20 | |
| | 26 | 28 | 25 | 25 | 104 | 21 | 23 | 21 | 21 | 86 | $X... = 190$ |

Note:
Treatment totals: $A = 42, B = 44, C = 50, D = 54$.
Treatment means: $A = 5.25, B = 5.50, C = 6.25, D = 6.75$.

$CF = (190)^2/32 = 1,128.13.$

$SSto = (7^2 + 5^2 + \cdots + 6^2 + 6^2) - CF = 53.87.$

$SSs = \{[(104)^2 + (86)^2]/16\} - CF = 10.12.$

$SStr = \{[(42)^2 + (44)^2 + (50)^2 + (54)^2]/8\} - CF = 11.37.$

$SSst = \{[(23)^2 + (25)^2 + (27)^2 + (29)^2 + (19)^2 + (19)^2 + (23)^2 + (25)^2]/4\} - CF - SSs - SStr = 0.38.$

$SSc : s = \{[(26)^2 + (28)^2 + (25)^2 + (25)^2 + (21)^2 + (23)^2 + (21)^2 + (21)^2]/4\} - \{[(104)^2 + (86)^2]/16\} = 2.25.$

$SSr : s = \{[(28)^2 + (23)^2 + (26)^2 + (27)^2 + (26)^2 + (16)^2 + (24)^2 + (20)^2]/4\} - \{[(104)^2 + (86)^2]/16\} = 18.25.$

$SSe = 53.87 - 10.12 - 11.37 - 0.38 - 2.25 - 18.25 = 11.50.$

**Table 4.21** Analysis of Variance for Tenderness Data in the Latin Square Design (Example 4.4-2)

| Source of Variance | DF | SS | MS | F Ratio |
|---|---|---|---|---|
| Total | 31 | 53.87 | | |
| Squares ($S$) | 1 | 10.12 | 10.12 | |
| Treatments ($T$) | 3 | 11.37 | 3.79 | 3.95* |
| $ST$ Interactions | 3 | 0.38 | 0.13 | |
| Columns: squares | 6 | 2.25 | 0.38 | |
| Rows: squares | 6 | 18.25 | 3.04 | |
| Error | 12 | 11.50 | 0.96 | |

*$p < .05$.

## Partially Replicated Latin Square

The additivity assumption of various effects in the LS designs is often inappropriate. The Latin square designs may be replicated to estimate the additivity of effects. But as in sensory evaluation, replication can be a difficult task owing to taste fatigue and adaptation. Therefore, partially replicated LS designs that minimize the number of treatments to be replicated are desired. Youden and Hunter (1955) proposed an extension of $m \times m$ Latin squares by including an $(m + 1)$th additional observation for each treatment; the additional observation should be taken only once from each row and each column. The following diagrams illustrate the manner in which the additional observation is taken from a $3 \times 3$ and $4 \times 4$ Latin square:

```
              Columns
            1    2    3
        1 │ A    B    CC
Rows    2 │ C    AA   B
        3 │ BB   C    A
```

```
              Columns
            1    2    3    4
        1 │ AA   B    C    D
Rows    2 │ B    D    A    CC
        3 │ D    BB   C    A
        4 │ C    A    DD   B
```

For example, in the $3 \times 3$ LS design, treatment $B$ is replicated only in column 1, row 3. The position of a treatment to be replicated is determined at random. These additional observations provide a measure of interaction or nonadditivity of effects and also of experimental error. For a Latin square of size $m$, such extension will result in $N = m(m + 1)$ observations.

Table 4.22 outlines the analysis of variance corresponding to the statistical model given here:

$$X_{ij(k)} = \mu + C_i + R_j + T_K + N_{ij(k)} + \varepsilon_{ij(k)}, \quad \begin{array}{l} i = 1, \ldots, m, \\ j = 1, \ldots, m, \\ k = 1, \ldots, m. \end{array} \quad (4.4\text{-}8)$$

The parameters $\mu$, $C_i$, $R_j$, and $T_k$ are as defined in (4.4-1). The interaction effect $N_{ij(k)}$ is a measure of nonadditivity of the rows, columns, and treatments. To calculate various sums of squares, let $X_{i..}$, $X_{.j.}$, $X_{..k}$, and $X_{...}$ denote the sum of the $i$th column, the $j$th row, the $k$th treatment, and the grand total, respectively. Define

$$c_i = X_{i..} - (R_i^* + X...)/m + 1, \quad (4.4\text{-}9)$$

where $R_i^*$ is the row total associated with the $i$th column in which the additional observation was taken. As a check, we must have $\Sigma c_i = 0$. Also define

$$t_k = X_{..k} - (R_k^* + C_k^* + X...)/m + 2, \quad (4.4\text{-}10)$$

**Table 4.22** Analysis of Variance for Partially Replicated Latin Square Designs

| Source of Variance | DF | SS | E(MS) |
|---|---|---|---|
| Total | $N - 1$ | SSto | |
| Rows | $m - 1$ | SSr | $\sigma_e^2 + m \sum_i R_i^2 / m - 1$ |
| Columns Adjusted for Rows | $m - 1$ | SSc | $\sigma_e^2 + m \sum_j C_j^2 / m - 1$ |
| Treatments Adjusted for Rows and Columns | $m - 1$ | SStr | $\sigma_e^2 + m \sum_k T_k^2 / m - 1$ |
| Interactions | $(m - 1)(m - 2)$ | SSn | $\sigma_e^2 + \sum_i \sum_j N_{ij(k)}^2 / (m - 1)(m - 2)$ |
| Error | $m$ | SSe | $\sigma_e^2$ |

where $R_k^*$ and $C_k^*$ are row and column totals, respectively, in which treatment $k$ is replicated. As a check, it should be that $\Sigma t_k = 0$. The calculations for the various sums of squares in Table 4.22 are

$$CF = \frac{X^2 \cdots}{N};$$

$$SSto = \sum_{i,j,(k)} X_{ij(k)}^2 - CF; \qquad (4.4\text{-}11)$$

$$SSr = \frac{\sum_j X^2 \cdot_j \cdot}{m+1} - CF; \qquad (4.4\text{-}12)$$

$$SSc(\text{adjusted for rows}) = \frac{m+1}{m(m+2)} \sum_i c_i^2; \qquad (4.4\text{-}13)$$

$$SStr(\text{adjusted for rows and columns}) = \frac{m+2}{m(m+3)} \sum_k t_k^2; \qquad (4.4\text{-}14)$$

$$SSn = SSto - SSr - SSc - SStr. \qquad (4.4\text{-}15)$$

To find $SSe$, let $d_i$ be the difference between the two observations for the $i$th treatment. Then

$$SSe = \sum_i d_i^2/2. \qquad (4.4\text{-}16)$$

The $k$th treatment mean $\overline{X}_k$ adjusted for row and column effects is

$$\overline{X}_k = \hat{\mu} + \bar{t}_k, \qquad (4.4\text{-}17)$$

where $\hat{\mu} = X\ldots/N$ and $\bar{t}_k = [(m+2)/m(m+3)]t_k$. The standard error of each $\overline{X}_k$ is estimated by

$$SE(\overline{X}_k) = \sqrt{[(m+2)/N]MSe}. \qquad (4.4\text{-}18)$$

---

**Example 4.4-3**

This example, taken from a paper by Youden and Hunter (1955), pertains to densities of briquettes formed from three sizes of particles (rows), compressed at three pressures (treatments), and fired at three temperatures (columns). Table 4.23 displays the data in a 3 × 3 Latin square where there is an additional observation for each treatment. The $SSto$ and $SSr$ were obtained using (4.4-11) and (4.4-12), respectively.

The sum of squares $SSc$ is obtained by first calculating $c_i$ from (4.4-9):

$$c_1 = 3587 - (3558 + 11047)/4 = -64.25,$$
$$c_2 = 3762 - (3692 + 11047)/4 = 77.25,$$
$$c_3 = 3698 - (3797 + 11047)/4 = -13.00.$$

**Table 4.23** Densities of Briquettes in a Partially Replicated 3 × 3 LS Design (Example 4.4-3)

| Particle Sizes | Temperature, °F | | | Row Totals |
| --- | --- | --- | --- | --- |
| | 1900 | 2000 | 2300 | |
| 5–10 | C: 964 | A: 933 | B: 942 958 | 3,797 |
| 10–15 | B: 996 | C: 905 943 | A: 908 | 3,692 |
| 15–20 | A: 842 845 | B: 981 | C: 890 | 3,558 |
| | 3587 | 3762 | 3698 | 11,047 |

Note:

Treatment totals: A = 3528, B = 3817, C = 3702.

$CF = (11,047)^2/12 = 10,169,684.08.$

$SSto = [(964)^2 + (936)^2 + \cdots + (890)^2] - CF = 21,512.92.$

$SSr = \{[(3797)^2 + (3692)^2 + (3558)^2]/4\} - CF = 7175.17.$

$SSc = (3 + 1)/3(3 + 2)[(-64.25)^2 + (77.25)^2 + (-13.00)^2] = 2737.23.$

$SStr = (3 + 2)/3(3 + 3)[(-110.40)^2 + (108.60)^2 + (1.80)^2] = 6662.60.$

$SSe = [(842 - 845)^2 + (905 - 943)^2 + (942 - 958)^2]/2 = 854.50.$

$SSn = 21,512.92 - SSr - SSc - SStr - SSe = 4,083.42.$

These values are substituted into (4.4-13) to obtain SSc, as shown at the bottom of Table 4.23. Likewise, for treatment sum of squares adjusted for row and column effects, we first calculate $t_k$ using (4.4-10):

$$t_1 = 3528 - 3638.40 = -110.40,$$
$$t_2 = 3817 - 3708.40 = 108.60,$$
$$t_3 = 3702 - 3700.00 = 1.80.$$

We obtain SStr by substituting these values into (4.4-14). Now the calculations of SSe and SSn follow from (4.4-15) and (4.4-16), respectively.

The analysis of variance is shown in Table 4.24. At 2 numerator and 3 denominator DF, the F ratio = 3331.30/284.83 = 11.69 is significant at the 5% level; thus, the null hypothesis of no treatment effect differences is rejected. The F ratio for interaction is significant at the 10% level and indicates the possibility of interactions. The presence of interactions

**Table 4.24** Analysis of Variance for Densities of Briquettes in a Partially Replicated LS Design (Example 4.4-3)

| Source of Variance | DF | SS | MS | F Ratio |
|---|---|---|---|---|
| Total | 11 | 21,512.92 | | |
| Particle Sizes (Rows) | 2 | 7,175.17 | | |
| Temperatures (Columns) | 2 | 2,737.23 | | |
| Pressure (Treatments) | 2 | 6,662.60 | 3,331.30 | 11.69* |
| Interactions | 2 | 4,083.42 | 2,041.71 | 7.17** |
| Error | 3 | 854.50 | 284.83 | |

*$p < .05$.
**$p < .10$.

would invalidate the assumption of additivity of effects in the LS designs. Using (4.4-17), the adjusted treatment means are estimated as follows:

$$A: \ \overline{X}_1 = 920.6 - 30.7 = 889.9,$$
$$B: \ \overline{X}_2 = 920.6 + 30.2 = 950.8,$$
$$C: \ \overline{X}_3 = 920.6 + \ \ 0.5 = 921.1.$$

The standard error of the adjusted treatment means using (4.4-18) is

$$SE(\overline{X}_k) = \sqrt{(5/12)284.83} = 10.89.$$

A multiple-comparison test may be done following Section 3.2

## 4.5 CROSS-OVER DESIGNS

The cross-over design, originally known as the changeover design, is an extension of the LS design. The cross-over design has the property that, if, say, treatment A follows treatment B in one sequence, then B follows A in another sequence, and so on for the other treatments. Such arrangement of treatments permits the estimation of carryover or residual effects of treatments. The cross-over designs were first used in dairy cattle feeding experiments, where the residual effects of feedings are considered significant. Indeed, feeding experiments are similar to sensory testings of foods; aftertaste or carryover effects of foods are known to exist in highly flavored and highly spiced products. Therefore, designs that would adjust for aftertaste effects are desirable. The cross-over designs serve this purpose.

In this section we consider balanced cross-over designs of the LS type that adjust for first-order carryover effects only, that is, the carryover effects that persist only in one period. Designs for the estimation of carryover effects extending to two periods are more complex. For these designs, the reader may refer to Williams (1950). The development of the cross-over designs is largely by Cochran et al. (1941), Patterson (1950), Lucas (1951), and Patterson and Lucas (1962). Consider a 4 × 4 Latin square where rows represent periods or orders of taste testing, columns are panelists or experimental units, and letters are treatments. A Latin square layout is shown here:

$$
\begin{array}{cc}
 & \text{Panelists} \\
 & \begin{array}{cccc} 1 & 2 & 3 & 4 \end{array} \\
\begin{array}{cc}
\text{Periods} & 1 \\
\text{or} & 2 \\
\text{testing} & 3 \\
\text{orders} & 4
\end{array} &
\left|\begin{array}{cccc}
A & B & C & D \\
B & C & D & A \\
C & D & A & B \\
D & A & B & C
\end{array}\right.
\end{array}
\qquad (4.5\text{-}1)
$$

An inspection of the layout with respect to tasting orders shows the following:

> $A$ follows $D$ in columns 2, 3, and 4;
> $B$ follows $A$ in columns 1, 3, and 4;
> $C$ follows $B$ in columns 1, 2, and 4;
> $D$ follows $C$ in columns 1, 2, and 3.

Notice that in the LS layout (4.5-1), carryover effects are present. For example, an observation on treatment $B$ from column 1 has in it the carryover effect of treatment $A$, and so on. To explain it further, let $X_a$, $X_b$, $X_c$, and $X_d$ denote observations from panelist (column) 1 for treatments $A$, $B$, $C$, and $D$, respectively. We notice that these observations coming from the same panelist are made up of effects as follows:

> $X_a =$ the effect of period 1, the effect of treatment $A$, and the random-error effect;
> $X_b =$ the effect of period 2, the effect of treatment $B$, the residual effect of treatment $A$, and the random-error effect;
> $X_c =$ the effect of period 3, the effect of treatment $C$, the residual effect of treatment $B$, and the random-error effect;

$X_d$ = the effect of period 4, the effect of treatment $D$, the residual effect of treatment $C$, and the random-error effect.

The observations from other panelists are also similarly confounded with residual effects. For the layout in (4.5-1), it is not possible to estimate the residual effects. An appropriate change in layout (4.5-1) to permit estimation of carryover effects can be made to result in a balanced changeover design, as shown in

$$
\begin{array}{c}
\text{Panelists} \\
\begin{array}{cc|cccc}
 & & 1 & 2 & 3 & 4 \\
\hline
 & 1 & A & B & C & D \\
\text{Testing} & 2 & B & C & D & A \\
\text{orders} & 3 & C & D & A & B \\
 & 4 & D & A & B & C \\
\end{array}
\end{array} \qquad (4.5\text{-}2)
$$

A cross-over design is said to be balanced for estimating the residual effects if each treatment in its layout is preceded or followed by every other treatment an equal number of times. For more details on cross-over designs, we may refer to Patterson and Lucas (1962). In layout (4.5-2) $B$ follows $A$ in column 1, but $A$ follows $B$ in column 4. Likewise, $C$ follows $B$ in column 1, $B$ follows $C$ in column 4, and so on. Thus, aside from possessing properties of a Latin square design, each treatment follows every other treatment once, satisfying conditions for balance.

Let us define parameters for a balanced cross-over design as follows:

$p$ = number of periods or tasting orders;
$t$ = number of treatments;
$c$ = number of experimental units in each square;
$s$ = number of squares;
$n$ = total number of experimental units.

Table 4.25 lists some balanced cross-over designs of the LS type useful in sensory evaluations. When $t > 4$, the incomplete block designs considered in Chapter 5 may be more suitable. A statistical model for a cross-over design balanced for first-order carryover effects is

$$
X_{ijk(m)} = \mu + S_i + P_j + (SP)_{ij} + C_{k:i} + T_m + \varepsilon_{ijk(m)} \qquad (4.5\text{-}3)
$$

where

$$
\begin{aligned}
i &= 1, \ldots, s; & j &= 1, \ldots, p; \\
k &= 1, \ldots, c; & m &= 1, \ldots, t;
\end{aligned}
$$

**Table 4.25** Balanced Cross-Over Designs Useful in Sensory Evaluation with $t = p = c^a$

| Design Number | Design Parameters | Treatment Arrangements of Complete Sets of Latin Squares | | |
|---|---|---|---|---|
| 2 | $t = p = c = 3$ | ABC | | ABC |
| | $s = 2$ | BCA | | CAB |
| | $n = 6$ | CAB | | BCA |
| | $\alpha = 5/24$ | | | |
| | $\alpha' = 1/6$ | | | |
| 5 | $t = p = c = 4$ | | ABCD | |
| | $s = 1$ | | BDAC | |
| | $n = 4$ | | CADB | |
| | $\alpha = 11/40$ | | DCBA | |
| | $\alpha' = \frac{1}{4}$ | | | |
| 6 | $t = p = c = 4$ | ABCD | ABCD | ABCD |
| | $s = 3$ | BADC | CDAB | DCBA |
| | $n = 12$ | CDAB | DCBA | BADC |
| | $\alpha = 11/120$ | DCBA | BADC | CDAB |
| | $\alpha' = 1/12$ | | | |

*aSource: Patterson and Lucas (1962).*

$X_{ijk(m)}$ is the observed response for the $i$th square, the $j$th row (order), the $k$th column, and the $m$th treatment; $\mu$ is the grand mean; $S_i$ is the effect of the $i$th square with $\Sigma_i S_i = 0$; $P_j$ is the effect of the $j$th order with $\Sigma_j C_{k:i} = 0$; $C_{k:i}$ is the effect of the $k$th column (panelist) nested within square with $\Sigma_k C_{k:i} = 0$; $T_m$ is the effect of the $m$th treatment including direct and carryover effects with $\Sigma_k C_{k:i} = 0$; and $\varepsilon_{ijk(m)}$ are random errors assumed to be independently and normally distributed.

Table 4.26 is the analysis of variance for the model in (4.5-3). The breakdown of the treatment sums of squares due to direct and carryover effects follows a procedure by Cochran et al. (1941) and Williams (1949). This procedure was illustrated by Patterson and Lucas

**Table 4.26** Analysis of Variance of a Balanced Cross-Over Design for Model (4.5-3)

| Source of Variance | DF | SS |
|---|---|---|
| Total | $np - 1$ | $SSto = \sum\limits_{i,j,k,(m)} X^2_{ijk(m)} - CF$ |
| Squares | $s - 1$ | $SSs = \left( \sum\limits_{i} X^2_{i...}/m^2 \right) - CF$ |
| Rows | $p - 1$ | $SSp = \left( \sum\limits_{j} X^2_{.j..}/sm \right) - CF$ |
| Squares $\times$ Rows | $(s - 1)(p - 1)$ | $SSsp = \left( \sum\limits_{i,k} X^2_{ij..}/m \right) - CF - SSs - SSp$ |
| Columns: Squares | $s(c - 1)$ | $SSc : s = \left( \sum\limits_{i,k} X^2_{i.k.}/m \right) - CF - SSs$ |
| Treatments | $2(t - 1)$ | $SStr$ [see (4.5-4)] |
| Direct Effects (Adjusted for Carryover Effects) | $t - 1$ | $SSd$ [see (4.5-6)] |
| Carryover Effects (Unadjusted for Direct Effects) | $t - 1$ | $SSr'$ [see (4.5-7)] |
| Direct Effects (Unadjusted for Carryover Effects) | $t - 1$ | $SSd'$ [see (4.5-9)] |
| Carryover Effects (Adjusted for Direct Effects) | $t - 1$ | $SSr$ [see (4.5-10)] |
| Error | $(n - s)(p - 1) - 2(t - 1)$ | $SSe$ (by difference) |

(1962) using dairy cattle milk production data and by Ferris (1957) using food taste testing data.

Following the notation of Patterson and Lucas, the steps in the calculations of treatment sums of squares are as follows:

1. Calculate the sums for rows, columns, and then the grand sum $G$.
2. Calculate the following quantities for each treatment:

$T$ = treatment sum;

$R$ = sum over columns of all treatments that immediately follow the treatment of interest in the same column (the sum of all these columns is denoted by $U_1$; if $p = c$, $U_1$ is not needed);

$U$ = sum of all columns that have the treatment of interest in the last row;

$P_1$ = sum of all first rows that include the treatment of interest;

$B$ = sum of all rows that include the treatment of interest.

**3.** Calculate $T'$ and $R'$ for each treatment by

$$T' = pT = U_1 - U, \qquad \sum T' = 0,$$
$$R' = pR - U_1 + (pP_1 - B)/c, \quad \sum R' = 0.$$

If $p = c$, use

$$T' = pT - B$$

and

$$R' = pR + U + P_1 - B(p+1)/p.$$

**4.** Calculate $\widehat{T}$ and $\widehat{R}$ for each treatment by

$$\widehat{T} = dT' + cR', \quad \sum \widehat{T} = 0,$$

and

$$\widehat{R} = T' + pR', \quad \sum \widehat{R} = 0,$$

where $d = c(p - 1) - 1$.

**5.** Calculate the treatment SS by

$$\text{SStr} = \alpha \left( \sum \widehat{T} T' + c \sum \widehat{R} R' \right) / p^2 d, \tag{4.5-4}$$

where

$$\alpha = pd(t-1)/n(p-1)(pd-c).$$

The SStr in (4.5-4) can be partitioned into two components, each with $(t - 1)$ DF: one due to the direct effects of treatments, denoted by SSd, and one due to first-order carryover effects of treatments denoted by SSr. We may write it as

$$
\begin{aligned}
\text{SStr} = &\ (\text{SS due to direct effects, adjusted for} \\
&\ \text{carryover effects}) + (\text{SS due to carryover} \\
&\ \text{effects, unadjusted for direct effects}) \\
= &\ \text{SSd} + \text{SSr}',
\end{aligned}
\tag{4.5-5}
$$

where

$$\text{SSd} = \alpha \sum \widehat{T}^2 / p^2 d^2, \tag{4.5-6}$$

and

$$\text{SSr}' = \text{SStr} - \text{SSd}. \tag{4.5-7}$$

Alternatively, we can have

$$SStr = (\text{SS due to direct effects, unadjusted for carryover effects}) + (\text{SS due to carryover effects, adjusted for direct effects}) \tag{4.5-8}$$

$$= SSd' + SSr,$$

where

$$SSd' = \alpha' \sum \widehat{T}^{12}/p^2, \tag{4.5-9}$$

$$SSr = c\alpha \sum \widehat{R}^2/p^3 d, \tag{4.5-10}$$

and

$$\alpha' = (t-1)/n(p-1).$$

Both (4.5-5) and (4.5-8) may also be used to check the computations.

The estimates of direct effects of treatments and carryover effects denoted by $\widehat{t}$ and $\widehat{r}$ are, respectively,

$$\widehat{t} = \alpha \widehat{T}/pd, \quad \sum \widehat{t} = 0, \tag{4.5-11}$$

with standard error

$$SE(\widehat{t}) = \sqrt{\alpha MSe}, \tag{4.5-12}$$

and

$$\widehat{r} = c\alpha \widehat{R}/pd, \quad \sum \widehat{r} = 0, \tag{4.5-13}$$

with standard error

$$SE(\widehat{r}) = \sqrt{(cp\alpha/d)MSe}. \tag{4.5-14}$$

The standard error of the sum $(\widehat{t}+\widehat{r})$ of both direct and carryover effects is

$$SE(\widehat{t}+\widehat{r}) = \sqrt{\frac{\alpha(d+cp+2c)}{d}MSe} \tag{4.5-15}$$

When the carryover effects are negligible, the estimates of direct effects of treatments are

$$\widehat{t}' = \alpha' T'/p, \tag{4.5-16}$$

with standard error given by

$$SE(\widehat{t}') = \sqrt{\alpha'\left[\frac{MSe + MSr}{(n-s)(p-1)}\right]}, \tag{4.5-17}$$

where MSe is the mean square error, and MSr is the carryover mean square (adjusted for direct effects).

The standard error of the differences between effects is obtained by multiplying the standard error of the effects by $\sqrt{2}$. On the basis of the estimates of $\hat{t}$ and $\hat{r}$, we may estimate each treatment mean $\overline{X}$, adjusted for carryover effects, by

$$\overline{X} = \hat{\mu} + \hat{t} + \hat{r}, \qquad\qquad (4.5\text{-}18)$$

where $\hat{\mu}$ is the estimate of the grand mean.

---

Table 4.27 displays a portion of sensory data from an experiment pertaining to color liking of a product. This portion forms two sets of 3 × 3 Latin squares, which is Design 2 in Table 4.25. The initial steps for calculating the SS are to obtain totals for rows and columns, as shown in Table 4.27. The SS for panelists within square SSc : s can be calculated in two ways. The first is illustrated in Table 4.27, and the second is to compute the SS for each square and then take the sum as follows:

**Example 4.5-1**

**Table 4.27** Color Liking Data to Illustrate a Balanced Cross-Over Design with $t = 3$, $p = 3$, $c = 3$, and $s = 2$ (Example 4.5-1)

| Squares: | | I | | | | II | | | | |
|---|---|---|---|---|---|---|---|---|---|---|
| Panelists: | | **1** | **2** | **3** | **Totals** | **4** | **5** | **6** | **Totals** | **Order Totals** |
| Orders | 1 | A: 5 | B: 7 | C: 8 | 20 | A: 3 | B: 4 | C: 6 | 13 | 33 |
| | 2 | B: 7 | C: 8 | A: 6 | 21 | C: 6 | A: 4 | B: 5 | 15 | 36 |
| | 3 | C: 8 | A: 6 | B: 8 | 22 | B: 5 | C: 6 | A: 5 | 16 | 38 |
| | | 20 | 21 | 22 | 63 | 14 | 14 | 16 | 44 | 107 |

*Note:*
$G = X.\!.\!.\!. = 107, \hat{\mu} = 5.94.$
$CF = (107)^2/18 = 636.0556.$
$SSto = (5^2 + 7^2 + \cdots + 5^2 + 5^2) - CF = 675.0000 - CF = 38.9444.$
$SSs = \{[(63)^2 + (44)^2]/9\} - CF = 656.1111 - CF = 20.0555.$
$SSp = \{[(33)^2 + (36)^2 + (38)]^2/6\} - CF = 638.1667 - CF = 2.1111.$
$SSsp = \{[(20)^2 + (21)^2 + \cdots + (15)^2 + (16)^2]/3\} - CF - SSs - SSp = 0.1111.$
$SSc : s = \{[(20)^2 + (21)^2 + \cdots + (14)^2 + (16)^2]/3\} - CF - SSs = 1.5556.$

$$\text{Square I}: \quad \{[(20)^2 + (21)^2 + (22)^2]/3\} - [(63)^2/9] = 0.6667,$$

$$\text{Square II}: \quad \{[(13)^2 + (15)^2 + (16)^2]/3\} - [(44)^2/9] = 0.8889.$$

Thus, SSc: $s = 0.6667 + 0.8889 = 1.5556$, as calculated in Table 4.27.

Table 4.28 gives calculations for the estimation of the direct and carryover effects of treatments. The totals for each treatment are obtained as in Section 4.4 for the LS design. The $R$ quantity, as defined earlier, involves columns 1, 3, 4, and 5. For treatment $A$,

$$R = 7 + 8 + 6 + 6 + = 27.$$

The sum of columns 1, 2, 4, and 5 is quantity $U_1$, as shown in Table 4.28. Note that $B$ and $p_1$ are the same for each treatment because the design is an LS and $B$ is equal to the grand total $G = 107$. Treatment $A$ is in the last row of columns 2 and 6; hence, $U = 21 + 16 = 37$. The preceding calculations are repeated for all treatments and are shown in Table 4.28.

Following steps 3 and 4, quantities $T'$, $R'$, $\widehat{T}$, and $\widehat{R}$ are obtained. Since $p = c$ in our example, using the appropriate formula, we have, for treatment $A$,

$$T' = 3(29) - 107 = -20,$$

$$R' = [3(27) + 37 + 33] - 107(3 + 1)/3 = 8.3333,$$

$$\widehat{T} = 5 - (20) + 3(8.3333) = -75.0001,$$

$$\widehat{R} = -20 + 3(8.3333) = 4.9999.$$

The estimates of the direct effects of treatment $A$ and carryover effects are calculated using (4.5-11) and (4.5-13), respectively:

$$\hat{t} = [5(-75.0001)/24]/15 = -1.0417,$$

$$\hat{r} = [3(5)(4.9999)/24]/15 = 0.2083.$$

Using (4.5-4), the treatment sum of squares is

$$\text{SStr} = [(5/24)(2763.0038 + 434.9883)]/45 = 14.8055,$$

which can be subdivided into SS owing to direct and carryover effects using (4.5-6), (4.5-7), (4.5-9), and (4.5-10) as follows:

$$\text{SSd} = [(5/24)(10,062.0396)]/225 = 0.3152;$$

$$\text{SS}'r = \text{SStr} - \text{SSd} = 14.8055 - 9.3152 = 5.4903;$$

$$\text{SS}'d = [(1/6)(762)]/9 = 14.1111;$$

$$\text{SSr} = [3(5/24)(149.9940)]/135 = 0.6944.$$

**Table 4.28** Estimation of Direct and Carryover Effects for the Color Liking Data (Example 4.5-1)

| Treatments | $T$ | $R$ | $B$ | $U_1$ | $U$ | $P_1$ | $T'$ | $R'$ | $\hat{T}$ | $\hat{R}$ | $\hat{t}$ | $\hat{r}$ |
|---|---|---|---|---|---|---|---|---|---|---|---|---|
| A | 29 | 27 | 107 | 70 | 37 | 33 | −20 | 8.3333 | −75.0001 | 4.9999 | −1.0417 | 0.2083 |
| B | 36 | 25 | 107 | 71 | 36 | 33 | 1 | 1.3333 | 8.9999 | 4.9999 | 0.1250 | 0.2083 |
| C | 42 | 22 | 107 | 73 | 34 | 33 | 19 | −9.6666 | 66.0002 | −9.9998 | 0.9167 | −0.4167 |
| | 107 | 74 | 321 | 214 | 107 | 99 | 0 | 0.0 | 0.0 | 0.0 | 0.0 | −0.0001 |

Note:

Treatment A:

$R = 7 + 8 + 6 + 6 = 27,$
$B = 20 + 21 + 22 + 13 + 15 + 16 = 107,$
$U_1 = 20 + 22 + 14 + 14 = 70,$
$U = 21 + 16 = 37,$
$P_1 = 20 + 13 = 33.$

Treatment B:

$R = 8 + 8 + 4 + 5 = 25,$
$B = 107,$
$U_1 = 20 + 21 + 14 + 16 = 71,$
$U = 22 + 14 = 36,$
$P_1 = 33.$

Treatment C:

$R = 6 + 6 + 5 + 5 = 22,$
$B = 107,$
$U_1 = 21 + 22 + 14 + 16 = 73,$
$U = 20 + 14 = 34,$
$P_1 = 33.$

As a check, note that $SSd + SSr' = 14.111 + 0.6944 = 14.8055 = SSd' + SSr$.

The analysis of variance is given in Table 4.29. The mean squares $MSd$, $MSr$, and $MSe$ due to direct treatment effects, carryover treatment effects, and the error, respectively, are obtained from their respective SS as given in the following:

**Table 4.29** Analysis of Variance of Color Liking Data (Example 4.5-1)

| Source of Variance | DF | SS | MS | F Ratio |
|---|---|---|---|---|
| Total | 17 | 38.9444 | | |
| Squares | 1 | 20.0555 | | |
| Tasting Orders | 2 | 2.1111 | | |
| Squares × Tasting Orders | 2 | 0.1111 | | |
| Panelists: Squares | 4 | 1.5556 | | |
| Treatments | 4 | 14.8055 | | |
| Direct Effects (Adjusted for Carryover Effects) | 2 | 9.3152 | 4.6576 | 60.96* |
| Carryover Effects (Unadjusted for Direct Effects) | 2 | 5.4903 | | |
| Direct Effects (Unadjusted for Carryover Effects) | 2 | 14.1111 | | |
| Carryover Effects (Adjusted for Direct Effects) | 2 | 0.6944 | 0.3472 | 4.54** |
| Error | 4 | 0.3056 | 0.0764 | |

*$p < .01$.
**$p < .10$.

**Table 4.30** Estimates of Adjusted Treatment Means and Their Standard Errors for the Color Liking Data ($\widehat{\mu} = 5.94$)

| Treatments | Direct Effect Means $\widehat{\mu} + \widehat{t}$ | Carryover Effects $\widehat{r}$ | Adjusted Means $\widehat{\mu} + \widehat{t} + \widehat{r}$ | Means Unadjusted for Carryover Effects $\widehat{\mu} + \widehat{t}$ |
|---|---|---|---|---|
| A | 4.90 | 0.2083 | 5.11 | 4.83 |
| B | 6.07 | 0.2083 | 6.28 | 6.00 |
| C | 6.86 | −0.4167 | 6.44 | 7.00 |
| Standard Errors | 0.13 | 0.17 | 0.25 | 0.14 |

$$MSd = 9.3152/2 = 4.6576,$$

$$MSr = 0.6944/2 = 0.3472,$$

$$MSe = 0.3056/4 = 0.0764.$$

The $F$ ratio for testing the significance of direct treatment effects adjusted for carryover effect is

$$F = MSd/MSe = 4.6576/0.0764 = 60.96,$$

which is significant at the 1% level. This result indicates that the color liking of products differs for at least one pair of treatments. For testing the significance of carryover effects adjusted for direct effects, the $F$ ratio is

$$F = MSr/MSe = 0.3472/0.0764 = 4.54,$$

which is significant at the 10% level. This suggests that the carryover effects may exist and should be accounted for. Table 4.30 contains treatment means adjusted and unadjusted for carryover effects and their standard errors.

## 4.6 SPLIT PLOT DESIGNS

Split plot designs originally developed for agricultural experiments where equally sized parcels (or blocks) of land were used to study the effect of various fertilizer experiments on crop yields. The blocks were divided into whole plots, and the whole plots further subdivided into smaller units called subplots. In the simplest such design, each whole plot receives treatments from Factor A, and each subplot within the whole plot receives treatments from Factor B. The essential feature of a split plot design is that experimental units (EUs) of different sizes are present. For example, say Factor A has $a = 4$ levels with 10 blocks. Each level of Factor A is applied in random order to the whole plots within a block. At this point, what we have is essentially a randomized block design with 10 blocks, each of size 4 (Table 4.31).

The split plot comes about by subdividing each whole plot EU into $b$ subplots, and treatment B levels are randomly assigned to the subplots within each main plot. In the preceding design, with $b = 2$, we could get the design shown in Table 4.32.

For Factor B, there exists 10*4 blocks of size 2 each. Every Factor A level in each block can be considered to be a block for Factor B. The experimental factor applied to the whole plot is also called the *whole plot* factor; the treatment applied to the smaller *subplot* is referred to as the *split plot factor*.

**Table 4.31** Whole Plot Factor

| Block | Factor A Levels by Whole Plots | | | |
|-------|-------|-------|-------|-------|
| 1 | $A_2$ | $A_1$ | $A_4$ | $A_3$ |
| 2 | $A_3$ | $A_4$ | $A_1$ | $A_2$ |
| ... | | | | |
| 10 | $A_4$ | $A_2$ | $A_3$ | $A_1$ |

**Table 4.32** Split Plot Factor

| Block | Whole Plot 1 | | Whole Plot 2 | | Whole Plot 3 | | Whole Plot 4 | |
|-------|------|------|------|------|------|------|------|------|
| 1 | $A_2B_1$ | $A_2B_2$ | $A_1B_2$ | $A_1B_1$ | $A_4B_2$ | $A_4B_1$ | $A_3B_1$ | $A_3B_2$ |
| 2 | $A_3B_2$ | $A_3B_1$ | $A_4B_2$ | $A_4B_1$ | $A_1B_1$ | $A_1B_2$ | $A_2B_1$ | $A_2B_2$ |
| ... | | | | | | | | |
| 10 | $A_4B_1$ | $A_4B_2$ | $A_2B_1$ | $A_2B_2$ | $A_3B_2$ | $A_3B_1$ | $A_1B_2$ | $A_1B_1$ |

Table 4.33 provides data from three production lines used to produce three products denoted as Control, Formula1, and Formula2 studying the effect of process temperature on the product where Temperature was applied according to a randomized scheme. Five trained sensory judges evaluated the exterior color of the products using a 5-point intensity rating scale where 1 = light and 5 = dark.

**Example 4.6-1**

In this experiment, blocks are the judges, the whole plots contain the Product (Factor A) as a treatment, and the subplots are the temperature applied to the product (Factor B). The model for this split plot design is

$$y_{ijk} = \mu + A_i + \delta_j + \gamma_{ij} \quad \text{\textit{Whole Plot part}}$$
$$+ B_k + AB_{ik} + \varepsilon_{ijk} \quad \text{\textit{Subplot part}}$$

where

$y_{ijk}$ = score given to $i$th level of Factor A at $k$th level of Factor B in block $j$,
$A_i$ = effect of $i$th level of Factor A,
$\delta_j$ = effect of $j$th block, $j = 1,2,\ldots,5$, corresponding to the 5 judges,
$\gamma_{ij}$ = whole plot error, $\sim N(0, \sigma_1^2)$, $i = 1,2,3$, corresponding to the 3 Products,
$B_k$ = effect of $k$th level of Factor B,
$AB_{ik}$ = effect of interaction of $i$th level of Factor A with $k$th level of Factor B,
$\varepsilon_{ijk}$ = subplot error, $\sim N(0, \sigma_2^2)$, $k = 1,2,3$, corresponding to 3 temperatures,
$\gamma_{ij}, \varepsilon_{ijk}$ are assumed independent.

**Table 4.33** Scores Given to 3 Products at 3 Different Process Temperatures by 5 Judges

| Product | Temperature | Judge | | | | |
|---------|-------------|-----|-----|-----|-----|-----|
| | | 1 | 2 | 3 | 4 | 5 |
| Control | Low | 2.5 | 3.0 | 2.5 | 3.5 | 3.0 |
| | Middle | 3.0 | 4.5 | 3.5 | 4.0 | 1.5 |
| | High | 4.0 | 4.0 | 4.0 | 3.5 | 2.5 |
| Formula1 | Low | 3.0 | 4.5 | 5.0 | 3.5 | 4.5 |
| | Middle | 3.0 | 5.0 | 5.0 | 4.5 | 4.0 |
| | High | 3.5 | 4.0 | 4.0 | 4.5 | 3.5 |
| Formula2 | Low | 5.0 | 4.5 | 5.0 | 4.5 | 4.5 |
| | Middle | 4.0 | 3.5 | 5.0 | 5.0 | 5.0 |
| | High | 4.0 | 4.5 | 4.5 | 5.0 | 5.0 |

**Table 4.34** Skeleton Analysis of Variance (ANOVA)

| Source | DF | Expected Mean Square |
|---|---|---|
| Blocks | $r - 1$ | $\sigma_2^2 + a\sigma_1^2 + ab\sigma_\delta^2$ |
| A | $a - 1$ | $\sigma_2^2 + b\sigma_1^2 + \phi^2(A)$ |
| Whole Plot Error | $(a - 1)(r - 1)$ | $\sigma_2^2 + b\sigma_1^2$ |
| B | $b - 1$ | $\sigma_2^2 + \phi^2(B)$ |
| AB | $(a - 1)(b - 1)$ | $\sigma_2^2 + \phi^2(AB)$ |
| Subplot Error | $a(b - 1)(r - 1)$ | $\sigma_2^2$ |

$\phi^2(A)$ represents the contribution of Factor A to its expected mean square in addition to the expected mean square of the whole plot error. Likewise, contributions to their expected mean squares of the subplot Factor B and the interaction component AB are $\phi^2(B)$ and $\phi^2(AB)$. It is clear from the analysis of variance Table 4.34 that the error term for the main effect of Factor A is the *block\*Factor A* treatment interaction, which involves both error terms defined in the model. The error term for Factor B is the sum of two interaction terms: (1) the Factor B by block interaction and (2) the *Factor A\*Factor B\*block* interaction and together they constitute the estimate of $\sigma_2^2$.

Table 4.35 shows the analysis of variance results of the data given in Table 4.33. The significant Product effect ($p < 0.001$) can be seen graphically in Figure 4.1

**Table 4.35** Analysis of Variance for Example 4.6-1

| Source | DF | Sum of Squares | Mean Square | F Value | Pr > F |
|---|---|---|---|---|---|
| Blocks | 4 | 3.86 | 0.96 | | |
| Product | 2 | 13.61 | 6.81 | 9.70 | <.001 |
| Block*Product (Error 1) | 8 | 5.61 | 0.70 | | |
| Temperature | 2 | 0.18 | 0.09 | 0.29 | 0.753 |
| Product*Temperature | 4 | 1.56 | 0.39 | 1.26 | 0.315 |
| Error 2 | 24 | 7.43 | 0.31 | | |
| Corrected Total | 44 | 32.24 | | | |

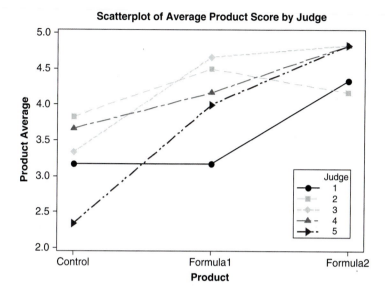

**Scatterplot of Average Product Score by Judge**

of the Product means by judge, where it is clear that Formulations 1 and 2 tend to exhibit a higher color score than Control.

Computations of relevant contrasts are not as straightforward as for purely factorial designs because of the presence of between and within block error. We can consider four types of contrasts: (1) across Factor A levels, (2) across Factor B levels, (3) across Factor A levels at a given level of Factor B, and (4) across Factor B levels at a given level of Factor A. A contrast among the Factor A levels involves a combination of the two error terms as shown in the skeleton ANOVA. The complication in carrying out significance testing of Factor A contrasts arises from the fact that the sampling distribution of the estimator of $\sigma_1^2 + \sigma_2^2$ is not a Chi square ($\chi^2$) (see Chapter 9) but a linear combination of $\chi^2$s. The estimator is a function of two SSs with different DF; consequently, some prefer to use Satterthwaite's formula to approximate the DF associated with the estimate of $\sigma_1^2 + \sigma_2^2$. The approximate DF are given by

$$q^* = \frac{[MS(Whole\ Plot\ Error) + (b-1)MS(Sub\ Plot\ Error)]^2}{\dfrac{MS(Whole\ Plot\ Error)^2}{(a-1)(r-1)} + \dfrac{MS(Sub\ Plot\ Error)^2}{a(b-1)(r-1)}}$$

and critical values from the Student's $t$ distribution with q* DF can be used to test hypotheses and construct confidence limits. For the sample dataset in Table 4.33, the analysis is carried out using the SAS system's MIXED procedure. Cochran and Cox (1957) provided another way to calculate a critical value based on weights associated with the critical Student's $t$ corresponding to the degrees of freedom for the estimates of $\sigma_1^2$ and $\sigma_2^2$.

The variance expressions for the four types of contrasts are as follows:

1. Contrast between levels of A or $\bar{\mu}_{i.}'s$

$$\tau = C_1\bar{\mu}_{1.} + C_2\bar{\mu}_{2.} + \ldots + C_a\bar{\mu}_{a.}$$

where $\sum C_i = 0$

$$\hat{\tau} = C_1\bar{y}_{1..} + C_2\bar{y}_{2..} + \ldots + C_b\bar{y}_{b..}$$
$$Var(\hat{\tau}) = (C_1^2 + C_2^2 + \ldots + C_b^2)(\sigma_2^2 + b\sigma_1^2)/br$$

2. Contrast between levels of B or $\bar{\mu}_{.k}'s$

$$\theta = C_1\bar{\mu}_{.1} + C_2\bar{\mu}_{.2} + \ldots + C_b\bar{\mu}_{.b}$$

where $\sum C_i = 0$

$$\hat{\theta} = C_1\bar{y}_{..1} + C_2\bar{y}_{..2} + \ldots + C_b\bar{y}_{..b}$$
$$Var(\hat{\theta}) = (C_1^2 + C_2^2 + \ldots + C_b^2)\sigma_2^2/ar$$

3. Contrast between levels of A (whole plot treatment) held at the same level of B (subplot treatment)

$$\gamma = C_1\bar{\mu}_{1k} + C_2\bar{\mu}_{2k} + \ldots + C_a\bar{\mu}_{ak}$$

where $\sum C_i = 0$

$$\hat{\gamma} = C_1\bar{y}_{1.k} + C_2\bar{y}_{2.k} + \ldots + C_a\bar{y}_{a.k}$$
$$Var(\hat{\gamma}) = (C_1^2 + C_2^2 + \ldots + C_b^2)(\sigma_1^2 + \sigma_2^2)/r$$

4. Contrast between levels of B (subplot treatment) held at the same level of A (whole plot treatment)

$$\delta = C_1\bar{\mu}_{i1} + C_2\bar{\mu}_{i2} + \ldots + C_b\bar{\mu}_{ib}$$

where $\sum C_i = 0$

$$\hat{\delta} = C_1\bar{y}_{i.1} + C_2\bar{y}_{i.2} + \ldots + C_b\bar{y}_{i.b}$$
$$Var(\hat{\delta}) = (C_1^2 + C_2^2 + \ldots + C_b^2)\sigma_2^2/r$$

The SAS code using the SAS MIXED procedure is given in Table 4.36, along with the statements requesting calculation of the relevant contrasts.

**Table 4.36** SAS Code MIXED Procedure for Example 4.6-1

Data A ;
input block SPLOT $ WPLOT $ Score ; /* block=Judge WPLOT=Product
    SPLOT=Temperature */
cards;

| | | | |
|---|---|---|---|
| 1 | Low | Control | 2.5 |
| 1 | Middle | Control | 3.0 |
| 1 | High | Control | 4.0 |
| 1 | Low | Formula1 | 3.0 |
| 1 | Middle | Formula1 | 3.0 |
| 1 | High | Formula1 | 3.5 |
| 1 | Low | Formula2 | 5.0 |
| 1 | Middle | Formula2 | 4.0 |
| 1 | High | Formula2 | 4.0 |
| 2 | Low | Control | 3.0 |
| 2 | Middle | Control | 4.5 |
| 2 | High | Control | 4.0 |
| 2 | Low | Formula1 | 4.5 |
| 2 | Middle | Formula1 | 5.0 |
| 2 | High | Formula1 | 4.0 |
| 2 | Low | Formula2 | 4.5 |
| 2 | Middle | Formula2 | 3.5 |
| 2 | High | Formula2 | 4.5 |
| 3 | Low | Control | 2.5 |
| 3 | Middle | Control | 3.5 |
| 3 | High | Control | 4.0 |
| 3 | Low | Formula1 | 5.0 |
| 3 | Middle | Formula1 | 5.0 |
| 3 | High | Formula1 | 4.0 |
| 3 | Low | Formula2 | 5.0 |
| 3 | Middle | Formula2 | 5.0 |
| 3 | High | Formula2 | 4.5 |
| 4 | Low | Control | 3.5 |
| 4 | Middle | Control | 4.0 |
| 4 | High | Control | 3.5 |
| 4 | Low | Formula1 | 3.5 |
| 4 | Middle | Formula1 | 4.5 |
| 4 | High | Formula1 | 4.5 |
| 4 | Low | Formula2 | 4.5 |
| 4 | Middle | Formula2 | 5.0 |
| 4 | High | Formula2 | 5.0 |

*(continued)*

**Table 4.36** SAS Code MIXED Procedure for Example 4.6-1—cont...

| 5 | Low | Control | 3.0 |
| 5 | Middle | Control | 1.5 |
| 5 | High | Control | 2.5 |
| 5 | Low | Formula1 | 4.5 |
| 5 | Middle | Formula1 | 4.0 |
| 5 | High | Formula1 | 3.5 |
| 5 | Low | Formula2 | 4.5 |
| 5 | Middle | Formula2 | 5.0 |
| 5 | High | Formula2 | 5.0 |

```
;
  proc mixed data=A method=reml;
    class block WPLOT SPLOT ;
    model SCORE = block WPLOT SPLOT WPLOT*SPLOT /ddfm=satterth;
    random block*WPLOT;
    contrast 'Product F1-C0 ' WPLOT -1 1 0 ;
    contrast 'Product F2-F1 ' WPLOT 0 -1 1 ;
    contrast 'Temp Low vs Mid ' SPLOT 0 -1 1 ; /* ordered high, low, mid */
    contrast 'Temp High vs Mid ' SPLOT 1 -1 0 ;

    /* Comparison of whole-plot means */
    estimate 'PRODUCT F1 VS C0 ' WPLOT -1 1 0 ;
    estimate 'PRODUCT F2 VS C0 ' WPLOT -1 0 1 ;
    estimate 'PRODUCT F2 VS F1 ' WPLOT 0 -1 1 ;

    /* Comparison of split-plot means */
    estimate 'Temp Mid vs Low ' SPLOT 0 -1 1 ; /* ordered high, low, mid */
    estimate 'Temp High vs Mid ' SPLOT 1 0 -1 ;

    /* Comparison of split-plot means at a fixed level of the whole-plot
       treatment */
    estimate 'Temp Mid-Low AT CONTROL' SPLOT 0 -1 1 WPLOT*SPLOT
            0 -1 1 0 0 0 0 0 0 ;
    estimate 'Temp Mid-Low AT FORMULA1' SPLOT 0 -1 1 WPLOT*SPLOT
            0 0 0 0 -1 1 0 0 0 ;
    estimate 'Temp Mid-Low AT FORMULA2' SPLOT 0 -1 1 WPLOT*SPLOT
            0 0 0 0 0 0 0 -1 1 ;
    estimate 'Temp High-Mid AT CONTROL' SPLOT 1 0 -1 WPLOT*SPLOT
            1 0 -1 0 0 0 0 0 0 ;
    estimate 'Temp High-Mid AT FORMULA1' SPLOT 1 0 -1 WPLOT*SPLOT
            0 0 0 1 0 -1 0 0 0 ;
    estimate 'Temp High-Mid AT FORMULA2' SPLOT 1 0 -1 WPLOT*SPLOT
            0 0 0 0 0 0 1 0 -1 ;

    /* Comparison of whole-plot means at a fixed level of the split-plot
       treatment */
```

*(continued)*

**Table 4.36** SAS Code MIXED Procedure for Example 4.6-1—cont...

estimate 'PRODUCT F1 VS C0 AT Low Temp' WPLOT -**1 1 0**
    WPLOT*SPLOT **0 -1 0 0 1 0 0 0 0**;
estimate 'PRODUCT F2 VS C0 AT Low Temp' WPLOT -**1 0 1**
    WPLOT*SPLOT **0 -1 0 0 0 0 1 0**;
estimate 'PRODUCT F1 VS C0 AT Mid Temp' WPLOT -**1 1 0**
    WPLOT*SPLOT **0 0 -1 0 0 1 0 0 0**;
estimate 'PRODUCT F2 VS C0 AT Mid Temp' WPLOT -**1 0 1**
    WPLOT*SPLOT **0 0 -1 0 0 0 0 0 1**;
estimate 'PRODUCT F1 VS C0 AT High Temp' WPLOT -**1 1 0**
    WPLOT*SPLOT -**1 0 0 1 0 0 0 0 0**;
estimate 'PRODUCT F2 VS C0 AT High Temp' WPLOT -**1 0 1**
    WPLOT*SPLOT -**1 0 0 0 0 0 1 0 0**;

/* Calculate least squares means */
lsmeans WPLOT SPLOT WPLOT*SPLOT ; /* / pdiff cl; */
**run;**

The MODEL statement contains the fixed effects terms, where WPLOT and SPLOT represent the whole plot and subplot effects. The RANDOM statement contains the random effects terms, and the SATTERTH option invokes Satterthwaite's procedure for approximating the degrees of freedom for the *F* tests. This applies to comparisons of whole plot means at a given (or different) level of the subplot factor. The SAS output is given in Table 4.37.

The standard errors calculated by PROC MIXED for various comparisons are as follows:

| Comparison | Formula |
| --- | --- |
| Among whole plot means | $\hat{\tau} = \bar{y}_{1..} - \bar{y}_{2..} = 3.27 - 4.10$ |
| | $Var(\hat{\tau}) = 2\left[\dfrac{\hat{\sigma}_2^2 + b\hat{\sigma}_1^2}{br}\right] = 0.306^2$ |
| Among subplot means | $\hat{\theta} = C_1\bar{y}_{..1} - C_2\bar{y}_{..2} = 3.9 - 4.10$ |
| | $Var(\hat{\theta}) = 2\hat{\sigma}_2^2/ar = 0.203^2$ |
| Among subplot means at a fixed level of whole plot factor | $\hat{\delta} = \bar{y}_{1.1} - \bar{y}_{1.2} = 2.9 - 3.3$ |
| | $Var(\hat{\delta}) = 2\hat{\sigma}_2^2/r = 0.352^2$ |
| Among whole plot levels at a fixed level of subplot factor | $\hat{\gamma} = \bar{y}_{1.1} - \bar{y}_{2.1} = 2.9 - 4.1$ |
| | $Var(\hat{\gamma}) = 2(\hat{\sigma}_1^2 + \hat{\sigma}_2^2)/r = 0.420^2$ |

**Table 4.37** SAS Output

Split-plot Design with 5 blocks (Judges), 3 Whole Plots (Product Regime) and 3 levels of the split-plot factor (Temperature)

The Mixed Procedure

Model Information

| | |
|---|---|
| Data Set | WORK.A |
| Dependent Variable | Score |
| Covariance Structure | Variance Components |
| Estimation Method | REML |
| Residual Variance Method | Profile |
| Fixed Effects SE Method | Model-Based |
| Degrees of Freedom Method | Satterthwaite |

Class Level Information

| Class | Levels | Values |
|---|---|---|
| block | 5 | 1 2 3 4 5 |
| wplot | 3 | Control Formula1 Formula2 |
| splot | 3 | High Low Middle |

Dimensions

| | |
|---|---|
| Covariance Parameters | 2 |
| Columns in X | 21 |
| Columns in Z | 15 |
| Subjects | 1 |
| Max Obs Per Subject | 45 |

Number of Observations

| | |
|---|---|
| Number of Observations Read | 45 |
| Number of Observations Used | 45 |
| Number of Observations Not Used | 0 |

Iteration History

| Iteration | Evaluations | −2 Res Log Like | Criterion |
|---|---|---|---|
| 0 | 1 | 83.76051341 | |
| 1 | 1 | 81.50901890 | 0.00000000 |

*(continued)*

**Table 4.37** SAS Output—cont...

Convergence criteria met.
Split-plot Design with 5 blocks (Judges), 3 Whole Plots (Product Regime) and 3 levels of the split-plot factor (Temperature)

The Mixed Procedure
Covariance Parameter
Estimates

| Cov Parm | Estimate |
|---|---|
| block*wplot | 0.1306 |
| Residual | 0.3097 |

Fit Statistics

| | |
|---|---|
| −2 Res Log Likelihood | 81.5 |
| AIC (smaller is better) | 85.5 |
| AICC (smaller is better) | 85.9 |
| BIC (smaller is better) | 86.9 |

Type 3 Tests of Fixed Effects

| Effect | Num DF | Den DF | F Value | Pr > F |
|---|---|---|---|---|
| block | 4 | 8 | 1.37 | 0.3245 |
| wplot | 2 | 8 | 9.70 | 0.0073 |
| splot | 2 | 24 | 0.29 | 0.7531 |
| wplot*splot | 4 | 24 | 1.26 | 0.3146 |

Estimates

| Label | Estimate | Standard Error | DF | t Value | Pr > \|t\| |
|---|---|---|---|---|---|
| PRODUCT F1 VS C0 | 0.8333 | 0.3058 | 8 | 2.73 | 0.0260 |
| PRODUCT F2 VS C0 | 1.3333 | 0.3058 | 8 | 4.36 | 0.0024 |
| PRODUCT F2 VS F1 | 0.5000 | 0.3058 | 8 | 1.64 | 0.1407 |
| Temp Mid vs Low | 0.1333 | 0.2032 | 24 | 0.66 | 0.5180 |
| Temp High vs Mid | 0 | 0.2032 | 24 | 0.00 | 1.0000 |

*(continued)*

**Table 4.37** SAS Output—cont...

| | | | | | |
|---|---|---|---|---|---|
| Temp Mid-Low AT CONTROL | 0.4000 | 0.3520 | 24 | 1.14 | 0.2670 |
| Temp Mid-Low AT FORMULA1 | 0.2000 | 0.3520 | 24 | 0.57 | 0.5752 |
| Temp Mid-Low AT FORMULA2 | −0.2000 | 0.3520 | 24 | -0.57 | 0.5752 |
| Temp High-Mid AT CONTROL | 0.3000 | 0.3520 | 24 | 0.85 | 0.4025 |
| Temp High-Mid AT FORMULA1 | −0.4000 | 0.3520 | 24 | -1.14 | 0.2670 |
| Temp High-Mid AT FORMULA2 | 0.1000 | 0.3520 | 24 | 0.28 | 0.7788 |
| PRODUCT F1 VS C0 AT Low Temp | 1.2000 | 0.4197 | 22.5 | 2.86 | 0.0090 |
| PRODUCT F2 VS C0 AT Low Temp | 1.8000 | 0.4197 | 22.5 | 4.29 | 0.0003 |
| PRODUCT F1 VS C0 AT Mid Temp | 1.0000 | 0.4197 | 22.5 | 2.38 | 0.0260 |
| PRODUCT F2 VS C0 AT Mid Temp | 1.2000 | 0.4197 | 22.5 | 2.86 | 0.0090 |
| PRODUCT F1 VS C0 AT High Temp | 0.3000 | 0.4197 | 22.5 | 0.71 | 0.4820 |
| PRODUCT F2 VS C0 AT High Temp | 1.0000 | 0.4197 | 22.5 | 2.38 | 0.0260 |

Split-plot Design with 5 blocks (Judges), 3 Whole Plots (Product Regime) and 3 levels of the split-plot factor (Temperature)

The Mixed Procedure

Contrasts

| Label | Num DF | Den DF | F Value | Pr > F |
|---|---|---|---|---|
| Product F1-C0 | 1 | 8 | 7.43 | 0.0260 |
| Product F2-F1 | 1 | 8 | 2.67 | 0.1407 |

*(continued)*

**Table 4.37** SAS Output—cont...

| Temp Low vs Mid | 1 | 24 | 0.43 | 0.5180 |
|---|---|---|---|---|
| Temp High vs Mid | 1 | 24 | 0.43 | 0.5180 |

| | | | Least Squares Means | | | | |
|---|---|---|---|---|---|---|---|
| | | | | Standard | | | |
| Effect | wplot | splot | Estimate | Error | DF | t Value | Pr > \|t\| |
| wplot | Control | | 3.2667 | 0.2162 | 8 | 15.11 | <.0001 |
| wplot | Formula1 | | 4.1000 | 0.2162 | 8 | 18.96 | <.0001 |
| wplot | Formula2 | | 4.6000 | 0.2162 | 8 | 21.27 | <.0001 |
| splot | | High | 4.0333 | 0.1713 | 22.5 | 23.54 | <.0001 |
| splot | | Low | 3.9000 | 0.1713 | 22.5 | 22.76 | <.0001 |
| splot | | Middle | 4.0333 | 0.1713 | 22.5 | 23.54 | <.0001 |
| wplot*splot | Control | High | 3.6000 | 0.2967 | 22.5 | 12.13 | <.0001 |
| wplot*splot | Control | Low | 2.9000 | 0.2967 | 22.5 | 9.77 | <.0001 |
| wplot*splot | Control | Middle | 3.3000 | 0.2967 | 22.5 | 11.12 | <.0001 |
| wplot*splot | Formula1 | High | 3.9000 | 0.2967 | 22.5 | 13.14 | <.0001 |
| wplot*splot | Formula1 | Low | 4.1000 | 0.2967 | 22.5 | 13.82 | <.0001 |
| wplot*splot | Formula1 | Middle | 4.3000 | 0.2967 | 22.5 | 14.49 | <.0001 |
| wplot*splot | Formula2 | High | 4.6000 | 0.2967 | 22.5 | 15.50 | <.0001 |
| wplot*splot | Formula2 | Low | 4.7000 | 0.2967 | 22.5 | 15.84 | <.0001 |
| wplot*splot | Formula2 | Middle | 4.5000 | 0.2967 | 22.5 | 15.16 | <.0001 |

An important advantage of using the SAS MIXED procedure for the statistical analysis of a split plot design is that the variance components are estimated directly. All significance tests are constructed correctly without having to specify error terms for any of the fixed factors. In addition, the correct standard errors are reported for all contrasts and estimates of the model parameters.

## EXERCISES

**4.1.** The Univex Fat Tester is a small electrical instrument used for the rapid determination of fat in hamburgers. Readings obtained by the Univex Tester and those obtained by the standard solvent extraction are as follows:

| Sample | Standard Extraction, % | Univex, % |
|--------|------------------------|-----------|
| 1 | 19.5 | 18.5, 19.0 |
| 2 | 25.8 | 24.1, 26.0 |
| 3 | 28.9 | 30.2, 27.9 |
| 4 | 31.0 | 31.9, 32.1 |
| 5 | 20.3 | 20.5, 19.0 |

1. Determine whether significant differences exist for fat readings between the Univex Tester and the solvent extraction method.
2. Construct the 95% confidence intervals for the difference between the mean readings of two methods.

**4.2.** Give the mean square error (MSe) for the analysis of data of Example 4.2-1. Estimate the means of chemical mixtures denoted by $T_1$ and $T_2$ in this example and call the estimates $\overline{X}_1$ and $\overline{X}_2$, respectively. Find the standard error of the difference $\overline{X}_1 - \overline{X}_2$ [i.e., $\text{SE}(\overline{X}_1 - \overline{X}_2)$].

**4.3.** The total sum of squares is defined as the squared deviation of observations from the grand mean. For instance, in an RCB design; we may define the SSto as follows:

$$\text{SSto} = \sum_{i=1}^{k} \sum_{j=1}^{n} (X_{ij} - \overline{X})^2.$$

Starting with the fact that

$$(X_{ij} - \overline{X}) = (X_{ij} - \overline{X}_{i\cdot} - \overline{X}_{\cdot j} + \overline{X}) + (\overline{X}_{i\cdot} - \overline{X}) + (\overline{X}_{\cdot j} - \overline{X}),$$

prove that Eq. (4.3-5) is true. Identify SStr, SSbl, and SSe.

**4.4.** Wooding (1969, 175) reported data from an irritation test study in which four concentrations of a substance contained in a base material were applied to the shaved skin of eight animals. The following data are the degrees of erythema (redness) of the skin on a subjective scale of 0 (no erythema) to 12 (extreme erythema). To compare the four concentrations, suggest a statistical model. Then prepare its analysis of variance. If the four concentrations are significantly different, use a multiple-comparison test to make further pairwise comparisons.

| Animal | Test Substance | | | |
|---|---|---|---|---|
| | **A** | **B** | **C** | **D** |
| 1 | 1.00 | 1.25 | 1.50 | 2.00 |
| 2 | 1.50 | 1.75 | 2.50 | 3.50 |
| 3 | 2.50 | 3.00 | 4.00 | 3.50 |
| 4 | 3.00 | 2.00 | 3.00 | 2.50 |
| 5 | 1.50 | 2.00 | 2.50 | 3.50 |
| 6 | 1.50 | 1.25 | 2.00 | 3.00 |
| 7 | 1.50 | 2.00 | 2.50 | 2.50 |
| 8 | 2.00 | 2.00 | 2.50 | 4.00 |

**4.5.** A panel of five judges scored five varieties of canned peas on a 0 to 10 flavor acceptability scale (Ferris, 1957, 251). Two Latin squares were necessary to yield a design balanced for first-order residual effects. The accompanying data are the total scores of five judges. Use the analysis of variance technique to test whether the five varieties differ significantly on the basis of flavor acceptability. Also estimate the mean flavor scores, adjusted for carryover effects, for each variety and their respective standard errors.

| Tasting Orders | First Week | | | | |
|---|---|---|---|---|---|
| | **M** | **Tu** | **W** | **Th** | **F** |
| 1 | D: 48.0 | A: 45.5 | C: 41.0 | E: 42.5 | B: 38.0 |
| 2 | A: 44.3 | C: 40.2 | E: 42.4 | B: 30.5 | D: 45.8 |
| 3 | E: 41.6 | B: 36.9 | D: 45.1 | A: 43.4 | C: 37.3 |
| 4 | C: 39.2 | E: 41.5 | B: 35.4 | D: 43.1 | A: 41.4 |
| 5 | B: 36.4 | D: 43.5 | A: 42.0 | C: 35.9 | E: 37.8 |

| Tasting Orders | Second Week | | | | |
|---|---|---|---|---|---|
| | **M** | **Tu** | **W** | **Th** | **F** |
| 1 | D: 46.4 | A: 43.5 | C: 39.0 | E: 40.5 | B: 36.4 |
| 2 | E: 40.2 | B: 37.5 | D: 44.4 | A: 41.9 | C: 37.8 |
| 3 | B: 36.0 | D: 44.3 | A: 40.6 | C: 36.1 | E: 38.4 |
| 4 | C: 37.8 | E: 38.1 | B: 33.6 | D: 42.3 | A: 39.0 |
| 5 | A: 40.8 | C: 35.5 | E: 37.8 | B: 38.7 | D: 40.0 |

# Incomplete Block Experimental Designs

In Chapter 4, we considered randomized complete block (RCB) designs in which the size of each block (row) equals the number of treatments being studied. If the number of treatments is large, it is often difficult to secure blocks consisting of like units to accommodate all $t$ treatments. In sensory testing, for instance, panelists are considered as blocks and the items to be tested as treatments. When the panelists are judging several food items, taste fatigue occurs and may produce biased responses. Hence, depending on the product being tested, it may not be desirable for a panelist to judge more than four items in one setting (block). For such situations, there are experimental designs in which the blocks are smaller in size than the number of treatments. Such designs are called incomplete block designs. This chapter considers an important class of incomplete block designs known as the balanced incomplete block (BIB) designs. Some extensions of the BIB designs are also considered.

## 5.1 BALANCED INCOMPLETE BLOCK DESIGNS

The BIB design was introduced by Yates (1936, 1940) in agricultural experimentation. A basic problem in agricultural experimentation is the heterogeneity of plots (blocks), particularly when they are large in size. When one has plots that are small in size, it may be possible to achieve homogeneity within plots and estimate the treatment differences with increased precision. To estimate all the treatment differences with equal precision, one should ensure the arrangement of treatments within blocks are "balanced." To study BIB designs, we introduce the following parameters.

## Parameters of Balanced Incomplete Block Designs

A BIB design is specified by its parameters:

$t$ = number of treatments;
$k$ = number of experimental units per block;
$r$ = number of replications of each treatment;
$b$ = number of blocks;
$\lambda$ = number of blocks in which each pair of treatments occurs together.

Although the experimental designs, such as the completely randomized (CR) and the RCB designs place no restriction on the number of replications per treatment, the BIB design must satisfy the restrictions

$$\lambda(t - 1) = r(k - 1),$$
$$tr = kb = N.$$

Note that $N$ is the total number of observations in the experiment and that $t, k, r, b,$ and $\lambda$ are all integers. The restrictions are that, for a given $t$ and $k$, the required $r$ and $b$ are fixed by the design instead of being specified by the researcher. This is an important consideration for researchers to be aware of in using this class of designs. If $t$ and $k$ are given, the possible choices of $b$, $r$, and $\lambda$ are (Yates, 1936)

$$b = \frac{t!}{k!(t - k)!},$$

$$r = \frac{(t - 1)!}{(k - 1)!(t - k)}, \tag{5.1-1}$$

$$\lambda = \frac{(t - 2)!}{(k - 2)!(t - k)!} = \frac{r(k - 1)}{t - 1},$$

where the symbol $t!$ (read as $t$ factorial) is defined to be $t! = t \times (t - 1) \times \cdots \times 3 \times 2 \times 1$. If $t = 5$, $t! = 5 \times 4 \times 3 \times 2 \times 1 = 120$. Thus, a BIB design with $t = 5$ treatments in blocks of size $k = 3$ would have

$$b = \frac{5!}{3!(5 - 3)!} = 10 \text{ blocks},$$

$$r = \frac{4!}{2!2!} = 6 \text{ replications},$$

$$\lambda = \frac{6(2)}{4} = 3.$$

A BIB design with these parameters is outlined in Table 5.1. In the table $\times$ denotes the presence of treatments in the block. Note that

**Table 5.1** A Basic Balanced Incomplete Block Design with
Parameters $t = 5$, $k = 3$, $r = 6$, $b = 10$, and $\lambda = 3$

| Blocks | Treatments | | | | |
|--------|---|---|---|---|---|
|  | 1 | 2 | 3 | 4 | 5 |
| 1 | × | × | × | | |
| 2 | × | × | | | × |
| 3 | × | | | × | × |
| 4 | | × | × | × | |
| 5 | | | × | × | × |
| 6 | × | × | | × | |
| 7 | × | | × | × | |
| 8 | × | | × | | × |
| 9 | | × | × | | × |
| 10 | | × | | × | × |

each pair of treatments occurs together in the same number of blocks.
For example, treatments 1 and 3 occur together in blocks 1, 7, and 8;
and treatments 1 and 4 occur together in blocks 3, 6, and 7.

A basic design, such as that in Table 5.1, may be repeated to secure
more observations. With $p$ repetitions of a basic design, the new parameters are

$$r^* = pr, \quad b^* = pb, \quad \lambda^* = p\lambda.$$

The simplest BIB design is with $k = 2$. Its remaining parameters are

$$r = t - 1, \quad b = \frac{t(t-1)}{2}, \quad \lambda = 1.$$

An extensive table of BIB designs was given by Cochran and Cox (1957).

## Intrablock Analysis

We first consider an intrablock analysis, where the block differences
are eliminated. In the intrablock analysis, comparisons among the
treatment effects are expressed in terms of comparisons between units
in the same block.

A statistical model for BIB designs is

$$X_{ij} = \mu + T_i + B_j + \varepsilon_{ij}, \quad \begin{matrix} i = 1, \ldots, t, \\ j = 1, \ldots, b, \end{matrix} \qquad (5.1\text{-}2)$$

where $X_{ij}$ is the observation for the $i$th treatment in the $j$th incomplete block; $\mu$ is the grand mean; $T_i$ is the effect of the $i$th treatment with $\Sigma T_i = 0$ (estimates of $T_i$ are denoted by $t_i$ which satisfy $\Sigma_i t_i = 0$); $B_j$ is the effect of the $j$th block with $\Sigma B_j = 0$ (estimates of $B_j$ are denoted by $b_j$, which satisfy $\Sigma_j b_j = 0$); and $\varepsilon_{ij}$ are random errors assumed to be independently and normally distributed with mean 0 and variance $\sigma_e^2$.

In the intrablock analysis, the block effects are considered fixed effects. In the interblock analysis to be considered later, the block effects are treated as random effects.

The model in (5.1-2) resembles that of RCB design with no interaction effects. But remember that we now have incomplete blocks and that the analysis is to be appropriately modified. Since the treatments are confounded with blocks, the treatment effects should be adjusted for the block effects. The reader may appreciate the rationale behind the adjustment if we examine an RCB design in which treatments and blocks are not confounded.

Consider the layout of an RCB design with parameters $t = 3$ and $b = 3$ in Table 5.2. The entries in the table represent the composition of the observations following Model (4.3-1) without the error terms. We note that the difference between any two columns in the table is

**Table 5.2** Composition of Observations in a Randomized Complete Block Design for Three Treatment and Three Blocks

| Blocks | Treatments | | | Totals |
|---|---|---|---|---|
| | $T_1$ | $T_2$ | $T_3$ | |
| $B_1$ | $\mu + T_1 + B_1$ | $\mu + T_2 + B_1$ | $\mu + T_3 + B_1$ | $3\mu + \sum_i T_i + 3B_1$ |
| $B_2$ | $\mu + T_1 + B_2$ | $\mu + T_2 + B_2$ | $\mu + T_3 + B_2$ | $3\mu + \sum_i T_i + 3B_2$ |
| $B_3$ | $\mu + T_1 + B_3$ | $\mu + T_2 + B_3$ | $\mu + T_3 + B_3$ | $3\mu + \sum_i T_i + 3B_3$ |
| | $3\mu + 3T_1 + \sum_j B_j$ | $3\mu + 3T_2 + \sum_j B_j$ | $3\mu + 3T_3 + \sum_j B_j$ | $12\mu + 3\sum_i T_i + 3\sum_j B_j$ |

free of block effects. For example, the difference between columns 1 and 2 is $3(T_1-T_2)$, which is free of block effects $B_j$. Likewise, differences between blocks are free of treatment effects. For instance, the difference between blocks 1 and 2 is $3(B_1-B_2)$, which is free of treatment effects.

Now let us consider a BIB design also with parameters $t = 3$ and $b = 3$. For these parameters, if we decide to have incomplete blocks of size $k = 2$, we must then have $r = 2$ and $\lambda = 1$. The comparison of observations without the error term from this BIB design, following Model (5.1-2), is shown in Table 5.3. The difference between the marginal totals of treatments 1 and 2 is

$$(2\mu + 2T_1 + B_1 + B_2) - (2\mu + 2T_2 + B_1 + B_3)$$
$$= (2T_1 - 2T_2) + (B_2 - B_3),$$

which is not free of block effects. Hence, in the analysis of the BIB designs, we must account for the block effects when estimating the treatment effects.

Let $X_{i\cdot}$ and $X_{\cdot j}$ denote the totals of observations corresponding to the $i$th treatment and the $j$th block, respectively. It has been shown in the literature (Yates, 1936; Rao, 1947; John, 1971, 1980) that the treatment effects in the BIB designs adjusted for the block effects are estimated by $t_i$ defined in (5.1-4). To understand the equations in (5.1-4), let $B_{(i\cdot)}$ be the total of all observations in blocks that contain the $i$th treatment and define

$$kQ_i = kX_{i\cdot} = B_{(i)}, \quad i = 1,\dots,t. \tag{5.1-3}$$

**Table 5.3** Layout of Observations of a Balanced Incomplete Block Design with Parameters $t = 3$, $b = 3$, $k = 2$, $r = 2$, and $\lambda = 1$, Following Model (5.1-2)

| Blocks | $T_1$ | $T_2$ | $T_3$ | Total $X_{\cdot j}$ |
|---|---|---|---|---|
| $B_1$ | $\mu + T_1 + B_1$ | $\mu + T_2 + B_1$ | | $2\mu + T_1 + T_2 + 2B_1$ |
| $B_2$ | $\mu + T_1 + B_2$ | | $\mu + T_3 + B_2$ | $2\mu + T_1 + T_3 + 2B_2$ |
| $B_3$ | | $\mu + T_2 + B_3$ | $\mu + T_3 + B_3$ | $2\mu + T_2 + T_3 + 2B_3$ |
| $X_{i\cdot}$ | $2\mu + 2T_1 + B_1 + B_2$ | $2\mu + 2T_2 + B_1 + B_3$ | $2\mu + 2T_3 + B_2 + B_3$ | $6\mu + 2\sum_i T_i + 2\sum_j B_j$ |

Estimate the effect of the $i$th treatment by $t_i$ as follows:

$$t_i = kQ_i/t\lambda, \quad i = 1, \ldots, t. \tag{5.1-4}$$

We may point out that $\Sigma kQ_i = 0$ and $B_{(i)}$ are called the *adjustment factors*. To understand the nature of $Q_i$, let us find $Q_1$ for the BIB design layout in Table 5.3. To find $Q_1$, we need $X_1.$ and $B_{(1)}$. Since treatment 1 occurs in blocks 1 and 2, $B_{(1)}$ is the total $X._1$ and $X._2$. Hence, from (5.1-3),

$$
\begin{aligned}
2Q_1 &= kX_1. - (X._1 + X._2) \\
&= 2(2\mu + 2T_1 + B_1 + B_2) - (4\mu + 2T_1 + T_2 + T_3 + 2B_1 + 2B_2) \\
&= 2T_1 - (T_2 + T_3),
\end{aligned}
$$

which is free of block effects. Similarly,

$$2Q_2 = 2T_2 - (T_1 + T_3),$$

which is also free of block effects, and so on. Table 5.4 may be used to facilitate calculations of $Q_i$ in the BIB designs.

**Table 5.4** A Balanced Incomplete Block Design Layout for $t = 4$, $k = 2$, $r = 3$, $b = 6$, and $\lambda = 1$, for Calculating $Q_i$ to Estimate Treatment Effects

| Blocks | Treatments | | | | $X._j$ |
|---|---|---|---|---|---|
| | **1** | **2** | **3** | **4** | |
| 1 | $X_{11}$ | $X_{21}$ | | | $X._1$ |
| 2 | | | $X_{32}$ | $X_{42}$ | $X._2$ |
| 3 | $X_{13}$ | | $X_{33}$ | | $X._3$ |
| 4 | | $X_{24}$ | $X_{34}$ | | $X._4$ |
| 5 | $X_{15}$ | | | $X_{45}$ | $X._5$ |
| 6 | | $X_{26}$ | | $X_{46}$ | $X._6$ |
| $X_{j.}$ | $X_1.$ | $X_2.$ | $X_3.$ | $X_4.$ | $X.$ |
| $kX_{j.}$ | $kX_1.$ | $kX_2.$ | $kX_3.$ | $kX_4.$ | $\hat{\mu} = X.. N$ |
| $B_{(i)}$ | $B_{(1)}$ | $B_{(2)}$ | $B_{(3)}$ | $B_{(4)}$ | |
| $kQ_i$ | $kQ_1$ | $kQ_2$ | $kQ_3$ | $kQ_4$ | $\sum_i kQ_i = 0$ |

**Table 5.5** Intrablock Analysis of Variance of Balanced Incomplete Block Designs Following Model (5.1-2)

| Source of Variance | DF | Sum of Squares |
|---|---|---|
| Total | $N - 1$ | $\text{SSto} = \sum_i \sum_j X_{ij}^2 - \text{CF}$ |
| Blocks (Unadjusted for Treatments) | $b - 1$ | $\text{SSbl}_{\text{unadj}} = \left( \sum_j X_{\cdot j}^2 / k \right) - \text{CF}$ |
| Treatments (Adjusted for Blocks) | $t - 1$ | $\text{SStr}_{\text{adj}} = k \sum_i Q_i^2 / t\lambda$ |
| Error | $N - b - t + 1$ | $\text{SSe} = \text{SSto} - \text{SSbl}_{\text{unadj}} - \text{SStr}_{\text{adj}}$ |

Estimates of the treatment means adjusted for block effects are given by

$$\overline{X}_i = \widehat{\mu} + t_i, \quad i = 1, \ldots, t, \tag{5.1-5}$$

where $\widehat{\mu} = X.../N$. The following expressions for standard errors are useful in making comparisons:

$$\text{SE}(t_i) = \sqrt{rk(t-1)\sigma^2/rt^2\lambda}, \tag{5.1-6}$$

$$\text{SE}(t_i - t_j) = \sqrt{2k\sigma^2/\lambda t}, \tag{5.1-7}$$

$$\text{SE}(\overline{X}_i) = \sqrt{k(t-1)\sigma^2/t(k-1)r}, \tag{5.1-8}$$

where $\sigma^2$ is the error variance estimated by the MSe from the intrablock analysis of variance table.

The intrablock analysis of variance is given in Table 5.5, with formulas for computing the various sums of squares.

---

**Example 5.1-1**

Four ($t = 4$) peanut butters were compared in a BIB design experiment having six ($b = 6$) blocks (panelists), each of size $k = 2$. The peanut butters were rated on a scale of 1 to 9. Their ratings appear in Table 5.6 along with $X_{i\cdot}$, $X_{\cdot j}$, $B_{(i)}$, and $Q_i$. Following the SS expressions in Table 5.5, their calculations are shown as footnotes to Table 5.6. The analysis of variance appears in Table 5.7. This table also lists estimates of the peanut butter brand effects from (5.1−4) and their adjusted means from (5.1−5).

To test the null hypothesis of all four brands being equally good, we use the $F$ ratio

$$F = \text{MStr}_{\text{adj}}/\text{MSe}$$

**Table 5.6** Peanut Butter Experiment (Example 5.1-1) with Design Parameters $t = 4$, $k = 2$, $r = 3$, $b = 6$, and $\lambda = 1$

| Panelists | Brands | | | | $X_{\cdot j}$ |
|---|---|---|---|---|---|
| | **1** | **2** | **3** | **4** | |
| 1 | 7 | 5 | | | 12 |
| 2 | | 2 | | 1 | 3 |
| 3 | | 2 | 4 | | 6 |
| 4 | 5 | | | 4 | 9 |
| 5 | 8 | | 6 | | 14 |
| 6 | | | 6 | 2 | 8 |
| $X_{i\cdot}$ | 20 | 9 | 16 | 7 | $X_{\cdot\cdot\cdot} = 52$ |
| $kX_{j\cdot}$ | 40 | 18 | 32 | 14 | $\overline{X} = 52/12 = 4.33$ |
| $B_{(i)}$ | 35 | 21 | 28 | 20 | |
| $kQ_i$ | 5 | $-3$ | 4 | $-6$ | $\sum_i kQ_i = 0$ |

Note:
$CF = (52)^2/12 = 225.3333.$
$SSto = (7^2 + 5^2 + \cdots + 4^2 + 2^2) - CF = 54.6667.$
$SSbl = [(12)^2 + 3^2 + \cdots + (14)^2 + 8^2]/2 - CF = 39.6667.$
$SStr_{adj} = [1/4(2)1][5^2 + (-3)^2 + 4^2 + (-6)^2] = 10.7500.$
$SSe = 54.6667 - 39.6667 - 10.7500 = 4.2500.$

**Table 5.7** Intrablock Analysis of Variance for the Peanut Butter Experiment (Example 5.1-1)

| Source of Variance | DF | SS | MS | F |
|---|---|---|---|---|
| Total | 11 | 54.6667 | | |
| Panelists | 5 | 39.6667 | 7.9333 | |
| Brands (Adjusted) | 3 | 10.7500 | 3.5833 | 2.53 |
| Error | 3 | 4.2500 | 1.4167 | |

| Brands | Brand Effects $\hat{t}$ | Adjusted Means $\overline{X}_{i\cdot}$ |
|---|---|---|
| 1 | 1.25 | 5.58 |
| 2 | $-0.75$ | 3.58 |
| 3 | 1.00 | 5.33 |
| 4 | $-1.50$ | 2.83 |

Note:
$$SE(\hat{t}_i) = \sqrt{\frac{3(2)(3)MSe}{3(16)1}} = 0.73. \quad SE(\overline{X}_i) = \sqrt{\frac{2(3)}{4(1)}\left(\frac{MSe}{3}\right)} = 0.84.$$

$$SE(\hat{t}_i - \hat{t}_j) = \sqrt{\frac{2(2)MSe}{4}} = 1.19.$$

and compare it with the tabled *F* value at the desired level of significance. In our example, the *F* ratio is found to be 2.53. The tabled *F* corresponding to 3 numerator and 3 denominator DF at the 5% level of significance is 9.28, which is much higher than the analysis of variance *F* ratio. Hence, the observed ratio is not significant, implying that all four brands are essentially equivalent.

## Interblock Analysis

Since not all treatments appear together within a block, differences between two blocks are not free of treatment effects. For example, refer to Table 5.3 and note that the difference between blocks 1 and 2 is confounded with $(T_2 - T_3)$. That is, the block differences contain information on comparisons between treatment effects. This treatment information contained in the block differences, when used properly to estimate treatment effects, is known as the *recovery of interblock information* (Yates, 1940). Yates pointed out that a second set of estimates of treatment effects, called *interblock estimates*, can be obtained. An underlying assumption in the recovery of interblock information is that block effects are random. That is, block effects $B_i$ are random variables with mean 0 and variance $\sigma_b^2$. It is also assumed in the recovery of interblock information that the block effects are uncorrelated with intrablock errors $\varepsilon_{ij}$. Yates argued that the interblock estimates of the treatment effects are

$$t_1' = (B_{(i)} - rk\widehat{\mu})/r - \lambda, \quad i = 1, \ldots, t. \tag{5.1-9}$$

We may check that $\Sigma t_i'$ must be zero. The interblock estimate of the treatment means are

$$\overline{X}_i' = \widehat{\mu} + t_i', \quad i = 1, \ldots, t. \tag{5.1-10}$$

## Combining Intrablock and Interblock Estimates

The two different estimates of the treatment effects may be combined to obtain improved estimates. Combining the two estimates is recommended when the block mean square adjusted for treatments is not larger than the intrablock mean square error (Rao, 1947; Federer, 1955). In this section we illustrate combining the two estimates by using the Yates (1940) method.

To use the Yates method for combining the two estimates, we need the mean square due to blocks adjusted for treatments. Remember that in the intrablock analysis, the SSto was partitioned into the block SS unadjusted for treatments (SSbl$_{\text{unadj}}$), the treatment SS adjusted for blocks (SStr$_{adj}$), and the error SS. An alternative is to partition SSto into block SS adjusted for treatments (SSbl$_{adj}$), treatment SS unadjusted for blocks (SStr$_{unadj}$), and the error SS. These two alternative analyses of variance are displayed in Table 5.8. The formulas for computing the various SS in Table 5.8 are as follows:

$$CF = \left(\sum_i \sum_j X_{ij}\right)^2 \Big/ N;$$

$$SSto = \sum_i \sum_j X_{ij}^2 - CF;$$

$$SStr_{unadj} = \sum_i X_{i.}^2/r - CF;$$

$$SSbl_{unadj} = \sum_j X_{.j}^2/k - CF; \tag{5.1-11}$$

$$SStr_{adj} = k \sum_i Q_i^2/t\lambda;$$

$$SSbl_{adj} = SSbl_{unadj} - SStr_{unadj} + SStr_{adj};$$

$$SSe = SSto - SStr_{unadj} - SSbl_{adj}$$
$$= SSto - SSbl_{unadj} - SStr_{adj}.$$

If the BIB basic design is repeated, say, $p$ times, we need to compute the SS due to repetitions denoted by SS$p$ with $(p - 1)$ DF. Let $R_k$ denote the totals of the $k$th repetition; then

$$SSp = \sum_k^p R_k^2/tr - \left(\sum_k^p R_k\right)^2 \Big/ trp. \tag{5.1-12}$$

The first step in the Yates method is to obtain quantities $W_i$, $i = 1, \ldots, t$, given by

**Table 5.8** Two Alternative Analyses of Variance for Recovery of Interblock Information

| Source of Variance | DF | SS | MS |
|---|---|---|---|
| Total | $N - 1$ | SSto | |
| Treatments (Unadjusted for Blocks) | $t - 1$ | SStr$_{unadj}$ | |
| Blocks (Adjusted for Treatments) | $b - 1$ | SSbl$_{adj}$ | $E_b$ |
| Intrablock Error | $N-t-b + 1$ | SSe | $E_e$ |
| Treatments (Adjusted for Block) | $t - 1$ | SStr$_{adj}$ | |
| Blocks (Unadjusted for Treatments) | $b - 1$ | SSbl$_{unadj}$ | |
| Intrablock Error | $N-t-b + 1$ | SSe | $E_e$ |

$$W_i = (t - k)X_{i.} - (t - 1)B_{(i)} + (k - 1)X... \qquad (5.1\text{-}13)$$

The next step is to compute a weighting factor $\theta$ using the intrablock mean square error $E_e$ and the block mean square adjusted for treatments $E_b$ as defined in analysis of variance Table 5.8. The factor $\theta$ is

$$\theta = \frac{(b - 1)(E_b - E_e)}{t(k - 1)(b - 1)E_b + (t - k)(b - t)E_e}. \qquad (5.1\text{-}14)$$

If the basic design is repeated $p$ times, expression (5.1-14) becomes

$$\theta = \frac{(b^* - p)(E_b - E_e)}{t(k - 1)(b^* - p)E_b + (t - k)(b^* - t - p + 1)E_e}, \qquad (5.1\text{-}15)$$

where $E_b$ now is given by

$$E_b = \frac{\text{Block adjusted for treatment SS} - \text{SS}p}{b^* - p}, \qquad (5.1\text{-}16)$$

where SS$p$ is computed from (5.1-12).

The combined estimates of the treatment effects are

$$t_i'' = \frac{X_{i.} + \theta W_i}{r} - \hat{\mu}, \quad i = 1, \ldots, t. \qquad (5.1\text{-}17)$$

In (5.1-17), $r$ should be replaced by $r^*$ when the basic design is repeated. The recovery of the interblock information is not necessary when $E_b > E_e$. The intrablock analysis will suffice. Other methods for combining the two estimates of treatment effects were reported by Seshadri (1963) and described by Cornell (1974b) for food research applications.

---

**Example 5.1-2**

This example was taken from the paper by Gacula and Kubala (1972). Four levels (treatments 1, 2, 3, 4) of tenderizing enzyme were injected into paired steaks. After 24 hours under refrigeration, the steaks were cooked and evaluated for tenderness by a six-member trained panel, using a 9-point rating scale. The sensory data and relevant calculations are shown in Table 5.9. In this example, the basic design consists of $t = 4$, $k = 2$, $r = 3$, $b = 6$, and $\lambda = 1$. With three repetitions of the basic design, the design parameters $r$, $b$, and $\lambda$ become

$$r^* = 3(3) = 9, \quad b^* = 3(6) = 18, \quad \lambda^* = 3(1) = 3.$$

The analysis of variance is given in Table 5.10. The $F$ ratio using intrablock information is $F = 1.75/1.28 = 1.36$ with 3 numerator and 15 denominator DF. This ratio is not significant at the 5% level. Hence, there is no evidence that the differences between treatments are statistically significant. The estimates of the adjusted treatment effects by intrablock and interblock analyses were obtained using expressions (5.1-4) and (5.1-9), respectively.

**Table 5.9** Steak Tenderness Data and Calculations (Example 5.1-2)

| Sessions (Repetitions) | Panelists | Treatments | | | | Panel Sums $X_{\cdot j}$ | Session Sums $R_k$ |
| --- | --- | --- | --- | --- | --- | --- | --- |
| | | 1 | 2 | 3 | 4 | | |
| I | 1 | 4 | 4 | | | 8 | 58 |
| | 2 | | | 7 | 5 | 12 | |
| | 3 | 7 | | 6 | | 13 | |
| | 4 | | 4 | | 6 | 10 | |
| | 5 | 4 | | | 5 | 9 | |
| | 6 | | 1 | 5 | | 6 | |
| II | 7 | 6 | 6 | | | 12 | 53 |
| | 8 | | | 3 | 4 | 7 | |
| | 9 | 6 | | 3 | | 9 | |
| | 10 | | 4 | | 5 | 9 | |
| | 11 | 3 | | | 5 | 8 | |
| | 12 | | 4 | 4 | | 8 | |
| III | 13 | 5 | 6 | | | 11 | 50 |
| | 14 | | | 7 | 6 | 13 | |
| | 15 | 3 | | 4 | | 7 | |
| | 16 | | 2 | | 4 | 6 | |
| | 17 | 4 | | | 5 | 9 | |
| | 18 | | 2 | 2 | | 4 | |
| $X_{j\cdot}$ | | 42 | 33 | 41 | 45 | $X_{\cdot\cdot} = 161$ | $\widehat{\mu} = 4.4722$ |
| $kX_{j\cdot}$ | | 84 | 66 | 82 | 90 | | |
| $B_{(i)}$ | | 86 | 74 | 79 | 83 | $\sum_i B_{(i)} = 322$ | |
| $kQ_i$ | $-2$ | $-8$ | 3 | 7 | 7 | $\sum_i kQ_i = 0$ | |
| $W_i$ | | $-13$ | 5 | 6 | 2 | $\sum_i W_i = 0$ | |

Note:

$CF = (161)^2/36 = 720.0278.$

$SSto = (4^2 + 7^2 + \cdots + 4^2 + 5^2) - CF = 78.9724.$

$SStr_{unadj} = \{[(42)^2 + (33)^2 + (41)^2 + (45)^2]/9\} - CF = 8.7500.$

$SSbl_{unadj} = \{[8^2 + (12)^2 + \cdots + 9^2 + 4^2]/2\} - CF = 54.4724.$

$SStr_{adj} = [(-2)^2 + (-8)^2 + 3^2 + 7^2]/24 = 5.2500.$

$SSbl_{adj} = 54.4724 - 8.7500 + 5.2500 = 50.9724.$

$SSe = 78.9724 - 8.7500 - 50.9724 = 19.2500.$

$SSp = \{[(58)^2 + (53)^2 + (50)^2]/3(4)\} - CF = 2.7224.$

**Table 5.10** Analysis of Variance for Steak Tenderness Data (Example 5.1-2)

| Source of Variance | DF | SS | MS |
|---|---|---|---|
| Total | 35 | 78.97 | |
| Treatments (Unadjusted for Blocks) | 3 | 8.75 | |
| Blocks (Adjusted for Treatment) | 17 | 50.97 | |
| Intrablock Error | 15 | 19.25 | |
| Treatments (Adjusted for Blocks) | 3 | 5.25 | 1.75 |
| Blocks (Unadjusted for Treatments) | 17 | 54.47 | |
| Repetitions | 2 | 2.72 | |
| Intrablock Error | 15 | 19.25 | 1.28 ($E_e$) |

For the combined estimates, we first calculate $E_b$ using (5.1-16). Thus,

$$E_b = (50.9724 - 2.7224)/(18 - 3) = 3.2167.$$

The weighting factor $\theta$ is

$$\theta = \frac{(18 - 3)(3.2167 - 1.2833)}{4(2 - 1)(18 - 3)3.2167 + (4 - 2)(14)1.2833} = 0.1267.$$

Now the combined estimates are computed from (5.1-17). Table 5.11 summarizes the intrablock, interblock, and combined estimates of the treatment effects. We obtain the adjusted treatment means by adding the grand mean $\hat{\mu} = 4.4722$ to the estimated effects. Hence, for treatment 1, we have

$$\overline{X}_{1 \text{ intra}} = 4.4722 - 0.1667 = 4.31,$$

$$\overline{X}_{1 \text{ inter}} = 4.4722 + 0.9167 = 5.39,$$

$$\overline{X}_{1 \text{ combined}} = 4.4722 + 0.0073 = 4.48.$$

We point out that in our analysis of the BIB designs, additivity is assumed. With reference to Model (5.1-1), additivity means that treatments and blocks do not interact. If, indeed, nonadditivity exists and is not accounted for in the model, the analysis would result in an upward bias in the error variance. A test for the presence of nonadditivity in BIB designs was given by Milliken and Graybill (1970). In Section 5.4, we include a design and its analysis that account for the interaction effects.

**Table 5.11** Estimates of Various Treatment Effects (Example 5.1-2)

| Treatment | Intrablock $t_i$ | Interblock $t_i'$ | Combined $t_i''$ |
|---|---|---|---|
| 1 | −0.1667 | 0.9167 | 0.0073 |
| 2 | −0.6667 | −1.0833 | −0.7335 |
| 3 | 0.2500 | −0.2500 | 0.1698 |
| 4 | 0.5833 | 0.4167 | 0.5566 |

## 5.2 BALANCED INCOMPLETE BLOCK DESIGNS AUGMENTED WITH A CONTROL

In some investigations, interest lies in comparing new treatments with the control treatment rather than new treatments with each other. For instance, a product may have established itself in a dominant position in the market over a period of time. Researchers in product development may present us with a number of promising new products for comparison with the established (control) product. Such situations are quite frequent in food and consumer research. Therefore, in this section we outline an augmented BIB design proposed by Basson (1959) and Pearce (1960) and its analysis. Basson discussed the design in the context of comparing new varieties with a standard variety in plant-breeding research. For an application of these designs to food research, we may refer to Gacula (1978).

Any BIB design can be augmented by having a reference (control) sample (treatment) in each block. When a BIB design is augmented with a control treatment, the parameters $t$ and $k$ of a BIB design become $(t + 1)$ and $(k + 1)$. The number of occurrences of the control with other treatments equals $r$, and the number of occurrences of pairs of other treatments equals $\lambda$. The layout of a BIB design, with $t = 4$, $k = 2$, $r = 3$, $b = 6$, and $\lambda = 1$, is given in Table 5.12, where × indicates the treatment appearing in a block. The augmented design is also shown in the table.

The intrablock model for statistical analysis is identical to the one in (5.1-2). For notational simplicity, the control treatment in the model is designated treatment 1. The other $t$ treatments are numbered 2, 3, ... , $(t + 1)$.

The restriction imposed on the treatment effects is $\sum_{i=2}^{t+1} T_i = 0$. We continue to denote by $X_i$. and $X_{.j}$, respectively, the totals of

**Table 5.12** A Balanced Incomplete Block Design and Its Augmented Design with Control in Every Block

| | Basic Design: | | | | |
|---|---|---|---|---|---|
| | Treatments | | | | |
| Blocks | 1 | 2 | 3 | 4 | Design Parameters |
| 1 | × | × | | | $t = 4$ |
| 2 | | | × | × | $r = 3$ |
| 3 | × | | × | | $\lambda = 1$ |
| 4 | | × | | × | $k = 2$ |
| 5 | × | | | × | $b = 6$ |
| 6 | | × | × | | $N = bk = 12$ |

| | Basic Design Augmented with Control: | | | | | |
|---|---|---|---|---|---|---|
| | Control | Treatments | | | | |
| Blocks | 1 | 2 | 3 | 4 | 5 | Design Parameters |
| 1 | × | × | × | | | $t + 1 = 5$ |
| 2 | × | | | × | × | $r = 3$ |
| 3 | × | × | | × | | $\lambda = 1$ |
| 4 | × | | × | | × | $k + 1 = 3$ |
| 5 | × | × | | | × | $b = 6$ |
| 6 | × | | × | × | | $N = b(k + 1) = 18$ |

the observations for the $i$th treatment and the $j$th block; $i = 1, \ldots, t + 1$; $j = 1, \ldots, b$. Also, $B_{(i)}$ denotes the total of all blocks having the $i$th treatment. Let $Q_i$ be defined as follows:

$$(k + 1)Q_i = (k + 1)X_i. - B_{(i)}, \quad i = 1, \ldots, t + 1. \qquad (5.2\text{-}1)$$

The intrablock estimates of the treatment effects are given by

$$t_1 = (k + 1)Q_1/bk, \qquad (5.2\text{-}2)$$

$$t_i = (k + 1)(Q_i + Q_1/t)/(rk + \lambda), \quad i = 2, \ldots, t + 1. \qquad (5.2\text{-}3)$$

For comparisons we require the following estimate of the standard errors of various statistics. For the control treatment,

$$\text{SE}(t_1) = \sqrt{(k + 1)\text{MSe}/bk}. \qquad (5.2\text{-}4)$$

For the other effects,

$$SE(t_i) = \sqrt{(t-1)^2(k+1)MSe/tr(kt-1)},\qquad(5.2\text{-}5)$$

$$SE(t_i - t_j) = \sqrt{2(k+1)(t-1)MSe/r(kt-1)},\qquad(5.2\text{-}6)$$

$$SE(t_i - t_1) = \sqrt{(k+1)(k+t-2)MSe/r(kt-1)},\qquad(5.2\text{-}7)$$

where $i \neq j = 2,\ldots,t$, and MSe is the mean square error from analysis of variance Table 5.13. Under SS the table lists the formulas for computing the SS quantities.

If we want to estimate the treatment means, our estimates for the control and other treatments are

$$\overline{X}_1 = \hat{\mu} + t_1\qquad(5.2\text{-}8)$$

and

$$\overline{X}_i = \hat{\mu} + t_i, \quad i = 2,\ldots,t+1,\qquad(5.2\text{-}9)$$

respectively.

**Table 5.13** Analysis of Variance Table for the Balanced Incomplete Block Design Augmented with Control Treatment in Every Block

| Source of Variance | DF[a] | SS[a] |
|---|---|---|
| Total | $N-1$ | $SSto = \sum_i \sum_j \sum_m X_{ijm}^2 - CF$ |
| Repetitions | $p-1$ | $SSp = \dfrac{\sum_m R_m^2}{b(k+1)} - CF$ |
| Blocks (Unadjusted for Treatment) within Repetitions | $p(b-1)$ | $SSbl : p = \dfrac{\sum_j X_{.j}^2}{k+1} - SSp - CF$ |
| Treatments (Adjusted for Blocks) | $t$ | $SStr_{adj} = \sum_i \hat{t}_i Q_i$ |
| Error | $N - pb + p + t$ | $SSe = SSto - SSp - SSbl{:}p - SStr_{adj}$ |

[a] $N = b(k+1)p$; $p$ = number of repetitions.

Note:

$X_{ijm}$ = the observation on the $i$th treatment, in the $j$th block and the $m$th repetition, $i = 1,\ldots, t+1; j = 1,\ldots, b, m = 1,\ldots, p$.

$R_m$ = the total of the $m$th repetition.

$CF = (\Sigma R_m)^2/N$.

All the other quantities are defined in Section 5.2.

The data in Table 5.14 are taken from files of the Armour Research Center pertaining to a nitrate-nitrite study. The table shows calculations of estimate of treatment effects, their standard errors, and the various SS quantities for the analysis of variance. The MSe required for calculating the standard error is taken from the analysis of variance in Table 5.15.

**Example 5.2-1**

With 4 numerator and 20 denominator DF, the ratio $F = 1.67$ is not significant at the 5% level. Had the $F$ ratio been found significant, it might have been necessary to carry out multiple-comparison tests (Chapter 3) to compare the new treatments with the control and also with each other, using estimates of their effects and standard errors.

**Table 5.14** Analysis of the Nitrate-Nitrite Study with Design Parameters $t + 1 = 5$, $k + 1 = 3$, $r^* = 6$, $b^* = 12$, $p = 2$, and $\lambda^* = 2$ (Example 5.2-1)

| Repetitions | Panelists | Control Treatment | | New Treatments | | | $X_{\cdot j}$ | $R_k$ |
|---|---|---|---|---|---|---|---|---|
| | | **1** | **2** | **3** | **4** | **5** | | |
| I | 1 | 4 | 5 | 4 | | | 13 | |
| | 2 | 5 | | | 3 | 6 | 14 | |
| | 3 | 5 | 7 | | 6 | | 18 | |
| | 4 | 4 | | 4 | 4 | | 12 | |
| | 5 | 5 | 6 | | | 5 | 16 | |
| | 6 | 4 | | 5 | | 3 | 12 | 85 |
| II | 7 | 5 | 6 | 3 | | | 14 | |
| | 8 | 3 | | | 4 | 4 | 11 | |
| | 9 | 5 | 6 | | 7 | | 18 | |
| | 10 | 4 | | 6 | 5 | | 15 | |
| | 11 | 4 | 7 | | | 5 | 16 | |
| | 12 | 6 | | 5 | | 2 | 13 | 87 |
| $X_{i\cdot}$ | | 54 | 37 | 27 | 29 | 25 | $X\ldots = 172$ | |
| $B_{(i)}$ | | 172 | 95 | 79 | 88 | 82 | | |
| $B_{(i)}/k + 1$ | | 57.3333 | 31.6667 | 26.3333 | 29.3333 | 27.3333 | | |
| $Q_i$ | | −3.3333 | 5.3333 | 0.6667 | −0.3333 | −2.3333 | $\sum_{1}^{5} Q_i = 0$ | |

*(continued)*

**Table 5.14** Analysis of the Nitrate-Nitrite Study with Design Parameters $t + 1 = 5$, $k + 1 = 3$, $r^* = 6$, $b^* = 12$, $p = 2$, and $\lambda^* = 2$ (Example 5.2-1)—*cont...*

| Repetitions | Panelists | Control Treatment | | New Treatments | | | $X_{\cdot j}$ | $R_k$ |
|---|---|---|---|---|---|---|---|---|
| | | **1** | **2** | **3** | **4** | **5** | | |
| $t_i$ | | −0.4167 | 0.9643 | −0.0357 | −0.2500 | −0.6786 | $\sum\limits_{2}^{5} t_i = 0$ | |
| $\overline{X}_i$ | | 4.50 | 5.88 | 4.88 | 4.67 | 4.24 | | |

*Note:*

$$CF = \frac{(172)^2}{36} = 821.7778.$$

$$SSto = (4^2 + 5^2 + \cdot + 5^2 + 2^2) - CF = 874.0000 - 821.7778 = 52.2222.$$

$$SSp = \frac{[(85)^2 + (87)^2]}{18} - CF - 821.8889 - 821.7778 = 0.1111.$$

$$SSbl : p = \frac{[(13)^2 + (14)^2 + \cdots + (16)^2 + (13)^2]}{3} - SSp - CF$$

$$= 841.3333 - 0.1111 - 821.7778 = 19.4444.$$

$$SStr_{adj} = (-3.3333)(-0.4167) + (5.3333)(0.9643) + \ldots + (-2.3333)(-0.6786) = 8.1748.$$

$$SSe = 52.2222 - 0.1111 - 19.4444 - 8.1748 = 24.2919.$$

*Standard errors of treatment effects from formulas (5.2-4) through (5.2-7) are*

$$SE(t_1) = \sqrt{3(1.2246)/(12)(2)} = 0.39;$$

$$SE(t_i) = \sqrt{\left[\frac{3^2(3)}{4(6)(8-1)}\right] 1.2246} = 0.44, \text{for each } i \neq j = 2, \ldots, t + 1;$$

$$SE(t_i - t_j) = \sqrt{2(3)(3)(1.2246)/6(8-1)} = 0.73, \text{for each } i = 2, \ldots, t + 1;$$

$$SE(t_i - t_1) = \sqrt{\left[\frac{3(6-2)}{6(8-1)}\right] 1.2246} = 0.59, \text{for each } i = 2, \ldots, t + 1.$$

**Table 5.15** Analysis of Variance (Example 5.2-1)

| Source of Variance | DF | SS | MS | F |
|---|---|---|---|---|
| Total | 35 | 52.2222 | | |
| Repetitions | 1 | 0.1111 | 0.1111 | |
| Blocks within Repetitions | 10 | 19.4444 | 1.9444 | |
| Treatments (Adjusted) | 4 | 8.1748 | 2.0437 | 1.67 |
| Error | 20 | 24.4919 | 1.2246 $(E_e)$ | |

## 5.3 DOUBLY BALANCED INCOMPLETE BLOCK DESIGNS

One of the assumptions underlying the intrablock analysis of the BIB designs using Model (5.1-1) is that the error terms $\varepsilon_{ij}$ are independently distributed. Calvin (1954) cited several studies of tasting experiments, in which scores are assigned to sample products (treatments), where there is evidence of scores being correlated within blocks. The score or rating that a particular sample (treatment) receives depends on the relative ratings of the other samples in the block. It is believed that in attempting to give an objective evaluation of a sample, the judge also makes comparisons with other samples in the block. Hence, the assumption that the errors are independently distributed is not valid. If such dependencies exist, one would like to account for them in statistical analyses. Calvin (1954) proposed a modification of Model (5.1-2) to account for dependent error terms in the model. His modified model is

$$X_{ih} = n_{ij}\left(\mu + T_i + B_h + \sum_{j \neq i} n_{jh} m_{ij} C_{ij} + \varepsilon_{ij}\right), \qquad (5.3\text{-}1)$$

where

$$h = 1, \ldots, t;$$

$$n_{ih} = \begin{cases} 1 & \text{if the } i\text{th treatment is in the } h\text{th block,} \\ 0 & \text{otherwise;} \end{cases}$$

$$n_{ij} = \begin{cases} 1 & \text{if } i < j, \\ -1 & \text{if } i > j; \end{cases}$$

and $X_{ih}$, $\mu$, $T_i$, $B_h$, and $\varepsilon_{ij}$ are defined in the usual way for BIB designs with $\Sigma_i T_i = \Sigma_h B_h = \Sigma_{i \neq j} m_{ij} C_{ij} = 0$. The quantity $C_{ij}$ is the effect common to the $i$th and $j$th treatments when they are both in the same block: $C_{ij} = C_{ji}$.

In addition to the parameters of a BIB design, one additional parameter is introduced. That is, each triplet of treatments occurs together in the same block $\lambda_1$ times, where

$$\lambda_1 = \frac{(t-3)!}{(k-3)!(t-k)!} = \frac{\lambda(k-2)}{t-2}. \qquad (5.3\text{-}2)$$

Thus, the BIB design is doubly balanced for pairs and triplets of treatments occurring together in the same block. Hence, the design is called DBIB. The additional condition given by (5.3-2) limits the use of DBIB designs to $t \geq 5$ and $k \geq 4$. In sensory evaluations, four treatments are the maximum that a judge can effectively evaluate in one setting (block).

There is no guarantee that a DBIB design exists for any $t$. Designs for $t = 5, 6, 7$, and $k = 4$ are shown in Table 5.16. It is apparent that the design size increases as the number of treatments increase. However, it is possible to reduce a DBIB design and yet retain its essential balancing property. Calvin (1954) listed reduced designs for all $t \leq 16$ and $k \leq 8$.

The treatment effects $t_i$ are estimated by

$$kt_i = kX_i. - B_{(i)}, \quad i = 1, \ldots, t, \tag{5.3-3}$$

where $X_i.$ and $B_{(i)}$ are as defined in the previous sections. We note that (5.3-3) is the same as in the BIB design analysis. The treatment means can be estimated by $\overline{X}_i = X.../N + t_i$, where $N = tr$. For comparisons between treatments, we need the following standard errors of their estimates:

$$SE(t_i) = \sqrt{k(t-1)^2 MSe/r(k-1)t^2}, \tag{5.3-4}$$

$$SE(\overline{X}_i) = \sqrt{k(t-1)^2 + t(k-1)MSe/rt^2(k-1)}, \tag{5.3-5}$$

$$SE(t_i - t_j) = \sqrt{2k(t-1)MSe/rt(k-1)}. \tag{5.3-6}$$

The MSe quantity is obtained from Table 5.17 corresponding to Model (5.3-1).

The breakdown of the total sum of squares as well as the formulas for CF, SSto, SSbl, SStr$_{adj}$, and SSe are shown in Table 5.17. The correlation sum of squares SSco adjusted for treatment and block effects is obtained by

$$SSco = \sum_i \sum_{j>i} (tC_{ij})^2 \Big/ 2t^2(\lambda - \lambda_1), \tag{5.3-7}$$

where

$$tC_{ij} = tA_{ij} - k(Q_i - Q_j),$$

$$A_{ij} = n_{ih}n_{jh} \sum_h (X_{ih} - X_{jh}),$$

$$kQ_i = kX_i. - B_{(i)},$$

$$\lambda_1 = \lambda(k-2)/(t-2).$$

**Table 5.16** Doubly Balanced Incomplete Block Designs for $t = 5, 6,$ and 7

| Design | $t$ | $k$ | $r$ | $b$ | $\lambda$ | $\lambda_1$ | Blocks | Treatments |
|---|---|---|---|---|---|---|---|---|
| 1 | 5 | 4 | 4 | 5 | 3 | 2 | 1 | 1 2 3 4 |
|   |   |   |   |   |   |   | 2 | 2 3 4 5 |
|   |   |   |   |   |   |   | 3 | 1 3 4 5 |
|   |   |   |   |   |   |   | 4 | 1 2 4 5 |
|   |   |   |   |   |   |   | 5 | 1 2 3 5 |
| 2 | 6 | 4 | 10 | 15 | 6 | 3 | 1 | 1 2 3 4 |
|   |   |   |   |   |   |   | 2 | 1 2 3 5 |
|   |   |   |   |   |   |   | 3 | 1 2 3 6 |
|   |   |   |   |   |   |   | 4 | 1 2 4 5 |
|   |   |   |   |   |   |   | 5 | 1 2 4 6 |
|   |   |   |   |   |   |   | 6 | 1 2 5 6 |
|   |   |   |   |   |   |   | 7 | 1 3 4 5 |
|   |   |   |   |   |   |   | 8 | 1 3 4 6 |
|   |   |   |   |   |   |   | 9 | 1 3 5 6 |
|   |   |   |   |   |   |   | 10 | 1 4 5 6 |
|   |   |   |   |   |   |   | 11 | 2 3 4 5 |
|   |   |   |   |   |   |   | 12 | 2 3 4 6 |
|   |   |   |   |   |   |   | 13 | 2 3 5 6 |
|   |   |   |   |   |   |   | 14 | 2 4 5 6 |
|   |   |   |   |   |   |   | 15 | 3 4 5 6 |
| 3 | 7 | 4 | 20 | 35 | 10 | 4 | 1 | 1 2 3 4 |
|   |   |   |   |   |   |   | 2 | 1 2 3 5 |
|   |   |   |   |   |   |   | 3 | 1 2 3 6 |
|   |   |   |   |   |   |   | 4 | 1 2 3 7 |
|   |   |   |   |   |   |   | 5 | 1 2 4 5 |
|   |   |   |   |   |   |   | 6 | 1 2 4 6 |
|   |   |   |   |   |   |   | 7 | 1 2 4 7 |
|   |   |   |   |   |   |   | 8 | 1 2 5 6 |

*(continued)*

**Table 5.16** Doubly Balanced Incomplete Block Designs for $t = 5$, 6, and 7—*cont...*

| Design | $t$ | $k$ | $r$ | $b$ | $\lambda$ | $\lambda_1$ | Blocks | Treatments |
|--------|-----|-----|-----|-----|-----------|-------------|--------|------------|
|        |     |     |     |     |           |             | 9      | 1 2 5 7    |
|        |     |     |     |     |           |             | 10     | 1 2 6 7    |
|        |     |     |     |     |           |             | 11     | 1 3 4 5    |
|        |     |     |     |     |           |             | 12     | 1 3 4 6    |
|        |     |     |     |     |           |             | 13     | 1 3 4 7    |
|        |     |     |     |     |           |             | 14     | 1 3 5 6    |
|        |     |     |     |     |           |             | 15     | 1 3 5 7    |
|        |     |     |     |     |           |             | 16     | 1 3 6 7    |
|        |     |     |     |     |           |             | 17     | 1 4 5 6    |
|        |     |     |     |     |           |             | 18     | 1 4 5 7    |
|        |     |     |     |     |           |             | 19     | 1 4 6 7    |
|        |     |     |     |     |           |             | 20     | 1 5 6 7    |
|        |     |     |     |     |           |             | 21     | 2 3 4 5    |
|        |     |     |     |     |           |             | 22     | 2 3 4 6    |
|        |     |     |     |     |           |             | 23     | 2 3 4 7    |
|        |     |     |     |     |           |             | 24     | 2 3 5 6    |
|        |     |     |     |     |           |             | 25     | 2 3 5 7    |
|        |     |     |     |     |           |             | 26     | 2 3 6 7    |
|        |     |     |     |     |           |             | 27     | 2 4 5 6    |
|        |     |     |     |     |           |             | 28     | 2 4 5 7    |
|        |     |     |     |     |           |             | 29     | 2 4 6 7    |
|        |     |     |     |     |           |             | 30     | 2 5 6 7    |
|        |     |     |     |     |           |             | 31     | 3 4 5 6    |
|        |     |     |     |     |           |             | 32     | 3 4 5 7    |
|        |     |     |     |     |           |             | 33     | 3 4 6 7    |
|        |     |     |     |     |           |             | 34     | 3 5 6 7    |
|        |     |     |     |     |           |             | 35     | 4 5 6 7    |

**Table 5.17** Analysis of Variance of Doubly Balanced Incomplete Block Designs [Model (5.3-1)]

| Source of Variance | DF | SS |
|---|---|---|
| Total | $N - 1$ | $SSto = \sum_i \sum_h X_{ih}^2 - CF$ |
| Blocks (Unadjusted for Treatments) | $b - 1$ | $SSbl = \sum_h X_{.h}^2/k - CF$ |
| Treatment (Adjusted) | $t - 1$ | $SStr_{adj} = \sum_i Q_i^2/kt\lambda$ |
| Correlations (Adjusted) | $(t-1)(t-2)/2$ | $SSco = \sum_i \sum_{j>i} (tC_{ij})^2/2t^2(\lambda - \lambda_1)$ |
| Error | By difference | $SSe = SSto - SSbl - SStr_{adj} - SSco$ |

Note that expression (5.3-7) can be rewritten simply as

$$\sum_i \sum_{j>i} \frac{C_{ij}^2}{2(\lambda - \lambda_1)},$$

but it is easier to work with $tC_{ij}$. Also note that for the $(i, j)$th treatment pair, $A_{ij}$ represents the sum of differences between treatments $i$ and $j$ over $\lambda$ blocks in which the pair occurs together.

---

**Example 5.3-1**

This example taken from Calvin (1954) pertains to an ice cream experiment. Six different amounts of vanilla ice cream samples were studied. The tasters were asked to score the samples from 0 to 5 in blocks of three as to the amount of vanilla in each sample. A score of 0 indicates no vanilla, and 5 indicates the highest amount. Each taster was a block, and no taster was used more than once in the experiment.

The design parameters are as follows:

$$t = 6, \quad r = 10, \quad \lambda = 6, \quad E = 0.9,$$
$$k = 4, \quad b = 15, \quad \lambda_1 = 3.$$

Table 5.18 displays the data and relevant calculations for the analysis. To facilitate the determination of SSco, one may record the differences for all pairs of treatments as in Table 5.19. For instance,

$$A_{12} = (2 - 4) + (0 - 4) + (0 - 3) + (1 - 3)$$
$$+ (0 - 2) + (0 - 1) = -14,$$

**Table 5.18** Vanilla Ice Cream Experiment (Example 5.3-1) with Design Parameters $t = 6$, $k = 4$, $r = 10$, $b = 15$, $\lambda = 6$, and $\lambda_1 = 3$

| Blocks | Treatments | | | | | | $X_{\cdot h}$ |
|---|---|---|---|---|---|---|---|
| | **1** | **2** | **3** | **4** | **5** | **6** | |
| 1 | 2 | 4 | 0 | 3 | | | 9 |
| 2 | 0 | 4 | 3 | | 5 | | 12 |
| 3 | 0 | 3 | 4 | | | 5 | 12 |
| 4 | 1 | 3 | | 4 | 5 | | 13 |
| 5 | 0 | 2 | | 4 | | 5 | 11 |
| 6 | 0 | 1 | | | 4 | 5 | 10 |
| 7 | 2 | | 3 | 1 | 4 | | 10 |
| 8 | 0 | | 0 | 3 | | 4 | 7 |
| 9 | 0 | | 2 | | 3 | 3 | 8 |
| 10 | 2 | | | 4 | 1 | 4 | 11 |
| 11 | | 0 | 2 | 3 | 1 | | 6 |
| 12 | | 1 | 1 | 2 | | 4 | 8 |
| 13 | | 2 | 3 | | 5 | 4 | 14 |
| 14 | | 1 | | 4 | 5 | 2 | 12 |
| 15 | | | 4 | 3 | 1 | 2 | 10 |
| $X_{i\cdot}$ | 7 | 21 | 22 | 31 | 34 | 38 | $X_{\cdot\cdot} = 153$ |
| $B_{(i)}$ | 103 | 107 | 96 | 97 | 106 | 103 | $\hat{\mu} = 2.55$ |
| $Q_i$ | −75 | −23 | −8 | 27 | 30 | 49 | $\sum_i Q_i = 0$ |
| $t_i$ | −2.08 | −0.64 | −0.22 | 0.75 | 0.83 | 1.36 | $\sum_i t_i = 0$ |
| $\overline{X}_i$ | 0.5 | 1.9 | 2.3 | 3.3 | 3.4 | 3.9 | |

Note:
$$CF = (153)^2/60 = 390.15.$$
$$SSto = (2^2 + 0^2 + \cdots + 2^2 + 2^2) - CF = 547 - 390.15 = 156.85.$$
$$SSbl_{unadj} = \{[9^2 + (12)^2 + \cdots + (10)^2]/4\} - CF = 408.25 - 390.15 = 18.10.$$
$$SStr_{adj} = [(-75)^2 + (-23)^2 + \cdots + (49)^2]/(4)(6)(6) = 10{,}248/144 = 71.17.$$
$$SSco = [(-32)^2 + (19)^2 + \cdots + (13)^2]/2(6)^2(6 - 3) = 5868/216 = 27.17.$$
$$SSe = 156.85 - 18.10 - 71.17 - 27.17 = 40.41.$$

**Table 5.19** Differences between Pairs of Treatments for Calculating SSco

| i–j | Treatment Pairs | | |
|-----|-----------------|------|------|
| | $A_{ij}$ | $tC_{ij}$ | $C_{ij}$ |
| 1–2 | −14 | −32 | −5.3 |
| 1–3 | − 8 | 19 | 3.2 |
| 1–4 | −12 | 30 | 5.0 |
| 1–5 | −17 | 3 | 0.5 |
| 1–6 | −24 | −20 | −3.3 |
| 2–3 | 1 | 21 | 3.5 |
| 2–4 | − 9 | − 4 | −0.7 |
| 2–5 | −14 | −31 | −5.2 |
| 2–6 | −15 | −18 | −3.0 |
| 3–4 | − 5 | 5 | 0.8 |
| 3–5 | − 2 | 26 | 4.3 |
| 3–6 | − 8 | 9 | 1.5 |
| 4–5 | 2 | 15 | 2.5 |
| 4–6 | − 1 | 16 | 2.7 |
| 5–6 | − 1 | 13 | 2.2 |

and so on. Further calculations are also given in Table 5.19 for computing SSco from (5.3-7). The analysis of variance is shown in Table 5.20. The $F$ ratio for differences between adjusted means is $F = 14.23/1.35 = 10.54$, which is statistically significant at the 1% level. Hence, the null hypothesis is rejected.

The $F$ ratio, $F = 2.72/1.35 = 2.01$, tests for the correlation effect. At 10 numerator and 30 denominator DF, the tabular $F$ at the 5% level is 2.16. Hence, the correlation effect is not considered significant at the 5% level. However, note that the correlation effect is significant at, say, the 10% level of significance. From (5.3-4) through (5.3-6), we get

**Table 5.20** Analysis of Variance of the Vanilla Experiment (Example 5.3-1)

| Source of Variance | DF | SS | MS | F |
|---|---|---|---|---|
| Total | 59 | 156.85 | | |
| Blocks (Unadjusted for Treatment) | 14 | 18.10 | | |
| Treatments (Adjusted) | 5 | 71.17 | 14.23 | 10.54* |
| Correlations (Adjusted) | 10 | 27.17 | 2.72 | 2.01 |
| Error | 30 | 40.41 | 1.35 | |

*$p < .01.$

$$SE(t_i) = \sqrt{\frac{4(5)^2(1.35)}{6(3)(6)^2}} = 0.46,$$

$$SE(\overline{X}_i) = \sqrt{\left[\frac{4-1}{4(10)(0.81)} + \frac{1}{(6)10}\right]1.35} = 0.38,$$

$$SE(t_i - t_j) = \sqrt{\frac{2(4)(5)(1.35)}{10(6)(3)}} = 0.55.$$

Since the hypothesis of the equality of treatments was rejected, we may carry out a multiple-comparison test.

## 5.4 COMPOSITE COMPLETE-INCOMPLETE BLOCK DESIGNS

The composite complete-incomplete block (CIB) designs, introduced by Cornell and Knapp (1972, 1974), Cornell (1974a), and Cornell and Schreckingost (1975) in the food sciences, belong to a family of designs known in the statistical literature as *extended complete block designs* (John, 1963). In Section 5.1, we pointed out the effects of nonadditivity or interaction on the analysis of BIB designs. The CIB designs may be used to separate the nonadditive effects from the experimental error so that what remains is the so-called pure error. Removal of the nonadditive effects (interaction effects) results in a more precise test of statistical significance.

A CIB design is formed by combining the complete and incomplete block designs with the same number of blocks $b$ and treatments $t$. As stated by Cornell and Knapp (1972), the construction of CIB designs requires only that the incomplete block portion be balanced. Consider the following layouts for complete and incomplete block designs with $t = 3$ and $b = 3$, where $\times$ denotes the particular treatment that is within the block:

Complete blocks : $k = t = 3$

| | | Treatments | | |
|---|---|---|---|---|
| | | 1 | 2 | 3 |
| | 1 | $\times$ | $\times$ | $\times$ |
| Blocks | 2 | $\times$ | $\times$ | $\times$ |
| | 3 | $\times$ | $\times$ | $\times$ |

Incomplete blocks : $k = 2$

| | | Treatments | | |
|---|---|---|---|---|
| | | 1 | 2 | 3 |
| | 1 | $\times$ | $\times$ | |
| Blocks | 2 | $\times$ | | $\times$ |
| | 3 | | $\times$ | $\times$ |

By combining the two layouts, we have the basic CIB design with blocks of size $(t + k)$, $1 \leq k < t$. The layout of the basic CIB design is

| | | Treatments | | |
|---|---|---|---|---|
| | | 1 | 2 | 3 |
| | 1 | $\times$ | $\times$ | $\times$ |
| | | $\times$ | $\times$ | |
| Blocks | 2 | $\times$ | $\times$ | $\times$ |
| | | $\times$ | | $\times$ |
| | 3 | $\times$ | $\times$ | $\times$ |
| | | | $\times$ | $\times$ |

For the preceding CIB design layout, $r = b(t + k)/t$. When the basic CIB design is repeated, the number of blocks $b$ is multiplied by the number of repetitions. Notice that in the basic CIB design there are cells with two observations on the same treatment. These cells provide an estimate of the pure error. Table 5.21 shows layouts of the CIB designs with $t = 3$, 4, and 5.

**Table 5.21** Some Composite Complete–Incomplete Block Designs

Design 1: $t = 3, b = 3, k = 1$

Incomplete blocks:

|  | Treatments | | |
|---|---|---|---|
| Blocks | 1 | 2 | 3 |
| 1 |  |  | x |
| 2 | x |  |  |
| 3 |  | x |  |

Resulting CIB design (special case):

|  | Treatments | | |
|---|---|---|---|
| Blocks | 1 | 2 | 3 |
| 1 | x | x | x |
|  |  |  | x |
| 2 | x | x | x |
|  | x |  |  |
| 3 | x | x | x |
|  |  | x |  |

Design 2: $t = 4, b = 6, k = 2$

Incomplete blocks:

|  | Treatments | | | |
|---|---|---|---|---|
| Blocks | 1 | 2 | 3 | 4 |
| 1 | x | x |  |  |
| 2 |  |  | x | x |
| 3 | x |  | x |  |
| 4 |  | x |  | x |
| 5 | x |  |  | x |
| 6 |  | x | x |  |

Resulting CIB design:

|  | Treatments | | | |
|---|---|---|---|---|
| Blocks | 1 | 2 | 3 | 4 |
| 1 | x | x | x | x |
|  | x | x |  |  |
| 2 | x | x | x | x |
|  |  |  | x | x |
| 3 | x | x | x | x |
|  | x |  | x |  |
| 4 | x | x | x | x |
|  |  | x |  | x |
| 5 | x | x | x | x |
|  | x |  |  | x |
| 6 | x | x | x | x |
|  |  | x | x |  |

*(continued)*

**Table 5.21** Some Composite Complete–Incomplete Block Designs—*cont...*

Design 3: $t = 5$, $b = 10$, $k = 2$
Resulting CIB design:

|  | Treatments |  |  |  |  |
|---|---|---|---|---|---|
| Blocks | 1 | 2 | 3 | 4 | 5 |
| 1 | x | x |  |  |  |
| 2 | x |  | x |  |  |
| 3 | x |  |  | x |  |
| 4 | x |  |  |  | x |
| 5 |  | x | x |  |  |
| 6 |  | x |  | x |  |
| 7 |  | x |  |  | x |
| 8 |  |  | x | x |  |
| 9 |  |  | x |  | x |
| 10 |  |  |  | x | x |

Resulting CIB design:

|  | Treatments |  |  |  |  |
|---|---|---|---|---|---|
| Blocks | 1 | 2 | 3 | 4 | 5 |
| 1 | x | x | x | x | x |
|   | x | x |  |  |  |
| 2 | x | x | x | x | x |
|   | x |  | x |  |  |
| 3 | x | x | x | x | x |
|   | x |  |  | x |  |
| 4 | x | x | x | x | x |
|   | x |  |  |  | x |
| 5 | x | x | x | x | x |
|   |  | x | x |  |  |
| 6 | x | x | x | x | x |
|   |  | x |  | x |  |
| 7 | x | x | x | x | x |
|   |  | x |  |  | x |
| 8 | x | x | x | x | x |
|   |  |  | x | x |  |
| 9 | x | x | x | x | x |
|   |  |  | x |  | x |
| 10 | x | x | x | x | x |
|   |  |  |  | x | x |

### Intrablock Analysis

Intrablock analysis of CIB designs follows a modified model used for analysis in Section 5.1. The modified model for the CIB design is

$$X_{ijp} = \mu + T_i + B_j + (TB)_{ij} + \varepsilon_{ijp}, \quad \begin{aligned} i &= 1,\ldots,t, \\ j &= 1,\ldots,b, \\ p &= 1, n_{ij}. \end{aligned} \quad (5.4\text{-}1)$$

The $X_{ijp}$ is the $p$th observation in the $(i, j)$th cell. The parameters $\mu$, $T_i$, $B_j$, and $\varepsilon_{ijp}$ are defined as in Model (5.1-2). The term $(TB)_{ij}$ is the interaction between the $i$th treatment and the $j$th block. Table 5.22 displays the intrablock analysis of variance of the CIB designs following Model (5.4-1). Note that in this table $n_{ij}$ is the number of observations in the $(i, j)$th cell, and its value is either 1 or 2. The quantity $d_{ij} = X_{ij1} - X_{ij2}$ is the difference between observations for each $(i, j)$th cell having two observations.

The adjusted treatment totals $Q_i$ for a CIB design are given by

$$Q_i = X_{i.} - \sum_j^b \frac{n_{ij} X_{.j}}{t + k}, \quad i = 1,\ldots,t, \quad (5.4\text{-}2)$$

where $X_{i.}$ and $X_{.j}$ are the treatment and block totals, respectively. Cornell and Knapp (1972) showed that the intrablock estimates of the treatment effects are

$$t_i = \frac{(t - 1)(t + k)}{b[(t + k)^2 - (t + 3k)]} Q_i, \quad i = 1,\ldots,t, \quad (5.4\text{-}3)$$

with the standard errors of $t_i$ and $t_i - t_j$ estimated by

$$SE(t_i) = \sqrt{\frac{(t + k)(t - 1)^2}{bt[(t + k)^2 - (t + 3k)]} MSe} \quad (5.4\text{-}4)$$

and

$$SE(t_i - t_j) = \sqrt{\frac{2(t - 1)(t + k)}{b[(t + k)^2 - (t + 3k)]} MSe}, \quad (5.4\text{-}5)$$

respectively, where MSe is the pure error mean square from Table 5.22. The treatment means are estimated by $\overline{X}_i = \hat{\mu} + t_i$, with standard error

$$SE(\overline{X}_i) = \sqrt{\left\{ \frac{1}{t + k} + \frac{(t + k)(t - 1)^2}{t[(t + k)^2 - (t + 3k)]} \right\} \frac{MSe}{b}}. \quad (5.4\text{-}6)$$

We may note that $SE(\overline{X}_i - \overline{X}_j)$ is given also by (5.4-5).

**Table 5.22** Intrablock Analysis of Variance for Composite Complete–Incomplete Block Designs Following Model (5.4–1)

| Source of Variance | DF | SS |
|---|---|---|
| Total | $N - 1$ | $SSto = \sum_i \sum_j \sum_p X_{ijp}^2 - CF$ |
| Blocks | $b - 1$ | $SSbl = \dfrac{\sum_i \sum_j X_{ij\cdot}^2}{n_{ij}} - \dfrac{t \sum_i X_{i\cdot\cdot}^2}{b(t+k)} - SStb$ |
| Treatments (Adjusted) | $t - 1$ | $SStr_{adj} = \dfrac{(t-1)(t+k)}{b\{(t+k)^2-(3t+k)\}} \sum_i Q_i^2$ |
| Interaction | $(b - 1)(t - 1)$ | $SStb = \dfrac{\sum_i \sum_j X_{ij\cdot}^2}{n_{ij}} - \dfrac{\sum_j X_{\cdot j}^2}{t+k} - SStr_{adj}$ |
| Pure Error | $bk$ | $SSe = \dfrac{\sum_i \sum_j d_{ij}^2}{2}$ |

The inter- and intrablock estimate of treatment effects for the CIB designs may also be combined; however, we do not discuss that here. The reader may refer to Seshadri (1963), Cornell and Knapp (1972), Trail and Weeks (1973), Cornell (1974b), and Cornell and Schreckingost (1975).

**Example 5.4-1**

Cornell and Schreckingost (1975) reported an example pertaining to the amount of fresh blueberry flavor in three pie fillings, denoted by *W*, *C*, and *M*. Each filling was evaluated for flavor on a $1-9$ flavor scale. The experimental data and the relevant calculations are shown in Table 5.23. The SS calculations in Table 5.23 follow the formulas given in Table 5.22. To obtain the interaction SS, one forms a treatment by block layout, as shown in Table 5.24. The $X_{ij\cdot}$'s that are totals of two observations ($n_{ij} = 2$) are underlined. The analysis of variance is shown in Table 5.25. The *F* ratios for testing the effects of treatments and interaction on pie flavor are, respectively,

$$F = 14.15/1.25 = 11.32,$$

with 2 numerator and 30 denominator DF, and

$$F = 4.09/1.25 = 3.27$$

with 28 numerator and 30 denominator DF. Both ratios are statistically significant at the 1% level, leading to the rejection of the equality of treatment effects or means and the rejection of the absence of interaction effects. Since interactions are significant, they cloud the interpretation and significance of the treatment effects. The analysis should be examined carefully in interpreting the results. The estimates of the standard errors are

**Table 5.23** Data from a Pie Filling Study Using a Composite Complete–Incomplete Block Design (Example 5.4-1) $t = 3$, $b = 15$, and $k = 2$

| Panelists (Blocks) | Pie Fillings W | Pie Fillings C | Pie Fillings M | $X_{\cdot j}$ |
|---|---|---|---|---|
| 1 | 8, 7 | 4, 5 | 6 | 30 |
| 2 | 6, 7 | 6, 6 | 4 | 29 |
| 3 | 5, 6 | 6, 4 | 5 | 26 |
| 4 | 7, 7 | 3, 2 | 6 | 25 |
| 5 | 5, 3 | 5, 7 | 3 | 23 |
| 6 | 3 | 6, 4 | 7, 8 | 28 |
| 7 | 8 | 3, 2 | 4, 7 | 24 |
| 8 | 7 | 7, 6 | 5, 5 | 30 |
| 9 | 5 | 2, 4 | 7, 5 | 23 |
| 10 | 9 | 7, 4 | 4, 5 | 29 |
| 11 | 3, 5 | 5 | 7, 6 | 26 |
| 12 | 7, 5 | 3 | 5, 5 | 25 |
| 13 | 6, 6 | 7 | 8, 5 | 32 |
| 14 | 8, 7 | 3 | 3, 5 | 26 |
| 15 | 7, 8 | 6 | 5, 5 | 31 |
| $X_{i\cdot}$ | 155 | 117 | 135 | $X_{\cdot\cdot} = 407$ |
| | | | | $\hat{\mu} = 5.43$ |
| $\sum_j n_{ij} X_{\cdot j}$ | 680 | 674 | 681 | |
| $Q_i$ | 19 | −17.8 | −1.2 | $\sum_i Q_i = 0$ |
| $t_i$ | 0.7917 | −0.7417 | 0.0500 | $\sum_i t_i = 0$ |
| $\overline{X}_i$ | 6.22 | 4.69 | 5.38 | |

Note:

$$CF = \frac{407^2}{75} = 2208.6533.$$

$$SSto = (8^2 + 7^2 + \cdots + 5^2 + 5^2) - CF = 2413.0000 - CF = 204.3467.$$

$$SStr_{adj} = \frac{2(5)}{15(25-9)}\left[(19.0)^2 + (-17.8)^2 + (-1.2)^2\right] = 28.3033.$$

$$SStb = \left[\frac{(15)^2}{2} + \frac{(13)^2}{2} + \cdots + \frac{8^2}{2} + \frac{10^2}{2}\right] - \left[\frac{(30)^2 + (29)^2 + \cdots + (26)^2 + (31)^2}{5}\right]$$
$$- SStr_{adj} = 2375.5000 - 2232.6000 - 28.3033 = 114.5967.$$

$$SSbl = \left[\frac{(15)^2}{2} + \frac{(13)^2}{2} + \cdots + \frac{8^2}{2} + \frac{(10)^2}{2}\right] - 3\left[\frac{(155)^2 + (117)^2 + (135)^2}{15(5)}\right] - SStb$$
$$= 2375.5000 - 2237.5600 - 114.5967 = 23.3433.$$

$$SSe = \frac{(8-7)^2 + \cdots + (5-5)^2}{2} = \frac{75.000}{2} = 37.5000.$$

**Table 5.24** Subtotals $X_{ij}$. for Calculating Interaction Sum of Squares[a]

| Blocks | W | C | M |
|---|---|---|---|
| 1 | 15 | 9 | 6 |
| 2 | 13 | 12 | 4 |
| 3 | 11 | 10 | 5 |
| 4 | 14 | 5 | 6 |
| 5 | 8 | 12 | 3 |
| 6 | 3 | 10 | 15 |
| 7 | 8 | 5 | 11 |
| 8 | 7 | 13 | 10 |
| 9 | 5 | 6 | 12 |
| 10 | 9 | 11 | 9 |
| 11 | 8 | 5 | 13 |
| 12 | 12 | 3 | 10 |
| 13 | 12 | 7 | 13 |
| 14 | 15 | 3 | 8 |
| 15 | 15 | 6 | 10 |

[a]The total of $n_{ij} = 2$ in the same cell are underlined.

**Table 5.25** Analysis of Variance of the Pie Fillings Data (Example 5.4-1)

| Source of Variance | DF | SS | MS | F Ratio |
|---|---|---|---|---|
| Total | 74 | 204.3467 | | |
| Treatments (Adjusted) | 2 | 28.3033 | 14.15 | 11.32* |
| Panelists | 14 | 23.3433 | 1.68 | |
| Interaction | 28 | 114.5967 | 4.09 | 3.27* |
| Pure Error | 30 | 37.5000 | 1.25 | |

*$p < .01$.

$$SE(t_i) = \sqrt{(5)(4)1.25/15(3)(25-9)} = 0.19,$$

$$SE(t_i - t_j) = \sqrt{2(2)(5)(1.25)/15(25-9)} = 0.32,$$

$$SE(\overline{X}_i) = \sqrt{\left[\frac{1}{5} + (5)(4)/3(25-9)\right](1.25/15)} = 0.23.$$

## EXERCISES

**5.1.** Ten methods of preparing dried whole eggs were scored for storage off-flavor using the scale 10 for no off-flavor and 0 for extreme off-flavor (Hanson et al., 1951, 9). The panel means from seven judges are reproduced as follows:

| Judging Period | Preparation Methods | | | | | | | | | |
|---|---|---|---|---|---|---|---|---|---|---|
| | 1 | 2 | 3 | 4 | 5 | 6 | 7 | 8 | 9 | 10 |
| 1 | 9.7 | 8.7 | | 5.4 | 5.0 | | | | | |
| 2 | | 9.6 | 8.8 | | | 5.6 | | | | 3.6 |
| 3 | | 9.0 | | 7.3 | | 3.8 | 4.3 | | | |
| 4 | 9.3 | | 8.7 | | 6.8 | | 3.8 | | | |
| 5 | 10.0 | | | 7.5 | | | | 4.2 | | 2.8 |
| 6 | | 9.6 | | | | | 5.1 | 4.6 | 3.6 | |
| 7 | | 9.8 | | 7.4 | | | | 4.4 | | 3.8 |
| 8 | | | | | 9.4 | | 6.3 | | 5.1 | 2.0 |
| 9 | 9.3 | 9.3 | 8.2 | | | | | | 3.3 | |
| 10 | | | | 8.7 | 9.0 | 6.0 | | | 3.3 | |
| 11 | 9.7 | | | | | 6.7 | 6.6 | | | 2.8 |
| 12 | | | 9.3 | 8.1 | | | | | 3.7 | 2.6 |
| 13 | 9.8 | | | | | 7.3 | | 5.4 | 4.0 | |
| 14 | | | 9.0 | 8.3 | | | 4.8 | 3.8 | | |
| 15 | | | 9.3 | | 8.3 | 6.3 | | 3.8 | | |

1. Write down a statistical model for analysis.
2. Prepare the analysis of variance table of your model and carry out a multiple-comparison test to divide the preparation methods into homogeneous groups.

**5.2.** Use the Yates method for combining intra- and interblock analyses for the data in Example 5.1-1. Compare the combined estimates of the treatment effects with those of the intrablock analysis in Example 5.1-1.

**5.3.** Sensory data pertaining to "texture liking" of low-sodium hot dogs are given in the following tabulation. Three treatments denoted by $A$, $B$, and $C$ and a control $R$ were evaluated following a BIB design with control in each block.

| Panelists (Blocks) | Sample Orders | Corresponding Texture Scores |
|---|---|---|
| 1 | R A B | 3, 5, 5 |
| 2 | C R A | 7, 6, 7 |
| 3 | B C R | 7, 7, 7 |
| 4 | R B A | 6, 7, 7 |
| 5 | A R C | 8, 7, 8 |
| 6 | C B R | 6, 6, 7 |
| 7 | A R B | 8, 2, 5 |
| 8 | C A R | 8, 8, 7 |
| 9 | R C B | 4, 7, 7 |
| 10 | A B R | 9, 9, 8 |
| 11 | R C A | 7, 8, 7 |
| 12 | C R B | 9, 7, 8 |

1. Carry out the analysis of variance and compute the adjusted treatment means, including the control.
2. Carry out a multiple-comparison test of your choice among the means for $A$, $B$, $C$, and $R$. Discuss your results.
3. Carry out a pairwise comparison between the treatment means and control mean.

**5.4.** A study was conducted on $t = 8$ soap bars to evaluate their lather characteristics. Data pertaining to the amount of lather are given in the following tabulation:

| Judges (Blocks) | Treatment Orders | Corresponding Lather Scores |
|---|---|---|
| 1 | 3, 4, 2, 1 | 9, 9, 8, 7 |
| 2 | 6, 7, 8, 5 | 8, 8, 6, 5 |
| 3 | 1, 2, 7, 8 | 4, 6, 5, 5 |
| 4 | 4, 5, 6, 3 | 8, 4, 9, 6 |
| 5 | 6, 1, 8, 3 | 7, 4, 7, 7 |
| 6 | 5, 7, 4, 2 | 8, 6, 4, 6 |
| 7 | 1, 4, 6, 7 | 2, 7, 7, 5 |
| 8 | 8, 5, 3, 2 | 6, 5, 8, 6 |

*(continued)*

| Judges (Blocks) | Treatment Orders | Corresponding Lather Scores |
|---|---|---|
| 9 | 5, 6, 2, 1 | 6, 6, 6, 3 |
| 10 | 3, 4, 7, 8 | 8, 4, 5, 7 |
| 11 | 3, 5, 7, 1 | 8, 7, 8, 3 |
| 12 | 8, 6, 4, 2 | 6, 6, 7, 5 |
| 13 | 4, 1, 5, 8 | 5, 2, 6, 4 |
| 14 | 2, 6, 3, 7 | 4, 8, 7, 6 |

1. Determine the design parameters $t$, $b$, $k$, $r$, $\lambda$, and $\lambda_1$, as defined in Section 5.3.
2. Carry out the analysis of variance of the lather scores following Model (5.3-1).
3. Carry out a multiple-comparison test among the adjusted treatment means.
4. In your model test the hypothesis that the correlation effect is zero. Interpret your results.

**5.5.** The accompanying data are taken from a CIB design (Pearce, 1960, Biometrika 47, 263). The purpose of the study was to find out whether the four weed killers would harm fruiting plants. The data represent the total spread in inches of 12 plants per plot approximately 2 months after the application of the weed killers (control data omitted).

| Blocks | Weed Killers | | | |
| | A | B | C | D |
|---|---|---|---|---|
| 1 | 166 | 166 | 107 | 133 |
|   | 163 |     |     |     |
| 2 | 117 | 139 | 103 | 132 |
|   |     | 118 |     |     |
| 3 | 130 | 147 | 95 | 103 |
|   |     |     | 109 |     |
| 4 | 136 | 152 | 104 | 119 |
|   |     |     |     | 132 |

1. Determine the CIB design parameters based on the layout of the data.
2. Carry out the intrablock analysis of variance and estimate the treatment effects.
3. Carry out the steps of a multiple-comparison test on means.

# Factorial Experiments

Factorial experiments can be attributed to Fisher (1960); they are used in studies involving factors where each factor is at more than one level. Some experimental areas in which factorial designs can be of use are

1. Food research, where factors may be ingredients in a product mix, formulas, amounts of spices, and types of flavor additives;
2. Chemical research, where factors may be temperature, pressure, reactants, and time;
3. Marketing research, where factors may be prices, product brands, locations, and seasons;
4. Sociological research, where factors may be age, sex, education, income, and racial groups.

There are several advantages in factorial designs, including the following:

1. The effects of varying all factors simultaneously can be evaluated over a wide range of experimental conditions.
2. All possible interactions among the factors can be estimated.
3. The estimates of main and interaction effects are obtained by optimal use of experimental materials.

In factorial experiments the factor levels are variously coded. If a factor $A$ has two levels, the levels may be designated low and high and coded 0 and 1 or $a_0$ and $a_1$, etc. If a factor $B$ has three levels, the levels may be designated low, medium, and high and coded 0, 1, and 2 or $b_0$, $b_1$, and $b_2$, etc. Depending on the coding scheme used in factorial experiments, each combination of factor levels can be appropriately denoted. For instance, we may examine Table 6.1 for some alternative codings and notations in factorial experiments with two factors $A$ and $B$; each at two levels. The notation in the last column of the table is most commonly used in factorial experiments when each factor is at two levels. This notation is fully explained in Section 6.1.

**Table 6.1** Some Alternative Notations in Two Factorial Experiments: Each Factor $A$ and $B$ at Two Levels

| Treatment Combination[a] | Five Alternative Notations for Treatment Combinations | | | | |
|---|---|---|---|---|---|
| | **1** | **2** | **3** | **4** | **5** |
| 1 | $a_1b_1$ | 11 | $a_0b_0$ | 00 | (1) |
| 2 | $a_1b_2$ | 12 | $a_0b_1$ | 01 | $b$ |
| 3 | $a_2b_1$ | 21 | $a_1b_0$ | 10 | $a$ |
| 4 | $a_2b_2$ | 22 | $a_1b_1$ | 11 | $ab$ |

[a]Definition of treatment combination:
 1: All factors at low levels.
 2: Factor A at low levels and factor B at high levels.
 3: Factor A at high levels and factor B at low levels.
 4: All factors at high levels.

## 6.1 THE $2^n$ FACTORIAL EXPERIMENTS

An experiment that consists of $n$ factors $A$, $B$, $C,\ldots$, each at two levels, is called a $2^n$ factorial experiment. In a $2^n$ factorial design, there are $2^n$ combinations of factor levels. In some analyses each combination is viewed as a treatment. In our presentation a symbol for a factorial combination includes each letter if the corresponding factor is at the high level in the combination. The symbol $ac$, for example, denotes the combination in which the factors $A$ and $C$ are at high levels and all the other factors at low levels. The combination with all the factors at low levels is denoted by $(1)$. Note that the last column of Table 6.1 follows this symbolism for the four combinations in $2^2$ factorial experiments. In a $2^3$ factorial experiment with factors $A$, $B$, and $C$, there are eight treatment combinations, which are as follows:

$$(1), \quad a, \quad b, \quad ab, \quad c, \quad ac, \quad bc, \quad abc.$$

In factorial experiments the symbols for factorial combinations are also used to denote the total experimental observations for the respective combinations. This simultaneous use of a symbol for a factorial combination and its corresponding experimental outcome does not cause any confusion.

The analysis of $2^n$ factorial experiments involves the partition of the total SS of the experimental outcomes into the SS due to the *main effects*, interaction effects, and experimental errors. In $2^n$ factorial experiments there are $n$ main effects, $\binom{n}{2} = n(n-1)/2$ two-factor interaction effects, $\binom{n}{3} = n(n-1)(n-2)/6$ three-factor interaction effects, and so on. In all, there are $2^n - 1$ effects or parameters that are to be estimated in addition to the grand mean $\mu$. The SS associated with each effect has 1 DF. There is 1 DF associated with the SS due to the grand mean called the correction factor (CF). Thus, to be able to estimate the error variance, one needs to repeat the $2^n$ factorial experiment $r \geq 2$ times. The breakdown of the sources of variation and the DF are shown in Table 6.2.

How do we define and estimate the effect parameters, as well as their respective SS, in $2^n$ factorial experiments? To answer the question, we consider one replication of a $2^3$ factorial experiment with factors $A$, $B$, and $C$. Since $n = 3$, there are three main effects and three two-factor and one three-factor interactions. Four independent estimates of the main effect of $A$ are given by the following contrasts among the outcomes of the factorial combinations:

1. Contrast $abc - bc$ estimates the main effect of $A$ when both $B$ and $C$ are at the high levels.
2. Contrast $ab - b$ estimates the main effect of $A$ when $B$ is at the high level and $C$ is at the low level.
3. Contrast $ac - c$ estimates the main effect of $A$ when $C$ is at the high level and $B$ is at the low level.
4. Contrast $a - (1)$ estimates the main effect of $A$ when both factors $B$ and $C$ are at low levels.

**Table 6.2** Sources of Variation with Degrees of Freedom in $2^n$ Factorial Experiments with $r$ Replications

| Source of Variance | DF |
|---|---|
| Grand mean | 1 |
| Main effects | $N$ |
| Two-factor interactions | $n(n-1)/2$ |
| Three-factor interactions | $n(n-1)(n-2)/6$ |
| Error | $2^n(r-1)$ |
| | $r2^n$ |

The average of these four independent estimates would be an overall best estimate of the main effect of A. That is,

$$\frac{1}{4}\left\{(abc - bc) + (ab - b) + (ac - c) + [a - (1)]\right\}$$

$$= \frac{1}{4}\left[abc + ab + ac + a - bc - b - c - (1)\right] \quad (6.1\text{-}1)$$

provides an overall estimate. Now remember that it would not be possible to estimate the error variance if the factorial combinations were used only once in the experiment. Hence, suppose that a $2^3$ factorial experiment consists of $r$ replications. For each replication (6.1-1) provides an estimate of the main effect of A. Once again we can average the $r$ estimates to obtain

$$\frac{abc + ab + ac + a - bc - b - c - (1)}{4r}, \quad (6.1\text{-}2)$$

which will yield "best" estimates of the main effect of A. Remember the interpretation of the symbols in (6.1-2). For example, $abc$ stands for the total of all the experimental outcomes for the combination $abc$ in the factorial experiment.

Following our arguments for estimating the main effect of A in $2^3$ factorial experiments consisting of $r$ replications, the estimates of the other two main effects are as follows. Estimate the main effects of B and C by the contrasts

$$\frac{abc + ab + bc + b - a - c - ac - (1)}{4r} \quad (6.1\text{-}3)$$

and

$$\frac{abc + ac + bc + c - a - b - ab - (1)}{4r}, \quad (6.1\text{-}4)$$

respectively. Now we can consider the estimation of two-factor interactions. Specifically, factors A and B interact if the estimates $abc - bc$ and $ab - b$ of the main effects of A when B is at the high level differ from the respective estimates $ac - c$ and $a - (1)$ when B is at the low level. Hence, an appropriate definition and estimate of the AB interaction is given by the contrast

$$\frac{(abc - bc) - (ac - c) + (ab - b) - [a - (1)]}{4r}$$

$$= \frac{abc + ab + c + (1) - bc - ac - a - b}{4r}. \quad (6.1\text{-}5)$$

Similarly, the other interaction effects $ac$, $bc$, and $abc$ are estimated by

$$\frac{abc + ac + b + (1) - bc - ab - a - c}{4r}, \qquad (6.1\text{-}6)$$

$$\frac{abc + bc + a + (1) - ab - ac - b - c}{4r}, \qquad (6.1\text{-}7)$$

and

$$\frac{abc + a + b + c - ab - ac - bc - (1)}{4r} \qquad (6.1\text{-}8)$$

contrasts, respectively.

In general, for $2^n$ factorial experiments, there are easy rules for writing down contrasts such as (6.1-2) through (6.1-8). Note that the denominator of these contrasts is the same for all the effects, which is $r2^{n-1}$. Verify that for $2^3$ factorial experiments the denominator is $r2^{3-1} = 4r$. The plus or minus sign of a factorial combination in an estimating contrast is determined as follows. For estimating the main effect of a factor, assign the plus sign to a combination if the factor is in it at the high level and the minus sign otherwise. For example, in the contrast for estimating the main effect of $A$, the combination $ac$ would receive the plus sign, whereas $bc$ would receive the minus sign. To determine the sign of a combination in a contrast for estimating an interaction effect, multiply the signs associated with the combination corresponding to the main effects of the factors in the interaction effect. Suppose we want the coefficient of treatment combination $ab$ for, say, the interaction effect $AC$. Obviously, the two factors involved in the interaction effect are $A$ and $C$. The coefficients of $ab$ for the main effects of $A$ and $C$ are 1 and $-1$, respectively, and hence the coefficient of $ab$ for the interaction effect $ac$ is $(1)(-1) = +1$. Table 6.3 can be verified by following these rules. The matrix of the table is called the *design matrix*. The columns of the design matrix in the table are orthogonal; that is, all the main effects and interaction effects are mutually independent contrasts.

Now we define the SS due to the main effects and the interaction effects. The word *contrast* in our definition of the SS refers only to the numerator of the estimating contrasts of the effects without the common divisor $r2^n$. Following this understanding, the SS for an effect in $2^n$ factorial experiments is defined as

$$SS = (\text{contrast})^2 / r2^n. \qquad (6.1\text{-}9)$$

An appropriate suffix is attached to the SS to indicate the source effect causing it. Thus, in $2^3$ factorial experiments, we may use $SSa$, $SSb$, $SSc$,

**Table 6.3** Design Matrix for $2^n$ Factorial Experiments

| Treatment Combination | Effects | | | | | | |
|---|---|---|---|---|---|---|---|
| | **A** | **B** | **AB** | **C** | **AC** | **BC** | **ABC** |
| (1) | −1 | −1 | 1 | −1 | 1 | 1 | −1 |
| a | 1 | −1 | −1 | −1 | −1 | 1 | 1 |
| b | −1 | −1 | −1 | 1 | −1 | 1 | 1 |
| ab | 1 | 1 | 1 | −1 | −1 | −1 | −1 |
| c | −1 | −1 | 1 | 1 | −1 | −1 | 1 |
| ac | 1 | −1 | −1 | 1 | 1 | −1 | −1 |
| bc | −1 | 1 | −1 | 1 | −1 | 1 | −1 |
| abc | 1 | 1 | 1 | 1 | 1 | 1 | 1 |
| ⋮ | | | | | | | |

SSab, SSac, SSbc, and SSabc to denote the SS associated with the effects. In the analysis of experiments where each factorial combination is treated as one treatment, the treatment SS is the total of all the SS due to all the $2^n − 1$ individual effects. Hence, for $2^3$ factorial experiments, the treatment SS is SSa + SSb + SSc + SSab + SSbc + SSabc with $2^3 − 1 = 7$ DF.

The total sum of squares, SSto, and the error sum of squares, SSe, are determined in the usual manner. The examples that follow illustrate two applications of the $2^2$ factorial experiments.

---

**Example 6.1-1**

The data in this example pertain to a $2^2$ factorial experiment taken from the files of the Armour Research Center. One of the factors is the source of meat, denoted by B. The two sources of meat are $B_1$ and $B_2$, designated as low and high, respectively. The other factor, A, refers to the type of meat additive. The two types of additives are denoted by $A_1$ and $A_2$ and designated as low and high, respectively. Thus, the four combinations are (1), a, b, and ab. This factorial experiment is repeated $r = 2$ times. The experimental data appear in Table 6.4 with all the pertinent calculations. The main effect of A, for example, is estimated by

$$\frac{ab + a − b − (1)}{4r} = \frac{69.2 − 67.7}{4r} = \frac{1.5}{4r},$$

**Table 6.4** Data of a $2^2$ Factorial Experiment and Calculations

| | B: Sources of Meat | | |
| | $B_1$ (Low) | $B_2$ (High) | Sums |
|---|---|---|---|
| A: Types of Additive | $A_1$ (low) = 18.5 | 15.7 | |
| | 16.6 | 16.9 | |
| | (1) = 35.1 | b = 32.6 | (1) + b = 67.7 |
| | $A_2$ (high) = 20.9 | 13.1 | |
| | 23.0 | 12.2 | |
| | a = 43.9 | ab = 25.3 | a + ab = 69.2 |
| Sums | (1) + a = 79.0 | b + ab = 57.9 | 136.9 |

| Treatment Combination | Sum | Effects | | |
| | | A | B | AB |
|---|---|---|---|---|
| (1) | 35.1 | −1 | −1 | +1 |
| a | 43.9 | +1 | −1 | −1 |
| b | 32.6 | −1 | +1 | −1 |
| ab | 25.3 | +1 | +1 | +1 |
| Sum of + | | 69.2 | 57.9 | 60.4 |
| Sum of − | | 67.7 | 79.0 | 76.5 |
| Contrasts (Numerator of Effects) | | 1.5 | −21.1 | −16.1 |

Note:
  $CF = (136.9)^2/8 = 2342.70$ with 1 DF.
 $SSto = [(18.5)^2 + \cdots + (12.2)^2] - CF = 93.47$ with 7 DF.
 $SSa = (1.5)^2/8 = 0.28$ with 1 DF.
 $SSb = (-21.1)^2/8 = 55.65$ with 1 DF.
 $SSab = (-16.1)^2/8 = 32.40$ with 1 DF.
 $SSe = 93.47 - (0.28 + 55.65 + 32.40) = 5.14$ with 4 DF.

and the SS due to $A$ is

$$SSa = \frac{1}{8}\left[(1.5)^2\right] = 0.28.$$

Table 6.5 summarizes the analysis of variance. Since the $ab$ interaction effect is found significant, the interpretation of the main effects is not clear. We may examine the graph in Figure 6.1 for the trend in the

**Table 6.5** Analysis of Variance for Example 6.1-1

| Source of Variance | DF | SS | MS | F Ratio |
|---|---|---|---|---|
| Total | 7 | 93.47 | | |
| Additives (A) | 1 | 0.28 | 0.28 | 0.2 |
| Meat Sources (B) | 1 | 55.65 | 55.65 | 43.1* |
| AB Interaction | 1 | 32.40 | 32.40 | 25.1* |
| Residual | 4 | 5.14 | 1.29 | |

*$p < .01$.

■**FIGURE 6.1** Plot of the $A \times B$ interaction for Example 6.1-1.

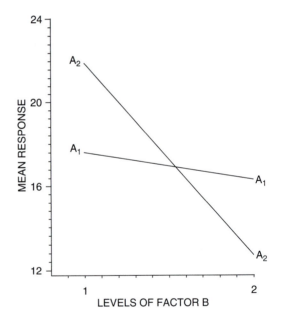

interaction effect. The graph suggests that the response varies with the factor combinations.

We consider one more example of a $2^2$ factorial experiment in a randomized complete block (RCB) design as follows.

A study on pricing was conducted by Barclay (1969) at the store level to answer the following:

**Example 6.1-2**

1. What would be the effect on total line profits of increasing the price of factor A by 4¢ per unit at retail?
2. What would be the corresponding effect of increasing the price of factor B by 4¢ per unit?
3. Is there an interaction between A and B? That is, would the effect of increasing one price depend on the level of the other price?

A $2^2$ factorial arrangement in an RCB design was used in the study with test stores considered as blocks and the factor combinations as treatments ($t = 4$). There were $b = 30$ blocks (stores). The treatment combinations, along with the total responses and the analysis of variance, are given in Table 6.6.

The estimate of the main effect of A is

$$\text{Main effect of} A = \frac{ab + a - b - (1)}{30(2)}$$
$$= \frac{88.323 + 89.589 - 90.765 - 90.384}{60}$$
$$= -0.0540.$$

**Table 6.6** Analysis of Pricing Study in Example 6.1-2

| Treatment Combination | Description of Package Price | | Total Response |
|---|---|---|---|
| | **Price of A** | **Price of B** | |
| (1) | Present (low) | Present (low) | 90.384 |
| a | Present + 4¢ (high) | Present (low) | 89.589 |
| b | Present (low) | Present + 4¢ (high) | 90.765 |
| ab | Present + 4¢ (high) | Present + 4¢ (high) | 88.323 |

| Analysis of Variance | | | |
|---|---|---|---|
| **Source of Variance** | **DF** | **SS** | **MS** |
| Blocks (Stores) | 29 | 3.923 | |
| Treatments | 3 | 0.117 | |
| Error | 87 | 2.639 | 0.030 |

Similarly,

$$\text{Main effect of } B = -0.0148,$$

$$\text{Interaction effect } AB = -0.0247.$$

It was decided to construct 90% confidence intervals for estimating the average changes in profits as a result of increasing the prices of items $A$ and $B$, and also for estimating the interaction effect. If we are estimating a parameter $\theta$ by a $(1 - \alpha)$ 100% confidence interval, then we calculate

$$\widehat{\theta} \pm [t_{DF, \alpha/2}] S_{\widehat{\theta}},$$

where $\widehat{\theta}$ is a point estimate of $\theta$, $t_{DF,\alpha}$ is the $\alpha$th percentile of the $t$ distribution at DF equal to the DF of SSe, and $SS_{\widehat{\theta}}$ is the estimated standard error of $\widehat{\theta}$. In our problem DF $= 87$. Since DF $> 30$, we may approximate the percentiles of the $t$ distribution by the percentiles of the standard normal distribution. Hence, for 90% confidence interval estimation, $t_{87, 0.05} \simeq Z_{0.05} = -1.645$.

Now let us determine $\widehat{S}_{\theta}$. Note that $\theta$ is the main effect of $A$ when we are estimating the average change in profit as a result of increasing the price of item $A$. If our interest is in estimating the average change in profit resulting from an increase in the price of $B$, then $\theta$ would be the main effect of $B$. Finally, $\theta$ may be the interaction effect of $A$ and $B$. In each of the three situations, $\theta$ is being estimated by a contrast among $(1)$, $a$, $b$, and $ab$. Remember that there are 30 replications (blocks) and each replication provides one contrast [refer to the arguments from (6.1-1) to (6.1-2)]. Hence, it follows that the standard error $\sigma_{\theta}$ of $\widehat{\theta}$, estimator of $\theta$ in either situation, is $\sigma/30$, where $\sigma$ is estimated by MSe. Hence, an estimator of $\sigma_{\widehat{\theta}}$ is $S_{\widehat{\theta}} = \sqrt{\text{MSe}/30} = \sqrt{0.03/30}$. Now returning to our expression for $(1 - \alpha)$ 100% confidence interval, a 90% confidence interval for $\theta$ is

$$\widehat{\theta} \pm 1.645\sqrt{0.03/30} \quad \text{or} \quad \widehat{\theta} \pm 0.0520.$$

Substituting the respective $\widehat{\theta}$ in each of the three situations, we have the following confidence intervals:

| Effect | 90% Confidence Interval |
| --- | --- |
| Price of A | $-0.0012$ to $-0.1068$ |
| Price of B | $-0.0676$ to $+0.0380$ |
| Interaction AB | $-0.0802$ to $+0.0254$ |

This means that increasing the price of item $A$ by 4¢ at retail would reduce total line profits. The loss in profits would range from 0.0012 to 0.1068 units per store. The confidence intervals for the average effect of $B$

and the interaction effect *AB* include zero. This suggests that increasing the price of item *B* would not significantly change the total profits. Likewise, the interaction between items *A* and *B* is not statistically significant. Remember that these conclusions are pertinent only at the $\alpha = 10\%$ level of significance.

---

**Example 6.1-3**

A portion of sensory data from a sodium hot dog study is shown in Table 6.7. The three factors, each at two levels, are NaCl, KCl, and dextrose:

| | | |
|---|---|---|
| *A*: NaCl | 2.5% (low) | 6.5% (high) |
| *B*: KCl | 2.0% (low) | 4.0% (high) |
| *C*: Dextrose | 2.0% (low) | 5.0% (high) |

Each factorial combination was compared to a control using a 9-point hedonic scale. The SSto can be split into the SS due to each effect with 1 DF. That is,

$$SSto = SSa + SSb + SSc + SSab + SSac + SSbc + SSabc$$

**Table 6.7** Data from a $2^3$ Factorial Experiment of Example 6.1-3

| | Factor Combinations | | | | | | | |
|---|---|---|---|---|---|---|---|---|
| **Replication** | **(1)** | ***a*** | ***b*** | ***ab*** | ***c*** | ***ac*** | ***bc*** | ***abc*** |
| 1 | 5 | 5 | 5 | 6 | 6 | 7 | 7 | 7 |
| 2 | 6 | 4 | 6 | 7 | 5 | 6 | 6 | 8 |
| 3 | 5 | 6 | 5 | 5 | 8 | 7 | 5 | 7 |
| 4 | 7 | 5 | 4 | 5 | 7 | 6 | 7 | 6 |
| 5 | 4 | 5 | 7 | 5 | 6 | 6 | 7 | 6 |
| | 27 | 25 | 27 | 28 | 32 | 32 | 32 | 34 |

Note:

$$CF = \frac{(237)^2}{40} = \frac{56,169}{40} = 1404.225 \text{ with 1 DF.}$$

$$SSto = (5^2 + 6^2 + \cdots + 6^2) - CF = 42.775 \text{ with 39 DF.}$$

$$SStr = \frac{[(27)^2 + (25)^2 + \cdots + (34)^2]}{5} - CF = 14.775 \text{ with 7 DF.}$$

where, for instance,

$$SSa = [abc + ab + ac + a - bc - b - c - (1)]^2/r2^3$$
$$= 1^2/5(8) = 0.025 \text{ with 1 DF,}$$

and so on. See Exercise 6.3.

---

## 6.2 THE $3^n$ FACTORIAL EXPERIMENTS

An experiment that consists of $n$ factors $A$, $B$, $C$,..., each at three levels, is called a $3^n$ factorial experiment. In a $3^n$ factorial design, there are $3^n$ combinations of factor levels. As mentioned before, the three levels of each factor may be designated as low, medium, and high. For a factor $A$, the three levels may be coded $a_0$, $a_1$, and $a_2$, and similarly for all other factors. Following this coding, in a $3^2$ factorial experiment with $A$ and $B$ factors, for example, the nine factorial combinations are

$$a_0b_0, \quad a_1b_0, \quad a_2b_0, \quad a_0b_1, \quad a_1b_1, \quad a_2b_1, \quad a_0b_2, \quad a_1b_2, \quad a_2b_2.$$

Each factorial combination symbol may also be used to denote the total of all its outcomes in the experiment.

Recall that in $2^n$ factorial experiments the effect of change for a factor from one level to the other is measured by the so-called main-effects parameter. When factors are at three levels, the effects of varying the levels (low to medium, low to high, and medium to high, e.g.) cannot be measured by just one parameter. Instead, when we treat one of the levels as origin, the linear and the quadratic-effects parameters are defined to measure the changes among the three levels. The linear and quadratic effects of factor $A$, for instance, are denoted by $A_L$ and $A_Q$, respectively. Both $A_L$ and $A_Q$ are appropriately defined contrasts, orthogonal to each other, among the factorial combinations. Hence, in $3^n$ factorial experiments, there are linear and quadratic effects for each factor, as well as their interaction effects.

Let us consider a $3^2$ factorial experiment with factors $A$ and $B$. Each replication of the experiment will provide $3^2 = 9$ observations, one observation per factorial combination. There will be $r3^2$ observations if there are $r$ replications. We need to estimate two linear effects parameters $A_L$ and $B_L$; two quadratic effects parameters $A_Q$ and $B_Q$; four interaction effects parameters $A_L B_L$, $A_L B_Q$, $A_Q B_L$, and $A_Q B_Q$; and the grand mean. The total variation in the experimental observations with $r3^2$ DF is partitioned into nine components, which are the SS due to effects, each with 1 DF, and the SS due to the grand mean (CF) with 1 DF. Hence, if $r = 1$, there would be no DF left for estimating the error variance component. So, whenever possible, the $3^2$ factorial experiment should be repeated $r \geq 2$ times. If that is not possible, however, then the SS due to interaction effects believed to be insignificant (higher order interactions, which are usually difficult to interpret and are insignificant) can be used in place of the SS due to error.

How do we define the SS in $3^n$ factorial experiments, due to various effects? As in $2^n$ factorial experiments, the effects in $3^n$ factorial experiments are also defined by means of appropriate contrasts among the factorial combinations. There are rules for determining the coefficients for each combination appearing in the contrasts for defining effects. Remember that the three levels of each factor are designated low, medium, and high or coded, for example, $a_0$, $a_1$, and $a_2$ for factor $A$, and so on. For instance, the coefficient of a factorial combination in the definition of a linear-effect contrast $A_L$ would be either $-1$, $0$, or $+1$, corresponding to the letters $a_0$, $a_1$, or $a_2$ appearing in the combination. The coefficient of a factorial combination in the definition of a quadratic effect contrast $A_Q$ would be $1$, corresponding to the letter $a_0$, or $a_2$ and $-2$, corresponding to the letter $a_1$ appearing in the combination. The coefficient of a factorial combination in the definition of an interaction-effect contrast would be the product of the coefficients of the combination for the factor effects that are in the interaction effect. Thus, for example, in a $3^2$ factorial, the coefficients of $a_0 b_1$ in the contrasts for $A_L$, $B_Q$, and $A_L B_Q$ are $-1$, $-2$, and $(-1)(-2) = 2$, respectively. Following these rules, the coefficient matrix (design matrix) is given in Table 6.8. The determination of the design matrix can be summarized by the coordinate system in Figure 6.2. The first coordinate corresponds to the levels of factor $A$, and the second coordinate to that of $B$. The two coordinate systems correspond to the linear and quadratic effects. For example, the coefficient of $a_0 b_1$ in the contrasts for the linear and the quadratic effects $B_L$ and $B_Q$, respectively, are $0$ and $-2$. Note also that the coefficient of $a_0 b_1$ for the interaction effect $A_L B_Q$ is $(-1)(-2) = 2$. We may point out that the coordinate system is most appropriate for the equally spaced factor levels.

**Table 6.8** Design or Coefficient Matrix of the Factorial Combinations in the Contrasts for Defining Effects in $3^2$ Factorial Experiments

| Treatment Combination | Effects | | | | | | | |
|---|---|---|---|---|---|---|---|---|
| | $A_L$ | $A_Q$ | $B_L$ | $B_Q$ | $A_L B_L$ | $A_L B_Q$ | $A_Q B_L$ | $A_Q B_Q$ |
| | 1 | 2 | 3 | 4 | 5 | 6 | 7 | 8 |
| $a_0 b_0$ or $T_1$ | −1 | 1 | −1 | 1 | 1 | −1 | −1 | 1 |
| $a_1 b_0$ or $T_2$ | 0 | −2 | −1 | 1 | 0 | 0 | 2 | −2 |
| $a_2 b_0$ or $T_3$ | 1 | 1 | −1 | 1 | −1 | 1 | −1 | 1 |
| $a_0 b_1$ or $T_4$ | −1 | 1 | 0 | −2 | 0 | 2 | 0 | −2 |
| $a_1 b_1$ or $T_5$ | 0 | −2 | 0 | −2 | 0 | 0 | 0 | 4 |
| $a_2 b_1$ or $T_6$ | 1 | 1 | 0 | −2 | 0 | −2 | 0 | −2 |
| $a_0 b_2$ or $T_7$ | −1 | 1 | 1 | 1 | −1 | −1 | 1 | 1 |
| $a_1 b_2$ or $T_8$ | 0 | −2 | 1 | 1 | 0 | 0 | −2 | −2 |
| $a_2 b_2$ or $T_9$ | 1 | 1 | 1 | 1 | 1 | 1 | 1 | 1 |
| $\sum_{i=1}^{9} c_{ij}^2$ | 6 | 18 | 6 | 18 | 4 | 12 | 12 | 36 |

■ **FIGURE 6.2** Coordinates of the factorial combinations for $3^2$ factorials.

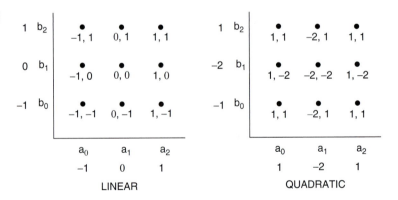

In general again, let us consider the $3^n$ factorial experiment. As in Table 6.8, one may denote the $3^n$ factorial combinations by $T_1$, $T_2$, $T_3$, etc., and the $(3^n - 1)$ effects may be numbered sequentially 1, 2, 3, etc. Thus, the $(i, j)$th element $c_{ij}$ of the design matrix is the coefficient of the factorial combination denoted by $T_i$ in the contrast for the effect numbered $j$. From Table 6.8, we see that the contrast for defining the linear effect $A_L$ is

$$-1(a_0b_0) + 0(a_1b_0) + \cdots + 0(a_1b_2) + 1(a_2b_2)$$
$$= a_2b_2 + a_2b_1 + a_2b_0 - a_0b_2 - a_0b_1 - a_0b_0$$
$$= T_9 + T_6 + T_3 - T_7 - T_4 - T_1.$$

The remaining contrasts of effects are obtained in a similar manner. Finally, the corresponding SS due to the effects are defined as

$$SS = (\text{contrast})^2 / r \sum_i c_{ij}^2.$$

An appropriate suffix is assigned to each SS to denote its source.

The analysis and computations of a $3^2$ factorial experiment may proceed following a two-way analysis of variance model as in Chapter 3 and Table 3.4. Since in factorial experiments the individual effects are of interest, the sums of squares SSa, SSb, and SSab in Table 2.4 are partitioned into the SS due to individual effects. Hence a modified version of Table 3.4 for displaying the analysis of $3^2$ factorial experiment is given by Table 6.9.

**Table 6.9** Analysis of Variance Table for a $3^2$ Factorial Experiment with Replications

| Source of Variance | SS | DF |
|---|---|---|
| Total (corrected) | SSto | $r3^2 - 1$ |
| A | SSa | 2 |
|   Linear, $A_L$ | $SSa_L$ | 1 |
|   Quadratic, $A_Q$ | $SSa_Q$ | 1 |
| B | SSb | 2 |
|   Linear, $B_L$ | $SSb_L$ | 1 |
|   Quadratic, $B_Q$ | $SSb_Q$ | 1 |
| AB | SSab | 4 |
|   Linear $\times$ linear, $A_L B_L$ | $SSa_L b_L$ | 1 |
|   Linear $\times$ quadratic, $A_L B_Q$ | $SSa_L b_Q$ | 1 |
|   Quadratic $\times$ linear, $A_Q B_L$ | $SSa_Q b_L$ | 1 |
|   Quadratic $\times$ quadratic, $A_Q B_Q$ | $SSa_Q b_Q$ | 1 |
|   Residual | SSe | $3^2(r - 1)$ |

| | |
|---|---|
| **Example 6.2-1** | Two compounds $A$ and $B$ are known to preserve the integrity of the moisture content in meats. Each compound was used at three levels in a $3^2$ factorial design concerning the water-binding capacity of meats. The experiment was repeated twice ($r = 2$), and the resulting data (in milliliters $H_2O$) are displayed in Table 6.10. The calculations for the sums of squares $SSa$, $SSb$, and $SSab$ follow from the two-way analysis of variance methods. These SS are broken down into SS with 1 DF each, as already discussed in this section. The calculations of the various SS are as follows: |

$$CF = (28.39)^2/18 = 44.7773 \quad \text{with} \quad 1 \text{ DF}$$

$$SSto = [(1.14)^2 + (1.05)^2 + \cdots + (0.10)^2 + (0.05)^2] - CF$$

$$= 14.3690 \quad \text{with} \quad 17 \text{ DF}.$$

SS due to factorial combinations:

$$SStr = [(2.19)^2 + (4.53)^2 + \cdots + (4.75)^2] + (0.15)^2/2 - CF$$

$$= 13.8289 \quad \text{with} \quad 8 \text{ DF}.$$

**Table 6.10** Data and Calculations of Sum of Squares for Example 6.2-1

| Factor $A$ | Factor B $b_0$ | Factor B $b_1$ | Factor B $b_2$ | Total for the A Levels |
|---|---|---|---|---|
| $a_0$ | 1.14 | 2.23 | 0.74 | 7.96 |
| | 1.05 | 2.30 | 0.50 | |
| $a_1$ | 1.87 | 3.13 | 1.43 | 12.03 |
| | 1.60 | 3.00 | 1.00 | |
| $a_2$ | 1.70 | 2.80 | 0.10 | 8.40 |
| | 1.80 | 1.95 | 0.05 | |
| Total for the $B$ Levels | 9.16 | 15.41 | 3.82 | 28.39 |

**Treatment Combination Totals**

| | | |
|---|---|---|
| $a_0b_0$ or $T_1 = 2.19$ | $a_0b_1$ or $T_4 = 4.53$ | $a_0b_2$ or $T_7 = 1.24$ |
| $a_1b_0$ or $T_2 = 3.47$ | $a_1b_1$ or $T_5 = 6.13$ | $a_1b_2$ or $T_8 = 2.43$ |
| $a_2b_0$ or $T_3 = 3.50$ | $a_2b_1$ or $T_6 = 4.75$ | $a_2b_2$ or $T_9 = 0.15$ |

SS due to factor $A$:

$$SSa = [(7.96)^2 + (12.03)^2 + (8.40)^2]/6 - CF$$
$$= 1.6631 \quad \text{with} \quad 2 \text{ DF.}$$

SS due to factor $B$:

$$SSb = [(19.16)^2 + (15.41)^2 + (3.82)^2]/6 - CF$$
$$= 11.2171 \quad \text{with} \quad 2 \text{ DF.}$$

SS due to $AB$ interaction:

$$SSab = SSto - SSa - SSb$$
$$= 0.9487 \quad \text{with} \quad 4 \text{ DF.}$$

For testing the significance of individual effects, $SSa$, $SSb$, and $SSab$ are partitioned into the SS due to each effect with 1 DF using the design matrix in Table 6.8. The linear effect contrast for $A$ and its SS are

$$A_L = -1(2.19) - 1(4.53) - 1(1.24) + 1(3.50) + 1(4.75) + 1(0.15)$$
$$= 0.44$$

and

$$SSa_L = (0.44)^2/2(6) = 0.0161,$$

respectively. The quadratic effect contrast for $A$ and its SS are

$$A_Q = 1(2.19) + 1(4.53) + 1(1.24) - 2(3.47) - 2(6.13) - 2(2.43)$$
$$+ 1(3.50) + 1(4.75) + 1(0.15) = -7.70$$

and

$$SSa_Q = (-7.70)^2/2(18) = 1.6469,$$

respectively.

One may check the accuracy of the calculations by verifying the fact that

$$SSa = SSa_L + SSa_Q.$$

Similarly, we compute

$$SSb_L = (-5.34)^2/2(6) = 2.3763,$$
$$SSb_Q = (-17.84)^2/2(18) = 8.8407,$$
$$SSa_L b_L = (-2.40)^2/2(4) = 0.7200,$$
$$SSa_L b_Q = (-0.22)^2/2(12) = 0.0020,$$
$$SSa_Q b_L = (-2.22)^2/2(12) = 0.2054,$$
$$SSa_Q b_Q = (1.24)^2/2(36) = 0.0214.$$

The analysis of variance is given in Table 6.11. Assuming the fixed-effects Model I, the $F$ ratio uses the residual mean square in the denominator. It is seen that factor $A$ is statistically significant ($F = 13.86^*$). This suggests that a difference in binding property exists among the three levels of factor $A$ averaged over all levels of factor $B$. Therefore, it is informative to divide $SSa$ into $SSa_L$ and $SSa_Q$. The result shows that only the quadratic effect is significant ($F = 27.45^*$). This finding indicates that the best result is obtained at the medium levels of factor $A$. We cannot, however, conclude whether this finding also holds for factor $B$ until we find that $A_QB_Q$ interaction is negligibly small. Figures 6.3 and 6.4 illustrate, respectively, the linear and quadratic effects of factor $A$. Each point on this graph represents a mean value for all three levels of factor $B$ at the indicated levels of factor $A$.

The effect of factor $B$ is also significant ($F = 93.48^*$). The statistical significance of $B$ is due to both the linear and quadratic effects, although

**Table 6.11** Analysis of Variance for the Data in Example 6.2-1

| Source of Variance | DF | SS | MS | F Ratio |
|---|---|---|---|---|
| Total (corrected) | 17 | 14.3690 | | |
| Treatment | 8 | 13.8289 | | |
| $A$ | 2 | 1.6631 | 0.8316 | 13.86* |
| $A_L$ | 1 | 0.0161 | 0.0161 | 0.27 |
| $A_Q$ | 1 | 1.6469 | 1.6469 | 27.45* |
| $B$ | 2 | 11.2171 | 5.6086 | 93.48* |
| $B_L$ | 1 | 2.3763 | 2.3763 | 39.61* |
| $B_Q$ | 1 | 8.8407 | 8.8407 | 147.35* |
| $AB$ | 4 | 0.9487 | 0.2372 | 3.95** |
| $A_LB_L$ | 1 | 0.7200 | 0.7200 | 12.00* |
| $A_LB_Q$ | 1 | 0.0020 | 0.0020 | 0.03 |
| $A_QB_L$ | 1 | 0.2054 | 0.2054 | 3.42 |
| $A_QB_Q$ | 1 | 0.0214 | 0.0214 | 0.36 |
| Residual | 9 | 0.5401 | 0.0600 | |

*$p < .01.$
**$p < .05.$

the quadratic effect contributes more, as seen by its larger mean square. This result indicates that the best binding property is obtained at the medium level of factor *B*. Graphically, these findings are illustrated in Figures 6.5 and 6.6, where each point on the graph represents the mean for all levels of factor *A*.

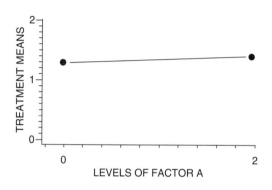

■ **FIGURE 6.3** Linear effect of factor *A* plotted against factor *B*.

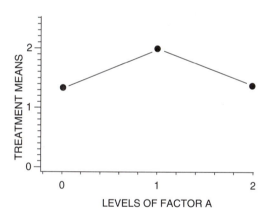

■ **FIGURE 6.4** Quadratic effect of factor *A* plotted against factor *B*.

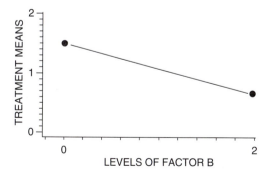

■ **FIGURE 6.5** Linear effect of factor *B* plotted against factor *A*.

■ **FIGURE 6.6** Quadratic effect of factor *B* plotted against factor *A*.

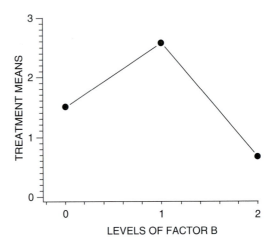

■ **FIGURE 6.7** Linear × linear interaction of factor *A* plotted against factor *B*: ●, low level of factor *B*; ▲, high level of factor *B*.

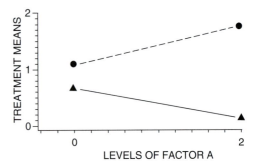

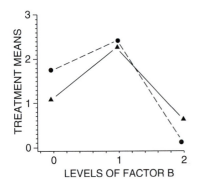

■ **FIGURE 6.8** Linear × quadratic interaction effect of factor *B* plotted against factor *A*: ▲, low level of factor *A*; ●, high level of factor *A*.

The *AB* interaction effect on binding property is significant at the 5% level. The reason for its significance is mostly associated with the $A_L B_L$ interaction ($F = 3.95^{**}$). The graphs of the $A_L B_L$ and the $A_L B_Q$ interactions are given in Figures 6.7 and 6.8, respectively. These graphs indicate that the binding property of meat at the low and high levels of factor *B* vary with the low and high levels of factor *A*. The $A_L B_Q$ interaction is not significant, since both the low and high levels of factor *A* behave similarly over all levels of factor *A*. The same argument is used to explain $A_Q B_L$ and $A_Q B_Q$ interactions in Figures 6.9 and 6.10, respectively. Since the $A_Q B_Q$ interaction effect is not significant, we can finally conclude that the best binding property is found at the medium levels of factors *A* and *B*.

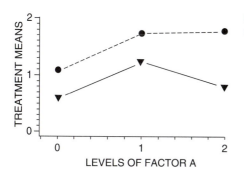

■ **FIGURE 6.9** Linear × quadratic interaction effect of factor $A$ plotted against factor $B$: ●, low level of factor $B$; ▼, high level of factor $B$.

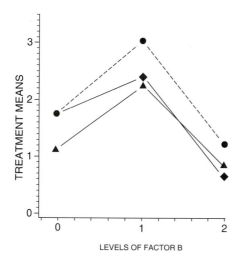

■ **FIGURE 6.10** Quadratic × quadratic interaction effect of factor $B$ plotted against factor $A$: ▲, low level of factor $A$; ●, medium level of factor $A$; ◆, high level of factor $A$.

## 6.3 THE $p \times q$ AND $p \times q \times k$ FACTORIAL EXPERIMENTS

In many instances, of course, not all the factors are at the same level. We consider only two such simple factorial experiments. The $p \times q$ factorial experiment consists of two factors $A$ and $B$: $A$ at $p$ levels and $B$ at $q$ levels. Table 6.12 provides a layout of a $p \times q$ factorial experiment in an RCB design. The first and second coordinates of treatment combination $(i, j)$ denote the $i$th level of $A$, $i = 1, \ldots, p$, and the $j$th level of $B$, $j = 1, \ldots, q$, respectively. There are several $pq$

**Table 6.12** $p \times q$ Factorial Experiment in a Randomized Complete Block Design Layout and Partition of Total Degrees of Freedom

| Levels of Factor A: | | 1 | | | | 2 | | | ... | | p | | |
|---|---|---|---|---|---|---|---|---|---|---|---|---|---|
| **Levels of Factor B:** | 1 | 2 | ... | q | 1 | 2 | ... | q | ... | 1 | 2 | ... | q |
| **Blocks** | | | | | | | | | | | | | |
| 1 | $X_{111}$ | $X_{121}$ | ... | $X_{1q1}$ | $X_{211}$ | $X_{221}$ | ... | $X_{2q1}$ | ... | $X_{p11}$ | $X_{p21}$ | ... | $X_{pq1}$ |
| 2 | $X_{112}$ | $X_{122}$ | ... | $X_{1q2}$ | $X_{212}$ | $X_{222}$ | ... | $X_{2q2}$ | ... | $X_{p12}$ | $X_{p22}$ | ... | $X_{pq2}$ |
| ⋮ | ⋮ | ⋮ | ⋮ | ⋮ | ⋮ | ⋮ | ⋮ | ⋮ | ⋮ | ⋮ | ⋮ | ⋮ | ⋮ |
| b | $X_{11b}$ | $X_{12b}$ | ... | $X_{1qb}$ | $X_{21b}$ | $X_{22b}$ | ... | $X_{2qb}$ | ... | $X_{p1b}$ | $X_{p2b}$ | ... | $X_{pqb}$ |

| Source of Variance | DF |
|---|---|
| Total | $pqb - 1$ |
| Blocks | $b - 1$ |
| Treatments | $pq - 1$ |
| A | $p - 1$ |
| B | $q - 1$ |
| AB | $(p - 1)(q - 1)$ |
| Error | $(pq - 1)(b - 1)$ |

combinations. Thus, one can view the layout in the table as that of an RCB design having $b$ blocks, each of size $pq$. Letting $X_{ijk}$ denote the observations in the $k$th block corresponding to the treatment combination $(i, j)$, we can, as in Section 4.3, compute the SS$b$ due to blocks, SStr due to treatments, etc. The SStr with $(pq - 1)$ DF can be broken down into SS$a$ due to factor $A$ with $(p - 1)$ DF, SS$b$ due to factor $B$ with $(q - 1)$ DF, and SS$ab$ due to interaction effects $AB$ with $(p - 1)(q - 1)$ DF. We illustrate the analysis for a $2 \times 3$ factorial experiment.

**Example 6.3-1**

The data in Table 6.13 concern a study on the effect of pumping methods (factor $A$) and curing length (factor $B$) on percent curing gain for ham muscle groups. Three data blocks (1, 2, 3) of this study are reproduced to illustrate the design and its analysis. The reader may follow the calculations in the footnote to this table.

**Table 6.13** Ham Curing Gain Data and Calculation of Sums of Squares for Example 6.3-1

| Levels of Factor A: | 1 | | | 2 | | | |
|---|---|---|---|---|---|---|---|
| **Levels of Factor B:** | **1** | **2** | **3** | **1** | **2** | **3** | **Block totals** |
| **Factorial Combination and Its Number:** | **(1,1) 1** | **(1,2) 2** | **(1,3) 3** | **(2,1) 4** | **(2,2) 5** | **(2,3) 6** | |
| Blocks | | | | | | | |
| 1 | 11.6 | 18.6 | 16.0 | 16.0 | 11.8 | 13.5 | 87.5 |
| 2 | 11.8 | 11.7 | 9.2 | 18.2 | 19.4 | 12.7 | 83.0 |
| 3 | 15.3 | 9.7 | 10.3 | 17.5 | 15.4 | 14.5 | 82.7 |
| | 38.7 | 40.0 | 35.5 | 51.7 | 46.6 | 40.7 | G = 253.2 |

| Levels of Factor B: | 1 | 2 | 3 | Totals |
|---|---|---|---|---|
| Levels of Factor A | | | | |
| 1 | 38.7 | 40.0 | 35.5 | 114.2 |
| 2 | 51.7 | 46.6 | 40.7 | 139.0 |
| | 90.4 | 86.6 | 76.2 | 253.2 |

Note:
  $CF = (253.2)^2/18 = 3561.68.$
  $SSto = [(11.6)^2 + \cdots + (14.5)^2] - CF = 169.92$ with 17 DF.
  $SSbl = [(87.5)^2 + (83.0)^2 + (82.7)^2]/6 - CF = 2.41$ with 2 DF.
  $SStr = [(38.7)^2 + \cdots + (40.7)^2]/3 - CF = 57.95$ with 5 DF.
  $SSa = [(114.2)^2 + (139.0)^2]/9 - CF = 34.17$ with 1 DF.
  $SSb = [(90.4)^2 + (86.6)^2 + (76.2)^2]/6 - CF = 18.01$ with 2 DF.
$SSab = SStr - SSa - SSb = 57.95 - 34.17 - 18.01 = 5.77$ with 2 DF.
  $SSe = SSto - SSbl - SStr = 169.91 - 2.41 - 57.95 = 109.56$ with 10 DF.

Table 6.14 summarizes the results of the analysis. None of the factor effects are found statistically significant. Therefore, we conclude that the percent curing gain is not affected by pumping methods and curing length.

Now let us consider a $p \times q \times k$ factorial experiment consisting of three factors $A$, $B$, and $C$ at $p$, $q$, and $k$ levels, respectively. A statistical model associated with the design is

$$X_{ijmn} = \mu + A_i + B_j + C_m + (AB)_{ij} + (AC)_{im}$$
$$+ (BC)_{jm} + (ABC)_{ijm} + \varepsilon_{ijmn},$$

**Table 6.14** Analysis of Variance for the Ham Curing Gain Data of Example 6.3-1

| Source of Variance | DF | SS | MS | F Ratio |
|---|---|---|---|---|
| Total | 17 | 169.92 | | |
| Blocks | 2 | 2.41 | 1.21 | |
| Treatments | 5 | 57.95 | | |
| A | 1 | 34.17 | 34.17 | 3.12 |
| B | 2 | 18.01 | 9.01 | 0.82 |
| AB | 2 | 5.77 | 2.89 | 0.26 |
| Residual | 10 | 109.56 | 10.96 | |

where

$$i = 1, \ldots, p; \quad j = 1, \ldots, q;$$

$$m = 1, \ldots, k; \quad n = 1, \ldots, r;$$

$X_{ijmn}$ is the observed response in the $n$th replication for the $i$th, $j$th, and $m$th levels of $A$, $B$, and $C$, respectively; $\mu$ is the grand mean; $A_i$ is the effect of the $i$th level of factor $A$, $\Sigma_i A_i = 0$; $B_j$ is the effect of the $j$th level of factor $B$, $\Sigma_j B_j = 0$; $C_m$ is the effect of the $m$th level of factor $C$, $\Sigma_m C_m = 0$; $(AB)_{ij}$ are the interaction effects between levels of $A$ and $B$, $\Sigma_i\Sigma_j(AB)_{ij} = 0$; $(AC)_{im}$ are the interaction effects between levels of $A$ and $C$, $\Sigma_i\Sigma_m(AC)_{im} = 0$; $(BC)_{jm}$ are the interaction effects between levels of $B$ and $C$, $\Sigma_j\Sigma_m(BC)_{jm} = 0$; $(ABC)_{ijm}$ are the interaction effects among levels of $A$, $B$ and $C$, $\Sigma_i\Sigma_j\Sigma_m(ABC)_{ijm} = 0$; and $\varepsilon_{ijmn}$ are random errors assumed to be independently and normally distributed with variance $\sigma_e^2$.

The formulas for calculating the SS are as follows:

$$CF = \frac{X^2 \cdots}{N} \quad \text{with 1 DF,} \quad \text{where } N = pqkr.$$

The total SS is

$$SSto = \sum_i \sum_j \sum_m \sum_n X_{ijmn}^2 - CF \text{ with } (N-1) \text{ DF.}$$

The SS due to A, B, and C factors are

$$SSa = \frac{\sum_i X_{i\cdots}^2}{qkr} - CF \text{ with } (p-1) \text{ DF,}$$

$$SSb = \frac{\sum_j X_{\cdot j\cdot\cdot}^2}{pkr} - CF \text{ with } (q-1) \text{ DF,}$$

and

$$SSc = \frac{\sum_m X_{\cdots m\cdot}^2}{pqr} - CF \text{ with } (k-1) \text{ DF,}$$

respectively. The SS due to interactions AB, AC, BC, and ABC are

$$SSab = \frac{\sum_i \sum_j X_{ij\cdot\cdot}^2}{kr} - SSa - SSb - CF \text{ with } (p-1)(q-1) \text{ DF,}$$

$$SSac = \frac{\sum_i \sum_m X_{i\cdot m\cdot}^2}{qr} - SSa - SSb - CF \text{ with } (p-1)(k-1) \text{ DF,}$$

$$SSbc = \frac{\sum_j \sum_m X_{\cdot jm\cdot}^2}{pr} - SSb - SSc - CF \text{ with } (q-1)(k-1) \text{ DF,}$$

$$SSabc = \frac{\sum_i \sum_j X_{ijm\cdot}^2}{r} - SSa - SSb - SSc - SSab - SSac - SSbc - CF$$

with $(p-1)(q-1)(k-1)$ DF.

The SSe is obtained by subtraction as usual.

---

**Example 6.3-2**

The effects of stress, manner of stuffing, and weight on fat stability (in milliliters of fat) of meat emulsion are investigated using a factorial design. The stress factor A is at two levels; the stuffing factor B and the weight factor C are at three levels each. The 2 × 3 × 3 factorial design was repeated three times.

Hence, $p = 2$, $q = 3$, $k = 3$, and $r = 3$. The data for this example appear in Table 6.15. Table 6.16 provides appropriate subtotals for the calculation of SS due to interactions. The analysis of variance without the SS column is given by Table 6.17. The F ratio indicates that there is a significant difference between the two stress levels. Also, significant differences exist among the weight groups. Since the interaction between factor A and factor C is

**Table 6.15** Fat Stability Data and Calculations of SS for Example 6.3-2

| Levels of A: | 1 | | | 2 | | | |
|---|---|---|---|---|---|---|---|
| **Levels of B:** | **1** | **2** | **3** | **1** | **2** | **3** | $X_{..k.}$ |
| **Levels of C:** | | | | | | | |
| 1 | 0.10 | 0.20 | 0.20 | 0.20 | 0.10 | 0.15 | 3.04 |
| | 0.14 | 0.20 | 0.20 | 0.15 | 0.20 | 0.10 | |
| | 0.20 | 0.15 | 0.20 | 0.20 | 0.15 | 0.20 | |
| 2 | 0.20 | 0.20 | 0.15 | 0.20 | 0.25 | 0.30 | 3.70 |
| | 0.20 | 0.20 | 0.15 | 0.30 | 0.25 | 0.20 | |
| | 0.20 | 0.15 | 0.15 | 0.20 | 0.25 | 0.15 | |
| 3 | 0.15 | 0.20 | 0.25 | 0.20 | 0.30 | 0.40 | 4.30 |
| | 0.20 | 0.20 | 0.20 | 0.25 | 0.25 | 0.30 | |
| | 0.30 | 0.20 | 0.20 | 0.25 | 0.25 | 0.20 | |
| $X_{.j..}$ | 1.69 | 1.70 | 1.70 | 1.95 | 2.00 | 2.00 | 11.04 |
| $X_{i...}$ | | 5.09 | | | 5.95 | | |

Note:

$CF = 11.04^2/54 = 2.257066$ with 1 DF.

$SSto = [(0.10)^2 + \cdots + (0.20)^2] - CF = 0.167533$ with 53 DF.

$SSa = [(5.09)^2 + (5.95)^2]/27 - CF = 0.013696$ with 1 DF.

$SSb = [(3.64)^2 + (3.70)^2 + (3.70)^3]/18 - CF = 0.000133$ with 2 DF.

$SSc = [(3.04)^2 + (3.70)^2 + (4.30)^2]/18 - CF = 0.044133$ with 2 DF.

$SSab = [(1.69)^2 + \cdots + (2.00)^2]/9 - SSa - SSb - CF = 0.000059$ with 2 DF.

$SSac = [(1.59)^2 + \cdots + (2.40)^2]/9 - SSa - SSc - CF = 0.015170$ with 2 DF.

$SSbc = [(0.99)^2 + \cdots + (1.55)^2]/6 - SSb - SSc - CF = 0.008267$ with 4 DF.

$SSabc = [(0.44)^2 + \cdots + (0.90)^2]/3 - SSa - SSb - SSc - SSab - SSac - SSbc - CF = 0.011007$ with 4 DF.

$SSe = SSto - SSa - SSb - SSc - SSab - SSac - SSbc - SSabc = 0.075067$ with 36 DF.

significant ($F = 3.64^*$), the researcher should examine these factors closely because their joint effect is different from the effect when these factors are not in combination with each other. Figure 6.11 illustrates the AC interaction. The interaction occurs at the lower level of C. If the lines for low and medium levels of A had been parallel to each other, the interaction effect would not have been significant. A nonsignificant interaction is illustrated in Figure 6.12.

**Table 6.16** Subtotals for the Calculation of the SS Due to Interactions

| $X_{ij..}$ for *AB* Interaction | | |
|---|---|---|
| **Levels of A:** | **1** | **2** |
| Levels of B: | | |
| 1 | 1.69 | 1.95 |
| 2 | 1.70 | 2.00 |
| 3 | 1.70 | 2.00 |

| $X_{i.m.}$ for *AC* Interaction | | |
|---|---|---|
| **Levels of A:** | **1** | **2** |
| Levels of C: | | |
| 1 | 1.59 | 1.45 |
| 2 | 1.60 | 2.10 |
| 3 | 1.90 | 2.40 |

| $X_{.jm.}$ for *BC* Interaction | | | |
|---|---|---|---|
| **Levels of B:** | **1** | **2** | **3** |
| Levels of C: | | | |
| 1 | 0.99 | 1.00 | 1.05 |
| 2 | 1.30 | 1.30 | 1.10 |
| 3 | 1.35 | 1.40 | 1.55 |

| $X_{ijm.}$ for *ABC* Interaction | | | | | | |
|---|---|---|---|---|---|---|
| **Levels of A:** | **1** | | | **2** | | |
| **Levels of B:** | **1** | **2** | **3** | **1** | **2** | **3** |
| Levels of C: | | | | | | |
| 1 | 0.44 | 0.55 | 0.60 | 0.55 | 0.45 | 0.45 |
| 2 | 0.60 | 0.55 | 0.45 | 0.70 | 0.75 | 0.65 |
| 3 | 0.65 | 0.60 | 0.65 | 0.70 | 0.80 | 0.90 |

**Table 6.17** Analysis of Variance for Fat Emulsion Stability Data (Example 6.3-2)

| Source of Variance | DF | MS | F Ratio |
|---|---|---|---|
| Total | 53 | | |
| Stresses (A) | 1 | 0.013696 | 6.57* |
| Stuffings (B) | 2 | 0.000067 | 0.03 |
| Weights (C) | 2 | 0.022067 | 10.58** |
| AB Interactions | 2 | 0.000030 | 0.01 |
| AC Interactions | 2 | 0.007585 | 3.64* |
| BC Interactions | 4 | 0.002067 | 1.00 |
| ABC Interactions | 4 | 0.002752 | 1.32 |
| Error | 36 | 0.002085 | |

*$p < .01$.
**$p < .05$.

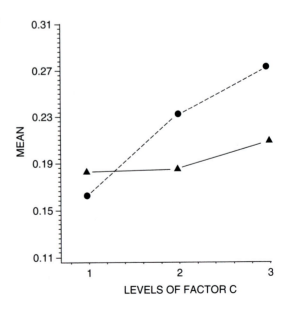

■ **FIGURE 6.11** An illustration of significant $A \times C$ interaction: ▲, low level of factor A; ●, high level of factor A.

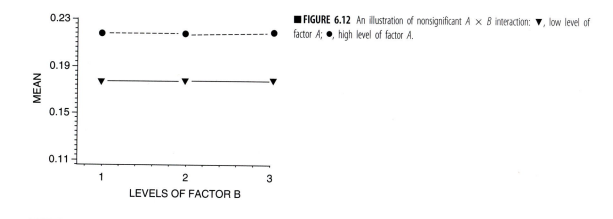

■**FIGURE 6.12** An illustration of nonsignificant $A \times B$ interaction: ▼, low level of factor $A$; ●, high level of factor $A$.

## 6.4 SIMPLE CONFOUNDING AND FRACTIONAL FACTORIAL EXPERIMENTS

A drawback of the factorial experiment is the large number of treatment combinations that result as the number of factors increases. In $2^n$ and $3^n$ factorial experiments with $n = 4$, for example, the number of factorial combinations is 16 and 81, respectively. If the experimental design is an RCB, one then requires blocks of sizes equal to the number of factorial combinations. However, large-size blocks, even when possible, create inefficient designs. Consequently, experiments are designed in blocks that are smaller in size than the number of factorial combinations. But when designed this way, the factorial experiments lead to one or more effects that are confounded with the block effects. That is, for estimating some effects, it is not possible to have the estimating contrasts free of the block effects. The seriousness of the confounding is minimized, however, by having factorial combinations so assigned within blocks that only predetermined higher order interaction effects are confounded with the blocks. The interaction effects to be confounded are those considered insignificant *a priori* by the researcher. It is through the confounding of insignificant effects only that minimal information is lost while having blocks of optimal sizes.

When all the factorial combinations are used in a replication, regardless of the design's having complete or incomplete blocks, it is called a *complete replication*. In many experiments it is prohibitive to use complete replications, particularly when the number of combinations is large. So the experiments are designed using only a *fraction of factorial combinations*. If, for instance, only half of the factorial combinations are used in an experiment, it is called a *½ replicate factorial*, and so on. From fractional factorial designs, however, it is not possible to estimate all the effects separately from each other. Any contrast among the factorial combinations used in the fractional factorial experiment will estimate a linear combination of two effects, and there is no way to separate them. Being aware of this drawback, one can use fractional factorial designs that permit the estimation of lower order effects (main effects, for instance) mixed with only the higher order interaction effects. If the higher order effects are indeed insignificant, then such designs will enable the researcher to have reasonable estimates of the main effects while using only a fraction of the complete replication. In this book we introduce the basic ideas only of fractional factorials and a complete factorial design resulting in confounded effects with the help of $2^n$ factorials. For details on these topics, we refer the reader to books on experimental designs. Some references are cited at the end of this section.

## Confounding in $2^n$ Factorial Experiments

We discuss the confounding of effects with blocks, using a $2^4$ factorial with factors $A$, $B$, $C$, and $D$. The generalizations to $2^n$ factorials are straightforward. The 16 factorial combinations are

| | | | |
|---|---|---|---|
| (1) | $d$ | $bc$ | $abd$ |
| $a$ | $ab$ | $bd$ | $acd$ |
| $b$ | $ac$ | $cd$ | $bcd$ |
| $c$ | $ad$ | $abc$ | $abcd$ |

Suppose we want to design the experiment so that each replication consists of two blocks, each of size eight. It is possible to arrange the 16 factorial combinations in two blocks so that any desired effect becomes confounded with the blocks, but as stated before, only some higher order effect or an effect of no interest should be confounded. The effect that is to be confounded is called the *defining effect*. Once we have a defining effect, the assignment of the factorial combinations within blocks is determined as follows. Define

$$L = 1_a X_a + 1_b X_b + 1_c X_c + 1_d X_d, \qquad (6.4\text{-}1)$$

where $1_a = 1$ if factor $A$ appears in the defining effect and 0 otherwise. Similarly, $1_b$, $1_c$, and $1_d$ are the indicator functions for factors $B$, $C$, and $D$, respectively. Thus, for example, the function $L$ for the defining effect $ACD$ is

$$L = X_a + X_c + X_d.$$

The $X$ variables are also indicator functions. They are indicators of the levels of factors in the factorial combinations. That is, $X_a = 1$ if $A$ appears at the high level in a factorial combination and 0 otherwise.

The $X_b$, $X_c$, and $X_d$ are similarly defined. Now evaluate the numerical value of the defining function $L$ for each factorial combination and reduce it to $L$ mod 2. By definition, $L$ mod 2 is the remainder when $L$ is divided by 2. Note that $L$ mod 2 is always 0 if $L$ is an even number and 1 if $L$ is an odd number. All the factorial combinations that lead to $L$ mod 2 = 0 are assigned to one block and the remaining combinations to the second block.

Suppose in a $2^4$ factorial experiment we wish to confound the *ABCD* interaction effect with the blocks. The defining function $L$ and its $L$ mod 2 values for each of the 16 combinations are given in Table 6.18. In the statistical analysis we cannot estimate the *ABCD* effect, and, consequently,

**Table 6.18** Confounding *ABCD* with Blocks in a $2^4$ Factorial

| Treatment Combination | $X_a$ | $X_b$ | $X_c$ | $X_d$ | Defining $L = 1(X_a) + 1(X_b) + 1(X_c) + 1(X_d)$ | $L$ mod $2^{a,b}$ |
|---|---|---|---|---|---|---|
| (1) | 0 | 0 | 0 | 0 | 1(0) + 1(0) + 1(0) + 1(0) | 0 |
| a | 1 | 0 | 0 | 0 | 1(1) + 1(0) + 1(0) + 1(0) | 1 |
| b | 0 | 1 | 0 | 0 | 1(0) + 1(1) + 1(0) + 1(0) | 1 |
| c | 0 | 0 | 1 | 0 | 1(0) + 1(0) + 1(1) + 1(0) | 1 |
| d | 0 | 0 | 0 | 1 | 1(0) + 1(0) + 1(0) + 1(1) | 1 |
| ab | 1 | 1 | 0 | 0 | 1(1) + 1(1) + 1(0) + 1(0) | 0 |
| ac | 1 | 0 | 1 | 0 | 1(1) + 1(0) + 1(1) + 1(0) | 0 |
| ad | 1 | 0 | 0 | 1 | 1(1) + 1(0) + 1(0) + 1(1) | 0 |
| bc | 0 | 1 | 1 | 0 | 1(0) + 1(1) + 1(1) + 1(0) | 0 |
| bd | 0 | 1 | 0 | 1 | 1(0) + 1(1) + 1(0) + 1(1) | 0 |
| cd | 0 | 0 | 1 | 1 | 1(0) + 1(0) + 1(1) + 1(1) | 0 |
| abc | 1 | 1 | 1 | 0 | 1(1) + 1(1) + 1(1) + 1(0) | 1 |
| abd | 1 | 1 | 0 | 1 | 1(1) + 1(0) + 1(1) + 1(1) | 1 |
| acd | 1 | 0 | 1 | 1 | 1(1) + 1(0) + 1(1) + 1(1) | 1 |
| bcd | 0 | 1 | 1 | 1 | 1(0) + 1(1) + 1(1) + 1(1) | 1 |
| abcd | 1 | 1 | 1 | 1 | 1(1) + 1(1) + 1(1) + 1(1) | 0 |

[a]$L$ mod 2 = 0: (1), ab, ac, ad, bc, bd, cd, abcd in block I.
[b]$L$ mod 2 = 1: a, b, c, d, abc, abd, acd, bcd in block II.

we cannot find the SS due to the $ABCD$ effect. Thus, the total SS is partitioned into the SS due to the factorial combinations with 14 DF, the SS due to blocking with 1 DF, and the SSe having the remaining DF.

| Example 6.4-1 | The data shown in Table 6.19 are from a $2^4$ factorial experiment on four factors: $A$, $B$, $C$, and $D$. The response variable is Vienna sausage weight expressed as percentage weight loss. The 16 factorial combinations were divided into two |

**Table 6.19** Experimental Data on Vienna Sausage—$ABCD$ Confounded with Blocks for Example 6.4-1

| Replicate 1 | | Replicate 2 | |
| --- | --- | --- | --- |
| **Block I** | **Block II** | **Block I** | **Block II** |
| (1): 4.79 | a: 2.37 | (1): 5.88 | a: 2.40 |
| ab: 3.27 | b: 2.09 | ab: 2.18 | b: 2.00 |
| ac: 2.32 | c: 1.23 | ac: 1.54 | c: 1.96 |
| ad: 2.80 | d: 1.00 | ad: 2.60 | d: 1.10 |
| bc: 1.74 | abc: 2.26 | bc: 2.61 | abc: 2.80 |
| bd: 4.07 | abd: 2.19 | bd: 4.36 | abd: 2.00 |
| cd: 1.44 | acd: 2.05 | cd: 1.61 | acd: 1.80 |
| abcd: 3.35 | bcd: 1.95 | abcd: 4.51 | bcd: 1.75 |
| | 38.92 | | 41.10 |

| | Replications | | |
| --- | --- | --- | --- |
| | **1** | **2** | **Total** |
| Block I | 23.78 | 25.29 | 49.07 |
| Block I | 15.14 | 15.81 | 30.95 |

Note:

$$CF = \frac{(80.02)^2}{32} = 200.10 \text{ with } 1 \text{ DF.}$$

$$SSto = [(4.79)^2 + (3.27)^2 + \cdots + (1.80)^2 + (1.75)^2] - CF$$

$$= 40.2210 \text{ with } 31 \text{ DF.}$$

$$SSrep = \frac{(38.92)^2 + (41.10)^2}{16} - CF = 0.1485 \text{ with } 1 \text{ DF.}$$

$$SSrep \times block = \frac{(23.78 - 15.14)^2 + (25.29 - 15.81)^2}{8[(1)^2 + (-1)^2]} - \frac{(49.07 - 30.95)^2}{32}$$

$$= 0.0220 \text{ with } 1 \text{ DF.}$$

blocks following Table 6.18. The experiment was replicated twice. The quantities at the bottom of this table are simply the sums of each block within replicates, which are used to calculate the replicate × block interaction.

The calculations of the SS are as in Section 6.1. These calculations of the SS are summarized in Table 6.20. For instance, SS$a$ is obtained by the contrast between factorial combinations with factor $A$ at a high level and those with factor $A$ at a low level. Thus, for example,

$$SSa = \frac{(\text{contrast})^2}{r2^2}$$

$$= \frac{(\text{sum of responses with} + \text{sign} - \text{sum of responses with} - \text{sign})^2}{2(2^4)}$$

$$= \frac{(40.44 - 39.58)^2}{32} = 0.0231.$$

All the other SS due to individual effects are found similarly and are given in the analysis of variance Table 6.21. One may examine the $F$ ratios for their statistical significances.

## Fractional Factorial

As stated at the beginning of this section, a fractional factorial consists of using only a fraction, such as ½ or ¼ of the total number of factorial combinations. Let us consider a ½ replicate factorial design of $2^5$ factorial. Out of the $2^5 = 32$ factorial combinations, we wish to use only 16 in an experiment. How do we select 16 combinations to be used? To answer this question, one must remember that in a $2^5$ factorial there are $2^5 - 1 = 31$ effect parameters, in addition to the grand mean $\mu$. If we design an experiment using only 16 combinations, no matter which ones, we cannot hope to estimate all 32 parameters separately, even if the experiment is repeated several times. But we can hope to estimate 16 independent linear combinations of the effects, each linear combination consisting of only 2 effects. The 2 effects in a combination are completely confounded because their effects cannot be separately estimated. In view of such a confounding, one may seek to design a ½ replicate so that one of the 2 confounded effects in each combination is insignificant. Consequently, if it can be done, each linear combination would really amount to its component effect, which is significant. However, as it turns out, the experimenter has only a limited control over the confounding pattern of effects.

**Table 6.20** Design Matrix and Some Calculations for the Sum of Squares Due to Effects of $2^4$ Factorials of Example 6.4-1

| Treatments: | (1) | a | b | ab | c | ac | bc | abc | d | ad | bd | abd | cd | acd | bcd | abcd | Sum | |
|---|---|---|---|---|---|---|---|---|---|---|---|---|---|---|---|---|---|---|
| Sum: | 10.67 | 4.77 | 4.09 | 5.45 | 3.19 | 3.86 | 4.35 | 5.06 | 2.10 | 5.40 | 8.43 | 4.19 | 3.05 | 3.85 | 3.70 | 7.86 | + | – |
| **Effect** | | | | | | | | | | | | | | | | | | |
| Total | + | + | + | + | + | + | + | + | + | + | + | + | + | + | + | + | 80.02 | 0 |
| A | – | + | – | + | – | + | – | + | – | + | – | + | – | + | – | + | 40.44 | 39.58 |
| B | – | – | + | + | – | – | + | + | – | – | + | + | – | – | + | + | 43.13 | 36.89 |
| AB | + | – | – | + | + | – | – | + | + | – | – | + | + | – | – | + | 41.57 | 38.45 |
| C | – | – | – | – | + | + | + | + | – | – | – | – | + | + | + | + | 34.92 | 45.10 |
| AC | + | – | + | – | – | + | – | + | + | – | + | – | – | + | – | + | 45.92 | 34.10 |
| BC | + | + | – | – | – | – | + | + | + | + | – | – | – | – | + | + | 43.91 | 36.11 |
| ABC | – | + | + | – | + | – | – | + | – | + | + | – | + | – | – | + | 41.85 | 38.17 |
| D | – | – | – | – | – | – | – | – | + | + | + | + | + | + | + | + | 38.58 | 41.44 |
| AD | + | – | + | – | + | – | + | – | – | + | – | + | – | + | – | + | 43.60 | 36.42 |
| BD | + | + | – | – | + | + | – | – | – | – | + | + | – | – | + | + | 46.67 | 33.35 |
| ABD | – | + | + | – | – | + | + | – | + | – | – | + | + | – | – | + | 34.27 | 45.75 |
| CD | + | + | + | + | – | – | – | – | – | – | – | – | + | + | + | + | 43.44 | 36.58 |
| ACD | – | + | – | + | + | – | + | – | + | – | + | – | – | + | – | + | 40.00 | 40.02 |
| BCD | – | – | + | + | + | + | – | – | + | + | – | – | – | – | + | + | 35.65 | 44.37 |
| ABCD | + | – | – | + | – | + | + | – | – | + | + | – | + | – | – | + | 49.07 | 30.95 |

**Table 6.21** Analysis of Variance of Weight Loss Data (Example 6.4-1)

| Source of Variance | DF | SS | MS | F Ratio |
|---|---|---|---|---|
| Total | 31 | 40.2210 | | |
| Replicates | 1 | 0.1485 | | |
| *ABCD* (Blocks) | 1 | 10.2604 | | |
| Replicates × Blocks | 1 | 0.0220 | | |
| Treatments | 14 | | | |
| *A* | 1 | 0.0231 | 0.0231 | 0.11 |
| *B* | 1 | 1.2168 | 1.2168 | 5.79* |
| *C* | 1 | 3.2385 | 3.2385 | 15.42** |
| *D* | 1 | 0.2556 | 0.2556 | 1.22 |
| *AB* | 1 | 0.3042 | 0.3042 | 1.45 |
| *AC* | 1 | 4.3660 | 4.3660 | 20.79** |
| *BC* | 1 | 1.9013 | 1.9013 | 9.05** |
| *AD* | 1 | 1.6110 | 1.6110 | 7.67* |
| *BD* | 1 | 5.5445 | 5.5445 | 25.93** |
| *CD* | 1 | 1.4706 | 1.4706 | 7.00* |
| *ABC* | 1 | 0.4232 | 0.4232 | 2.02 |
| *ABD* | 1 | 4.1185 | 4.1185 | 19.61** |
| *ACD* | 1 | 0.0000 | 0.0000 | — |
| *BCD* | 1 | 2.3762 | 2.3762 | 11.32** |
| Residual | 14 | 2.9406 | 0.2100 | |

*$p < .05$.
**$p < .01$.

A ½ replication is determined as soon as we have decided to confound the grand mean $\mu$ with one of the 31 effects (usually a higher order interaction effect believed to be insignificant). To understand this fact, suppose we decide to confound the *ABCDE* interaction with $\mu$. If we use the equality sign to denote the confounding and $I$ to denote $\mu$ in the equation, we then have a so-called defining equation:

$$I = ABCDE.$$

The defining equation, representing the confounding of $\mu$ with the *ABCDE* effect, determines uniquely all the other pairs whose components, called *aliases*, are confounded. The alias of effect *A*, for example, is obtained from the defining equation as follows. Multiplying both sides of the defining equation by *A*, we find

$$A = A^2 BCDE, \tag{6.4-2}$$

wherein the multiplication operation *I* is treated as unity. Also, the multiplication is always further simplified by letting $A^2 = I$, $B^2 = I$, and so on. Hence, Eq. (6.4-2) becomes

$$A = BCDE,$$

which means that effect *A* is confounded with *BCDE* interactions. All the aliases corresponding to the defining equation $I = ABCDE$ are listed in Table 6.22.

**Table 6.22** Aliases and a ½ Replicate of a $2^5$ Factorial Corresponding to Defining Equation $I = ABCDE$

| Factorial Combinations with $L$ mod $2 = 0$ | Pairs of Aliases |
|---|---|
| (1) | $\mu$, *ABCDE* |
| *ab* | *AB, CDE* |
| *ac* | *AC, BDE* |
| *bc* | *BC, ADE* |
| *ad* | *AD, BCE* |
| *bd* | *BD, ACE* |
| *cd* | *CD, ABE* |
| *abcd* | *ABCD, E* |
| *ae* | *AE, BCD* |
| *be* | *BE, ACD* |
| *ce* | *CE, ABD* |
| *abce* | *ABCE, D* |
| *de* | *DE, ABC* |
| *abde* | *ABDE, C* |
| *acde* | *ACDE, B* |
| *bcde* | *BCDE, A* |

The defining equation determines not only all the pairs of aliases but also the fractional replication. The effect that is confounded with the mean $\mu$ is called, as before, the *defining effect*. For the defining effect we can write function $L$, following (6.4-1). All the factorial combinations with $L$ mod $2 = 0$ constitute a ½ replication and all those with $L$ mod $2 = 1$ constitute the remaining ½ replication. It is up to the experimenter to choose which of the two ½ replications will be used for the experiment. For our example $ABCDE$ is the defining effect and its defining equation is

$$L = X_a + X_b + X_c + X_d + X_e.$$

Table 6.23 lists $L$ mod 2 for all 32 combinations in our $2^5$ factorial.

Suppose we desire to have only a ¼ replication of a $2^5$ factorial. Obviously, more than 2 effects would be confounded with each other, because with a ¼ replication of a $2^5$ factorial, we can estimate only ¼ of $2^5$, namely, 8 independent linear combinations—each combination of 4 parameters out of 32 (31 effect parameters and one grand mean $\mu$). Hence, for a ¼ replicate we need to decide 3 effects that are to be confounded with the mean $\mu$. The mean and the 3 effects confounded with each other now constitute the defining combination.

Indeed, the fact is that having chosen 2 effects confounded with $\mu$, the experimenter has no control over the choice of the third effect in the defining combination. In our example suppose we decided to confound the $BCE$ and $ABCDE$ effects with each other and with $\mu$. Then the effect resulting from the multiplication

$$BCE \times ABCDE = AB^2C^2DE^2 = AD$$

is also confounded with $\mu$. Furthermore, as in a ¼ replication, the defining combination determines the other combinations of the confounded effects, as well as a ¼ replication. To partition $2^5$ combinations into four ¼ replications, we use the defining effects $BCE$ and $ABCDE$ of our example to write down two defining equations, $L_1$ and $L_2$, respectively, from (6.4-1). They are

$$L_1 = X_b + X_c + X_e,$$
$$L_2 = X_a + X_b + X_c + X_d + X_e.$$

Note that for each factorial combination, $(L_1$ mod 2, $L_2$ mod 2) is either (0, 0), (0, 1), (1, 0), or (1, 1). Following these 4 possible pairs for the functions $L_1$ and $L_2$, all 32 combinations are partitioned in Table 6.24 into 4 sets, each constituting a ¼ replication. Also the bottom of the table lists the corresponding 8 combinations of the confounding effects.

**Table 6.23** Construction of a ½ Replicate of a $2^5$ Factorial with *ABCDE* as the Defining Effect[a]

| Treatment Combination | $X_a$ | $X_b$ | $X_c$ | $X_d$ | $X_e$ | $L$ | $L$ mod 2 |
|---|---|---|---|---|---|---|---|
| (1) | 0 | 0 | 0 | 0 | 0 | 0 | 0 |
| a | 1 | 0 | 0 | 0 | 0 | 1 | 1 |
| b | 0 | 1 | 0 | 0 | 0 | 1 | 1 |
| ab | 1 | 1 | 0 | 0 | 0 | 2 | 0 |
| c | 0 | 0 | 1 | 0 | 0 | 1 | 1 |
| ac | 1 | 0 | 1 | 0 | 0 | 2 | 0 |
| bc | 0 | 1 | 1 | 0 | 0 | 2 | 0 |
| abc | 1 | 1 | 1 | 0 | 0 | 3 | 1 |
| d | 0 | 0 | 0 | 1 | 0 | 1 | 1 |
| ad | 1 | 0 | 0 | 1 | 0 | 2 | 0 |
| bd | 0 | 1 | 0 | 1 | 0 | 2 | 0 |
| abd | 1 | 1 | 0 | 1 | 0 | 3 | 1 |
| cd | 0 | 0 | 1 | 1 | 0 | 2 | 0 |
| acd | 1 | 0 | 1 | 1 | 0 | 3 | 1 |
| bcd | 0 | 1 | 1 | 1 | 0 | 3 | 1 |
| abcd | 1 | 1 | 1 | 1 | 0 | 4 | 0 |
| e | 0 | 0 | 0 | 0 | 1 | 1 | 1 |
| ae | 1 | 0 | 0 | 0 | 1 | 2 | 0 |
| be | 0 | 1 | 0 | 0 | 1 | 2 | 0 |
| abe | 1 | 1 | 0 | 0 | 1 | 3 | 1 |
| ce | 0 | 0 | 1 | 0 | 1 | 2 | 0 |
| ace | 1 | 0 | 1 | 0 | 1 | 3 | 1 |
| bce | 0 | 1 | 1 | 0 | 1 | 3 | 1 |
| abce | 1 | 1 | 1 | 0 | 1 | 4 | 0 |
| de | 0 | 0 | 0 | 1 | 1 | 2 | 0 |
| ade | 1 | 0 | 0 | 1 | 1 | 3 | 1 |
| bde | 0 | 1 | 0 | 1 | 1 | 3 | 1 |
| abde | 1 | 1 | 0 | 1 | 1 | 4 | 0 |

(continued)

**Table 6.23** Construction of a ½ Replicate of a $2^5$ Factorial with *ABCDE* as the Defining Effect[a]—cont...

| Treatment Combination | $X_a$ | $X_b$ | $X_c$ | $X_d$ | $X_e$ | $L$ | $L$ mod 2 |
|---|---|---|---|---|---|---|---|
| cde | 0 | 0 | 1 | 1 | 1 | 3 | 1 |
| acde | 1 | 0 | 1 | 1 | 1 | 4 | 0 |
| bcde | 0 | 1 | 1 | 1 | 1 | 4 | 0 |
| abcde | 1 | 1 | 1 | 1 | 1 | 5 | 1 |

[a]$L = X_a + X_b + X_c + X_d + X_e$.

**Table 6.24** Four Sets of ¼ Replication Corresponding to Defining Effects *BCE* and *ABCDE*[a]

| ($L_1$ mod 2, $L_2$ mod 2) | | | |
|---|---|---|---|
| **(0, 0)** | **(0, 1)** | **(1, 0)** | **(1, 1)** |
| (1) | a | ab | b |
| bc | abc | ac | c |
| ad | d | bd | abd |
| abcd | bcd | cd | acd |
| be | abe | ae | e |
| ce | ace | abce | bce |
| abde | bde | de | ade |
| acde | cde | bcde | abcde |
| **Confounding Combinations** | | | |

Defining combinations: $\mu$, *BCE*, *ABCDE*, *AD*

| | |
|---|---|
| Seven other combinations that follow from the defining combinations | $\begin{cases} A, ABCE, BCDE, D \\ B, CE, ACDE, ABD \\ AB, ACE, CDE, BD \\ C, BE, ABDE, ACD \\ AC, ABE, BDE, CD \\ BC, E, ADE, ABCD \\ ABC, AE, DE, BCD \end{cases}$ |

[a]$L_1 = X_b + X_c + X_e$; $L_2 = X_a + X_b + X_c + X_d + X_e$.

We have covered briefly some basics of confounding and fractional factorial experiments. We have not attempted to give a numerical example. The reader is referred to books by Cochran and Cox (1957), Davies (1967), Peng (1967), and papers by Finney (1945, 1946), Plackett and Burman (1946), Kempthorne (1947), and Davies and Hay (1950).

## EXERCISES

**6.1.** Two levels of colorant $A$ and two of colorant $B$ were formulated for ground meat. Four observations were available for each $A$- and $B$-level combination.

  **1.** Set up the layout of the data for calculating the effects, their SS, and the error SS.
  **2.** Set up the analysis of variance table and the test methods for testing the significance of the effects.

**6.2.** Write down the design matrix for $2^4$ factorial experiments. How would you calculate all the SS due to each effect and the error? Prepare the analysis of variance table.

**6.3.** Refer to Example 6.1-3 and compute $SSb$, $SSc$, $SSab$, $SSac$, $SSbc$, $SSabc$, and $SSe$. Prepare the analysis of variance table. Test at the $\alpha = 1\%$ level of significance if any effect is significant.

**6.4.** Refer to Example 6.1-3. Estimate the main effects of $A$ and $B$ by 99% confidence intervals. Use these intervals to test the significance of the $A$ and $B$ effects at $\alpha = 1\%$ level of significance. Do your conclusions agree with the analysis of variance table in Exercise 6.3?

**6.5.** An experiment was conducted to study the Browning reaction in frankfurters. Meat emulsions consisting of three levels of sucrose (0%, 1.5%, 3.0%) and three levels of $Na_2SO_2$ (0%, 1.5%, 3.0%) were made. The following data are for percentage moisture:

| | Dextrose Level, % | | |
|---|---|---|---|
| **Na₂SO₂ Level, %** | **0** | **1.5** | **3.0** |
| 0 | 53.9 | 54.0 | 55.0 |
| | 53.0 | 53.8 | 56.0 |
| 1.5 | 52.0 | 56.0 | 55.2 |
| | 53.0 | 55.4 | 56.5 |
| 3.0 | 54.0 | 55.3 | 55.8 |
| | 52.5 | 54.0 | 54.1 |

**1.** Prepare the analysis of variance.

**2.** Plot the linear and the quadratic effects associated with sucrose and $Na_2SO_2$ and their interactions.

**6.6.** Design a $2^4$ factorial in blocks of size eight, confounding the *ABC* interaction effect with blocks. How many effects are measurable? Set up an analysis of variance table.

**6.7.** Consider a $2^6$ factorial. Construct a ½ replication and also a ¼ replication using the highest order interactions as the defining effects. Also list the combinations of the confounding effects.

# Response Surface Designs and Analysis

There is considerable interest in designs for fitting response surfaces and determining optimum operating conditions. Response surface designs and their analyses are used to find combinations of a number of experimental factors that will lead to optimum responses. An optimum response can be a maximum or a minimum, depending on its nature. The methodology was first developed and described by Box and Wilson (1951). Later, Bradley (1958) wrote an article explaining the mathematical and statistical tools used in the response surface methodology. Following Bradley, Hunter (1958, 1959) wrote a series of articles discussing statistical analyses of some response surface design models. There are a number of books dealing with this topic, such as those by Cochran and Cox (1957), Davies (1967), and Myers (1971). MacDonald and Bly (1963) used the response surface methodology to determine optimal levels of several emulsifiers in cake mix shortenings, and Nielsen et al. (1972) used its analysis techniques on a protein denaturation study. Other applications in food research areas are to be found in Kissell and Marshall (1962) and Jedlicka et al. (1975).

In this chapter we discuss the statistical analysis of response surface designs and some guidelines for its use. A workable knowledge of elementary calculus is helpful. We have included some mathematical details that may be of help to readers and have tried to portray concepts through graphs.

## 7.1 GENERAL CONCEPTS

We begin by considering

$$Y = f(X_1, X_2, \ldots, X_n) + \varepsilon, \tag{7.1-1}$$

where $Y$ refers to the observed response best known as the *dependent variable*; $f$ is the response function of $X_1, X_2, \ldots, X_n$, which are quantitative

**247**

variables known as the *independent variables,* and $\varepsilon$ is the random error term. Although the precise form of the response function $f$ is always unknown, experience has shown that it can usually be approximated by suitable linear or quadratic functions of the independent variables. A linear regression relationship,

$$Y = \beta_0 + \beta_1 X_1 + \beta_2 X_2 + \cdots + \beta_n X_n + \varepsilon, \qquad (7.1\text{-}2)$$

is the simplest and is known as the *first-order model* or equation. The *second-order model* is the quadratic regression relationship

$$Y = \beta_0 + \beta_1 X_1 + \cdots + \beta_n X_n + \beta_{11} X_1^2 + \cdots + \beta_{nn} X_n^2$$
$$+ \beta_{12} X_1 X_2 + \cdots + \beta_{n-1,n} X_{n-1} X_n + \varepsilon. \qquad (7.1\text{-}3)$$

The parameters of these equations are usually not known and, therefore, must be estimated from the experimental outcomes. The physical meanings of the parameters are as follows:

$\beta_0 =$ the intercept (grand mean), and its estimate is denoted by $b_0$;

$\beta_i =$ the linear effect of $X_i$, and its estimate is denoted by $b_i$, $i = 1, \ldots, n$;

$\beta_{ii} =$ the quadratic effect of $X_i$, and its estimate is denoted by $b_{ii}$, $i = 1, \ldots, n$;

$\beta_{ij} =$ the interaction effect of $X_i$ and $X_j$, and its estimate is denoted by $b_{ij}$, $i < j$, $i = 1, \ldots, n-1$ and $j = 1, \ldots, n$.

The error terms $\varepsilon_i$ are assumed to be normally distributed with mean 0 and variance $\sigma^2$. The variation as quantified by $\sigma^2$ in the distribution of responses is essentially due to the differences among the design points and uncontrollable experimental errors. The variation due to differences among the design points can be explained by the response function if it is known. The true response function is seldom known, and hence must be estimated. Consequently, the contribution to the error variation is due not only to the experimental errors alone but also to the inadequacy or the so-called lack of fit of the estimated model. If the estimated model is adequate in predicting responses, then the sum of squares due to errors (SSe) is essentially due to the experimental errors. To judge the adequacy of an estimated model, one partitions SSe into the sum of squares due to experimental errors and the sum of squares due to lack of fit of the model. This partition is explained fully in Section 7.2.

The statistical principle of *least squares* is used to estimate the parameters of a hypothesized regression response function. When the parameters of the hypothesized response function are replaced by their estimates, the resulting function is said to be a *fitted response function*. Thus, $\widehat{Y}$ given by

$$\widehat{Y} = b_0 + b_1 X_1 + \cdots + b_n X_n$$

is a fitted linear response function that can be used to predict responses for desired values of the independent variables. When the fitted response function $\widehat{Y}$ is graphed as a function of independent variables, the resulting graph is called a *response surface plot* or *contour map*.

Contour maps of responses are an attractive feature of response surface analysis. Let us introduce some commonly encountered surface plots generated by the first- and second-order fitted response functions. Figure 7.1 is the response surface of a first-order equation. The center of the graph $C_0$ plays an important role and is discussed in Section 7.2. In this figure the response $Y$ increases with an increasing $MgCl_2$ level, the effect of $CaCl_2$ being nil. Figures 7.2 and 7.3 depict response surfaces generated by second-order models. The curvilinear plots result from quadratic and interaction effects in the model. The profile of the plot in Figure 7.2 illustrates the effectiveness of added nitrite in controlling *Staphylococcus* growth. The plot in Figure 7.3 illustrates the dominant contribution of the quadratic and interaction terms to the model. This means that the residual nitrite depends on the added nitrate and nitrite. The response surface in Figure 7.4 is *saddle shaped*. It is a complex response surface with

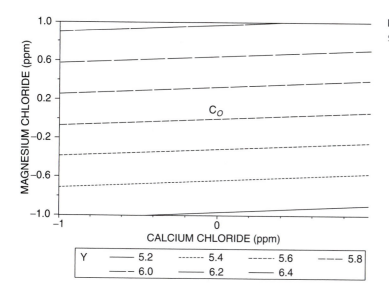

■ **FIGURE 7.1** Response surface plots of emulsion stability as a function of $CaCl_2$ and $MgCl_2$ levels.

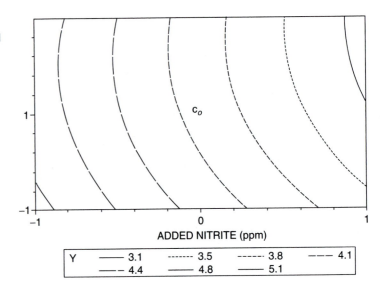

■ **FIGURE 7.2** Response surface plots of *Staphylococcus* count in pepperoni as functions of added nitrate and nitrite.

| Y | ——— 3.1 | ------- 3.5 | ------ 3.8 | ---- 4.1 |
|---|---------|-------------|------------|----------|
|   | —— 4.4  | ——— 4.8     | ——— 5.1    |          |

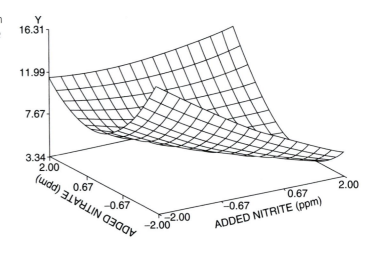

■ **FIGURE 7.3** Response surface plots of residual nitrite in pepperoni stored in a vacuum as functions of added nitrite and nitrate.

maximum and minimum values encountered at various combinations of the independent variables. Figure 7.5 illustrates a response surface that is a *hill* or *basin*. This form is seen when the optimum response is near the center of the design. In this figure the optimum response is a maximum value. As we go further away from the center of the design, the response decreases. From the foregoing we see that by constructing surface plots we may be able to locate and characterize optimum responses. The analytical steps required to construct response surfaces are discussed next.

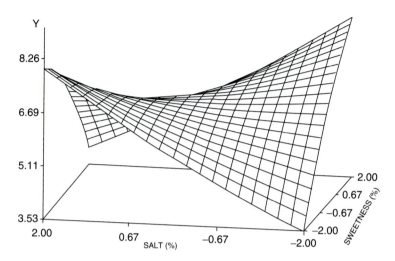

■ **FIGURE 7.4** A response contour depicting a saddle-shaped response surface.

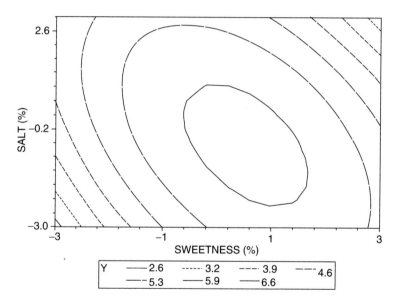

■ **FIGURE 7.5** Response contour depicting a hill-shaped response surface.

## Construction of Response Surfaces

Let us consider a first-order fitted response model in two variables:

$$\widehat{Y} = b_0 + b_1 X_1 + b_2 X_2. \tag{7.1-4}$$

For a given $\widehat{Y}$, (7.1-4) is the equation of a straight line in coordinates $X_1$ and $X_2$. The line can be plotted if we have the coordinates of any two points on it. Note that if $X_1 = 0$, then $X_2 = \widehat{Y} - b_0/b_2$; and if $X_2 = 0$, $X_1 = \widehat{Y} - b_0/b_1$. Thus, two points on the line are $(0, \widehat{Y} - b_0/b_2)$ and $(\widehat{Y} - b_0/b_1, 0)$. Since everywhere on the line joining these two points the predicted response is $\widehat{Y}$, the line can be viewed as a contour. For each given $\widehat{Y}$, there is a line contour. Thus, if (7.1-4) is a "good fit" response model, one can use it to obtain a collection of contour lines for mapping the response surface and finding the coordinates $(X_1, X_2)$ corresponding to the optimum response.

The task of empirically mapping a second-order response surface is a bit involved. Consider a fitted second-order model in two variables:

$$\widehat{Y} = b_0 + b_1 X_1 + b_2 X_2 + b_{11} X_1^2 + b_{22} X_2^2 + b_{12} X_1 X_2. \tag{7.1-5}$$

For a given $\widehat{Y}$, (7.1-5) is the equation of a circle, an ellipse, a hyperbola, or a parabola, depending on the signs and magnitudes of the regression coefficients $b_0$, $b_1$, $b_2$, $b_{11}$, $b_{22}$, and $b_{12}$. To plot contours of (7.1-5), we need to trace all points $(X_1, X_2)$ that correspond to a given $\widehat{Y}$. How do we determine the coordinates $X_1$ and $X_2$ of points on a contour? To provide a formula, we first rewrite (7.1-5) as follows:

$$b_{11}X_1^2 + (b_1 + b_{12}X_2)X_1 + (b_{22}X_2^2 + b_2X_2 + b_0 - \widehat{Y}) = 0. \tag{7.1-6}$$

Let

$$a = b_{11}, \; b = (b_1 + b_{12}X_2), \\ c = b_{22}X_2^2 + b_2X_2 + b_0 - \widehat{Y}. \tag{7.1-7}$$

Using $a$, $b$, and $c$ as just defined, we see that (7.1-6) is simply a quadratic in $X_1$. That is,

$$aX_1^2 + bX_1 + c = 0. \tag{7.1-8}$$

Remember that $a$, $b$, and $c$ are defined through $X_2$ and $\widehat{Y}$, in addition to the regression coefficients $b_0$, $b_1$, $b_2$, $b_{11}$, $b_{22}$, and $b_{12}$. The two roots (coordinates of $X_1$) satisfying (7.1-8) are given by

$$X_1 = \frac{-b + \sqrt{b^2 - 4ac}}{2a} \quad \text{and} \quad X_1 = \frac{-b - \sqrt{b^2 - 4ac}}{2a} \tag{7.1-9}$$

In short, to plot a contour map of (7.1-5) for a fixed $\widehat{Y}$ value, first choose the $X_2$ coordinate and then find two $X_1$ coordinates from

(7.1-9). This provides us with two points $(X_1, X_2)$, with the $X_2$ coordinate being the same. Next we can choose a different $X_2$ coordinate and once again find the two corresponding $X_1$ coordinates, thus obtaining two more points on the contour for the fixed $\widehat{Y}$. Repeating this process, we can find as many points as necessary to draw a contour. We can repeat the whole process to draw a contour for another $\widehat{Y}$ value, and so on.

We note again that the contours of the surface given by (7.1-5) depend on the regression coefficients. In mapping the second-order equations, the coefficients of the squared terms influence the direction of the curvature of the plots. The fitted model may lead to negative or extremely large responses that have no practical meaning. As a result, the extrapolation of the responses beyond the experimental range of the independent variables may not be valid. It is, therefore, important to design experiments appropriately to allow for exploration of the fitted response surfaces over the desired range of experimental variables. Once we have the plots, such as those in Figures 7.1 through 7.5, we can examine the changes occurring on the response surface. For most practical applications, visual examination of the surface plots would suffice to isolate the regions of independent variables corresponding to optimal responses. As an aid to graphing the surface plots and finding optimal responses, we briefly introduce certain tools from the differential calculus.

Let us consider a function $Y(X)$ in one variable $X$. One approach to finding a maximum or minimum of $Y(X)$ is to differentiate it with respect to $X$ and equate the derivative to zero; that is,

$$dY(X)/d(X) = 0.$$

Any solution $X_0$ of the preceding differential equation is usually such that $Y(X_0)$ is either a maximum or minimum value of $Y(X)$ in the region around $X_0$. If the second derivative $d^2Y(X)/dX^2$, evaluated at $X_0$, is positive (negative), then $Y(X_0)$ is a minimum (maximum). When the second derivative at $X_0$ is zero, then $Y(X_0)$ is neither a maximum nor a minimum.

The calculus of partial differentiation is used for finding the optimum of a multivariable function. Consider a two-variable function in (7.1-5), which is a second-order fitted response equation:

$$\widehat{Y} = b_0 + b_1 X_1 + b_2 X_2 + b_{11} X_1^2 + b_{22} X_2^2 + b_{12} X_1 X_2. \qquad (7.1\text{-}10)$$

When $X_2$ is treated as fixed, the partial derivative of $\widehat{Y}$ with respect to $X_1$, when equated to zero, gives

$$\partial\widehat{Y}/\partial X_1 = b_1 + 2b_{11}\,X_1 + b_{12}\,X_2 = 0.$$

Similarly, the partial derivative of $\widehat{Y}$ with respect to $X_2$ gives

$$\partial\widehat{Y}/\partial X_2 = b_2 + 2b_{22}X_2 + b_{12}\,X_1 = 0.$$

The solutions of the preceding equations are called *stationary points*. Let $X_{1,0}$ and $X_{2,0}$ denote the $X_1$ and $X_2$ coordinates, respectively, of a stationary point. Also let $\widehat{Y}_0$ denote the predicted response when, in (7.1-10), $X_1$ is replaced by $X_{1,0}$ and $X_2$ by $X_{2,0}$. Though not immediately obvious, it turns out that

$$\widehat{Y}_0 = b_0 + \tfrac{1}{2}(b_1 X_{1,0} + b_2 X_{2,0}), \tag{7.1-11}$$

and $\widehat{Y}_0$ is a candidate for a maximum or minimum response. The stationary point leading to $\widehat{Y}_0$ may be outside the experimental range of the independent variables. Figure 7.5 is a case in which $\widehat{Y}_0$ corresponds to a point within the experimental range, whereas in Figure 7.3 it corresponds to a stationary point outside the range.

It is usually easier to visualize the nature of a response surface when it is expressed in a so-called canonical form. The determination of a stationary point is the first step in writing a second-order response equation in canonical form. The second step involves changing the center of the $X_1$ and $X_2$ coordinates to the stationary point by letting

$$U_1 = X_1 - X_{1,0}, \qquad U_2 = X_2 - X_{2,0}.$$

This transformation reduces $\widehat{Y} - \widehat{Y}_0$ to

$$\widehat{Y} - \widehat{Y}_0 = b_{11}U_1^2 + b_{22}U_2^2 + b_{12}U_1U_2. \tag{7.1-12}$$

Finally, the cross-product term in (7.1-12) is removed, and new coefficients are obtained for new variables $V_1$ and $V_2$ such that

$$\widehat{Y} - \widehat{Y}_0 = \lambda_1 V_1^2 + \lambda_2 V_2^2, \tag{7.1-13}$$

where $\lambda_1$ and $\lambda_2$ are the characteristic roots of a symmetric matrix $B$:

$$B = \begin{bmatrix} b_{11} & \tfrac{1}{2}b_{12} \\ \tfrac{1}{2}b_{12} & b_{22} \end{bmatrix}. \tag{7.1-14}$$

Equation (7.1-13) is said to be in canonical form. The characteristic roots of $B$ are obtained by equating the determinant

$$\begin{vmatrix} b_{11} - \lambda & \frac{1}{2}b_{12} \\ \frac{1}{2}b_{12} & b_{22} - \lambda \end{vmatrix}$$

to zero and solving the resulting equation in $\lambda$. That is, we need to solve

$$(b_{11} - \lambda)(b_{22} - \lambda) - \tfrac{1}{4}b_{12}^2 = 0$$

$$\lambda^2 - \lambda(b_{11} + b_{22}) + b_{11}b_{22} - \tfrac{1}{4}b_{12}^2 = 0 \qquad (7.1\text{-}15)$$

or

$$\lambda^2 - \lambda(b_{11} + b_{22}) + b_{11}b_{22} - \tfrac{1}{4}b_{12}^2 = 0$$

for $\lambda$. Note that (7.1-15) is a quadratic form of (7.1-8) in $\lambda$ with $a = 1$, $b = -(b_{11} + b_{22})$, and $c = b_{11}b_{22} - \tfrac{1}{4}b_{12}^2$. Hence, we can find the two roots $\lambda_1$ and $\lambda_2$ of (7.1-15) using (7.1-9).

The following conclusions can be reached on the basis of the signs and the magnitudes of the $\lambda$s:

1. If both $\lambda_1$ and $\lambda_2$ are negative, the response surface has a maximum value $\widehat{Y}_0$, corresponding to the stationary point $(X_{1,0}, X_{2,0})$.
2. If both $\lambda_1$ and $\lambda_2$ are positive, the response surface has a minimum value $\widehat{Y}_0$ corresponding to the stationary point.

In both cases 1 and 2, the experimenter may conduct further confirmatory experiments with independent variables near the stationary point to converge to an improved optimum combination of the independent variables:

3. If $\lambda_1$ and $\lambda_2$ are of opposite signs, then the surface portrays saddle points. Additional experiments need to be designed for optimum responses.
4. A large (small) value of $\lambda_1$ is indicative of rapid (slow) changes in the responses along the $X_1$ axis. Similar conclusions hold for $\lambda_2$.

If the stationary point is far removed from the range of experimental values of the independent variables, the fitted response equation is not adequate for depicting responses near the stationary point.

Further experimentation with independent variables is desirable near the stationary point. But in food and consumer research, the researcher may not be able to expand the range since the levels of independent variables are often restricted. For instance, the addition of nitrite to food products to ensure microbiological safety is restricted by government regulations. The restriction of the independent variables to maintain nutritional balance and cost is another example. In such studies, therefore, the exploration of the response surface is limited to the allowable range. In studies where no restriction is placed on the range of the independent variables, researchers can plan their experiments freely to locate optimum responses. For further details, we refer to works by Box and associates (1951, 1954, 1955), Anderson (1953), and books by Davies (1967) and Myers (1971).

## 7.2 **FITTING OF RESPONSE SURFACES AND SOME DESIGN CONSIDERATIONS**

We have assumed that the response function $f(X_1, X_2, \ldots, X_n)$ can be approximated by either a first-order equation, as in (7.1-2), or a second-order equation, as in (7.1-3). Notice that the relationship in (7.1-2) is a special case of the one in (7.1-3) if the coefficients of all the second-order terms, that is, $\beta_{11}, \ldots, \beta_{nn}$, and $\beta_{12}, \ldots, \beta_{n-1,n}$, are equal to zero. We discuss the estimation for the second-order equation. The estimation for the first-order equation can be pursued similarly.

Let $Y_i$ denote the response variable $Y$ corresponding to $X_1$ at $X_{1i}$, $X_2$ at $X_{2i}$, and $X_n$ at $X_{ni}$. Hence, from (7.1-3),

$$\widehat{Y}_i = \beta_0 + \beta_1 X_{1i} + \cdots + \beta_n X_{ni} + \beta_{11} X_{1i}^2 + \cdots + \beta_{nn} X_{ni}^2$$
$$+ \beta_{12} X_{1i} X_{2i} + \cdots + \beta_{n-1,n} X_{n-1i} X_{ni} + \varepsilon_i, \quad i = 1, \ldots, N. \tag{7.2-1}$$

If $b_i$, $b_{ii}$, and $b_{ij}$ estimate $\beta_i$, $\beta_{ii}$, and $\beta_{ij}$, then, corresponding to the design point $(X_{1i}, X_{2i}, \ldots, X_{ni})$, the predicted response $\widehat{Y}_i$ is

$$\widehat{Y}_i = b_0 + b_1 X_{1i} + \cdots + b_n X_{ni} + b_{11} X_{1i}^2 + \cdots + b_{nn} X_{ni}^2$$
$$+ b_{12} X_{1i} X_{2i} + \cdots + b_{n-1,n} X_{n-1i} X_{ni}.$$

The estimates $b_i$, $b_{ii}$, and $b_{ij}$ are called the least squares estimates if they are such that $\Sigma_{i=1}^{N} (Y_i - \widehat{Y}_i)^2$ is as small as possible. Leaving the mathematical details aside, we need the so-called normal equations to find the least squares estimates. To this end, let

$$Y = \begin{bmatrix} Y_1 \\ \vdots \\ Y_n \\ \vdots \\ Y_{nn} \\ \vdots \\ Y_n \end{bmatrix}, \quad b_1 = \begin{bmatrix} b_1 \\ \vdots \\ b_n \\ \vdots \\ b_{11} \\ \vdots \\ b_{n-1,n} \end{bmatrix},$$

and

$$X = \begin{bmatrix} 1 & X_{11} \cdots X_{n1} & X_{11}^2 \cdots X_{n1}^2 & X_{11}X_{21} \cdots X_{n-1,1}X_{n1} \\ 1 & X_{12} \cdots X_{n2} & X_{12}^1 \cdots X_{n2}^2 & X_{12}X_{22} \cdots X_{n-1,2}X_{n2} \\ \vdots & \vdots \quad \vdots & \vdots \quad \vdots & \vdots \quad \vdots \\ 1 & X_{1N} \cdots X_{nN} & X_{nN}^2 \cdots X_{nN}^2 & X_{1N}X_{2N} \cdots X_{n-1,N}X_{nN} \end{bmatrix}.$$

Obviously, $X$ is a matrix of $N$ rows. The $i$th row of the matrix $X$ corresponds to the $i$th design point, $Y$ is the vector of responses, and $b$ denotes the vector of estimates of regression coefficients. For $b$ to be the vector of least squares estimates, the following normal equation must be satisfied:

$$(X'X)b = X'Y, \tag{7.2-2}$$

where $X'$ is a matrix with the $i$th column being the $i$th row of $X$, and $X'X$ is the matrix whose element in the $i$th row and $j$th columns is the inner product of the $i$th row of $X'$ with the $j$th columns of $X$. For instance, if $U = (U_1, \ldots, U_n)$ is a row vector and

$$V = \begin{bmatrix} V_1 \\ V_2 \\ \vdots \\ V_n \end{bmatrix}$$

is a column vector, then their inner product is $U_1V_1 + U_2V_2 + \cdots + U_nV_n$.

The inverse (generalized inverse, if necessary) of $X'X$ is denoted by $(X'X)^{-1}$. In this notation the least squares solution $b$ can be expressed as

$$b = (X'X)^{-1}X'Y. \tag{7.2-3}$$

The variance of $b_i$, denoted by $\text{var}(b_i)$, is known to be the following:

$$\text{var}(b_i) = \{i\text{th diagonal element of}(X'X)^{-1}\}\sigma^2, \quad i = 1, \ldots, n. \tag{7.2-4}$$

Consistent with (7.2-4), the variance of any other regression coefficient is equal to $\sigma^2$ multiplied by the corresponding diagonal element of $(X'X)^{-1}$.

Usually, experiments are designed with observations repeated for each design point. Suppose we take $r_i$ observations corresponding to the $i$th design point. Let $Y_{ij}$ denote the $j$th observation for the $i$th design point, $j = 1, \ldots, r_i$, and $M = \sum_{j=1}^{N} r_i$. That is, in all, the experiment would yield $M$ observations. We continue using (7.2-3) for computing $b$, where now

$$
Y = \begin{bmatrix} Y_{11} \\ \vdots \\ Y_{1r1} \\ \vdots \\ Y_{21} \\ \vdots \\ Y_{2r2} \\ \vdots \\ Y_{N1} \\ \vdots \\ Y_{Nr_N} \end{bmatrix}, \quad
X = \begin{array}{c} \left.\begin{array}{c} \text{same} \\ r_1 \\ \text{rows} \end{array}\right\} \\ \left.\begin{array}{c} \text{same} \\ r_2 \\ \text{rows} \end{array}\right\} \\ \\ \left.\begin{array}{c} \text{same} \\ r_N \\ \text{rows} \end{array}\right\} \end{array}
\begin{bmatrix}
1 & X_{11} \cdots X_{n1} & X_{11}^2 \cdots X_{n-1,1} & X_{n1} \\
- & - & - & - \\
1 & X_{11} \cdots X_{n1} & X_{11}^2 \cdots X_{n-1,1} & X_{n1} \\
1 & X_{12} \cdots X_{n2} & X_{12}^2 \cdots X_{n-1,2} & X_{n2} \\
- & - & - & - \\
1 & X_{12} \cdots X_{n2} & X_{12}^2 \cdots X_{n-1,2} & X_{n2} \\
\vdots & \vdots & \vdots & \vdots \\
1 & X_{1N} \cdots X_{nN} & X_{1N}^2 \cdots X_{n-1,N} & X_{nN} \\
- & - & - & - \\
1 & X_{1N} \cdots X_{nN} & X_{1N}^2 \cdots X_{n-1,N} & X_{nN}
\end{bmatrix}
$$

We may point out that several excellent statistical packages are widely available for computers for solving the normal equations and much more. So the researcher need not be concerned with the tasks of obtaining $X'X$, inverting it, and then calculating $(X'X)^{-1}X'Y$. All that is required of the researcher is preparation of data cards having $Y_i$ identified with its corresponding row of the $X$ matrix and some control cards necessary for the use of an appropriate statistical program. Some well-known packages are Design-Expert (Stat-Ease, Inc., Suite 191, 2021 East Hennepin Ave., Minneapolis, MN 55413-9827), SAS (SAS Institute, Inc., SAS Campus Drive, Cary, NC 27513), and SPSS (SPSS, Inc., 233 S. Wacker Drive, 11th Floor, Chicago, IL 60606).

After a response model has been fitted for $\widehat{Y}$ and if the model is a good fit, it can then be used to study response surfaces for optimum design points (combinations of independent variables). The goodness of fit of a fitted model is measured by the closeness of the fitted responses to the corresponding observed responses. Let $\widehat{Y}_{ij}$ and $Y_{ij}$ denote the fitted and the observed responses, respectively, for the $j$th repetition of the $i$th design point. Then a statistical measure of closeness is given by SSe, the so-called error sum of squares, which is

$$
\text{SSe} = \sum_i \sum_j (Y_{ij} - \widehat{Y}_{ij})^2.
$$

Note that the SSe tends to be small (good fit) if the differences $(Y_{ij} - \widehat{Y}_{ij})$ due to errors tend to be small. If the predicted responses $\widehat{Y}_{ij}$ are not in agreement with the observed responses $Y_{ij}$, then the experimenter may want to know whether the disagreement is due to the experimental errors alone or to a combination of experimental errors and the inappropriateness of the response model being fitted. The contribution to the SSe due to experimental errors alone is obtained from the observations of repeated design points as follows. The contribution to the SSe from the $i$th design point is

$$\sum_{j=1}^{r_j} Y_{ij}^2 - r_i \overline{Y}_i^2$$

with $(r_i - 1)$ DF. Hence, the total contribution to SSe due only to the experimental errors from all the design points, denoted by SSexp, is

$$\text{SSexp} = \sum_{i=1}^{N} \left[ \sum_{j=1}^{r_i} Y_{ij}^2 - r_i \overline{Y}_i^2 \right] \qquad (7.2\text{-}5)$$

with $\left( \sum_{i=1}^{N} r_i - N \right)$ DF. The difference between SSe and SSexp is due to a combination of experimental errors and an inadequate fitted model. This sum of squares is usually called the sum of squares due to the "lack of fit" and is denoted by SSf. Thus, a measure of the model's inappropriateness is

$$\text{SSf} = \text{SSe} - \text{SSexp},$$

with its DF obtained by the difference in the corresponding DF for SSe and SSexp. The SSexp, when divided by its DF, that is, MSexp, provides an estimate of $\sigma^2$. If SSf is essentially due to the experimental errors, then SSf divided by its DF, that is, MSf, also provides an estimate of $\sigma^2$. One therefore uses the $F$ ratio

$$F = \text{MSf}/\text{MSexp}$$

to test whether the two estimates of $\sigma^2$ are significantly different from each other. If the two estimates are judged to be significantly different, then we may conclude that the hypothesized model is inappropriate for describing the responses.

Now let us consider some designs for fitting response surfaces. Consider a simple design for two quantitative factors $A$ and $B$. Suppose each factor is used at a low and a high level. Thus, we have $2^2$ factorial combinations: $(1)$, $a$, $b$, $ab$. Let $Y_{1j}$, $Y_{2j}$, $Y_{3j}$, and $Y_{4j}$ denote the experimental observations in the $j$th replication of the design points, which are the factorial combinations $(1)$, $a$, $b$, and $ab$, respectively. Let $X_1 = 1$ if factor $A$ is at the high level and $-1$ otherwise.

Such standardization is always possible through a proper change of scale. Similarly, define $X_2$ to be the standardized variable for factor B. Following Model (7.2-1), the observations in the $j$th replication can be represented as

$$Y_{ij} = \beta_0 + \beta_1 X_{1i} + \beta_2 X_{2i} + \beta_{12} X_{1i} X_{2i} + \varepsilon_{ij}, \qquad \begin{aligned} i &= 1, 2, 3, 4, \\ j &= 1, \ldots, r. \end{aligned} \qquad (7.2\text{-}6)$$

Note that in Model (7.2-6) there are no quadratic effect parameters since each factor appears only at two levels. Even if we were to rewrite Model (7.2-1) with quadratic effect terms in it, the model would reduce to (7.2-6), combining the quadratic effects with the grand mean $\beta_0$ since $X_{1i}^2 = X_{2i}^2 = 1$. Now it follows that

$$Y = \begin{bmatrix} Y_{11} \\ \vdots \\ Y_{1r} \\ Y_{21} \\ \vdots \\ Y_{2r} \\ Y_{31} \\ \vdots \\ Y_{3r} \\ Y_{41} \\ \vdots \\ Y_{4r} \end{bmatrix}, \quad X = \begin{bmatrix} 1 & -1 & -1 & 1 \\ \vdots & \vdots & \vdots & \vdots \\ 1 & -1 & -1 & 1 \\ 1 & 1 & -1 & -1 \\ \vdots & \vdots & \vdots & \vdots \\ 1 & 1 & -1 & -1 \\ 1 & -1 & 1 & -1 \\ \vdots & \vdots & \vdots & \vdots \\ 1 & -1 & 1 & -1 \\ 1 & 1 & 1 & 1 \\ \vdots & \vdots & \vdots & \vdots \\ 1 & 1 & 1 & 1 \end{bmatrix} \begin{matrix} \left.\vphantom{\begin{matrix}1\\1\\1\end{matrix}}\right\} r \text{ rows} \\ \left.\vphantom{\begin{matrix}1\\1\\1\end{matrix}}\right\} r \text{ rows} \\ \left.\vphantom{\begin{matrix}1\\1\\1\end{matrix}}\right\} r \text{ rows} \\ \left.\vphantom{\begin{matrix}1\\1\\1\end{matrix}}\right\} r \text{ rows} \end{matrix}$$

The matrix $X$ is called the *design matrix*. For the reader who is familiar with matrix operations, it is easy to verify that

$$X'X = \begin{bmatrix} 4r & 0 & 0 & 0 \\ 0 & 4r & 0 & 0 \\ 0 & 0 & 4r & 0 \\ 0 & 0 & 0 & 4r \end{bmatrix}, \quad (X'X)^{-1} = \frac{1}{4r} \begin{bmatrix} 1 & 0 & 0 & 0 \\ 0 & 1 & 0 & 0 \\ 0 & 0 & 1 & 0 \\ 0 & 0 & 0 & 1 \end{bmatrix},$$

and the vector $b$ of least squares estimates from (7.2-3) is

$$b = \begin{bmatrix} b_0 \\ b_1 \\ b_2 \\ b_{12} \end{bmatrix} = (X'X)^{-1}X'Y$$

$$= \frac{1}{4r} \begin{bmatrix} Y_{..} \\ Y_{2.} + Y_{4.} - Y_{1.} - Y_{3.} \\ Y_{3.} + Y_{4.} - Y_{1.} - Y_{2.} \\ Y_{1.} + Y_{4.} - Y_{2.} - Y_{3.} \end{bmatrix}.$$

It is interesting to note that $b_0$, the least squares estimate of $\beta_0$, is the grand mean $\overline{Y}$; $b_1$ and $b_2$ are estimates of the main effects of $A$ and $B$, respectively; and $b_{12}$ estimates the interaction effect. In the preceding example, had we assumed that the interaction effect were insignificant, then the Model (7.2-6) would have been a first-order model given by (7.1-2) with $n = 2$ factors.

An experimenter initially uses simple $2^n$ factorial designs to fit models such as (7.2-6) with main effects and two-factor interactions. If the main effects are found to be large compared to the two-factor interactions, the experimenter would then try new factor levels that are changed in the direction of optimum response in the initial experiments. Box and Wilson (1951) explored with further new experiments the path of *steepest ascent*. In the new experiments each factor is varied proportionally with its unit effect in the initial experiments. This process is continued until the first-order effects make no further significant contribution to the search for the optimum response. Indeed, the experimenter may use a $2^n$ factorial experiment to be sure that in this region the first-order effects are small and that no new path needs to be followed in search of the optimum response. Having reached the region where the first-order effects make no further significant contribution, the experimenter can conduct some additional experiments specifically designed to estimate the quadratic and interaction effects. To fit the second-order model in $n$ variables, the experimenter may use $3^n$ factorial designs or some suitably augmented $2^n$ factorial designs.

Specifically, let us consider a $3^2$ factorial whose nine design points are shown in Figure 7.6; the design matrix for fitting the second-order model using only one replicate is shown in Table 7.1. Remember that the actual levels were standardized so that the coordinates of low, medium, and high levels are $-1$, $0$, and $1$, respectively. In interpreting the results, the standardized levels are, of course, changed back to the actual levels.

Note that not every inner product of two columns of the design matrix in Table 7.1 is zero. For instance, the inner product of the constant column and the $X_{1i}^2$ column is not zero. Hence, the design is nonorthogonal and, consequently, the $X'X$ matrix will have some off-diagonal elements that are not zero. As a result of the nonorthogonality, it is not possible to compute the sums of squares due to the quadratic effects without first adjusting for the linear effects in the model. Sometimes one may orthogonalize the design matrix and adjust the parameters of the model accordingly. A design matrix like that in Table 7.1 can be orthogonalized rather easily if we subtract

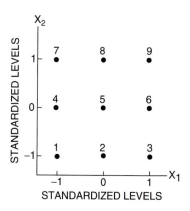

■ **FIGURE 7.6** Design points of a $3^2$ factorial with low, medium, and high levels, standardized at $-1$, $0$, and $1$, respectively.

**Table 7.1** Design Matrix for a $3^2$ Factorial Experiment with One Replication to Fit a Second-Order Model

| Design Point | Constant $\beta_0$ | $X_{1i}$ | $X_{2i}$ | $X_{1i}^2$ | $X_{2i}^2$ | $X_{1i}X_{2i}$ |
|---|---|---|---|---|---|---|
| 1 | 1 | −1 | −1 | 1 | 1 | 1 |
| 2 | 1 | 0 | −1 | 0 | 1 | 0 |
| 3 | 1 | 1 | −1 | 1 | 1 | −1 |
| 4 | 1 | −1 | 0 | 1 | 0 | 0 |
| 5 | 1 | 0 | 0 | 0 | 0 | 0 |
| 6 | 1 | 1 | 0 | 1 | 0 | 0 |
| 7 | 1 | −1 | 1 | 1 | 1 | −1 |
| 8 | 1 | 0 | 1 | 0 | 1 | 0 |
| 9 | 1 | 1 | 1 | 1 | 1 | 1 |

**Orthogonalized Form of the Preceding Matrix**

| Design Point | Constant $\beta_0$ | $X_{1i}$ | $X_{2i}$ | $X_{1i}^2$ | $X_{2i}^2$ | $X_{1i}X_{2i}$ |
|---|---|---|---|---|---|---|
| 1 | 1 | −1 | −1 | $\frac{1}{3}$ | $\frac{1}{3}$ | 1 |
| 2 | 1 | 0 | −1 | $-\frac{2}{3}$ | $\frac{1}{3}$ | 0 |
| 3 | 1 | 1 | −1 | $\frac{1}{3}$ | $\frac{1}{3}$ | −1 |
| 4 | 1 | −1 | 0 | $\frac{1}{3}$ | $-\frac{2}{3}$ | 0 |
| 5 | 1 | 0 | 0 | $-\frac{2}{3}$ | $-\frac{2}{3}$ | 0 |
| 6 | 1 | 1 | 0 | $\frac{1}{3}$ | $-\frac{2}{3}$ | 0 |
| 7 | 1 | −1 | 1 | $\frac{1}{3}$ | $\frac{1}{3}$ | −1 |
| 8 | 1 | 0 | 1 | $-\frac{2}{3}$ | $\frac{1}{3}$ | 0 |
| 9 | 1 | 1 | 1 | $\frac{1}{3}$ | $\frac{1}{3}$ | 1 |

$$C = \frac{\sum_{i-1}^{M} X_{1i}^2}{M} = \frac{\sum_{i=1}^{M} X_{2i}^2}{M} \tag{7.2-7}$$

from each element of $X_{1i}^2$ and $X_{2i}^2$ columns, where $M$ is the total number of rows in the design matrix. For the design matrix in Table 7.1, $C = \frac{2}{3}$, and the resulting orthogonal matrix is the lower half of the table.

Having found $b_0$, $b_1$, $b_{11}$, $b_{22}$, and $b_{12}$, the only adjustment needed as a result of the orthogonalization of the $3^2$ factorial design matrix just discussed is as follows. The least squares estimate $b_0$ of the constant term $\beta_0$ is always $\overline{Y}$ when the design matrix is used in the orthogonal form. Had we used the matrix in its original form rather than the orthogonal one, the least squares estimate of $\beta_0$ would have been

$$b_0 - \bar{Y} - Cb_{11} - C_{22}, \qquad (7.2\text{-}8)$$

where $C$ is the constant used in the orthogonalization.

## 7.3 ILLUSTRATIONS OF FITTINGS OF FIRST- AND SECOND-ORDER MODELS

First-order models and the designs used for fitting them are useful in exploratory work. The designs most commonly used to fit the first-order models are $2^n$ factorial designs. In addition to the $2^n$ design points, a center point is also frequently incorporated into the design to permit improved estimation of the variation due to experimental errors. The center point in the design may correspond to the absence of all the factors. For a $2^2$ factorial, with factors $A$ and $B$, for example, four design points corresponding to the four factorial combinations and one additional center point are shown in Figure 7.7. An analysis of variance for a $2^n$ factorial with one center point is given in Table 7.2. There are $(N + 1)$ design points, the $i$th design point being replicated $r_i$ times, where $N = 2^n$.

The formulas for calculating various sums of squares follow the table.

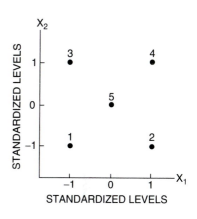

■ **FIGURE 7.7** Design points of a $2^2$ factorial with factors $A(X_1)$ and $B(X_2)$; low and high levels are standardized at $-1$ and $1$, respectively, and there is a center point.

**Table 7.2** Analysis of Variance of $N = 2^n$ Factorial with $(N + 1)$ Design Points, the $i$th Point Replicated $r_i$ Times[a]

| Source of Variance | DF | SS | MS |
|---|---|---|---|
| Total | $M - 1$ | SSto | |
| Linear Regression in $n$ Variables | $n$ | SSr | MSr |
| Error | $M - n - 1$ | SSe $=$ SSto $-$ SStr | MSe |
| Lack of Fit | $N - n$ | SSf $=$ SSe $-$ SSexp | MSf |
| Experimental Error | $M - N - 1$ | SSexp | |

[a]The total number of observations is $M = \sum_{i=1}^{N+1} r_i$.

Note:

$SSto = \Sigma_i \, \Sigma_j \, Y_{ij}^2 - (Y\ldots)^2/M.$

$SSr = b'X'Y = \sum_{i=1}^{n} b_i g_i,$ where $b_i$ is the $i$th component of $b = (X'X)^{-1}X'Y$ and $g_i$ is the $i$th component of $g = X'Y.$

$SSexp$ is found from (7.2-5): $SSexp = \sum_{i=1}^{N+1} [\Sigma_j Y_{ij}^2 - Y_i^2/r_i].$

To test the appropriateness of the fitted model, one computes

$$F = MSf/MSe$$

and compares it with the tabulated $F$ distribution values.

**Example 7.3-1**  An exploratory study was conducted to find the effect of $CaCl_2$ ($X_1$) and $MgCl_2$ ($X_2$) on water stability ($Y$) of meat emulsion. A part of the data in the study is reproduced in Table 7.3. The standardized and experimental levels for both the independent variables are

| Standardized levels: | −1 | 0 | 1 |
|---|---|---|---|
| **Experimental levels of $X_1$:** | 50 | 125 | 200 |
| **Experimental levels of $X_2$:** | 50 | 125 | 200 |

**Table 7.3** Design Points and Their Responses (Example 7.3-1)

| Design Point | $\beta_0$ | $X_{1i}$ | $X_{2i}$ | $Y_{ij}$ | $Y_{i\cdot}$ | $\bar{Y}_{i\cdot}$ |
|---|---|---|---|---|---|---|
| 1 | 1 | −1 | −1 | 6.4 | | |
| | 1 | −1 | −1 | 6.2 | 18.1 | 6.03 |
| | 1 | −1 | −1 | 5.5 | | |
| 2 | 1 | 1 | −1 | 5.7 | | |
| | 1 | 1 | −1 | 4.6 | 15.2 | 5.07 |
| | 1 | 1 | −1 | 4.9 | | |
| 3 | 1 | −1 | 1 | 6.6 | | |
| | 1 | −1 | 1 | 6.6 | 19.0 | 6.33 |
| | 1 | −1 | 1 | 5.8 | | |
| 4 | 1 | 1 | 1 | 7.1 | | |
| | 1 | 1 | 1 | 6.6 | 21.4 | 7.13 |
| | 1 | 1 | 1 | 7.7 | | |
| 5 | 1 | 0 | 0 | 4.8 | | |
| | 1 | 0 | 0 | 5.1 | | |
| | 1 | 0 | 0 | 5.0 | 25.2 | 5.04 |
| | 1 | 0 | 0 | 5.4 | | |
| | 1 | 0 | 0 | 4.9 | | |

Note:
$\Sigma_i \Sigma_j Y_{ij}^2 = 588.55$; $Y.. = 98.90$.

The first four points in Table 7.3 constitute a $2^2$ factorial, which is also known as the "cube" portion of the design. The fifth point is the center point. The cube portion is replicated three times, whereas the center point is replicated five times.

The design matrix $X$, response vector $Y$, and some calculations in the matrix form are illustrated as follows:

$$X = \begin{bmatrix} 1 & -1 & -1 \\ 1 & -1 & -1 \\ 1 & -1 & -1 \\ 1 & 1 & -1 \\ 1 & 1 & -1 \\ 1 & 1 & -1 \\ 1 & -1 & 1 \\ 1 & -1 & 1 \\ 1 & -1 & 1 \\ 1 & 1 & 1 \\ 1 & 1 & 1 \\ 1 & 1 & 1 \\ 1 & 0 & 0 \\ 1 & 0 & 0 \\ 1 & 0 & 0 \\ 1 & 0 & 0 \\ 1 & 0 & 0 \end{bmatrix}, \quad Y = \begin{bmatrix} 6.4 \\ 6.2 \\ 5.5 \\ 5.7 \\ 4.6 \\ 4.9 \\ 6.6 \\ 6.6 \\ 5.8 \\ 7.1 \\ 6.6 \\ 7.7 \\ 4.8 \\ 5.1 \\ 5.0 \\ 5.4 \\ 4.9 \end{bmatrix},$$

$$X' = \begin{bmatrix} 1 & 1 & 1 & 1 & 1 & 1 & 1 & 1 & 1 & 1 & 1 & 1 & 1 & 1 & 1 & 1 & 1 \\ -1 & -1 & -1 & 1 & 1 & 1 & -1 & -1 & -1 & 1 & 1 & 1 & 0 & 0 & 0 & 0 & 0 \\ -1 & -1 & -1 & -1 & -1 & -1 & -1 & 1 & 1 & 1 & 1 & 1 & 0 & 0 & 0 & 0 & 0 \end{bmatrix},$$

$$(X'X) = \begin{bmatrix} 17 & 0 & 0 \\ 0 & 12 & 0 \\ 0 & 0 & 12 \end{bmatrix}, \quad (X'X)^{-1} = \begin{bmatrix} \dfrac{1}{17} & 0 & 0 \\ 0 & \dfrac{1}{12} & 0 \\ 0 & 0 & \dfrac{1}{12} \end{bmatrix},$$

$$X'Y = \begin{bmatrix} 98.9000 \\ -0.5000 \\ 7.1000 \end{bmatrix} = \begin{bmatrix} g_0 \\ g_1 \\ g_2 \end{bmatrix}.$$

Now the normal equations (7.2-2) can be written as

$$\begin{bmatrix} 17 & 0 & 0 \\ 0 & 12 & 0 \\ 0 & 0 & 12 \end{bmatrix} \begin{bmatrix} b_0 \\ b_1 \\ b_2 \end{bmatrix} = \begin{bmatrix} 98.9000 \\ -0.5000 \\ 7.1000 \end{bmatrix}.$$

It follows that

$$b = \begin{bmatrix} b_0 \\ b_1 \\ b_2 \end{bmatrix} = \begin{bmatrix} 5.8176 \\ -0.0417 \\ 0.5917 \end{bmatrix}.$$

Hence, the fitted first-order response equation is

$$\hat{Y} = 5.8176 - 0.0471X_1 + 0.5917X_2.$$

If we need to test hypotheses or find confidence intervals for the parameters, we then need their variances. From (7.2-4),

$$\text{var}(b_0) = \sigma^2/17, \quad \text{var}(b_1) = \sigma^2/12, \quad \text{var}(b_2) = \sigma^2/12.$$

Since $\sigma^2$ is unknown, it is estimated by MSe if the estimated model is a good fit, and by MSexp if the model is not a good fit. In our example, we will see that the first-order fitted model does not fit the data adequately. Hence, $\sigma^2$ is estimated by MSexp, which is 0.1949.

Let us carry the calculations for the various sums of squares as follows:

$$\begin{aligned} \text{SSto} &= \sum_i \sum_j Y_{ij}^2 - 17\bar{Y}^2 \\ &= 588.55 - 17(5.82)^2 \\ &= 13.1847 \quad \text{with} \quad 16 \text{ DF.} \end{aligned}$$

$$\begin{aligned} \text{SSr} &= \sum_{i=1}^{2} b_i g_i \\ &= (-0.5000)(-0.0417) + (7.1000)(0.5917) \\ &= 4.2220 \quad \text{with} \quad 2 \text{ DF.} \end{aligned}$$

$$\begin{aligned} \text{SSe} &= \text{SSto} - \text{SSr} \\ &= 8.9627 \quad \text{with} \quad (16 - 2) = 14 \text{ DF.} \end{aligned}$$

From (7.2-5),

$$\begin{aligned} \text{SSexp} &= (6.4)^2 + (6.2)^2 + (5.5)^2 - (18.1)^2/3 \\ &\quad + (5.7)^2 + (4.6)^2 + (4.9)^2 - (15.2)^2/3 \\ &\quad + (6.6)^2 + (6.6)^2 + (5.8)^2 - (19.0)^2/3 \\ &\quad + (7.1)^2 + (6.6)^2 + (7.7)^2 - (21.4)^2/3 \\ &\quad + (4.8)^2 + (5.1)^2 + (5.0)^2 + (5.4)^2 \\ &\quad + (4.9)^2 - (25.2)^2/5 \\ &= 2.3388 \quad \text{with} \quad 12 \text{ DF.} \end{aligned}$$

| Table 7.4 Analysis of Variance of the Data in Example 7.3-1 | | | | |
|---|---|---|---|---|
| **Source of Variance** | **DF** | **SS** | **MS** | **F Ratio** |
| Total | 16 | 13.1847 | | |
| Linear Regression Model in Two Variables | 2 | 4.2220 | 2.1110 | |
| Error | 14 | 8.9627 | 0.6402 | |
| Lack of Fit | 2 | 6.6239 | 3.3119 | $16.99^a$ |
| Experimental Error | 12 | 2.3388 | 0.1949 | |

$^a p < .001.$

Finally

$$SSf = SSe - SSexp = 8.9627 - 2.3388$$
$$= 6.6239 \quad \text{with} \quad 2 \text{ DF}.$$

These results are summarized in Table 7.4. We notice that the hypothesis of the linear model's being adequate is rejected even at the significance level 0.001, since MSf/MSexp = 16.9928 is too large.

Even though we know that the first-order fitted equation is very inadequate, let us try to use it to map the response surface. We examine a contour corresponding to $\widehat{Y} = 5.0$, which is well within the range of the observed responses. For $\widehat{Y} = 5.0$, we see that the fitted model is a line contour; that is,

$$0.0417X_1 - 0.5917X_2 = 0.8176.$$

If $X_2 = -1$, we notice that $X_1 = 33.80$, a coordinate that is far beyond the range of the experimental design points. This again confirms that the fitted equation is not a good one for mapping the response surface.

As we just noticed in Example 7.3-1, the first-order models are often inadequate and provide a poor description of the geometric shape of the response surfaces. For better description, the model must include the quadratic and the interaction effects. The estimation of the quadratic and the interaction effects requires designs with three or more levels to be run for each factor. The three quantitative levels are designated low, medium, and high with standardized coefficients $-1$, 0, and 1, respectively. In subsequent sections we discuss several other designs suitable for fitting second-order models. Here we illustrate the process of fitting second-order models with $3^2$ factorial designs.

**Example 7.3-2**

We refer to the $3^2$ factorial experiment and its outcomes considered in Example 6.2-1. Two compounds $A$ and $B$, each at three numerical levels, were used to study their effects on the water-binding capacity of meats. Factor $A$ was used at three levels: 62, 64, and 66 ppm, whereas factor $B$ was at 7, 9, and 11 ppm. The design consisted of $r = 2$ replications. For the statistical analysis the numerical levels were standardized to $-1, 0, 1$. Had each design point been repeated only once, the design matrix in both the orthogonal and the nonorthogonal forms for the fitting of the second-order model would have been the one given in Table 7.1. In this example, however, the design consists of two replications. Therefore, there are two observations for each design point, and, corresponding to each design point, there are two identical rows in the design matrix. The design points, their matrix $X$ in orthogonal form, and the response vector $Y$ are as shown:

| Design point $i$ | Constant $\beta_0$ | $X_{1i}$ | $X_{2i}$ | $X_{2i}^2$ | $X_{1i}^2$ | $X_1X_{2i}$ | | | |
|---|---|---|---|---|---|---|---|---|---|
| 1 | 1 | $-1$ | $-1$ | $\frac{1}{3}$ | $\frac{1}{3}$ | 1 | | 1.14 | $Y_{11}$ |
|  | 1 | $-1$ | $-1$ | $\frac{1}{3}$ | $\frac{1}{3}$ | 1 | | 1.05 | $Y_{12}$ |
| 2 | 1 | 0 | $-1$ | $-\frac{2}{3}$ | $\frac{1}{3}$ | 0 | | 1.87 | $Y_{21}$ |
|  | 1 | 0 | $-1$ | $-\frac{2}{3}$ | $\frac{1}{3}$ | 0 | | 1.60 | $Y_{22}$ |
| 3 | 1 | 1 | $-1$ | $\frac{1}{3}$ | $\frac{1}{3}$ | $-1$ | | 1.70 | $Y_{31}$ |
|  | 1 | 1 | $-1$ | $\frac{1}{3}$ | $\frac{1}{3}$ | $-1$ | | 1.80 | $Y_{32}$ |
| 4 | 1 | $-1$ | 0 | $\frac{1}{3}$ | $-\frac{2}{3}$ | 0 | | 2.23 | $Y_{41}$ |
|  | 1 | $-1$ | 0 | $\frac{1}{3}$ | $-\frac{2}{3}$ | 0 | | 2.30 | $Y_{42}$ |
| 5 | 1 | 0 | 0 | $-\frac{2}{3}$ | $-\frac{2}{3}$ | 0 | $X =$ | 3.13 | $Y_{51}$ |
|  | 1 | 0 | 0 | $-\frac{2}{3}$ | $-\frac{2}{3}$ | 0 | | 3.00 | $Y_{52}$ |
| 6 | 1 | 1 | 0 | $\frac{1}{3}$ | $-\frac{2}{3}$ | 0 | | 2.80 | $Y_{61}$ |
|  | 1 | 1 | 0 | $\frac{1}{3}$ | $-\frac{2}{3}$ | 0 | | 1.95 | $Y_{62}$ |
| 7 | 1 | $-1$ | 1 | $\frac{1}{3}$ | $\frac{1}{3}$ | $-1$ | $, \; Y =$ | 0.74 | $Y_{71}$ |
|  | 1 | $-1$ | 1 | $\frac{1}{3}$ | $\frac{1}{3}$ | $-1$ | | 0.50 | $Y_{72}$ |
| 8 | 1 | 0 | 1 | $-\frac{2}{3}$ | $\frac{1}{3}$ | 0 | | 1.43 | $Y_{81}$ |
|  | 1 | 0 | 1 | $-\frac{2}{3}$ | $\frac{1}{3}$ | 0 | | 1.00 | $Y_{82}$ |
| 9 | 1 | 1 | 1 | $\frac{1}{3}$ | $\frac{1}{3}$ | 1 | | 0.10 | $Y_{91}$ |
|  | 1 | 1 | 1 | $\frac{1}{3}$ | $\frac{1}{3}$ | 1 | | 0.05 | $Y_{92}$ |

Again, for those who are familiar with matrix operations, it is easy to see that

$$(X'X) = \begin{bmatrix} 18 & & & & & 0 \\ & 12 & & & & \\ & & 12 & & & \\ & & & 4 & & \\ & & & & 4 & \\ 0 & & & & & 8 \end{bmatrix}, \quad (X'X)^{-1} = \begin{bmatrix} \frac{1}{18} & & & & & 0 \\ & \frac{1}{12} & & & & \\ & & \frac{1}{12} & & & \\ & & & \frac{1}{4} & & \\ & & & & \frac{1}{4} & \\ 0 & & & & & \frac{1}{8} \end{bmatrix}.$$

To set up the normal equations, we compute

$$X'Y \begin{bmatrix} 28.3900 \\ 0.4400 \\ -5.3400 \\ -2.5676 \\ -5.9475 \\ -2.4000 \end{bmatrix} = \begin{bmatrix} g_0 \\ g_1 \\ g_2 \\ g_{11} \\ g_{22} \\ g_{12} \end{bmatrix}.$$

Now the normal equations $(X'X)\, b = X'Y$ yield

$$b = \begin{bmatrix} b_0 \\ b_1 \\ b_2 \\ b_{11} \\ b_{22} \\ b_{12} \end{bmatrix} = \begin{bmatrix} 1.5772 \\ 0.0367 \\ -0.4450 \\ -0.6417 \\ -1.4867 \\ -0.3000 \end{bmatrix}.$$

Note that $b_0 = \bar{Y}$. Hence, the second-order fitted model, using the orthogonalized form of the design matrix, is

$$\hat{Y} = 1.5772 + 0.0367X_1 - 0.4450X_2 - 0.6417X_1^2$$
$$- 1.4867X_2^2 - 0.3000X_1X_2.$$

To obtain the fitted second-order equation corresponding to the design matrix in the nonorthogonalized form, we need to incorporate the adjustment given by (7.2-8). That is, instead of $b_0$ being 1.5772, it is

$$b_0 = 1.5772 - \frac{2}{3}(-0.6417) - \frac{2}{3}(-1.4867) = 2.9962,$$

and the fitted second-order model is

$$\hat{Y} = 2.9962 + 0.0367X_1 - 0.4450X_2 - 0.6417X_1^2$$
$$- 1.4867X_2^2 - 0.3000X_1X_2. \tag{7.3-1}$$

One should keep in mind that the essential purpose of the orthogonalization is to be able to carry out the statistical computations easily. But this is really a minor consideration in this age of computers and in view of universally available statistical software consisting of excellent programs for all sorts of statistical analyses, and regression analyses in particular.

From (7.2-4) and the diagonal matrix $(X'X)^{-1}$ corresponding to the orthogonalized form of $X$, we have

$$\text{var}(\overline{Y}) = \sigma^2/18, \quad \text{var}(b_1) = \sigma^2/12, \quad \text{var}(b_2) = \sigma^2/12,$$
$$\text{var}(b_{11}) = \sigma^2/4, \quad \text{var}(b_{22}) = \sigma^2/4, \quad \text{var}(b_{12}) = \sigma^2/8.$$

Following adjustment (7.2-8) for $b_0$, one can verify that

$$\text{var}(b_0) = \text{var}(\overline{Y}) + \left(\frac{2}{3}\right)^2 \text{var}(b_{11}) + \left(\frac{2}{3}\right)^2 \text{var}(b_{22})$$

$$= \sigma^2 \left[\frac{1}{18} + \frac{1}{9} + \frac{1}{9}\right] = 5\sigma^2/18.$$

Here again, $\sigma^2$ is unknown and can be estimated by MSe if the estimated model is a good fit and by MSexp if the model is not a good fit.

The calculations for the sum of squares are as follows:

$$\text{SSto} = \sum_i \sum_j Y_{ij}^2 - \frac{Y_{..}^2}{18} = 14.3890 \quad \text{with} \quad 17 \text{ DF},$$

$$\text{SSr} = b_1g_1 + b_2g_2 + b_{11}g_{11} + b_{22}g_{22} + b_{12}g_{12}$$
$$= 13.6021 \quad \text{with} \quad 5 \text{ DF}.$$

We may note that SSr can be partitioned into components due to linear and quadratic effects, written as SSrl and SSrq, respectively. They are

$$\text{SSrl} = b_1g_1 + b_2g_2$$
$$= 0.0151 + 2.3763 = 2.3924 \quad \text{with} \quad 2 \text{ DF},$$

$$\text{SSrq} = b_{11}g_{11} + b_{22}g_{22} + b_{12}g_{12}$$
$$= 1.6476 + 8.8421 + 0.7200 = 11.2097 \quad \text{with} \quad 3 \text{ DF},$$

$$\text{SSe} = \text{SSto} - \text{SSr}$$
$$= 14.3690 - 13.6021 = 0.7669 \quad \text{with} \quad (17 - 5) = 12 \text{ DF}.$$

From (7.2-5),

$$SSexp = 2.4021 - (2.19)^2/2 + 6.0569 - (3.47)^2/2 + 6.1300$$
$$-(3.50)^2/2 + 10.2629 - (4.53)^2/2 + 18.7969$$
$$-(6.13)^2/2 + 11.6425 - (4.75)^2/2 + 0.7976 - (1.24)^2/2$$
$$+3.0449 - (2.43)^2/2 + 0.0125 - (0.15)^2/2$$

$$= 0.0041 + 0.0365 + 0.0050 + 0.0025 + 0.0085$$
$$+0.3613 + 0.0288 + 0.0925 + 0.0013$$

$$= 0.5405 \quad \text{with} \quad 9 \text{ DF.}$$

Finally,

$$SSf = SSe - SSexp$$
$$= 0.7669 - 0.5405 = 0.2264 \quad \text{with} \quad 3 \text{ DF.}$$

To test the adequacy of the second-order model, we calculate the $F$ ratio,

$$F = MSf/MSexp = 0.0755/0.0600 = 1.2583,$$

and note that this $F$ ratio is much less than even the 90th percentile of the $F$ distribution with 3 numerator and 9 denominator DF. Hence, the second-order model is judged to be adequate. The breakdown of the SSto into various components is summarized in Table 7.5.

**Table 7.5** Analysis of Variance of the Data in Example 7.3-2

| Source of Variance | DF | SS | MS | F Ratio |
|---|---|---|---|---|
| Total | 17 | 14.3690 | | |
| Model | 5 | 13.6021 | | |
| Linear | 2 | 2.3924 | 1.1962 | 18.72 |
| $b_1$ | 1 | 0.0151 | | |
| $b_2$ | 1 | 2.3763 | | |
| Quadratic and Interaction | 3 | 11.2097 | 3.7366 | 58.48 |
| $b_{11}$ | 1 | 1.6476 | | |
| $b_{22}$ | 1 | 8.8421 | | |
| $b_{12}$ | 1 | 0.7200 | | |
| Error | 12 | 0.7669 | 0.0639 | |
| Lack of Fit | 3 | 0.2264 | 0.0755 | 1.26[a] |
| Experimental Error | 9 | 0.5404 | 0.0600 | |

[a]*Insignificant F ratio.*

Since the fitted second-order model provides a good fit, it may be used to search for optimum levels. Treating $\widehat{Y}$ as a function of $X_1$ and $X_2$, we take partial derivatives $\partial\widehat{Y}/\partial X_1$ and $\partial\widehat{Y}/\partial X_2$ and equate them with zero. That is,

$$\partial\widehat{Y}/\partial X_1 = 0.0367 - 1.2834X_1 - 0.3000X_2 = 0;$$
$$\partial\widehat{Y}/\partial X_2 = -0.4450 - 2.9734X_2 - 0.3000X_1 = 0.$$

Solving the two equations for $X_1$ and $X_2$, we get the stationary point

$$(X_{1,0}X_{2,0}) = (0.0651, -0.1562).$$

The optimum response, if it exists, will correspond to the stationary point. The predicted response at the stationary point is

$$\begin{aligned}\widehat{Y}_0 = {}& 2.9962 + (0.0367)(0.0651) + (-0.4450)(-0.1562)\\ & -(0.6417)(0.0042) + (-1.4867)(0.0234)\\ & +(-0.3000)(-0.0099)\end{aligned}$$

$$= 3.04.$$

The stationary point $(0.0651, -0.1562)$ is located near the center of the experimental region. To find out whether $\widehat{Y}_0$ is a maximum or minimum response, we calculate the characteristic roots of matrix $B$ defined in (7.1-14). Following Eq. (7.1-15), the characteristic roots are given by

$$\begin{vmatrix} -0.6417 - \lambda & \dfrac{1}{2}(-0.3000) \\[2mm] \dfrac{1}{2}(-0.3000) & -1.4867 - \lambda \end{vmatrix} = 0,$$

or

$$(-0.6417 - \lambda)(-1.4867 - \lambda) - \left[\dfrac{1}{2}(-0.3000)\right]^2 = 0,$$

or

$$\lambda^2 + 2.1284\lambda + 0.9315 = 0.$$

The two roots of the quadratic equation can be obtained using (7.1-9) with $a = 1$, $b = 2.1284$, and $c = 0.9315$. The two roots are $\lambda_1 = -0.6159$ and $\lambda_2 = -1.5126$. As a check on the calculations, we may note that $\lambda_1 + \lambda_2 = b_{11} + b_{22}$ is always true. Since both roots are negative, the response surface has a maximum $\widehat{Y} = 3.04$, corresponding to the stationary point. The fitted response model in the canonical form is

$$\widehat{Y} = 3.04 - 0.6159V_1^2 - 1.5126V_2^2.$$

We may follow the procedure outlined in Section 7.1 to sketch the contours of the fitted response model. For instance, if $\widehat{Y} = 2.50$ and $X_1 = 0$, the fitted response model leads to

$$2.50 = 2.9961 - 0.4450X_2 - 1.4867X_2^2.$$

This equation can be rewritten as

$$1.4867X_2^2 + 0.4450X_2 - 0.4961 = 0,$$

and from (7.1-9), two coordinates of $X_2$ satisfying the quadratic equation are

$$X_2 = \frac{-0.4450 + 1.7743}{2.9734} = 0.447$$

and

$$X_2 = \frac{-0.4450 - 1.7743}{2.9734} = -0.746.$$

These coordinates provide two points $(0, 0.447)$ and $(0, -0.746)$ on the response surface contour for $\widehat{Y} = 2.50$. This procedure can be repeated for several combinations of $\widehat{Y}$, $X_1$, and $X_2$ values to permit the plotting of the surface contours. One can use existing computer software to carry out arithmetic calculations to plot the response surface contours. For our example, we plotted the contours in Figure 7.8 for $\widehat{Y} = 0.50$ and $\widehat{Y} = 2.50$.

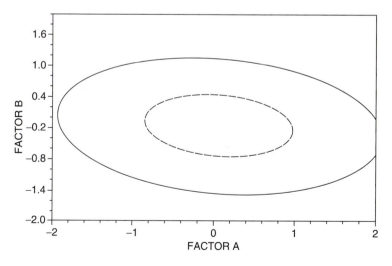

■ **FIGURE 7.8** Response surface contour corresponding to $\widehat{Y} = 0.50$ (——) and $\widehat{Y} = 2.50$ (- - -) of the fitted response Model (7.3-1) in Example 7.3-2.

Note that the surface has a peak at $\widehat{Y} = 3.04$, which corresponds to the actual numerical levels of $X_1$ and $X_2$ in the neighborhood of 6.4 and 9 ppm, respectively.

The use of $3^n$ factorial designs in fitting the second-order models has its drawbacks. It is a fact, though not immediately clear, that the fitted response models from $3^n$ factorial experiments do not predict responses with equal power in all directions from the center of the design. An experimental design having equal predicting powers in all directions at a constant distance from the center of the design is called a *rotatable design*. It is true that $2^n$ factorial designs are rotatable when used to fit the first-order models. Hence, instead of $3^n$ factorial designs, as one may suspect, it is possible to augment $2^n$ suitable factorial or fractional factorial designs to obtain rotatable designs for fitting the second-order models. An augmented $2^2$ factorial design appears in Figure 7.9. Points 1–4 in the figure correspond to $2^2$ factorial combinations. Points 5–8 are called *axial points*, and point 9 is called the *center point* of the design. The distance $\alpha$ of the axial points from the center point can be chosen so that the design is rotatable. The matrix for estimating the second-order

■ **FIGURE 7.9** Design points of an augmented $2^2$ factorial design.

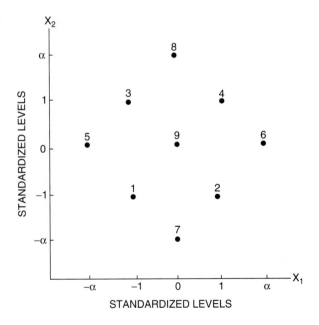

**Table 7.6** Design Matrix of a $2^2$ Factorial Augmented Design for Fitting the Second-Order Model

| Design Point $i$ | Constant $\beta_0$ | $X_{1i}$ | $X_{2i}$ | $X_{1i}^2$ | $X_{2i}^2$ | $X_{1i}X_{2i}$ |
|---|---|---|---|---|---|---|
| 1 | 1 | $-1$ | $-1$ | 1 | 1 | 1 |
| 2 | 1 | 1 | $-1$ | 1 | 1 | $-1$ |
| 3 | 1 | $-1$ | 1 | 1 | 1 | $-1$ |
| 4 | 1 | 1 | 1 | 1 | 1 | 1 |
| 5 | 1 | $-\alpha$ | 0 | $\alpha^2$ | 0 | 0 |
| 6 | 1 | $\alpha$ | 0 | $\alpha^2$ | 0 | 0 |
| 7 | 1 | 0 | $-\alpha$ | 0 | $\alpha^2$ | 0 |
| 8 | 1 | 0 | $\alpha$ | 0 | $\alpha^2$ | 0 |
| 9 | 1 | 0 | 0 | 0 | 0 | 0 |

model for the design in Figure 7.9 is given in Table 7.6. The reader can see in the section to follow that the augmented $2^2$ factorial design is a member of an important class of *composite designs* by Box and Wilson (1951). We discuss some rotatable designs in Section 7.5. For a detailed discussion of rotatability, the reader is referred to Hunter (1958a).

## 7.4 **COMPOSITE DESIGNS**

In addition to nonrotatability, $3^n$ factorial designs also suffer from another drawback. In studies involving more than two independent variables, the use of $3^n$ factorials would result in a large number of design points and therefore in excessive experimentation. To reduce the number of design points, Box and Wilson (1951) introduced so-called composite designs. To form a composite design, we may start with a $2^n$ factorial design and add enough extra design points to enable it to fit the second-order models. These designs not only have fewer points compared with three-level factorial designs, but they can also be performed in stages. For instance, a $2^n$ factorial

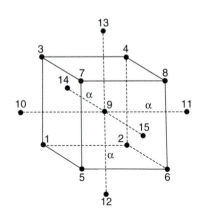

**FIGURE 7.10** The 15 design points of a three-variable composite design.

design may be used to fit the first-order model repeatedly to reach a near-stationary point for the optimum response. Once a near-stationary region has been isolated, additional $(2n + 1)$ points can be added to the $2^n$ factorial design points to fit a second-order model searching for the optimum response. If $n = 3$, we would have $2^n + (2n + 1) = 15$ points in a composite design compared with $3^n = 27$ points in a three-level factorial design. The extra $(2n + 1)$ points are added as follows.

To the $2^n$ points of two-level factorial designs, $(2n + 1)$ extra points are added so that one of them is at the origin (center) and the remaining $2n$ at a distance of $\alpha$ units from the origin, spread equally in pairs along the $n$ axis of the design. For $n = 3$, a three-dimensional representation of the 15 points of a composite design is shown in Figure 7.10. These 15 points along with a column of constant terms (ones) are also displayed in Table 7.7.

Let us continue by considering a composite design with $N = 2^n + (2n + 1)$ points. Suppose there are $m$ observations from the basic $2^n$ factorial design points and that, of the extra $(2n + 1)$ points, each is replicated $r$ times to yield $r(2n + 1)$ more observations. Thus, there are $M = m + r(2n + 1)$ observations in all, and they can be used to fit the second-order Model (7.1-3). For any composite design of this sort, the design matrix $X$ for fitting the second-order model is such that its columns are orthogonal to each other except for the non-orthogonality of any two columns corresponding to the quadratic effects. However, the design matrix can be made completely orthogonal in two steps. First, choose the distance $\alpha$ of the axial points from the center point so that

$$\alpha = [\rho m/(4r^2)]^{1/4}, \tag{7.4-1}$$

where

$$\rho = [M^{1/2} - m^{1/2}]^2.$$

Next, rewrite the matrix to obtain a new orthogonal matrix as follows. All the columns in the design matrix corresponding to the constant $\beta_0$, the linear effects, and the interaction effects remain unchanged. But from each column of quadratic effects, subtract its average $C$ to obtain a new quadratic-effect column. For algebraic definition of $C$, refer to (7.2-7). The new matrix so obtained is orthogonal. Here again, the only adjustment needed as a result of using the orthogonal form of the design matrix instead of the original form is given by (7.2-8).

**Table 7.7** Coordinates of Points of a Three-Variable Composite Design along with the Column of Constant Terms (Design Matrix)

| Design Point $i$ | Constant $\beta_0$ | Independent Variables | | | |
|---|---|---|---|---|---|
| | | $X_{1i}$ | $X_{2i}$ | $X_{3i}$ | |
| 1 | 1 | −1 | −1 | −1 | |
| 2 | 1 | 1 | −1 | −1 | |
| 3 | 1 | −1 | 1 | −1 | |
| 4 | 1 | 1 | 1 | −1 | Eight two-level factorial design points |
| 5 | 1 | −1 | −1 | 1 | |
| 6 | 1 | 1 | −1 | 1 | |
| 7 | 1 | −1 | 1 | 1 | |
| 8 | 1 | 1 | 1 | 1 | |
| 9 | 1 | 0 | 0 | 0 | Center point |
| 10 | 1 | −$\alpha$ | 0 | 0 | |
| 11 | 1 | $\alpha$ | 0 | 0 | |
| 12 | 1 | 0 | −$\alpha$ | 0 | Six axial points at $\alpha$ distance from the center point |
| 13 | 1 | 0 | $\alpha$ | 0 | |
| 14 | 1 | 0 | 0 | −$\alpha$ | |
| 15 | 1 | 0 | 0 | $\alpha$ | |

We may point out that the distance $\alpha$ of the axial points from the center, as determined by (7.4-1), measures the quantitative level of the factors on the standardized scale. The corresponding experimental level can be easily determined through the respective scale and location transformations used for standardizing the experimental levels of each factor.

As an example, let us examine the matrix of a composite design for fitting second-order Model (7.1-3) in $n = 3$ variables. Let us assume here that the basic $2^3$ factorial design and the extra $(2n + 1) = 7$ points are replicated only once; hence, $m = 8$. So there are only $M = 8 + 7 = 15$ observations, one for each design point. To write the design matrix in its orthogonal form, we find from (7.4-1) that $\alpha = 1.215$, and from (7.2-7) that $C = 0.730$. Consequently, the entries in the quadratic-effects columns are either $1 - 0.730 = 0.270$ or $0 - 0.730 = -0.730$. The design matrix $X$ in the orthogonal form is given in Table 7.8. Since $X$ is orthogonal, $X'X$ has nonzero elements only on its diagonal.

**Table 7.8** Design Matrix $X$ in Orthogonal Form for Fitting the Second-Order Model Using $M = 2^n + 2n + 1$ Points of a Composite Design with $n = 3$ and $r = 1$

| Design Point $i$ | $\beta_0$ | $X_{1i}$ | $X_{2i}$ | $X_{3i}$ | $X_{1i}^2$ | $X_{2i}^2$ | $X_{3i}^2$ | $X_{1i}X_{2i}$ | $X_{1i}X_{3i}$ | $X_{2i}X_{3i}$ |
|---|---|---|---|---|---|---|---|---|---|---|
| 1 | 1 | −1 | −1 | −1 | 0.270 | 0.270 | 0.270 | 1 | 1 | 1 |
| 2 | 1 | 1 | −1 | −1 | 0.270 | 0.270 | 0.270 | −1 | −1 | 1 |
| 3 | 1 | −1 | 1 | −1 | 0.270 | 0.270 | 0.270 | −1 | 1 | −1 |
| 4 | 1 | 1 | 1 | −1 | 0.270 | 0.270 | 0.270 | 1 | −1 | −1 |
| 5 | 1 | −1 | −1 | 1 | 0.270 | 0.270 | 0.270 | 1 | −1 | −1 |
| 6 | 1 | 1 | −1 | 1 | 0.270 | 0.270 | 0.270 | −1 | 1 | −1 |
| 7 | 1 | −1 | 1 | 1 | 0.270 | 0.270 | 0.270 | −1 | −1 | 1 |
| 8 | 1 | 1 | 1 | 1 | 0.270 | 0.270 | 0.270 | 1 | 1 | 1 |
| 9 | 1 | 0 | 0 | 0 | −0.730 | −0.730 | −0.730 | 0 | 0 | 0 |
| 10 | 1 | −1.215 | 0 | 0 | 0.746 | −0.730 | −0.730 | 0 | 0 | 0 |
| 11 | 1 | 1.215 | 0 | 0 | 0.746 | −0.730 | −0.730 | 0 | 0 | 0 |
| 12 | 1 | 0 | −1.215 | 0 | −0.730 | 0.746 | −0.730 | 0 | 0 | 0 |
| 13 | 1 | 0 | 1.215 | 0 | −0.730 | 0.746 | −0.730 | 0 | 0 | 0 |
| 14 | 1 | 0 | 0 | −1.215 | −0.730 | −0.730 | 0.746 | 0 | 0 | 0 |
| 15 | 1 | 0 | 0 | 1.215 | −0.730 | −0.730 | 0.746 | 0 | 0 | 0 |
| Diagonal Elements of $X'X$ | | | | | | | | | | |
| | 15 | 10.952 | 10.952 | 10.952 | 4.361 | 4.361 | 4.361 | 8 | 8 | 8 |
| Diagonal Elements of $(X'X)^{-1}$ | | | | | | | | | | |
| | 0.0677 | 0.0913 | 0.0913 | 0.0913 | 0.2293 | 0.2293 | 0.2293 | 0.1250 | 0.1250 | 0.1250 |

**Example 7.4-1**

A study was conducted to investigate the levels of sucrose ($X_1$), dextrose ($X_2$), and corn syrup ($X_3$) with respect to their effects on the microbial population and sensory characteristics of a food product. Initially, a $2^3$ factorial design was planned with each factor in amounts of 4 oz and 12 oz per 100 lb of mix. Later on, however, the design was augmented to have a composite design with 6 axials and 1 center point in addition to the $2^3$ factorial points. For the center point, each factor was at 8 oz per 100 lb of mix. For each factor, the actual amounts of 4, 8, and 12 oz were standardized to $-1$, 0, and 1 levels, respectively. The actual levels 3.14 and 12.86 of each factor for the axial points were determined corresponding to $\alpha = -1.215$ and $\alpha' = 1.215$, respectively, so the design matrix of the second-order response model is orthogonal, as given in Table 7.8.

The response variable was the flavor characteristic. It was nearly impossible for any panelist to accurately judge the flavor characteristic simultaneously for the 15 food samples, one for each design point. Hence, to reduce the number of food samples for simultaneous evaluation, a balanced incomplete block (BIB) design arrangement was used where each panelist judged only 3 food samples. There were 35 panelists, and each sample was evaluated 7 times in the BIB arrangement. Leaving aside details that are of no interest here, we record the means (adjusted for panelist) as observations $Y_1, Y_2, \ldots, Y_{15}$ for the 15 design points, respectively, in the order of design points listed in Table 7.8. The 15 observations are

6.8, 6.0, 4.0, 6.0, 7.0, 6.0, 3.8, 5.9, 7.1, 4.3, 3.1, 3.0, 4.0, 6.9, 6.1.

Note that $\Sigma_i^N Y_i = 80.0$ and $\Sigma_i^N Y_i^2 = 456.8200$. The matrices $(X'X)$ and $(X'X)^{-1}$ are diagonal. Their diagonal elements appear at the bottom of Table 7.8. The following calculations were carried out:

$$
g = X'Y =
\begin{bmatrix}
g_0 \\
g_1 \\
g_2 \\
g_3 \\
g_{11} \\
g_{22} \\
g_{33} \\
g_{12} \\
g_{13} \\
g_{23}
\end{bmatrix}
=
\begin{bmatrix}
80.0000 \\
0.8420 \\
-4.8850 \\
-1.0720 \\
-1.9776 \\
-2.5680 \\
6.2880 \\
5.9000 \\
-0.1000 \\
-0.5000
\end{bmatrix},
$$

$$
b = (X'X)^{-1}X'Y = \begin{bmatrix} b_0 \\ b_1 \\ b_2 \\ b_3 \\ b_{11} \\ b_{22} \\ b_{33} \\ b_{12} \\ b_{13} \\ b_{23} \end{bmatrix} = \begin{bmatrix} 5.3333 \\ 0.0769 \\ -0.4460 \\ -0.0979 \\ -0.4566 \\ -0.5921 \\ 1.4404 \\ 0.7375 \\ -0.0125 \\ -0.0625 \end{bmatrix}.
$$

If SSrl, SSrq, and SSri denote the sums of squares due to the linear, quadratic, and interaction effects, then

$$
\begin{aligned}
\text{SSrl} &= b_1 g_1 + b_2 g_2 + b_3 g_3 \\
&= (0.0769)(0.8420) + (-0.4460) + (-4.8850) + (-0.0979)(1.0720) \\
&= 2.3483 \quad \text{with} \quad 3 \text{ DF},
\end{aligned}
$$

$$
\begin{aligned}
\text{SSrq} &= b_{11} g_{11} + b_{22} g_{22} + b_{33} g_{33} \\
&= 11.4807 \quad \text{with} \quad 3 \text{ DF},
\end{aligned}
$$

$$
\begin{aligned}
\text{SSri} &= b_{12} g_{12} + b_{13} g_{13} + b_{23} g_{23} \\
&= 4.3839 \quad \text{with} \quad 3 \text{ DF},
\end{aligned}
$$

Hence,

$$
\begin{aligned}
\text{SSr} &= \text{SSrl} = \text{SSrq} + \text{SSri} \\
&= 18.2129 \quad \text{with} \quad 9 \text{ DF},
\end{aligned}
$$

Also,

$$
\text{SSto} = \sum_i Y_i^2 - \frac{Y_{\cdot\cdot}^2}{15} = 456.82 - \frac{(80.0)^2}{15}
$$

$$
= 30.1533 \quad \text{with} \quad 14 \text{ DF}.
$$

Hence,

$$
\begin{aligned}
\text{SSe} &= \text{SSto} - \text{SSr} = 30.1533 - 18.2129 \\
&= 11.9404 \quad \text{with} \quad (14 - 9) = 5 \text{ DF}.
\end{aligned}
$$

The analysis of variance is summarized in Table 7.9.

Now to obtain the fitted equation, we first need to find $b_0$, incorporating the adjustment in (7.2-8):

$$
\begin{aligned}
b_0 &= 5.3333 - 0.7300(-0.4566) - 0.7300(-0.5921) \\
&\quad - 0.7300(1.4404) \\
&= 5.0473.
\end{aligned}
$$

**Table 7.9** Analysis of Variance for the Data of the Composite Design in Example 7.4-1

| Source of Variance | DF | SS | MS |
|---|---|---|---|
| Total | 14 | $SSto = 30.1533$ | |
| Model | 9 | $SSr = 18.2129$ | 2.0236 |
| Linear | 3 | $SSrl = 2.3483$ | 0.7828 |
| Quadratic | 3 | $SSrq = 11.4807$ | 3.8269 |
| Interaction | 3 | $SSri = 4.3839$ | 1.4613 |
| Error | 5 | $SSe = 11.9404$ | 2.3881 |

Hence, the fitted second-order model is

$$\widehat{Y} = 5.0473 + 0.0769X_1 - 0.4460X_2 - 0.0979X_3 - 0.4566X_1^2$$
$$- 0.5921X_2^2 + 1.4404X_3^2 + 0.7375X_1X_2 - 0.0125X_1X_3 - 0.0625X_2X_3.$$

Response surface plots are constructed following Section 7.1. Differentiating $\widehat{Y}$ with respect to each independent variable and equating the derivative to zero, we get the following equations:

$$\partial\widehat{Y}/\partial X_1 = 0.0769 - 0.9321X_1 + 0.7375X_2 - 0.0125X_3 = 0;$$

$$\partial\widehat{Y}/\partial X_2 = -0.4460 - 1.1842X_2 + 0.7375X_1 - 0.0625X_3 = 0;$$

$$\partial\widehat{Y}/\partial X_3 = -0.0979 + 2.8808X_3 - 0.0125X_1 - 0.0625X_2 = 0.$$

Solving for $X_1$, $X_2$, and $X_3$, we get $(X_{1,0}, X_{2,0} X_{3,0}) = (0.7750, 0.8564, 0.0559)$ as the stationary point. The corresponding predicted response $\widehat{Y}_0$ is a candidate for the optimum (maximum or minimum) response. We find that $\widehat{Y}_0 = 4.50$. To represent the fitted response surface in its canonical form, we determine the characteristic roots of

$$B = \begin{pmatrix} b_{11} & \frac{1}{2}b_{12} & \frac{1}{2}b_{13} \\ \frac{1}{2}b_{12} & b_{22} & \frac{1}{2}b_{23} \\ \frac{1}{2}b_{13} & \frac{1}{2}b_{23} & b_{33} \end{pmatrix}.$$

To this end, the characteristic equation of $B$ is

$$\begin{vmatrix} -0.4566 - \lambda & 0.3688 & -0.0063 \\ 0.3688 & -0.5921 - \lambda & 0.0313 \\ -0.0063 & 0.0313 & 1.4404 - \lambda \end{vmatrix} = 0.$$

The characteristic equation is a cubic; therefore, there are three roots. They are found to be $\lambda_1 = -0.1497$, $\lambda_2 = 0.8987$, and $\lambda_3 = 1.4401$. As a check, note that

$$\sum_{i-1}^{3} \lambda_i = \sum_{i-1}^{3} b_{ii}.$$

Since the characteristic roots are of different signs, the response surface is saddle shaped (see Figure 7.4), where a maximum or minimum $\hat{Y}_0$ is found at various combinations of the independent variables.

## Composite Designs from Fractional Factorials

Even for composite designs the experimental design points increase rapidly as $n$ increases. For instance, for $n = 5$, 6, and 7, the composite designs would have $M = 43$, 77, and 143 experimental points. Hence, the composite designs desired are those in which the $2^n$ factorial is replaced by a fractional replicate of $2^n$. In keeping with the aims of composite designs, fractional replications should be designed to permit the estimation of all the linear and quadratic effects and some, if not all, the interaction effects in the second-order models. To accomplish this, we need to employ those fractional replicates in which no main effect is confounded with the grand mean. We may recall from Section 6.4 the ways of constructing a fractional replication. From such fractional replicate composite designs, it is always possible to estimate $\beta_0$, all linear effects $\beta_i$, all the quadratic effects $\beta_{ii}$, and one of the interaction effects selected from each of the alias sets. We consider the examples of such a design given by Hartley (1959). Westlake (1965) considered designs based on fractions of factorials, which require even fewer experimental points than those of Hartley.

## A ½ Replicate of $2^4$

The commonly used ½ replicate design of $2^4$ has a defining equation of $ABCD = I$. With this design all the main effects are clear of two-factor interactions. The two-factor interactions are paired in the following three alias pairs:

$$AB = CD, \qquad AC = BD, \qquad AD = BC.$$

Hence, only three of the six interaction effects $\beta_{ij}$ can be estimated if this design is to be used. An attractive ½ fractional replicate design results if $I = ABC$ is the design's defining equation. Then the alias pairs are

$$A = BC, \qquad B = AC, \qquad C = AB, \qquad D = ABCD,$$
$$AD = BCD, \qquad BD = ACD, \qquad CD = ABD.$$

Note that all the two-factor interactions occur in different alias pairs so that all the six $\beta_{ij}$ interaction effects can be estimated, along with all the linear effects $\beta_i$, all the quadratic effects $\beta_{ii}$, and $\beta_0$. In correspondence with the interaction effect $ABC$ in the defining equation, we can evaluate

$$L = X_a + X_b + X_c$$

for each factorial combination (see Section 6.4). All the factorial combinations with $L$ mod $= 0$ constitute a ½ replicate, and the remaining ½ replicate corresponds to $L$ mod $2 = 1$. The two ½ replicates, corresponding to the defining equation $I = ABC$, are

$$L \bmod 2 = 0: \quad (1) \quad ab \quad ac \quad bc \quad d \quad abd \quad acd \quad bcd,$$
$$L \bmod 2 = 1: \quad a \quad b \quad c \quad abc \quad ad \quad bd \quad cd \quad abcd.$$

The ½ replicate of $2^4$ factorial corresponding to $L$ mod $2 = 1$ when augmented to be a composite design is shown in Table 7.10.

## A ¼ Replicate of $2^6$

Again the desired ¼ replication of $2^6$ factorial is the one that would keep all the main effects clear from the two-factor interactions. In such a replicate, however, some two-factor interactions are confounded with other two-factor interactions. Such a replicate when augmented to be a composite design would not permit the estimation of all the two-factor interactions. However, a ¼ replicate of $2^6$ factorial corresponding to the following defining equations does permit the estimation of all the two-factor interactions:

$$I = ABC, \qquad I = DEF.$$

Following the procedure outlined in Section 6.4, we can partition $2^6$ factorial combinations into four ¼ replicates. Table 7.11 lists 16 points of a ¼ replicate of $2^6$ factorial. These 16 points can be augmented by $2n = 12$ axial points and 1 center point to have a composite design.

The ½ replicate of a $2^5$ factorial given in Table 7.12 is such that, when augmented by $(2n + 1) = 11$ points to form a composite design, it

**Table 7.10** The Design Matrix of a $\frac{1}{2}$ Replicate of $2^4$ Factorial for Augmented Composite Design

| Factorial Combination | Design Point | Constant $\beta_0$ | $X_{1i}$ | $X_{2i}$ | $X_{3i}$ | $X_{4i}$ | |
|---|---|---|---|---|---|---|---|
| a | 1 | 1 | 1 | −1 | −1 | −1 | |
| b | 2 | 1 | −1 | 1 | 1 | −1 | |
| c | 3 | 1 | −1 | −1 | 1 | −1 | |
| abc | 4 | 1 | 1 | 1 | 1 | −1 | $\frac{1}{2}$ replicate |
| ad | 5 | 1 | 1 | −1 | −1 | 1 | |
| bd | 6 | 1 | −1 | 1 | −1 | 1 | |
| cd | 7 | 1 | −1 | −1 | 1 | 1 | |
| abcd | 8 | 1 | 1 | 1 | 1 | 1 | |
| | 9 | 1 | 0 | 0 | 0 | 0 | Center point |
| | 10 | 1 | −$\alpha$ | 0 | 0 | 0 | |
| | 11 | 1 | $\alpha$ | 0 | 0 | 0 | |
| | 12 | 1 | 0 | −$\alpha$ | 0 | 0 | |
| | 13 | 1 | 0 | $\alpha$ | 0 | 0 | |
| | 14 | 1 | 0 | 0 | −$\alpha$ | 0 | Axial points |
| | 15 | 1 | 0 | 0 | $\alpha$ | 0 | |
| | 16 | 1 | 0 | 0 | 0 | −$\alpha$ | |
| | 17 | 1 | 0 | 0 | 0 | $\alpha$ | |

**Table 7.11** A $\frac{1}{4}$ Replicate of $2^6$ Factorial Resulting in 16 Experimental Points and the Constant Column

| Factorial Combination | Design Point $i$ | Constant $\beta_0$ | $X_{1i}$ | $X_{2i}$ | $X_{3i}$ | $X_{4i}$ | $X_{5i}$ | $X_{6i}$ |
|---|---|---|---|---|---|---|---|---|
| ad | 1 | 1 | 1 | −1 | −1 | 1 | −1 | −1 |
| bd | 2 | 1 | −1 | 1 | −1 | 1 | −1 | −1 |
| cd | 3 | 1 | −1 | −1 | 1 | 1 | −1 | −1 |
| abcd | 4 | 1 | 1 | 1 | 1 | 1 | −1 | −1 |
| ae | 5 | 1 | 1 | −1 | −1 | −1 | 1 | −1 |
| be | 6 | 1 | −1 | 1 | −1 | −1 | 1 | −1 |

*(continued)*

**Table 7.11** A $\frac{1}{4}$ Replicate of $2^6$ Factorial Resulting in 16 Experimental Points and the Constant Column—*cont...*

| Factorial Combination | Design Point $i$ | Constant $\beta_0$ | $X_{1i}$ | $X_{2i}$ | $X_{3i}$ | $X_{4i}$ | $X_{5i}$ | $X_{6i}$ |
|---|---|---|---|---|---|---|---|---|
| ce | 7 | 1 | −1 | −1 | 1 | −1 | 1 | −1 |
| abce | 8 | 1 | 1 | 1 | 1 | −1 | 1 | −1 |
| af | 9 | 1 | 1 | −1 | −1 | −1 | −1 | 1 |
| bf | 10 | 1 | −1 | 1 | −1 | −1 | −1 | 1 |
| cf | 11 | 1 | −1 | −1 | 1 | −1 | −1 | 1 |
| abcf | 12 | 1 | 1 | 1 | 1 | −1 | −1 | 1 |
| adef | 13 | 1 | 1 | −1 | −1 | 1 | 1 | 1 |
| bdef | 14 | 1 | −1 | 1 | −1 | 1 | 1 | 1 |
| cdef | 15 | 1 | −1 | −1 | 1 | 1 | 1 | 1 |
| abcdef | 16 | 1 | 1 | 1 | 1 | 1 | 1 | 1 |

**Table 7.12** A $\frac{1}{2}$ Replicate of $2^5$ Factorial Resulting in 16 Experimental Points and the Constant Column

| Factorial Combination | Point $i$ | Constant $\beta_0$ | $X_{1i}$ | $X_{2i}$ | $X_{3i}$ | $X_{4i}$ | $X_{5i}$ |
|---|---|---|---|---|---|---|---|
| a | 1 | 1 | 1 | −1 | −1 | −1 | −1 |
| b | 2 | 1 | −1 | 1 | −1 | −1 | −1 |
| c | 3 | 1 | −1 | −1 | 1 | −1 | −1 |
| abc | 4 | 1 | 1 | 1 | 1 | −1 | −1 |
| d | 5 | 1 | −1 | −1 | −1 | 1 | −1 |
| abd | 6 | 1 | 1 | 1 | −1 | 1 | −1 |
| acd | 7 | 1 | 1 | −1 | 1 | 1 | −1 |
| bcd | 8 | 1 | −1 | 1 | 1 | 1 | −1 |
| e | 9 | 1 | −1 | −1 | −1 | −1 | 1 |
| abe | 10 | 1 | 1 | 1 | −1 | −1 | 1 |
| ace | 11 | 1 | 1 | −1 | 1 | −1 | 1 |
| bce | 12 | 1 | −1 | 1 | 1 | −1 | 1 |
| ade | 13 | 1 | 1 | −1 | −1 | 1 | 1 |
| bde | 14 | 1 | −1 | 1 | −1 | 1 | 1 |
| cde | 15 | 1 | −1 | −1 | 1 | 1 | 1 |
| abcde | 16 | 1 | 1 | 1 | 1 | 1 | 1 |

will permit estimation of $\beta_0$ and all the linear, quadratic, and interaction effects. For statistical analysis, the design matrices of the composite designs resulting from fractional replications of $2^n$ factorials can be orthogonalized as outlined in the earlier sections.

## 7.5 ROTATABLE DESIGNS

The predictive power of a fitted response model at various points in the experimental region depends on the design being used. As stated in Section 7.3, it may be desirable to have designs that predict uniformly at all constant distances from their center points. Such designs are called *rotatable designs*. For rotatable designs the variances and covariances of the estimated coefficients in the fitted model remain unchanged when the design points are rotated about its center. The first-order design for estimating a first-order model with a center point shown in Figure 7.7 is rotatable. The simplest second-order rotatable designs for estimating second-order models in two variables are provided by the vertices of a pentagon, a hexagon, and an octagon, each with a center point, as illustrated in Figure 7.11. To provide a good predictability toward the center of the design, the center point is replicated more frequently than the peripheral points. Table 7.13 lists the coordinates of points for the pentagon, hexagon, and octagon designs. As shown by these coordinates, the peripheral points are equidistant from the design center to satisfy rotatability. Also note that the number of levels of the independent variables is fixed by the design. For instance, the hexagon design requires five levels of variable $X_1$ and three of variable $X_2$. The usefulness of these designs lies in the fact that they have a relatively small number of experimental points, which allows the fitting of the second-order models and also provides a measure of experimental error.

■ **FIGURE 7.11** Three two-factor rotatable designs.

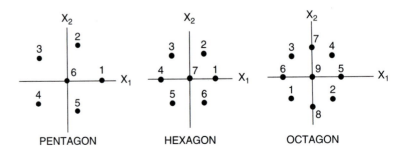

PENTAGON HEXAGON OCTAGON

**Table 7.13** Experimental Design Points of Some Two-Factor Second-Order Rotatable Designs

| Design Point | $X_1$ | $X_2$ | |
|---|---|---|---|
| Pentagon Design | | | |
| 1 | 1.000 | 0 | Five levels of each variable |
| 2 | 0.309 | 0.951 | |
| 3 | −0.809 | 0.588 | |
| 4 | −0.809 | −0.588 | |
| 5 | 0.309 | −0.951 | |
| 6 | 0 | 0 | |
| Hexagon Design | | | |
| 1 | 1.000 | 0 | Five levels of $X_1$ and three levels of $X_2$ |
| 2 | 0.500 | 0.866 | |
| 3 | −0.500 | 0.866 | |
| 4 | −1.000 | 0 | |
| 5 | −0.500 | −0.886 | |
| 6 | 0.500 | −0.866 | |
| 7 | 0 | 0 | |
| Octagon Design | | | |
| 1 | −1 | −1 | Five levels of each variable |
| 2 | 1 | −1 | |
| 3 | −1 | 1 | |
| 4 | 1 | 1 | |
| 5 | 1.4142 | 0 | |
| 6 | −1.4142 | 0 | |
| 7 | 0 | 1.4142 | |
| 8 | 0 | −1.4142 | |
| 9 | 0 | 0 | |

**Example 7.5-1**

This example is taken from a review paper by Hunter (1959). The purpose of the experiment is to explore the relationship of the yield $Y$ of a chemical process to the temperature $X_1$ of the reacting mass and to the rate $X_2$ of the addition of reagents. The area of interest is in a temperature interval of 15 to 75°C, and a time interval of 1 to 3 hours. Using a hexagon design, the actual and the standardized levels for each independent variable are shown as follows:

| Temperature (°C), $X_1$: | 15 | 30 | 45 | 60 | 75 |
|---|---|---|---|---|---|
| Standardized levels of $X_1$: | −1 | −0.5 | 0 | 0.5 | 1 |
| Time (hour), $X_2$: | | 1.134 | 2 | 2.866 | |
| Standardized levels of $X_2$: | | −0.866 | 0 | 0.866 | |

The center point is replicated four times, whereas all the other points of the hexagon design are replicated only once. For fitting a second-order response model, the matrix of this design is written as follows. We subtract $C$, that is,

$$C = \sum_i \frac{X_{1i}^2}{M} = \sum_i \frac{X_{2i}^2}{M},$$

from each $X_{1i}^2$ and $X_{2i}^2$ to obtain the columns corresponding to the quadratic effects, where $M$ is the number of rows of the design matrix. This matrix, in the form of a hexagon design for fitting a second-order model, is given in Table 7.14. We may point out that the matrix is not completely orthogonal since the inner product of the $(X_{1i}^2 - C)$ and $(X_{2i}^2 - C)$ columns is not zero.

The last column of the table lists the observed responses, one for each design point.

The following calculations lead to estimates of the coefficients and the various sums of squares:

$$X'X = \begin{bmatrix} 10 & 0 & 0 & 0 & 0 & 0 \\ 0 & 3 & 0 & 0 & 0 & 0 \\ 0 & 0 & 3 & 0 & 0 & 0 \\ 0 & 0 & 0 & 1.35 & 0.15 & 0 \\ 0 & 0 & 0 & 0.15 & 1.35 & 0 \\ 0 & 0 & 0 & 0 & 0 & 0.75 \end{bmatrix},$$

**Table 7.14** Design Matrix of a Second-Order Rotatable Hexagon Design with Center Point Replicated Four Times[a]

| Point $i$ | Constant $\beta_0$ | $X_{1i}$ | $X_{2i}$ | $(X_{1i}^2 - C)$ | $(X_{2i}^2 - C)$ | $X_{1i}X_{2i}$ | $Y_i$ |
|---|---|---|---|---|---|---|---|
| 1 | 1 | 1.0 | 0 | 0.70 | −0.30 | 0 | 96.0 |
| 2 | 1 | 0.5 | 0.8667 | −0.05 | 0.45 | 0.4335 | 78.7 |
| 3 | 1 | −0.5 | 0.8667 | −0.05 | 0.45 | −0.4335 | 76.7 |
| 4 | 1 | −1.0 | 0 | 0.70 | −0.30 | 0 | 54.6 |
| 5 | 1 | −0.5 | −0.8667 | −0.05 | 0.45 | 0.4335 | 64.8 |
| 6 | 1 | 0.5 | −0.8667 | −0.05 | 0.45 | −0.4335 | 78.9 |
| 7 | 1 | 0 | 0 | −0.30 | −0.30 | 0 | 97.4 |
| 8 | 1 | 0 | 0 | −0.30 | −0.30 | 0 | 90.5 |
| 9 | 1 | 0 | 0 | −0.30 | −0.30 | 0 | 93.0 |
| 10 | 1 | 0 | 0 | −0.30 | −0.30 | 0 | 86.3 |

[a]*Total number of observations is M = 10.*
*Note:*
$\Sigma_i Y_i = 816.90;\ \Sigma_{i\ x1i}^2 = \Sigma_i X_{2i}^2 = 1.35;\ \Sigma_i Y_i^2 = 68{,}471.69.$
$C = \Sigma_i X_{1i}^2/10 = \Sigma_i X_{2i}^2/10 = 0.30.$

$$(X'X)^{-1} = \begin{bmatrix} 0.10 & 0 & 0 & 0 & 0 & 0 \\ 0 & 0.3334 & 0 & 0 & 0 & 0 \\ 0 & 0 & 0.3334 & 0 & 0 & 0 \\ 0 & 0 & 0 & 0.7500 & 0.0834 & 0 \\ 0 & 0 & 0 & 0.0834 & 0.7500 & 0 \\ 0 & 0 & 0 & 0 & 0 & 1.3334 \end{bmatrix},$$

$$g = X'Y = \begin{bmatrix} g_0 \\ g_1 \\ g_2 \\ g_{11} \\ g_{22} \\ g_{12} \end{bmatrix} = \begin{bmatrix} 816.9000 \\ 49.4500 \\ 10.1322 \\ -19.6950 \\ -20.7450 \\ -5.2393 \end{bmatrix},$$

$$b = (X'X)^{-1}X'Y = \begin{bmatrix} b_0 \\ b_1 \\ b_2 \\ b_{11} \\ b_{22} \\ b_{12} \end{bmatrix} = \begin{bmatrix} 81.6900 \\ 16.4833 \\ 3.3776 \\ -16.500 \\ -17.2000 \\ -6.9862 \end{bmatrix}.$$

The various sums of squares are as follows:

$$SSto = \Sigma Y_i^2 - \frac{(\Sigma_i Y_i)^2}{M} = 68,471.69 - \frac{(816.90)^2}{10}$$

$$= 1739.1290 \quad \text{with} \quad 9 \text{ DF}.$$

Since only the center point is replicated four times, we can obtain the sum of squares due to the experimental errors from the corresponding four responses. That is,

$$SSexp = (97.4)^2 + (90.5)^2 + (93.0)^2 + (86.3)^2 - \tfrac{1}{4}(367.20)^2$$

$$= 64.74 \quad \text{with} \quad 3 \text{ DF}.$$

The sums of squares due to the linear, quadratic, and interaction components of the fitted model are

$$SSrl = b_1 g_1 + b_2 g_2 = 849.3217 \quad \text{with} \quad 2 \text{ DF},$$
$$SSrq = b_{11} g_{11} + b_{22} g_{22} = 681.7815 \quad \text{with} \quad 2 \text{ DF},$$
$$SSri = b_{12} g_{12} = 36.6028 \quad \text{with} \quad 1 \text{ DF},$$

respectively. The sum of squares due to the regression model is

$$SSr = SSrl + SSrq + SSri$$
$$= 1567.7060 \quad \text{with} \quad 5 \text{ DF}.$$

The error sum of square is

$$SSe = SSto - SSr = 171.4230 \quad \text{with} \quad 4\text{DF},$$

and the sum of squares due to the lack of fit is

$$SSf = SSe - SSexp = 106.6830 \quad \text{with} \quad 1 \text{ DF}.$$

The analysis of variance of this example is given in Table 7.15.

To test the adequacy of the model, we compute the $F$ ratio

$$F = Msf/MSexp$$
$$= 106.6830/21.5800 = 4.94,$$

which is not statistically significant at the 5% level. Following this information, if the second-order model is considered adequate to describe the response surface, then the tests for the linear, quadratic, and interaction effects may be further carried out.

**Table 7.15** Analysis of Variance of Responses Resulting from the Design in Table 7.14

| Source of Variance | DF | SS | MS | F Ratio |
|---|---|---|---|---|
| Total | 9 | SSto = 1,739.1290 | | |
| Model | 5 | SSr = 1,567.7060 | | |
| Linear ($b_1$, $b_2$) | 2 | SSrl = 849.3217 | | |
| Quadratic ($b_{11}$, $b_{22}$) | 2 | SSrq = 681.7815 | | |
| Interaction ($b_{12}$) | 1 | SSri = 36.6028 | | |
| Error | 4 | SSe = 171.4230 | 42.8558 | |
| Lack of Fit | 1 | SSf = 106.6830 | 106.6830 | 4.94 |
| Experimental Error | 3 | SSexp = 64.7400 | 21.5800 | |

Now let us write the fitted equation. Since we wrote the matrix columns for the quadratic effects as $(X_{1i}^2 - C)$ and $(X_{2i}^2 - C)$, where $C = 0.30$, we incorporate the adjustment in (7.2-8) to find $b_0$. Hence,

$$b_0 = 81.69 - 0.30(-16.5000) - 0.30(-17.2000) = 91.80.$$

The fitted second-order model is

$$\widehat{Y} = 91.80 + 16.4833X_1 + 3.3776X_2 - 16.5000X_1^2$$
$$- 17.2000X_2^2 - 6.9862X_1X_2.$$

When we use MSexp $= 21.58$ as an estimate of $\sigma^2$, the variances and the covariances of the coefficients of the regression equation are given by (7.2-4). Thus, var($b_1$) $=$ var($b_2$) $= \sigma^2(0.3334)$ is estimated by (21.5800) (0.3334) $= 7.1926$. Also, var($b_{11}$) and var($b_{22}$) are estimated by (21.5800) (0.7500) $= 16.1850$ and 21.5800(1.3334) $= 28.7748$, respectively.

To determine an optimum time–temperature combination, if it exists, where the response is stationary (maximum or minimum), we differentiate the fitted equation with respect to $X_1$ and $X_2$ and equate to zero. That is,

$$\frac{\partial \widehat{Y}}{\partial X_1} = 16.4833 - 33.000X_1 - 6.9862X_2 = 0,$$

$$\frac{\partial \widehat{Y}}{\partial X_2} = 3.3776 - 6.9862X_1 - 34.4000X_2 = 0.$$

The solution of the preceding equations is $X_{1,0} = 0.50$ and $X_{2,0} = 0.01$. Hence, the possible optimum response is obtained at the stationary point

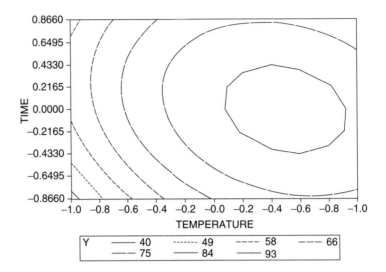

■ **FIGURE 7.12** Response surface contours of the second-order fitted response model of Example 7.5-1.

(0.50, 0.01). In terms of the actual levels, (0.50, 0.01) translates into (60°C, 2 hours). Further calculus methods may be used to determine whether (60°C, 2 hours) is a maximum or minimum point, or neither. For this example, we chose to draw the response surface contours to study the response surface. The response surface contours are constructed as shown in Figure 7.12 for $\hat{Y} = 40, 49, \ldots, 93$. The plots clearly indicate that the maximum response is obtained at 60°C and at 2-hour intervals of reagent addition.

## Composite Rotatable Designs

The most widely used class of designs for estimating second-order response surfaces is that of composite designs discussed in Section 7.4. These designs can be made rotatable by an appropriate choice of $\alpha$, the distance of the axial points from the center of the design. The octagon design in Figure 7.11 is an example of a composite rotatable design with $\alpha = \sqrt{2} = 1.4142$ to satisfy rotatability. Let us consider a composite design with its points as follows:

1. There are $n_c$ points: $(\pm 1, \pm 1, \ldots, \pm 1)$, which follow the familiar $2^n$ factorial or a suitable fractional replicate of the $2^n$ factorial.
2. There are $2n$ axial points: $(\pm \alpha, 0, 0, \ldots, 0)$, $(0, \pm \alpha, 0, \ldots, 0), \ldots, (0, 0, 0, \ldots, \pm \alpha)$.
3. There is a center point: $(0, 0, \ldots, 0)$, which is replicated $n_0$ times. Hence, there are $M = n_c + 2n + n_0$ total design points.

For the preceding composite design to be rotatable, it is necessary that $\alpha = n_c^{1/4}$. Remember that the experimental levels are so standardized that the $n_c$ points of the $2^n$ factorial (or a suitable fraction thereof) have coordinates equal to $+1$ or $-1$. If $X_{ij}$ denotes the actual $i$th coordinate of the $j$th point, $i = 1,\ldots,n$, and $j = 1,\ldots,M$, then the standardized coordinate $U_{ij}$ is

$$U_{ij} = (X_{ij} - \overline{X}_{i.})/S_i,$$

where

$$S_i = \left[\frac{\sum_{j=1}^{M}(X_{ij} - \overline{X}_{i.})^2}{n_c + 2(n_c)^{1/2}}\right]^{1/2}.$$

When fitting a second-order response model, one can make the resulting matrix of the composite rotatable designs orthogonal if the center point is repeated a certain number of times. Table 7.16 lists some second-order composite rotatable designs satisfying orthogonality. For instance, if a ½

**Table 7.16** Second-Order Composite Rotatable Designs Satisfying Orthogonality[a]

| Number of Variables $n$ | Factorial Points $n_c$ | Axial Points $2n$ | Center Point Repeated $n_0$ Times | Total Number of Observations $M$[b] | $\alpha = n_c^{1/4}$ |
|---|---|---|---|---|---|
| 2 | 4 | 4 | 8 | 16 | 1.414 |
| 3 | 8 | 6 | 9 | 23 | 1.682 |
| 4 | 16 | 8 | 12 | 36 | 2.000 |
| 5 | 32 | 10 | 17 | 59 | 2.378 |
| 5 ($\frac{1}{2}$ fraction) | 16 | 10 | 10 | 36 | 2.000 |
| 6 | 64 | 12 | 24 | 100 | 2.828 |
| 6 ($\frac{1}{2}$ fraction) | 32 | 12 | 15 | 59 | 2.378 |
| 7 | 124 | 14 | 35 | 177 | 3.364 |
| 7 ($\frac{1}{2}$ fraction) | 64 | 14 | 22 | 100 | 2.828 |
| 8 | 256 | 16 | 52 | 324 | 4.000 |
| 8 ($\frac{1}{2}$ fraction) | 128 | 16 | 33 | 177 | 3.364 |
| 8 ($\frac{1}{4}$ fraction) | 64 | 16 | 20 | 100 | 2.828 |

[a]Source: Box and Hunter (1957).
[b]$M = n_c + 2n + n_0$.

replicate of a $2^5$ factorial is to be augmented to have a second-order composite rotatable design with orthogonal matrix, then $n_c = \frac{1}{2}\,(2^5) = 16$, $2n = 10$, $n_0 = 10$, $M = n_c + 2n + n_0 = 36$, and $\alpha = 16^{1/4} = 2$.

## Arrangements of Composite Designs in Blocks

For greater accuracy and also for practical reasons, the designs are arranged in blocks that are smaller in size than the number of experimental design points. A block, for example, may consist of only experimental points that can be performed in only one trial of the experiment. The composite designs can be easily carried out in orthogonal blocking arrangements (DeBaun, 1956; Box and Hunter, 1957). The orthogonal blocking arrangements provide for the coefficients of the response model to be independent of the block differences. The composite designs consist of $n_c$ points at the vertices design of a cube, corresponding to a $2^n$ factorial or a suitable fraction thereof, together with $2n$ axial points and a center point replicated $n_0$ times. The set of design points at the vertices of the cube and the set of axial points are each a first-order rotatable design. These two sets of points of the composite designs are assigned to different blocks. The center point is repeated in each block as often as is necessary to have orthogonal blocking. Let $n_{c0}$ and $n_{a0}$ denote the number of replications of the center point assigned to the cubic part and the axial part, respectively. Then, for orthogonal blocking of the composite designs, it is required that (see Box and Hunter, 1957)

$$\alpha = \left[ \frac{n_c(2n + n_{a0})}{2(n_c + n_{c0})} \right]^{1/2}.$$

Now remember that for rotatability of the composite designs, we must have $\alpha = n_c^{1/4}$. Hence, for a composite design to be rotatable and also for it to be arranged in orthogonal blocks, it is required that

$$n_c^{1/2} = 2\left[ \frac{(n_c + n_{c0})}{(2n + n_{a0})} \right] \tag{7.5-1}$$

hold. Because $n_c$, $n_{c0}$, $2n$, and $n_{a0}$ are all integers, (7.5-1) will not be true, in general. Hence, rotatability and, at the same time, orthogonality between the block effects and the coefficients of the second-order model are not always possible to obtain. One can, however, choose $n_0$ and subdivide it into $n_{c0}$ and $n_{a0}$ so that either rotatability or orthogonality holds exactly and the other holds as nearly as possible. In practice, it is simpler to have $\alpha$ for exact orthogonality and to get as close as possible to its value for rotatability.

One cannot further arrange the axial points into smaller blocks without destroying the rotatability. But the cube points of $2^n$ factorials or fractions, of $2^n$ factorials can be further arranged in smaller blocks following a confounding system that confounds only third- or higher order interaction effects with the blocks without destroying rotatability. If, following a desired confounding system, the cube points are arranged in smaller blocks, each of the same size, then the center point is replicated in each block an equal number of times to maintain orthogonality.

Table 7.17 shows blocking arrangements for some composite designs. For instance, if $n = 2$, there are four cubic points and four axial points, together with $n_0 = 6$ replications of the center point. The center point is replicated an equal number of times in the cubic and axial blocks so that restriction (7.5-1) for rotatability of the design and also orthogonal blocking is satisfied. When (7.5-1) holds, $\alpha = 1.4142$ for rotatability coincides with the $\alpha$ value for orthogonal blocking. Table 7.18 shows the blocking of the points of a composite design when $n = 2$.

**Table 7.17** Blocking Arrangements for Rotatable and Near-Rotatable Composite Designs[a]

| Number of Variables | 2 | 3 | 4 | 5 | 5 ($\frac{1}{2}$ fraction) | 6 | 6 ($\frac{1}{2}$ fraction) | 7 | 7 ($\frac{1}{2}$ fraction) |
|---|---|---|---|---|---|---|---|---|---|
| Cubic Points of $2^n$ (or Suitable Fraction) Factorial | | | | | | | | | |
| $n_c$ | 4 | 8 | 16 | 32 | 16 | 64 | 32 | 128 | 64 |
| Number of blocks | 1 | 2 | 2 | 4 | 1 | 8 | 2 | 16 | 8 |
| Number of points in blocks | 4 | 4 | 8 | 8 | 16 | 8 | 16 | 8 | 8 |
| Number of added center points, $n_{c0}$ | 3 | 4 | 4 | 8 | 6 | 8 | 8 | 16 | 8 |
| Total number of points in blocks, $n_c + n_{c0}$ | 7 | 6 | 10 | 10 | 22 | 9 | 20 | 9 | 9 |

*(continued)*

**Table 7.17** Blocking Arrangements for Rotatable and Near-Rotatable Composite Designs[a]—cont...

| Number of Variables | 2 | 3 | 4 | 5 | 5 ($\frac{1}{2}$ fraction) | 6 | 6 ($\frac{1}{2}$ fraction) | 7 | 7 ($\frac{1}{2}$ fraction) |
|---|---|---|---|---|---|---|---|---|---|
| Axial block | | | | | | | | | |
| $2n$ | 4 | 6 | 8 | 10 | 10 | 12 | 12 | 14 | 14 |
| Number of added points, $n_{a0}$ | 3 | 2 | 2 | 4 | 1 | 6 | 2 | 11 | 4 |
| Total number of points | 7 | 8 | 10 | 14 | 11 | 18 | 14 | 25 | 18 |
| Grand total points in the design, $M$ | 14 | 20 | 30 | 54 | 33 | 90 | 54 | 169 | 80 |
| Value of $\alpha$ for orthogonal blocking | 1.4142 | 1.6330 | 2.0000 | 2.3664 | 2.0000 | 2.8284 | 2.3664 | 3.3636 | 2.8284 |
| Value of $\alpha$ for rotatability | 1.4142 | 1.6818 | 2.0000 | 2.3784 | 2.0000 | 2.8284 | 2.3784 | 3.3333 | 2.8284 |

[a]Source: Box and Hunter (1959).

**Table 7.18** Orthogonal Blocking of Points of a Composite Rotatable Design with Two Factors

| Blocks | Points | $X_1$ | $X_2$ | |
|---|---|---|---|---|
| I (Cubic Block) | 1 | −1 | −1 | |
| | 2 | 1 | −1 | $n_c$ |
| | 3 | −1 | 1 | |
| | 4 | 1 | 1 | |
| | 5 | 0 | 0 | |
| | 6 | 0 | 0 | $n_{c0}$ |
| | 7 | 0 | 0 | |
| | 8 | −$\alpha$ | 0 | |
| II (Axial Block) | 9 | $\alpha$ | 0 | $2n$ |
| | 10 | 0 | −$\alpha$ | |
| | 11 | 0 | $\alpha$ | |
| | 12 | 0 | 0 | |
| | 13 | 0 | 0 | $n_{a0}$ |
| | 14 | 0 | 0 | |

We also consider the blocking arrangements for the composite design when $n = 3$. In this design there are 8 cubic points and 6 axial points. For this design it is not possible to find values $n_{ao}$ and $n_{co}$ so that (7.5-1) holds. Hence, it is not possible to achieve orthogonal blocking and at the same time exact rotatability. However, exact orthogonal blocking and near-rotatability can be attained if $n_{co} = 4$ and $n_{ao} = 2$. For these choices of $n_{co}$ and $n_{ao}$, the $\alpha$ value for orthogonal blocking is 1.6330, which is close to its value 1.6818 for exact rotatability. The $n_c + n_{co} = 8 + 4 = 12$ points in the cubic set are arranged in two blocks of equal size six, confounding the highest order interaction effect $ABC$ with the blocks. The axial block consists of $2n + n_{ao} = 6 + 2 = 8$ points. There are two blocks of equal size, six for the cubic portion, and one block of size eight for the axial portion (see Table 7.19).

The block arrangements of the composite designs for $n > 3$ factors can be developed similarly. If a design is arranged in $b$ blocks, then the sum of squares due to the blocks is

$$\text{SSbl} = \sum_{i=1}^{b} \frac{B_i^2}{n_i} - \text{CF}, \qquad (7.5\text{-}2)$$

where $B_i$ and $n_i$ represent the total of the $i$th block and its size, respectively. Let $Z_{i1}, Z_{i2}, \ldots, Z_{in_{i0}}$ denote the $Y_{ij}$ observations corresponding to the $n_{i0}$ center points replicated in the $i$th block. Note that

$$\sum_{i=1}^{b} n_{i0} = n_0.$$

Then the sum of squares due to the experimental error is given by

$$\text{SSexp} = \sum_{i=1}^{b} \left[ \sum_{j=1}^{n_{i0}} Z_{ij}^2 - n_0 \overline{Z}_{..}^2 \right]. \qquad (7.5\text{-}3)$$

For the analysis of variance of these designs, see Table 7.20.

**Table 7.19** Orthogonal Blocking of a Three-Factor Composite Design with *ABC* Interaction Confounded with Blocks I and II

| Blocks | Constant $\beta_0$ | $X_{1i}$ | $X_{2i}$ | $X_{3i}$ | |
|--------|--------------------|----------|----------|----------|---|
| I | 1 | −1 | −1 | −1 | |
| | 1 | 1 | 1 | −1 | $n_c$ |
| | 1 | 1 | −1 | 1 | |
| | 1 | −1 | 1 | 1 | |
| | 1 | 0 | 0 | 0 | Two center points in block I, $n_{c0}$ |
| | 1 | 0 | 0 | 0 | |
| II | 1 | 1 | −1 | −1 | |
| | 1 | −1 | 1 | −1 | $n_c$ |
| | 1 | −1 | −1 | 1 | |
| | 1 | 1 | 1 | 1 | |
| | 1 | 0 | 0 | 0 | Two center points in block II, $n_{c0}$ |
| | 1 | 0 | 0 | 0 | |
| III | 1 | −α | 0 | 0 | |
| | 1 | α | 0 | 0 | |
| | 1 | 0 | −α | 0 | $2n$ |
| | 1 | 0 | α | −0 | |
| | 1 | 0 | 0 | −α | |
| | 1 | 0 | 0 | α | |
| | 1 | 0 | 0 | 0 | Two center points in axial block, $n_{a0}$ |
| | 1 | 0 | 0 | 0 | |

**Table 7.20** Analysis of Variance of the Composite Designs Arranged in $b$ Blocks for Fitting Second-Order Response Model

| Source of Variance | DF | SS |
|---|---|---|
| Total | $M - 1$ | $SSto = \sum_i \sum_j Y_{ij}^2 - CF$ |
| Model | $n(n + 3)/2$ | $SStr = SSrl + SSrq +$ $SSri$ |
| Linear | $n$ | $SSrl = \sum_i g_i b_i$ |
| Quadratic | $n$ | $SSrq = \sum_i g_{ii} b_{ii}$ |
| Interaction | $n(n - 1)/2$ | $SSri = \sum_i g_{ij} b_{ij}$ |
| Error | $M - n(n + 3)/2$ | $SSe = SSto - SStr$ |
| Blocks | $b - 1$ | SSbl from (7.5-2) |
| Lack of Fit | By difference | $SSf = SSe - SSbl -$ $SSexp$ |
| Experimental Error | $n_0 - b$ | SSexp from (7.5-3) |

**Example 7.5-2**

Hot dogs must have a particular color for consumer acceptance. Three colorants A, B, and C, known to impart a desirable color, were studied, using a composite rotatable design. For logistic reasons the design was arranged in orthogonal blocks following the arrangement given in Table 7.17 for three variables. Three variables $X_1$, $X_2$, and $X_3$ denote the levels of factors A, B, and C, respectively. Note that each variable has five levels, $\pm 1$, $\pm \alpha$, and 0, corresponding to the cubic, axial, and center points. The $\alpha$ value for the axial point is 1.633. The design matrix for fitting the second-order model is given by Table 7.21. The quadratic-effects columns were made orthogonal to the constant column by subtracting $C$ from $X_{1i}^2$, $X_{2i}^2$, and $X_{3i}^2$ for each $i$, where

$$C = \sum_{i=1}^{20} \frac{X_{1i}^2}{M} = \sum_{i=1}^{20} \frac{X_{2i}^2}{M} = \sum_{i=1}^{20} \frac{X_{3i}^2}{M} = \frac{13.3334}{20} = \frac{2}{3}.$$

The table also gives the average score $Y$ of 12 color experts. An 8-point hedonic scale was used with "1 = dislike extremely" and "8 = like extremely." The color evaluation for each experimental point within a block was done at random.

**Table 7.21** Design Matrix and Observation Vector Y for Example 7.5-2

| Blocks | Constant $\beta_0$ | $X_{1i}$ | $X_{2i}$ | $X_{3i}$ | $X_{1i}^2 - c$ | $X_{2i}^2 - c$ | $X_{3i}^2 - c$ | $X_{1i}X_{2i}$ | $X_{1i}X_{3i}$ | $X_{2i}X_{3i}$ | $Y_i$ |
|---|---|---|---|---|---|---|---|---|---|---|---|
| I | 1 | -1 | -1 | -1 | $\frac{1}{3}$ | $\frac{1}{3}$ | $\frac{1}{3}$ | 1 | 1 | 1 | 5.1 |
| | 1 | 1 | 1 | -1 | $\frac{1}{3}$ | $\frac{1}{3}$ | $\frac{1}{3}$ | 1 | -1 | -1 | 4.7 |
| | 1 | 1 | -1 | 1 | $\frac{1}{3}$ | $\frac{1}{3}$ | $\frac{1}{3}$ | -1 | 1 | -1 | 4.0 |
| | 1 | -1 | 1 | 1 | $\frac{1}{3}$ | $\frac{1}{3}$ | $\frac{1}{3}$ | -1 | -1 | 1 | 6.6 |
| | 1 | 0 | 0 | 0 | $-\frac{2}{3}$ | $-\frac{2}{3}$ | $-\frac{2}{3}$ | 0 | 0 | 0 | 6.0 |
| | 1 | 0 | 0 | 0 | $-\frac{2}{3}$ | $-\frac{2}{3}$ | $-\frac{2}{3}$ | 0 | 0 | 0 | 5.9 |
| II | 1 | 1 | -1 | -1 | $\frac{1}{3}$ | $\frac{1}{3}$ | $\frac{1}{3}$ | -1 | -1 | 1 | 5.0 |
| | 1 | -1 | 1 | -1 | $\frac{1}{3}$ | $\frac{1}{3}$ | $\frac{1}{3}$ | -1 | 1 | -1 | 5.4 |
| | 1 | -1 | -1 | 1 | $\frac{1}{3}$ | $\frac{1}{3}$ | $\frac{1}{3}$ | 1 | -1 | -1 | 4.9 |
| | 1 | 1 | 1 | 1 | $\frac{1}{3}$ | $\frac{1}{3}$ | $\frac{1}{3}$ | 1 | 1 | 1 | 3.9 |
| | 1 | 0 | 0 | 0 | $-\frac{2}{3}$ | $-\frac{2}{3}$ | $-\frac{2}{3}$ | 0 | 0 | 0 | 6.2 |
| | 1 | 0 | 0 | 0 | $-\frac{2}{3}$ | $-\frac{2}{3}$ | $-\frac{2}{3}$ | 0 | 0 | 0 | 6.3 |
| III | 1 | -1.633 | 0 | 0 | 2.0 | $-\frac{2}{3}$ | $-\frac{2}{3}$ | 0 | 0 | 0 | 6.3 |
| | 1 | 1.633 | 0 | 0 | 2.0 | $-\frac{2}{3}$ | $-\frac{2}{3}$ | 0 | 0 | 0 | 3.9 |
| | 1 | 0 | -1.633 | 0 | $-\frac{2}{3}$ | 2.0 | $-\frac{2}{3}$ | 0 | 0 | 0 | 7.0 |
| | 1 | 0 | 1.633 | 0 | $-\frac{2}{3}$ | 2.0 | $-\frac{2}{3}$ | 0 | 0 | 0 | 6.8 |
| | 1 | 0 | 0 | -1.633 | $-\frac{2}{3}$ | $-\frac{2}{3}$ | 2.0 | 0 | 0 | 0 | 4.5 |
| | 1 | 0 | 0 | 1.633 | $-\frac{2}{3}$ | $-\frac{2}{3}$ | 2.0 | 0 | 0 | 0 | 4.8 |
| | 1 | 0 | 0 | 0 | $-\frac{2}{3}$ | $-\frac{2}{3}$ | $-\frac{2}{3}$ | 0 | 0 | 0 | 5.0 |
| | 1 | 0 | 0 | 0 | $-\frac{2}{3}$ | $-\frac{2}{3}$ | $-\frac{2}{3}$ | 0 | 0 | 0 | 5.5 |
| | 20 | 0 | 0 | 0 | 0 | 0 | 0 | 0 | 0 | 0 | 107.8 |
| Sums of Squares | 20 | 13.3334 | 13.3334 | 13.3334 | 13.3334 | 13.3334 | 13.3334 | 8 | 8 | 8 | 598.66 |

The following calculations were carried out:

$$
g = X'Y =
\begin{bmatrix}
\Sigma Y_i \\
g_1 \\
g_2 \\
g_3 \\
g_{11} \\
g_{22} \\
g_{33} \\
g_{12} \\
g_{13} \\
g_{23}
\end{bmatrix}
=
\begin{bmatrix}
107.8000 \\
-8.3192 \\
1.2754 \\
-0.3101 \\
-5.0699 \\
4.5302 \\
-7.4699 \\
-2.4000 \\
-2.8000 \\
-1.6000
\end{bmatrix},
\quad
b = (X'X)^{-1}X'Y =
\begin{bmatrix}
b_0 \\
b_1 \\
b_2 \\
b_3 \\
b_{11} \\
b_{22} \\
b_{33} \\
b_{12} \\
b_{13} \\
b_{23}
\end{bmatrix}
=
\begin{bmatrix}
-5.3800 \\
-0.6239 \\
0.0955 \\
-0.0233 \\
-0.3995 \\
0.2755 \\
-0.5683 \\
-0.3000 \\
-0.3500 \\
0.2000
\end{bmatrix},
$$

The calculations for the sum of squares are as follows:

$$
\text{SSto} = \sum Y_i^2 - \frac{(\Sigma Y_i)^2}{M} = 598.66 - \frac{(107.8)^2}{20}
$$

$$
= 17.6180 \quad \text{with} \quad 19 \text{ DF};
$$

$$
\text{SSrl} = g_1 b_1 + g_2 b_2 + g_3 b_3
$$
$$
= (-8.3192)(-0.6239) + (1.2754)(0.0955)
$$
$$
+ (-0.3101)(-0.0233)
$$
$$
= 5.3194 \quad \text{with} \quad 3 \text{ DF};
$$

$$
\text{SSrq} = g_{11} b_{11} + g_{22} b_{22} + g_{33} b_{33}
$$
$$
= (-5.0699)(-0.3995) + (4.5302)(0.2755)
$$
$$
+ (-7.4699)(-0.5683)
$$
$$
= 7.5186 \quad \text{with} \quad 3 \text{ DF};
$$

$$
\text{SSri} = g_{12} b_{12} + g_{13} b_{13} + g_{23} b_{23}
$$
$$
= (-2.4000)(-0.3000) + (-2.8000)(-0.35000)
$$
$$
+ (1.60000)(0.2000)
$$
$$
= 2.0200 \quad \text{with} \quad 3 \text{ DF}.
$$

Hence,

$$
\text{SSto} = \sum Y_i^2 - \frac{(\Sigma Y_i)^2}{M} = 598.66 - \frac{(107.8)^2}{20}
$$

$$
= 17.6180 \quad \text{with} \quad 19 \text{ DF};
$$

$$
\text{SSrl} = g_1 b_1 + g_2 b_2 + g_3 b_3
$$
$$
= (-8.3192)(-0.6239) + (1.2754)(0.0955)
$$
$$
+ (-0.3101)(-0.0233)
$$
$$
= 5.3194 \quad \text{with} \quad 3 \text{ DF};
$$

$$SSrq = g_{11}b_{11} + g_{22}b_{22} + g_{33}b_{33}$$
$$= (-5.0699)(-0.3995) + (4.5302)(0.2755)$$
$$+ (-7.4699)(-0.5683)$$
$$= 7.5186 \quad \text{with} \quad 3 \text{ DF};$$
$$SSri = g_{12}b_{12} + g_{13}b_{13} + g_{23}b_{23}$$
$$= (-2.4000)(-0.3000) + (-2.8000)(-0.35000)$$
$$+ (1.60000)(0.2000)$$
$$= 2.0200 \quad \text{with} \quad 3 \text{ DF}.$$

Hence,

$$SSr = SSrl + SSrq + SSri = 14.8580 \quad \text{with} \quad 9 \text{ DF},$$

and

$$SSe = SSto - SSr = 17.6180 - 14.8580 = 2.7600 \quad \text{with} \quad 10 \text{ DF}.$$

Furthermore, from (7.5-2),

$$SSbl = \left[ \frac{(32.30)^2}{6} + \frac{(31.70)^2}{6} + \frac{(43.80)^2}{8} \right] - \frac{(107.8)^2}{20}$$
$$= 0.1264 \quad \text{with} \quad 2 \text{ DF},$$

and from (7.5-3),

$$SSexp = [(6.0)^2 + (5.9)^2 - 2(5.95)^2] + [(6.2)^2 - 2(6.25)^2]$$
$$+ [(5.0)^2 + (5.5)^2 - 2(5.25)^2]$$
$$= 0.1350 \quad \text{with} \quad 3 \text{ DF}.$$

Hence,

$$SSf = SSe - SSbl - SSexp = 2.4986 \quad \text{with} \quad 5 \text{ DF}.$$

The analysis of variance is summarized by Table 7.22. Since $F = Msf / MSexp = 11.104$ exceeds the 95th percentile of the $F$ distribution with 5 numerator DF and 3 denominator DF, it appears that the model is inadequate. Hence, we use MSexp to estimate $\sigma^2$, and we can estimate the variances of various coefficients by the respective diagonal elements of $(X'X)^{-1}$ multiplied by MSexp following (7.2-4).

Remember that the columns corresponding to the quadratic effects of the design were made orthogonal to the constant column by subtracting $C = \frac{2}{3}$. Therefore, we use (7.2-8) to adjust for estimating $\beta_0$ by

$$b_0 = 5.38 - \tfrac{2}{3}(-0.3995) - \tfrac{2}{3}(0.2755) - \tfrac{2}{3}(0.5683)$$
$$= 5.8415.$$

**Table 8.1** Termination Times for 17 Samples of a Food Product and Hazard Values of Failure Times

| (1) Reverse Rank $k$ | (2) Age at Termination, Days | (3) Hazard $h(X)$, 100/$k$ | (4) Cumulative Hazard $H(X)$, % |
|---|---|---|---|
| 17 | 21 | 5.88 | 5.88 |
| 16 | 23 | 6.25 | 12.13 |
| 15 | 25 | 6.67 | 18.80 |
| 14 | 38 | 7.14 | 25.94 |
| 13 | 43 | 7.69 | 33.63 |
| 12 | 43 | 8.33 | 41.96 |
| 11 | 52 | 9.09 | 51.05 |
| 10 | 56 | 10.00 | 61.05 |
| 9 | 61 | 11.11 | 72.16 |
| 8 | 63 | 12.50 | 84.66 |
| 7 | 67 | 14.29 | 98.95 |
| 6 | 69[a] | — | — |
| 5 | 69 | 20.20 | 118.95 |
| 4 | 70 | 25.00 | 143.95 |
| 3 | 75[a] | — | — |
| 2 | 86[a] | — | — |
| 1 | 107 | 100.00 | 243.95 |

[a]*Terminated owing to lack of sample.*

failure and censored times of 17 samples from different plants producing a food product. Each sample was judged on a 7-point off-flavor rating scale by a panel of judges. An average panel score of 3.5 was considered indicative of the product's failure. The failure and censored times were recorded in days from the date of production. The data consist of 14 failure and 3 censored times. The censored times correspond to samples that had not failed but were removed from the experiment because insufficient amounts remained for testing. In general, steps in hazard plotting are as follows:

1. Arrange the termination (failure as well as censored) times in increasing order, regardless of whether the observed times are failure or censored times. Table 8.1 lists the termination times in increasing order in column 2.

2. Assign reverse ranks to the termination times. The lowest termination time receives the highest rank. The reverse ranks for our example appear in column 1 of the table. In general, if there are $n$ termination times, the reverse rank of the $i$th smallest termination time is $(n - i + 1)$.

3. Obtain the hazard value for each failure time as follows. If a termination time is a failure time, then its hazard value (expressed as percentage) is 100 divided by its reverse rank $k$. That is, if the $i$th termination time is a failure time, then its hazard value is $100/(n - i + 1)$. For our example, the hazard values are given in column 3 of the table.

4. For each failure time, calculate the corresponding cumulative hazard value, which is the sum of its hazard value and the hazard values of all the preceding failure times. Column 4 of the table lists the cumulative hazard values for the termination times.

5. Choose a theoretical life distribution and use its hazard graph paper to plot cumulative hazard values versus their failure times. Plot each failure time on the vertical axis against its corresponding cumulative hazard value on the horizontal axis of the hazard graph paper. The resulting plot is called a hazard plot. Hazard graph papers may be obtained from Technical and Engineering Aids for Management (TEAM, P.O. Box 25, Tamworth, NH 03886).

As will become clear in the following section, a hazard plot should be nearly a straight line when plotted on the hazard graph paper of an appropriate life distribution. If a hazard plot does not appear to be close to a straight line, one should do the plotting again but on the hazard paper of some other likely life distribution. It is possible that hazard plotting on two different hazard papers will yield reasonable straight lines on each paper. This would suggest that both graphs are appropriate within the ranges of the observed data and that either paper may be used, keeping in mind other practical considerations.

## 8.2 SHELF LIFE MODELS

A fundamental assumption underlying statistical analyses of life testing experiments is that the life distribution of a consumer product belongs to a family of probability distributions. Parameters of a probability distribution relate directly or indirectly to such physical aspects as location and scale values. Parameters of life distributions are estimated by use of life testing experimental data. Once the parameters of a life model have been estimated, the model can be used to predict probabilities of various events, such as future failures; see Nelson (1969, 1970). This section considers some frequently used shelf life statistical models and related inferences.

## Normal Distribution

The normal probability distribution was introduced in Chapter 1 through its pdf, given by (1.1-5). The mean and variance parameters were denoted by $\mu$ and $\sigma^2$, respectively. It is apparent from Figure 1.1 that the curves of the pdf's $f(X)$ of normal distributions approach zero for $X$ values greater than $\mu + 3\sigma$ and also for values less than $\mu - 3\sigma$. It means that, for all practical purposes, the space of a normally distributed random variable $X$ is from $\mu - 3\sigma$ to $\mu + 3\sigma$. Thus, for a normally distributed random variable $X$ to denote the life of a product, we must assume that $\mu - 3\sigma$ is nonnegative. This must be so assumed because the lifetime of a product is never negative. In other words, for a normal distribution with mean $\mu$ and variance $\sigma^2$ to serve as a life model, it is necessary that $\mu > 3\sigma$. With this understanding, suppose we use the normal distribution and want to study its hazard function and hazard plotting for making statistical inferences.

A mathematically oriented reader would have no difficulty in finding hazard functions $h(X)$ knowing that $h(X)\Delta X$ denotes the conditional probability of a product failing in the time interval $(X, X + \Delta X)$ when the product has not failed up to time $X$. If we want to study the hazard function of the normal probability distribution, as in Chapter 1, $\phi(X)$ and $\Phi(X)$ denote, respectively, the pdf and the cdf of the standard normal distribution with mean 0 and variance 1. In this notation, the hazard function $h(X)$ in (8.2-1) of the normal distribution with mean $\mu$ and variance $\sigma^2$ is given by

$$h(X) = \frac{(1/\sigma)\phi[(X - \mu)/\sigma]}{1 - \Phi[(X - \mu)/\sigma]}.$$

When we use the techniques of integral calculus, the cumulative hazard function up to time $X$, denoted by $H(X)$, is

$$H(X) = - \ln\{1 - \Phi[(X - \mu)/\sigma]\}, \qquad (8.2-1)$$

where the symbol "ln" stands for the logarithm to the base $e$ and is called the natural logarithm. It is possible to rewrite (8.2-1) so that failure time $X$ can be expressed in terms of its cumulative hazard $H(X)$. That is,

$$X = \mu + \sigma U(H), \qquad (8.2\text{-}2)$$

where

$$U(H) = \Phi^{-1}[1 - \exp(-H)], \qquad (8.2\text{-}3)$$

and $\Phi^{-1}$ is the inverse function of $\Phi$. Once we have the cumulative hazard value $H(X)$ for failure time $X$, we can find the corresponding $U(H)$ through (8.2-3). Notice that the failure time $X$ is related to $U(H)$ linearly with the slope $\sigma$ and the intercept $\mu$. Therefore, this relationship can be used not only for estimating parameters $\mu$ and $\sigma$ but also for judging the appropriateness of the normality assumption. Curves of the pdf $f(X)$, the cdf $F(X)$, the hazard function $h(X)$, and the cumulative hazard function $H(X)$ for the normally distributed random variable appear in Figures 8.2, 8.3, 8.4, and 8.5, respectively.

## Lognormal Distribution

The life $X$ of a product is said to follow a lognormal distribution if its pdf and cdf are

$$f(X) = \frac{0.4343}{X\sigma} \phi\left(\frac{\log X - \mu}{\sigma}\right), \quad X > 0, \qquad (8.2\text{-}4)$$

**■ FIGURE 8.2** Normal probability density function. *Source:* Nelson (1970). Copyright by John Wiley & Sons, Inc.

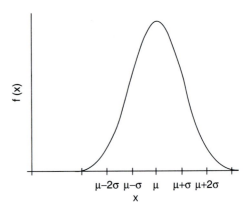

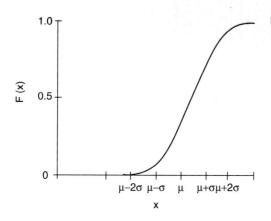

■ **FIGURE 8.3** Normal cumulative distribution function. *Source:* Nelson (1969). Copyright American Society for Quality Control, Inc. Reprinted by permission.

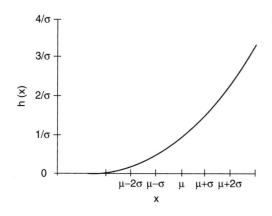

■ **FIGURE 8.4** Normal hazard function. *Source:* Nelson (1969). Copyright American Society for Quality Control, Inc. Reprinted by permission.

and

$$F(X) = \Phi\left(\frac{\log X - \mu}{\sigma}\right), \qquad (8.2\text{-}5)$$

respectively. In the preceding expressions, the functions $\phi$ and $\Phi$ are, as before, the pdf and cdf of the standard normal distribution. The parameters $\mu$ and $\sigma$ are the mean and the standard deviation of the distribution of log $X$, which is normally distributed. We note that log $X$ is a base 10 logarithm. The hazard function of the lognormal distribution is

$$h(X) = \frac{(0.4343/X\sigma)\phi[(\log X - \mu)/\sigma]}{1 - \Phi[(\log X - \mu)/\sigma]}, \quad X > 0. \qquad (8.2\text{-}6)$$

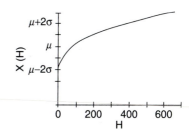

■ **FIGURE 8.5** Normal cumulative hazard function. *Source:* Nelson (1969). Copyright American Society for Quality Control, Inc. Reprinted by permission.

The hazard function is zero at time zero. It first increases to a maximum and then decreases to zero with increasing time $X$. This behavior of the hazard function may render the lognormal model unsuitable for products that continue to deteriorate with increasing time.

The log of the failure time $X$ is related to the cumulative hazard function $H(X)$ of the lognormal distribution through $U(H)$ in a linear manner. That is,

$$\log X = \mu + \sigma U(H), \tag{8.2-7}$$

where $U(H)$ is as defined by (8.2-3). Equation (8.2-7) suggests that normal hazard paper will also serve as lognormal hazard paper if the vertical time scale is made a logarithmic scale. We note here also that (8.2-7) can be used to estimate $\sigma$ and $\mu$ from failure data plotted on lognormal hazard paper. The intercept $\mu$ corresponds to $U(H) = 0$, which in turn is true if $H = \ln 2$. Figures 8.6, 8.7, 8.8, and 8.9 show, respectively, sketches of pdf, cdf, hazard, and cumulative hazard functions for the lognormal distribution. For lognormal applications as a life model, we refer to Goldthwaite (1961) and Barlow and Proschan (1965). Gacula and Kubala (1975) used the lognormal for modeling the shelf life of deteriorating food products.

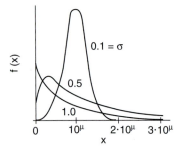

■ **FIGURE 8.6** Lognormal probability density function. *Source:* Nelson (1970). Copyright by John Wiley & Sons, Inc.

## Weibull Distribution

The life $X$ of a product is governed by the Weibull distribution if its pdf and cdf are

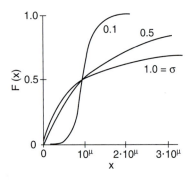

■ **FIGURE 8.7** Lognormal cumulative distribution function. *Source:* Nelson (1970). Copyright by John Wiley & Sons, Inc.

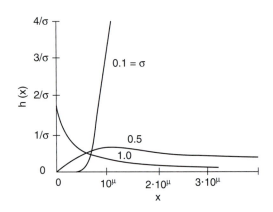

■ **FIGURE 8.8** Lognormal hazard function. *Source:* Nelson (1970). Copyright by John Wiley & Sons, Inc.

$$f(X) = (\beta/\sigma^\beta)X^{\beta-1} \exp[-(X/\alpha)^\beta], \quad X > 0, \qquad (8.2\text{-}8)$$

and

$$F(X) = 1 - \exp[-(X/\alpha)^\beta], \qquad (8.2\text{-}9)$$

respectively. Both parameters $\alpha$ and $\beta$ are nonnegative. The Weibull distribution is a versatile distribution because of its shape parameter $\beta$. Depending on parameter $\beta$, the Weibull can have an increasing, decreasing, or even a constant hazard function. Because of this property, the Weibull is favored over the lognormal for modeling the shelf life of food products. We note that when $\beta = 1$, the Weibull distribution is the well-known exponential probability distribution.

The Weibull distribution is skewed. The mean and variance of the distribution depend on both parameters $\alpha$ and $\beta$. For large $\beta$, the Weibull pdf tends to be symmetric around the median. The hazard function of the Weibull distribution is

$$h(X) = (\beta/\alpha^\beta)X^{\beta-1}, \quad X > 0. \qquad (8.2\text{-}10)$$

The hazard function is a decreasing function of time $X$ if $\beta < 1$ and increasing if $\beta > 1$. For $\beta = 1$, $h(X) = 1/\alpha$, which simply does not change with time $X$. That is, the hazard function of the exponential distribution ($\beta = 1$) is constant.

The cumulative hazard function is

$$H(X) = (X/\alpha)^\beta, \quad X > 0. \qquad (8.2\text{-}11)$$

We can rewrite (8.2-11) to express time $X$ as a function of the cumulative hazard function $H$; that is,

$$X = \alpha[H(X)]^{1/\beta},$$

or in log units of time $X$,

$$\log X = \log \alpha + (1/\beta) \log H(X). \qquad (8.2\text{-}12)$$

Curves of various functions for the Weibull distribution are given in Figures 8.10 through 8.13.

From (8.2-12), one can see that $\log X$ is a linear function of $\log H(X)$ with slope $1/\beta$ and intercept $\log \alpha$. Thus, the hazard plotting of the Weibull distribution is a straight line on log–log graph paper. From the hazard plot on the Weibull hazard paper, it is possible to estimate the slope $1/\beta$ of the line in (8.2-12) and, in turn, $\beta$.

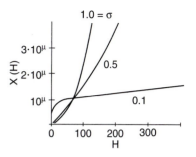

■ **FIGURE 8.9** Lognormal cumulative hazard function. *Source:* Nelson (1969). Copyright American Society for Quality Control, Inc. Reprinted by permission.

■ **FIGURE 8.10** Weibull probability density function. *Source:* Nelson (1969). Copyright American Society for Quality Control, Inc. Reprinted by permission.

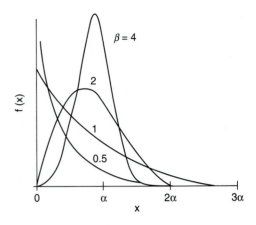

■ **FIGURE 8.11** Weibull probability distribution function. *Source:* Nelson (1969). Copyright American Society for Quality Control, Inc. Reprinted by permission.

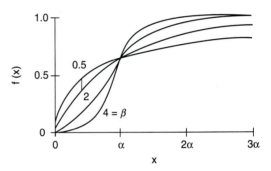

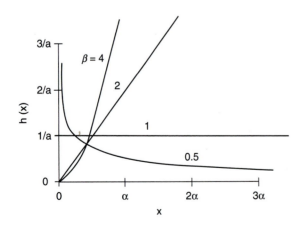

■ **FIGURE 8.12** Weibull hazard function. *Source:* Nelson (1969). Copyright American Society for Quality Control, Inc. Reprinted by permission.

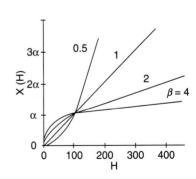

■ **FIGURE 8.13** Weibull cumulative hazard function. *Source:* Nelson (1969). Copyright American Society for Quality Control, Inc. Reprinted by permission.

For $H = 1$, $\log H = 0$, and in that case $\log X = \log \alpha$ or $X = \alpha$. This fact can be used to estimate $\alpha$ from the hazard plot on the Weibull hazard paper. That is, we can estimate $\alpha$ by the failure time $X$ corresponding to $\log H(X) = 0$ or $H(X) = 1$.

Let us plot the time $X$ and its cumulative hazard values given in Table 8.1 on the Weibull hazard paper. We know from (8.2-12) that on the Weibull hazard paper, time $X$ is on the log $X$ scale and also that its cumulative hazard $H(X)$ is on the log $H(X)$ scale. The resulting plot in Figure 8.14 appears to be quite close to a straight line. To use the plot for estimating $\beta$, start at the dot on the upper left corner of the hazard paper and draw a line parallel to the fitted line. Extend the line until it meets the scale of values for the shape parameter $\beta$. We notice from the plot that the parallel line meets the scale for the slope parameter at $\beta = 2.45$. Hence, $\beta$ would be estimated by 2.45 from the termination-time data in Table 8.1, assuming that the failures were indeed governed by the Weibull distribution. To obtain $\alpha$ from the hazard plot, remember that $\alpha$ corresponds to a failure time $X$ so that $H(X) = 1$ or 100%. From the 100% point on the cumulative hazard scale on the horizontal axis, draw a vertical line to where it intersects the plotted line. From the point of intersection, draw a horizontal line to the left to where it intersects the ordinate axis. The point of intersection on the ordinate is an estimate of the $X$ value. For our example, $\alpha$ is estimated to be nearly 59 days.

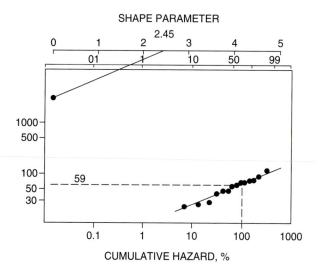

■ **FIGURE 8.14** Hazard plot of the data in Table 8.1 on Weibull hazard paper.

| Example 8.2-1 | Nine samples of a food product were tested over time for off-flavor development. Ten trained panelists were used to evaluate each sample for off-flavor development on a 7-point rating scale beginning with category 1, indicating no off-flavor, to category 7, indicating very strong off-flavor. On this scale, an average score of 2.5 or more was considered a high enough off-flavor score for the product to have failed. Since no off-flavor development was expected in the first few weeks, the following design was used (Gacula, 1975). At the end of 3 weeks (21 days), only three samples out of nine were tested to save the time and cost of evaluating all nine samples. Indeed, out of three samples, none was found to have failed by the end of the third week. So no sample was evaluated during the next week because it was considered still too early for off-flavor development. At the end of the fifth week, only four samples out of the nine were evaluated. Out of the four samples, one sample was found to have failed with an average off-flavor score of 3.5. At the end of the sixth week, five samples of the remaining eight were evaluated. Of the five samples tested, two were judged to have failed with average scores of 2.5 and 2.6. This suggested that off-flavor development was beginning to accelerate. Hence, all remaining six samples were tested at the end of the seventh week. At this time, two more samples were found to have failed with average off-flavor scores of 3.1 and 3.4. From this time on, the testing of the remaining four samples was conducted at intervals of 3 days instead of a week. On the 52nd day, two more samples had failed with average scores of 4.0 and 3.2. On the 55th day, neither of the two remaining samples had failed, but one of the two samples contained an insufficient amount of the product for subsequent evaluation. Thus, on day 55, there was a censored observation. On the 58th day, the one remaining sample was evaluated and found to have failed with an average off-flavor score of 4.1. The resulting data of the design described appear in Table 8.2, with hazard and cumulative hazard values in columns 4 and 5, respectively. For example, hazard and cumulative hazard values at $X = 35$ and $X = 42$ are as follows: |
|---|---|

$$h(35) = 100/9 - 11.1,$$
$$h(42) = 100/8 = 12.5,$$
$$H(35) = h(35),$$
$$H(42) = h(35) + h(42)$$
$$= 11.1 + 12.5 = 23.6.$$

The next step is to plot columns 3 versus 5 of Table 8.2 on an appropriate hazard probability paper. The Weibull hazard paper is used in this example (see Figure 8.15). The points seem to lie along a line, which suggests that

**Table 8.2** Off-Flavor Average Scores for Example 8.2-1

| (1) Reverse Rank *k* | (2) Mean Score | (3) Age | (4) Hazard $h(X)$, 100/k | (5) Cumulative Hazard $H(X)$, % |
|---|---|---|---|---|
| 9 | 3.5 | 35 | 11.1 | 11.1 |
| 8 | 2.6 | 42 | 12.5 | 23.6 |
| 7 | 2.5 | 42 | 14.3 | 37.9 |
| 6 | 3.1 | 49 | 16.7 | 54.6 |
| 5 | 3.4 | 49 | 20.0 | 74.6 |
| 4 | 4.0 | 52 | 25.0 | 99.6 |
| 3 | 3.2 | 52 | 33.3 | 132.9 |
| 2 | 2.4[a] | *55 | — | — |
| 1 | 4.1 | 58 | 100.0 | 232.9 |

[a]Unfailed unit.

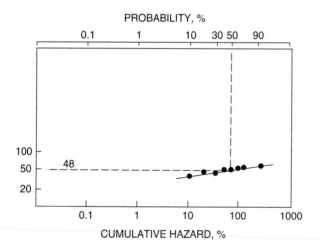

■ **FIGURE 8.15** Hazard plot of the data in Example 8.2-1.

the failure times follow the Weibull distribution. The median of the failure distribution is obtained by drawing a line from the intersection of the 50th point on the probability scale and the "fitted" line to the ordinate axis. From Figure 8.15 the median is estimated to be 48 days. On the rating scale, using 2.5 as a cutoff point for the failure criterion, the analysis indicates that, by the 48th day, chances are 50% that the product will fail.

The probability that the product will go bad or become unacceptable by time $X$ is called the *probability of sensory failure* and can be obtained by cdf $F(X)$. The cdf expresses the probability of failures by time $X$ for future samples. To determine the cdf from Figure 8.15, draw a horizontal line from the selected time to the fitted line. At the point of intersection, draw an upward line and read the cdf. Using this procedure for several values of time $X$, we obtain the following cdf values from Figure 8.15:

| Age ($X$) | cdf $F(X)$ | $R(X) = 1 - F(X)$ |
|---|---|---|
| 35 | 0.08 | 0.92 |
| 40 | 0.18 | 0.82 |
| 45 | 0.35 | 0.65 |
| 50 | 0.60 | 0.40 |

Thus, by time $X = 35$ days, the cdf is 8%. Likewise, by 50 days we estimate that 60% of the product should have failed or become unacceptable.

A practical guide to use in determining expiration dates is a statistic $R(X) = 1 - F(X)$. Since $F(X)$ expresses the proportion of samples that will fail on or before age $X$, $R(X)$ defines the proportion of the items surviving at age $X$. Thus, with an expiration date of 35 days from the production date, the reliability is 92% that the product will remain acceptable for consumption, and so on.

Instead of using hazard plots for estimating the cdf and parameters of a life distribution, one can use other methods of estimation. Methods for estimating the parameters of normal and lognormal distributions are rather well known and illustrated in nearly all statistics texts. We describe briefly two methods for estimating the parameters of the Weibull distribution. One of the methods is due to Dubey (1967). Let $Y_p$ denote the $p$th percentile of the observed failure times. For $p_1$ and $p_2$, such that $0 < p_1 < p_2 < 1$, Dubey proposed the following estimator:

$$\widehat{\beta} = \frac{\ln[-\ln(1 - p_1)] - \ln[-\ln(1 - p_2)]}{\ln Y_{p1} - \ln Y_{p2}}. \qquad (8.2\text{-}13)$$

Dubey argued that a choice of $p_1 = 0.17$ and $p_2 = 0.97$ leads to an "efficient" estimator $\widehat{\beta}$. For $p_1 = 0.17$ and $p_2 = 0.97$, the numerator of (8.2-13) is $-2.9349$, and therefore,

$$\widehat{\beta} = 2.9349/(\ln Y_{0.97} - \ln Y_{0.17}).$$

The two steps for calculating $\beta$ are as follows. First, record the lifetimes $X_1 \leq X_2 \leq \cdots \leq X_n$ in increasing order. Second, calculate $Y_{0.17}$ and $Y_{0.97}$ using an approximate rule that, for $0 < p < 1$, the $p$th percentile $Y_p$ is the $p(n+1)$th observation in the ordered arrangement $X_1 \leq X_2 \leq \cdots \leq X_n$.

We illustrate Dubey's $\beta$ calculations for the failure data in Table 8.3. Note that $n = 14$. Now $Y_{0.17}$ is the $[(0.17)(15)]$th $= (2.6)$th or the 3rd observation.

Hence, $Y_{0.17} = 25$. Similarly, $Y_{0.97} = 107$, which is the $[(0.97)(15)$th $= (14.6)$th or 15th observation. Hence,

$$\ln Y_{0.97} - \ln Y_{0.17} = \ln 107 - \ln 25 = 4.67283 - 3.21888 = 1.45395.$$

Therefore,

$$\widehat{\beta} = 2.9349/1.45395 = 2.02.$$

| Table 8.3 Sample of 14 Observations from a Weibull Distribution | | |
|---|---|---|
| **Observation Number** | **Age to Failure $X$, Days** | **$Y = \ln X$** |
| 1 | 21 | 3.04452 |
| 2 | 23 | 3.13549 |
| 3 | 25 | 3.21888 |
| 4 | 38 | 3.63759 |
| 5 | 43 | 3.76120 |
| 6 | 43 | 3.76120 |
| 7 | 52 | 3.95124 |
| 8 | 56 | 4.02535 |
| 9 | 61 | 4.11087 |
| 10 | 63 | 4.14313 |
| 11 | 67 | 4.20469 |
| 12 | 69 | 4.23411 |
| 13 | 70 | 4.24850 |
| 14 | 107 | 4.67283 |

Note:
$\Sigma_i Y_i = 54.1495; \; \overline{Y} = 3.8678; \; \Sigma_i Y_i^2 = 212.3596.$
$S^2 = \{212.3596 - [(54.1495)^2/14]\}/13 = 0.2445; \; S = 0.4738.$
$\widetilde{\beta} = 1.282/0.4738 = 2.71; \; \widetilde{\alpha} = \exp[3.8678 + (0.5772/2.71)] = 60.$

A quick and efficient estimator of $\alpha$, given by Dubey, is

$$\hat{\alpha} = Y_{0.17}(0.183)^{-1/\beta}. \qquad (8.2\text{-}14)$$

Again for the data in Table 8.3, we find that

$$\hat{\alpha} = 25(0.183)^{-1/2.02} = 57.95 \simeq 58.$$

Menon (1963) also provided estimators of $\alpha$ and $\beta$. Following his method, estimate $\beta$ as follows. Suppose $X_1, X_2, \ldots, X_n$ is a sample of $n$ independent observations from a Weibull distribution. Define $Y_i = \ln X_i$ for $i = 1, \ldots, n$ and compute the variance of $Y_1, Y_2, \ldots, Y_n$ values. That is,

$$S^2 = \Sigma(Y_i - \overline{Y})^2/(n-1),$$

and its square root is $S$. Menon proposed to estimate $\alpha$ and $\beta$ by

$$\tilde{\beta} = 1.282/S \quad \text{and} \quad \tilde{\alpha} = \exp(\overline{Y} + 0.5772/\tilde{\beta}).$$

To illustrate the calculations of $\tilde{\beta}$ and $\tilde{\alpha}$, let us assume again that the failure times of a food product shown in Table 8.3 constitute a sample of 14 independent observations from a Weibull distribution. This sample of 14 observations and other calculations leading to $\tilde{\beta}$ and $\tilde{\alpha}$ are shown in Table 8.3.

---

## Graphical Test for Failure Distributions

For hazard plotting we must have hazard probability paper. To use the appropriate hazard paper, we have to know the life distribution of failure times. Consider the sensory failure times for a product, shown in Table 8.4, taken from Gacula and Kubala (1975). On a 7-point-intensity off-flavor scale, all 26 experimental samples failed. The hazard plots for three failure distribution models are shown in Figures 8.16 through 8.18 by graphing columns 2 versus 4 of Table 8.4. The lines in these figures were visually drawn. The hazard plots of the failure times on both lognormal and Weibull paper are straight lines, suggesting good data fits to both distributions. To differentiate between the Weibull and the lognormal, one needs more data at the tail ends.

Assuming that the hazard plot is a "good" straight-line fit on the Weibull hazard paper, we can estimate $\beta$ from the plot. Note that from the Weibull hazard plot in Figure 8.18, $\beta$ is estimated by 4.17, which, by its magnitude, indicates that the failure distribution is approximately symmetric. For symmetric shelf life distributions, their median

**Table 8.4** Time to Failure and Hazard Data for a Product

| (1)<br>Number of<br>Terminations $\geq X$<br>(Reverse Rank $k$) | (2)<br>Age at<br>Termination $X$,<br>Days | (3)<br>Hazard $h$<br>$(X)$, 100/k | (4)<br>Cumulative<br>Hazard $H$<br>$(X)$, % |
|---|---|---|---|
| 26 | 24 | 3.85 | 3.85 |
| 25 | 24 | 4.00 | 7.85 |
| 24 | 26 | 4.17 | 12.02 |
| 23 | 26 | 4.35 | 16.37 |
| 22 | 32 | 4.55 | 20.92 |
| 21 | 32 | 4.76 | 25.68 |
| 20 | 33 | 5.00 | 30.68 |
| 19 | 33 | 5.26 | 35.94 |
| 18 | 33 | 5.56 | 41.50 |
| 17 | 35 | 5.88 | 47.38 |
| 16 | 41 | 6.25 | 53.63 |
| 15 | 42 | 6.67 | 60.30 |
| 14 | 43 | 7.14 | 67.44 |
| 13 | 47 | 7.69 | 75.13 |
| 12 | 48 | 8.33 | 83.46 |
| 11 | 48 | 9.09 | 92.55 |
| 10 | 48 | 10.00 | 102.55 |
| 9 | 50 | 11.11 | 113.66 |
| 8 | 52 | 12.50 | 126.16 |
| 7 | 54 | 14.29 | 140.45 |
| 6 | 55 | 16.67 | 157.12 |
| 5 | 57 | 20.00 | 177.12 |
| 4 | 57 | 25.00 | 202.12 |
| 3 | 57 | 33.33 | 235.45 |
| 2 | 57 | 50.00 | 285.45 |
| 1 | 61 | 100.00 | 385.45 |

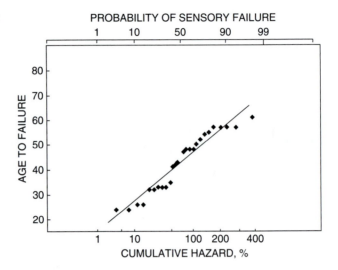

■**FIGURE 8.16** Normal hazard plot for the data in Table 8.4. *Source:* Gacula and Kubala (1975).

■**FIGURE 8.17** Lognormal hazard plot for the data in Table 8.4. *Source:* Gacula and Kubala (1975).

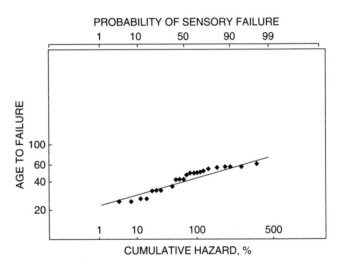

■**FIGURE 8.18** Weibull hazard plot for the data in Table 8.4. *Source:* Gacula and Kubala (1975).

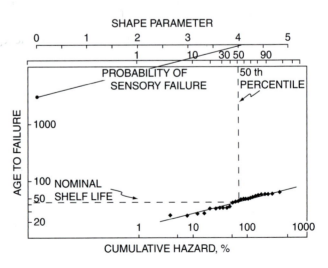

may be used to estimate the mean time $X$. The median of a distribution is in that value $X$ for which $F(X) = \frac{1}{2}$.

## 8.3  REGRESSION ANALYSIS

Regression analysis may be used to analyze life testing experiments in which one of the aims has been to estimate the rate of change in the quality of a product as a function of time and to estimate the product's life. We will introduce only simple regression analysis methods that can play important roles in analyzing shelf life testing experiments.

Consider the following simple statistical linear relationship between two variables $X$ and $Y$:

$$Y = \beta_0 + \beta_1 X + \varepsilon. \tag{8.3-1}$$

In (8.3-1) $\beta_0$ and $\beta_1$ are the intercept and slope parameters, respectively, of the linear relationship $Y = \beta_0 + \beta_1 X$ between $Y$ and $X$. The component $\varepsilon$ in the relationship quantifies all the random fluctuations in the experimental outcomes of the $Y$ variable from the expected outcomes $\beta_0 + \beta_1 X$ corresponding to each level of variable $X$. The component $\varepsilon$ is commonly referred to as the error term in the model and is assumed to be normally distributed with mean 0 and unknown variance $\sigma^2$. For each fixed level of variable $X$, there is a range of all possible $Y$ values. When the error term $\varepsilon$ is assumed to be normally distributed with mean 0 and variance $\sigma^2$, then $Y$ is normally distributed with mean $\beta_0 + \beta_1 X$ and variance $\sigma^2$. In statistical terminology, $Y$ is called the dependent and $X$ the independent variable. Also, the mean of $Y$ for any fixed value of the $X$ variable, namely, $\beta_0 + \beta_1 X$, is called the simple linear regression of $Y$ on $X$. If there are two or more independent variables, say $X_1, X_2, \ldots, X_p$, then the mean of the dependent variable $Y$, given fixed levels of $X_1, X_2, \ldots, X_p$, is called the multiple regression of $Y$ on $X_1, X_2, \ldots, X_p$. The multiple regression is said to be a linear multiple regression if all the parameters appear linearly in the regression. In regression analyses, both the independent and dependent variables may be viewed as having an appropriate joint probability distribution. Usually, however, the independent variables are treated as mathematical variables whose levels can be suitably and optimally chosen by the experimenter. Then the dependent variable is observed through experimentation for various chosen levels of independent variables. That is why dependent variables are treated as random variables. Thus, in a simple linear regression, an experimenter may decide to observe dependent variable $Y$ for several values of independent variable $X$ fixed at $X = X_1, X = X_2, \ldots, X = X_k$. A representation of the simple linear

■ **FIGURE 8.19** Representation of simple linear regression.

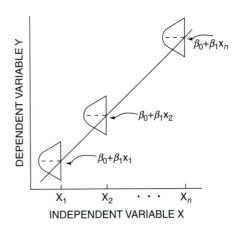

regression is given in Figure 8.19, where $Y$ is shown to have a normal probability distribution with mean $\beta_0 + \beta_1 X_i$ at each $X = X_i$, $i = 1, \ldots, k$.

We are interested only in outlining the regression analysis for the simple linear regression model of $Y$ on $X$ given by (8.3-1) under the assumption that $\varepsilon$ is normally distributed with mean 0 and variance $\sigma^2$. It is implicit in our assumption that the variance $\sigma^2$ does not depend on any particular level of $X$. We further assume that all the experimental observations on $Y$ are statistically independent. With the simple linear regression model having been hypothesized, experiments are designed to observe dependent variable $Y$ for selected levels of independent variable $X$. The data may be recorded as in Table 8.5. For instance, with $X$ fixed at $X_1$, the experiment was conducted $r_1$ times to observe $Y$ values denoted by $Y_{11}, Y_{12}, \ldots, Y_{1r1}$. Note that it is not necessary to observe an equal number of $Y$ values for each $X$ level. For some $X$ values, the experiment may be conducted only once to observe one $Y$ value. For other $X$ values, $Y$ may be observed more than once. In stability studies concerned with the length of time

**Table 8.5** Layout of Data for a Simple Linear Regression of $Y$ on $X$

| Independent Variable $X$ | Dependent Variable $Y$ |
|---|---|
| $X = X_1$ | $Y_{11}, Y_{12}, \ldots, Y_{1r1}$ |
| $X = X_2$ | $Y_{21}, Y_{22}, \ldots, Y_{2r2}$ |
| $\vdots$ | $\vdots$ |
| $X = X_k$ | $Y_{k1}, Y_{k2}, \ldots, Y_{krk}$ |

required for a food product to become unfit for consumption, levels $X_1, X_2, \ldots, X_k$ of the independent variable may denote $k$ different successive times at which judges in the panel are asked to judge whether the food has remained good for consumption. The experimenter decides the successive times at which to seek the panel's judgment. We note that it is not necessary to require the same number of judges every time that evaluations are made.

Having conducted a life testing experiment and obtained data like those given in Table 8.5, we can format the regression analysis as shown in columns 1 and 2 of Table 8.6. There are, in all, $\Sigma_{i=1}^{k} r_i$ observations on the independent variable appearing in Table 8.5. These $n$ observations are denoted in a single subscript notation by $Y_1, Y_2, \ldots, Y_n$ in Table 8.6. Of these $n$ observations, $r_1$ observations correspond to independent variable $X = X_1$, $r_2$ observations to $X = X_2$, and so on. This format permits a one-to-one correspondence between $Y_i$ and $X_i$, $i = 1, \ldots, n$. The one-to-one correspondence between $Y_i$ and $X_i$ is purely artificial and is created only to have the experimental data in a standard format for regression analysis. Otherwise, we do know that of the $X_1, X_2, \ldots, X_k$ in column 1 of Table 8.6, only $k$ are distinct.

Now Table 8.6 can be used to calculate the following quantities for the regression analysis:

$$\overline{X} = \sum X/n, \overline{Y} = \sum Y/n,$$
$$S_{XX} = \sum X^2 - n\overline{X}^2, \ S_{YY} = \sum Y^2 - n\overline{Y}^2, \qquad (8.3\text{-}2)$$
$$S_{XY} = \sum XY - n\overline{X}\overline{Y}.$$

**Table 8.6** Data in a Format for Regression Analysis and Related Calculations

| (1) Independent Variable $X_i$ | (2) Dependent Variable $Y_i$ | (3) $X_i^2$ | (4) $Y_i^2$ | (5) $X_i Y_i$ |
|---|---|---|---|---|
| $X_1$ | $Y_1$ | $X_1^2$ | $Y_1^2$ | $X_1 Y_1$ |
| $X_n$ | $Y_2$ | $X_2^2$ | $Y_2^2$ | $X_2 Y_2$ |
| $\vdots$ | $\vdots$ | $\vdots$ | $\vdots$ | $\vdots$ |
| $X_n$ | $Y_n$ | $X_n^2$ | $Y_n^2$ | $X_n Y_n$ |
| $\sum X$ | $\sum Y$ | $\sum X^2$ | $\sum Y^2$ | $\sum XY$ |

Following computations of expressions in (8.3-2), $\beta_1$ and $\beta_0$ are estimated by

$$b_1 = S_{XY}/S_{XX} \quad \text{and} \quad b_0 = \overline{Y} - b_1\overline{X}, \qquad (8.3\text{-}3)$$

respectively. The estimates obtained from $b_0$ and $b_1$ are used in place of $\beta_0$ and $\beta_1$, respectively, to have a statistically optimum estimate of the mean of the $Y$ values corresponding to the values of independent variable $X$. That is,

$$\widehat{Y} = b_0 + b_1 X. \qquad (8.3\text{-}4)$$

Thus, the mean of the $Y$ values, for instance, $\beta_0 + b_1 X_i$ when $X = X_i$, is estimated by

$$\widehat{Y}_i = b_0 + b_1 X_i.$$

Equation (8.3-4) is said to be a "fitted" simple linear regression of $Y$ on $X$. We point out that, although (8.3-1) is a hypothetical model, when it is fitted to the experimental data, we get (8.3-4). That is why (8.3-4) is termed a fitted regression model. If the hypothesized model is inadequate, the fitted model would also very likely be inadequate. So before accepting a fitted simple linear regression, we usually ask the following question: Does the fitted simple linear regression (8.3-4) provide a good fit in the sense that it can be reliably used to predict $Y$ values for specified $X$ values? A preliminary answer to the question can be given through a statistic $r$, called the correlation coefficient, defined by

$$r = S_{XY}/\sqrt{S_{XX}S_{YY}}. \qquad (8.3\text{-}5)$$

The square of the correlation coefficient denoted by $r^2$ is called the *index of determination*, which measures the portion of the total variation in $Y$ values accounted for by the fitted regression of $Y$ on $X$. It is a mathematical fact and follows from the definition that $r$ is always a number between $-1$ and $+1$. The closer $r$ is to $+1$ or $-1$, the stronger is the linear relationship between $X$ and $Y$. If indeed the relationship between $X$ and $Y$ is linear, then $r$ should be very close to $+1$ or $-1$, depending on the increasing or decreasing linear relationship. Equivalently, $r^2$ is always between 0 and 1, and the closer it is to 1, the stronger is the linear relationship between $X$ and $Y$. Hence, $r^2$ can be used as a measure of the goodness of fit of the simple regression model to the observed data. Figure 8.20 illustrates a scatter diagram of points resulting from plotting data like those in Table 8.5. The line passing through the scatter diagram represents a fitted regression equation given by (8.3-4). A fitted regression equation is considered satisfactory if the points in the

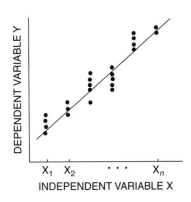

■ **FIGURE 8.20** Scatter diagram for data like those in Table 8.5 and the graph of a fitted simple linear regression equation $\widehat{Y} = b_0 + b_1 X$.

scatter diagram are judged to be in close proximity to the fitted regression line. An overall measure of the departure of the points from its fitted equation in a scatter diagram is given by

$$\text{SSe} = \sum_{i=1}^{n} (Y_i - \widehat{Y}_i)^2, \tag{8.3-6}$$

where $\widehat{Y}_i$ and $Y_i$ are the predicted and the observed values, respectively, of the dependent variable $Y$ corresponding to independent variable $X$ equal to $X_i$. If the fitted regression is a good fit, we may conclude that the hypothesized model is an adequate one, and then SSe is essentially the sum of squares due to the experimental error component $\varepsilon$ with $(n - 2)$ DF. Hence, provided the fitted regression is a good fit, we can estimate $\sigma^2$ (the variance of the distribution of the experimental error component $\sigma$) by

$$\text{MSe} = \text{SSe}/(n - 2), \tag{8.3-7}$$

where SSe is as defined in (8.3-6). The MSe in (8.3-7) provides an estimate of $\sigma^2$, and quite often, for the sake of simplicity, the square root of MSe is denoted by $S$, which is an estimate of $\sigma$.

---

**Example 8.3-1**

A study was conducted to evaluate the effect of nitrite levels on the storage stability of vacuum-packed bacon. For each nitrite level, sensory evaluations were made initially and then every week thereafter. So the independent variable $X$ was 0 when the initial sensory evaluations were made and its subsequent values were 7, 14, 21, etc. Sensory evaluations of off-flavor were made on a 7-point scale. At one end of the scale, a score of 1 meant no off-flavor, and at the other end a score of 7 meant a very strong off-flavor. For each nitrite level, off-flavor scores were observed at each time $X$. Thus, there was a dependent variable associated with each nitrite level. Of the six nitrite levels in the original study, only three levels $A$, $B$, and $C$ are being considered in this example. Let $Y$ be the dependent variable. There is one dependent variable $Y$ for each of the three levels $A$, $B$, and $C$. For each $X$, the observed $Y$ scores for the three nitrite levels are given in Table 8.7. We may point out that for each time $X$, the scores of the dependent variables given in Table 8.7 are in fact the average scores of a panel of judges who made sensory evaluations of the off-flavor development. In using average scores only, we have lost some information. Had we retained the off-flavor score for each judge in the panel, the resulting data for a nitrite level would have appeared as in Table 8.5 or in the format of Table 8.6 for regression analysis. A regression analysis using the scores of each judge in the panel would be

**Table 8.7** Stability Data for Vacuum-Packed Bacon for Three Nitrite Levels A, B, and C

| Evaluation Time X (Independent Variable) | Average Off-Flavor Scores at Nitrite Levels A, B, and C | | |
|---|---|---|---|
| | $Y_A$ | $Y_B$ | $Y_C$ |
| 0 | 2.09 | 1.31 | 1.34 |
| 7 | 1.89 | 1.18 | 1.19 |
| 14 | 2.26 | 1.98 | 1.69 |
| 21 | 2.35 | 1.80 | 1.78 |
| 28 | 3.25 | 2.82 | 1.56 |
| 35 | 3.05 | 2.17 | 1.74 |
| 42 | 3.76 | 2.67 | 2.37 |
| 49 | 4.56 | 3.39 | 3.03 |
| 56 | — | 2.90 | 3.00 |
| 63 | — | 3.70 | 2.60 |

a more reliable analysis than one using only average scores. In using the average score, one is using less informative data, because the average of a set of data has no information about its variance. However, we will illustrate regression analysis using the average scores in Table 8.7.

For each nitrite level, we can follow Table 8.6 to calculate $\sum X$, $\sum Y$, $\sum X^2$, $\sum Y^2$, and $\sum XY$. For nitrite level A, $n = 8$, and following the formulas in (8.3-2) we get

$$\sum X = 196, \qquad \sum Y = 23.21;$$
$$\bar{X} = 24.5, \qquad \bar{Y} = 2.90;$$
$$\sum X^2 = 6860, \qquad \sum Y^2 = 73.37;$$
$$\sum XY = 673.33,$$
$$S_{XX} = 2058, \qquad S_{YY} = 6.03;$$
$$S_{XY} = 104.69.$$

For nitrite level $B$, $n = 10$, we get

$$\sum X = 315, \qquad \sum Y = 23.92;$$
$$\overline{X} = 31.5, \qquad \overline{Y} = 2.39;$$
$$\sum X^2 = 13,965, \qquad \sum Y^2 = 63.65;$$
$$\sum XY = 902.44,$$
$$S_{XX} = 4042.50, \qquad S_{YY} = 6.43;$$
$$S_{XY} = 148.96.$$

For nitrite level $C$, $n = 10$, we get

$$\sum X = 315, \qquad \sum Y = 20.30;$$
$$\overline{X} = 31.5, \qquad \overline{Y} = 2.03;$$
$$\sum X^2 = 13,965, \qquad \sum Y^2 = 45.26;$$
$$\sum XY = 753.76,$$
$$S_{XX} = 4042.50, \qquad S_{YY} = 4.05;$$
$$S_{XY} = 114.31.$$

Now we may use (8.3-3) to find $b_0$ and $b_1$ to fit a simple linear regression for each nitrite level. For level $A$, we calculate

$$b_1 = \frac{104.69}{2058} = 0.051 \quad \text{and} \quad b_0 = 2.90 - (24.5)(0.051) = 1.65.$$

Hence, the fitted simple linear regression (8.3-4) for off-flavor score $Y$ corresponding to nitrite level $A$ is, on days $X$,

$$\widehat{Y}_A = 1.65 + 0.051X. \tag{8.3-8}$$

The correlation coefficient $r_A$ from (8.3-5) is

$$r_A = \frac{104.69}{\sqrt{(2058)(6.03)}} = 0.913,$$

and $r_A^2 = 0.83$.

Similarly, for nitrite levels $B$ and $C$, we find

$$\widehat{Y}_B = 1.23 + 0.037X,$$
$$r_B = 0.922, r_B^2 = 0.85, \tag{8.3-9}$$

$$\widehat{Y}_C = 1.14 + 0.028X,$$
$$r_C = 0.894, r_C^2 = 0.80. \tag{8.3-10}$$

We may now illustrate the calculations for SSe and MSe following (8.3-6) and (8.3-7), respectively. For nitrite levels $A$, $B$, and $C$, consider Table 8.8.

**Table 8.8** Calculations of SSe for Data in Table 8.7

| (1) X | (2) $Y_A$ | (3) $\widehat{Y}_A$ | (4) $Y_A - \widehat{Y}_A$ | (5) $(Y_A - \widehat{Y}_A)^2$ | (6) $Y_B$ | (7) $\widehat{Y}_B$ | (8) $Y_B - \widehat{Y}_B$ | (9) $(Y_B - \widehat{Y}_B)^2$ | (10) $Y_C$ | (11) $\widehat{Y}_C$ | (12) $Y_C - \widehat{Y}_C$ | (13) $(Y_C - \widehat{Y}_C)^2$ |
|---|---|---|---|---|---|---|---|---|---|---|---|---|
| 0 | 2.09 | 1.65 | 0.44 | 0.194 | 1.31 | 1.23 | 0.08 | 0.006 | 1.34 | 1.14 | 0.24 | 0.058 |
| 7 | 1.89 | 2.01 | −0.12 | 0.014 | 1.18 | 1.49 | −0.31 | 0.096 | 1.19 | 1.34 | −0.25 | 0.062 |
| 14 | 2.26 | 2.36 | −0.10 | 0.010 | 1.98 | 1.75 | 0.23 | 0.053 | 1.69 | 1.53 | 0.16 | 0.026 |
| 21 | 2.35 | 2.72 | −0.37 | 0.137 | 1.80 | 2.01 | −0.21 | 0.044 | 1.78 | 1.70 | 0.08 | 0.006 |
| 28 | 3.25 | 3.08 | 0.17 | 0.029 | 2.82 | 2.27 | 0.55 | 0.303 | 1.56 | 1.92 | −0.36 | 0.130 |
| 35 | 3.05 | 3.44 | −0.39 | 0.152 | 2.17 | 2.53 | −0.36 | 0.123 | 1.74 | 2.12 | −0.38 | 0.144 |
| 42 | 3.76 | 3.79 | −0.03 | 0.001 | 2.67 | 2.78 | −0.11 | 0.012 | 2.37 | 2.32 | 0.05 | 0.003 |
| 49 | 4.56 | 4.15 | 0.41 | 0.170 | 3.39 | 3.04 | 0.35 | 0.123 | 3.03 | 2.51 | 0.52 | 0.270 |
| 56 | — | — | — | — | 2.90 | 3.30 | −0.40 | 0.160 | 3.00 | 2.71 | 0.29 | 0.084 |
| 63 | — | — | — | — | 3.70 | 3.56 | 0.14 | 0.020 | 2.60 | 2.90 | −0.30 | 0.090 |
| | | | | 0.707 | | | | 0.940 | | | | 0.873 |

Column 3 follows from the fitted equation (8.3-8), and the resulting SSe is 0.707. Column 7 follows from the fitted regression equation (8.3-9), and the resulting SSe is 0.940. Column 11 follows from the fitted regression equation (8.3-10), and the resulting SSe is 0.873.

The MSe for the three levels are denoted by $S_A^2$, $S_B^2$, and $S_C^2$; they are obtained from (8.3-7):

$$S_A^2 = \tfrac{1}{6}(0.707) = 0.118, \quad S_A = 0.34,$$

$$S_B^2 = \tfrac{1}{8}(0.940) = 0.118, \quad S_B = 0.34,$$

$$S_C^2 = \tfrac{1}{8}(0.873) = 0.109, \quad S_C = 0.33.$$

Having fitted a regression equation of $Y$ on $X$, we may use it to predict $Y$ values corresponding to desired $X$ values, provided the fit is a good one. Also, a simple fitted linear regression of $Y$ on $X$ can be easily inverted to represent $X$ in terms of $Y$ values. Thus, with our notation, a simple linear regression $\widehat{Y} = b_0 + b_1 X$ can be inverted and written as

$$\widehat{X} = (Y - b_0)/b_1 \tag{8.3-11}$$

to estimate $X$ for any $Y$ value.

Instead of using (8.3-11) to estimate $X$, one may wish to fit a simple linear regression of $X$ on $Y$. That is, $X$ may be treated as a dependent variable and $Y$ as an independent one to fit a simple linear regression

$$\widehat{X} = b_0 + b_1 Y, \tag{8.3-12}$$

where

$$b_1 = S_{XY}/S_{XY}, \quad \text{and} \quad b_0 = \overline{X} - b_1 \overline{Y}. \tag{8.3-13}$$

Both (8.3-11) and (8.3-12) will give the same estimate of $X$ when $Y = \overline{Y}$. Otherwise, the choice between them for estimating $X$ is not clear. For $Y < \overline{Y}$, (8.3-12) will provide inflated estimates compared to those provided by (8.3-11), and vice versa. The two alternatives were compared by Krutchkoff (1967) by means of Monte Carlo methods, and his findings seem to support (8.3-12) over (8.3-11).

---

We illustrate the estimation of $X$ from (8.3-11) and (8.3-12) using the data in Example 8.3-1 for nitrite level $A$ only. Note that for nitrite level $A$, the fitted regression equation of $Y$ on $X$ is given by (8.3-8). So $b_0 = 1.65$ and $b_1 = 0.051$. With these constants, expression (8.3-11) becomes

**Example 8.3-2**

$$\widehat{X} = (Y - 1.65)/0.051. \tag{8.3-14}$$

On the other hand, when we fit a simple linear regression of $X$ on $Y$, $b_0$ and $b_1$, given by (8.3-12), are

$$b_1 = 17.37 \quad \text{and} \quad b_0 = -25.88.$$

Hence, the fitted regression (8.3-12) of $X$ on $Y$ is

$$\widehat{X} = -25.88 + 17.37Y. \tag{8.3-15}$$

Suppose that whenever the off-flavor score is 3, the product becomes unfit for store distribution and in that respect has failed. According to our analyses, how long would it take for the product to fail? If we substitute $Y = 3$ in (8.3-14), $\widehat{X} = 27$. From (8.3-15) also, $\widehat{X} = 27$ when $Y = 3$. This should not be surprising because the two estimates of $X$ are always equal when $Y = \overline{Y}$. Notice that $\overline{Y} = 2.90$ is practically equal to 3 and that is why the two estimates are identical. If we wanted to estimate the time it takes to develop a moderate off-flavor score $Y = 4$, then from (8.3-14),

$$\widehat{X} = (4 - 1.65)/0.051 = 46 \text{ days,}$$

whereas from (8.3-15),

$$\widehat{X} = -25.88 + 17.37(4) = 44 \text{ days}.$$

The two estimates are not identical.

## Confidence Intervals

In regression analysis, the slope of a regression equation determines the strength of the linear relationship. Indeed, in simple linear regression the slope is zero if and only if the correlation coefficient is zero. Therefore, it may be of interest to test hypotheses about $\beta$ or find confidence intervals for estimating it. To these ends, we need to know that under the assumptions of our model, $b_1$, estimator of $\beta_1$, is normally distributed with mean $\beta_1$ and standard error

$$S_{b-1} = S/\sqrt{S_{XX}}, \tag{8.3-16}$$

where $S = \sqrt{MSe}$. This fact may be used to construct confidence intervals and tests of hypotheses about $\beta_1$. A$(1 - \alpha)$ 100% confidence interval for $\beta_1$ is given by

$$b_1 \pm (t_{\alpha/2, n-2}) S_{b_1}, \tag{8.3-17}$$

where $t_{\alpha/2, n-2}$ is the $[(\alpha/2)100]$th percentile of the $t$ distribution with $(n - 2)$ DF. We can use confidence intervals to test hypotheses about $\beta_1$. For instance, we may not reject at the $\alpha$% level of significance the null hypothesis that $\beta_1 = \beta_1^0 = 0$ if $\beta_1^0$ is found contained in an observed $(1 - \alpha)100$% confidence interval given by (8.3-17). We note that, in particular, when $\beta_1^0 = 0$, the null hypothesis becomes $\beta_1 = 0$, which is equivalent to a null hypothesis about the correlation coefficient being zero.

For constructing confidence intervals for $\beta_0$, one needs the standard error of its estimate $b_0$, which is given by

$$S_{b_0} = S\sqrt{\sum X^2 / S_{XX}}. \tag{8.3-18}$$

A $(1 - \alpha)100$% confidence interval for $\beta_0$ can be obtained by

$$b_0 \pm (t_{\alpha/2, n-2}) S_{b_0}. \tag{8.3-19}$$

In many applications the researcher may wish to estimate the mean $\beta_0 + \beta_1 X$ of the $Y$ values as a function of the independent variable $X$. An estimate of $\beta_0 + \beta_1 X$ is $b_0 + b_1 X$, as given by (8.3-4). The

$(1 - \alpha)100\%$ confidence bands on the mean $\beta_0 + \beta_1 X$ for any $X$ are given by

$$(b_0 + b_1 X) \pm S \left\{ \sqrt{2 F_{1-\alpha,2,n-2} \left[ \frac{1}{n} + \frac{(X - \bar{X})^2}{S_{XX}} \right]} \right\} \qquad (8.3\text{-}20)$$

where $F_{1-\alpha,2,n-2}$ is the $[(1 - \alpha)100]$th percentile of the $F$ distribution with 2 and $n - 2$ DF of the numerator and the denominator, respectively. The theoretical details on confidence bands in (8.3-20) are given in a book by Scheffé (1959).

---

**Example 8.3-3**

For the data in Table 8.7 corresponding to nitrite level $A$, we have fitted a simple linear regression for off-flavor score $Y$ on the number of days $X$ in Example 8.3-1. The fitted simple linear regression

$$\hat{Y} = 1.65 + 0.051X$$

is equivalent to stating that $\beta_0 + \beta_1 X$ is estimated by

$$b_0 + b_1 X = 1.65 + 0.051X.$$

In Example 8.3-1, $S$ was denoted by $S_A$, which is 0.34; $S_{XX} = 2058$, and $\bar{X} = 24.5$. With these quantities, a 95% confidence interval on $b_1$ can be obtained from (8.3-17) with $b_1 = 0.051$ and $t_{0.025,6} = 2.45$ from the table of the $t$ distribution (Table A.2).

From (8.3-16),

$$S_{b_1} = 0.34/\sqrt{2058} = 0.008.$$

Substituting these quantities in (8.3-17), we get

$$0.051 \pm (2.45)(0.008).$$

Hence, a 95% confidence interval for $\beta_1$ is

$$0.031 \leq \beta_1 \leq 0.071.$$

We may also use (8.3-19) to set up a confidence interval on $\beta_0$. From Example 8.3-1, we have $\Sigma X^2 = 6860$ and from (8.3-18),

$$S_{b_0} = (0.34)\sqrt{6830/2058} = 0.62.$$

Hence, a 95% confidence interval on $\beta_0$ is

$$1.65 \pm (2.45)(0.62)$$

or

$$0.13 \leq \beta_0 \leq 3.17.$$

Now suppose we want a 95% confidence bands on $\beta_0 + \beta_1 X$. For this purpose, to use (8.3-20), we need to have $F_{0.95,2,6}$. From the $F$ distribution table (Table A.4), one can see that $F_{0.95,2,6} = 5.14$. We have all the quantities needed in (8.3-20) for setting up 95% degree of confidence bands on $\beta_0 + \beta_1 X$. The bands are given by

$$1.654 + 0.051X \pm 0.34 \sqrt{2(5.14)\left[\frac{1}{8} + \frac{(X - 24.5)^2}{2058}\right]}$$

For instance, when $X = 7$, values of lower and upper bands are given by

$$1.654 + 0.051(7) \pm 0.34 \sqrt{10.28\left[0.125 + \frac{(7 - 24.5)^2}{2058}\right]}.$$

Thus, at $X = 7$, the lower and upper bands are 1.44 and 2.59, respectively. Other calculations are given in Table 8.9 and shown graphically in Figure 8.21.

From Figure 8.21 we notice that graphs of the confidence bands given by (8.3-20) are curvilinear. A practical implication, which becomes clear from the graphs, is that predictions about the mean function $\beta_0 + \beta_1 X$ are subject to increasing variation as one moves away from the center coordinates $(\overline{X}, \overline{Y})$ of the observed data.

There are also straight-line confidence bands for estimating the mean function $\beta_0 + \beta_1 X$. Straight-line bands have been considered in the literature by Acton

**Table 8.9** Curvilinear and Straight-Line 95% Confidence Bands for $b_0 + b_1 X$

| X | Lower Bands | | Upper Bands | |
|---|---|---|---|---|
| | **Curvilinear** | **Straight Line** | **Curvilinear** | **Straight Line** |
| 7 | 1.44 | 1.27 | 2.59 | 2.75 |
| 14 | 1.90 | 1.78 | 2.83 | 2.95 |
| 21 | 2.35 | 2.29 | 3.12 | 3.15 |
| 28 | 2.68 | 2.65 | 3.48 | 3.51 |
| 35 | 2.97 | 2.85 | 3.90 | 4.02 |
| 42 | 3.22 | 3.05 | 4.37 | 4.53 |
| 49 | 3.44 | 3.26 | 4.86 | 5.04 |

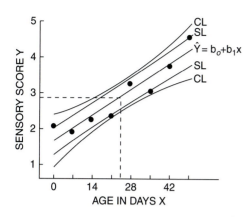

(1959), Folks and Antle (1967), Graybill and Bowden (1967), Dunn (1968), and others. To obtain $(1 - \alpha)100\%$ degree of confidence straight-line bands, Graybill and Bowden suggested the following:

$$b_0 + b_1X \pm Sd_\alpha \left[ \frac{1}{\sqrt{n}} + \frac{|X - \bar{X}|}{\sqrt{S_{XX}}} \right], \tag{8.3-21}$$

where $d_\alpha$ is found in Table 8.10 and $|X - \bar{X}|$ means that the difference $(X - \bar{X})$ is always taken to be positive. For example, if $(X - \bar{X}) = -3$, then $|X - \bar{X}| = 3$. We may point out that Bowden and Graybill (1966) provided an extensive table for $d_\alpha$. For the reader wishing to have the straight-line bands given by (8.3-21), it will be necessary to go to the extensive table for $d_\alpha$ values.

Continuing with our Example 8.3-3, we want 95% degree of confidence straight-line bands, and we refer to Table 8.10 for $\alpha = 0.05$. Recall that, since $n = 8$, $n - 2 = 6$. From Table 8.10, $d_{0.05} = 2.92$. All the other entries in (8.3-21) have been already identified in Example 8.3-3. Their substitution into (8.3-21) leads to

$$1.65 + 0.051X \pm (0.34)(2.92) \left[ \frac{1}{\sqrt{8}} + \frac{|X - 24.5|}{\sqrt{2058}} \right]$$

or

$$1.65 + 0.051X \pm 0.99 \left[ 0.35 + \frac{|X - 24.5|}{45.37} \right]. \tag{8.3-22}$$

For $X = 7$, we have from (8.3-22) the lower band $= 1.27$ and the upper band $= 2.75$. Table 8.9 lists the lower and the upper bands for other $X$ values. The plot of the straight-line bands in Table 8.9 is given in Figure 8.22. Since both the lower and upper bands are straight lines, we require only two points for each band to plot them as shown in Figure 8.22.

**Table 8.10** Abbreviated Table for $d_\alpha$ Values[a]

| $n - 2$ | α | |
|---|---|---|
| | **0.05** | **0.10** |
| 4 | 3.38 | 2.66 |
| 6 | 2.92 | 2.39 |
| 8 | 2.72 | 2.26 |
| 10 | 2.61 | 2.19 |
| 12 | 2.54 | 2.15 |
| 14 | 2.49 | 2.12 |
| 16 | 2.46 | 2.10 |
| 20 | 2.41 | 2.07 |
| 24 | 2.38 | 2.05 |
| 30 | 2.35 | 2.03 |
| 40 | 2.32 | 2.01 |
| 50 | 2.30 | 1.99 |

[a]*Source: Bowden and Graybill (1966). With permission of the American Statistical Association.*

■ **FIGURE 8.22** Straight-line 95% confidence bands for $b_0 + b_1X$ corresponding to the data in Example 8.3-3 and the estimated shelf lives.

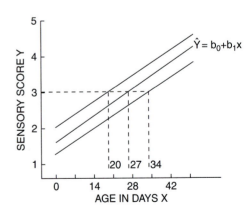

Once we have a plot of the bands, it can be used to construct an interval of possible $X$ values that may correspond to a given $Y$. This may be of interest in shelf life experiments where, for example, $Y$ may denote the off-flavor score of a product as a straight-line function of time $X$. We may wish to specify the time interval during which the product is likely to develop an off-flavor score $Y$ of a specified magnitude. From the specified $Y$ score in the vertical scale, we may extend a horizontal line until it intersects the graphs of the lower and upper bands. From the points of intersections, draw vertical lines downward until each line meets the horizontal scale of time $X$. The two resulting $X$ values on the time scale will specify an interval estimate for $X$. The procedure is illustrated in Figure 8.22 for estimating $X$ when $Y = 3$.

## Nonlinear Relationships

Sometimes a fitted linear regression line of $Y$ on $X$ or $X$ on $Y$ is not adequate for making predictions. This may happen when the true relationship between $Y$ and $X$ is not linear. The linearity can be judged by the magnitude of the correlation coefficient. We discuss briefly only two types of nonlinear relations between $Y$ and $X$, which satisfactorily approximate many other nonlinear relationships.

First, consider a regression model of $Y$ on $X$ given by

$$Y = \beta_0 + \beta_1 X + \beta_2 X^2 + \varepsilon. \qquad (8.3\text{-}23)$$

It is a quadratic model because of the $X^2$ term in the model. Even though it is a quadratic model, we still have only two variables $Y$ and $X$. Experimental data appear as in Table 8.6. Let $X_i$ denote the $i$th value of the independent variable $X$ and $Y_i$ be the corresponding value of the dependent variable $Y$, $i = 1, \ldots n$. Thus, $(X_1, Y_1)$, $(X_2, Y_2), \ldots, (X_n, Y_n)$ is a sample of $n$ pairs.

Define

$$Y = \begin{bmatrix} Y_1 \\ Y_2 \\ \vdots \\ Y_n \end{bmatrix}, \quad X = \begin{bmatrix} 1 & X_1 & X_1^2 \\ 1 & X_2 & X_2^2 \\ \vdots & \vdots & \vdots \\ 1 & X_n & X_n^2 \end{bmatrix},$$

and compute

$$X'X = \begin{bmatrix} n & \sum X_i & \sum X_i^2 \\ \sum X_i & \sum X_i^2 & \sum X_i^3 \\ \sum X_i^2 & \sum X_i^3 & \sum X_i^4 \end{bmatrix}.$$

Then the estimate of

$$\begin{bmatrix} \beta_0 \\ \beta_1 \\ \beta_2 \end{bmatrix}$$

is given by

$$\begin{bmatrix} b_0 \\ b_1 \\ b_2 \end{bmatrix} = (X'X)^{-1} \begin{bmatrix} \sum Y_i \\ \sum X_i Y_i \\ \sum X_i^2 Y_i \end{bmatrix}, \qquad (8.3\text{-}24)$$

where $b_0, b_1$, and $b_2$ are given by (8.3-24). Then

$$\hat{Y} = b_0 + b_1 X + b_2 X^2, \qquad (8.3\text{-}25)$$

When estimating shelf life $X$ for a given $Y$, we rewrite Eq. (8.3-25) as

$$b_2 X^2 + b_1 X + (b_0 - Y) = 0. \qquad (8.3\text{-}26)$$

Equation (8.3-26) is a quadratic in $X$ and can be rewritten as

$$aX^2 + bX + c = 0, \qquad (8.3\text{-}27)$$

with $a = b_2$, $b = b_1$, and $c = b_0 - Y$. The two roots of (8.3-27) are given by

$$X = (-b \pm \sqrt{b^2 - 4ac})/2a. \qquad (8.3\text{-}28)$$

It can be verified that, in Example 8.3-1, the second-order fitted regression equation $Y$ on $X$ corresponding to level $A$ is

$$\hat{Y} = 2.00 + 0.002X + 0.0001X^2.$$

When the fitted equation is rewritten as in (8.3-27), $a = 0.001$, $b = 0.002$, and $c = (2.00 - Y)$. Hence, the two roots of $X$ for any $Y$ can be found from formula (8.3-28). For $Y = 3$, we find that one of the two roots is $X = 31$.

One other frequently fitted nonlinear relationship is the exponential relationship of the following form:

$$Y = \alpha \exp(\beta X).$$

Every exponential relation of this form can be rewritten as

$$\ln Y = \ln \alpha + \beta X. \qquad (8.3\text{-}29)$$

Letting $Z = \ln Y$, $\beta_0 = \ln \alpha$, and $\beta_1 = \beta$, Eq. (8.3-29) can be expressed as

$$Z = \beta_0 + \beta_1 X,$$

which is a linear relationship between $Z$ and $X$. We can now follow linear regression analysis methods to fit a simple linear regression of $Z$ and $X$ and make statistical decisions keeping in mind that $Z = \ln Y$. Indeed, for the data in Example 8.3-1, when we use $Z = \ln Y$ scores to fit a simple linear regression of $Z$ on $X$, it is found that

$$Z = b_0 + b_1 X = 0.601 + 0.016X, \qquad (8.3\text{-}30)$$

where now $b_1$ and $b_0$ are computed as follows. If

$$S_{XZ} = \sum (X - \bar{X})(Z - \bar{Z}),$$

then

$$b_1 = S_{XZ}/S_{XX}, \quad \text{and} \quad b_0 = \bar{Z} - b_1 \bar{X}.$$

In terms of $Y$, (8.3-30) is

$$\ln Y = 0.601 + 0.016X$$

or

$$Y = \exp(0.601) \exp(0.016X) = 1.82 \exp(0.016X). \qquad (8.3\text{-}31)$$

One can use (8.3-31) to estimate $X$ (the shelf life) corresponding to different off-flavor scores. For example, when $Y = 3$, the $X$ that satisfies (8.3-31) must satisfy

$$\exp(0.016X) = 3/1.82 = 1.648. \qquad (8.3\text{-}32)$$

Since the exponential function is widely tabulated, it can be seen that for (8.3-32) to hold, we must have

$$0.016X = 0.499,$$

and hence $X \simeq 31$ days.

It is rather interesting to note that the quadratic model also estimates $X = 31$ days, corresponding to $Y = 3$ for the off-flavor data used in the illustrations.

## EXERCISES

**8.1.** The following are ages to failure for a refrigerated food product:

| Sample | Age to Failure, Days | Sample | Age to Failure, Days |
|--------|---------------------|--------|---------------------|
| 1 | 72 | 10 | 55 |
| 2 | 48 | 11 | 34 |
| 3 | 36 | 12 | 65 |
| 4 | 42 | 13 | 73 |
| 5 | 41 | 14 | 43 |
| 6 | 33 | 15 | 51 |
| 7 | 49 | 16 | 64 |
| 8 | 58 | 17 | 29 |
| 9 | 43 | 18 | 69 |

1. Compute the hazard and cumulative hazard functions. Then carry out the hazard plotting for the normal, lognormal, and Weibull distributions. Which distribution gives the best fit to the observed failures?
2. Estimate the median shelf life of the product using the Weibull distribution.
3. Estimate the Weibull shape parameter $b$ graphically and by using appropriate formulas.
4. Is the failure rate increasing or decreasing?

**8.2.** Construct the hazard function for the normal and lognormal distributions when $\mu = 2$ and $\sigma = 1$.

**8.3.** Off-flavor data for a refrigerated food product are tabulated in the following, where each value (based on a 7-point scale) is an average of 24 observations (3 packages $\times$ 8 panelists):

| Testing Periods, Days (X) | Mean Scores (Y) |
|---------------------------|-----------------|
| 0 | 1.80 |
| 4 | 1.92 |
| 8 | 2.50 |
| 12 | 3.00 |
| 16 | 2.90 |
| 20 | 3.45 |
| 24 | 3.55 |
| 28 | 4.00 |

**1.** Fit a simple linear regression of $Y$ on $X$ and also of $X$ on $Y$. Estimate $X$ from the fitted regressions when $Y = 3$.

**2.** Determine the 95% straight-line confidence bands for $b_0 + b_1 X$ and graph them.

**8.4.** The following are off-flavor scores $Y$ corresponding to three storage temperatures $X$. For storage temperatures of 100, 70, and 40°F, the independent variable $X$ was coded to be 2, 1, and 0, respectively.

**1.** Fit a simple linear regression of $Y$ on $X$.

**2.** Determine the 95% confidence limits on $b_1$, $b_0$, $b_0 + b_1$, and $b_0 + 2b_1$. What physical meaning do these parameters have in this problem?

| Temperature X | Off-flavor Y |
|---|---|
| 0 | 4.50 |
| 0 | 4.25 |
| 0 | 4.75 |
| 0 | 3.00 |
| 0 | 3.75 |
| 0 | 3.00 |
| 0 | 4.25 |
| 0 | 3.50 |
| 1 | 6.00 |
| 1 | 5.00 |
| 1 | 5.75 |
| 1 | 4.75 |
| 1 | 3.50 |
| 1 | 5.50 |
| 1 | 5.25 |
| 1 | 3.75 |
| 2 | 5.75 |
| 2 | 5.00 |
| 2 | 5.00 |
| 2 | 4.00 |
| 2 | 6.25 |
| 2 | 6.00 |
| 2 | 4.50 |
| 2 | 4.75 |

# Nonparametric Statistical Methods

The theoretical validity of statistical methods rests on underlying statistical assumptions. For instance, statistical methods in the analysis of experimental designs require that observations be normally and independently distributed. Statistical methods that depend on specific distributional assumptions are called *parametric methods*. Widely used parametric methods have been proven to be robust in the sense that they remain valid when used to analyze large sets of data even if the underlying distributional assumptions are not true. But for small quantities of data, parametric methods can be misleading when some underlying distributional assumptions are not met. Owing to this and other practical considerations, there have also been developed statistical methods that do not rest upon specific distributional assumptions. Such methods are called *nonparametric methods*. In this chapter we consider a number of important nonparametric statistical methods that are useful in the analyses of data obtained in biological, food, and consumer research studies.

## 9.1  SOME METHODS FOR BINARY DATA

### Tests for Independent Proportions

#### *The Binomial Test*

The binomial test is a one-proportion test. It is a test for parameter $p$ in a binomial model for given $x$, $n$, and $p_0$, where $x$ is the number of successes in total trials, $n$ is the number of trials, and $p_0$ is a specified value $0 < p_0 < 1$. The assumptions for the binomial test are

- The outcome of each trial is binary, i.e., yes/no, 1/0, success/failure, present/absent, etc.
- The probability of a success, $p$, is the same for each trial.
- The $n$ trials are independent.

The null hypothesis is

$$H_0: p = p_0.$$

The alternative hypothesis for a two-sided test is

$$H_a: p \neq p_0.$$

For a one-sided test, it is

$$H_a: p > p_0 \text{ for upper tail test, or}$$
$$H_a: p < p_0 \text{ for lower tail test.}$$

The test statistic is the observed number of successes $c$. Reject $H_0$, if (9.1-1) or (9.1-2) holds; otherwise, do not reject $H_0$.

For a two-sided test,

$$c > c_{1-\alpha/2} \text{ or } c < c_{\alpha/2}, \tag{9.1-1}$$

for a one-sided test

$$c > c_{1-\alpha} \text{ or } c < c_{\alpha}, \tag{9.1-2}$$

where $c_{1-\alpha}, c_\alpha, c_{1-\alpha/2}, c_{\alpha/2}$ are the percentile points of a binomial distribution. Most statistical software contains built-in programs for the percentile points and the test, e.g., *qbinom* and *binom.test* in S-PLUS. It is noted that the critical values in Table A.25 are $c_{1-\alpha} + 1$ for a one-sided test and $c_{1-\alpha/2} + 1$ for a two-sided test.

When a sample size, $n$, is large, e.g., $np_0 \geq 5$ and $n(1 - p_0) \geq 5$, an asymptotic normal statistic, without or with continuity correction, can be used. It is

$$Z = \frac{|x/n - p_0|}{\sqrt{p_0(1 - p_0)/n}} \tag{9.1-3}$$

or

$$Z = \frac{|x/n - p_0| - 1/(2n)}{\sqrt{p_0(1 - p_0)/n}}. \tag{9.1-4}$$

---

**Example 9.1-1**

A new beverage with less sugar was produced. In order to determine whether consumers can detect the difference between the new beverage (B) and the current beverage (A) in sweetness, researchers conducted a sensory difference test using the 3-alternative forced-choice (3-AFC) method (Green and Swets, 1966). A significance level $\alpha = 0.05$ was selected for the test.

One hundred consumer panelists participated in the test. A set of three samples was presented to each of the panelists. The sample sets were AAB, ABA, and BAA. The panelists were asked to select the sample less sweet in a sample set. They had to select one sample even if they could not identify the one less sweet. If the consumers could not detect the difference, the probability of selecting B as less sweet should have been 1/3. Of the 100 panelists, 45 selected beverage B as less sweet. Using an S-PLUS built-in function for a one-sided binomial test, we can get $c_{1-\alpha}$, i.e., $c_{0.95} = 41$,

```
> qbinom(0.95,100,1/3)
```

```
[1] 41
```

Because $45 > 41$, we can conclude at a $\alpha = 0.05$ level that there was a significant difference between beverage A and beverage B in perceptive sweetness. We can also do the test directly using an S-PLUS program as follows:

```
> binom.test(45,100,1/3,alternative = "greater")
        Exact binomial test
data: 45 out of 100
number of successes = 45, n = 100, p-value = 0.01
alternative hypothesis: true p is greater than 0.33333
```

Because the $P$-value $= 0.01 < 0.05$, we can reach the same conclusion. Using normal approximation in (9.1-3) and (9.1-4), we get $Z = 2.475$ and $Z = 2.369$, with the associated $P$-values 0.007 and 0.009, respectively. For the binomial test in this case, we can also use Table A.25 to find a critical value. It is 42 for $n = 100$, $\alpha = 0.05$, and $p_0 = 1/3$. The null hypothesis is rejected because the observed number of correct responses $c = 45$ is larger than the corresponding critical value 42.

---

### The z Test for Two Proportions

For comparison of two independent proportions, i.e., two unknown success probabilities, $p_1$ and $p_2$ from two independent binomial samples, the commonly used approach is the normal $z$ test. The null hypothesis is $H_0: p_1 = p_2$, and the alternative hypothesis is $H_a: p_1 \neq p_2$ (for a two-sided test) or $H_a: p_1 > p_2$ or $H_a: p_1 < p_2$ (for a one-sided test). The test statistic, without or with continuity correction, is

$$Z = \frac{|\hat{p}_1 - \hat{p}_2|}{\sqrt{\bar{p}(1 - \bar{p})(1/n_1 + 1/n_2)}} \tag{9.1-5}$$

or

$$Z = \frac{|\hat{p}_1 - \hat{p}_2| - 0.5(1/n_1 + 1/n_2)}{\sqrt{\bar{p}(1 - \bar{p})(1/n_1 + 1/n_2)}}, \tag{9.1-6}$$

where $\bar{p}$ is a weighted mean of $\hat{p}_1$ and $\hat{p}_2$, $\bar{p} = \dfrac{n_1 \hat{p}_1 + n_2 \hat{p}_2}{n_1 + n_2}$. It follows approximately the standard normal distribution.

| Example 9.1-2 | A consumer preference test for the current product and a new product was conducted in two regions with 100 and 120 consumer panelists, respectively. The investigator wanted to know if there was a significant difference between the two regions in preference. A two-sided test with a significance level $\alpha = 0.05$ was selected. There were 80 and 75 consumers who preferred the new product in the two regions. The proportions are $\hat{p}_1 = 0.8$ and $\hat{p}_2 = 0.625$, respectively. The weighted mean of the two proportions is $\bar{p} = \dfrac{100 \times 0.8 + 120 \times 0.625}{100 + 120}$ $= 0.7045$, and the value of the test statistic is $Z = 2.8328$ according to (9.1-5). Because $2.8328 > 1.96 = z_{1-\alpha/2} = z_{0.975}$, we can conclude at a 0.05 significance level that the two preference proportions are significantly different. |
|---|---|

### The Chi-square Test for $2 \times 2$ Table

Data for two independent binomial samples can be presented in a $2 \times 2$ table like that shown in Table 9.1, where rows represent samples and columns represent "present" and "absent." The test statistic in (9.1-7) follows a $\chi^2$ distribution with 1 DF. The critical values of a $\chi^2$ distribution with 1 DF for one- and two-sided tests are given in Table 9.2. The $P$-value of a one-sided test should be half of the $P$-value of the two-sided test. It can be demonstrated that the square of the statistic (9.1-5), $Z^2$, is equal to

$$X^2 = \frac{(n_{11}n_{22} - n_{21}n_{12})^2 n}{n_{1+}n_{2+}n_{+1}n_{+2}} \tag{9.1-7}$$

The S-PLUS program *prop.test* can be used to test two or multiple independent proportions.

**Table 9.1** $2 \times 2$ Table for Two Independent Binomial Samples

|  | **Present** | **Absent** |  |
|---|---|---|---|
| Sample 1 | $n_{11}$ | $n_{1+} - n_{11}$ | $n_{1+}$ |
| Sample 2 | $n_{+1} - n_{11}$ | $n_{2+} - n_{+1} + n_{11}$ | $n_{2+}$ |
|  | $n_{+1}$ | $n_{+2}$ | $n$ |

**Table 9.2** Critical Values for Chi-square Test with 1 DF

| $\alpha$-Significance Level | 0.10 | 0.05 | 0.01 |
|---|---|---|---|
| Two-Sided Test | 2.706 | 3.841 | 6.635 |
| One-Sided Test | 1.642 | 2.706 | 5.412 |

**Example 9.1-3**

The data in Example 9.1-2 can be presented in a 2 × 2 table, as shown in Table 9.3. According to (9.1-7), we calculate

$$X^2 = \frac{(80 \times 45 - 75 \times 20)^2 \times 220}{100 \times 120 \times 155 \times 65} = 8.0248.$$

Because $8.0248 > 3.841 = \chi^2_{1-\alpha,1}$ (from Table 9.2), we can claim that the two proportions are significantly different. Using the S-PLUS program *prop.test*, we can obtain the same results. The reader should note that $Z^2 = 2.8328^2 = 8.0248$; the statistic (9.1-7) is equivalent to (9.1-5). If we have some prior information about the preference and believe that the preference proportion in one region is larger than that in the other region, the test should be one-sided. In this situation the critical value is 2.706 instead of 3.841. The *P*-value should be $0.0046/2 = 0.0023$.

```
> prop.test(c(80,75),c(100,120),correct = F, alter =
  "two.sided")

2-sample test for equality of proportions without
  continuity correction

data: c(80, 75) out of c(100, 120)
X-square = 8.0248, df = 1, p-value = 0.0046
alternative hypothesis: two.sided
95 percent confidence interval:
0.05817028 0.29182972
sample estimates:
prop'n in Group 1 prop'n in Group 2
          0.80              0.625
```

**Table 9.3** 2 × 2 Table for Two Independent Binomial Samples (Example 9.1-3)

|  | **Preference** | **Nonpreference** | |
|---|---|---|---|
| Region 1 | $n_{11} = 80$ | $n_{12} = 20$ | $n_{1+} = 100$ |
| Region 2 | $n_{21} = 75$ | $n_{22} = 45$ | $n_{2+} = 120$ |
|  | $n_{+1} = 155$ | $n_{+2} = 65$ | $n = 220$ |

### Fisher's Exact Test

For comparison of two independent binomial samples with small sample sizes, Fisher's (1934) exact test is appropriate. For a given row and column marginal total in Table 9.1, $n_{11}$, the number of successes in sample 1 determines the other three cell counts. The number of $n_{11}$ follows a hypergeometric distribution with parameters $n_{1+}$, $n_{2+}$ and $n_{+1}$. The probability of a particular value $n_{11}$ and the other three cell counts is given as follows:

$$\Pr(n_{11}) = \frac{\binom{n_{1+}}{n_{11}} \binom{n_{2+}}{n_{+1} - n_{11}}}{\binom{n_{1+} + n_{2+}}{n_{+1}}}. \tag{9.1-8}$$

The range of possible values of $n_{11}$ is $m_- \leq n_{11} \leq m_+$, where $m_- = \max(0, n_{+1} - n_{2+})$ and $m_+ = \min(n_{+1}, n_{1+})$.

The basic idea behind Fisher's exact test is that if the two proportions are homogeneous (or two characteristics of a sample are independent) in the $2 \times 2$ table and if we consider the marginal totals as given, the test is easy to determine both theoretically and computationally. The practice is to enumerate all possible outcomes and add up the probabilities of those tables more extreme than the one observed. For the null hypothesis $H_0: p_1 = p_2$ versus the alternative hypothesis $H_a: p_1 > p_2$, or $H_a: p_1 < p_2$, the exact $P$-value is

$$P = \sum_{n_{11}=x_0}^{m_+} \Pr(n_{11}), \tag{9.1-9}$$

or

$$P = \sum_{n_{11}=m_-}^{x_0} \Pr(n_{11}), \tag{9.1-10}$$

where $x_0$ is an observed number of $n_{11}$. If the value of (9.1-9) or (9.1-10) is smaller than the specified Type I error $\alpha$, the null hypothesis should be rejected.

For the two-sided test with $H_0: p_1 = p_2$ versus the alternative hypothesis $H_a: p_1 \neq p_2$, the exact $P$-value is usually defined as the two-tailed sum of the probabilities of tables no more likely than the observed table. To calculate it, we first calculate all the values of $\Pr(n_{11})$, $m_- \leq n_{11} \leq m_+$ (the sum of the values should be 1), then select the values that are smaller than or equal to the value of $\Pr(x_0)$. The sum of the values is the $P$-value for the two-sided test. If the $P$-value is smaller than the specified Type I error $\alpha$, the null hypothesis should be rejected.

Fisher's exact test can also be used for the chi-square test of independence between the row and column variable of the $2 \times 2$ table. The practice and result are the same as the two-sided test of homogeneity. Many statistical software packages contain functions of the hypergeometric distribution and Fisher's exact test.

**Example 9.1-4**

R. A. Fisher (1935) designed a tea-tasting experiment for the exact test. This is an interesting example that maybe we can regard as a kind of sensory discrimination test. The objective of the example was to test a lady's claim that she could tell, when drinking tea, whether the milk had been poured before or after the tea. In this experiment the lady was offered randomly eight cups of tea, in four of which milk was added first, and was asked to decide which of these had the milk added first and which last. She was told that there were four cups of each type. The data are given in Table 9.4. This is a somewhat rare example of the generation of a $2 \times 2$ table in which both margins are determined in advance.

The possible numbers of $n_{11}$, i.e., responses of "Milk first" for the cups in which milk was added first, are 0, 1, 2, 3, and 4. The number of $n_{11}$ follows a hypergeometric distribution with parameters (4, 4, 4). The probabilities of $n_{11} = 0$, 1, 2, 3, and 4 are 0.0142857, 0.2285714, 0.5142857, 0.2285714, and 0.0142857, respectively. Using the S-PLUS built-in function *dhyper*, we can get the values easily. For example, for $n_{11} = 3$, the probability is 0.2285714.

```
> dhyper(3,4,4,4)
[1] 0.2285714
```

For a two-sided test with $H_0: p_1 = p_2$ versus the alternative hypothesis $H_a$: $p_1 \neq p_2$, where $p_1$ and $p_2$ are the probabilities of response "Milk first" for the two types of cups, the *P*-value should be $2 \times (0.2286 + 0.0143) = 0.486$. This result suggests that, based on the small data, the lady's claim cannot be confirmed. Using an S-PLUS program, we can obtain the same results:

**Table 9.4** Fisher's Tea-Tasting Experiment (Example 9.1-4)

| Poured First | Guess Poured First | | Total |
|---|---|---|---|
| | **Milk** | **Tea** | **Total** |
| Milk | 3 | 1 | 4 |
| Tea | 1 | 3 | 4 |
| Total | 4 | 4 | |

*Source: Based on Fisher's experiment (1935).*

```
> fisher.test(a)

    Fisher's exact test

data: a
p-value = 0.4857
alternative hypothesis: two.sided
> a
    [,1] [,2]
[1,] 3 1
[2,] 1 3
```

---

**Example 9.1-5**

In order to compare purchase intent for a product between male and female consumers, researchers selected 25 female and 20 male consumers. For the question "How likely would you be to buy this product?" there were five possible responses: 5 = Definitely would buy it; 4 = Probably would buy it; 3 = Might or might not buy it; 2 = Probably would not buy it; 1 = Definitely would not buy it. We found that 10 of the 25 female consumers and 3 of the 20 male consumers gave the answers: "Definitely would buy it" or "Probably would buy it." The results are presented in Table 9.5.

We want to know if male and female consumers have the same probability of selecting 5 or 4, i.e., the top two boxes. The null hypothesis is $H_0: p_1 = p_2$ versus the alternative hypothesis $H_a: p_1 > p_2$; i.e., the probability for female consumers is larger than that for male consumers. Type I error $\alpha = 0.1$ is selected. The observed number of $n_{11}$ is $x_0 = 10$, $m_+ = \min(13, 25) = 13$ in this case. According to (9.1-9), $P = \sum_{n_{11}=10}^{13} \text{Pr}(n_{11}) = 0.05104 + 0.01160 + 0.00142 + 0.00007 = 0.06414$. Because $P = 0.06414 < 0.1$, we can conclude at a 0.1 significance level that the probability of selecting the top two boxes for female consumers is larger than that for male

**Table 9.5** Purchase Intent Data (Example 9.1-5)

|  | Buy | Not Buy |  |
|---|---|---|---|
| Female | $n_{11} = 10$ | $n_{1+} - n_{11} = 15$ | $n_{1+} = 25$ |
| Male | $n_{+1} - n_{11} = 3$ | $n_{2+} - n_{+1} + n_{11} = 17$ | $n_{2+} = 20$ |
|  | $n_{+1} = 13$ | $n_{+2} = 32$ | $n = 45$ |

consumers. For a two-sided test, because $m_- = \max(0, 13\text{-}20) = 0$, $m_+ = 13$, there are 14 possible values of $\Pr(n_{11})$ for $n_{11} = 0, 1, 2, \ldots, 13$. They are 0.00000, 0.00004, 0.00069, 0.00582, 0.02910, 0.09167, 0.18805, 0.25521, 0.22969, 0.13558, 0.05104, 0.01160, 0.00142, and 0.00007. The sum of the extreme $\Pr(n_{11})$ values, which are smaller than or equal to $\Pr(x_0) = \Pr(10) = 0.05104$, is 0.0998. Because $0.0998 < 0.1$, we can draw the same conclusion for the two-sided test. Using the S-PLUS built-in function *fisher.test*, we can obtain the result for the two-sided test quickly:

```
> fisher.test(a)

  Fisher's exact test

data: a
p-value = 0.0998
alternative hypothesis: two.sided
> a
    [,1] [,2]
[1,] 10 15
[2,]  3 17
```

## Tests for Dependent Proportions

### The McNemar Test

For a randomized complete block design, which is often used in practice, correlated and matched samples are involved. For comparison of two correlated and matched binomial samples, McNemar's (1947) test can be used. The data for two correlated and matched binomial samples can be presented in a $2 \times 2$ table, as shown in Table 9.6.

**Table 9.6** $2 \times 2$ Table for Two Matched Binomial Samples

| | | Sample 1 | |
|---|---|---|---|
| | | **Present** | **Absent** |
| Sample 2 | Present | *a* | *b* |
| | Absent | *c* | *d* |
| | | | *n* |

The proportions of the two samples are $p_1 = \frac{a+c}{n}$ and $p_2 = \frac{a+b}{n}$, respectively. The difference of the two proportions is $d = p_2 - p_1 = \frac{b-c}{n}$. Under the null hypothesis $H_0: p_1 = p_2$, the standard error of the difference is $\frac{\sqrt{b+c}}{n}$; hence, the McNemar test statistic can be derived as

$$Z = \frac{b-c}{\sqrt{b+c}},$$ (9.1-11)

or

$$X^2 = \frac{(b-c)^2}{b+c}.$$ (9.1-12)

It is an approximate standard normal test statistic (9.1-11) or an approximate $\chi^2$ statistic with 1 DF (9.1-12).

---

**Example 9.1-6**

In order to test possible brand name effect for purchase intent for a product, researchers selected 100 consumer panelists. A product with and without brand name was presented to each of the panelists. For the same question as that in Example 9.1-5, the numbers of the panelists whose responses fell into the top two boxes are given in Table 9.7. This is a one-sided test with $H_0$: $p_1 = p_2$ versus $H_a: p_1 < p_2$. A significance level $\alpha = 0.05$ was selected. Using (9.1-11), we get

$$Z = \frac{23 - 11}{\sqrt{23 + 11}} = 2.058 > 1.645 = z_{0.95},$$

where $z_{0.95}$ can be found from Table A.1. Using (9.1-12), we get

$$X^2 = \frac{(23 - 11)^2}{23 + 11} = 4.235 > 2.706 = \chi^2_{0.95,1},$$

where $\chi^2_{0.9,1} = z^2_{0.95}$ can be found from Table 9.2. Hence, we reject $H_0: p_1 = p_2$ in favor of $H_a : p_1 < p_2$ at $\alpha = 0.05$. The conclusion is that brand name

**Table 9.7** Responses of Purchase Intent for a Product With and Without Brand Name (Example 9-1.6)

|  |  | Without Brand Name | | |
|---|---|---|---|---|
|  |  | **Buy** | **Not Buy** |  |
| With Brand Name | Buy | 26 | 23 | 49 |
|  | Not Buy | 11 | 40 | 51 |
|  |  | 37 | 63 | 100 |

effect exists for the product. We can obtain the same results by using the S-PLUS built-in function *mcnemar.test*. Note that the *P*-value should be 0.0396/2 = 0.019799 for the one-sided test.

```
> mcnemar.test(a,corr = F)

McNemars chi-square test without continuity correction

data: a
McNemar's chi-square = 4.2353, df = 1, p-value = 0.0396
> a
   [,1] [,2]
[1,] 26 23
[2,] 11 40
```

### Cochran's Q Test

For comparison of more than two correlated and matched binomial samples, Cochran's $Q$ statistic (1950) can be used. It is an extension of McNemar's test to multiple samples. The data can be arranged in a two-way table with $n$ rows and $m$ columns, in which each column represents a sample (treatment) and each row represents a matched group. The data in the two-way table are either $X_{ij} = 1$ or $X_{ij} = 0$, which represent "success" or "failure," or any other dichotomization of the possible treatment results. The statistic is given as

$$Q = \frac{m(m-1)\sum_{j=1}^{m}(T_j - \overline{T})^2}{m\sum_{i=1}^{n}u_i - \sum_{i=1}^{n}u_i^2}, \qquad (9.1\text{-}13)$$

or

$$Q = \frac{m(m-1)\sum_{j=1}^{m}T_j^2 - (m-1)S^2}{mS - \sum_{i=1}^{n}u_i^2}, \qquad (9.1\text{-}14)$$

where $u_i = \sum_{j=1}^{m}X_{ij}$ is the total number of successes in the $i$th row; $T_j = \sum_{i=1}^{n}X_{ij}$ is the total number of successes in the $j$th sample (column); $S = \sum_{i=1}^{n}u_i = \sum_{j=1}^{m}T_j$ is the total number of successes in all samples; $\overline{T} = \frac{S}{m}$ is the average number of successes in a sample.

If the true probability of success is the same in all samples, the limiting distribution of Q, when the number of rows is large, is the $\chi^2$ distribution with $(m - 1)$ DF.

**Example 9.1-7**

For this example, there are four prototypes. The investigator wanted to know if there is a significant difference among the prototypes on consumer preference to a current product. One hundred consumer panelists were selected. Each panelist evaluated four pairs of the samples (i.e., each of the four prototypes with a current product) with the response 1 = prefer the prototype and 0 = prefer the current one. The responses of the panelists are listed in Table 9.8. The preference numbers for the four prototypes are

**Table 9.8** Preference Data (Example 9.1-7)

| No. | $P_1$ | $P_2$ | $P_3$ | $P_4$ | $u_i$ | No. | $P_1$ | $P_2$ | $P_3$ | $P_4$ | $u_i$ |
|---|---|---|---|---|---|---|---|---|---|---|---|
| 1 | 0 | 1 | 1 | 0 | 2 | 18 | 1 | 1 | 0 | 0 | 2 |
| 2 | 0 | 1 | 1 | 0 | 2 | 19 | 1 | 1 | 1 | 1 | 4 |
| 3 | 1 | 1 | 0 | 1 | 3 | 20 | 0 | 0 | 0 | 0 | 0 |
| 4 | 1 | 1 | 1 | 1 | 4 | 21 | 1 | 0 | 0 | 0 | 1 |
| 5 | 1 | 1 | 0 | 1 | 3 | 22 | 1 | 0 | 0 | 1 | 2 |
| 6 | 1 | 1 | 1 | 1 | 4 | 23 | 1 | 1 | 1 | 0 | 3 |
| 7 | 1 | 1 | 1 | 1 | 4 | 24 | 1 | 1 | 1 | 1 | 4 |
| 8 | 1 | 1 | 1 | 0 | 3 | 25 | 0 | 1 | 1 | 1 | 3 |
| 9 | 0 | 0 | 1 | 1 | 2 | 26 | 1 | 1 | 1 | 1 | 4 |
| 10 | 1 | 1 | 1 | 1 | 4 | 27 | 0 | 1 | 0 | 1 | 2 |
| 11 | 0 | 1 | 1 | 1 | 3 | 28 | 0 | 1 | 1 | 1 | 3 |
| 12 | 1 | 1 | 1 | 1 | 4 | 29 | 1 | 1 | 1 | 1 | 4 |
| 13 | 1 | 1 | 1 | 1 | 4 | 30 | 0 | 0 | 1 | 1 | 2 |
| 14 | 1 | 1 | 1 | 1 | 4 | 31 | 0 | 1 | 1 | 1 | 3 |
| 15 | 0 | 1 | 1 | 1 | 3 | 32 | 1 | 0 | 1 | 0 | 2 |
| 16 | 0 | 1 | 1 | 1 | 3 | 33 | 1 | 1 | 1 | 0 | 3 |
| 17 | 1 | 1 | 1 | 1 | 4 | 34 | 1 | 0 | 1 | 1 | 3 |

*(continued)*

**Table 9.8** Preference Data (Example 9.1-7)—*cont...*

| No. | $p_1$ | $p_2$ | $p_3$ | $p_4$ | $u_i$ | No. | $p_1$ | $p_2$ | $p_3$ | $p_4$ | $u_i$ |
|-----|-----|-----|-----|-----|-----|-----|-----|-----|-----|-----|-----|
| 35 | 1 | 0 | 0 | 1 | 2 | 62 | 1 | 1 | 0 | 1 | 3 |
| 36 | 1 | 0 | 1 | 1 | 3 | 63 | 1 | 1 | 1 | 1 | 4 |
| 37 | 1 | 0 | 0 | 1 | 2 | 64 | 1 | 0 | 0 | 1 | 2 |
| 38 | 1 | 1 | 1 | 0 | 3 | 65 | 1 | 0 | 0 | 1 | 2 |
| 39 | 1 | 1 | 1 | 1 | 4 | 66 | 1 | 1 | 1 | 1 | 4 |
| 40 | 0 | 0 | 1 | 1 | 2 | 67 | 1 | 1 | 1 | 1 | 4 |
| 41 | 1 | 1 | 1 | 1 | 4 | 68 | 1 | 0 | 0 | 1 | 2 |
| 42 | 1 | 1 | 1 | 1 | 4 | 69 | 1 | 0 | 1 | 0 | 2 |
| 43 | 0 | 1 | 1 | 1 | 3 | 70 | 0 | 0 | 0 | 0 | 0 |
| 44 | 0 | 1 | 1 | 1 | 3 | 71 | 1 | 1 | 0 | 1 | 3 |
| 45 | 1 | 1 | 1 | 1 | 4 | 72 | 0 | 1 | 1 | 1 | 3 |
| 46 | 1 | 1 | 1 | 1 | 4 | 73 | 1 | 0 | 1 | 1 | 3 |
| 47 | 1 | 0 | 0 | 1 | 2 | 74 | 1 | 0 | 1 | 1 | 3 |
| 48 | 0 | 1 | 1 | 1 | 3 | 75 | 1 | 0 | 1 | 1 | 3 |
| 49 | 1 | 1 | 1 | 1 | 4 | 76 | 0 | 1 | 1 | 1 | 3 |
| 50 | 0 | 0 | 0 | 0 | 0 | 77 | 1 | 1 | 1 | 1 | 4 |
| 51 | 1 | 0 | 1 | 1 | 3 | 78 | 0 | 0 | 1 | 1 | 2 |
| 52 | 1 | 1 | 1 | 1 | 4 | 79 | 1 | 1 | 1 | 1 | 4 |
| 53 | 1 | 1 | 1 | 1 | 4 | 80 | 0 | 0 | 0 | 0 | 0 |
| 54 | 1 | 1 | 1 | 1 | 4 | 81 | 1 | 0 | 1 | 0 | 2 |
| 55 | 1 | 0 | 1 | 1 | 3 | 82 | 1 | 0 | 1 | 1 | 3 |
| 56 | 1 | 1 | 1 | 1 | 4 | 83 | 1 | 1 | 1 | 1 | 4 |
| 57 | 1 | 1 | 1 | 1 | 4 | 84 | 1 | 1 | 1 | 1 | 4 |
| 58 | 1 | 1 | 1 | 1 | 4 | 85 | 1 | 1 | 1 | 0 | 3 |
| 59 | 1 | 0 | 1 | 1 | 3 | 86 | 1 | 0 | 1 | 1 | 3 |
| 60 | 1 | 0 | 1 | 1 | 3 | 87 | 0 | 1 | 1 | 1 | 3 |
| 61 | 1 | 1 | 0 | 1 | 3 | 88 | 1 | 0 | 1 | 0 | 2 |

*(continued)*

**Table 9.8** Preference Data (Example 9.1-7)—*cont...*

| No. | $p_1$ | $p_2$ | $p_3$ | $p_4$ | $u_i$ | No. | $p_1$ | $p_2$ | $p_3$ | $p_4$ | $u_i$ |
|-----|-------|-------|-------|-------|-------|-----|-------|-------|-------|-------|-------|
| 89 | 0 | 0 | 1 | 1 | 2 | 95 | 1 | 1 | 1 | 1 | 4 |
| 90 | 0 | 0 | 0 | 0 | 0 | 96 | 1 | 1 | 1 | 1 | 4 |
| 91 | 1 | 1 | 0 | 1 | 3 | 97 | 0 | 0 | 0 | 0 | 0 |
| 92 | 1 | 1 | 1 | 1 | 4 | 98 | 1 | 1 | 1 | 1 | 4 |
| 93 | 1 | 1 | 1 | 1 | 4 | 99 | 0 | 1 | 1 | 1 | 3 |
| 94 | 0 | 0 | 0 | 0 | 0 | 100 | 1 | 1 | 1 | 1 | 4 |
| | | | | | | $T_j$ | 72 | 65 | 77 | 80 | $S = 294$ |

72, 65, 77, and 80, respectively. According to (9.1-14), the $Q$ statistic is computed as

$$Q = \frac{(4)(4-1)(72^2 + 65^2 + 77^2 + 80^2) - (4-1)294^2}{(4)(294) - (2^2 + 2^2 + 3^2 + \cdots + 4^2 + 3^2 + 4^2)} = 8.234.$$

Because $8.234 > 7.815 = \chi^2_{0.95,3}$, which can be found from Table A.3, we can conclude at a 0.05 significance level that the four prototypes differ significantly in preference. The associated $P$-value is 0.041. Using an S-PLUS code, we can obtain the results quickly:

```
> cochqtest(cochqdat4)
[1]] 8.234 0.041
> cochqdat4
numeric matrix: 100 rows, 4 columns.
   [,1] [,2] [,3] [,4]
[1,] 0 1 1 0
[2,] 0 1 1 0
...
```

## Combining 2 × 2 Contingency Tables

### *Some Possible Approaches*

Suppose that we are interested in comparing probabilities of the same event for two populations. For example, consider two independent samples and their outcomes in $2 \times 2$ tables. Suppose further that the sampling is repeated $r$ times. Thus, the total experimental data consist of a series of $2 \times 2$ tables such as the following for the $i$th stage sample:

| | **Frequencies of Events** | | |
|---|---|---|---|
| **Samples** | **Occurring** | **Not Occurring** | **Sample Sizes** |
| Population I | $O_{11}^{(i)} = X_i$ | $O_{12}^{(i)} = n_i - X_i$ | $O_{1+}^{(i)} = n_i$ |
| Population II | $O_{21}^{(i)} = Y_i$ | $O_{22}^{(i)} = m_i - Y_i$ | $O_{2+}^{(i)} = m_i$ |
| | $O_{+1}^{(i)} = X_i + Y_i$ | $O_{+2}^{(i)} = n_i + m_i - X_i - Y_i$ | $O_{++}^{(i)} = n_i + m_i$ |

Note that sample sizes $n_i$ and $m_i$ need not be equal at any stage. We want a test for comparing probabilities of occurrence of the same event in two populations. Cochran (1954) discussed three possible statistics:

1. Combine all the tables in one $2 \times 2$ table and then compute the $X^2$ statistic. This statistic is valid if there is reason to believe that the probability of the event does not change from one stage to another.
2. Compute the $X^2$ for each $2 \times 2$ table from (9.1-7) and then add them all. Thus, the $X^2$ statistic for all the $i$th stage samplings is $X^2 = \sum_{i=1}^{k} X_i^2$ with $k$ DF. This statistic does not account for the direction of the difference between the two probabilities of the same event.
3. For each table, compute

$$w_i = \frac{n_i m_i}{n_i + m_i}, \quad d_i = \frac{X_i}{n_i} - \frac{Y_i}{m_i}, \quad p_i = \frac{X_i + Y_i}{n_i + m_i}, \quad q_i = 1 - p_i \quad (9.1\text{-}15)$$

and then a test statistic

$$Z = \sum_{i=1}^{k} w_i d_i \Big/ \sqrt{\sum_{i=1}^{k} w_i p_i q_i}. \quad (9.1\text{-}16)$$

Note that $Z$ takes into account the sign of each difference $d_i$. Also, $Z$ has approximately the standard normal distribution. If the two samples at each stage are of the same size $n$, then (9.1-15) simplifies to

$$w_i = n/2, \quad d_i = (X_i - Y_i)/n, \quad w_i d_i = (X_i - Y_i)/2, \quad p_i = (X_i + Y_i)/2n. \quad (9.1\text{-}17)$$

| | |
|---|---|
| **Example 9.1-8** | A study was conducted on two types of casing for use in sausage manufacturing. The basis of comparison was the number of sausages that burst during cooking. Fifteen samples of each type, properly identified, were cooked in a pan together for 5 minutes, and the number of sausages that burst was obtained for each type. This procedure was repeated three times using the same sample size. Table 9.9 shows the experimental outcomes. The $X^2$ statistics for each experiment are shown in the footnote to the table. At $\alpha = 0.05$, $\chi^2_{0.95,1} = 3.84$, which is greater than any of the three computed statistics. If we follow Cochran's second suggestion, then our statistic is $X^2 = X_1^2 + X_2^2 + X_3^2 = 1.429 + 0.682 + 0.159 = 2.27$ with 3 DF. Since $2.27 < 7.81 = \chi^2_{0.95,3}$, there appears to be no significant difference between the test and the control casings across experiments. |

Should one follow Cochran's first suggestion, then the pooled $2 \times 2$ table is as shown in Table 9.10. The $X^2$ statistic computed from the pooled table is $X^2 = 0.216$ with 1 DF. No significant difference is found.

**Table 9.9** Three $2 \times 2$ Tables and Their $X^2$ Statistics (Examples 9.1-8 and 9.1-11)

| Experiment Number | Type of Casing | Frequencies | | Marginal Totals, $n_{i.}$ | Percentage Burst | $X^2$ |
|---|---|---|---|---|---|---|
| | | **Burst** | **No Burst** | | | |
| 1 | Test | 3 | 12 | 15 | 20.0 | |
| | Control | 6 | 9 | 15 | 40.0 | |
| | Marginal totals $n_{.j}$ | 9 | 21 | $n = 30$ | $p_1 = 0.30$ | 1.429 |
| 2 | Test | 5 | 10 | 15 | 33.0 | |
| | Control | 3 | 12 | 15 | 20.0 | |
| | Marginal totals $n_{.j}$ | 8 | 22 | $n = 30$ | $p_2 = 0.27$ | 0.682 |
| 3 | Test | 4 | 11 | 15 | 26.7 | |
| | Control | 5 | 10 | 15 | 33.3 | |
| | Marginal totals $n_{.j}$ | 9 | 21 | $n = 30$ | $p_3 = 0.30$ | 0.159 |

*Note:*
*Calculation of $X^2$ statistics for each experiment from (9.1-7):*

*Experiment 1:* $X_1^2 = \dfrac{[(3)(9) - (6)(12)]^2 (30)}{(9)(21)(15)(15)} = 60{,}750/42{,}525 = 1.429$ *with 1 DF.*

*Experiment 2:* $X_2^2 = \dfrac{[(5)(12) - (3)(10)]^2 (30)}{(8)(22)(15)(15)} = 27{,}000/39{,}600 = 0.682$ *with 1 DF.*

*Experiment 3:* $X_3^2 = \dfrac{[(4)(10) - (5)(11)]^2 (30)}{(9)(21)(15)(15)} = 6{,}750/42{,}525 = 0.159$ *with 1 DF.*

**Table 9.10** Pooled Data (Example 9.1-8)

|  | **Burst** | **Did Not Burst** | **Total** |
|---|---|---|---|
| Test | 12 | 33 | 45 |
| Control | 14 | 31 | 45 |
|  | 26 | 64 | 90 |

**Table 9.11** Calculations for Computing the $Z$ Statistic Defined in (9.1-15 and 9.1-16)[a]

| Experiments | $d_i = \frac{x_i}{n_i} - \frac{y_i}{m_i}$ | $w_i d_i$ | $p_i$ | $q_i = 1 - p_i$ | $w_i$ | $w_i p_i q_i$ |
|---|---|---|---|---|---|---|
| 1 | −3/15 | −3/2 | 9/30 = 0.30 | 0.70 | 15/2 | 1.575 |
| 2 | 2/15 | 1 | 8/30 = 0.27 | 0.73 | 15/2 | 1.468 |
| 3 | −1/15 | −1/2 | 9/30 = 0.30 | 0.70 | 15/2 | 1.575 |
|  |  | −1.0 |  |  |  | 4.618 |

[a]Since $n_i = m_i$, use (9.1-17) instead of (9.1-15).
Note:
$\sqrt{\sum_i w_i p_i q_i} = 2.15$; $Z = -1/2.15 = -0.465$; $X_1^2 = Z^2 = 0.216$ with 1 DF.

Suppose we calculate a test statistic following Cochran's third suggestion. The necessary calculations are given in Table 9.11. Note that $n_i = m_i = 15$ for each sample. For example, note that $d_1 = (3/15) - (6/15) = 0.20$, $w_1 = 15/2$, $p_1 = 9/30$, and so on. Since the sample size for each experiment is the same, the weighting factor $w_1 = 15/2$ is the same for all three experiments. At $\alpha = 0.05$ level, $z_\alpha = -1.645$, which is less than the observed $Z = -0.465$. Hence, we cannot reject the null hypothesis of no association between sausage casing and sausage bursting in cooking.

### The Cochran-Mantel-Haenszel Test

Mantel and Haenszel (1959) and Mantel (1963) proposed another approach to combining $2 \times 2$ tables. This approach differs only slightly from Cochran's statistic (9.1-16) in the multiplying factors $\frac{n_i + m_i - 1}{n_i + m_i}$ (Radhakrishna, 1965). When the sample sizes in the $k$ $2 \times 2$ tables are all large, the difference is trivial. For this reason the

Mantel-Haenszel method is often called the Cochran-Mantel-Haenszel *(CMH)* test. The null hypothesis for the test is

$$H_0: \ p_1^{(1)} = p_2^{(1)}, p_1^{(2)} = p_2^{(2)}, \ldots, p_1^{(k)} = p_2^{(k)},$$

and the alternative hypothesis is

$$H_a: \text{either } p_1^{(i)} > p_2^{(i)} \text{ for some } i, \text{or } p_1^{(i)} < p_2^{(i)} \text{ for some } i, \text{but not both.}$$

The *CMH* test statistic is

$$CMH = \frac{\left[\sum_{i=1}^{k}(O_{11}^{(i)} - E_i)\right]^2}{\sum_{i=1}^{k} V_i}, \tag{9.1-18}$$

where $O_{rc}^{(i)}, (r, c = 1, 2)$ denotes the observed number in a $2 \times 2$ table for the *i*th-stage sample; $E_i = \dfrac{O_{1+}^{(i)} O_{+1}^{(i)}}{O_{++}^{(i)}}$; $V_i = \dfrac{O_{1+}^{(i)} O_{2+}^{(i)} O_{+1}^{(i)} O_{+2}^{(i)}}{O_{++}^{(i)2}(O_{++}^{(i)} - 1)}$. The *CMH* statistic is distributed asymptotically $\chi^2$ with 1 DF.

---

**Example 9.1-9**

For the data shown in Table 9.9 for Example 9.1-8, by using (9.1-18), we get $E_1 = E_3 = (15 \times 9)/30 = 4.5$; $E_2 = (15 \times 8)/30 = 4$; $V_1 = V_3 = (15 \times 15 \times 9 \times 21)/[30^2 \times (30 - 1)] = 1.6293$; $V_2 = (15 \times 15 \times 8 \times 22)/[30^2 \times (30 - 1)] = 1.5172$; hence,

$$CMH = \frac{(3 - 4.5 + 5 - 4 + 4 - 4.5)^2}{2 \times 1.6293 + 1.5172} = 0.2094.$$

Because $0.2094 < 3.841 = \chi^2_{0.95,1}$, we cannot reject the null hypothesis of no association between sausage casing and sausage bursting in cooking. Using the S-PLUS built-in function *mantelhaen.test*, we can obtain the same results:

```
> mantelhaen.test(cmhdat4,correct = F)

Mantel-Haenszel  chi-square  test  without  continuity
    correction

data: cmhdat4
Mantel-Haenszel chi-square = 0.2094, df = 1, p-value =
    0.6472
> cmhdat4

, , 1
  [,1][,2]
[1,] 3 12
[2,] 6  9
```

```
, , 2
  [,1] [,2]
[1,] 5 10
[2,] 3 12

, , 3
  [,1] [,2]
[1,] 4 11
[2,] 5 10
```

## Measure of Association for 2 × 2 Tables

One commonly used measure of association in $2 \times 2$ tables for two independent binomial populations is the odds ratio introduced first by Cornfield (1951) as a measure of relative risk. In $2 \times 2$ tables, the odds for population 1 are defined as $\theta^{(1)} = \pi_{11}/\pi_{12}$; the odds for population 2 are $\theta^{(2)} = \pi_{21}/\pi_{22}$. The odds ratio can be defined as the ratio of the two odds in (9.1-19), which is also referred to as the *cross-products ratio* (Fisher, 1962):

$$\theta = \frac{\theta^{(1)}}{\theta^{(2)}} = \frac{\pi_{11}\pi_{22}}{\pi_{21}\pi_{12}} \qquad (9.1\text{-}19)$$

where $\pi_{ij}$ $(i,j = 1,2)$ is the cell probability. The sample odds ratio, i.e., the estimate of the population odds ratio, is

$$\widehat{\theta} = \frac{n_{11}n_{22}}{n_{12}n_{21}}. \qquad (9.1\text{-}20)$$

The odds ratio can be any number between zero and infinity. The value $\theta = 1$ implies that the two probabilities of "Success" of the two populations in a $2 \times 2$ table are identical, or the two characteristics in a $2 \times 2$ table are independent. The values of $\theta = k$ and $\theta = 1/k$ represent the same level of association. The logarithm of the odds ratio, which is symmetric about zero, is often used. The asymptotic standard error of the logarithm of the sample odds ratio is

$$SD(\ln(\widehat{\theta})) = \sqrt{\frac{1}{n_{11}} + \frac{1}{n_{12}} + \frac{1}{n_{21}} + \frac{1}{n_{22}}} \qquad (9.1\text{-}21)$$

or the adjusted version,

$$SD(\ln(\widehat{\theta})) = \sqrt{\frac{1}{n_{11} + 0.5} + \frac{1}{n_{12} + 0.5} + \frac{1}{n_{21} + 0.5} + \frac{1}{n_{22} + 0.5}}. \qquad (9.1\text{-}22)$$

A large-sample confidence interval for the logarithm of the sample odds ratio is

$$\ln(\widehat{\theta}) \pm z_{1-\alpha/2} SD(\ln(\widehat{\theta})). \qquad (9.1\text{-}23)$$

---

**Example 9.1-10**

For the data shown in Table 9.3 for Example 9.1-3, the estimated odds ratio is $\widehat{\theta} = \dfrac{80 \times 45}{20 \times 75} = 2.4$ from (9.1-20), and the logarithm of the odds ratio is 0.88. The standard error of the logarithm of the sample odds ratio is $SD(\ln(\widehat{\theta})) = \sqrt{\dfrac{1}{80} + \dfrac{1}{20} + \dfrac{1}{75} + \dfrac{1}{45}} = 0.3131$ from (9.2-21), and a 0.95 confidence interval for the logarithm of the sample odds ratio is then $0.88 \pm 1.96 \times 0.3131 = (0.2617, 1.4892)$ from (9.1-23).

---

## 9.2 SOME METHODS BASED ON RANKS

Most nonparametric methods use ranks assigned to experimental observations in decision-making rules. There are advantages and some disadvantages in using ranks rather than the observations themselves. A significant advantage with ranks is that they are least disturbed by a few gross errors and outliers in observed data sets. Also, the use of ranks in decision-making rules requires no specific distributional assumptions about sampled data. Moreover, numerical calculations tend to be easy when ranks are used. But a main drawback is that, if ranks only are used, a certain amount of information provided by the actual magnitude of the observations is lost. Also, there arises a practical difficulty in assigning ranks when two or more observations are of the same magnitude. If two or more observations are of the same magnitude, they are then tied for the same rank. How should the ties be resolved? There are various alternative suggestions in the literature given, for example, by Putter (1955), Gridgeman (1959), Pratt (1959), Lehman (1961), Odesky (1967), Bradley (1968), and Draper et al. (1969). Some of the suggestions for assigning ranks to tied observations are as follows:

1. When two or more observations in a sample are tied for the same rank, retain only one observation from among those tied and delete the rest from consideration. This suggestion leads to a reduced sample size and is criticized for ignoring information provided by ties in a sample. That is why this suggestion is not widely accepted.

2. When two or more observations are tied, rank them randomly to assign ranks. For instance, if three observations are tied, we may rank them randomly to assign the three ranks for which they are tied. This suggestion is easy to follow but complicates the distributional theory of test statistics.

3. A more commonly used suggestion is: Assign observations of the same magnitude to the rank that is the average of all the ranks for which the observations are tied. The average of ranks is sometimes called the *midrank*.

4. In preference studies involving only "for" or "against," "like" or "dislike," and "win" or "loss" type responses, a no preference response is a tie. In such studies, ties may be proportionally resolved. For instance, consider a consumer preference study consisting of 248 panelists who compared two food products $X$ and $Y$. Suppose 98 preferred brand $X$ over brand $Y$, whereas 50 preferred brand $Y$ over brand $X$. Thus, there were in all $98 + 50 = 148$ expressed preferences, with 100 panelists giving no preference for one food over the other. The two brands were tied in 100 responses. The 100 ties may be artificially changed into preferences in proportion to the expressed preferences. That is, out of 100 ties, $(98/148) \times 100 = 66$ ties resolved in favor of brand $X$. The remaining $100 - 66 = 34$ ties are automatically resolved as preferences for brand $Y$.

## Rank Tests for Single and Paired Samples

### The Sign Test for Comparing Two Populations

In many situations, observations are subject to natural pairing. That is, observations come in pairs $(X_i, Y_i)$, with one observation in each pair from population $X$ and the other from population $Y$. For instance, in taste testing experiments, each judge may be asked to state preference scores for two competing food products $X$ and $Y$. In general, observations are naturally paired when two treatments are compared in a randomized block design with blocks of size two. Each block consists of the same two treatments. Thus, the sample consists of as many pairs as there are blocks. When the observations on two treatments are paired, the sign test is an easy statistical tool for judging the significance of the difference between the two treatments. To carry out the sign test, associate a plus $(+)$ sign to each pair $(X_i, Y_i)$ if the difference $(X_i - Y_i)$ is positive and a minus $(-)$ sign otherwise. Since there are $n$ pairs of observations, the total of the $+$ and $-$ signs is $n$. If the two populations or treatments being compared are essentially the same, one expects a nearly equal number of $+$ and $-$ signs. Hence, the null hypothesis

$$H_0: \text{No difference between two treatments}$$

would be rejected in favor of an appropriate alternative if there are too few signs of one kind. Let $R^+$ and $R^-$ denote the number of $+$ and $-$ signs, respectively. Remember that $R^+ + R^- = n$; hence, once we know $R^+$, we also know $R^-$. Let $R$ denote the minimum of $R^+$ and $R^-$. Although $R$ is always less than or at most equal to $n/2$, we expect it to be closer to $n/2$ when $H_0$ is true. If $H_0$ is false, $R$ should be observed significantly smaller than $n/2$. Hence, the sign test simply stated is: reject $H_0$ if $R \leq r_0$, where the critical value $r_0$ can be found corresponding to the specified level of significance. Since $R$ can take only integer values with positive probability, it may not be possible to find a critical value $r_0$ that would correspond to a specified level of significance. But this is not really a handicap, because a level of significance, whether it be 0.05, 0.01, or any other value, is really an arbitrary choice. Hence, instead of finding $r_0$ for a specified level of significance, we denote the observed value of $R$ to be $r_0$ and find the significance probability

$$P(r_0) = P(R \leq r_0). \tag{9.2-1}$$

If the observed significance probability $P(r_0)$ is less than the specified level of significance $\alpha$, we will reject the null hypothesis at the specified level in favor of an appropriate alternative. Indeed, it is more informative to know the significance probability $P(r_0)$ in reaching an appropriate decision. Table A.18 provides significance probabilities $P(r_0)$ for $n = 10$ through $n = 100$ and for various values of $r_0$ for each $n$.

It is of practical significance to know that the sign test can be carried out using the normal approximation through the $Z$ statistic

$$Z = 2(R - n)/\sqrt{n}. \tag{9.2-2}$$

If $H_0$ is true, for large $n$, $Z$ in (9.2-2) follows approximately the standard normal distribution. Indeed, the approximation is quite good even for $n$ as small as 20. Using this result, we may reject the null hypothesis of the equality of two distributions in favor of the two-sided alternative if $Z < z_{\alpha/2}$ or $Z > z_{1-\alpha/2}$ at the $\alpha$ level of significance. If the alternative is one-sided, $Z$ must be compared either with $z_\alpha$ or $z_{1-\alpha}$, depending on the direction of the alternative.

Statistical properties of the sign test have been thoroughly studied. Gibbons (1964) showed that the sign test compares well with the $t$ test and that there is almost no difference in power in the case of leptokurtic distribution. For sensory data, Gacula et al. (1972) showed that the sign test compares favorably with the paired test in terms of power and Type I error for establishing statistical significance. Other studies show that the sign test is robust against the basic parametric assumptions (Stewart, 1941; Walsh, 1946; Dixon, 1953).

To summarize, the steps in carrying out the sign test are as follows:

1. Obtain the difference $d_i$, $d_i = X_i - Y_i$, $i = 1, 2, \ldots, n$, and assign it a $+$ sign if $d_i > 0$ and $-$ sign if $d_i < 0$.
2. In the case of ties ($d_i = 0$), assign no sign and reduce the sample size $n$ by the number of ties.
3. Observe the test statistic $R$ that is the smaller of $R^+$ and $R^-$, the number of $+$ and the number of $-$ signs, respectively. Let $r_0$ denote this observed value of $R$.
4. Use Table A.18 to find significance probability $P(r_0)$ and reject $H_0$ in favor of an appropriate alternative if $P(r_0)$ is less than the desired level of significance.

---

**Example 9.2-1**

There were two treatments involved in a study of beef tenderness. An objective evaluation using the Armour Tenderometer meat tenderness tester was performed. The higher the numerical value, the less tender is the meat. For our purpose the first 25 pairs of observations are given in Table 9.12. Following Steps 1–4 just outlined, the test statistic $R$ is observed to be the number of $+$ signs, and its observed value is $r_0 = 6$. Because of one tie, the sample size $n$ is $25 - 1 = 24$. Table A.18 shows that for $r_0 = 6$ and $n = 24$, the significance probability $P(r_0)$ is 0.0227 or about 2%. Thus, we conclude that the treated group is significantly more tender than the control at any $\alpha$ level greater than 0.023. Hence, the two groups may also be judged to be significantly different at the $\alpha = 0.05$ level of significance.

We would like to point out that the sign test is also applicable to testing the null hypothesis that the median $\mu_X$ of population $X$ is known equal to $\mu_0$ against an appropriate alternative. Obviously, in this case there is only one population to be observed. Let $X_1, X_2, \ldots, X_n$ denote the random sample. We may carry out all the steps of the sign test with differences $d_i = X_i - \mu_0$, $i = 1, \ldots, n$. Actually, the sign test is just a special case (with $p_0 = 0.5$) of the binomial test discussed in Section 9.1. Because the sign test is famous and the oldest nonparametric method, it deserves special consideration.

**Table 9.12** Meat Tenderness Readings for Illustration of the Sign Test (Example 9.2-1)

| Treated, X | Control, Y | Sign, X – Y |
|---|---|---|
| 18.9 | 18.4 | + |
| 16.0 | 19.3 | − |
| 17.3 | 21.1 | − |
| 16.5 | 17.9 | − |
| 18.6 | 20.3 | − |
| 18.7 | 16.8 | + |
| 15.1 | 15.4 | − |
| 15.4 | 15.4 | 0 |
| 15.9 | 17.9 | − |
| 14.7 | 15.3 | − |
| 14.8 | 15.4 | − |
| 14.9 | 15.7 | − |
| 14.3 | 17.4 | − |
| 13.9 | 14.1 | − |
| 21.5 | 19.2 | + |
| 25.0 | 24.4 | + |
| 18.4 | 17.0 | + |
| 13.1 | 23.0 | − |
| 20.5 | 21.5 | − |
| 21.2 | 23.4 | − |
| 20.6 | 20.7 | − |
| 16.9 | 18.6 | − |
| 18.1 | 20.0 | − |
| 18.7 | 21.6 | − |
| 24.5 | 21.5 | + |

*Note:*
*Sample size $n = 25$; + signs $R^+ = 6$; − signs $R^- = 18$; the reduced value of n due to tied observations $= 24$; the test statistic $R = r_0 = 6$; and the significance probability from Table A.18 is $P(r_0) = 0.0227$.*

Using the S-PLUS built-in function *binom.test*, for the data from Example 9.2-1 shown in Table 9.12, we can obtain the following results:

```
> binom.test(6,24,1/2)

Exact binomial test

data: 6 out of 24
number of successes = 6, n = 24, p-value = 0.0227
alternative hypothesis: true p is not equal to 0.5
```

### The Wilcoxon Signed Rank Test

The signed rank test is a modification of the sign test in that it takes into account the magnitude of the differences $d_i$ in addition to the signs. The test statistic $T$ is computed as follows:

1. Assign ranks to the absolute differences $|d_i|$ in order of magnitude. Thus, the smallest $|d_i|$ receives rank 1, and so on. If a $d_i = 0$, drop it and reduce $n$ by the number of zero differences. In case of ties while ranking $|d_i|$, the average rank is assigned.
2. Associate the sign of the original difference $d_i$ with the rank of $|d_i|$.
3. Compute $T^+$ and $T^-$, the sums of positive and negative ranks, respectively. It is a fact that $T^+ + T^- = n(n + 1)/2$.
4. The signed rank test statistic $T$ is the minimum of $T^+$ and $T^-$. If $T < T_\alpha$ we reject in favor of the appropriate alternative at the $\alpha$ level of significance.

For $n = 5$ through $n = 50$, Table A.19 provides the closest percentiles to $T_\alpha$ as critical values for testing one- and two-sided alternatives to the null hypothesis of no difference at a desired $\alpha$ level. For each $n$, the first entry in the table corresponds to a true $\alpha$ smaller than and closest to the desired $\alpha$ level, while the second entry corresponds to a true $\alpha$ larger than and closest to the desired $\alpha$ level. Each column is headed by the desired $\alpha$ level of significance. Here again, note that the desired level is not achieved because the distribution of $T$ is also discrete. That is why the table lists only the closest percentiles. As an example, suppose $n = 10$ and we desire to test the null hypothesis against the two-sided alternative at the $\alpha = 0.05$ level of significance. First, locate the column under $\alpha = 0.05$ for the two-sided alternative. Then for $n = 10$, the second entry from the table is $T_{0.0322} = 9$, which is the $(0.0322)100$th percentile and the one larger than and closest to $\alpha/2 = 0.05$.

The signed rank test statistic $T$ also can be converted into the $Z$ statistic having the approximate standard normal distribution under $H_0$. That is,

$$Z = (T - \mu)/\sigma, \qquad (9.2\text{-}3)$$

where

$$\mu = n(n+1)/4 \quad \text{and} \quad \sigma^2 = n(n+1)(2n+1)/24. \qquad (9.2\text{-}4)$$

If there are ties among the nonzero $|d_i|$, one should use the average of the integer ranks of the ties group. When applying the large-sample approximation, one should use a correct factor in the null variance. In the situation,

$$\sigma^2 = n(n+1)(2n+1)/24 - \sum_{i=1}^{g} t_i(t_i-1)(t_i+1)/48, \qquad (9.2\text{-}5)$$

where $g$ is the number of tie groups, $t_i$ is the number of ties in the $i$th tie group.

---

**Example 9.2-2**

A study was conducted to compare the percent shrinkage of beef carcasses under cryogenic refrigeration. One side of the carcass was under control refrigeration and the other under cryogenic refrigeration. Let $\mu_X$ and $\mu_Y$ denote the medians of the percent shrinkage under cryogenic and controlled refrigeration. The null and the alternative hypotheses are

$$H_0:\ \mu_X = \mu_Y;$$
$$H_a:\ \mu_X < \mu_Y.$$

The calculations are shown in Table 9.13. Since difference $d_2 = 0$, $n$ reduces to 14. For $\alpha = 0.05$, the tabulated value from Table A.19 for the one-sided test is $T_{0.045} = 25$. Since the observed value of $T$ is 8.5, which is less than 25, the hypothesis of no difference between the two treatments is rejected. Hence, the researcher may conclude that cryogenic refrigeration causes significantly less shrinkage than the control.

From (9.2-5) with $n = 14$, $g = 1$, $t_1 = 2$,

$$\mu = 52.5, \quad \sigma^2 = 253.58.$$

Hence,

$$Z = (8.5 - 52.5)/\sqrt{253.58} = -2.76.$$

Since $Z = -2.76$ is approximately the first percentile, that is, $Z_{0.01}$, of the standard normal distribution, we reject $H_0$ in favor of $\mu_X < \mu_Y$.

Using the S-PLUS built-in function *wilcox.test* for the data, we found the results shown next. The reader should note that the results of the normal approximation used in the program are not exactly the same as $Z = -2.76$. This is due to the different methods for handling zero differences. In the program the zero values are not discarded, but some adjustments for the zero values are made in the estimations of $\mu$ and $\sigma$ based on Lehmann (1975, 130). The practice to discard the zeros was suggested by

**Table 9.13** Percentage Shrinkage of Beef Carcasses under Control and Cryogenic Conditions (Example 9.2-2)

| Carcass Number | Control, Y (%) | Cryogenic, X (%) | $d_i$ | Rank (+) | Rank (−) |
|---|---|---|---|---|---|
| 1 | 1.17 | 0.98 | 0.19 | 4 | |
| 2 | 0.83 | 0.83 | 0.00 | — | — |
| 3 | 0.83 | 0.53 | 0.30 | 8 | |
| 4 | 1.63 | 0.72 | 0.91 | 13 | |
| 5 | 2.05 | 0.78 | 1.27 | 14 | |
| 6 | 1.44 | 0.58 | 0.86 | 12 | |
| 7 | 0.85 | 1.09 | −0.24 | | 5.5 |
| 8 | 1.02 | 0.74 | 0.28 | 7 | |
| 9 | 1.29 | 1.05 | 0.24 | 5.5 | |
| 10 | 1.16 | 1.17 | −0.01 | | 1 |
| 11 | 1.01 | 0.97 | 0.04 | 3 | |
| 12 | 1.31 | 0.78 | 0.53 | 10 | |
| 13 | 1.25 | 0.61 | 0.64 | 11 | |
| 14 | 1.21 | 1.23 | −0.02 | | 2 |
| 15 | 1.31 | 0.96 | 0.35 | 9 | |
| Sum of Ranks | | | | 96.5 | 8.5 |
| Test Statistic *T* | | | | | 8.5 |
| *Z* Score of *T* | | | | | −2.75 |
| Significance Probability of Z Score | | | | | 0.013 |

Wilcoxon (1949). According to Conover (1973), each method of handling ties at zero is more powerful than the other in some situations. Hence, there is little reason to prefer one to the other (Conover, 1999, 360).

```
> wilcox.test(wilsrdat[,1],wilsrdat[,2],alternative =
  "greater",paired = T,exact = F,correct = F)

Wilcoxon signed-rank test

data: wilcodat[, 1] and wilcodat[, 2]
signed-rank normal statistic without correction Z =
  2.6989, p-value = 0.0035
```

```
alternative hypothesis: true mu is greater than 0
> wilsrdat
   [,1] [,2]
[1,] 1.17 0.98
[2,] 0.83 0.83
[3,] 0.83 0.53
[4,] 1.63 0.72
......
```

## Rank Tests for Two Independent Samples

### *The Wilcoxon Rank Sum Test*

Wilcoxon (1945, 1946, 1947) developed tests that use ranks in testing the equality of two populations. Investigations into the optimal statistical properties of his tests have been written up in many research papers in the literature. One may, for instance, refer to Wilcoxon and Wilcox (1964) and Bradley (1968).

The Wilcoxon rank sum test is applicable when two random samples $X_1$, $X_2, \ldots, X_n$ from $X$ population and $Y_1, Y_2, \ldots, Y_m$ from $Y$ population are taken independently of each other. The two samples need not be of equal size. The null hypothesis is that *both X and Y populations are alike*. The null and the alternative hypotheses are sometimes stated in terms of the location parameters $\mu_X$ and $\mu_Y$ of the $X$ and $Y$ populations, respectively. That is, if two populations are alike in all respects except perhaps the location, then the null hypothesis may be stated as follows:

$$H_0: \ \mu_X = \mu_Y.$$

The alternative could be a one- or two-sided hypothesis, depending on the nature of inquiry. For example, it might be of interest to know that the treated population $X$ has a higher median value than the untreated population $Y$. In this case the alternative would be $H_a: \mu_X > \mu_Y$.

The Wilcoxon rank sum method requires the ranking of each of the $(m + n)$ observations in both samples in order of magnitude. The rank sum test statistic $W$ is the sum of ranks of observations in the smaller of the two samples. One can always formulate the setup without any loss of generality so that $m \leq n$. If the samples are of equal size $(m = n)$, one may choose $W$ for either sample. It can be seen that if $W$ is the sum of $m$ ranks, then $W$ can be as small as

$$1 + 2 + \cdots + m = m(m + 1)/2, \tag{9.2-6}$$

or as large as

$$(n + 1) + (n + 2) + \cdots + (n + m) = m(2n + m + 1)/2. \qquad (9.2\text{-}7)$$

In between these extremes lies the distribution of $W$. If $H_0$ is false, then one expects the observed value of $W$ to lie at or near the extremes. That is, the tail values in the distribution of $W$ constitute the rejection region of $H_0$. For the two-sided alternative, $H_0$ is rejected whenever $W$ is observed as "large" or "small" in either of the two tails. For $H_a: \mu_X < \mu_Y$, reject $H_0$ in favor of $H_a$ when $W$ is large in the right tail, and for $H_a: \mu_X > \mu_Y$, reject $H_0$ in favor of $H_a$ if $W$ is in the left tail. The specification of the tails for the rejection of $H_0$ depends on the desired level of significance $\alpha$. Since $W$ can assume only integer values, it is not always possible to specify the tails of the distribution of $W$ for the rejection region that will necessarily correspond to the exact desired $\alpha$ level. However, one can use Table A.15 to specify for the rejection of $H_0$ the tail values of $W$ having a significance probability closest to the desired significance level $\alpha$. To illustrate, suppose $m = 8$ and $n = 10$. In Table A.15, $m \leq n$ always. Remember that $W$ corresponds to the smaller sample of size $m$. Suppose we want to carry out a two-sided test at the $\alpha = 0.01$ significance level. Referring to Table A.15 under $P = 0.01$ (two sided) for $m = 8$ and $n = 10$, we find the following numbers:

| | |
|---|---|
| 47, 105 | 0.0043 |
| 48, 104 | 0.0058 |

The interpretation of these entries is the following:

$$P(W \leq 47) = P(W \geq 105) = 0.0043,$$

$$P(W \leq 48) = P(W \geq 104) = 0.0058.$$

If the rejection of $H_0$ occurs in tails $W \leq 47$ and $W \geq 105$, the significance probability is $2 \times 0.0043 = 0.0086$, which is less than the specified 0.01 level. On the other hand, if we use tails $W \leq 48$ and $W \geq 104$, the significance probability is $2 \times 0.0058 = 0.0116$, which is just greater than the 0.01 level. Neither of the rejection regions in the tails meets the exact 0.01 level of significance. Depending on the experimenter's assessment of the risk involved in wrongly rejecting $H_0$, either one of the two rejection regions (tails) can be used for decision making.

In short, Table A.15 provides two rejection regions (tails) with associated significance probabilities $P$ closest to the desired $\alpha$ level. For one rejection region, the significance probability $P$ is less than $\alpha$, whereas for the other, $P$ is greater than $\alpha$. The table is reproduced only for $3 \leq m \leq n \leq 10$ from a more extensive table by Wilcoxon et al. (1970).

Many statistical software packages provide functions for the distribution of the Wilcoxon rank sum statistic. For example, S-PLUS provides the four functions *rwilcox, pwilcox, dwilcox,* and *qwilcox,* which can generate random numbers, calculate cumulative probabilities, compute densities, and return quantiles for the distribution, respectively. For example, for $m = 8$, $n = 10$, the cumulative probability of $W$ from 0 to 48 is 0.00583; the cumulative probability of $W$ larger than 103 is also 0.00583.

```
> pwilcox(48,8,10)
[1] 0.005827506
> 1-pwilcox(103,8,10)
[1]] 0.005827506
```

When $m$ and $n$ are both large ( $>10$), we may use the fact that the distribution of $W$ is approximately normal with mean and variance given, respectively, by

$$\mu = m(m+n+1)/2 \quad \text{and} \quad \sigma^2 = mn(m+n+1)/12. \qquad (9.2\text{-}8)$$

When there are ties, the null mean of $W$ is unaffected, but the null variance is reduced to

$$\sigma^2 = mn(m+n+1)/12 - \left\{ mn \sum_{i=1}^{g}(t_i^3 - t_i)/(12mn(m+n-1)) \right\}, \qquad (9.2\text{-}9)$$

where $g$ is the number of tied groups and $t_i$ is the size of the $i$th tied group. For carrying out the large sample test, we may use

$$Z = (W - \mu)/\sigma, \qquad (9.2\text{-}10)$$

where the mean $\mu$ and the standard deviation $\sigma$ are given by (9.2-8) or (9.2-9). For a two-sided alternative, reject $H_0$ at the $\alpha$ level of significance if the computed $Z$ value from (9.2-10) is either greater than $z_{1-\alpha/2}$ or less than $z_{\alpha/2}$. For a one-sided alternative, the decision rule would reject $H_0$ in favor of alternative $\mu_X > \mu_Y$ provided that $Z < z_\alpha$ and in favor of $\mu_X < \mu_Y$ provided that $Z > z_{1-\alpha}$. Remember that symbol $z_\alpha$ always stands for the $\alpha$th percentile of the standard normal distribution.

Before considering an example, let us decide the manner in which the tied observations will be handled. Whenever two or more observations in a combined sample of $(m + n)$ observations are found to be of the same magnitude, all those observations will be assigned the same midrank, which is the average of all ranks for which the observations were tied.

**Example 9.2-3**

The data in Table 9.14 pertain to percent moisture in beef patties for control and test groups. Note that $m = n = 10$ and that all the observations have been ranked according to their magnitude. The sum of ranks $W$ is obtained for the smaller sample. Since both samples in this example are of equal size, $W$ can be used for either sample. From Table A.15 we notice that, corresponding to $m = 10, n = 10$,

$$P(W \leq 78) = P(W \geq 132) = 0.0216,$$
$$P(W \leq 79) = P(W \geq 131) = 0.0262.$$

Note that for the rejection region in the tails $W \leq 78$ and $W \geq 132$, and the significance probability is $2(0.0216) = 0.0432 < 0.05$; whereas for the rejection region in the tails $W \leq 79$ and $W \geq 131$, and the significance probability is $2(0.0262) = 0.0524 > 0.05$. Since $m = n = 10$, we can use either 79 or 131 for observed values of the rank sum statistic $W$. Hence, for an $\alpha = 0.05$ level of significance, it is a borderline case as to whether $H_0$ will be accepted or rejected. Had we chosen $\alpha = 0.1$, then we would have rejected $H_0$. The fact is

**Table 9.14** Percentage Moisture in Beef Patties for Test and Control Groups (Examples 9.2-3 and 9.2-4)

| Control | Rank | Test | Rank |
|---------|------|------|------|
| 55.6 | 1.5 | 55.6 | 1.5 |
| 57.2 | 12.0 | 58.2 | 17.0 |
| 56.8 | 10.0 | 56.4 | 7.5 |
| 55.7 | 3.0 | 57.5 | 13.0 |
| 56.4 | 7.5 | 59.8 | 20.0 |
| 56.2 | 6.0 | 58.8 | 19.0 |
| 57.0 | 11.0 | 58.4 | 18.0 |
| 56.1 | 5.0 | 57.8 | 15.0 |
| 56.5 | 9.0 | 55.9 | 4.0 |
| 57.6 | 14.0 | 58.0 | 16.0 |
| Sum of Ranks $W$ | 79.0 | | 131.0 |
| $Z$ Score of $W$ | $-1.96$ | | |

that, on the basis of these data, we can reject $H_0$ at any significance level $\alpha$ greater than 0.0524 but not at any significance level $\alpha$ less than 0.0526. In view of this fact, we emphasize that it is the experimenter's task to reach an appropriate decision by taking into account all risks and any nonstatistical considerations. This example clearly demonstrates that the specification of the significance level $\alpha$ should be properly viewed as being arbitrary. There is no firm basis on which $\alpha$ is necessarily 0.05 or 0.01, etc.; it is a matter of convenience and the traditional use of these values. Hence, instead of simply reporting the rejection or acceptance of $H_0$ at an arbitrarily specified significance level, it is more informative to report the significance probability, under the null hypothesis, of obtaining a value as extreme as or more extreme than the observed value of the test statistic.

Even though, in this example, both $m$ and $n$ are 10 and cannot be considered large with regard to the normal approximation, we illustrate it anyhow. Let us find $Z$ in (9.2-10) for $\mu$ and $\sigma^2$ as computed from (9.2-8) or (9.2-9). We find $\mu = 105$, $\sigma^2 = 175$, and $\sigma = 13.22$. Hence, corresponding to the rank sum $W = 79$, $Z = (79 - 105)/13.22 = -1.96$. If the test is two-sided at $\alpha = 0.05$, then $z_{\alpha/2} = z_{0.025} = -1.96$. Since the observed $Z$ and the tabulated $z_{0.025}$ are equal, the decision is a borderline case, as indeed it was earlier. It is interesting to note that even for $m$ and $n$ values as small as 10, the use of the approximate normal distribution for $W$ leads to the same conclusion. This suggests that the normal approximation is quite good, even for small sample sizes. Using the S-PLUS built-in function *wilcox.test* for the data, we obtain the following results:

```
> wilcox.test(wildat[,1],wildat[,2],paired = F,exact =
    F,correct = F)

Wilcoxon rank-sum test

data: wildat[, 1] and wildat[, 2]
rank-sum normal statistic without correction Z = −1.9669,
    p-value = 0.0492
alternative hypothesis: true mu is not equal to 0
> wildat
    [,1] [,2]
[1,] 55.6 55.6
[2,] 57.2 58.2
[3,] 56.8 56.4
 . . . . . . . . . . .
```

To illustrate the case of unequal sample sizes, we eliminate the first and third observations from the test group data in Table 9.14. The sum of ranks for the smaller sample ($m = 8$) is now $W = 107$. From Table A.15 we observe, for the two-sided test, that

**Example 9.2-4**

$$P(W \le 53) = P(W \ge 99) = 0.0217.$$

Suppose the rejection region is specified by the tails $W \le 53$ and $W \ge 99$. Since $W = 107$ is greater than 99, we would reject $H_0$ at any level $\alpha$ greater than $2(0.0217) = 0.0434$.

We illustrated the Wilcoxon rank sum test for two independently taken random samples—one from each population. When the word *population* is used for treatments, the design for selecting two independent random samples, one from each population, is equivalent to having a completely randomized experimental design with two treatments. Thus, the use of the rank sum test is valid when the designs used are completely randomized with two treatments. In many practical situations, the blocking of treatments is desirable. For instance, in sensory testing, for various reasons, the panel session cannot be completed in one sitting. A provision is made, therefore, to conduct the experiment at different times (blocks). It is known that panel scores vary with judges over intervals of time, creating panel and time interactions. The interaction effects can be removed by ranking the observations within blocks. Therefore, we next consider a variant of the rank sum test when the experimental design is the randomized block design with two treatments and $b$ blocks. The distribution of the rank sum statistic $W$ for samples resulting from a randomized block design is given by Nelson (1970). The critical values of this distribution are reproduced in Table A.16. The critical values tabulated are such that the listed significance levels in the table are never exceeded. The critical values with superscript $+$ can be increased by one to give a significance probability closer to (but not less than) the stated value. The tables are given for treatments repeated the same number of times ($n$) within a block.

To the null hypothesis of equality of two treatments or populations, we compute the rank sum statistic within each block for both treatments. Then the rank sums for all the blocks are added to obtain a grand rank sum statistic $W$ for each treatment. Thus, two grand rank sums are obtained, one for each treatment.

For two-sided alternatives we may use the smaller of the two grand rank sums. On the other hand, if the alternative is one-sided, say $\mu_X < \mu_Y$, then the correct statistic to use is the grand rank sum associated with the sample from population $X$. If the observed value of the grand rank sum statistic used for testing is smaller than the tabulated value corresponding to a desired $\alpha$ level of significance, we reject $H_0$. Remember that Table A.16 provides critical values for testing whether the grand rank sum statistic is too small for the null hypothesis to be true.

Wilcoxon (1947) suggested that the distribution of the rank sums approaches normality as either $n$, the number of replicates or the number of blocks, gets large. For the significance level $\alpha$, the critical values for the two-sided and one-sided tests are

$$W_{\alpha/2} = bt + Z_{\alpha/2}\sqrt{bnt/6} \tag{9.2-11}$$

and

$$W_\alpha = bt + Z_\alpha\sqrt{bnt/6}, \tag{9.2-12}$$

respectively. In the preceding expressions, $b$ is the number of blocks, $n$ is the number of replicates within each block, $t = 2n(2n + 1)/4$, and $z_{\alpha/2}$ and $z_\alpha$ are the $(\alpha/2)$ 100th and $(\alpha)$ 100th percentiles of the standard normal distribution, respectively. Remember that the smaller of the grand rank sums is to be compared with the appropriate critical value.

---

**Example 9.2-5**

Table 9.15 shows texture scores for control and test products based on an 8-point hedonic scale. The ties are resolved within blocks by assigning the average of the ranks for which the observations were tied. The grand rank sums for the two blocks are 105.5 for the control and 166.5 for the test. Our test statistic is the smaller rank sum. For $\alpha = 0.05$, the tabulated values in Table A.16 for the one-sided and two-sided tests are 109 and 113, respectively. Both are greater than the smaller of the two grand rank sums. Hence, we may reject the null hypothesis in favor of $\mu_X \neq \mu_Y$ and conclude that the median texture scores of the control and test groups are different. It is instructive to note that for $\alpha = 0.05$, the critical values from the normal approximations, (9.2-11) and (9.2-12), are $W_{0.025} = 109.64$ and $W_{0.05} = 113.94$ for the two-sided and one-sided alternatives, respectively.

**Table 9.15** Texture Scores of Control and Test Products Based on 8-Point Hedonic Scale (Example 9.2-5)

| Blocks | Control | | Test | |
|---|---|---|---|---|
| | **Scores** | **Ranks** | **Scores** | **Ranks** |
| a.m. | 4 | 6.5 | 4 | 6.5 |
| | 3 | 2.5 | 4 | 6.5 |
| | 5 | 12.0 | 6 | 15.5 |
| | 4 | 6.5 | 5 | 12.0 |
| | 4 | 6.5 | 5 | 12.0 |
| | 5 | 12.0 | 5 | 12.0 |
| | 2 | 1.0 | 4 | 6.5 |
| | 3 | 2.5 | 6 | 15.5 |
| Rank Sum | | 49.5 | | 86.5 |
| p.m. | 4 | 4.5 | 5 | 8.5 |
| | 4 | 4.5 | 4 | 4.5 |
| | 5 | 8.5 | 7 | 15.0 |
| | 6 | 12.0 | 6 | 12.0 |
| | 3 | 1.5 | 5 | 8.5 |
| | 3 | 1.5 | 4 | 4.5 |
| | 5 | 8.5 | 6 | 12.0 |
| | 7 | 15.0 | 7 | 15.0 |
| Rank Sum | | 56.0 | | 80.0 |
| Grand Rank Sum | | 105.5 | | 166.5 |
| Number of Replicates Within Blocks (*n*) | | 8 | | 8 |

*Note:*
*Tabulated critical value (from Table A.16) = 109.*
$W_{0.025}$ *(from 9.2-11) = 109.64.*
$W_{0.05}$ *(from 9.2-12) = 113.94.*

### The Mann-Whitney U Statistic

A very important version of the Wilcoxon rank sum test for two independent samples is the Mann-Whitney $U$ statistic proposed by Mann and Whitney (1947). The statistic $U_{XY}$ is defined as the number of pairs $X_i, Y_j$ $(i = 1, \ldots, n_X; j = 1, \ldots, n_Y)$ with $X_i < Y_j$; $U_{YX}$ is defined as the number of pairs $X_i, Y_j$ with $X_i > Y_j$, where $n_X, n_Y$ denote the sample sizes of $X$ and $Y$, respectively. When there are ties, $U$ is computed by

$$U_{XY} = [\text{number of pairs } (X_i, Y_j) \text{ with } X_i < Y_j] \\ + [\text{number of pairs } (X_i, Y_j) \text{ with } X_i = Y_j]/2, \tag{9.2-13}$$

$$U_{YX} = [\text{number of pairs } (X_i, Y_j) \text{ with } X_i > Y_j] \\ + [\text{number of pairs } (X_i, Y_j) \text{ with } X_i = Y_j]/2. \tag{9.2-14}$$

The relationship between $U_{XY}$ and $U_{YX}$ is

$$U_{XY} + U_{YX} = n_X n_y. \tag{9.2-15}$$

The Mann-Whitney $U$ statistic is equivalent to the Wilcoxon $W$ statistic in the sense that one is a linear function of the other. Because of the relationship, the statistics $U$ and $W$ are both referred to as the Mann-Whitney-Wilcoxon statistic (MWW). The relationship between the $W$ and $U$ is

$$U_{YX} = W_X - n_X(n_X + 1)/2, \tag{9.2-16}$$

$$U_{XY} = W_Y - n_Y(n_Y + 1)/2. \tag{9.2-17}$$

For the summarized rating frequency data in a $2 \times k$ table, the Mann-Whitney $U$ statistic can be calculated from

$$U_{XY} = \sum_{i=1}^{k-1} b_i \sum_{j=i+1}^{k} a_j + \sum_{i=1}^{k} a_i b_i / 2, \tag{9.2-18}$$

$$U_{YX} = n_X n_Y - U_{XY}, \tag{9.2-19}$$

where $a = (a_1, a_2, \ldots a_k)$, $b = (b_1, b_2, \ldots b_k)$ denote the frequency vectors of $k$-point scale ratings for two independent samples $X$ (with frequency numbers $a$) and $Y$ (with frequency numbers $b$), $a_1, b_1$ denote the frequencies of samples $X$ and $Y$ for the $k$-th category, and $a_k, b_k$ denote the frequencies of samples $X$ and $Y$ for the first category.

Either one of the statistics $U_{XY}$ and $U_{YX}$ can be used for a test. If the value of $U_{XY}$ is larger than that of $U_{YX}$, it suggests that $X$ tends to be less than $Y$. For large-sample $(n_X, n_Y \geq 8)$, the test statistic is

$$Z = |U - \mu| / \sigma, \tag{9.2-20}$$

where

$$\mu = n_X n_Y / 2,$$

$$\sigma^2 = \frac{n_X n_Y (n_X + n_Y + 1)}{12} - \frac{n_X n_Y}{12(n_X + n_Y)(n_X + n_Y - 1)} \sum_{i=1}^{g} (t_i^3 - t_i),$$

where $g$ is the number of tied groups and $t_i$ is the size of the $i$th tied group.

In order to investigate purchase intent of male and female consumers for a product, researchers randomly drew 25 male and 20 female consumers. A 5-point scale was used with 5 = Definitely would buy it to 1 = Definitely would not buy it. The ratings given by the 45 consumers are displayed and summarized in Table 9.16. The ranks of the 45 ratings are also listed in Table 9.16. The rank sum for the male consumers is $W_X = 479.5$ and $W_Y = 555.5$ for the female. Hence, $U_{YX} = 479.5 - 25 \times 26/2 = 154.5$ and $U_{XY} = 555.5 - 20 \times 21/2 = 345.5$ from (9.2-16) and (9.2-17). From (9.2-18) and (9.2-19), we can also obtain $U_{XY} = 345.5$ and $U_{YX} = 25 \times 20 - 345.5 = 154.5$. Because the value of $U_{XY}$ is larger than that of $U_{YX}$,

**Example 9.2-6**

**Table 9.16** Rating Data for Purchase Intent (Example 9.2-6)

| ID | Gender | Rating | Rank | ID | Gender | Rating | Rank | ID | Gender | Rating | Rank |
|----|--------|--------|------|----|--------|--------|------|----|--------|--------|------|
| 1 | 1 | 5 | 40.5 | 16 | 1 | 2 | 10.5 | 31 | 2 | 5 | 40.5 |
| 2 | 1 | 1 | 3.5 | 17 | 1 | 4 | 30 | 32 | 2 | 4 | 30 |
| 3 | 1 | 1 | 3.5 | 18 | 1 | 2 | 10.5 | 33 | 2 | 2 | 10.5 |
| 4 | 1 | 3 | 19.5 | 19 | 1 | 2 | 10.5 | 34 | 2 | 4 | 30 |
| 5 | 1 | 4 | 30 | 20 | 1 | 2 | 10.5 | 35 | 2 | 4 | 30 |
| 6 | 1 | 5 | 40.5 | 21 | 1 | 1 | 3.5 | 36 | 2 | 4 | 30 |
| 7 | 1 | 2 | 10.5 | 22 | 1 | 5 | 40.5 | 37 | 2 | 5 | 40.5 |
| 8 | 1 | 4 | 30 | 23 | 1 | 4 | 30 | 38 | 2 | 4 | 30 |
| 9 | 1 | 3 | 19.5 | 24 | 1 | 1 | 3.5 | 39 | 2 | 5 | 40.5 |
| 10 | 1 | 3 | 19.5 | 25 | 1 | 3 | 19.5 | 40 | 2 | 3 | 19.5 |
| 11 | 1 | 1 | 3.5 | 26 | 2 | 5 | 40.5 | 41 | 2 | 3 | 19.5 |
| 12 | 1 | 3 | 19.5 | 27 | 2 | 4 | 30 | 42 | 2 | 3 | 19.5 |
| 13 | 1 | 3 | 19.5 | 28 | 2 | 5 | 40.5 | 43 | 2 | 4 | 30 |
| 14 | 1 | 5 | 40.5 | 29 | 2 | 3 | 19.5 | 44 | 2 | 2 | 10.5 |
| 15 | 1 | 2 | 10.5 | 30 | 2 | 5 | 40.5 | 45 | 2 | 1 | 3.5 |

**Table 9.17** Frequencies of Ratings (Example 9.2-6)

|  | "5" | "4" | "3" | "2" | "1" |  |
|---|---|---|---|---|---|---|
| Male | 4 | 4 | 6 | 6 | 5 | 25 |
| Female | 6 | 7 | 4 | 2 | 1 | 20 |
|  | 10 | 11 | 10 | 8 | 6 | 45 |

it suggests that male consumers tend to be lower than female consumers in purchase intent.

We can calculate $\mu = 25 \times 20/2 = 250$ from (9.2-20). And we can find from Table 9.17 that there are five tied groups with the sizes 10, 11, 10, 8, and 6, respectively. Hence, from (9.2-20),

$$\sigma^2 = \frac{25 \times 20 \times 46}{12} - \frac{25 \times 20}{12(25 + 20)(25 + 20 - 1)} \times [(10^3 - 10)$$
$$+ (11^3 - 11) + \cdots (6^3 - 6)] = 1832.2.$$

The value of the test statistic is

$$Z = \frac{345.5 - 250}{\sqrt{1832.2}} = -2.231,$$

with the associated $P$-value $= 0.026$ for the two-sided test. We can conclude at an $\alpha = 0.05$ significance level that female consumers have higher purchase intent than males for that product. The reader should note that exactly the same results can be obtained by using $U_{YX}$. In this situation,

$$Z = \frac{154.5 - 250}{\sqrt{1832.2}} = -2.231.$$

The reader also should note that exactly the same results can be obtained by using the $W$ statistic, either $W_X$ or $W_Y$. For $W_X = 479.5$, $\mu_X = 25 \times (25 + 20 + 1)/2 = 575$,

$$Z = \frac{|479.5 - 575|}{\sqrt{1832.2}} = 2.231.$$

For $W_Y = 555.5$, $\mu_Y = 20 \times (25 + 20 + 1)/2 = 575$,

$$Z = \frac{|555.5 - 460|}{\sqrt{1832.2}} = 2.231.$$

Using the S-PLUS built-in function *wilcox.test* for the data in Table 9.16, we obtain the following results:

```
> wilcox.test(udat6[udat6[,1]==1,2],udat6[udat6[,1]=
  =2,2],correct=F)

Wilcoxon rank-sum test

data: udat6[udat6[,1]==1,2] and udat6[udat6[,1]==2,2]
rank-sum normal statistic without correction Z = -2.2311,
  p-value = 0.0257
alternative hypothesis: true mu is not equal to 0
> udat6
   x y
[1,] 1 5
[2,] 1 1
[3,] 1 1
[4,] 1 3
. . . . .
```

## Rank Tests for Multiple Independent Samples

### The Kruskal–Wallis H Test

The Kruskal–Wallis $H$ test is a generalization of the Wilcoxon rank sum test to more than two populations (Kruskal and Wallis, 1952). For instance, one may be evaluating four methods ($k = 4$) of roasting turkey or testing the effects of five treatments ($k = 5$) on the control of bacterial growth. In such cases the experimental design is the one-way analysis of variance model. If the parametric assumptions for the analysis of variance model are in doubt, then the $H$ test provides an alternative method of analysis. The Kruskal–Wallis $H$ test is a nonparametric version of a one-way analysis of variance.

We may state for the null hypothesis that all the $k$ populations are identical with respect to their mean or median values. That is,

$$H_0: \ \mu_1 = \mu_2 = \ldots = \mu_k.$$

The alternative is that at least one member of the pair of the $k$ populations is different from each other. That is,

$$H_a: \ \mu_i \neq \mu_j \text{ for at least one pair } (i \neq j), \ i, j = 1, \ldots, k.$$

The $H$ test statistic, when there are no tied observations, is

$$H = \frac{12}{N(N+1)} \sum_{i=1}^{k} \frac{R_i^2}{n_i} - 3(N+1), \qquad (9.2\text{-}21)$$

where $N$ is the total number of observations from all the $k$ samples, $n_i$ is the number of observations in the $i$th sample, and $R_i$ is the sum of the ranks for the $i$th sample. For large samples, Kruskal and Wallis have shown that the $H$ statistic approximately follows the $\chi^2$ distribution with $k - 1$ DF. We may use the $\chi^2$ distribution tables to find the critical values. For the $\alpha$ level of significance, find the $(1 - \alpha)$ 100th percentile $\chi^2_{1-\alpha,k-1}$ of the $\chi^2$ distributed with $(k - 1)$ DF and reject $H_0$ at the $\alpha$ level if $H > \chi^2_{1-\alpha,k-1}$. In the case of tied observations, the observations are replaced by the average of the ranks for which they are tied, and a modification is incorporated in computing $H$. The statistic $H$ is replaced by

$$H' = \frac{H}{1 - \sum_{i=1}^{g}(t_i^3 - t_i)/(N^3 - N)}, \qquad (9.2\text{-}22)$$

where $g$ is the number of tied groups and $t_i$ is the size of the $i$th tied group. When there is no tie, $H'$ reduces to $H$.

**Example 9.2-7**

To illustrate the $H$ test, let us consider the Instron readings obtained in three locations on frankfurters. The readings taken on both ends (lower and upper) and the center portion of each frankfurter are displayed in Table 9.18.

**Table 9.18** Three Instron Readings from the Lower, Center, and Upper Portions of Frankfurters (Example 9.2-7)

| | Lower, $T_1$ | | Center, $T_2$ | | Upper, $T_3$ |
|---|---|---|---|---|---|
| **Readings** | **Ranks** | **Readings** | **Ranks** | **Readings** | **Ranks** |
| 2.58 | 12 | 2.46 | 5 | 2.08 | 1 |
| 2.57 | 10 | 2.58 | 12 | 2.30 | 3 |
| 2.58 | 12 | 2.49 | 7 | 2.21 | 2 |
| 2.76 | 14 | 2.87 | 15 | 2.47 | 6 |
| 2.32 | 4 | 2.56 | 9 | 2.50 | 8 |
| $n_i$ | 5 | | 5 | | 5 |
| $\sum_i R_i$ | 52 | | 48 | | 20 |
| $\sum_i R_i^2/n_i$ | 540.8 | | 460.8 | | 80 |

The first step is to combine all the observations and rank them in order from 1 to $N$, the tied observations receiving the average of their ranks. Note that $N = \Sigma_i^k n_i = 15$. There were three tied observations, each receiving an average rank of 12. For these data, the $H$ statistic is

$$H = \frac{12}{15(16)}(540.8 + 460.8 + 80) - 3(16) = 6.08.$$

Because there is only one tied group with three 2.58, $g = 1$, the size of the group is 3; hence,

$$H' = \frac{6.8}{1 - (3^3 - 3)/(15^3 - 15)} = 6.1237 \text{ with 2 DF.}$$

For the $\alpha = 0.05$ level of significance, the critical value of $\chi^2_{0.95,2} = 5.99$, which is exceeded by the computed value of $H' = 6.1237$. Therefore, the null hypothesis is rejected, and we conclude that there are differences in readings among the three locations at the 5% level of significance. Using the S-PLUS built-in function *kruskal.test*, we can obtain the same results:

```
> kruskal.test(krusdat[,2],krusdat[,1])

Kruskal-Wallis rank sum test

data: krusdat[, 2] and krusdat[, 1]
Kruskal-Wallis chi-square = 6.1237, df = 2, p-value =
   0.0468
alternative hypothesis: two.sided
> krusdat
  [,1] [,2]
[1,] 1 2.58
[2,] 1 2.57
[3,] 1 2.58
[4,] 1 2.76
...........
```

In rejecting the null hypothesis of the equality of populations, the $H$ test does not isolate the populations that are different from each other. We consider a multiple-comparison test by Steel (1961). Steel's test requires the same number of observations from each population. The test is explained with the help of our example and Table 9.19. The following steps are involved:

1. Jointly rank the $T_1$ and $T_2$ readings, $T_1$ and $T_3$ readings, and $T_2$ and $T_3$ readings. Thus, we will have two sets of ranks for the $T_1$ readings when they are ranked with $T_2$ and $T_3$, respectively. Similarly, we should have two sets of ranks for the $T_2$ and $T_3$ readings (see Table 9.19).

**Table 9.19** Illustration of Steel's Rank Sum Multiple-Comparison Test Using Data from Table 9.18 (Example 9.2-7)

| $T_1$ Readings | Ranks of $T_1$ with | | $T_2$ Readings | Ranks of $T_2$ with | | $T_3$ Readings | Ranks of $T_3$ with | |
| --- | --- | --- | --- | --- | --- | --- | --- | --- |
| | $T_2$ | $T_3$ | | $T_1$ | $T_3$ | | $T_1$ | $T_2$ |
| 2.58 | 6.5 | 8.5 | 2.46 | 2 | 4 | 2.08 | 1 | 1 |
| 2.57 | 5 | 7 | 2.58 | 6.5 | 9 | 2.30 | 3 | 3 |
| 2.58 | 8 | 8.5 | 2.49 | 3 | 6 | 2.21 | 2 | 2 |
| 2.76 | 9 | 10 | 2.87 | 10 | 10 | 2.47 | 5 | 5 |
| 2.32 | 1 | 4 | 2.56 | 4 | 8 | 2.50 | 6 | 7 |
| $R_{ij}$ | 29.5 | 38 | | 25.5 | 37 | | 17 | 18 |
| $R_{ji}$ | 25.5 | 17 | | 29.5 | 18 | | 38 | 37 |
| $R'_{ij} = \min(R_{ij} + R_{ji})$ | 25.5 | 17 | | 25.5 | 18 | | 17 | 18 |

*Note:*
*Test statistics are $R_{12}' = 25.5$, $R_{13}' = 17$, and $R_{23}' = 18$; at the $\alpha = 0.05$ level, $R_{0.05, \, 3.5} = 16$ from Table A.17.*

2. Add the ranks of $T_1$ to get $R_{12}$ and $R_{13}$. Similarly, find $R_{21}$, $R_{23}$, $R_{31}$, and $R_{32}$. As a check, note that $R_{ij} + R_{ji} = n(2n + 1)$ for all pairs $(i, j)$; therefore, $R_{ij}$ is called the conjugate of $R_{ji}$.
3. Compare $R_{ij}'$, the minimum of $R_{ij}$ and $R_{ji}$, for each pair with a critical value $R_{\alpha,k,n}$ found from Table A.17 for a predetermined $\alpha$ level. Populations $T_i$ and $T_j$ are judged significantly differently from each other if $R_{ij}' \leq R_{\alpha,k,n}$ at an overall significance level $\alpha$.

In our example, $n = 5$ and $k = 3$; suppose that we want an overall significance level no greater than $\alpha = 0.05$. From Table A.17, $R_{0.05,3,5} = 16$, and our test statistics are $R_{12}' = 25.5$, $R_{13}' = 17$, and $R_{23}' = 18$. On the basis of these data and according to Steel's method, we cannot declare significant differences for any of the three pairs.

### The Jonckheere–Terpstra Test for Ordered Alternatives

In some practical situations, some prior information is available about the order of the treatments for comparison. If the null hypothesis of no difference is rejected, the natural order should be confirmed. Terpstra (1952) and Jonckheere (1954) independently proposed a test, which is more powerful than the Kruskal–Wallis test in this situation, i.e., for comparison of $k$ independent samples with an ordered alternative hypothesis

$$H_a: \mu_1 \leq \mu_2 \leq \ldots \leq \mu_k, \text{with at least one strict inequality.}$$

The Jonckheere–Terpstra statistic, *JT*, is constructed by adding all Mann–Whitney $U$ statistics for samples $i$ and $j$, where $i, j = 1, 2, \ldots, k$ with $i < j$, i.e.,

$$JT = \sum_{i<j} U_{ij}. \tag{9.2-23}$$

For large values of $N$, $(N = n_1 + n_2 + \ldots n_k)$, the *JT* statistic is approximately normally distributed with expectation and variance $\mu$ and $\sigma^2$, i.e.,

$$Z = \frac{JT - \mu}{\sigma}, \tag{9.2-24}$$

where

$$\mu = \frac{N^2 - \sum_{i=1}^{k} n_i^2}{4} \quad \text{and} \quad \sigma^2 = V_1 + V_2 + V_3 \tag{9.2-25}$$

where

$$V_1 = \frac{1}{72} \left[ N(N-1)(2N+5) - \sum_{i=1}^{k} n_i(n_i - 1)(2n_i + 5) \right. $$

$$\left. - \sum_{i=1}^{g} t_i(t_i - 1)(2t_i + 5) \right];$$

$$V_2 = \frac{1}{36N(N-1)(N-2)} \left[ \sum_{i=1}^{k} n_i(n_i - 1)(n_i - 2) \right] \left[ \sum_{i=1}^{g} t_i(t_i - 1)(t_i - 2) \right];$$

$$V_3 = \frac{1}{8N(N-1)} \left[ \sum_{i=1}^{k} n_i(n_i - 1) \right] \left[ \sum_{i=1}^{g} t_i(t_i - 1) \right].$$

See Hollander and Wolfe (1999, 204). We should emphasize that it is not justified to specify the hypothesis of ordered alternatives and to use the Jonckheere–Terpstra test after the observations are obtained.

| **Example 9.2-8** | For Example 9.2-7, if the experience tells us that the Instron readings obtained in the three locations have a natural order—$\mu_1 \geq \mu_2 \geq \mu_3$, i.e., the readings should be larger at the lower location—the Jonckheere–Terpstra test can be used in this case. According to the definition of the Mann–Whitney $U$ statistic, $U_{12} = 11$, $U_{13} = 2$, $U_{23} = 3$; hence, $JT = U_{12} + U_{13} + U_{23} = 11 + 2 + 3 = 16$. We get from (9.2-25) that |
|---|---|

$$\mu = \frac{15^2 - (5^2 + 5^2 + 5^2)}{4} = 37.5;$$

$$V_1 = \frac{1}{72}[15(15 - 1)(30 + 5) - 3 \times 5(5 - 1)(10 + 5) - 3(3 - 1)(6 + 5)] = 88.667;$$

$$V_2 = \frac{1}{36 \times 15(15 - 1)(15 - 2)}[3 \times 5(5 - 1)(5 - 2)][3(3 - 1)(3 - 2)] = 0.011;$$

$$V_3 = \frac{1}{8 \times 15(15 - 1)}[3 \times 5(5 - 1)][3(3 - 1)] = 0.214.$$

Hence,

$$\sigma^2 = 88.667 + 0.011 + 0.214 = 88.89.$$

The value of the $Z$ statistic is then $Z = \dfrac{16 - 37.5}{\sqrt{88.89}} = -2.28$ with the associated $P$-value $= 0.011$. We can reject the null hypothesis of no difference and accept the alternative hypothesis at a 0.05 significance level. It is noted that although using the Kruskal–Wallis test, we can draw the same conclusion: the smaller $P$-value suggests the Jonckheere–Terpstra test is more powerful. We should mention that we could also use the statistic (9.2.23),

$$JT = U_{21} + U_{31} + U_{32} = 14 + 23 + 22 = 59.$$

Hence,

$$Z = \frac{59 - 37.5}{\sqrt{88.89}} = 2.28.$$

The same conclusion can be drawn. Because $U_{21} > U_{12}$, $U_{31} > U_{13}$, and $U_{32} > U_{23}$, it confirms that the order of the treatments in the alternative hypothesis is correct. There are built-in functions in some statistical software, e.g., SAS, SPSS, and statXact, for the Jonckheere–Terpstra test. One can easily write a program for the test using S-PLUS, for example:

```
> jttest(krusdat,2,1)
[1]   16.00000000  37.50000000  88.89194139  -2.28038021
       0.01129257
> krusdat
   [,1] [,2]
[1,] 1 2.58
[2,] 1 2.57
```

```
[3,] 1 2.58
[4,] 1 2.76
[5,] 1 2.32
[6,] 2 2.46
...
```

## Rank Tests for Multiple Related Samples

### The Friedman Rank Sum Test

The Friedman rank sum test proposed first by Friedman (1937, 1940) is probably the best-known rank test for a complete random block design. When there is a two-way layout with one blocking variable and one treatment variable, for example, each of $n$ panelists evaluating $k$ products with ranking, one can use the Friedman rank sum test to test the null hypothesis that there is no product effect; i.e., $H_0: \mu_1 = \mu_2 = \ldots = \mu_k$ versus the alternative hypothesis $H_a: \mu_1, \mu_2, \ldots, \mu_k$ are not all equal. The Friedman test is a nonparametric version of a two-way analysis of variance. The test statistic is

$$F = \frac{12}{nk(k+1)} \sum_{j=1}^{k} \left[ R_j - n(k+1)/2 \right]^2$$

$$= \left[ \frac{12}{nk(k+1)} \sum_{j=1}^{k} R_j^2 \right] - 3n(k+1). \tag{9.2-26}$$

For a large sample, the test statistic follows asymptotically a chi-square distribution with $(k-1)$ DF. In the case of ties, the average ranks should be used, and the test statistic should be adjusted as

$$F' = F/[1 - S/nk(k^2 - 1)], \tag{9.2-27}$$

where $S = \sum_{j=1}^{n} \sum_{i=1}^{g_j} (t_{ij}^3 - t_{ij})$, where $g_j$ is the number of tied groups in the $j$th block and $t_{ij}$ denotes the size of the $i$th group in the $j$th block.

When the null hypothesis of no difference is rejected, multiple comparisons based on rank sums can be made. The comparison procedure using the least significant difference in (9.2-28) was developed by Nemenyi (1963) and discussed by Hochberg and Tamhane (1987) and Hollander and Wolfe (1999). Decide $\mu_i \neq \mu_j$ if

$$|R_i - R_j| \geq Q_{k,\infty}^{\alpha} \left[ \frac{nk(k+1)}{12} \right]^{0.5}; \tag{9.2-28}$$

otherwise, decide $\mu_i = \mu_j$, where $1 \leq i < j \leq k$. One can find the value of $Q_{k,\infty}^{\alpha}$ in Table 9.20 or Table A.6.

**Table 9.20** Upper $\alpha$ Point of the Studentized Range Distribution with Parameter $k$ and Degrees of Infinite ($Q^\alpha_{k,\infty}$)

| | $\alpha \leq 0.1$ | | | | $\alpha \leq 0.05$ | | | | $\alpha \leq 0.01$ | | | |
|---|---|---|---|---|---|---|---|---|---|---|---|---|
| | $k$ | | | | $k$ | | | | $k$ | | | |
| | 3 | 4 | 5 | 6 | 3 | 4 | 5 | 6 | 3 | 4 | 5 | 6 |
| $n \to \infty$ | 2.90 | 3.24 | 3.48 | 3.66 | 3.31 | 3.63 | 3.86 | 4.03 | 4.12 | 4.40 | 4.60 | 4.76 |

**Example 9.2-9**

In a consumer test for overall liking for five brands of a product A, B, C, D, and E, 30 consumer panelists were selected randomly and $\alpha = 0.1$ was chosen. Each panelist was asked to rank the five products from the most liked (1) to the least liked (5). The ranks are listed in Table 9.21. The rank sums for the five brands are 77.5, 97, 77.5, 84, and 114. For $n = 30$ and $k = 5$, from (9.2-26), the value of the $F$ statistic is

$$F = 12(77.5^2 + 97^2 + 77.5^2 + 84^2 + 114^2)/(30 \times 5 \times 6) - 3 \times 30 \times 6 = 12.98.$$

**Table 9.21** Ranking Test for Brands (Example 9.2-9)

| ID | A | B | C | D | E | ID | A | B | C | D | E | ID | A | B | C | D | E |
|---|---|---|---|---|---|---|---|---|---|---|---|---|---|---|---|---|---|
| 1 | 2 | 4 | 3 | 1 | 5 | 11 | 4 | 3 | 2 | 1 | 5 | 21 | 1 | 5 | 2 | 3 | 4 |
| 2 | 1 | 3 | 2 | 4 | 5 | 12 | 2 | 3.5 | 3.5 | 1 | 5 | 22 | 2 | 4 | 3 | 1 | 5 |
| 3 | 2 | 4 | 1 | 3 | 5 | 13 | 5 | 2 | 4 | 3 | 1 | 23 | 3 | 2 | 4 | 5 | 1 |
| 4 | 4 | 1 | 4 | 2 | 4 | 14 | 1 | 5 | 2 | 3 | 4 | 24 | 4 | 2 | 3 | 5 | 1 |
| 5 | 4 | 2 | 1 | 4 | 4 | 15 | 1 | 4 | 2 | 3 | 5 | 25 | 2 | 1 | 4 | 3 | 5 |
| 6 | 2 | 4 | 1 | 3 | 5 | 16 | 2 | 2 | 2 | 4 | 5 | 26 | 1 | 4 | 2 | 3 | 5 |
| 7 | 2 | 5 | 1 | 3 | 4 | 17 | 1 | 3 | 2 | 5 | 4 | 27 | 4 | 3 | 5 | 1 | 2 |
| 8 | 2 | 4 | 3 | 1 | 5 | 18 | 2 | 3 | 1 | 5 | 4 | 28 | 5 | 3 | 4 | 1 | 2 |
| 9 | 3.5 | 1.5 | 3.5 | 5 | 1.5 | 19 | 4 | 5 | 1.5 | 3 | 1.5 | 29 | 2 | 4 | 3 | 1 | 5 |
| 10 | 5 | 2 | 4 | 3 | 1 | 20 | 1 | 4 | 2 | 3 | 5 | 30 | 3 | 4 | 2 | 1 | 5 |
| | | | | | | | | | | | | | 77.5 | 97 | 77.5 | 84 | 114 |

We found that Panelist 4 had a tie group with size 3, panelist 5 also had a tie group with size 3, panelist 9 had two tie groups with size 2, panelist 12 also had a tie group with size 2, panelist 16 had a tie group with size 3, and panelist 19 had a tie group with size 2.

According to (9.2-27),

$$S = (3^3 - 3) + (3^3 - 3) + 2(2^3 - 2) + (2^3 - 2) + (3^3 - 3) + (2^3 - 2) = 96.$$

Hence,

$$F' = \frac{12.98}{1 - 96/[30 \times 5(5^2 - 1)]} = 13.34.$$

Because $13.34 > 7.779 = \chi^2_{0.9,4}$ from Table A.3, we can conclude at $\alpha = 0.1$ significance level that the panelists showed significant difference for the brands in overall liking, with the most overall liking for brands A and C and the least overall liking for brand E. For $\alpha = 0.1$, $k = 5$ and $n = 30$, the least significant difference is $3.48 \times \sqrt{\dfrac{30 \times 5 \times 6}{12}} = 30.12$. Hence, the results of multiple comparisons for the rank sums of the products in Table 9.1 are as follows:

A    C    D    B    E

The S-PLUS program *friedman.test* can be used for the test for a specified data form:

```
> friedman.test(fmdat2[,3],fmdat2[,2],fmdat2[,1])

Friedman rank sum test

data: fmdat2[, 3] and fmdat2[, 2] and fmdat2[, 1]
Friedman chi-square = 13.3356, df = 4, p-value = 0.0097
alternative hypothesis: two.sided

> fmdat2[1:5,]

id prod rank
1 1 1 2
2 2 1 1
3 3 1 2
4 4 1 5
...........
```

### The Page Test for Ordered Alternatives

The alternative hypothesis for the Friedman test is that the treatment effects are not all the same; it is a two-sided alternative. In some practical situations, the ordered alternative is of interest. The page test can be used for this situation. Just as the Jonckheere–Terpstra test for ordered alternatives is more powerful than the Kruskal–Wallis test in a completely random design, i.e., for multiple independent samples, the page test with ordered alternatives is more powerful than the Friedman test in a randomized complete block design, i.e., for multiple correlated samples. For the null hypothesis $H_0: \mu_1 = \mu_2 = \ldots = \mu_k$ and the alternative hypothesis $H_a: \mu_{1'} \geq \mu_{2'} \geq \ldots \geq \mu_{k'}$ with at least one strict inequality, Page (1963) proposed a test statistic based on a weighted sum of rank sums of treatments. It is

$$L = \sum_{i=1}^{k} iR_i. \tag{9.2-29}$$

The weight $i$ is the hypothetical rank of the treatment; i.e., $R_1$ is the least rank sum predicted by the alternative hypothesis, and $R_k$ is the largest rank sum predicted by the alternative hypothesis. Tables of some exact critical values of the statistic are given in Page (1963) and Odeh (1977). Page (1963) also gave a formula for normal-deviate critical values of $L$,

$$L_{critical} = \frac{n(k^3 - k)}{12} \left[ \frac{Z_{critical}}{\sqrt{n(k-1)}} + \frac{3(k+1)}{k-1} \right] \tag{9.2-30}$$

where $Z_{critical}$ represents the $(1 - \alpha)100$th percentile of the standard normal distribution. The normal-deviate critical values are very good approximations of the exact critical vales of $L$. The critical values $(L_0)$ based on (9.2-30) are calculated and given in Table A.26 for $n$ to 100, $k$ to 7, and for $\alpha = 0.05$ and 0.1, respectively.

For large values of $n$ and $k$, the statistic (9.2-30) follows asymptotically a normal distribution with $\mu$ and $\sigma^2$, where

$$\mu = nk(k + 1)^2/4 \quad \text{and} \quad \sigma^2 = n(k - 1)k^2(k + 1)^2/144.$$

Hence, the $Z$ statistic

$$Z = \frac{L - \mu}{\sigma} \tag{9.2-31}$$

follows asymptotically the standard normal distribution.

**Example 9.2-10**

In order to investigate perceived sweetness of samples A, B, and C, researchers randomly selected 50 consumer panelists. The samples were presented to each of the panelists, who were asked to rank the samples from the most sweet (1) to the least sweet (3). The null hypothesis is $H_0: \mu_A = \mu_B = \mu_C$, and the alternative hypothesis is $H_a: \mu_A \geq \mu_B \geq \mu_C$ with at least one strict inequality. The data are listed in Table 9.22. The rank sums for the samples A, B, and C are 90, 98, and 112, respectively. Because the assumed most sweet sample with the lowest rank sum is sample A and the assumed least sweet sample with the highest rank sum

**Table 9.22** Ranks for Three Samples (Example 9.2-10)

| ID | A | B | C | ID | A | B | C |
|----|---|---|---|----|---|---|---|
| 01 | 2 | 3 | 1 | 21 | 1 | 2 | 3 |
| 02 | 2 | 3 | 1 | 22 | 2 | 1 | 3 |
| 03 | 2 | 3 | 1 | 23 | 2 | 1 | 3 |
| 04 | 1 | 3 | 2 | 24 | 1 | 2 | 3 |
| 05 | 1 | 3 | 2 | 25 | 2 | 1 | 3 |
| 06 | 3 | 1 | 2 | 26 | 1 | 2 | 3 |
| 07 | 1 | 3 | 2 | 27 | 1 | 2 | 3 |
| 08 | 3 | 1 | 2 | 28 | 2 | 1 | 3 |
| 09 | 1 | 3 | 2 | 29 | 2 | 1 | 3 |
| 10 | 1 | 3 | 2 | 30 | 1 | 2 | 3 |
| 11 | 1 | 3 | 2 | 31 | 1 | 2 | 3 |
| 12 | 1 | 3 | 2 | 32 | 1 | 2 | 3 |
| 13 | 1 | 3 | 2 | 33 | 1 | 2 | 3 |
| 14 | 1 | 3 | 2 | 34 | 1 | 2 | 3 |
| 15 | 3 | 1 | 2 | 35 | 1 | 2 | 3 |
| 16 | 1 | 3 | 2 | 36 | 2 | 1 | 3 |
| 17 | 1 | 3 | 2 | 37 | 2 | 1 | 3 |
| 18 | 1 | 2 | 3 | 38 | 2 | 1 | 3 |
| 19 | 1 | 2 | 3 | 39 | 1 | 2 | 3 |
| 20 | 2 | 1 | 3 | 40 | 3 | 2 | 1 |

*(continued)*

**Table 9.22** Ranks for Three Samples (Example 9.2-10)—*cont...*

| ID | A | B | C | ID | A | B | C |
|----|---|---|---|----|---|---|---|
| 41 | 3 | 1 | 2 | 46 | 3 | 2 | 1 |
| 42 | 3 | 2 | 1 | 47 | 3 | 1 | 2 |
| 43 | 3 | 2 | 1 | 48 | 3 | 2 | 1 |
| 44 | 3 | 2 | 1 | 49 | 3 | 2 | 1 |
| 45 | 3 | 1 | 2 | 50 | 3 | 1 | 2 |
|    |   |   |   |    | 90 | 98 | 112 |

is sample C, the statistic value is then $L = 90 + 2(98) + 3(112) = 622$. From Table A.26, we find for $\alpha = 0.05$, $n = 50$, and $k = 3$, the critical value is 617. Because $L = 622 > l_0 = 617$, the null hypothesis is rejected, and the alternative hypothesis is accepted at $\alpha = 0.05$ level. Using large-sample approximation in (9.2-31), we get $\mu = 50 \times 3 \times (3+1)^2/4 = 600$, $\sigma^2 = 50 \times 2 \times 3^2 \times 4^2/144 = 100$, and $Z = 2.2$ with the associated $P$-value $= 0.014$. We can obtain the same conclusion. However, if using the Friedman tests for the same data, we get $P$-value $= 0.084$. It shows that the page test with the ordered alternative hypothesis is more powerful.

```
> pagetest(pagedat2)
[1]] 2.200 0.014
> friedman(pagedat2)
Friedman test with adjustment for ties:
p-value = 0.084
```

### The Anderson Statistic

Anderson (1959) proposed a test statistic that is different from the Friedman statistic for ranking data in a randomized complete block design. In consumer preference studies, it is common to seek a ranking of treatments or items. Suppose the number $t$ of treatment $T_1$, $T_2, \ldots, T_t$ is not large, so they may be ranked simultaneously. If this is so, each consumer is asked to rank the $t$ treatments by assigning ranks from 1 to $t$. Suppose $n$ consumers (judges) are asked to assign ranks to $t$ items. The results of the ranking may be summarized in a $t \times t$ table, such as Table 9.23. In the table, $n_{ij}$ denotes the number of consumers who assigned the $j$th rank to the $i$th treatment.

**Table 9.23** Layout of Simultaneous Rankings of $t$ Treatments by Each of $n$ Judges, Where $n_{ij}$ Denotes the Number of Judges Assigning the $j$th Rank to the $i$th Treatment

| Treatments | 1 | 2 | 3 | ... | $t$ | Totals |
|---|---|---|---|---|---|---|
| | | | **Ranks** | | | |
| $T_1$ | $n_{11}$ | $n_{12}$ | $n_{13}$ | ... | $n_{1t}$ | $n$ |
| $T_2$ | $n_{21}$ | $n_{22}$ | $n_{23}$ | ... | $n_{2t}$ | $n$ |
| $\vdots$ | $\vdots$ | $\vdots$ | $\vdots$ | ... | $\vdots$ | $\vdots$ |
| $T_t$ | $n_{t1}$ | $n_{t2}$ | $n_{t3}$ | ... | $n_{tt}$ | $n$ |
| Total | $n$ | $n$ | $n$ | ... | $n$ | |

Note that the total of each row and also of each column must be $n$. Hence, the cell frequencies $n_{ij}$ cannot be considered to constitute a random sample. Therefore, the conventional $X^2$ statistic is no longer appropriate for the analysis of such tables.

The null hypothesis of interest is that the probability $p_{ij}$ of the $i$th treatment receiving the $j$th rank is $1/t$. That is, the treatments are ranked randomly by the judges. The null hypothesis can be stated as

$$H_0: \ p_{ij} = 1/t.$$

Anderson (1959) developed a $X^2$ statistic for testing $H_0$. His statistic is

$$X^2 = \left(\frac{t-1}{n}\right) \sum_{i=1}^{t} \sum_{j=1}^{t} \left(n_{ij} - \frac{n}{t}\right)^2, \qquad (9.2\text{-}32)$$

which has approximately the $\chi^2$ distribution with $(t-1)^2$ DF. It is noted that the Anderson statistic, $A$, contains the Friedman statistic, $F$, in that it can be decomposed into the sum of the Friedman statistic (the location effect) and other terms, $A' = A - F$ often referred to as the variance effects, skewness effects, etc. See Anderson (1959) and Best (1993). The statistic $A'$ follows approximately the chi-square distribution with $(t-1)(t-2)$ DF. A significant result of $A'$ may suggest possible consumer segmentation.

---

Anderson (1959) reported a consumer preference study at Mississippi State College involving three varieties $T_1$, $T_2$, and $T_3$ of snap beans. The three varieties were displayed in retail stores, and consumers were asked to rank the beans according to first, second, and third choices. The data

**Example 9.2-11**

**Table 9.24** Rank Frequencies for Three Varieties (Example 9.2-11)[a]

| Varieties | Ranks | | | Totals |
| | 1 | 2 | 3 | |
|---|---|---|---|---|
| $T_1$ | 42 ($n_{11}$) | 64 ($n_{12}$) | 17 ($n_{13}$) | 123 |
| $T_2$ | 31 ($n_{21}$) | 16 ($n_{22}$) | 76 ($n_{23}$) | 123 |
| $T_3$ | 50 ($n_{31}$) | 43 ($n_{32}$) | 30 ($n_{33}$) | 123 |
| | 123 | 123 | 123 | 369 |

[a]From a study by Anderson (1959).

obtained in one store are presented in Table 9.24. From (9.2-32), $X^2 = 53.04$ with $(t - 1)^2 = 4$ DF. At the $\alpha = 0.01$ level, $\chi^2_{0.99,4} = 13.28$ from Table A.3. The observed value of the statistic is

$$X^2 = \frac{2}{123}\left[\sum_{i=1}^{3}\sum_{j=1}^{3}(n_{ij} - 41)^2\right] = 53.04 \text{ with 4 DF.}$$

This value is highly significant. Therefore, we conclude that at least one of the varieties is a preferred one. The experimenter, on inspecting the highest ranking, will find that $T_3$ has the largest number of first ranks, followed by $T_1$ and $T_2$. If the null hypothesis of random rankings is rejected, one may wish to further examine the contrasts among the $n_{ij}$ following Anderson (1959).

The Friedman statistic without ties can also be calculated using Anderson's types of data in Table 9.24. In this case, $n = 123$, $k = t = 3$ and $R_1 = 42 + 2 \times 64 + 3 \times 17 = 221$; $R_2 = 31 + 2 \times 16 + 3 \times 76 = 291$ and $R_3 = 50 + 2 \times 43 + 3 \times 30 = 226$; hence

$$F = \frac{12}{123 \times 3(3 + 1)}(221^2 + 291^2 + 226^2) - 3 \times 123(3 + 1) = 24.797,$$

and $A' = A - F = 53.04 - 24.797 = 28.243$. Both $F$ and $A'$ follow a chi-square distribution with 2 DF, and both $P$-values of the two statistics are smaller than 0.001. We can conclude that both of the location effects and the other effects are significant.

```
> andersontest(andersondat)
Location: df = 2 , SS = 24.797 , p-value = 0
Spread: df = 2 , SS = 28.244 , p-value = 0
Total: df = 4 , SS = 53.041 , p-value = 0
```

### The Durbin Test for Balanced Incomplete Block Designs

In this section we have so far considered a situation in which the number of treatments $t$ is not large so that all the treatments can be ranked by each judge simultaneously and the rankings recorded in a $t \times t$ table, such as Table 9.23. But as $t$ increases, it becomes progressively more confusing for judges to rank all $t$ treatments simultaneously. When the simultaneous ranking of more than two treatments becomes confusing and undesirable, the incomplete block design experimental arrangement suggests itself. For details about the incomplete block design, refer to Chapter 5. Instead of having judges compare and rank all the $t$ objects simultaneously, one may ask judges to compare and rank only $k$ objects at a time, where $k$ is a desired number smaller than $t$.

Durbin (1951) discussed the ranking experiments in a randomized balanced incomplete block (BIB) design and developed a Friedman-type statistic. In the BIB designs, there are $n$ blocks (panelists) and $t$ treatments (products). There are $k < t$ treatments ranked within each of the $n$ blocks, every treatment appears in $r$ blocks $(r < n)$, and every treatment appears with every other treatment exactly $\lambda$ times. The parameters in the BIB designs satisfy the restriction

$$\lambda(t - 1) = r(k - 1). \tag{9.2-33}$$

The Durbin statistic is

$$D = \left[\frac{12(t-1)}{rt(k^2-1)}\right] \sum_{j=1}^{t} \left\{ R_j - \frac{r(k+1)}{2} \right\}^2, \tag{9.2-34}$$

which follows asymptotically a chi-square distribution with $(t-1)$ DF.

---

**Example 9.2-12**

For illustration of the calculation of the $D$ statistic in (9.2-34), we consider the data in Table 9.25. There are seven panelists in a ranking test for seven products with a BIB design. For $t = 7$, $k = 3$, $n = 7$, $r = 3$, $\lambda = 1$ and $R_1 = 3$, $R_2 = 4$, $R_3 = 7$, $R_4 = 8$, $R_5 = 9$, $R_6 = 3$, $R_7 = 8$, then

$$\frac{12(t-1)}{rt(k^2-1)} = \frac{12(6)}{3(7)(8)} = 0.429,$$

$$\frac{r(k+1)}{2} = \frac{3(4)}{2} = 6.$$

**Table 9.25** Ranked Data in a Balanced Incomplete Block Design ($t = 7$, $k = 3$, $n = 7$, $r = 3$, $\lambda = 1$) (Example 9.2-12)

| | Treatments (Products) | | | | | | |
|---|---|---|---|---|---|---|---|
| | **A** | **B** | **C** | **D** | **E** | **F** | **G** |
| **Blocks (Panelists)** | | | | | | | |
| 1 | 1 | 2 | | 3 | | | |
| 2 | 1 | | 2 | | 3 | | |
| 3 | | | 2 | 3 | | 1 | |
| 4 | 1 | | | | | 2 | 3 |
| 5 | | 1 | 3 | | | | 2 |
| 6 | | 1 | | | 3 | | 2 |
| 7 | | | | 2 | 3 | 1 | |
| $R_i$ | 3 | 4 | 7 | 8 | 9 | 4 | 7 |

Hence,

$$D = 0.429 \times [(3 - 6)^2 + (4 - 6)^2 + (7 - 6)^2 + (8 - 6)^2 + (9 - 6)^2$$
$$+ (4 - 6)^2 + (7 - 6)^2]$$
$$= 13.73.$$

Because $D = 13.73 > 12.59$, the 95th percentile of a chi-square distribution with $7 - 1 = 6$ DF, we conclude at a 0.05 significance level that the seven products for comparison are significantly different.

```
> durbin2(7,3,3,c(3,4,7,8,9,4,7))
[1]  13.71428571  0.03299579
```

## Measures of Association for Ranking Data

### The Spearman Rank Correlation Coefficient

Spearman's rank correlation is named as such because it uses ranks of observations (instead of observations) in the expression of Spearman's product moment correlation measure. Another statistic for measuring dependence between two quantities is the quadrant sum statistic, but we will not consider it here. The quadrant sum statistic was proposed by Olmstead and Tukey (1947), and the interested reader is advised to refer to their paper and to the book by Steel and Torrie (1980).

Let $(X_1, Y_1), (X_2, Y_2), \ldots, (X_n, Y_n)$ be random samples. Rank the $X$ values, assigning ranks from 1 through $n$. Tied observations may be assigned the average of their ranks. Similarly, rank the $Y$ values. Let $R_i$ and $S_i$ denote the ranks of $X_i$ and $Y_i$, respectively, $i = 1, \ldots, n$. For each pair we compute the difference $d_i = R_i - S_i$. Then calculate statistics $D$ and $r_S$ as follows:

$$D = \sum_{i=1}^{n} d_i^2, \qquad (9.2\text{-}35)$$

and

$$r_S = 1 - 6D/(n^3 - n). \qquad (9.2\text{-}36)$$

The statistic $r_S$ should be adjusted for ties. It is

$$r_S^* = \frac{r_S - C_1}{C_2}, \qquad (9.2\text{-}37)$$

where

$$C_1 = \frac{\sum_{i=1}^{g_1}(s_i^3 - s_i) + \sum_{j=1}^{g_2}(t_j^3 - t_j)}{2n(n^2 - 1)} \quad \text{and} \quad C_2 = \sqrt{\left[1 - \frac{\sum_{i=1}^{g_1}(s_i^3 - s_i)}{n(n^2 - 1)}\right]\left[1 - \frac{\sum_{j=1}^{g_2}(t_j^3 - t_j)}{n(n^2 - 1)}\right]}$$

where $g_1$ and $g_2$ are the numbers of tied groups for $X$ and $Y$, respectively, and $s_i$ and $t_j$ are the sizes of the $i$th and $j$th tied groups, $i = 1, \ldots, g_1; j = 1, \ldots, g_2$ (Hollander and Wolfe, 1999, 396).

The statistic $r_S$ (or $r_S^*$) just defined is called the rank correlation and measures the extent of the linear dependence between the two quantities $X$ and $Y$. The distribution of $r_S$ ranges from $-1$ to 1. When $r_S$ is observed in either tail of its distribution, the degree of linear dependence is considered significant. Two quantities with rank correlation $r_S$ equal to 1 or $-1$ will be considered to have perfect linear relationships. The linear relationship is increasing when $r_S$ is positive and decreasing when $r_S$ is negative. There is no significant linear dependence between two quantities if their rank correlation measure is not significantly different from zero. Thus, a statistical hypothesis of independence between two quantities is rejected when their rank correlation statistic $r_S$ is observed in either tail of its distribution. The exact specification for the tails to reject the null hypothesis of independence rests upon the desired level of significance.

We note that statistics $r_S$ and $D$ are equivalent in the sense that, given either one, we can find the other from relationship (9.2-36). Hence, we may use $D$ instead of $r_S$ for testing independence if we so wish. To be able to carry out tests, we need their distributions under the null hypothesis. We may point out here that the exact as well as the asymptotic distributions of our test statistics tend to be somewhat complicated whenever ties occur. That is why here we consider and use distributions of $D$ and $r_S$ under the assumption of no ties, even though in reality there may be tied observations when assigning ranks for calculating $D$ and $r_S$.

Note from (9.2-36) that $r_S = 1$, if and only if $D = 0$. Also, $r_S = -1$ if and only if $D$ takes its maximum possible value $(n^3 - n)/3$. Under the assumption of independence, the distribution of $r_S$ is symmetric about zero and that of $D$ is symmetric about $(n^3 - n)/6$. Hence, for any $d$ in the space of statistic $D$,

$$P(D \le d) = P\left[D \ge \frac{2(n^3 - n)}{6} - d\right]. \qquad (9.2\text{-}38)$$

For instance, if $n = 5$, the space of $D$ consists of all integers from 0 through 40, and over this space the distribution of $D$ (under the null hypothesis of independence) is symmetric about $(n^3 - n)/6 = 20$. Because of this symmetry, $P(D \le 0) = P(D \ge 40)$, $P(D \ge 1) = P(D \ge 39)$, and so on. Probabilities $P(D \le d)$ for all $d$ in the space of the distribution of $D$, under the null hypothesis of independence and no ties, are tabulated in Table A.20 for $n = 2$ through $n = 11$. If the number of pairs, that is, sample size $n$, is 11 or less, we may use Table A.20 for testing the null hypothesis of independence to find the significance probabilities of rejection regions.

For large $n$, one may use the $T$ statistic defined as follows:

$$T = r_S \sqrt{(n - 2)/(1 - r_S^2)}. \qquad (9.2\text{-}39)$$

Note that $T$ is computed from $r_S$, which is defined in (9.2-36). It has been shown that under the null hypothesis of independence, the asymptotic distribution of the $T$ statistic follows the $t$ distribution with $(n - 2)$ DF. The use of the asymptotic distribution is believed to be satisfactory if $n > 10$. The decision rule, following the asymptotic distribution, is to reject the null hypothesis of independence if $T$ is observed in the tails of its distribution. That is, at the $\alpha$ level of significance, reject the null hypothesis of independence in favor of the two-sided alternative if $|T| > t_{1-\alpha/2,n-2}$, where $|T|$ denotes the absolute value of observed $T$. If the alternative is believed to be one-sided, then the rejection region is defined by the appropriate tail of the $t$ distribution.

It can be shown that the Spearman rank correlation coefficient $r_S$ is equal to the conventional correlation coefficient, i.e., Pearson product-moment correlation coefficient with ranks substituted for the observations $(X, Y)$. Hence, the original variables can be converted to ranks, and the ordinary Pearson correlation then gives the value of the Spearman rank correlation coefficient.

---

The off-flavor scores and plate counts for five treatments with various levels of nitrogen and carbon dioxide are shown in Table 9.26. The panel scores are averages of 10 judges using a scale of 1 to 7 with $1 =$ none and $7 =$ very strong off-flavor. The investigator wished to determine the degree of association between the two response variables. The plate

**Example 9.2-13**

**Table 9.26** Illustration of Spearman Rank Correlation for Testing Independence (Example 9.2-13)

| Treatment | Off-Flavor Score, $X_i$ | Rank of $X_i$, $R_i$ | Log Plate Count, $Y_i$ | Rank of $Y_i$, $S_i$ | $d_i = R_i - S_i$ | $d_i^2$ |
|---|---|---|---|---|---|---|
| $T_1$ | 1.4 | 1 | 6.36 | 5 | −4 | 16 |
| $T_2$ | 3.5 | 5 | 2.45 | 1 | 4 | 16 |
| $T_3$ | 2.8 | 4 | 5.43 | 3 | 1 | 1 |
| $T_4$ | 2.0 | 3 | 3.63 | 2 | 1 | 1 |
| $T_5$ | 1.7 | 2 | 5.75 | 4 | −2 | 4 |
| | | | | | | $D = 38$ |

Note:
Significance probability $= P(D \geq 38) = 0.042$ from Table A.20.

counts are averages of two determinations. The investigator believed that, if the linear relationship is significant, it must be an inverse relationship in the sense that off-flavor scores increase with decreasing plate counts. Thus, the alternative is one-sided, and the rejection region for the null hypothesis of independence in favor of the one-sided alternative will be specified by large values of $D$.

From our calculations in Table 9.26, $D = 38$. From Table A.20, the probability of observing $D \geq 38$ is 0.042. Hence, at any $\alpha$ level greater than 0.042, say 0.05, we reject the hypothesis of independence. Even though $n = 5 < 10$, let us compute $T$ from (9.2-39) and use it for testing. Since $r_S = -0.90$, we find $T = -3.57$. Because $t_{\alpha,n-2} = t_{0.05,3} = -2.35$ is greater than the observed $T = -3.57$, we conclude that low off-flavor scores are associated with high total plate counts and that the relationship is inverse since $r_S$ is found negative. Notice that both the exact and approximate tests in this example lead to the same conclusion, even though $n = 5$.

---

**Example 9.2-14**

Three candidate judges were screened for their ability to detect levels of smoke in a product. The null hypothesis of interest is that each judge assigns ranking at random against the alternative that each is capable of correct ranking. Six smoke levels of a test product were presented to each candidate. They were instructed to rank the product samples according to the level of smoke perceived, giving a rank of 1 to the lowest level and 6 to the product sample with the highest smoke level. As there is a true ranking of levels of smoke, ties were not allowed. The results are shown in Table 9.27.

We can verify that when $n = 6$ and $D = 4$, $r_S = 0.89$. From Table A.20, we have $P(D \leq 4) = 0.017$. Hence, even at a significance level of less than 5%, we conclude that the rankings made by candidates $A$ and $B$ could not be simply a result of random consideration, but of their ability to perceive the amount of smoke present in the sample. For candidate $C$, $D = 14$ and $r_S = 0.60$ with $P(D \leq 14) = 0.121$. Therefore, at $\alpha = 0.05$, or even at $\alpha = 0.10$, rankings made by candidate $C$ are far from consistent with the true rankings. Thus, candidate $C$ may not be included in the panel of judges for this type of study.

**Table 9.27** An Illustration of the Use of Spearman Rank Correlation for Selecting Judges (Example 9.2-14)

| Candidate | True Ranking $R_i$ | Candidate's Ranking $S_i$ | $d_i = R_i - S_i$ | $d_i^2$ |
|---|---|---|---|---|
| A | 1 | 2 | −1 | 1 |
| | 2 | 1 | 1 | 1 |
| | 3 | 3 | 0 | 0 |
| | 4 | 4 | 0 | 0 |
| | 5 | 6 | −1 | 1 |
| | 6 | 5 | 1 | 1 |
| | | | | $D = 4$ |
| | | Significance probability $= 0.017$ | | |
| B | 1 | 1 | 0 | 0 |
| | 2 | 3 | −1 | 1 |
| | 3 | 2 | 1 | 1 |
| | 4 | 5 | −1 | 1 |
| | 5 | 4 | 1 | 1 |
| | 6 | 6 | 0 | 0 |
| | | | | $D = 4$ |
| | | Significance probability $= 0.017$ | | |
| C | 1 | 3 | −2 | 4 |
| | 2 | 1 | 1 | 1 |
| | 3 | 2 | 1 | 1 |
| | 4 | 6 | −2 | 4 |
| | 5 | 5 | 0 | 0 |
| | 6 | 4 | 2 | 4 |
| | | | | $D = 14$ |
| | | Significance probability $= 0.121$ | | |

**Example 9.2-15**

Two panelists evaluated 20 samples with different intensity of a sensory characteristic using a 5-point scale. The data are given in Table 9.28. In order to assess consistency of performance between the two panelists, we can use the Spearman correlation coefficient. We note that

$$D = (3 - 8.5)^2 + (7.5 - 18.5)^2 + \cdots + (13.5 - 18.5)^2 = 542.5 \text{ from } (9.2\text{-}35),$$

hence,

$$r_S = 1 - \frac{6 \times 542.5}{20(20^2 - 1)} = 0.592 \text{ from } (9.2\text{-}36).$$

We also note that there are four tied groups with sizes 5, 4, 8, and 2 for the first panelist and five tied groups with sizes 3, 3, 4, 6, and 4; hence,

$$C_1 = \frac{(5^3 - 5) + \cdots + (2^3 - 2) + (3^3 - 3) + \cdots + (4^3 - 4)}{2 \times 20 \times (20^2 - 1)} = 0.068,$$

$$C_2 = \sqrt{\left[1 - \frac{(5^3 - 5) + \cdots + (2^3 - 2)}{2 \times 20 \times (20^2 - 1)}\right]\left[1 - \frac{(3^3 - 3) + \cdots + (4^3 - 4)}{2 \times 20 \times (20^2 - 1)}\right]}$$

$$= 0.9317.$$

The adjusted Spearman correlation coefficient is

$$r_S^* = \frac{0.592 - 0.068}{0.9317} = 0.562 \text{ from } (9.2\text{-}37).$$

**Table 9.28** Evaluation Results of Two Panelists for 20 Samples Using a 5-Point Scale (Example 9.2-15)

|      | 1   | 2    | 3    | 4    | 5   | 6    | 7   | 8   | 9    | 10  |
|------|-----|------|------|------|-----|------|-----|-----|------|-----|
| X    | 2   | 3    | 4    | 4    | 2   | 4    | 2   | 2   | 5    | 3   |
| R(X) | 3   | 7.5  | 13.5 | 13.5 | 3   | 13.5 | 3   | 3   | 19   | 7.5 |
| Y    | 3   | 5    | 4    | 4    | 1   | 1    | 1   | 3   | 4    | 2   |
| R(Y) | 8.5 | 18.5 | 13.5 | 13.5 | 2   | 2    | 2   | 8.5 | 13.5 | 5   |
|      | 11  | 12   | 13   | 14   | 15  | 16   | 17  | 18  | 19   | 20  |
| X    | 5   | 4    | 2    | 4    | 5   | 3    | 3   | 4   | 4    | 4   |
| R(X) | 19  | 13.5 | 3    | 13.5 | 19  | 7.5  | 7.5 | 13.5| 13.5 | 13.5|
| Y    | 4   | 4    | 3    | 2    | 5   | 3    | 2   | 4   | 5    | 5   |
| R(Y) | 13.5| 13.5 | 8.5  | 5    | 18.5| 8.5  | 5   | 13.5| 18.5 | 18.5|

The value of the approximate $T$ statistic is

$$T = 0.562\sqrt{(20-2)/(1-0.562^2)} = 2.88$$

with the associated *P*-value $= 0.01$. Using the S-PLUS built-in function *cor. test* for the ranks of $X$ and $Y$, we can obtain the Spearman correlation coefficient. It is the same as the conventional correlation coefficient, i.e., Pearson's product-moment correlation.

```
> cor.test(rank(spdat3[,1]),rank(spdat3[,2]))

    Pearson's product-moment correlation

data: rank(spdat3[, 1]) and rank(spdat3[, 2])
t = 2.8861, df = 18, p-value = 0.0098
alternative hypothesis: true coef is not equal to 0
sample estimates:
    cor
0.562462
```

## The Kendall Coefficient of Concordance

For three or more sets of ranks, Kendall's coefficient of concordance, $W$, introduced independently by Kendall and Babington-Smith (1939) and Wallis (1939), provides a measure of concordance among the sets of ranks (see Kendall 1962, 1970 for details). For example, there are $n$ blocks representing $n$ independent panelists, each one assigning ranks to the same set of $k$ samples, and Kendall's $W$ measures the extent to which the $n$ panelists agree on their ranks of the $k$ samples. Kendall's $W$ statistic is defined as

$$W = \frac{12S}{n^2 k(k^2 - 1)}, \tag{9.2-40}$$

where $S = \sum_{j=1}^{k} [R_j - n(k+1)/2]^2$, $R_j, j = 1, 2, \ldots, k$ is rank sum for the $j$th sample. For ties, the $W$ statistic should be adjusted as

$$W^* = \frac{W}{1 - A/[nk(k^2 - 1)]}, \tag{9.2-41}$$

where

$$A = \sum_{j=1}^{n} \sum_{i=1}^{g_j} (t_{ij}^3 - t_{ij}),$$

where $g_j$ is the number of tied groups in the $j$th block (panelist) and $t_{ij}$ denotes the size of the $i$th group in the $j$th block (panelist). The value of $W$ (or $W^*$) ranges between 0 and 1. When there is complete agreement among all the $n$ sets of ranks, the value will equal 1 while there is no pattern of agreement among all the $n$ sets of ranks; i.e., if the ranks assigned to the $k$ samples are completely random for each of the $n$ panelists, the rank sum is the same for the $k$ samples, and the $W$ value will equal 0.

Kendall's coefficient of concordance, $W$ (or $W^*$), is related to the Friedman statistic, $F$ (or $F^*$), as

$$W = \frac{F}{n(k-1)} \qquad (9.2\text{-}42)$$

or

$$W^* = \frac{F^*}{n(k-1)}. \qquad (9.2\text{-}43)$$

The $W$ statistic can also be used as a test statistic for the null hypothesis of no agreement or no association among the rankings. The alternative hypothesis is that association or agreement exists. For large $n$, the statistic $n(k-1)W$ or $n(k-1)W^*$ follows approximately a chi-square distribution with $(k-1)$ DF. It is noted that the hypotheses in the Friedman test and in the test for $W$ are different. One is for treatments and the other is for blocks (panelists). However, the test results are identical for the same data.

Kendall's coefficient of concordance is related to the Spearman rank correlation coefficient as

$$\bar{r}_S = \frac{nW-1}{n-1}, \qquad (9.2\text{-}44)$$

where $\bar{r}_S$ denotes the average of the $\binom{n}{2}$ Spearman rank correlation coefficients calculated for every possible pair consisting of two sets of ranks.

| **Example 9.2-16** | The data in Table 9.29 are used for illustration of Kendall's coefficient of concordance. Each of the four panelists evaluated the samples A, B, and C using a 5-point scale. The corresponding rank or midrank is also given in parentheses. We calculate $S = [5 - 4 \times (3+1)/2]^2 + [7.5 - 4(3+1)/2]^2 + [11.5 - 4(3+1)]^2 = 21.5$; hence, from (9.2-40), |

**Table 9.29** Evaluation Results of Four Panelists for Three Samples Using a 5-Point Scale (Example 9.2-16)

| Panelist | Sample | | |
|---|---|---|---|
| | **A** | **B** | **C** |
| 1 | 2 (1) | 4 (2) | 5 (3) |
| 2 | 1 (1) | 4 (2.5) | 4 (2.5) |
| 3 | 1 (1) | 3 (2) | 4 (3) |
| 4 | 3 (2) | 1 (1) | 5 (3) |
| R | (5) | (7.5) | (11.5) |

*Note:*
*The number in parentheses is rank.*

$$W = 12 \times 21.5/[4^2 \times 3(3^2 - 1)] = 0.6718.$$

Because there is one tied group with size of 2, $A = 2^3 - 2 = 6$. Then from (9.2-41), the adjusted Kendall's coefficient of concordance is

$$W^* = 0.6718/[1 - 6/(4 \times 3 \times 8)] = 0.717.$$

It indicates there is a moderate degree of association among the four sets of ranks. From (9.2-43), the value of the statistic $F^*$ is

$$F^* = n(k - 1)W^* = 4 \times (3 - 1) \times 0.717 = 5.733,$$

with an associated $P$-value $= 0.057$ for a $\chi^2$ distribution with 2 DF. We can also confirm the relationship between Kendall's coefficient of concordance and the Spearman rank correlation coefficient. We can calculate $r_S = 0.866$, 1, 0.5, 0.866, 0, and 0.5 for the pairs of ranks (1,2), (1,3), (1,4), (2,3), (2,4), (3,4), respectively. Hence,

$$\bar{r}_S = (0.866 + 1 + 0.5 + 0.866 + 0 + 0.5)/6 = 0.62.$$

It is equal to $(4 \times 0.717 - 1)/(4 - 1) = 0.62$ from (9.2-44).

Kendall's coefficient of concordance may be requested in some statistical software, e.g., the SPSS. In fact, one can use any software that can conduct the Friedman test to calculate and test the Kendall coefficient of concordance because of the relationship between the $W$ and $F$ statistics. For the data in Table 9.29, using the S-PLUS program *friedman.test*, we obtain $F^* = 5.733$; hence, $W^* = 5.733/[4 \times (3 - 1)] = 0.717$ with the associated $P$-value $= 0.057$.

```
> friedman.test(x[,3],x[,2],x[,1])

    Friedman rank sum test

data: x[, 3] and x[, 2] and x[, 1]
Friedman chi-square = 5.7333, df = 2, p-value = 0.0569
alternative hypothesis: two.sided
   id sample rating
1  1   1    2
2  2   1    1
3  3   1    1
4  4   1    3
....
```

## 9.3 SOME METHODS FOR CATEGORICAL DATA

### Contingency Tables and Chi-square Statistics

Categorical data refer to counts or frequencies with respect to various categories. When categorical data are classified in a two-way table, the rows of the table may correspond to one category and the columns to the second. Some examples of experimental studies giving rise to categorical data are as follows:

1. Experiments designed for investigating relationships between two or more attributes; for example, consumer surveys carried out for relating occupation and purchasing decisions; genetic studies designed for investigating a possible association between the color of eyes and the color of hair.
2. Experiments designed for assessing the effectiveness of treatments where responses are on a nominal scale; for example, dead or alive; preferred or not preferred; excellent, good, or poor.
3. Experiments whose outcomes are quantitative but that are categorized in frequency distribution forms for summarizing and other statistical analyses.

Usually, the layout of categorical data shown in Table 9.30 is called an $r \times c$ contingency table. The table consists of $r$ rows and $c$ columns. The $n_{ij}$, $i = 1, \ldots, r$, $j = 1, \ldots, c$, is the cell frequency and pertains to the number of observations (counts) falling in the $(i, j)$th cell of the table. Depending on the design and nature of the inquiry,

**Table 9.30** An $r \times c$ Contingency Table for Categorical Data

| Rows | Columns | | | | |
|------|---------|---------|-----|---------|---------|
|      | **1**   | **2**   | **...** | **c**   | **Totals** |
| 1    | $n_{11}$ | $n_{12}$ | ... | $n_{1c}$ | $n_{1+}$ |
| 2    | $n_{21}$ | $n_{22}$ | ... | $n_{2c}$ | $n_{2+}$ |
| $\vdots$ | $\vdots$ | $\vdots$ | | $\vdots$ | $\vdots$ |
| r    | $n_{r1}$ | $n_{r2}$ | ... | $n_{rc}$ | $n_{r+}$ |
|      | $n_{+1}$ | $n_{+2}$ | ... | $n_{+c}$ | $n$ |

the marginal totals $n_{i+}$ and $n_{+j}$ in the table may or may not be known before observing and classifying the data.

The contingency tables that we will discuss in this section consist of $1 \times c$ (or $r \times 1$) contingency tables, $r \times c$ contingency tables, $r \times r$ contingency tables, and multidimensional tables. The $2 \times 2$ table that we discussed in Section 9.1 is a special contingency table when both $r$ and $c$ are 2. Because of the many rows and columns in the generalized contingency tables, the one-sided hypotheses for the $2 \times 2$ tables are no longer appropriate for the tables.

For the $r \times c$ contingency tables with ordered categories, singly ordered tables or doubly ordered tables (i.e., the categories of one or both of the factors in the $r \times c$ contingency tables are ordered), the methods based on ranks in Section 9.2 may be more powerful than the tests just for categorical data.

For categorical data analyses, the chi-square statistics are very important. By a chi-square statistic, we mean a sample quantity, which, for large sample sizes, has approximately the $\chi^2$ distribution. The Pearson chi-square statistic $X^2$ and the likelihood ratio chi-square statistic $G^2$ are introduced in this section with their various applications.

### The Pearson Chi-square Statistic

First, consider Table 9.30 with $r$ rows and $c$ columns. There are, in all, $n$ observations classified into $r$ rows and $c$ columns. In the $(i, j)$th cell of the $i$th row and $j$th column, there are $n_{ij}$ observations out of $n$. Although $n_{ij}$ is the observed frequency of the $(i, j)$th cell, we can also compute its expected frequency under the null hypothesis of interest.

If the observed and expected frequencies are "close," we may not reject the null hypothesis. An overall statistical measure of closeness between the observed and expected frequencies for each and every cell is given by

$$X^2 = \sum_{i=1}^{r} \sum_{j=1}^{c} \frac{(n_{ij} - e_{ij})^2}{e_{ij}}, \qquad (9.3\text{-}1)$$

where $e_{ij}$ denotes the expected frequency of the $(i, j)$th cell under the null hypothesis of interest. The distribution of the $X^2$ statistic in (9.3-1) is invariably approximated by the $\chi^2$ distribution with $(r - 1)(c - 1)$ DF, and the null hypothesis is rejected at the $\alpha$ level of significance if $X^2 \geq \chi^2_{1-\alpha,(r-1)(c-1)}$.

For the second version of the Pearson chi-square statistic, we consider a sample of $n$ items categorized in a frequency distribution table with $k$ categories (classes). Let the observed frequency of the $i$th class be denoted $n_i$. Also, let $e_i$ denote the expected frequency of the $i$th class under the null hypothesis of interest. For testing the null hypothesis, the $X^2$ statistic is

$$X^2 = \sum_{i=1}^{k} \frac{(n_i - e_i)^2}{e_i}. \qquad (9.3\text{-}2)$$

The distribution of the $X^2$ statistic in (9.3-2) is approximated by the $\chi^2$ distribution with $(k - 1)$ DF. The null hypothesis is rejected if $X^2 \geq \chi^2_{1-\alpha,k-1}$ at the $\alpha$ level of significance.

For better approximations to the distributions of the statistics in (9.3-1) and (9.3-2), it is suggested that the expected frequencies $e_{ij}$ for each cell and $e_i$ for each class be large. For this reason, many statisticians, as a rule of thumb, recommend that no expected frequency be less than 5 when using the $\chi^2$ statistics just defined. They suggest that neighboring classes be combined if this requirement is not met. The incorporation of the suggestion will yield larger expected frequencies. We wish to add that the suggestion is controversial and not universally accepted. The cause of controversy lies in the fact that the grouping of neighboring classes further suppresses information from the sample, since the grouping leads to fewer classes. Some classes, having lost their separate identities, can no longer provide all the information contained in them. Hence, when expected frequencies are small, an effect may be a loss of power of the tests based on $X^2$ statistics. Indeed, critics of the grouping have shown that exact distributions of the $X^2$ statistics are satisfactorily approximated by the $\chi^2$ distribution, even if expected frequencies are small. In conclusion, although some statisticians recommend grouping and others do not, it may be desirable to have larger ($>5$) expected frequencies if, by grouping further, one does not overly summarize the data into too few classes.

There is yet another consideration in using $\chi^2$ tests when the expected frequencies are small. Remember that the categorical data are discrete but that the $\chi^2$ distribution is continuous. Hence, it is also suggested that, if the expected frequencies are small, a so-called Yates correction for continuity should be introduced into the computation of the $X^2$ statistics in (9.3-1) and (9.3-2) as follows:

$$X^2 = \sum_{i=1}^{r} \sum_{j=1}^{c} \frac{(|n_{ij} - e_{ij}| - 0.5)^2}{e_{ij}}, \tag{9.3-3}$$

and

$$X^2 = \sum_{i=1}^{k} \frac{(|n_i - e_i| - 0.5)^2}{e_i}. \tag{9.3-4}$$

Some statisticians suggest that the $X^2$ statistics be corrected for continuity when the expected frequencies are between 5 and 10. Again, there is no consensus in the literature as to whether or not the continuity correction should be applied. Snedecor and Cochran (1967) recommended its use, whereas others have argued against on the grounds that it overcorrects and leads to much smaller values of the statistics. Conover (1974) proposed the use of the correction only when the marginal totals are fixed and nearly equal. Mantel (1976) also supported the use of the continuity correction. The reader who is interested in this controversy may refer to works by Grizzle (1967), Katti (1968), Bennett and Underwood (1970), Detre and White (1970), and Pirie and Hamdan (1972). In summary, statistical tests based on corrected $X^2$ statistics are conservative in that they tend to reject the null hypotheses less frequently. Uncorrected $X^2$ statistics, on the other hand, tend to be liberal in rejecting the null hypotheses.

### The Likelihood Ratio Chi-square Statistic

The likelihood ratio chi-square statistic is an alternative to the Pearson chi-square statistic. The statistic is based on the famous likelihood ratio theory attributed to Wilks (1935, 1938). For the $r \times c$

contingency table data, the likelihood ratio chi-square statistic can be written as

$$G^2 = 2 \sum_{i=1}^{r} \sum_{j=1}^{c} n_{ij} \log(n_{ij}/e_{ij}), \qquad (9.3\text{-}5)$$

where "log" denotes a natural logarithm. Like the Pearson $X^2$ statistic, $G^2$ is distributed approximately as a $\chi^2$ distribution with $(r-1)(c-1)$ DF. For one sample having categorical data with $k$ categories, the $G^2$ statistic is

$$G^2 = 2 \sum_{i=1}^{k} n_i \log(n_i/e_i), \qquad (9.3\text{-}6)$$

which is distributed approximately as a $\chi^2$ distribution with $(k-1)$ DF.

There are some advantages of the $G^2$ statistic over other statistics. As pointed out by Lawal (2003), one advantage of $G^2$ relates to its partitioning property, which allows us to decompose the overall $G^2$ into small components, and another advantage of $G^2$ is that it simplifies the process of comparing one model against another. However, Agresti (1990) indicated that the chi-square approximation is usually poor for $G^2$ when $n/rc < 5$. Although there are some differences between the two statistics, they share many properties and commonly produce the same conclusions.

## Goodness of Fit Tests

The goodness of fit tests judge whether an observed sample has been taken from some hypothesized population. Here, we consider two examples of goodness of fit tests. One of them concerns a population whose items can be classified in one of the predetermined $k$ mutually exclusive classes. The numbers of observations in each class are presented in a $1 \times k$ contingency table. Let $p_i$ denote the proportion of the population in the $i$th class, $i = 1, \ldots, k$. In this example the researcher wanted to test whether proportions $p_i$ are equal to $p_{i0}$, $i = 1, \ldots, k$. The null and alternative hypotheses are

$$H_0: \ p_i = p_{i0}, \quad i = 1, \ldots, k,$$

$$H_a: \ p_i \neq p_{i0}, \quad \text{for at least one } i.$$

For testing these hypotheses, we observe a random sample of size $n$ and count the number $n_i$ of items out of $n$ belonging to the $i$th class. On the other hand, if $H_0$ is true, we expect $e_i = np_{i0}$ items in the $i$th class. Hence, from (9.3-2), we can calculate the $X^2$ statistic and compare it with the tabulated $\chi^2_{1-\alpha,k-1}$ value for the $\alpha$ level of significance test.

In genetic studies it has been hypothesized that the proportions of black, brown, and wild-type plumage color patterns in fowl are $\frac{1}{2}$, $\frac{1}{4}$, and $\frac{1}{4}$, respectively. We wanted to test this hypothesis. Hence, our null hypothesis is

$$H_0: \; p_1 = \frac{1}{2}, \quad p_2 = \frac{1}{4}, \quad p_3 = \frac{1}{4},$$

where $p_1$, $p_2$, and $p_3$ denote proportions of black, brown, and wild-type plumage color patterns, respectively. In a sample of $n$ offspring, when $H_0$ is true, one expects $e_1 = n/2$, $e_2 = n/4$, and $e_3 = n/4$ offspring of black, brown, and wild-type plumage color patterns, respectively. We use the numerical data from a study by Smyth (1965) for testing the preceding null hypothesis. The following results of genotype matings were observed and are listed in Table 9.31. The expected frequencies under the null hypothesis appear alongside the observed frequencies.

The $X^2$ goodness of fit statistic from (9.3-2) is

$$X^2 = \frac{(173 - 166.50)^2}{166.50} + \frac{(78 - 83.25)^2}{83.25} + \frac{(82 - 83.25)^2}{83.25} = 0.6036.$$

This value of $X^2$ is to be compared with $\chi^2_{1-\alpha,2}$. At $\alpha = 0.05$, $\chi^2_{0.95,2} = 5.99$. Since the observed $X^2 = 0.6036$ is much less than 5.99, we may conclude that the observed distribution of the plumage color patterns seems to be in agreement with the hypothesized distribution in $H_0$. If the correction for continuity is applied, then the test statistic is

$$X^2 = \frac{(|173 - 166.50| - 0.5)^2}{166.50} + \frac{(|78 - 83.25| - 0.5)^2}{83.25} + \frac{(|82 - 83.25| - 0.5)^2}{83.25}$$

$$= 2.162 + 0.2710 + 0.0068 = 0.4940.$$

In this example, the use of the continuity correction does not alter the final conclusion. Note, however, that the corrected $X^2$ value is less than the

**Example 9.3-1**

**Table 9.31** Numerical Data from a Study by Smyth (1965) (Example 9.3-1)

|  | Black | Brown | Wild Type | Total |
|---|---|---|---|---|
| Observed | 173 | 78 | 82 | 333 |
| Expected | 166.50 | 83.25 | 83.25 | 333 |

uncorrected one. This is always the case. That is why the corrected $X^2$ statistic is less likely to reject $H_0$ than the uncorrected one. Using the $G^2$ statistic from (9.3-6), we computed

$$G^2 = 2 \times [173 \times \log(173/166.5) + 78 \times \log(78/83.25)$$
$$+ 82 \times \log(82/83.25)] = 0.6076.$$

The same conclusion as above can be drawn.

Our second example concerns the null hypothesis for testing that a random sample is taken from a normally distributed population. In other words, on the basis of information from a random sample of some quantity (random variable), $X$, we want to test a null hypothesis that $X$ is normally distributed with mean $\mu$ and variance $\sigma^2$. For using the $X^2$ goodness of fit statistic, it is necessary that the theoretical range of $X$ be divided into $k$ mutually exclusive classes in some optimum way. Observations in the sample are classified into these classes. Let $n_i$ be the number of sample observations with magnitude such that they are classified in the $i$th class. We can also calculate the expected frequencies under the null hypothesis, as explained in the example to follow. Again, the $X^2$ goodness of fit statistic defined in (9.3-2) is used to test the null hypothesis.

---

**Example 9.3-2**   We wanted to test whether an observed sample is taken from a normally distributed population. For this purpose consider columns 1 and 2 of Table 9.32, which summarize 200 Tenderometer readings of beef carcasses in a frequency distribution form. All 200 observations are classified into $k = 16$ classes in column 1. Observed frequencies $n_i$ are in column 2. The histogram of the frequency distribution is given in Figure 9.1. We wanted to test the following null hypothesis:

$H_0$ : The Tenderometer reading $X$ follows a normal distribution with mean $\mu$ and variance $\sigma^2$.

Since we did not specify parameters $\mu$ and $\sigma^2$ in $H_0$, they are estimated by the mean $\overline{X} = 21.0$ lb, and the variance $S^2 = (2.25)^2$, respectively, of the frequency distribution. Next, we need to compute the expected frequencies $e_i$ for each class. Let $p_i$ be the probability that a reading is in

**Table 9.32** An Illustration of $\chi^2$ Goodness of Fit Test for the Normal Distribution (Example 9.3-2)

| (1) Class Intervals | (2) Observed Frequencies $n_i$ | (3) Probabilities $p_i$ | (4) Expected Frequencies $e_i = np_i$ | (5) $(n_i - e_i)^2/e_i$ |
|---|---|---|---|---|
| <14.5 | 1 | 0.0019 | 0.39 | 0.97 |
| 14.5–15.5 | 5 | 0.0053 | 1.06 | 14.56 |
| 15.5–16.5 | 7 | 0.0155 | 3.10 | 4.91 |
| 16.5–17.5 | 15 | 0.0372 | 7.43 | 7.71 |
| 17.5–18.5 | 18 | 0.0734 | 14.67 | 0.76 |
| 18.5–19.5 | 24 | 0.1192 | 23.85 | 0.00 |
| 19.5–20.5 | 31 | 0.1596 | 31.92 | 0.03 |
| 20.5–21.5 | 26 | 0.1759 | 35.17 | 2.39 |
| 21.5–22.5 | 25 | 0.1596 | 31.92 | 1.50 |
| 22.5–23.5 | 17 | 0.1192 | 23.85 | 1.97 |
| 23.5–24.5 | 8 | 0.0734 | 14.67 | 3.03 |
| 24.5–25.5 | 11 | 0.0372 | 7.43 | 1.71 |
| 25.5–26.5 | 8 | 0.0155 | 3.10 | 7.75 |
| 26.5–27.5 | 2 | 0.0053 | 1.06 | 0.82 |
| 27.5–28.5 | 1 | 0.0015 | 0.30 | 1.63 |
| >28.5 | 1 | 0.0004 | 0.09 | 9.74 |

Note:
$n = 200$; $\overline{X} = 21.0$ lb; $S = 2.25$ lb; $X^2 = 60.40$ with 13 DF.

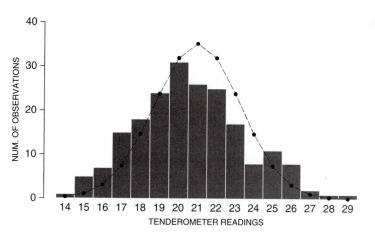

■ **FIGURE 9.1** Histogram of Tenderometer readings of 200 beef carcasses with normal curve superimposed.

the $i$th class when indeed $H_0$ is true. Then $e_i = np_i$, where $n = 200$. By changing the $X$ into $Z$ scores, we can find $p_1, p_2, \ldots .p_k$ as follows:

$$p_1 = P(X \leq 14.5)$$

$$= P\left(Z \leq \frac{14.5 - \overline{X}}{S}\right)$$

$$= P(Z \leq -2.89) = 0.0019,$$

$$p_2 = P(14.5 \leq X \leq 15.5)$$

$$= P\left(\frac{14.5 - \overline{X}}{S} \leq Z \leq \frac{15.5 - \overline{X}}{S}\right)$$

$$= P(-2.89 \leq Z \leq 2.44) = 0.0054,$$

and so on, for the remaining $p_3, p_4, \ldots, p_{16}$. Table 9.32 lists all these probabilities in column 3 and the expected frequencies in column 4. We can now calculate the $X^2$ statistic in (9.3-2), which is $X^2 = 60.40$. As stated, the $X^2$ statistic in (9.3-2) has a $\chi^2$ distribution with $(k - 1)$ DF, where $k$ is the number of classes. But when using the $X^2$ statistic for a goodness of fit test, one always reduces its DF parameter by the number of estimated parameters of the hypothesized distribution. In our case two parameters $\mu$ and $\sigma^2$ are estimated. Hence, DF $= (k - 1) - 2 = 13$. At the $\alpha = 0.01$ level of significance, $\chi^2_{1-\alpha,13} = \chi^2_{0.99,13} = 26.22$, which is much less than the observed $X^2 = 60.40$. Therefore, we may reject the null hypothesis that the Tenderometer readings $X$ follow a normal distribution.

Figure 9.1 shows a histogram of the Tenderometer readings of 200 beef carcasses. The normal curve obtained by plotting column 4 is superimposed. The disagreement between the histogram and normal curve is quite apparent.

## Analysis of $r \times c$ Contingency Tables

### Chi-square Tests for Independence

In many situations we want to judge the independence of one or more attributes from another. In these types of investigations, observations are classified in a contingency table of rows and columns according to levels of attributes. An $r \times c$ contingency table is given is Table 9.30, where $n$ observations in a random sample are classified in $r$ rows and $c$ columns. The total number of the sample, $n$, is fixed, but neither row total nor column total is fixed. In fact the test for independence is essentially an extension of the previous test of goodness of fit. The random sample of size $n$ falls into the $r \times c$ mutually exclusive classes with the cell probability $p_{ij}$, $i = 1, 2, \ldots, r$ and $j = 1, 2, \ldots, c$. The sample follows a multinomial distribution with parameters $n$ and $\mathbf{p}$, where $\mathbf{p}$ is a vector, $\mathbf{p} = (p_{11}, \ldots, p_{rc})$. If an observation is taken at random, the probability of its being classified in the $(i, j)$th cell is

$$p_{ij} = \frac{n_{ij}}{n}, \quad i = 1, \ldots, r; \quad j = 1, \ldots, c. \tag{9.3-7}$$

Also the probability of a randomly selected observation being in the $i$th row, regardless of the column, is

$$p_{i+} = n_{i+}/n, \quad i = 1, \ldots, r. \tag{9.3-8}$$

Similarly, the probability of an observation being in the $j$th column is

$$p_{+j} = n_{+j}/n, \quad j = 1, \ldots, c. \tag{9.3-9}$$

We are interested in the following problem: Do the two attributes (rows and columns) behave independently? The term *independence* means that the distribution of one attribute does not change with the distribution of the second. Statistically stated, the null hypothesis of the independence of one of two attributes from the other is

$$H_0: \ p_{ij} = (p_{i+})(p_{+j}) \tag{9.3-10}$$

for all cells $(i, j)$ in a contingency table.

In order to test the hypothesis in (9.3-10), we wish to calculate the expected frequencies $e_{ij}$ for each and every cell when $H_0$ is true. If $H_0$ is true, then for each and every cell,

$$p_{ij} = (n_i/n)(n_j/n).$$

Thus, when $H_0$ is true, we expect

$$e_{ij} = np_{ij} = n(n_{i+}/n)(n_{+j}/n) = (n_{i+})(n_{+j})/n \tag{9.3-11}$$

observations, as opposed to $n_{ij}$ in the $(i, j)$th cell. An overall measure of the closeness of the observed and the expected frequencies for each and every cell is given by the chi-square statistic $X^2$ in (9.3-1) or $G^2$ in (9.3-5). Since $(r - 1) + (c - 1)$ parameters, i.e., $(r - 1)$ row totals

$n_{1+}, \ldots, n_{r-1+}$ and $(c-1)$ column totals $n_{+1}, \ldots, n_{+c-1}$, have been estimated, the chi-square statistic $X^2$ or $G^2$ has $(rc-1) - (r-1) - (c-1)$ $= (r-1)(c-1)$ DF.

---

**Example 9.3-3**

The data given in Table 9.33 for illustration are only part of a much larger set of data. The sampling design is as follows: 105 respondents were each given a sample of the product to evaluate for its texture quality and juiciness. Hence, $n = 105$ is the sample size. We want to judge the relationship between these two variables—for example, whether a fine or coarse texture is associated with a high or low amount of juiciness.

The $X^2$ statistic for this $2 \times 2$ table, when computed from (9.1-7), is the following:

$$X^2 = \frac{[(31)(42) - (12)(20)]^2 \, 105}{(43)(62)(51)(54)} = 16.13$$

with 1 DF. At $\alpha = 0.01$, $\chi^2_{1-\alpha,1} = \chi^2_{0.99,1} = 6.63$, which is less than 16.13. Therefore, the observed value 16.13 is significant at this level, and we conclude that texture quality for this product is related to the amount of juiciness.

```
> chisq.test(ch9dat1,correct = F)

Pearson's  chi-square  test  without  Yates'  continuity
    correction

data: ch9dat1
X-square = 16.1293, df = 1, p-value = 0.0001
> ch9dat1
   [,1] [,2]
[1,] 31 20
[2,] 12 42
```

**Table 9.33** Data for Evaluation of Texture and Juiciness (Example 9.3-3)

|  | Fine Texture | Coarse Texture | Total |
|---|---|---|---|
| High Juiciness | 31 | 20 | 51 |
| Low Juiciness | 12 | 42 | 54 |
|  | 43 | 62 | 105 |

Table 9.34 is part of a data set from a survey relating to the introduction of a new product. After tasting the product, respondents were asked to complete a questionnaire. The results for one of the questions are contained in this table. This particular question seeks to relate occupational level with purchasing decision. Because of the small number of respondents in the last three purchasing categories, these categories were combined to obtain a $4 \times 3$ table. To obtain the $X^2$ statistic, we need expected cell frequencies under the null hypothesis of no association. They are obtained from (9.3-11) as follows:

**Example 9.3-4**

$$e_{11} = (169)(163)/297 = 92.75,$$

$$e_{12} = (169)(94)/297 = 53.29,$$

and so on. The expected cell frequencies are also given in Table 9.34 within the parentheses. The $X^2$ statistic from (9.3-1) is

$$X^2 = \frac{(87 - 92.75)^2}{92.75} + \frac{(57 - 53.49)^2}{53.49} + \cdots + \frac{(1 - 1.27)^2}{1.27}$$
$$= 0.3565 + 0.2303 + \cdots + 0.0574 = 3.43$$

with $(4-1)(3-1) = 6$ DF. Since $\chi^2_{0.95,6} = 12.59 > 3.43$, it may be concluded at the $\alpha = 0.05$ significance level that there is no association between the occupational level and the purchasing decision with regard to the new

**Table 9.34** Frequencies Relating Purchasing Decisions and Occupations (Example 9.3-4)

| Purchasing Decision | Occupation | | | Totals |
| --- | --- | --- | --- | --- |
| | White Collar[a] | Blue Collar[a] | Other[a] | |
| Definitely Would Buy | 87 (92.75) | 54 (47.20) | 22 (23.05) | 163 |
| Probably Would Buy | 57 (53.49) | 22 (27.22) | 15 (13.29) | 94 |
| Might or Might Not Buy | 19 (17.64) | 8 (8.98) | 4 (4.38) | 31 |
| Probably Would Not Buy | 2 ⎫ | 2 (2.61) | 1 (1.27) | 5 ⎫ |
| Definitely Would Not Buy | 1 ⎬ 6 (5.12) | – | – | 1 ⎬ 9 |
| Don't Know | 3 ⎭ | – | – | 3 ⎭ |
| | 169 | 86 | 42 | 297 |

[a]The numbers within parentheses are the expected cell frequencies.

product. Using the $G^2$ test statistic, we computed $G^2 = 3.46$. The same conclusion can be obtained as follows:

```
> chisq.test(ch9dat2,correct = F)

    Pearson's chi-square test without Yates' continuity
correction

data: ch9dat2
X-square = 3.4299, df = 6, p-value = 0.7533
> ch9dat2
  [,1][,2][,3]
[1,] 87 54 22
[2,] 57 22 15
[3,] 19 8  4
[4,] 6  2  1
```

### Chi-square Tests for Homogeneity

The independence test discussed in the previous section assumes that the row and column totals are not fixed in advance. However, some experimental situations have a fixed number of observations in either each row or each column. For example, in order to compare purchase intent for a specified product of consumers in three different regions A, B, C, researchers randomly drew 300 consumers from each of the three regions. A 5-point scale was used for 1 = Definitely would not buy to 5 = Definitely would buy. See Table 9.35. With this

**Table 9.35** Homogeneity Test with Fixed Marginal (Row or Column) Totals (Example 9.3-5)

| Purchase Intent | Region | | | $\widehat{p}_0$ |
|---|---|---|---|---|
| | **A** | **B** | **C** | |
| Definitely Would Buy | 96 (65) | 63 (65) | 36 (65) | 0.22 |
| Probably Would Buy | 75 (54) | 42 (54) | 45 (54) | 0.18 |
| Might or Might Not Buy | 81 (106) | 103 (106) | 129 (106) | 0.35 |
| Probably Would Not Buy | 33 (53) | 54 (53) | 72 (53) | 0.18 |
| Definitely Would Not Buy | 15 (22) | 53 (22) | 18 (22) | 0.07 |
| Total | 300 | 300 | 300 | 1 |

sampling plan, we would have three multinomial populations with parameters $\mathbf{p}_j = (p_{1j}, \ldots, p_{5j})$ and $n_{+j}$, $j = 1, 2, 3$, where $n_{+j} = 300$ in this case. The interest is to test if the three consumer populations have the same cell response probabilities, i.e., $H_0: \mathbf{p}_1 = \mathbf{p}_2 = \mathbf{p}_3 = \mathbf{p}_0$. This is the test of homogeneity of several multinomial populations. Under this null hypothesis, the estimate of the common probability vector can be shown to be $\hat{\mathbf{p}}_0 = (n_{1+}, n_{2+}, \ldots, n_{5+})/n$, where $n = \sum_{j=1}^{3} n_{+j} = 900$ in this case. The expected cell number under null hypothesis is then $e_{ij} = n_{+j}n_{i+}/n$. We can see that the expected cell number is the same as that in a test of independence. It can be demonstrated that either the $X^2$ or $G^2$ statistic with $(r - 1)(c - 1)$ degrees of freedom can be used to do the test.

Although the tests for independence and homogeneity are computationally identical, the two types of tests are different in sampling plans, hypotheses, distributions, and result interpretations. For the test of independence, only the overall sample size is fixed in advance, whereas for the test of homogeneity, one set of marginal totals is fixed. In the test of independence, the null hypothesis is that the cell probability equals the product of the corresponding marginal probabilities, whereas in the test of homogeneity, the null hypothesis is that probability vectors are the same for the populations. For the test of independence, the sample of size $n$ follows a multinomial distribution with $r \times c$ categories, whereas for the test of homogeneity, the set of independent samples follows the product multinomial probability model. In the test of independence, both categorical variables are responses. The test evaluates whether or not the two categorical response variables are independent. In the test of homogeneity, one is considered as a response variable, and the other is regarded as fixed factor. The test evaluates whether or not the distribution of the response variable is the same across each factor level.

**Example 9.3-5**

Table 9.35 shows the data from a homogeneity test. The objective of the test was to verify whether the consumers in different regions have the same response proportions of purchase intent for a product. The estimated common response proportion vector is displayed in the last column of the table. The expected cell numbers are listed in the parentheses. The value of the Pearson chi-square statistic is 73.87. The associated P-value < 0.001 for a chi-square distribution with $(5 - 1)(3 - 1) = 8$ DF. This value suggests that the consumers in the different regions differed significantly in purchase intent for the product.

```
> chisq.test(chdat35)
     Pearson's chi-square test without Yates' continuity
     correction

data: chdat35
X-square = 73.8744, df = 8, p-value = 0
> chdat35
       [,1] [,2] [,3]
[1,]   96  63  36
[2,]   75  42  45
[3,]   81 108 129
[4,]   33  54  72
[5,]   15  33  18
```

### Exact Tests for Small Samples

Fisher's exact test for a $2 \times 2$ table was discussed in Section 9.1. The test can be generalized to data presented in an $r \times c$ contingency table. The generalized Fisher exact test is based on multivariate hypergeometric (for one sample case) or product hypergeometric distribution (for several sample cases) under the assumption that both the row and column totals are considered to be fixed. Fisher's exact test is feasible for analysis of the $r \times c$ contingency table data with small cell numbers.

Unlike the simplicity of Fisher's exact test for a $2 \times 2$ table, there are $(r-1)(c-1)$ pivot cells rather than only one pivot cell ($n_{11}$) in an $r \times c$ contingency table. Once the values of $(r-1)(c-1)$ pivot cells are fixed, the values of the other $(r+c-1)$ cells can be completely determined immediately from the fixed row and column totals. Under the assumption that the row and column totals, $n_{i+}, n_{+j}$, are fixed, for either sampling model the probability of a possible cell counts $\{n_{ij}\}$ is

$$P(n_{ij}|n_{+j}; n_{j+}) = \frac{(\prod_i n_{i+}!)(\prod_j n_{+j}!)}{n! \prod_i \prod_j n_{ij}!} \qquad (9.3\text{-}12)$$

Because there are no unknown parameters in (9.3-12), exact inference can be conducted based on the calculated probabilities of all possible outcomes of cell counts. In fact, the situations that both row and column totals are determined in advance are rare in practice. Fisher's exact tests can still be used reasonably when both margins of the table are not naturally fixed with the possible price of conservativeness. Exact tests are permutation tests, which involve the procedures for determining statistical significance directly from data rather than some particular sampling distribution; these tests are computer-intensive and require appropriate statistical software. Many statistical software packages include Fisher's exact tests for $r \times c$ contingency table data with a small number of total counts (usually fewer than 200). The advances in computational power and efficiency of algorithms have made exact methods feasible for a wider variety of inferential analyses and for a larger collection of table sizes and sample sizes. As pointed out by some authors, e.g., Agresti (1992), in the future, statistical practice for categorical data should place more emphasis on "exact" methods and less emphasis on methods using possibly inadequate large-sample approximations.

---

**Example 9.3-6**

In order to investigate a possible relationship between age groups and overall liking for a product, researchers randomly selected 24 consumers. There were 10, 8, and 6 consumers, respectively, in the three age groups: young, middle-aged, and old. The responses in a 9-point rating scale with 1 = Dislike extremely to 9 = Like extremely are displayed in Table 9.36. Because there are many cells with 0 or 1, the conventional chi-square tests are not suitable for the data. Fisher's exact test should be used. It is obvious that both margins of this table were not fixed in the sampling scheme. However, we will assume that the margins are fixed as the observed values. The S-PLUS program *fisher.test* is used for analysis of the data. The small *P*-value (0.0181) suggests that there is a close relationship between age groups and overall liking for the product. In other words, consumers in different age groups have significantly different overall liking for the product at a 0.05 significance level.

**Table 9.36** Data for Evaluation of the Relationship between Age Groups and Overall Liking (Example 9.3-6)

|  | 9 | 8 | 7 | 6 | 5 | 4 | 3 | 2 | 1 | Total |
|---|---|---|---|---|---|---|---|---|---|---|
| Young | 4 | 3 | 2 | 1 | 0 | 0 | 0 | 0 | 0 | 10 |
| Middle-Aged | 0 | 1 | 2 | 0 | 3 | 0 | 0 | 1 | 1 | 8 |
| Old | 0 | 0 | 0 | 0 | 1 | 1 | 1 | 1 | 2 | 6 |
| Total | 4 | 4 | 4 | 1 | 4 | 1 | 1 | 2 | 3 | 24 |

```
> fisher.test(fisherdat)

     Fisher's exact test

data: fisherdat
p-value = 0.0181
alternative hypothesis: two.sided
> fisherdat
  9 8 7 6 5 4 3 2 1
Y 4 3 2 1 0 0 0 0 0
M 0 1 2 0 3 0 0 1 1
  0 0 0 0 0 1 1 1 1 2
```

## Measures of Association for $r \times c$ Contingency Tables

There are many different measures of associations for $r \times c$ tables according to different criteria. See, for example, Bishop et al. (1975, Section 11.3) and Liebetrau (1983). Here, we will discuss only the measures based on the Pearson chi-square statistic and Goodman and Kruskal's lambda measures. The reasonable interpretation of the type of measures based on the Pearson chi-square statistic is that each measure departs from zero when the data in the table depart from independence. The reasonable interpretation of Goodman and Kruskal's lambda measures is the predictability, i.e., the proportional reduction in error (PRE) in predicting the category of the row (or column) variable from the category of the column (or row) variable by the information on the column (or row) category.

### Measures Based on Chi-square

### Pearson's Coefficient of Mean Square Contingency, $\phi^2$

The definition of $\phi^2$ is

$$\phi^2 = \sum_{i=1}^{r} \sum_{j=1}^{c} \frac{(p_{ij} - p_{i+}p_{+j})^2}{p_{i+}p_{+j}}. \tag{9.3-13}$$

The maximum likelihood estimation of $\phi^2$ is

$$\hat{\phi}^2 = \sum_{i=1}^{r} \sum_{j=1}^{c} \frac{(\hat{p}_{ij} - \hat{p}_{i+}\hat{p}_{+j})^2}{\hat{p}_{i+}\hat{p}_{+j}} = \frac{X^2}{N}, \tag{9.3-14}$$

where $X^2$ is Pearson's chi-square statistic for $r \times c$ tables. The range of $\phi^2$ is $0 \le \phi^2 \le \min\{r, c\} - 1$. In the case of independence, $\phi^2 = 0$ and in the case of perfect association, $\phi^2 \le \min\{r, c\} - 1$. The weakness of the measure is that it depends on the dimensions of the tables.

### Pearson's Contingency Coefficient, C

To overcome the deficiency of $\phi^2$ Pearson (1904) proposed a measure called the (Pearson) contingency coefficient as follows:

$$C = \sqrt{\frac{\phi^2}{1 + \phi^2}}. \tag{9.3-15}$$

The estimation of $C$ is

$$\widehat{C} = \sqrt{\frac{X^2}{N + X^2}}. \tag{9.3-16}$$

Since the maximum value of $C$ is $\sqrt{[\min(r, c) - 1]/\min(r, c)}$ in the perfect association, the range of $C$ still depends on the dimensions of the tables.

### Cramer's Contingency Coefficient, V

Cramer (1946) proposed the measure, $V$,

$$V = \sqrt{\frac{\phi^2}{\min(r, c) - 1}}. \tag{9.3-17}$$

The estimation of $V$ is

$$\widehat{V} = \sqrt{\frac{X^2}{[\min(r, c) - 1]N}}. \tag{9.3-18}$$

The measure $V$ is in the range of 0 and 1. In the case of perfect association, $V = 1$ for all $r \times c$ tables and $V = 0$ in the case of independence.

---

**Example 9.3-7**

For the data in Table 9.34, the $X^2 = 3.43$, $N = 297$, $\min(4,3) - 1 = 2$; hence, $\widehat{\phi}^2 = 3.43/297 = 0.0115$ from (9.3-14), $\widehat{C} = \sqrt{3.43/(297 + 3.43)} = 0.107$ from (9.3-16), and $\widehat{V} = \sqrt{3.43/(297 \times 2)} = 0.076$ from (9.3-18). The measures and asymptotic variances of the measures can be obtained by using the computer program *rctabm*. The asymptotic variances are $\widehat{\sigma}_\infty^2(\widehat{\phi}^2) = 0.0001499$, $\widehat{\sigma}_\infty^2(\widehat{C}) = 0.00314$, and $\widehat{\sigma}_\infty^2(\widehat{V}) = 0.00162$.

```
> rctabm(ch9dat2)
[1]] 0.0115483458 0.1068480417 0.0759879785 0.0001498997
0.0031351707 0.0016225240
> ch9dat2
     [,1] [,2] [,3]
[1,]  87  54  22
[2,]  57  22  15
[3,]  19   8   4
[4,]   6   2   1
```

### Proportional Reduction in Error (PRE) Measures

Goodman and Kruskal (1954) proposed the lambda measures for a contingency table with a nominal scale. $\lambda_{C/R}$ is designed for predictions of a column variable from a row variable; $\lambda_{R/C}$ is designed for predictions of a row variable from a column variable, and $\lambda$ is designed for the symmetric situation in which neither variable is reasonably to be assumed as a predicted variable. Under the multinomial sampling model, the maximum likelihood estimations of the measures are, respectively,

$$\widehat{\lambda}_{C/R} = \frac{\sum_{i=1}^{r} \max_j(n_{ij}) - \max_j(n_{+j})}{N - \max_j(n_{+j})}, \tag{9.3-19}$$

where $\max_j(n_{ij})$ is the largest entry in the $i$th row of the table, and $\max_j(n_{+j})$ is the largest of the column marginal totals.

$$\widehat{\lambda}_{R/C} = \frac{\sum_{j=1}^{c} \max_i(n_{ij}) - \max_i(n_{i+})}{N - \max_i(n_{i+})}, \tag{9.3-20}$$

where $\max_j(n_{ij})$ is the largest entry in the $j$th column of the table, and $\max_i(n_{i+})$ is the largest of the row marginal totals.

$$\widehat{\lambda} = \frac{\sum_{i=1}^{r} \max_j(n_{ij}) + \sum_{j=1}^{c} \max_i(n_{ij}) - \max_j(n_{+j}) - \max_i(n_{i+})}{2N - \max_j(n_{+j}) - \max_i(n_{i+})}. \tag{9.3-21}$$

The values of the lambda measures are always in the range of 0 and 1, and $\lambda_{C/R}$ and $\lambda_{R/C}$ values are usually not the same. When the row and column variable are independent from each other, the lambda values should be zero, but not vice versa; i.e., the lambda value may be zero without statistical independence holding.

For the data in Table 9.37, $\sum_{i=1}^{3} \max_j(n_{ij}) = 11 + 15 + 18 = 44$, $\max_j(n_{+j}) =$ 35, $N = 100$; hence,

Example 9.3-8

$$\widehat{\lambda}_{C/R} = (44 - 35)/(100 - 35) = 0.138 \text{ from } (9.3\text{-}19),$$

Since $\sum_{j=1}^{4} \max_i(n_{ij}) = 8 + 15 + 18 + 9 = 50$, $\max_i(n_{i+}) = 37$; hence,

$$\widehat{\lambda}_{R/C} = (50 - 37)/(100 - 37) = 0.2063 \text{ from } (9.3\text{-}20),$$

and

$$\widehat{\lambda} = (44 + 50 - 35 - 37)/(200 - 35 - 37) = 0.1719 \text{ from } (9.3\text{-}21).$$

Using the S-PLUS code *rctabm2*, we can obtain the estimates and their asymptotical variances:

```
> rctabm2(rctabm2dat6)
[[1]]  0.138461538  0.206349206  0.171875000  0.005505690
0.010198083 0.005027682
> rctabm2dat6
     [,1] [,2] [,3] [,4]
[1,]  8   11    8  9
[2,]  7   15    4  1
[3,]  2    9   18  8
```

**Table 9.37** Data of a 3 × 4 Contingency Table (Example 9.3-8)

|      | C1 | C2 | C3 | C4 | Total |
|------|----|----|----|----|-------|
| R1   | 8  | 11 | 8  | 9  | 36    |
| R2   | 7  | 15 | 4  | 1  | 27    |
| R3   | 2  | 9  | 18 | 8  | 37    |
| Total| 17 | 35 | 30 | 18 | 100   |

## Analysis of Square ($r \times r$) Contingency Tables

In this section, we discuss a particular type of contingency table; i.e., the $r \times r$ square tables in which row and column variables have the same number of categories and the category definitions are the same for each variable. The square tables may arise in different ways, for example,

1. Each individual in a sample evaluates two products using an $r$-point scale or evaluates one product at different times or conditions.
2. A sample of individuals is cross-classified according to two essentially similar categorical variables, e.g., grade of vision of left and right eye.
3. A sample of pairs of matched individuals, such as husbands and wives or fathers and sons, is each classified according to some categorical variable of interest, e.g., purchase intent for a product for husbands and wives.
4. Two raters (panelists) independently evaluate a sample of subjects to a set of categories.

The analysis of square tables involves the analysis for two dependent samples with categorical scales. The interest of the analysis is to test symmetry and marginal homogeneity. In a $2 \times 2$ table, symmetry and marginal homogeneity are equivalent, but they are not equivalent in an $r \times r$ square table with $r > 2$. Complete symmetry is more restrictive than marginal homogeneity. Complete symmetry implies marginal homogeneity, but marginal homogeneity does not imply complete symmetry.

### Testing Symmetry

A square table is symmetric if

$$H_0: \ p_{ij} = p_{ji} \quad \text{for all} \quad i \neq j, \ i, \ j = 1, 2, \ldots, r, \tag{9.3-22}$$

where $p_{ij}$ is the probability of an observation belonging to cell $(i, j)$. Under the null hypothesis in (9.3-22), the estimated expected cell frequency for cell $(i, j)$ is $\widehat{e}_{ij} = \dfrac{n_{ij} + n_{ji}}{2}$ for $i \neq j$ and $\widehat{e}_{ii} = n_{ii}$ for $i = j$.

The test statistics for symmetry are

$$X^2 = \sum_{i>j} (n_{ij} - n_{ji})^2 / (n_{ij} + n_{ji}), \tag{9.3-23}$$

$$G^2 = 2 \sum_{i \neq j} n_{ij} \log \frac{2n_{ij}}{n_{ij} + n_{ji}}, \tag{9.3-24}$$

which follow asymptotically a chi-square distribution with $r(r-1)/2$ DF under the null hypothesis in (9.3-22).

---

**Example 9.3-9**

In order to investigate possible price effect for consumer's purchase intent for a product, researchers randomly selected 200 consumers. Each consumer was asked if he or she would like to buy the product with price 1 or price 2. The responses of the consumers for purchase intent under the conditions of two prices are presented in Table 9.38. We can calculate

$$X^2 = \frac{(7-26)^2}{7+26} + \frac{(2-4)^2}{2+4} + \frac{(9-12)^2}{9+12} = 12.035.$$

**Table 9.38** Data for Price and Purchase Intent (Example 9.3-9)

| Price 1 | Price 2 | | | Total |
|---------|-----|--------|---------|-------|
| | **Buy** | **Unsure** | **Not Buy** | **Total** |
| Buy | 65 | 26 | 4 | 95 |
| Unsure | 7 | 56 | 12 | 75 |
| Not Buy | 2 | 9 | 19 | 30 |
| Total | 74 | 91 | 35 | 200 |

The associated *P*-value $= 0.007$ for a chi-square distribution with $3(3 - 1)/2 = 3$ DF. We can also calculate

$$G^2 = 2 \times \left( 7 \times \log\frac{2 \times 7}{7 + 26} + 2 \times \log\frac{2 \times 2}{2 + 4} + 9 \times \log\frac{2 \times 9}{9 + 12} \right.$$
$$\left. + 26 \times \log\frac{2 \times 26}{7 + 26} + 4 \times \log\frac{2 \times 4}{2 + 4} + 12 \times \log\frac{2 \times 12}{9 + 12} \right) = 12.752.$$

The associated *P*-value is 0.005. We should reject the null hypothesis of symmetry and conclude that the effect of price is significant for consumers' purchase intent for the product. We can use the following S-PLUS code for the test:

```
> symtest(smdat5)
[1] 12.035 0.007 12.752 0.005
> smdat5
    [,1] [,2] [,3]
[1,]  65   26    4
[2,]   7   56   12
[3,]   2    9   19
```

### Testing Marginal Homogeneity

Marginal homogeneity of a square table is defined by

$$H_0: \quad p_{i+} = p_{+j}, \quad i = 1, 2, \ldots, r.$$

Stuart (1955) and Maxwell (1970) gave a suitable statistic for the marginal homogeneity test as

$$X^2 = \mathbf{d}'\widehat{\mathbf{V}}^{-1}\mathbf{d}, \tag{9.3-25}$$

where $\mathbf{d}$ is a column vector of any $(r-1)$ of the differences $d_1, d_2, \ldots, d_r$ (one of the $d_i$'s should be dropped and the same statistic value should be obtained when any one $d_i$ value is omitted), where $d_i = n_{i+} - n_{+i}$ and $\widehat{V}$ is the $(r-1)(r-1)$ sample covariance matrix of $\mathbf{d}$ with elements,

$$\widehat{v}_{ij} = -(n_{ij} + n_{ji}) \quad \text{for} \quad i \neq j \quad \text{and} \quad \widehat{v}_{ii} = -(n_{ij} + n_{ji}).$$

Under the null hypothesis, the statistic $X^2$ follows a chi-square distribution with $(r-1)$ DF. Fleiss and Everitt (1971) showed that when $r = 3$, the test statistic can be written as

$$X^2 = \frac{\bar{n}_{23}d_1^2 + \bar{n}_{13}d_2^2 + \bar{n}_{12}d_3^2}{2(\bar{n}_{12}\bar{n}_{23} + \bar{n}_{12}\bar{n}_{13} + \bar{n}_{13}\bar{n}_{23})}, \tag{9.3-26}$$

where $\bar{n}_{ij} = (n_{ij} + n_{ji})/2$.

---

**Example 9.3-10**

For the data in Table 9.38, $\mathbf{d} = (21, -16)$, $\widehat{V} = \begin{bmatrix} 39 & -33 \\ -33 & 54 \end{bmatrix}$, we can calculate the statistic value by using a built-in program in some statistical software and get $X^2 = 11.43$.

```
> t(d)%*%solve(v)%*%d
[1,] 11.42773
> d
[1] 21 -16
> v
    [,1] [,2]
[1,]   39  -33
[2,]  -33   54
```

Because $r = 3$, we can also use (9.3-26) for $d_1 = 21$, $d_2 = -16$, $d_3 = -5$, $\bar{n}_{12} = 16.5$, $\bar{n}_{13} = 3$, $\bar{n}_{23} = 10.5$,

$$X^2 = \frac{10.5 \times 21^2 + 3 \times (-16)^2 + 16.5 \times (-5)^2}{2 \times (16.5 \times 10.5 + 16.5 \times 3 + 3 \times 10.5)} = 11.43.$$

The associated $P$-value is 0.003 for a chi-square distribution with 2 DF. We can use the following S-PLUS code for the test:

```
> smtest(smdat5)
[1]] 11.428 0.003
```

### Testing Marginal Homogeneity for Ordered Categories

For a square table with ordered categories, Agresti (1990, 363) discussed another more powerful large-sample statistic for testing marginal homogeneity. It is

$$Z = \frac{\hat{\tau}}{\hat{\sigma}(\hat{\tau})}, \tag{9.3-27}$$

where

$$\hat{\tau} = \log\left(\frac{\sum\sum_{i<j} n_{ij}}{\sum\sum_{i>j} n_{ij}}\right), \quad \hat{\sigma}^2(\hat{\tau}) = \left(\sum\sum_{i<j} n_{ij}\right)^{-1} + \left(\sum\sum_{i>j} n_{ij}\right)^{-1}.$$

The statistic $Z$ in (9.3-27) follows asymptotically the standard normal distribution.

---

For the data in Table 9.38, we can regard the three categories "Buy," "Unsure," and "Not Buy" as ordered categories for purchase intent. We get $\hat{\tau} = \log\left(\frac{42}{18}\right) = 0.847$, $\hat{\sigma}^2(\hat{\tau}) = \frac{1}{42} + \frac{1}{18} = 0.079$; hence, $Z = \frac{0.847}{\sqrt{0.079}} = 3.01$ from (9.3-27). The associated *P*-value is 0.001. Using the following S-PLUS code, we can get the results quickly:

**Example 9.3-11**

```
> mhortest(smdat5)
[1] 0.847 0.079 3.008 0.001
```

---

### Measures of Agreement

It is often of interest, for example, to measure agreement of performance of panelists. Agreement is a specific type of association. Goodman and Kruskal (1954) described this problem as one of measuring the reliability between two observers. We distinguish between measuring agreement and measuring association because there can be high association along with either low or high agreement. Cohen (1960) suggested a measure such as (9.3-28), which is generally known as the kappa coefficient. It is defined by

$$\kappa = \frac{p_0 - p_e}{1 - p_e}, \tag{9.3-28}$$

where $p_0 = \sum_{i=1}^{r} p_{ii}$ denotes the probability of agreement; i.e., two observers assign an item to the same category of the $r$ categories; and $p_e = \sum_{i=1}^{r} p_{i+}p_{+j}$ denotes the probability of agreement by chance. Hence, the kappa coefficient is a measure of agreement above chance. The values of $\kappa$ can range from $-p_e/(1-p_e)$ to 1 for a given set of marginal totals. It equals 1 when there is perfect agreement. It equals 0 when the agreement equals that expected by chance in the situation of independence. Negative values of $\kappa$ can occur if agreement is weaker than expected by chance, but this is rare. The maximum likelihood estimator of $\kappa$ under the multinomial sampling is

$$\hat{\kappa} = \frac{n \sum_{i=1}^{r} n_{ii} - \sum_{i=1}^{r} n_{i+}n_{+i}}{n^2 - \sum_{i=1}^{r} n_{i+}n_{+i}}. \tag{9.3-29}$$

The asymptotic variance of $\hat{\kappa}$ was given by Everitt (1968) and Fleiss et al. (1969). We can use the S-PLUS code *kappaest* to get $\hat{\kappa}$, asymptotic variance of $\hat{\kappa}$, $\hat{\sigma}_{\infty}^2(\hat{\kappa})$, and confidence interval of $\kappa$.

---

**Example 9.3-12**

For the data in Table 9.38, we can calculate $\hat{p}_0 = (65 + 56 + 19)/200 = 0.7$, $\hat{p}_e = (95 \times 74 + 75 \times 91 + 30 \times 35)/200^2 = 0.3726$; hence, $\hat{\kappa} = (0.7 - 0.3726)/(1 - 0.3726) = 0.52$ from (9.3-29). We can also get $\hat{\sigma}_{\infty}^2(\hat{\kappa}) = 0.00255$ and the 95% confidence interval of $\kappa$, [0.439, 0.605], by using the following S-PLUS code:

```
> kappaest(smdat5)
[1] 0.521817095 0.002549632 0.438762044 0.604872146
```

---

## 9.4 MULTIDIMENSIONAL CONTINGENCY TABLES

### Introduction

When more than two attributes are under investigation, the researcher may wish to classify the sample observations according to each attribute to study the various dependencies and relationships. In this section, we introduce some aspects of three-dimensional contingency tables. For details and analyses of multidimensional tables, we refer to Darroch (1962), Lewis (1962), Birch (1963), Cox (1970), Bishop et al. (1975), Upton (1978), Everitt (1992), and Agresti (1996, 2002).

Following a straightforward extension of the two-dimensional tables, we let $n_{ijk}$ and $e_{ijk}$ denote, respectively, the observed and the expected frequencies of the $i$th level of attribute $A$, the $j$th level of attribute $B$, and the $k$th level of attribute $C$; $i = 1,\ldots, a$; $j = 1,\ldots, b$; $k = 1,\ldots, c$. Hence, in all there are $(abc)$ cells and $n = \sum_i \sum_j \sum_k n_{ijk}$ observations in the $a \times b \times c$ contingency table. If the observations constitute a random sample of size $n$, then the probability of an observation being in the $(i, j, k)$th cell is

$$p_{ijk} = n_{ijk}/n, \quad i = 1,\ldots, a; \quad j = 1,\ldots, b; \quad k = 1,\ldots, c. \quad (9.4\text{-}1)$$

Let the marginal probability of an observation being in the $i$th level of $A$ and the $j$th level of $B$ regardless of $C$ be denoted by $p_{ij+}$. The marginal probability $p_{ij+}$ is obtained by summing $p_{ijk}$ over the subscript $k$. That is

$$p_{ij+} = \sum_{k=1}^{c} p_{ijk} = n_{ij+}/n.$$

All other marginal probabilities can be found similarly.

## Simpson's Paradox

Sometimes one needs to collapse a three-dimensional table into a two-dimensional marginal table since the collapsed table has larger cell frequencies and fewer parameters and is easier to interpret. However, we cannot expect that collapsing is always valid. Simpson's paradox (Simpson, 1951) reveals the fact that invalidly collapsing a three-dimensional table into a two-dimensional table may lead to a misleading conclusion. This paradox is also called the Yule–Simpson paradox because Yule (1903) first discovered this phenomenon. If a variable in a three-dimensional table is a lurking variable (or confounding variable)—i.e., the variable that has an important effect on the other variables—collapsing (summing) over the variable is not valid. Simpson's paradox is viewed as association reversal that is of fundamental importance in statistical practice. For more discussions on the topic, see, for example, Blyth (1972), Bishop et al. (1975), Whittemore (1978), Ducharme and Lepage (1986), Wermuth (1987), Samuels (1993), and Dong (1998). The following example describes Simpson's paradox.

**Example 9.4-1**

In order to investigate whether or not consumers have higher purchase intent for a new product B than the current product A, 100 consumer panelists evaluated A, and another 100 panelists evaluated B. The responses of the panelists are displayed in Table 9.39. It is obvious that consumers have higher purchase intent for the new product B (56%) than the current product (48%). However, if the consumers are classified into "Users" and "Non-Users" groups for the type of products, the data can be presented as shown in Table 9.40. We find that the purchase intent for the current product (A) is higher than the new product (B) in both "Users" and "Non-Users" groups (70% versus 65% and 33% versus 23%)! This contradiction is Simpson's paradox. Simpson's paradox can occur because "Users" is a lurking variable. We can see that "Users" for these types of products have higher purchase intent for both A and B than the "Non-Users" (70% and 65% versus 33% and 23%). This outcome suggests that the binary variable ("Users", "Non-Users") should not be ignored when comparing purchase intent for new and current products.

**Table 9.39** Total Responses for Products (Example 9.4-1)

| | Response | |
|---|---|---|
| **Product** | **Would Buy** | **Would Not Buy** |
| A | 48 | 52 |
| B | 56 | 44 |

**Table 9.40** The Responses by User Groups (Example 9.4-1)

| | Users | | Non-Users | |
|---|---|---|---|---|
| **Product** | **Would Buy** | **Would Not Buy** | **Would Buy** | **Would Not Buy** |
| A | 28 (70%) | 12 | 20 (33%) | 40 |
| B | 51 (65%) | 27 | 5 (23%) | 17 |

## Tests for Independence

In a two-dimensional table, there is only one model of primary interest: independence of rows and columns. However, with multidimensional tables, there are different types of independence, including complete (mutual) independence, joint independence, pairwise (marginal) independence, and conditional independence. Hence, there are different tests for independence.

### Mutual Independence

Given a random sample of $n$ observations in a three-dimensional contingency table, for mutual independence we first test the following null hypothesis:

$$H_0: \; p_{ijk} = (p_{i++})(p_{+j+})(p_{++k}), \tag{9.4-2}$$

where $p_{i++} = n_{i++}/n$, $p_{+j+} = n_{+j+}/n$, and $p_{++k} = n_{++k}/n$. Hence, the expected frequency of the $(i, j, k)$th cell under $H_0$ is

$$e_{ijk} = np_{i++}p_{+j+}p_{++k} = (n_{i++})(n_{+j+})(n_{++k})/n^2. \tag{9.4-3}$$

The expected frequencies under $H_0$ given by (9.4-3) are to be compared with the observed frequencies $n_{ijk}$ for each and every cell by means of the following $X^2$ statistic:

$$X^2 = \sum_i \sum_j \sum_k (n_{ijk} - e_{ijk})^2 / e_{ijk}. \tag{9.4-4}$$

Note that (9.4-4) is a straightforward extension of (9.3-1) for two-dimensional tables. The $X^2$ statistic in (9.4-4) has approximately the $\chi^2$ distribution with $(abc - a - b - c + 2)$ DF. Even if the null hypothesis of complete independence in (9.4.2) is rejected, pairwise independence may still hold. It may also be that while any two of the three classifications are interacting, the third one remains independent of them. So there are three hypotheses, since any one of the three classifications could be independent of the remaining two.

### Joint Independence

To test whether classification $A$ is jointly independent of the $B$ and $C$ classifications, we set up the following null hypothesis:

$$H_0: \; p_{ijk} = (p_{i++})(p_{+jk}). \tag{9.4-5}$$

A $X^2$ statistic for testing $H_0$ is given by

$$X^2 = \sum_i \sum_j \sum_k (n_{ijk} - e_{ijk})^2 / e_{ijk}, \tag{9.4-6}$$

where $e_{ijk} = (n_{i++})(n_{+jk})/n$. The $X^2$ statistic in (9.4-6) has approximately the $\chi^2$ distribution with $(a - 1)(bc - 1)$ DF. The two other analogous hypotheses can be similarly set up and tested.

### Pairwise (Marginal) Independence

For testing the pairwise independence of the three classifications, a null hypotheses for the independence (or interaction) of the $A$ and $B$ classifications is

$$H_0: \; p_{ij+} = (p_{i++})(p_{+j+}). \tag{9.4-7}$$

A $X^2$ statistic with $(a - 1)(b - 1)$ DF for testing the preceding null hypothesis is

$$X^2 = \sum_i \sum_j (n_{ij+} - e_{ij+})^2/e_{ij+}, \tag{9.4-8}$$

where $e_{ij+} = (n_{i++})(n_{+j+})/n$.

Whenever $H_0$ in (9.4-7) is true, it means that the $AB$ interaction is zero for each level of the $C$ classification. The null hypotheses for testing the $AC$ and $BC$ interactions equal to zero can be similarly postulated and tested.

### Conditional Independence

Sometimes it is of interest to test, for example, the hypothesis that, for a given level of $C$ classification, the $A$ and $B$ classifications are independent. That is, for a chosen $k$th level of $C$, we want to test

$$H_0: \; p_{ijk} = (p_{i+k})(p_{+jk})/p_{++k}. \tag{9.4-9}$$

A $X^2$ test statistic with $(a - 1)(b - 1)$ DF for testing $H_0$ in (9.4-9) is

$$X_k^2 = \sum_i \sum_j (n_{ijk} - e_{ijk})^2/e_{ijk}, \tag{9.4-10}$$

where $e_{ijk} = (n_{i+k})(n_{+jk})/n_{++k}$. For an overall test of the conditional independence of the $A$ and $B$ classifications, given the $C$ classification at any level, one may compute the $X^2$ statistics in (9.4-10) for each level of $C$. Thus, we would have $c\ X^2$ statistics $X_k^2$, one for each level of $C$. Then $X^2 = \sum_{k=1}^{c} X_k^2$, with $c(a - 1)(b - 1)$ DF, can be used to test the overall conditional independence of the $A$ and $B$ classifications,

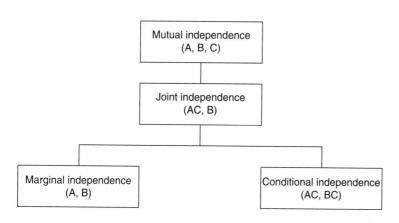

■ **FIGURE 9.2** Relationships among types of A, B, C independence.

given $C$. Similarly, the overall conditional independences of the $A$ and $C$ and also $B$ and $C$ classifications can be tested.

The relationships among the four types of independence are summarized in Figure 9.2.

## Numerical Example for Tests for Independence

Table 9.41 lists the attributes used in Table 9.42, which summarizes data from a marketing study of a new product.

**Example 9.4-2**

A random sample of $n = 296$ respondents was classified according to these attributes, and observed frequencies $n_{ijk}$ were obtained. We illustrate various $\chi^2$ tests.

**Table 9.41** Attributes of the Marketing Study Data in Table 9.42

| Attributes | Symbol | Description |
|---|---|---|
| Occupation: $A$ | $A_1$ | White collar |
| | $A_2$ | Blue collar |
| Purchasing decision: $B$ | $B_1$ | Yes |
| | $B_2$ | No |
| Sex: $C$ | $C_1$ | Male |
| | $C_2$ | Female |

**Table 9.42** Three-Dimensional Contingency Table of Responses from a Marketing Study (Example 9.4-2)

| C: | | Male, $C_1$ | | Female, $C_2$ | | Marginal Totals |
|---|---|---|---|---|---|---|
| B: | | Yes, $B_1$ | No, $B_2$ | Yes, $B_1$ | No, $B_2$ | |
| A: | White collar ($A_1$) | 19 | 12 | 102 | 34 | $n_{1++} = 167$ |
| | Blue collar ($A_2$) | 9 | 3 | 79 | 38 | $n_{2++} = 129$ |
| Marginal Totals | | $n_{+11} = 28$ | $n_{+21} = 15$ | $n_{+12} = 181$ | $n_{+22} = 72$ | $n = 296$ |

*Note:*
*Other marginal totals are*
$n_{+1+} = 28 + 181 = 209;$
$n_{+2+} = 15 + 72 = 87;$
$n_{++1} = 28 + 15 = 43;$
$n_{++2} = 181 + 72 = 253.$

### Null Hypothesis (9.4-2): Independence of A, B, and C

The marginal totals are given in Table 9.42. Calculations for expected frequencies are shown in Table 9.43. The observed $X^2$ is

$$X^2 = \frac{(19 - 17.13)^2}{17.13} + \frac{(9 - 13.23)^2}{13.23} + \cdots + \frac{(34 - 41.95)^2}{41.95} + \frac{(38 - 32.41)^2}{32.41}$$

$$= 8.53 \text{ with 4 DF.}$$

**Table 9.43** Expected Frequencies for Testing Independence of $A$, $B$, and $C$ from Formula (9.4-3)

| Cell, $ijk$ | $n_{ijk}$ | $n_{i++}n_{+j+}n_{++k}/n^2$ | $e_{ijk}$ |
|---|---|---|---|
| 111 | 19 | $n_{1++}n_{+1+}n_{++1}/n^2 = (167)(209)(43)/(296)^2 =$ | 17.13 |
| 211 | 9 | $n_{2++}n_{+1+}n_{++1}/n^2 = (129)(209)(43)/(296)^2 =$ | 13.23 |
| 121 | 12 | $n_{1++}n_{+2+}n_{++1}/n^2 = (167)(87)(43)/(296)^2 =$ | 7.13 |
| 221 | 3 | $n_{2++}n_{+2+}n_{++1}/n^2 = (129)(87)(43)/(296)^2 =$ | 5.51 |
| 112 | 102 | $n_{1++}n_{+1+}n_{++2}/n^2 = (167)(209)(253)/(296)^2 =$ | 100.79 |
| 212 | 79 | $n_{2++}n_{+1+}n_{++2}/n^2 = (129)(209)(253)/(296)^2 =$ | 77.85 |
| 122 | 34 | $n_{1++}n_{+2+}n_{++2}/n^2 = (167)(87)(253)/(296)^2 =$ | 41.95 |
| 222 | 38 | $n_{2++}n_{+2+}n_{++2}/n^2 = (129)(87)(253)/(296)^2 =$ | 32.41 |

For the level of significance $\alpha = 0.05$, $\chi^2_{0.95,4} = 9.49$. Since $8.53 < 9.49$, the null hypothesis of independence is not rejected at the $\alpha = 0.05$ level. By using the S-PLUS built-in program *loglin*, we can obtain the values of both the likelihood ratio chi-square statistic and Pearson chi-square statistic; they are $G^2 = 8.41$ and $X^2 = 8.53$ with 4 DF, respectively. This suggests that occupation, purchasing decision, and sex may not be independent. If the null hypothesis is not rejected, one does not proceed to test other hypotheses. For the sake of illustration, however, let us consider the following hypotheses.

```
> loglin(multabdat,list(1,2,3))
2 iterations: deviation 0
$lrt:
[1] 8.406467
$pearson:
[1] 8.529655
$df:
[1] 4
```

### Null Hypothesis (9.4-5): Classification A Is Independent of B and C

From (9.4-6),

$$e_{111} = n_{+11}n_{1++}/n = (28)(167)/296 = 15.80,$$

$$e_{211} = n_{+11}n_{2++}/n = (28)(129)/296 = 12.20,$$

and so on. Then

$$X^2 = \frac{(19 - 15.80)^2}{15.80} + \frac{(9 - 12.20)^2}{12.20} + \cdots + \frac{(38 - 31.38)^2}{31.38}$$

$$= 7.36 \text{ with 3 DF.}$$

Since $\chi^2_{0.95,3} = 7.81$ at the $\alpha = 0.05$ level, the null hypothesis is not rejected. By using the S-PLUS built-in program *loglin*, we can obtain the values of both the likelihood ratio chi-square statistic and the Pearson chi-square statistic, which are $G^2 = 7.69$ and $X^2 = 7.36$ with 3 DF, respectively. We may conclude that occupation is independent of the joint distribution of purchasing decision and sex. It follows that pairwise interactions $AB$ and $AC$ are not significant. But again for the sake of illustration, we proceed to test interaction hypotheses.

```
> loglin(multabdat,list(1,c(2,3)))
2 iterations: deviation 0
$lrt:
[1]] 7.694894
$pearson:
[1]] 7.359211
$df:
[1] 3
```

### Null Hypothesis (9.4-7): AB Interaction

The required marginal totals summed over the $C$ classification are shown in Table 9.44. From (9.4-8), expected cell frequencies and the $X^2$ statistic are

$$e_{11+} = n_{1++}n_{+1+}/n = (167)(209)/296 = 117.92,$$

$$e_{21+} = n_{2++}n_{+1+}/n = (129)(209)/296 = 91.08,$$

$$e_{12+} = n_{1++}n_{+2+}/n = (167)(87)/296 = 49.08,$$

$$e_{22+} = n_{2++}n_{+2+}/n = (129)(87)/296 = 37.92,$$

$$X^2 = \frac{(121-117.92)^2}{117.92} + \frac{(88-91.08)^2}{91.08} + \frac{(46-49.08)^2}{49.08} + \frac{(41-37.92)^2}{37.92}$$

$$= 0.628$$

with 1 DF. The observed $X^2$ value is not significant. By using the S-PLUS built-in program *loglin*, we can obtain the values of the

**Table 9.44** Marginal Totals of $2 \times 2 \times 2$ Table 9.42 for *AB, AC,* and *BC* Interactions

| | **AB Interaction** | | | | **AC Interaction** | | |
|---|---|---|---|---|---|---|---|
| | $B_1$ | $B_2$ | Totals | | $C_1$ | $C_2$ | Totals |
| $A_1$ | $121(n_{11+})$ | $46(n_{12+})$ | $167(n_{1++})$ | $A_1$ | $31(n_{1+1})$ | $136(n_{1+2})$ | $167(n_{1++})$ |
| $A_2$ | $88(n_{21+})$ | $41(n_{22+})$ | $129(n_{2++})$ | $A_2$ | $12(n_{2+1})$ | $117(n_{2+2})$ | $129(n_{2++})$ |
| | $209(n_{+1+})$ | $87(n_{+2+})$ | $296(n)$ | | $43(n_{++1})$ | $253(n_{++2})$ | $296(n)$ |
| | | | | | **BC Interaction** | | |
| | | | | | $C_1$ | $C_2$ | Totals |
| | | | | $B_1$ | $28(n_{+11})$ | $181(n_{+12})$ | $209(n_{+1+})$ |
| | | | | $B_2$ | $15(n_{+21})$ | $72(n_{+22})$ | $87(n_{+2+})$ |
| | | | | | $43(n_{++1})$ | $253(n_{++2})$ | $296(n)$ |

**Table 9.45** Observed and Expected Cell Frequencies for Testing Second-Order Interactions and the Corresponding $X^2$ Statistic

| Hypotheses for | | | | | |
|---|---|---|---|---|---|
| **AB Interaction** | | **AC Interaction** | | **BC Interaction** | |
| $n_{ij+}$ | $e_{ij+}$ | $n_{i+k}$ | $e_{i+k}$ | $n_{+jk}$ | $e_{+jk}$ |
| 121 | 117.92 | 31 | 24.26 | 28 | 30.36 |
| 88 | 91.08 | 12 | 18.74 | 15 | 12.64 |
| 46 | 49.08 | 131 | 142.74 | 181 | 178.64 |
| 41 | 37.92 | 117 | 110.26 | 72 | 74.36 |
| $X^2$ | 0.628 | | 5.674 | | 0.730 |
| DF | 1 | | 1 | | 1 |

likelihood ratio chi-square statistic and Pearson chi-square statistic, which are $G^2 = 0.63$ and $X^2 = 0.63$ with 1 DF, respectively. Therefore, the null hypothesis of insignificant $AB$ interaction is not rejected. The result indicates that a decision as to whether or not one buys the product is not influenced by the occupational level of the buyer. For the null hypotheses of $AC$ and $BC$ interactions, the $X^2$ statistics are calculated in Table 9.45.

```
> loglin(multabdat[,,1]+multabdat[,,2],list(1,2))
2 iterations: deviation 0
$lrt:
[1] 0.6280417
$pearson:
[1] 0.6298857
$df:
[1] 1
```

Finally, let us consider a null hypothesis of the conditional independence of $A$ and $B$ classifications, given classification $C$. Other null hypotheses of conditional independence can be similarly tested.

## Null Hypothesis (9.4-9): Conditional Independence of A and B Given C

Compute the marginal totals for the $A$ and $B$ classifications for each level of $C$. These marginal totals are shown in Table 9.46. From these totals expected cell frequencies are computed using (9.4-10). They are

**Table 9.46** Marginal Totals of $A$ and $B$ for Each Level of $C$

|  | $B_1$ | $B_2$ | Totals |
|---|---|---|---|
| Level $C_2$ |  |  |  |
| $A_1$ | 19 | 12 | $31(n_{1+1})$ |
| $A_2$ | 9 | 3 | $12(n_{2+1})$ |
|  | $28(n_{+11})$ | $15(n_{+21})$ | $43(n_{++1})$ |
| Level $C_1$ |  |  |  |
| $A_1$ | 102 | 34 | $136(n_{1+2})$ |
| $A_2$ | 79 | 38 | $117(n_{2+2})$ |
|  | $181(n_{+12})$ | $72(n_{+22})$ | $253(n_{++2})$ |

$$e_{111} = \frac{(31)(28)}{43} = 20.186, \quad e_{112} = 97.296,$$

$$e_{121} = \frac{(31)(15)}{43} = 10.814, \quad e_{122} = 38.704,$$

$$e_{211} = \frac{(28)(12)}{43} = 7.814, \quad e_{212} = 83.704,$$

$$e_{221} = \frac{(15)(12)}{43} = 4.186, \quad e_{222} = 33.296.$$

Two $X^2$ statistics computed for levels $C_1$ and $C_2$, respectively, are $X_1^2 = 0.716$ and $X_2^2 = 1.728$. For an overall conditional test, we use $X^2 = X_1^2 + X_2^2 = 2.444$ with 2 DF. By using the S-PLUS built-in program *loglin*, we can obtain the values of both likelihood ratio chi-square statistic and Pearson chi-square statistic, which are $G^2 = 2.47$ and $X^2 = 2.44$ with 2 DF, respectively. At the $\alpha = 0.05$ level, $\chi^2_{0.95,2} = 5.99$. Hence, one cannot reject the null hypothesis of overall conditional independence of occupation from purchasing decision within sexes.

```
> loglin(multabdat,list(c(2,3),c(1,3)))
2 iterations: deviation 0
$lrt:
[1] 2.466134
$pearson:
[1] 2.443582
$df:
[1] 2
```

## EXERCISES

**9.1.** The following data give the number of days at termination of sensory tests for items from five production plants. An investigator wishes to determine whether age at termination varies among plants.

| | | Plants | | |
|---|---|---|---|---|
| **1** | **2** | **3** | **4** | **5** |
| 33 | 27 | 40 | 49 | 29 |
| 40 | 33 | 41 | 53 | 27 |
| 35 | 25 | 81 | 55 | 42 |
| 30 | 21 | 75 | 53 | 30 |

**1.** Carry out the Kruskal–Wallis $H$ test to test the null hypothesis of no difference among plants with respect to the age at termination.

**2.** If the null hypothesis is rejected, carry out Steel's rank sum test for pairwise significance.

**9.2.** The following data are taken from a routine dog food preference study. In this study 10 dogs were used and two products (control and test) were offered to each dog simultaneously over a 10-day feeding period. The total food consumption for each dog during the entire feeding period is as follows:

| Dog | Control, gm | Test, gm |
|---|---|---|
| 1 | 4240 | 3435 |
| 2 | 5275 | 4680 |
| 3 | 5280 | 4240 |
| 4 | 4260 | 5070 |
| 5 | 5015 | 680 |
| 6 | 3805 | 5508 |
| 7 | 3325 | 3315 |
| 8 | 4310 | 4135 |
| 9 | 4110 | 1855 |
| 10 | 4755 | 2360 |

Test whether there is a significant difference in the dog's preferences between the control and the test foods using the Wilcoxon signed-rank test.

**9.3.** An in-laboratory consumer test was conducted on a newly developed poultry product. Forty-eight respondents, who were contacted prior to the test, ranked the three products according to their preferences. The numbers given in the following table are the number of first, second, and third choices for each product:

| | Rank | | |
|---|---|---|---|
| **Product** | **1** | **2** | **3** |
| A | 28 | 14 | 6 |
| B | 12 | 18 | 18 |
| C | 8 | 16 | 24 |

1. Why does the total of ranks for each product equal 48?
2. Use the Anderson statistic to test whether a significant difference in preferences exists.

**9.4.** The following data in the $2 \times 3 \times 2$ table were taken from a product usage study. For this $2 \times 3 \times 2$ table, follow various tests in Section 9.4 and state your conclusions at the $\alpha = 0.05$ level of significance.

| | Usage | | | | | |
|---|---|---|---|---|---|---|
| | **Once per Month** | | | **Two or Three Times per Month** | | |
| **Brand:** | **X** | **Y** | **Z** | **X** | **Y** | **Z** |
| Income Level | | | | | | |
| 1 | 26 | 20 | 36 | 50 | 37 | 15 |
| 2 | 13 | 15 | 10 | 16 | 5 | 2 |

**9.5.** The following liver weight in grams data were extracted from a dose toxicity study (90 days of inclusion of a chemical in the diet of rats) reported by Weil and Gad (1980):

| **Sex** | **Rat Number** | **0.25% in Diet** | **2.0% in Diet** |
|---|---|---|---|
| Male | 1 | 19.27 | 20.01 |
| | 2 | 14.98 | 15.68 |
| | 3 | 15.26 | 21.51 |
| | 4 | 10.63 | 18.34 |
| | 5 | 16.88 | 15.85 |

| Sex | Rat Number | 0.25% in Diet | 2.0% in Diet |
|-----|-----------|---------------|--------------|
| Female | 1 | 10.14 | 9.27 |
| | 2 | 8.84 | 13.84 |
| | 3 | 7.74 | 10.55 |
| | 4 | 8.50 | 9.93 |
| | 5 | 9.60 | 10.66 |

1. Test the null hypothesis of no chemical effects on the liver weight of rats using a nonparametric procedure.
2. Test the null hypothesis using a randomized complete block model and compare it with the result of the nonparametric procedure used in (1). Interpret your statistical analyses.

**9.6.** Two hundred consumer panelists participated in a consumer preference test using a paired comparison method for two products A and B, and a significance level of 0.05 was selected. Determine the critical values for a one-sided test and two-sided test using a computer program for the binomial distribution.

**9.7.** The following data were given in a consumer preference test for the current product A and new product B. Test whether there is a significant difference between users and non-users of the product in preferences by using a $z$ test, chi-square test, and Fisher's exact test, respectively.

| | Preferring A | Preferring B | |
|-----|-----|-----|-----|
| Users | 56 | 24 | 80 |
| Non-Users | 72 | 48 | 120 |
| | 128 | 72 | 200 |

**9.8.** In a test for consumer purchase intent for two products, each of 125 consumers evaluated the two products using a 5-point scale. Test whether consumers have the same purchase intent for the two products in terms of the proportions of the top two boxes. The top two boxes are 5 = Definitely would buy it and 4 = Probably would buy it. The response data are summarized in the following table. Use the McNemar test to analyze the data.

| | | Product B | |
|---|---|---|---|
| | | **Top Two Boxes** | **Other Boxes** |
| Product A | Top two boxes | 72 | 21 |
| | Other boxes | 9 | 23 |
| | | | 125 |

**9.9.** A panel consisting of 12 panelists evaluates four prototypes to see whether the prototypes have similar and acceptable sensory characteristics compared with the control sample. The data are given in the following table with 1 = acceptable and 0 = unacceptable. Use Cochran's Q statistic to analyze the data.

| | Prototype | | | | |
|---|---|---|---|---|---|
| **Panelist** | **1** | **2** | **3** | **4** | $u_i$ |
| 1 | 1 | 1 | 0 | 1 | 3 |
| 2 | 1 | 1 | 0 | 1 | 3 |
| 3 | 1 | 0 | 0 | 1 | 2 |
| 4 | 0 | 1 | 1 | 1 | 3 |
| 5 | 1 | 1 | 0 | 0 | 2 |
| 6 | 1 | 0 | 1 | 1 | 3 |
| 7 | 1 | 1 | 0 | 0 | 2 |
| 8 | 0 | 0 | 0 | 0 | 0 |
| 9 | 1 | 1 | 0 | 1 | 3 |
| 10 | 1 | 1 | 1 | 1 | 4 |
| 11 | 1 | 0 | 0 | 1 | 2 |
| 12 | 1 | 1 | 0 | 0 | 2 |
| $T_j$ | 10 | 8 | 3 | 8 | 29 |

**9.10.** Rosner (1986) reported the data produced in a cancer study, as shown in the following table. In the table, "Case" refers to an individual who had cancer, and "Control" refers to an individual who did not have cancer. Use the Cochran–Mantel–Haenszel (CMH) statistic to conduct an overall test of independence between cancer status and passive smoking status.

|  |  | Passive Smoker | Not a Passive Smoker |
|---|---|---|---|
| Nonsmoker | | | |
| | Case | 120 | 111 |
| | Control | 80 | 155 |
| Smoker | | | |
| | Case | 161 | 117 |
| | Control | 130 | 124 |

**9.11.** Best (1995) provided data of Japanese and Australian consumers for ratings of sweetness of chocolate using a 7-point scale with 7 = Like Extremely to 1 = Dislike Extremely. The data are as follows:

| | Sweetness Liking Score | | | | | | |
|---|---|---|---|---|---|---|---|
| **Consumers** | **1** | **2** | **3** | **4** | **5** | **6** | **7** |
| Australian | 2 | 1 | 6 | 1 | 8 | 9 | 6 |
| Japanese | 0 | 1 | 3 | 4 | 15 | 7 | 1 |

**1.** Use the Mann–Whitney $U$ statistic to analyze the data.
**2.** Use the Pearson chi-square statistic to analyze the data.

**9.12.** Twenty panelists evaluated all of five products for sweetness and ranked the products from "Most Sweet" to "Least Sweet." The ranks are as follows:

| | Samples | | | | | | Samples | | | | |
|---|---|---|---|---|---|---|---|---|---|---|---|
| **ID** | **P1** | **P2** | **P3** | **P4** | **P5** | **ID** | **P1** | **P2** | **P3** | **P4** | **P5** |
| 1 | 1 | 2 | 3 | 5 | 4 | 11 | 1 | 2 | 5 | 4 | 3 |
| 2 | 3 | 2 | 5 | 1 | 4 | 12 | 1 | 3 | 2 | 5 | 4 |
| 3 | 5 | 1 | 3 | 4 | 2 | 13 | 2 | 3 | 1 | 4 | 5 |
| 4 | 4 | 3 | 1 | 2 | 5 | 14 | 4 | 5 | 1 | 2 | 3 |
| 5 | 2 | 3 | 1 | 4 | 5 | 15 | 4 | 3 | 1 | 2 | 5 |
| 6 | 1 | 3 | 2 | 4 | 5 | 16 | 1 | 3 | 2 | 5 | 4 |
| 7 | 1 | 3 | 2 | 5 | 4 | 17 | 1 | 4 | 2 | 3 | 5 |
| 8 | 1 | 2 | 4 | 3 | 5 | 18 | 2 | 1 | 4 | 3 | 5 |
| 9 | 4 | 1 | 3 | 2 | 5 | 19 | 4 | 3 | 2 | 5 | 1 |
| 10 | 1 | 2 | 5 | 4 | 3 | 20 | 4 | 1 | 2 | 3 | 5 |
| | | | | | | | 47 | 50 | 51 | 70 | 82 |

1. Conduct a Friedman test for the designs.
2. If a significant difference is found, conduct multiple comparisons based on the ranks.

**9.13.** For the data in Exercise 9.12, if there is some prior information about the order of the alternative hypothesis, i.e., $H_a$: P1 > P2 > P3 > P4 > P5 with at least one strict inequality, use the page test to analyze the data ($\alpha = 0.05$).

**9.14.** For the data in Exercise 9.12, calculate Kendall's coefficient of concordance, $W$.

**9.15.** For the data in Exercise 9.1, if there is some prior information about the order of the ages in the alternative hypothesis, i.e., $H_a$ : P1' > P2' > P3' > P4' > P5', where P1', P2', P3', P4', P5', denote the production plants 3, 4, 1, 5 and 2, respectively. Use the Jonckheere–Terpstra test to analyze the data.

**9.16.** In a central location consumer test, each of a total 120 panelists evaluated two products (A and B) for saltiness using a 5-point just-about-right (JAR) intensity scale from $1$ = Not at all salty enough to $3$ = Just about right to $5$ = Much too salty. The cross-classified data of the responses are as follows:

|  |  | Sample B | | | | | |
|---|---|---|---|---|---|---|---|
|  |  | **1** | **2** | **3** | **4** | **5** | **Total** |
|  | 1 | 0 | 0 | 1 | 1 | 2 | 4 |
|  | 2 | 0 | 4 | 2 | 5 | 2 | 13 |
| Sample A | 3 | 1 | 2 | 6 | 6 | 2 | 17 |
|  | 4 | 0 | 1 | 2 | 31 | 11 | 45 |
|  | 5 | 0 | 0 | 2 | 8 | 31 | 41 |
|  | Total | 1 | 7 | 13 | 51 | 48 | 120 |

Analyze the data by using
1. A symmetry test;
2. A marginal homogeneity test; and
3. A marginal homogeneity test for ordered categories.

**9.17.** For the data in Exercise 9.4, can we collapse validly the $2 \times 3 \times 2$ table (A, B, C) over USEGE (C) into a $2 \times 3$ table (A, B)?

**9.18.** Use Simpson's paradox to interpret the necessity of consumer segmentation.

# Sensory Difference and Similarity Tests

## 10.1 SENSORY DIFFERENCE TESTS

In Chapter 2 we considered some methods for constructing sensory scales to quantify sensory judgments. In this section we discuss six commonly used sensory difference tests. They are called the triangle, duo–trio, paired comparison (also called the 2-alternative forced-choice, or 2-AFC), 3-alternative forced-choice (3-AFC), A–Not–A, and same-different tests. The sensory difference tests, also known as discriminative difference tests, have long been used for judging and establishing the difference between products without reporting the magnitude or direction of difference. There is a long list of publications that describe the application of difference tests, such as those by Jellinek (1964), Amerine et al. (1965), Ellis (1966), Abbott (1973), Amerine and Roessler (1976), Larmond (1977), Moskowitz (1983, 1988), Stone and Sidel (1985, 2004), O'Mahony (1986), Meilgaard et al. (1991), Chambers and Wolf (1996), Lawless and Heymann (1998), and Bi (2006).

### Triangle Test

#### *Conventional Triangle Test*

The triangle (triangular) test is a well-known and widely used test for detecting differences in product quality and screening potential judges in sensory evaluation. The test is discussed in many published articles, such as those by Peryam and Swartz (1950), Lockhart (1951), and Byer and Abrams (1953). The ISO (ISO 4120-1983) standard and the ASTM (E1885-2004) standard describe the test.

The experimental procedure in conducting a triangle test is as follows. Three properly coded samples are presented simultaneously to the judges. Two of the samples are identical, and the remaining one is odd (different). The order of sample presentation must be balanced throughout the entire experiment. A sample questionnaire is shown in Figure 10.1. The task of each judge is to pick the odd one from the three samples. The probability of the correct selection, say $P_T$, purely by chance, is $\frac{1}{3}$. If the judges are discriminating, then the probability of the correct selection must be significantly greater than $\frac{1}{3}$. Hence, the null and alternative hypotheses may be stated as follows:

$$H_0: \quad P_T = \frac{1}{3}; \quad H_a: \quad P_T > \frac{1}{3} \qquad (10.1\text{-}1)$$

Note that $H_a$ is a one-sided hypothesis. If the null hypothesis cannot be rejected, then either the odd sample is not really different or the judges are not sufficiently discriminating. Indeed, one may argue that judges cease to be effective discriminators when samples are nearly identical.

An experimental design for the triangle test is shown in Table 10.1. In the table, let $A$ designate the control or standard sample and $B$ the test sample (treatment). The position of the odd sample in the order of presentation of the three samples to panelists is underlined, and these six arrangements of the three samples are called the *basic design*. The basic design is repeated to obtain additional replications

■ **FIGURE 10.1** Questionnaire for a triangle test.

Name _____Date_____

This is a Triangle Test. You must make a choice.

Two of the three samples are the same, one is different (odd).

Write down the number of the odd sample _____.

Comments: _____
_____
_____

**Table 10.1** Basic Design for a Triangle Test[a]

| Panelists | Odd Sample | Orders | Three-Digit Code |
|---|---|---|---|
| 1 | A | ABB | 101, 475, 178 |
| 2 | | BAB | 946, 557, 721 |
| 3 | | BBA | 453, 412, 965 |
| 4 | B | BAA | 876, 633, 798 |
| 5 | | ABA | 321, 613, 197 |
| 6 | | AAB | 154, 995, 668 |

[a]Number of unknown samples, 3; probability of correct selection by chance under $H_0$, $P_T = \frac{1}{3}$.

and balance for the order effect. For balancing, both the A and B samples are designated odd in the design an equal number of times. Also, sample A appears in a particular position the same number of times as sample B. Panelists are randomly assigned to the different orders of sample presentation.

There is experimental evidence that the outcomes of the triangle tests are biased by the designation of the odd sample. Grim and Goldblith (1965), using data on off-flavor from an irradiation study of whole-egg magma, observed that panelists who received two irradiated samples and the control sample as the odd one were less able to correctly identify the odd sample than those who received the irradiated sample as the odd one. A study by Wasserman and Talley (1969) on smoked frankfurters supports the findings of the irradiation study. Wasserman and Talley noted that a judge was able to determine a significant difference between smoked and unsmoked frankfurters when the smoked frankfurter was designated the odd sample. But a judge was less likely to detect significant differences when the unsmoked sample was the odd one. The design shown in Table 10.1 takes into consideration such bias, as it is balanced with respect to the designation of the odd and identical samples.

There are other considerations when using the triangle test in sensory evaluation. It is important that panel members have a clear understanding of the test criterion. Judgment should be based only on the characteristic or the response called for in the questionnaire. If the detection of smoke is the response variable, incidental differences in, for example, size, color, or appearance should be discounted. Also the identical or like samples should be homogeneous. If, in a triangle test, the difference between like samples is larger than the difference to be detected between the odd and like samples, then the null hypothesis of no difference tends to be falsely accepted. Let $X_1$ and $X_2$ be, respectively, the psychological responses generated by the two like samples and $Y$ the response evoked by the odd sample. A correct response in the triangle test is obtained if we have the following relationships:

$$|X_1 - X_2| < |X_1 - Y| \quad \text{and} \quad |X_1 - X_2| < |X_2 - Y|. \tag{10.1-2}$$

See Ura (1960) and David and Trivedi (1962). When products to be evaluated vary greatly from sample to sample, the preceding relationships may not hold, and the triangle test may no longer apply and should be used discretely.

The important statistical assumptions underlying the triangle test are that panelists make their selections independently of each other. Let $N$ denote the number of panelists used in the triangle test. The number of correct selections, $m$ out of $N$ comparisons, follows the binomial probability distribution with parameters $P_T$ and $N$. Tables A.21 and A.25 provide critical values corresponding to various significance levels and the number of panelists or judgments under the null hypothesis of $P_T = \frac{1}{3}$.

When $N$ is large, $N \geq 30$, we may use the $Z$ score for the number $m$ of correct selections to carry out the test. The $Z$ statistic or score for $m$ is

$$Z = (3m - N)/\sqrt{2N}, \tag{10.1-3}$$

which has the standard normal distribution. The $Z$ score should be compared with $z_{1-\alpha}$ at the $\alpha$ level of significance. Reject the null hypothesis in favor of the one-sided alternative $P_T > \frac{1}{3}$ if $Z > z_{1-\alpha}$.

---

**Example 10.1-1**

A study was carried out for evaluating the flavor of frankfurters prepared with and without sodium nitrite. To eliminate the incidental differences due to color, the expert panelists were blindfolded. Some of the experimental outcomes for this example are displayed in Table 10.2.

As one can see in this table, the number of times the odd sample was correctly selected is $m = 4$ out of $N = 12$ comparisons. At the 5% significance level, the critical value from Table A.21 is 8. For $m$ to be significant, it should be equal to or greater than the tabular value. In this

**Table 10.2** Design of a Triangle Test for Evaluating the Flavor of Frankfurters With and Without Sodium Nitrite

| Panelists | Odd Sample | Orders | Selection | |
|---|---|---|---|---|
| | | | Correct | Incorrect |
| 1 | A (nitrite) | ABB | A | — |
| 2 | | BAB | — | B |
| 3 | | BBA | A | — |
| 4 | B (no nitrite) | BAA | — | A |
| 5 | | ABA | — | A |
| 6 | | AAB | B | — |
| 7 | A (nitrite) | ABB | — | B |
| 8 | | BAB | — | B |
| 9 | | BBA | A | — |
| 10 | B (no nitrite) | BAA | — | A |
| 11 | | ABA | — | A |
| 12 | | AAB | — | A |
| Total | | | $m = 4$ | 8 |

case $m = 4 < 8$; hence, it is not significant at the 5% level of significance. We may conclude, therefore, that there is no detectable flavor difference between the two types of frankfurters.

## Modified Triangle Test

The triangle test has been modified to extract additional information from a sensory difference test. The modified triangle test is due to Bradley (1964) and Bradley and Harmon (1964). Another modification of the triangle test is due to Gridgeman (1964, 1970a), who called it the two-stage triangle test. The first stage pertains only to the identification of the odd sample and, hence, is the usual triangle test. The second stage pertains to the quantification of the difference between the odd and like samples. A questionnaire for the modified triangle test is shown in Figure 10.2. In this questionnaire a nondirectional degree of difference is provided by item (2) in the first stage, whereas item (3) in the second stage provides a directional selection.

■ **FIGURE 10.2** Questionnaire for the modified triangle test. *Source:* Larmond (1967).

Date_____Taster_____

Product _____

Instructions: Here are three samples for evaluation. Two of these samples are duplicates. Separate the odd sample for difference only.

(1)        <u> Sample </u>                             <u>Check odd sample</u>

               <u>  314  </u>                                 _____

               <u>  628  </u>                                 _____

               <u>  542  </u>                                 _____

(2)        Indicate the degree of difference between the duplicate samples and the odd sample.

               Slight _____      Much_____

               Moderate _____      Extreme_____

(3)        Acceptability:

               Odd sample more acceptable_____

               Duplicate samples more acceptable_____

(4)        Comments: _____
_____
_____

The question often arises as to the psychological effect of the first stage on the judgment made in the second. Schutz and Bradley (1953) reported that preference judgments in the second stage show a definite bias against the odd sample. Therefore, the preference outcomes in the second stage are of doubtful validity. Several experiments were conducted by Gregson (1960), and his findings support those of Schutz and Bradley that the second stage is biased against the odd sample. The main conclusions of these experiments are the following:

1. The bias against the odd sample in the second stage is not related to the items being tested but is intrinsic to the method.
2. The bias does not always occur. It is likely to occur when the items being compared are physically or hedonically very similar.

3. The direction and the extent of the bias depend on the information given to panelists about the differences they are asked to detect.
4. The bias can be balanced out by using a balanced experimental design.

There are various suggestions for analyzing the experimental outcomes of a two-stage triangle test. A suggestion due to Amerine et al. (1965) requires the use of only the outcomes from panelists who were able to correctly identify the odd sample in the first stage. The obvious drawback of this approach is that not all the outcomes are used; hence, any statistical analysis can be considered conditional only on the basis of the outcomes of the first stage. Another suggestion is to use the second-stage outcomes only when results of the first stage are judged statistically significant. A third suggestion is to use all the outcomes from both stages, regardless of the first-stage outcomes. Statistical analyses following these suggestions have been discussed by, among others, Bradley and Harmon (1964) and Gridgeman (1964, 1970a). In the following, we consider the statistical analyses by Bradley and Harmon and by Gridgeman.

### Bradley–Harmon Model

The Bradley–Harmon model requires that in the second stage respondents assign scores that measure the degree of difference between the sample selected as odd and the remaining pair of like samples on the basis of a scoring scale. Scoring scales may differ, but a typical scale according to Bradley and Harmon (1964) would be as follows: $0 =$ no difference; $2 =$ very slight difference; $4 =$ slight difference; $6 =$ moderate difference; $8 =$ large difference; $10 =$ very large difference. When the odd sample is correctly selected, the score on the degree of difference scale is a measure of

$$R = |Y - 0.5(X_1 + X_2)| \qquad (10.1\text{-}4)$$

where $R$ is the absolute difference between the sensation $Y$ evoked by the odd sample and the average of sensations $X_1$ and $X_2$ evoked by the like samples. When the odd sample is not correctly selected, the score on the degree of difference scale is a measure of

$$W = |X_1 - 0.5(Y + X_2)| \quad \text{or} \quad W = |X_2 - 0.5(Y + X_1)| \qquad (10.1\text{-}5)$$

We may emphasize that both $R$ and $W$ are measured directly on a degree of difference scoring scale. The experimental procedure does not require measurements of sensations $Y$, $X_1$, and $X_2$. That is, for the Bradley–Harmon model, the panelists provide measurements on $R$ and $W$ and not on sensations $Y$, $X_1$, and $X_2$.

When $N$ comparisons are conducted for a triangle test, let $m$ denote the number of correct identifications of the odd sample in the first stage and $R_1$, $R_2$,..., $R_m$ the corresponding degree of difference scores in the second stage. Corresponding to $n = N - m$ incorrect identifications in the first stage are the degree of difference scores denoted by $W_1$, $W_2$,..., $W_n$. The degree of difference scores in the second stage provide valuable information (Davis and Hanson, 1954; Gridgeman, 1964, 1970a).

Bradley and Harmon (1964) derived appropriate conditional distribution of scores $R$ and $W$ and gave a likelihood ratio statistic for the modified triangle method. The natural logarithm of the likelihood function is a function of parameters $\delta$ and $\sigma$ [note that Bradley and Harmon (1964) use $\theta = \mu/\sigma$ for $\delta$]. Parameter $\delta$ denotes a standard distance between two products for comparison in the Thurstonian scale, and $\sigma$ is the standard variation of sensory sensation in the scale. The log likelihood with parameters $\delta$ and $\sigma$ is

$$\ln L = f(\delta, \sigma) = k - N \ln \sigma - S/3\sigma^2 - (4m + n)\delta^2/12$$
$$+ \sum_{i=1}^{m} \ln \cosh(2\delta R_i/3\sigma) + \sum_{i=1}^{n} \ln \cosh(\delta W_i/3\sigma)$$
$$+ \sum_{i=1}^{m} \ln I(\sqrt{2}R_i/3\sigma) + \sum_{i=1}^{n} \ln[I_{+,i} + I_{-,i}], \qquad (10.1\text{-}6)$$

where

$$k = 2N \ln 2 - 0.5N \ln (3\pi),$$

$$S = \sum_{i=1}^{m} R_i^2 + \sum_{i=1}^{n} W_i^2,$$

$$\cosh(u) = (e^u + e^{-u})/2.$$

In (10.1-6), $I$ denotes the incomplete standard normal distribution function defined as

$$I(u) = \frac{1}{\sqrt{2\pi}} \int_0^u e^{-t^2/2} dt,$$

$$I_{+,i} = I[(\sqrt{2}W_i/3\sigma) + (\delta/\sqrt{2})],$$

$$I_{-,i} = I[(\sqrt{2}W_i/3\sigma) - (\delta/\sqrt{2})].$$

The usual null hypothesis of the triangle test is $H_0: P_T = 1/3$ versus the alternative hypothesis, $H_a: P_T > 1/3$. The test is equivalent to $H_0: \delta = 0$ versus $H_a: \delta > 0$. These hypotheses remain for the modified triangle test. The likelihood ratio test statistic for the modified triangle test is

$$X_1^2 = -2 \ln \lambda = -2[Max \ln L|_{H_0} - Max \ln L|_{H_a}] \qquad (10.1\text{-}7)$$

where $Max \ln L|_{H_0}$ is the maximum of (10.1-6) in the null hypothesis, i.e., the maximum of $\ln L$ at $\sigma = \tilde{\sigma}$ when $\delta = 0$; and $Max \ln L|_{H_a}$ is the maximum of (10.1-6) in the alternative hypothesis, i.e., the maximum of $\ln L$ at $\delta = \hat{\delta}$ and $\sigma = \hat{\sigma}$. The test statistic $X_1^2$ follows asymptotically a chi-square distribution with 1 DF. The S-PLUS code *moditr* can be used to estimate the parameters $\hat{\delta}$, $\hat{\sigma}$ (non-null standard variance), $\tilde{\sigma}$ (null hypothesis standard variance), and the value of the likelihood ratio statistic $X_1^2$, the associated P-value, as well as the covariance matrix of the estimators $\hat{\delta}$ and $\hat{\sigma}$, which can be used for calculation of the confidence interval of $\delta$ (or $\sigma$),

$$\hat{\delta} \pm z_{1-\alpha/2}\hat{\sigma}. \qquad (10.1\text{-}8)$$

One should note that the zero score should be replaced by a very small nonzero value (e.g., 0.01) in estimations of the parameters by using the S-PLUS code.

---

**Example 10.1-2**

Bradley and Harmon (1964) gave a numerical example involving a taste test for cereal. Data for $N = 44$ modified triangle tests together with scored degree of difference are shown in Table 10.3. The test with $H_0$: $\delta = 0$ ($P_T = \frac{1}{3}$) against the one-sided alternative $H_a$: $\delta > 0$ ($P_T > \frac{1}{3}$). The scoring scale used was 0, 2, 4, 6, 8, and 10, where $0 =$ No difference to $10 =$ Very large difference. Note that $m = 20$ and $n = 24$ in this case.

By using the S-PLUS code *moditr*, we get $\hat{\delta} = 1.08$, $\hat{\sigma} = 2.11$, $\tilde{\sigma} = 2.49$, and $X_1^2 = 4.13$ with associated P-value $= 0.042$. Hence, the null hypothesis should be rejected, and the alternative hypothesis should be accepted. The conclusion is that the experimental cereal is significantly different from the control cereal. By using the S-PLUS code *modtr*, we also get the covariance matrix of $\hat{\delta}$ and $\hat{\sigma}$,

**Table 10.3** Scores on Degree of Difference on Cereal Tests (Example 10.1-2)

|  | Scores |
|---|---|
| R (Scores for Correct Tests) | 2,4,2,8,6,6,2,2,6,2,2,4,8,0,4,8,6,2,3,4 |
| W (Scores for Incorrect Tests) | 4,6,6,6,2,4,2,2,0,0,2,2,0,5,4,2,2,2,2,2,0,6,4,6 |

$$\begin{pmatrix} 0.1045 & -0.0505 \\ -0.0505 & 0.0493 \end{pmatrix}.$$

Hence, the 95% confidence interval of $\delta$ is $1.08 \pm 1.96 \times \sqrt{0.1045}$, i.e., [0.45,1.71].

```
> moditr(rr,ww)
1.08085501024766    2.10691779896738    2.48596939351806
-142.738695183547  -144.804501656304   4.13161294551401
0.0420892161744655
   [,1]   [,2]
[1,] 0.10454583 -0.05050535
[2,] -0.05050535 0.04933321
> rr
[1] 2.00 4.00 2.00 8.00 6.00 6.00 2.00 2.00 6.00 2.00 2.00
    4.00 8.00 0.01 4.00 8.00 6.00 2.00 3.00 4.00
> ww
[1] 4.00 6.00 6.00 6.00 2.00 4.00 2.00 2.00 0.01 0.01 2.00
    2.00 0.01 5.00 4.00 2.00 2.00 2.00 2.00 0.01 6.00
    4.00 6.00
```

An alternative statistic, suggested by Harmon (1963), is

$$Z = \sqrt{N}[\overline{H}/0.7694S - 1.9053], \qquad (10.1\text{-}9)$$

where

$$\overline{H} = \left(\sum_i^m R_i + \sum_j^n W_j\right)\Big/N,$$
$$S^2 = \left(\sum_i^m R_i^2 + \sum_j^n W_j^2 - N\overline{H}^2\right)\Big/N - 1.$$

Notice that $\overline{H}$ is the average of the $R$ and $W$ scores. For large $N$, the $Z$ statistic has the standard normal distribution, and we may use Table A.1 to obtain the desired significance probability. Reject the null hypothesis $P_T = \frac{1}{3}$ if $Z > Z_{1-\alpha}$ at the $\alpha$ level of significance.

Let us consider briefly a drawback of the suggestion that uses only the second-stage data $R_1, R_2, \ldots, R_m$ from those panelists who were correct in the first stage. In the literature it has been suggested that $t = \overline{R}/S_R$ should be used as a test statistic for testing $H_0$: $\delta = 0$ against $H_a$: $\delta \neq 0$, where $\overline{R}$ and $S_R$ are, respectively, the mean and standard deviation of $R_1$, $R_2, \ldots, R_m$. Since $m$ itself is a random variable, the statistic $t = \overline{R}/S_R$ may have the $t$ distribution with $(m-1)$ DF only conditionally, given that the number of correct responses in the first stage is $m$. It is indeed very difficult to work out the unconditional distribution, since $m$ was not really known before the experiment was performed.

---

Table 10.4 pertains to a study comparing a test steak sauce with a control. A 6-point degree of difference scale was used where $0 =$ no difference and $6 =$ very large difference. We wanted to test the null hypothesis that there is no difference between the odd and identical samples; that is, $H_0: \delta = 0$ $(P_T = \frac{1}{3})$ against the one-sided alternative $H_a: \delta > 0$ $(P_T > \frac{1}{3})$.

**Example 10.1-3**

Since $\bar{H} = 1.83$, $S = 1.53$, when we use (10.1-9), the $Z$ statistic is

$$Z = \sqrt{29}\{1.83/0.7694(1.53)] - 1.9053\} = -1.89$$

which is significant at the 0.029 level. Hence, we may reject $H_0$ and conclude that there is a significant difference between the odd and identical samples.

To carry out the usual triangle test, note that $N = 29$, $m = 16$, and $n = 13$. That is, in the first stage there were 16 correct identifications out of 29. Hence, an estimate of $P_T$ is $\widehat{P}_T = \frac{16}{29} = 0.55$, and the $Z$ score of 16 from (10.1-3) is

$$Z = (48 - 29)/\sqrt{2(29)} = 2.49.$$

Since the alternative is one-sided, we may reject the null hypothesis that $P_T = \frac{1}{3}$ at the $\alpha = 0.006$ level (Table A.1) in favor of $P_T > \frac{1}{3}$. Note that the same conclusion was reached using the Harmon test statistic. Suppose we want to estimate $P_T$ by a confidence interval. A $(1 - \alpha)$ 100% degree of confidence interval estimate for $P_T$ is given by

$$\widehat{P}_T \pm z_{\alpha/2}\sqrt{\widehat{P}_T(1 - \widehat{P}_T)/N}. \qquad (10.1\text{-}10)$$

In our example, the 95% confidence interval for $P_T$ is

$$0.55 \pm 1.96\sqrt{(0.55)(0.45)/29,}$$

which estimates $P_T$ within $(0.37, 0.73)$ with a 95% degree of confidence. We can use the S-PLUS code *bhmod* to get the results: $\bar{H}$, $S^2$, $Z$, P-value, and the 95% confidence interval for $P_T$.

**Table 10.4** Degree-of-Difference Scores on Test Steak Sauce against a Control in Stage 2 (Example 10.1-3)

| R (Scores for Correct Selection in Stage 1) | | W (Scores for Incorrect Selection in Stage 1) | |
|---|---|---|---|
| 2 | | 5 | |
| 1 | | 3 | |
| 0 | | 1 | |
| 2 | | 5 | |
| 2 | | 1 | |
| 3 | | 0 | |
| 3 | | 2 | |
| 4 | | 1 | |
| 3 | | 1 | |
| 0 | | 2 | |
| 4 | | 0 | |
| 4 | | 0 | |
| 0 | | 1 | |
| 1 | | | |
| 1 | | | |
| 1 | | | |
| $m$ | 16.00 | $n$ | 13.00 | $N = 29$ |
| $\Sigma R_i$ | 31.00 | $\Sigma W_i$ | 22.00 | |
| $\overline{R}$ | 1.94 | $\overline{W}$ | 1.69 | |
| $\sum R_i^2$ | 91.00 | $\sum W_i^2$ | 72.00 | |

*Note: $\overline{H} = (31.0 + 22.0)/29 = 1.83$; $S^2 = [91.0 + 72.0 - 29(1.83)^2]/29 - 1 = 2.35$; $S = 1.53$.*

```
> bhmod(bhdat1,bhdat2)
[1] 1.82758621 2.36206897 −1.93738311 0.02634926
  0.37072239 0.73272588
> bhdat1
[1] 2 1 0 2 2 3 3 4 3 0 4 4 0 1 1 1
> bhdat2
[1] 5 3 1 5 1 0 2 1 1 2 0 0 1
```

### Gridgeman Two-Stage Model

Let us briefly examine the data in Table 10.4 when they are classified in a $2 \times 2$ table, as shown in Table 10.5. The degree of difference scores in the second stage are classified either under "no difference" or "difference." A score of 0 is classified under "no difference" and a score greater than 0 under "difference." To the 16 correct selections of the odd sample in the first stage, 13 panelists gave degree of difference scores ranging from 1 to 4. Of the 13 panelists who made incorrect selections in the first stage, 10 gave degree of difference scores ranging from 1 to 5. These results suggest that both groups of panelists were fairly consistent with regard to their respective decisions in the second stage. This led Gridgeman (1964, 1970a) to consider a model to account for situations in which panelists with incorrect selections at the first stage continue to make meaningful nonrandom informative second-stage judgments.

Recall that in the modified triangle test, the first stage involves identifying the odd sample from two identical samples. In the second stage a degree of difference score, which measures the extent of the difference between the odd and identical samples, is obtained. As an alternative to the Bradley–Harmon model for quantifying the degree of difference in the second stage, Gridgeman proposed dichotomized responses, either $S = 0$ or $S > 0$, indicating, respectively, "no difference" or "difference." Thus, there are four classes underlying the Gridgeman model, as shown in Table 10.6. The null and alternative hypotheses tested by Gridgeman can be stated as follows:

$H_0$: Panelists make random judgments in both stages.

$H_a$: Panelists make nonrandom judgments in both selection and scoring.

**Table 10.5**  Summary of the Data of Table 10.4 When the Degree of Difference Is Dichotomized into "No Difference" and "Difference" Categories

| Stage 1, Selection | Stage 2, Degree of Difference | | Total |
|---|---|---|---|
| | **No Difference** | **Difference** | |
| Correct | 3 | 13 | 16 |
| Incorrect | 3 | 10 | 13 |
| | 6 | 23 | 29 |

**Table 10.6** Classification of Two-Stage Triangle Test Outcomes for the Gridgeman Model

| | Stage 2, Degree of Difference $S$ | | |
|---|---|---|---|
| **Stage 1, Selection of Odd Sample** | $S = 0$ | $S > 0$ | **Total** |
| Correct | $n_{11}$ | $n_{12}$ | $n_{1.}$ |
| Incorrect | $n_{21}$ | $n_{22}$ | $n_{2.}$ |
| | $n_{.1}$ | $n_{.2}$ | $N$ |

To test the preceding hypotheses, the Gridgeman model uses a suitable probability distribution of weights $W$ on the four classes stated in Table 10.6. A probability distribution of weights can be easily seen when $H_0$ is indeed true; that is, each panelist makes decisions completely randomly in both stages. In the first stage, the probability of correctly selecting the odd sample if the selection is made at random is $P_T = \frac{1}{3}$. In the second stage, again under the assumption of random decisions, $P(S = 0) = P(S > 0) = \frac{1}{2}$. With this explanation, the probability distribution of weights under $H_0$, given in Table 10.7, can be verified.

Obviously, the decision as to the selection of weights $W_{11}$, $W_{21}$, $W_{22}$, and $W_{12}$ has to be subjective rather than statistical in nature. For any set of weights we can find the mean and variance of its probability distribution under $H_0$ as given by

$$\mu_W = (W_{11} + 2W_{21} + 2W_{22} + W_{12})/6 \qquad (10.1\text{-}11)$$

$$\sigma_W^2 = (W_{11}^2 + 2W_{21}^2 + 2W_{22}^2 + W_{12}^2)/6 - \mu_W^2 \qquad (10.1\text{-}12)$$

**Table 10.7** Frequency and Probability Distributions of Weights under $H_0$ for the Gridgeman Model

| Classes | Weights $W$ | Observed Frequencies $n_{ij}$ | Probability Distribution of Weights | A Simple Set of Weights |
|---|---|---|---|---|
| First Stage Correct and $S = 0$ (11) | $W_{11}$ | $n_{11}$ | $\frac{1}{6}$ | 0 |
| First Stage Incorrect and $S = 0$ (21) | $W_{21}$ | $n_{21}$ | $\frac{2}{6}$ | 1 |
| First Stage Incorrect and $S > 0$ (22) | $W_{22}$ | $n_{22}$ | $\frac{2}{6}$ | 2 |
| First Stage Correct and $S > 0$ (12) | $W_{12}$ | $n_{12}$ | $\frac{1}{6}$ | 3 |
| | | $N$ | 1 | |

Gridgeman considered various possible choices for the weights. A simple and reasonable set of weights is given in the last column of Table 10.7. For this choice of weights, the mean and variance of their distribution under $H_0$ are $\mu_W = \frac{9}{6} = \frac{3}{2}$ and $\sigma_W^2 = \left(\frac{19}{6}\right) - \left(\frac{9}{4}\right) = \frac{11}{12}$, respectively.

Let $n_{ij}$ be the number of responses in the $(i, j)$th class and $N$ the total number of responses. If we use the simple set of weights, then we can calculate the total weight $S$ from

$$S = 0(n_{11}) + 1(n_{21}) + 2(n_{22}) + 3(n_{12}). \tag{10.1-13}$$

When the null hypothesis is true and the panelists respond independently of each other, then the mean $\mu_S$ and the variance $\sigma_S^2$ of the distribution of $S$ are obtained as follows:

$$\mu_S = N\mu_W = 3N/2, \tag{10.1-14}$$

$$\sigma_S^2 = N\sigma_W^2 = 11N/12. \tag{10.1-15}$$

For large $N$ we may use the $Z$ score of $S$, which approximates the standard normal distribution. Thus, we have

$$Z = (S - \mu_S)/\sigma_S, \tag{10.1-16}$$

where $\sigma_S = \sqrt{11N/12}$.

---

Gridgeman (1964) reported results of a two-stage triangle test on the relative tenderness of two poultry meats $A$ and $B$ from seven experiments. The data are summarized in Table 10.8.

**Example 10.1-4**

In our notation $A > B$ means that $A$ is more tender than $B$, and $A < B$ means $B$ is more tender than $A$. Of $N = 246$ panelists, 92 were able to pick the odd sample, and the rest did not. In the second stage, there was a clear

**Table 10.8** Data Summary for a Two-Stage Triangle Test on the Relative Tenderness of Poultry Meats (Example 10.1-4)

| Stage 1 | Stage 2 | | |
|---|---|---|---|
| | $A > B$ | $A < B$ | Total |
| Correct Selection | 56 | 36 | 92 |
| Incorrect Selection | 95 | 59 | 154 |
| | 151 | 95 | 246 |

separation between the odd and identical samples, sample $A$ being more tender. We want to test the following null and alternative hypotheses:

$H_0$:  Judgments are random (meats $A$ and $B$ are equally tender).
$H_a$:  Meats $A$ and $B$ are not equally tender.

Notice that $H_a$ is two-sided. We first compute $S$ from (10.1-13). That is,

$$S = 0(56) + 1(95) + 2(59) + 3(36) = 321.$$

Then from (10.1-14, 10.1-15)

$$\mu_S = 3(246)/2 = 369 \quad \text{and} \quad \sigma_S = \sqrt{11(246)/12} = 15.02.$$

Now from (10.1-16), the $Z$ score of $S$ is

$$Z = (321 - 369)/15.02 = -3.2$$

which corresponds to a small significance probability equal to $2(0.0007) = 0.0014$. Therefore, we conclude that the two meats are not equally tender. It follows from the experimental outcomes that meat $A$ is more tender than $B$ (151 versus 95).

```
> gridgeman(matrix(c(56,95,36,59),2,2))
[1] 246.0000 321.0000 369.0000 225.5000 -3.1965 0.0014
```

**Example 10.1-5**

Let us also consider a case in which $N$ is small; consequently, we cannot use the $Z$ statistic. The critical values of the $S$ statistic are given for $N = 2, 4, 6, 8$, and so on, up to $N = 30$ in Table A.22. For illustration we use the data shown in Table 10.9 and find that

$$S = 0(12) + 1(10) + 2(4) + 3(2) = 24.$$

**Table 10.9** Data Summary for a Two-Stage Triangle Test (Example 10.1-5)

| Stage 1 | Stage 2 | | |
| --- | --- | --- | --- |
| | $A > B$ | $A < B$ | Total |
| Correct Selection | 12 | 2 | 14 |
| Incorrect Selection | 10 | 4 | 14 |
| | 22 | 6 | 28 |

for the one-sided alternative of the duo–trio test, the critical values are 14, 15, and 16, corresponding to levels of significance 0.10, 0.05, and 0.01, respectively. Hence, at the $\alpha = 0.05$ significance level, we can reject the hypothesis of no difference between the control and test samples in favor of the alternative that there is a significant difference. Note that the null hypothesis is not rejected at a significance level of $\alpha = 0.01$ or less. We can also use Table A.25. The critical value is 15 for the duo–trio test for sample size 20 and $\alpha = 0.05$. Because the observed number equals the critical value, the null hypothesis is rejected.

Suppose we use $Z$ from (10.1-19) for our test statistic. We find

$$Z = (30 - 20)/\sqrt{20} = 2.25.$$

Since at $\alpha = 0.05$, $z_{1-\alpha} = z_{0.95} = 1.645$, we can reject the null hypothesis. But note again that at $\alpha = 0.01$, $z_{1-\alpha} = z_{0.95} = 2.32$, which is greater than 2.25. Hence, at a significance level of 0.01 or less, the null hypothesis cannot be rejected. The same conclusion was reached before, when using Table A.23, based on the binomial distribution.

The sign test (Chapter 9) can also be used for testing the null hypothesis of the duo–trio test. For the sign test, the test statistic is the smaller one of the numbers of correct and incorrect selections. Hence, the sign test statistic for the data of our example is observed to be 5. From Table A.18, when $n = 20$ and the critical value $r_0 = 5$, the significance level for the one-sided alternative is $0.0414/2 = 0.0207$.

## Pair Difference Test (2-AFC)

The pair difference test is a paired comparison test. This test is sometimes referred to as a 2-alternative force choice (2-AFC) test. The ISO standard (ISO 5495-1983) and ASTM standard (ASTM E2164-2001) describe the test. The pair test for detecting difference (nondirectional difference selection) is also used for detecting the direction of difference (directional difference selection). Hence, for its valid applications, it is important that the framing of the questionnaire be clear and understood by members of the panel. In the pair difference test, only two samples are compared. When there are more than two samples to be compared in pairs, the method of paired comparison is used. This method is discussed in Chapter 11.

The experimental procedure of a pair difference test is as follows: two properly coded samples are presented simultaneously or successively in random order. The panelists are instructed to pick the sample that more closely meets a specified criterion. Examples of questionnaires for nondirectional and directional pair difference tests are shown in Figures 10.4 and 10.5, respectively. To make the pair test effective, all the members of the panel must understand the selection criterion to mean the same thing. That is, words like *saltiness, sweetness, spoiled flavor, hardness,* or some other familiar criterion must be uniformly understood. Unlike the duo–trio test, the pair difference test does not have reference standard sample. Table 10.11 shows the basic design. Note that the basic design is balanced for the order effect. The basic design may be repeated $r$ times to obtain $N = 2r$ comparisons. If there are $r$ panelists, each panelist may be asked to provide two comparisons of the basic design.

■ **FIGURE 10.4** Questionnaire for a nondirectional pair difference test. *Source:* Amerine et al. (1965).

Set_____

Date_____

Name_____

| Pair | Check one | |
|:---:|:---:|:---:|
| | There is a difference | There is no difference |
| A | | |
| B | | |
| C | | |

■ **FIGURE 10.5** Questionnaire for a directional pair difference test.

Name_____          Date_____

Product_____

Circle the sample number within each pair that you prefer in <u>saltiness</u>.

| Pair | Sample number | |
|:---:|:---:|:---:|
| A | _____ | _____ |
| B | _____ | _____ |
| C | _____ | _____ |
| D | _____ | _____ |

**Table 10.11** Basic Design for the Pair Difference Test

| Sample Set | Order of Unknowns |
|:---:|:---:|
| 1 | A, B |
| 2 | B, A |

Also in the pair test, the probability of correct selection by chance alone is $\frac{1}{2}$. Therefore, the null hypothesis is

$$H_0: \quad P = 1/2. \qquad (10.1\text{-}20)$$

If, in a pair test, the panel members are asked to state whether a "difference" or "no difference" exists (Figure 10.4), then the alternative hypothesis would be one-sided; that is,

$$H_a: \quad P > 1/2. \qquad (10.1\text{-}21)$$

If, on the other hand, a pair test is designed to have panelists state only, for example, their preferences for one object over the other, then the alternative hypothesis is directional (Figure 10.5). For such pair tests the alternative hypothesis is two-sided; that is,

$$H_a: \quad P \neq 1/2. \qquad (10.1\text{-}22)$$

If the panel members are making comparisons with respect to the amount of a certain attribute, then the test statistic is the number $m$ of correct selections for testing the null hypothesis against the one-sided alternative. When the panel members are making preference comparisons, the test statistic is the number $m$ of preferences for one object over the other for testing the null hypothesis against the two-sided alternative. Regardless of the alternative hypothesis, the underlying statistical model of $m$ is the binomial with parameters $P$ and $N$. We can also use the $Z$ score of $m$ as defined by (10.1-19) for the test statistic, which has approximately the standard normal distribution.

---

**Example 10.1-7**

The data in Table 10.12 are used to illustrate a directional pair difference test. The instruction given to panelists may be as follows: "Please taste the two samples and state your preferences by writing down the code number of the sample you prefer in the space provided." We note that $m = 6$, which is less than the critical value 9 found from Table A.23 for the $\alpha = 0.05$ significance level. We cannot reject the null hypothesis of no preferences at a significance level of 0.05 or less. We may use the $Z$ score, which is

**Table 10.12**  Results of a Pair Difference Test (Example 10.1-7)

| Panelists | Orders | Judgments |
|-----------|--------|-----------|
| 2 | AB | A |
| 9 | BA | A |
| 8 | AB | A |
| 4 | BA | B |
| 3 | AB | B |
| 5 | BA | A |
| 1 | AB | B |
| 6 | BA | B |
| 7 | AB | A |
| 10 | BA | A |

*Number of judgments preferring A: m = 6.*
*Number of judgments preferring B: 10 − m = 4.*

$$Z = (12 - 10)/\sqrt{10} = 0.63,$$

and find that its significance probability is $2(0.264) = 0.528$. Hence, $H_0$ cannot be rejected at any level less than 0.528. We can also find from Table A.25 that the critical value is 9 for $n = 10$, $\alpha = 0.05$ for a one-sided 2-AFC test. Hence, we cannot reject the null hypothesis based on the data.

**Example 10.1-8**  We may also use the data of the preceding example to illustrate a nondirectional difference test. A questionnaire for a nondirectional pair difference test is shown in Figure 10.4. When the pair difference test is nondirectional, the test statistic is the number $m$ of agreeing judgments for differentiation. If, in Example 10.1-7, we treat the number of judgments preferring sample $A$ as agreeing judgments that a difference exists, then $m = 6$. We can proceed to test the null hypothesis as before, except that for a nondirectional pair difference test, the alternative is two-sided. Hence, the observed number $m$ of agreeing judgments is compared with the critical values obtained from Table A.23 or Table A.25 corresponding to the two-sided alternative.

## Three-Alternative Forced-Choice Test (3-AFC)

The 3-alternative forced-choice (3-AFC) test is developed in the theory of signal detection (Green and Swets, 1966). There are three stimuli: *A, B,* and *B;* or *A, A,* and *B,* where *A* is stronger than *B.* The panelist is presented with one of the sets *ABB, BAB,* or *BBA* with the instruction to select the strongest stimulus of the three. Or the panelist is presented with one of the sets *AAB, ABA,* or *BAA* with the instruction to select the weakest stimulus of the three. The panelist has to select a stimulus even if he or she cannot identify the one with the strongest or weakest stimulus. The design of the 3-AFC method is the same as that of the conventional triangle method, but the instructions are different (see Figure 10.6). One is to select the odd sample; the other is to select the "strongest" or "weakest" sample. The different instructions may affect profoundly the panelist's performances and testing powers of the two tests. For a specified sensory intensity, i.e., a perceived magnitude of a stimulus, the proportion of correct responses in a 3-AFC test will be larger than that in a triangle test. The problem can be best interpreted by the Thurstonian model, which we will discuss in Chapter 11.

The test statistic for the 3-AFC test is the number $m$ of correct selections out of $N$ comparisons. If all comparisons are being made independently, then the test statistic follows the binomial distribution with parameters $P$ and $N$. The null hypothesis is

$$H_0: \quad P = 1/3, \tag{10.1-23}$$

and the alternative hypothesis is

$$H_a: \quad P > 1/3. \tag{10.1-24}$$

Name _____ Date _____

This is a 3-AFC Test. You must make a choice.

One of the three samples is the most sweet (bitter, etc.).

Write down the number of the sample _____.

Comments: _____
_____
_____

■ **FIGURE 10.6** Questionnaire for a 3-AFC test.

The probability of correct responses in a null hypothesis is 1/3 because, if samples A and B have the same sensory intensity, the probability of selecting A as "stronger" for the sets ABB, BAB, and BBA or selecting B as "weaker" for the sets AAB, ABA, and BAA is just the chance probability 1/3. The test is always one-sided because the probability of correct responses cannot be smaller than the chance probability. The minimum number of correct responses for the 3-AFC test can be found in Table A.25. For a given significance level $\alpha$ and sample size $N$, the critical number can also be obtained by using, for example, the S-PLUS program $qbinom (1 - \alpha, N, 1/3) + 1$.

---

**Example 10.1-9**

A new beverage was produced with decreased sugar. In order to investigate whether consumers could detect the difference in sweetness between the new beverage and the current one, researchers conducted a difference test using the 3-AFC. One hundred panelists were drawn randomly, and a significance level of 0.05 was specified. Forty-six panelists of the total 100 panelists correctly selected the new product as a less sweet beverage. The minimum number of correct responses for the test with $\alpha = 0.05$ and $N = 100$ is 42 from Table A.25. Because $46 > 42$, we conclude at $\alpha = 0.05$ significance level that consumers perceived the new beverage is less sweet than the current product. We can also obtain the critical value using the S-PLUS program $qbinom$ as follows:

```
> qbinom(0.95,100,1/3) + 1
[1] 42
```

---

## A–Not A Test

The A–Not A method in sensory analysis (Peryam, 1958) is also called the yes-no method in the theory of signal detection (Green and Swets, 1966). The ISO standard (ISO 8588-1987) describes the test. The two samples are A and Not A. One sample, which is either A or Not A, is presented to each panelist. The panelist is asked whether the received sample is "A" or "Not A."

The A–Not A method is a response-biased method. This means that the panelists may have different criteria and probabilities of response "A" for sample A and different criteria and probabilities of response "A" for sample Not A even if they have the same sensitivity to the same sensory difference. The Thurstonian model and the theory of signal detection rescue the method as a difference test. The model and theory demonstrate that response bias in the method is independent of sensitivity to sensory difference.

In fact, if we regard the responses "A" and "Not A" as "1" and "0" or "yes" and "no," any experiment for comparison of two samples (A and Not A) with binary responses ("1" and "0") for each of the samples can be treated as a generalized A–Not A test. For the conventional A–Not A method, the test is one-sided because the proportion of response "A" for sample A should always be larger than or equal to the proportion of response "A" for sample Not A. However, for the generalized A–Not A method, the test can be either one-sided or two-sided.

There are different designs for the A–Not A test. One is the monadic design, in which each panelist evaluates one sample that is A or Not A. A questionnaire for the monadic A–Not A test is shown in Figure 10.7. The total numbers of panelists who evaluate sample A and sample Not A are fixed in advance. This design involves comparisons of two independent proportions of response "1" for the two samples. If two samples for testing are the same—i.e., the two samples have a similar sensory characteristic—the probability of response "A" for sample A, $P_A$, and the probability of response "A" for sample Not A, $P_N$, should be the same. Hence, the null hypothesis of the test is

$$H_0: \quad P_A = P_N \qquad\qquad (10.1\text{-}25)$$

Assessor Number _____ Date _____    ■ **FIGURE 10.7** Questionnaire for A–Not A test.

Instructions:

1. Before taking this test, familiarize yourself with the flavor (taste) of the sample "A" and "Not A", which are available from the attendant.

2. You will receive a sample of either "A" or "Not A". Evaluate the samples and indicate whether the sample is "A" or "Not A".

|         Code          |          "A"          |         "Not A"        |
| --------------------- | --------------------- | --------------------- |
|                       |                       |                       |

Comments: _____
_____
_____

versus the alternative hypothesis:

$$H_a: \quad P_A > P_N. \tag{10.1-26}$$

The data for the test should be given in a $2 \times 2$ table with columns of samples A and Not A and rows of responses "A" and "Not A." This is a two-proportion homogeneity test. Because $P_A$ should always be larger than $P_N$ if the null hypothesis is rejected, the test is one-sided. Pearson's chi-square statistic should be used for the test:

$$X^2 = \sum \frac{(O - E)^2}{E}, \tag{10.1-27}$$

where $O$ is the observed number of each cell in a $2 \times 2$ table and $E$ is the expected value for each cell. The expected values can be calculated by $E_{ij} = \frac{r_i c_j}{N}$, where $r_i$ is the total for row $i$ and $c_j$ is the total for column $j$. With Yates' continuity correction, the statistic is

$$X^2 = \sum \frac{(|O - E| - 0.5)^2}{E}. \tag{10.1-28}$$

The test statistic follows asymptotically a chi-square distribution with 1 DF. The critical values of the one-sided and two-sided tests can be found in Table 9.2. Some built-in programs (e.g., *chisq.test*, *fisher.test*) in S-PLUS can be used for the test. However, one should note that the P-value of the output of the programs is for a two-sided test. The P-value should be divided by 2 for a one-sided test.

Another alternative is the mixed design, in which total sample size is fixed, but the sample sizes for samples A and Not A are not fixed in advance. A questionnaire for the mixed designed A–Not A is the same as that in Figure 10.7. Each panelist randomly draws a sample—either A or Not A—from a sample pool. The number of samples in the sample pool should be much larger than the number of panelists. In this design, both the sample and response are random variables. The null hypothesis in the design is $H_0$: sample and response are independent of each other. The alternative hypothesis is $H_a$: sample and response are not independent. This is an independence test. However, the practice of statistical analysis for the homogeneity test and the independence test is similar. The independence test should be a two-sided test.

If each panelist evaluates both sample A and Not A, the evaluation involves a comparison of two correlated proportions; in this case, McNemar's test should be used. See, for example, Meilgaard et al. (1991) and Bi (2006, Section 2.3.3) for the A–Not A test in this design.

Assessor Number _____ Date _____

Instructions:

1. Before taking this test, familiarize yourself with the flavor (taste) of the sample "A" and "Not A", which are available from the attendant.

2. You will receive two samples. Evaluate the samples in the order that they are presented from left to right and indicate whether the sample evaluated is "A" or "Not A".

| Code | "A" | "Not A" | Code | "A" | "Not A" |
|------|-----|---------|------|-----|---------|
| ____ | ____ | ____ | ____ | ____ | ____ |

Comments: _____

_____

_____

However, one should note that, for the conventional A–Not A method, the paired design is valid only in a situation in which panelists are not aware that the pair of samples is certainly different. A questionnaire for the paired designed A–Not A is shown in Figure 10.8. If the panelists know that they receive a pair of samples, A and Not A, they will give the response "A" for a sample; then they will certainly give a "Not A" response for the other sample. In this situation, the method becomes a forced-choice method without response bias.

For comparison of purchase intent for two products, the 5-point scale is widely used: 5 = Definitely would buy, 4 = Probably would buy, 3 = Maybe/maybe not buy 2 = Probably would not buy, 1 = Definitely would not buy. An often-used procedure is to collapse the 5-point scale data into binary data. For example, the responses 5 and 4 (i.e., the top two boxes) are assigned 1, and other responses are assigned 0. The comparison of purchase intent for the two products involves comparison of two proportions of 1. If each panelist evaluates both of the products, the two proportions are correlated. Analysis of this kind of data can be regarded as analysis of paired-designed A–Not A test data. The test statistic is

$$Z = \frac{x_{10} - x_{01}}{\sqrt{x_{10} + x_{01}}} \qquad (10.1\text{-}29)$$

where $x_{10}$ is the number of panelists who give response "A" for sample A and response "Not A" for sample Not A; $x_{01}$ is the number of panelists who give response "Not A" for sample A and response "A" for sample Not A. In the generalized A–Not A test, $x_{10}$ is the number of panelists who give response "1" or "yes" for sample A and response "0" or "no" for sample Not A; $x_{01}$ is the number of panelists who give response "0" or "no" for sample A and response "1" or "yes" for sample Not A. The test statistic (10.1-29) follows asymptotically the standard normal distribution.

**Example 10.1-10**

Some ingredients of a product were changed. The manufacturer wanted to know whether the flavor of the product decreased. A difference test using the A–Not A method was conducted with 200 consumers who were users of the product. The Type I error $\alpha = 0.05$ was selected. The testing data are shown in Table 10.13, where A is the current product and Not A is the changed product.

The estimated proportions are $\widehat{P}_A = 0.6$ and $\widehat{P}_B = 0.45$. The chi-square statistic can be calculated as follows:

$$X^2 = \frac{(60 - 100 \times 105/200)^2}{100 \times 105/200} + \frac{(40 - 100 \times 95/200)^2}{100 \times 95/200}$$

$$+ \frac{(45 - 100 \times 105/200)^2}{100 \times 105/200} + \frac{(55 - 100 \times 95/200)^2}{100 \times 95/200}$$

$$= 4.51.$$

With Yates' continuity correction, the statistic value is 3.93. Because 4.51 (or 3.93) > 2.706, we can conclude at $\alpha = 0.05$ level of significance that the flavor of the new product significantly changed. The same value can be obtained by using the S-PLUS built-in program *chisq.test*. The P-value for a one-sided test should be $0.0337/2 = 0.0168$. When we use Pearson's chi-square test with Yates' continuity correction, the P-value for a one-sided test should be $0.0474/2 = 0.0237$. When we use Fisher's exact test *fisher.test*, the P-value should be $0.0472/2 = 0.0236$ for a one-sided test.

**Table 10.13**  Data for an A–Not A Test (Example 10.1-10)

| | | Sample | | |
|---|---|---|---|---|
| | | **A** | **Not A** | **Total** |
| Response | "A" | 60 | 45 | 105 |
| | "Not A" | 40 | 55 | 95 |
| | Total | 100 | 100 | 200 |

```
> chisq.test(matrix(c(60,40,45,55),2,2),correct=F)
    Pearson's chi-square test without Yates' continuity
      correction
data: matrix(c(60, 40, 45, 55), 2, 2)
X-square = 4.5113, df = 1, p-value = 0.0337
> chisq.test(matrix(c(60,40,45,55),2,2),correct=T)
  Pearson's chi-square test with Yates' continuity
    correction
data: matrix(c(60, 40, 45, 55), 2, 2)
X-square = 3.9298, df = 1, p-value = 0.0474
> fisher.test(matrix(c(60,40,45,55),2,2))
  Fisher's exact test
data: matrix(c(60, 40, 45, 55), 2, 2)
p-value = 0.0472
alternative hypothesis: two.sided
```

One hundred consumer panelists participated in a test of purchase intent for two products A and B. Each panelist evaluated both of the products. The 5-point purchase intent scale was used, and the data for the responses in the top two boxes ("5" and "4", i.e., "Buy" responses) and the responses in the other boxes ("3", "2," and "1," i.e., "Not buy") are shown in Table 10.14. This test can be regarded as a paired A–Not A test. The estimated purchase intent in terms of the proportions of the top two boxes are 0.64 and 0.48, respectively, for products A and B. The value of the test statistic (10.1-29) is

**Example 10.1-11**

$$Z = \frac{24 - 8}{\sqrt{24 + 8}} = 2.828.$$

The associated *P*-value = 0.0047 for the two-sided test. A significant difference between the two products was found for purchase intent. Using the S-PLUS built-in function *mcnemar.test*, we can obtain the same result:

**Table 10.14** Data Summary for a Paired A–Not A Test (Example 10.1-11)

|  |  | Product B | | |
|---|---|---|---|---|
|  |  | "Buy" | "Not Buy" | Total |
| Product A | "Buy" | 40 | 24 | 64 |
|  | "Not Buy" | 8 | 28 | 36 |
|  | Total | 48 | 52 | 100 |

```
> mcnemar.test(cbind(c(40,8),c(24,28)),correct=F)
McNemar's chi-square test without continuity correction
data: cbind(c(40, 8), c(24, 28))
McNemar's chi-square = 8, df = 1, p-value = 0.0047
```

## Same-Different Test

The same-different test method is sometimes referred to as the simple-difference test (e.g., Meilgaard et al., 1991); see Figure 10.9. The ASTM Standard (ASTM E2139-2005) describes the test. For comparison of samples $A$ and $B$, there are two concordant pairs—$AA$, $BB$—and two discordant pairs—$AB$ and $BA$. Each panelist is presented either a concordant pair or a discordant pair and asked if the pair is "concordant" or "discordant." The same-different method is also a response-biased method. This means that the panelists may have different criteria and probabilities of response "Same" for the concordant pair ($P_S$) and different criteria and probabilities of response "Same" for the discordant pair ($P_D$) even if they have the same sensitivity to the same sensory difference.

The null hypothesis of the test is

$$H_0: \quad P_S = P_D \tag{10.1-30}$$

versus the alternative hypothesis:

$$H_a: \quad P_S > P_D \tag{10.1-31}$$

■ **FIGURE 10.9** Questionnaire for a same-different test. *Source:* ASTM E 2139-05.

Assessor Number _____ Date _____

Instructions: You will receive two samples. These samples may be the same or they may be different.

Evaluate the samples in the order that they are presented from left to right.

Indicate whether the samples are the same or different by checking the appropriate line below

The two samples are the same _____

The two samples are different _____

Comments: _____

_____

_____

The designs and statistical analyses for the same-different test are the same as those for the A–Not A test. However, one should keep in mind that the two tests involve different cognitive strategies and decision rules. For a same sensory difference, the same-different test has lower power than the A–Not A test to detect the difference. The same-different test should be used in a situation in which there is a large sensory difference or a large sample size is available.

---

For a monadic same-different test with 200 panelists, each panelist received a sample pair either concordant or discordant. The data for this example are shown in Table 10.15. The estimated proportions are $\widehat{P}_S = 0.7$ and $\widehat{P}_D = 0.6$. The chi-square statistic value with Yates' continuity correction is 1.78, and the P-value for a one-sided test is $0.1821/2 = 0.091$ when we use the S-PLUS program *chisq.test*. The P-value for a one-sided test is $0.1819/2 = 0.091$ when we use the S-PLUS program *fisher.test*.

**Example 10.1-12**

```
> chisq.test(matrix(c(70,30,60,40),2,2))
    Pearson's chi-square test with Yates' continuity
      correction
data: matrix(c(70, 30, 60, 40), 2, 2)
X-square = 1.7802, df = 1, p-value = 0.1821
> fisher.test(matrix(c(70,30,60,40),2,2))
        Fisher's exact test
data: matrix(c(70, 30, 60, 40), 2, 2)
p-value = 0.1819
alternative hypothesis: two.sided
```

**Table 10.15**  Data for a Same-Different Test (Example 10.1-12)

|  |  | Sample Pair | | |
|---|---|---|---|---|
|  |  | **Concordant Pair** | **Discordant Pair** | **Total** |
| Response | "Same" | 70 | 60 | 230 |
|  | "Different" | 30 | 40 | 170 |
|  | Total | 100 | 100 | 200 |

---

## Comparisons of Difference Test Designs

Depending on our purpose and experimental considerations, if we wanted to use a sensory test, we would use one of the tests discussed in the preceding sections of this chapter. When nonstatistical considerations allow the use of one of six difference tests, we may choose the one that is optimum statistically. To evaluate the statistical optimality of a test, we use its power function. The power function of a test gives the probabilities of rejecting the null hypothesis when indeed it is false. When the power function of a test dominates the power function of any other test, it is optimum statistically.

### Comparisons of the Forced-Choice Methods

Hopkins and Gridgeman (1955) studied the power of the triangle, duo–trio, and pair tests for the detection of flavor intensity. The results of their study are that for sensory detection of flavor intensity, the triangle test is the most powerful, followed by the duo–trio and pair difference tests, in that order. Ura (1960) considered the power functions of the three tests when they are used for selecting judges. His findings ranked the three difference tests in the same order as reported in the Hopkins and Gridgeman study. David and Trivedi (1962) studied the power functions of the three difference tests for detecting and quantifying differences between average sensations when the probability distribution of the resulting sensations from a sample is either normal or rectangular. In this case the pair difference test is the most powerful, followed by the triangle test and the duo–trio test, in that order. See also Byer and Abrams (1953), Gridgeman (1955a), and Filipello (1956). Bradley (1963) also compared the triangle test with the duo–trio test. He found that the triangle test would need fewer comparisons than the duo–trio test for obtaining the same amount of power. It can be summarized that for testing the recognition of differences, for example, testing the ability of judges, the triangle test is more powerful than the duo–trio and pair tests. However, for quantifying the degree of difference, the pair test is the most powerful.

We may also examine some relationships between $P_T$ and $P_D$, the probabilities of correct selection in the triangle and duo–trio tests. Consider two very different samples so that the probability of recognizing the difference sensorily is $Q$. Hopkins and Gridgeman (1955a) showed that $P_T$ and $P_D$ are related through $Q$ as follows:

$$P_T = (1 + 2Q)/3, \tag{10.1-32}$$

$$P_D = (1 + Q)/3. \tag{10.1-33}$$

From (10.1-32) and (10.1-33), Bradley (1963) showed that the value of $P_D$ for a given $P_T$ is

$$P_{DH} = (1 + 3P_T)/4. \qquad (10.1\text{-}34)$$

Since the preceding relationship follows from the reasoning of Hopkins and Gridgeman, the resulting $P_D$ in the relationship for a given $P_T$ is denoted $P_{DH}$.

David and Trivedi (1962) also argued for a relationship between $P_D$ and $P_T$. Their relationship $P_D$ for a given $P_T$ is denoted by $P_{DD}$. The relationship is

$$P_{DD} = 2P_T/(1 + P_T). \qquad (10.1\text{-}35)$$

In addition to these relationships between $P_T$ and $P_D$, there are relationships between these probabilities and $\theta$, the standardized degree of difference between the odd and identical samples. Table A.24 is reproduced from Bradley (1963) to show these relationships. Values of $P_{DH}$ and $P_{DD}$ were obtained using (10.1-34) and (10.1-35). An application of this table is as follows. Suppose, in a triangle test, $m$ correct judgments were observed out of $N$ comparisons. Then $P_T$ is estimated by $\widehat{P}_T = m/N$. For each $P_T$ value in the table, one finds $\widehat{\theta}$, an estimated value of $\theta$ corresponding to $\widehat{P}_T$. Obviously, large values of $\theta$ correspond to large values of $P_T$ and $P_D$. When there is no difference between the odd and identical samples, $\theta = 0.0$, and the probabilities of correct selection $P_T$ and $P_D$ are $\frac{1}{3}$ and $\frac{1}{2}$, respectively. When $\theta \geq 7.0$, the probabilities of correct selection in the triangle and the duo–trio tests are very close and approach 1.0.

Ennis (1993) compared four forced-choice methods based on a same sensory difference in terms of Thurstonian $\delta$ (or $d'$). The meaning of the notation $\delta$ (or $d'$) is the same as $\theta$ in Table A.24. We will discuss Thurstonian $\delta$ (or $d'$) in Chapter 11. Bi (2006, Chapter 3) gave a more detailed discussion on the topic of statistical power analysis for discrimination tests, including the forced-choice methods and the methods with response bias. The power of a discrimination test using a forced-choice method can be calculated with

$$\text{Power} = \Pr[Z > \lambda] = 1 - \Phi(\lambda), \qquad (10.1\text{-}36)$$

where $\Phi(\lambda)$ is the distribution function of the standard normal distribution;

$$\lambda = \frac{\sigma_0 z_{1-\alpha}/\sqrt{n} - p_1 + p_0}{\sigma_1/\sqrt{n}},$$

$$\sigma_0 = \sqrt{p_0(1-p_0)};$$

$$\sigma_1 = \sqrt{p_1(1-p_1)};$$

and $p_0 = 1/2$ for the 2-AFC and duo–trio methods and $p_0 = 1/3$ for the 3-AFC and triangle methods. Here, $p_1$ is an assumed probability of correct responses in a discrimination test using a forced-choice method corresponding to an assumed sensory difference in terms of Thurstonian $\delta$ (or $d'$). Table 10.16 gives $p_1$ values corresponding to

**Table 10.16** $p_c$ Values Corresponding to Thurstonian $\delta$ for Forced-Choice Methods

| $\delta$ | 2-AFC | DT | 3-AFC | TRI | $\delta$ | 2-AFC | DT | 3-AFC | TRI |
|---|---|---|---|---|---|---|---|---|---|
| 0.30 | 0.58 | 0.51 | 0.42 | 0.34 | 1.20 | 0.80 | 0.61 | 0.69 | 0.45 |
| 0.35 | 0.60 | 0.51 | 0.44 | 0.34 | 1.25 | 0.81 | 0.62 | 0.70 | 0.46 |
| 0.40 | 0.61 | 0.51 | 0.45 | 0.35 | 1.30 | 0.82 | 0.63 | 0.72 | 0.47 |
| 0.45 | 0.62 | 0.52 | 0.47 | 0.35 | 1.35 | 0.83 | 0.64 | 0.73 | 0.48 |
| 0.50 | 0.64 | 0.52 | 0.48 | 0.36 | 1.40 | 0.84 | 0.65 | 0.74 | 0.49 |
| 0.55 | 0.65 | 0.53 | 0.50 | 0.36 | 1.45 | 0.85 | 0.65 | 0.75 | 0.50 |
| 0.60 | 0.66 | 0.53 | 0.51 | 0.37 | 1.50 | 0.86 | 0.66 | 0.77 | 0.51 |
| 0.65 | 0.68 | 0.54 | 0.53 | 0.37 | 1.55 | 0.86 | 0.67 | 0.78 | 0.52 |
| 0.70 | 0.69 | 0.54 | 0.54 | 0.38 | 1.60 | 0.87 | 0.68 | 0.79 | 0.53 |
| 0.75 | 0.70 | 0.55 | 0.56 | 0.38 | 1.65 | 0.88 | 0.69 | 0.80 | 0.54 |
| 0.80 | 0.71 | 0.55 | 0.57 | 0.39 | 1.70 | 0.89 | 0.70 | 0.81 | 0.55 |
| 0.85 | 0.73 | 0.56 | 0.59 | 0.40 | 1.75 | 0.89 | 0.71 | 0.82 | 0.56 |
| 0.90 | 0.74 | 0.57 | 0.60 | 0.40 | 1.80 | 0.90 | 0.71 | 0.83 | 0.57 |
| 0.95 | 0.75 | 0.58 | 0.62 | 0.41 | 1.85 | 0.90 | 0.72 | 0.84 | 0.58 |
| 1.00 | 0.76 | 0.58 | 0.63 | 0.42 | 1.90 | 0.91 | 0.73 | 0.85 | 0.59 |
| 1.05 | 0.77 | 0.59 | 0.65 | 0.43 | 1.95 | 0.92 | 0.74 | 0.86 | 0.59 |
| 1.10 | 0.78 | 0.60 | 0.66 | 0.43 | 2.00 | 0.92 | 0.75 | 0.87 | 0.60 |
| 1.15 | 0.79 | 0.61 | 0.68 | 0.44 | | | | | |

Note: 2-AFC: 2-alternative forced-choice method.
3-AFC: 3-alternative forced-choice method.
DT: Duo-trio method.
TRI: Triangle method.

Thurstonian $\delta$ from 0.30 to 2.00 with 0.05 increments. For example, for $\delta = 1$, $p_1 = 0.76$, 0.58, 0.63, and 0.42, respectively, for the 2-AFC, duo–trio, 3-AFC, and triangle methods.

The sample sizes for a specified power can be calculated as

$$n = \frac{(\sigma_1 z_{power} + \sigma_0 z_{1-\alpha})^2}{(p_1 - p_0)^2} \tag{10.1-37}$$

For continuity correction, the effective sample size should be adjusted by $n_e$:

$$n_e = \frac{2}{p_1 - p_0} \tag{10.1-38}$$

Specifically, $n - n_e$ should be used for estimating power and $n + n_e$ for estimating sample size. We can use the S-PLUS built-in program *binomial. sample.size* to get the results with or without continuity correction.

The comparisons of the four forced-choice methods are shown in Figure 10.10. One should note that, for a same sensory difference, the largest power is for the 3-AFC test and the smallest power is for the duo–trio test.

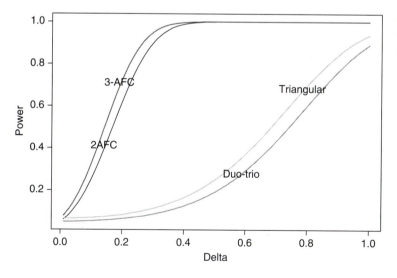

■ **FIGURE 10.10** Powers of difference tests using the four forced-choice methods, $n = 300$, $\alpha = 0.05$.

**Example 10.1-13**

For a discrimination test using the conventional triangle method with sample size $n = 100$, $\alpha = 0.05$, we want to know how much the power can be reached for a specified sensory difference $\delta = 1$, i.e., $p_1 = 0.42$ from Table 10.16 or Table A.24. According to (10.1-36), we get

$$\lambda = \frac{\sqrt{2/9} \times 1.64/\sqrt{100} - 0.42 + 1/3}{\sqrt{0.42 \times 0.58}/\sqrt{100}} = -0.1848;$$

hence, the power is $1 - \Phi(-0.1848) = 0.57$. For continuity correction,

$$n_e = \frac{2}{0.42 - 1/3} = 23,$$

the effective sample size is $n = 100 - 23 = 77$, and

$$\lambda = \frac{\sqrt{2/9} \times 1.64/\sqrt{77} - 0.42 + 1/3}{\sqrt{0.42 \times 0.58}/\sqrt{77}} = 0.0303;$$

the power is then $1 - \Phi(0.0303) = 0.49$. Using the S-PLUS program *binomial.sample.size*, we can obtain the same result:

```
> binomial.sample.size(p = 1/3, p.alt = 0.42,n1 = 100,
alternative = "great",correct = F)
    p.null p.alt delta alpha power n1
1 0.3333333 0.42 0.08666667 0.05 0.5733592 100
> binomial.sample.size(p = 1/3, p.alt = 0.42,n1 = 100,
alternative = "great",correct = T)
    p.null p.alt delta alpha power n1
1 0.3333333 0.42 0.08666667 0.05 0.4876556 100
```

If we want to know the sample size needed to reach the power of 0.8 in this case, we can calculate it from (10.1-37). Since $z_{0.8} = 0.842$, the sample size is

$$n = \frac{(\sqrt{0.42 \times 0.58} \times 0.842 + \sqrt{2/9} \times 1.64)^2}{(0.42 - 1/3)^2} = 188.12.$$

We therefore can say that it is 189 for the sample size. For continuity correction, the effective sample size should be $189 + 23 = 212$. We can obtain the same results by using the S-PLUS program *binomial.sample. size*:

```
> binomial.sample.size(p = 1/3, p.alt = 0.42,power = 0.8,
alternative = "great",correct = F)
    p.null p.alt delta alpha power n1
1 0.3333333 0.42 0.08666667 0.05 0.8 189
```

```
> binomial.sample.size(p = 1/3, p.alt = 0.42, power = 0.8,
alternative = "great", correct = T )
      p.null p.alt delta alpha power n1
 1 0.3333333 0.42 0.08666667 0.05 0.8 212
```

### Comparisons of the Methods with Response Bias

Statistically, the force-choice methods involve a comparison of a proportion with a specified value (1/2 for the 2-AFC and duo–trio methods and 1/3 for the 3-AFC and triangle methods), while the methods with response bias involve a comparison of two proportions, e.g., $\widehat{p}_A$ and $\widehat{p}_N$; i.e., the proportion of response "A" for sample A and the proportion of response "A" for sample Not A. Bi and Ennis (2001) discussed the power of the monadic A–Not A test. Bi (2006, Section 3.3) discussed the power of the A–Not A test in different designs. It is the same for the same-different method statistically.

For a specified proportion $p_N$ (or $p_D$ in the same-different method), which relates with response bias, and a specified $\delta$, which represents a sensory difference, we can determine the proportion $p_A$ (or $p_S$ in the same-different method); see Table 10.17. The test power for a monadic A–Not A test with the null hypothesis $H_0: p_A = p_N$ and the alternative hypothesis $H_a: p_A = p_N + d$, when a sensory difference is $\delta$, is

$$\text{Power} = 1 - \Phi(\lambda) \qquad (10.1\text{-}39)$$

where $\lambda = \dfrac{\sigma_0 z_{1-\alpha} - d}{\sigma_1}$, $\sigma_0^2$ is the variance of $\widehat{p}_A - \widehat{p}_N$ under the null hypothesis, and $\sigma_1^2$ is the variance of $\widehat{p}_A - \widehat{p}_N$ under the alternative hypothesis,

$$\sigma_0^2 = p_0(1 - p_0)\left(\frac{1}{N_A} + \frac{1}{N_N}\right);$$

$$\sigma_1^2 = \frac{p_A(1 - p_A)}{N_A} + \frac{p_N(1 - p_N)}{N_N},$$

and the best estimate of $p_0$ is the weighted mean of $\widehat{p}_A$ and $\widehat{p}_N$:

$$\widehat{p}_0 = \frac{h\widehat{p}_A + \widehat{p}_N}{1 + h},$$

**Table 10.17**   The Probability $p_A$ in an A–Not A Method and the Probability $p_S$ in a Same-Different Method for a Given Response Bias ($p_N$ and $p_D$) and Sensory Difference ($\delta$)

| | The A-Not A Method | | | | | The Same-Different Methods | | | | |
|---|---|---|---|---|---|---|---|---|---|---|
| | $p_N$ | | | | | $p_D$ | | | | |
| | 0.10 | 0.20 | 0.30 | 0.40 | 0.50 | 0.10 | 0.20 | 0.30 | 0.40 | 0.50 |
| $\delta$ | | | | | | | | | | |
| 0.30 | 0.163 | 0.294 | 0.411 | 0.519 | 0.618 | 0.102 | 0.204 | 0.306 | 0.408 | 0.510 |
| 0.35 | 0.176 | 0.312 | 0.431 | 0.538 | 0.637 | 0.103 | 0.206 | 0.309 | 0.411 | 0.513 |
| 0.40 | 0.189 | 0.329 | 0.450 | 0.558 | 0.655 | 0.104 | 0.208 | 0.312 | 0.415 | 0.517 |
| 0.45 | 0.203 | 0.348 | 0.470 | 0.578 | 0.674 | 0.105 | 0.210 | 0.315 | 0.419 | 0.522 |
| 0.50 | 0.217 | 0.366 | 0.490 | 0.597 | 0.692 | 0.106 | 0.213 | 0.318 | 0.423 | 0.527 |
| 0.55 | 0.232 | 0.385 | 0.510 | 0.617 | 0.709 | 0.108 | 0.215 | 0.322 | 0.428 | 0.533 |
| 0.60 | 0.248 | 0.404 | 0.530 | 0.636 | 0.726 | 0.109 | 0.218 | 0.327 | 0.434 | 0.539 |
| 0.65 | 0.264 | 0.424 | 0.550 | 0.654 | 0.742 | 0.111 | 0.222 | 0.331 | 0.440 | 0.546 |
| 0.70 | 0.280 | 0.444 | 0.570 | 0.672 | 0.758 | 0.113 | 0.225 | 0.337 | 0.446 | 0.553 |
| 0.75 | 0.298 | 0.464 | 0.589 | 0.690 | 0.773 | 0.115 | 0.229 | 0.342 | 0.453 | 0.561 |
| 0.80 | 0.315 | 0.483 | 0.609 | 0.708 | 0.788 | 0.117 | 0.234 | 0.348 | 0.461 | 0.569 |
| 0.85 | 0.333 | 0.503 | 0.628 | 0.725 | 0.802 | 0.120 | 0.238 | 0.355 | 0.469 | 0.578 |
| 0.90 | 0.351 | 0.523 | 0.646 | 0.741 | 0.816 | 0.122 | 0.243 | 0.362 | 0.477 | 0.588 |
| 0.95 | 0.370 | 0.543 | 0.665 | 0.757 | 0.829 | 0.125 | 0.249 | 0.370 | 0.486 | 0.597 |
| 1.00 | 0.389 | 0.563 | 0.683 | 0.772 | 0.841 | 0.128 | 0.255 | 0.378 | 0.496 | 0.608 |
| 1.05 | 0.408 | 0.582 | 0.700 | 0.787 | 0.853 | 0.131 | 0.261 | 0.386 | 0.506 | 0.619 |
| 1.10 | 0.428 | 0.602 | 0.718 | 0.801 | 0.864 | 0.135 | 0.268 | 0.396 | 0.517 | 0.630 |
| 1.15 | 0.448 | 0.621 | 0.734 | 0.815 | 0.875 | 0.139 | 0.275 | 0.405 | 0.528 | 0.641 |
| 1.20 | 0.468 | 0.640 | 0.750 | 0.828 | 0.885 | 0.143 | 0.282 | 0.416 | 0.540 | 0.653 |
| 1.25 | 0.487 | 0.658 | 0.766 | 0.840 | 0.894 | 0.147 | 0.290 | 0.426 | 0.552 | 0.665 |
| 1.30 | 0.507 | 0.677 | 0.781 | 0.852 | 0.903 | 0.152 | 0.299 | 0.438 | 0.565 | 0.678 |
| 1.35 | 0.527 | 0.694 | 0.796 | 0.864 | 0.912 | 0.157 | 0.308 | 0.450 | 0.578 | 0.690 |
| 1.40 | 0.547 | 0.712 | 0.809 | 0.874 | 0.919 | 0.162 | 0.318 | 0.462 | 0.591 | 0.703 |

*(continued)*

**Table 10.17** The Probability $p_A$ in an A–Not A Method and the Probability $p_S$ in a Same-Different Method for a Given Response Bias ($p_N$ and $p_D$) and Sensory Difference ($\delta$)—cont...

| | The A-Not A Method | | | | | The Same-Different Methods | | | | |
| | $p_N$ | | | | | $p_D$ | | | | |
| | **0.10** | **0.20** | **0.30** | **0.40** | **0.50** | **0.10** | **0.20** | **0.30** | **0.40** | **0.50** |
|---|---|---|---|---|---|---|---|---|---|---|
| 1.45 | 0.567 | 0.728 | 0.823 | 0.884 | 0.926 | 0.168 | 0.328 | 0.475 | 0.605 | 0.716 |
| 1.50 | 0.586 | 0.745 | 0.835 | 0.894 | 0.933 | 0.174 | 0.339 | 0.488 | 0.619 | 0.728 |
| 1.55 | 0.606 | 0.761 | 0.848 | 0.903 | 0.939 | 0.180 | 0.350 | 0.502 | 0.633 | 0.741 |
| 1.60 | 0.625 | 0.776 | 0.859 | 0.911 | 0.945 | 0.187 | 0.362 | 0.517 | 0.647 | 0.754 |
| 1.65 | 0.644 | 0.791 | 0.870 | 0.919 | 0.950 | 0.195 | 0.375 | 0.531 | 0.661 | 0.766 |
| 1.70 | 0.662 | 0.805 | 0.880 | 0.926 | 0.955 | 0.203 | 0.388 | 0.547 | 0.676 | 0.778 |
| 1.75 | 0.680 | 0.818 | 0.890 | 0.933 | 0.960 | 0.211 | 0.402 | 0.562 | 0.690 | 0.790 |
| 1.80 | 0.698 | 0.831 | 0.899 | 0.939 | 0.964 | 0.220 | 0.416 | 0.578 | 0.704 | 0.802 |
| 1.85 | 0.715 | 0.843 | 0.908 | 0.945 | 0.968 | 0.230 | 0.431 | 0.593 | 0.718 | 0.813 |
| 1.90 | 0.732 | 0.855 | 0.916 | 0.950 | 0.971 | 0.240 | 0.447 | 0.609 | 0.732 | 0.824 |
| 1.95 | 0.748 | 0.866 | 0.923 | 0.955 | 0.974 | 0.251 | 0.463 | 0.625 | 0.746 | 0.834 |
| 2.00 | 0.764 | 0.877 | 0.930 | 0.960 | 0.977 | 0.262 | 0.479 | 0.641 | 0.759 | 0.844 |

where $h = N_A/N_N$. The sample size can be estimated by using

$$N_N = \left[ \frac{\sqrt{p_N(1 - p_N) + p_A(1 - p_A)/h}z_{power} + \sqrt{p_0(1 - p_0)(1 + 1/h)}z_{1-\alpha}}{p_A - p_N} \right]^2.$$

$$(10.1\text{-}40)$$

A continuity correction factor is

$$n_e = (h + 1)/[h(p_A - p_N)]. \qquad (10.1\text{-}41)$$

The comparisons of the two methods with response bias are shown in Figure 10.11. One should note that for the same sensory difference in terms of Thurstonian $\delta$ and the specified proportion of false alarm, i.e., $p_N$ (or $p_D$), the A–Not A method has a larger power than the same-different method.

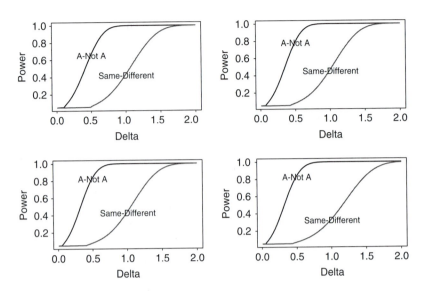

■ **FIGURE 10.11** Powers of difference tests using the A–Not A and the same-different methods, $N_A = N_N = N_S = N_D = 100$, $\alpha = 0.05$, and $p_N = p_D = 0.1, 0.2, 0.3,$ and $0.4$, respectively.

---

**Example 10.1-14**  For a monadic A–Not A test with $\alpha = 0.05$ and sample sizes $N_A = N_N = 100$, we want to know the test power if it is assumed that $p_N = 0.2$, $p_A = 0.4$, which corresponds to about $\delta = 0.6$ from Table 10.17. Because $h = 1$,

$$\widehat{p}_0 = \frac{0.4 + 0.2}{1 + 1} = 0.3,$$

$$\sigma_0^2 = 0.3(1 - 0.3)\left(\frac{1}{100} + \frac{1}{100}\right) = 0.0042,$$

$$\sigma_1^2 = \frac{0.4(1 - 0.4)}{100} + \frac{0.2(1 - 0.2)}{100} = 0.004;$$

hence,

$$\lambda = \frac{\sqrt{0.0042} \times 1.64 - 0.2}{\sqrt{0.004}} = -1.478.$$

The power is $1 - \Phi(-1.478) = 0.93$ from (10.1-39). The continuity correction factor is

$$n_e = (1 + 1)/[(0.4 - 0.2)] = 10$$

from (10.1-41). Using the effective sample sizes $N_A = N_N = 100 - 10 = 90$, we can get the power 0.91. Using the built-in S-PLUS program *binomial. sample.size*, with and without continuity correction, we can obtain the following results:

```
> binomial.sample.size(p = 0.2, p2 = 0.4, alternative =
  "greater", n1 = 100,correct = F)
  p1 p2 delta alpha power n1 n2 prop.n2
1 0.2 0.4 0.2 0.05 0.9301359 100 100 1
> binomial.sample.size(p = 0.2, p2 = 0.4, alternative =
  "greater", n1 = 90,correct = F)
  p1 p2 delta alpha power n1 n2 prop.n2
1 0.2 0.4 0.2 0.05 0.9056655 90 90 1
> binomial.sample.size(p = 0.2, p2 = 0.4, alternative =
  "greater", n1 = 100,correct = T)
  p1 p2 delta alpha power n1 n2 prop.n2
1 0.2 0.4 0.2 0.05 0.9056655 100 100 1
```

For sample sizes for a specified power of 0.8, we can get from (10.1-40) that

$$N_N = \left[ \frac{\sqrt{0.2(1 - 0.2) + 0.4(1 - 0.4)/1} \times 0.84 + \sqrt{0.3(1 - 0.3)(1 + 1/1)} \times 1.64}{0.4 - 0.2} \right]^2$$

$$= 63.86.$$

Hence, $N_A = N_N = 64$. For continuity correction, the sample sizes should be $N_A = N_N = 64 + 10 = 74$. We can obtain the same results by using the S-PLUS program *binomial.sample.size:*

```
> binomial.sample.size(p = 0.2, p2 = 0.4,power = 0.8,
  alternative = "greater",correct = F)
    p1 p2 delta alpha power n1 n2 prop.n2
1 0.2 0.4 0.2 0.05 0.8 64 64 1
> binomial.sample.size(p = 0.2, p2 = 0.4,power = 0.8,
  alternative = "greater",correct = T)
    p1 p2 delta alpha power n1 n2 prop.n2
1 0.2 0.4 0.2 0.05 0.8 74 74 1
```

## 10.2 **SENSORY SIMILARITY (EQUIVALENCE) TESTS**

### Introduction

Similarity tests, which are also known as equivalence tests or noninferiority tests (one-sided equivalence tests), have been well developed and widely applied in some fields, for example, pharmaceutical and clinical fields. See, for example, Metzler and Haung (1983), Chow and Liu (1992), and Wellek (2003) for an overview. Bi (2005, 2007ab) discussed the applications of the methodology in sensory and consumer research.

In many situations in sensory and consumer research, the objective is not to demonstrate difference but to demonstrate similarity or equivalence. Proof of exact equality is theoretically impossible. Similarity testing merely demonstrates statistically that the difference between two products for comparison is smaller than the allowed difference in intensity or preference. It is widely acknowledged that the conventional hypothesis testing used for difference testing is inappropriate in the context of similarity. The basic difficulty is that the null hypothesis of no difference can never be proved or established but can be possibly disproved according to the logic of statistical hypothesis testing (see, e.g., Blackwelder, 1982). Absence of evidence of difference is not evidence of similarity (Altman and Bland, 1995). Lack of significance may merely be the result of inadequate sample size, while a trivial difference may be statistically significant with large sample size. In order to conduct similarity testing, we need some new ways of thinking about statistical hypothesis testing and some new statistical models.

One so-called power approach has been used for similarity testing. In this approach, a large power value is selected for a specified allowed difference. A sample size is then determined to ensure the large power to detect the difference. If the null hypothesis of no difference is not rejected, similarity is then concluded. This approach is based on the logic that if a difference is larger than a specified allowed difference, the difference should likely be detected and the null hypothesis of no difference should likely be rejected. On the other hand, if a difference is smaller than a specified allowed difference, the null hypothesis should likely not be rejected. At one time, the "power approach" was a standard method in bioequivalence testing. However, due to its unsuitability, the approach was finally abandoned. Some authors, for example, Schuirmann (1987), have shown in a detailed examination that the power approach is quite inadequate for similarity testing. One reason is that this approach contorts the logic of hypothesis testing. According to the logic, we cannot prove and accept the null hypothesis in any situation. Furthermore, one weakness of this method is that for a large sample size and a small measurement error, it is unlikely to draw a conclusion of similarity even for a slight difference but effective equivalence.

In this section we will discuss some similarity testing methods used for sensory and consumer research. These methods are based on interval hypothesis testing. In this approach the null hypothesis is that the difference of two treatments is larger than a specified allowed difference, and the alternative hypothesis is that the difference is smaller than the specified value. If the null hypothesis is rejected, the alternative hypothesis (i.e., similarity between the two products for comparison) can be concluded.

## Similarity Tests Using Forced-Choice Methods

The forced-choice methods used conventionally for discrimination testing can also be used for similarity testing, but a different statistical testing model is needed. First, we need to specify an allowed or ignorable difference in terms of the proportion or probability of "discriminators," $p_{do}$. The probability of correct responses, $p_{co}$, corresponding to $p_{do}$ is then calculated, $p_{co} = p_{do} + p_0(1 - p_{do})$, where $p_0$ is a guessing probability, $p_0 = 1/2$ for the 2-AFC and duo–trio tests, and $p_0 = 1/3$ for the 3-AFC and triangle tests.

The null and alternative hypotheses of the similarity testing are

$$H_0: \quad p_c \geq p_{co}$$

$$H_a: \quad p_0 \leq p_c < p_{co}. \qquad (10.2\text{-}1)$$

This test is one-sided because $p_0 \leq p_c$ is usually assumed for forced-choice methods. The test statistic $c$ is the number of correct responses in a similarity test with sample $n$. The critical number $c_0$ is the maximum value satisfying

$$\sum_{x=0}^{c_0} \binom{n}{x} p_{co}^x (1 - p_{co})^{n-x} < \alpha. \qquad (10.2\text{-}2)$$

If the observed number of correct responses $c$ is smaller than or equal to a critical number $c_0$, the null hypothesis is rejected and the alternative hypothesis is accepted at a significance level $\alpha$. It means that similarity can be concluded.

Tables A.27 and A.28 show the critical numbers for similarity testing using the 2-AFC, duo–trio, 3-AFC, and triangle methods for $\alpha = 0.05$ and 0.1, $p_{do} = 0.1$ to 0.5 with a step of 0.1, and for sample size $n = 5$ to 100. We can use the S-PLUS code *simforce0* to calculate the critical values for a given sample size $n$, method $m$ ($m = 2$ for 2-AFC and duo–trio; $m = 3$ for 3-AFC and triangle), similarity limit $p_{do}$, and significance level $\alpha$.

---

**Example 10.2-1**

One hundred panelists were selected for a similarity test using the 3-AFC method for sweetness of two product brands. The allowed proportion of "discriminators" for the method was selected as $p_{do} = 0.2$, and the significance level was $\alpha = 0.05$. The observed number of correct responses in the test was 35. We can find from Table A.28 that $c_0 = 37$, which is the maximum value for

$$\sum_{x=0}^{37} \binom{100}{x} \times 0.4667^x (1 - 0.4667)^{100-x} < 0.05,$$

where $p_{c0} = 0.2 + 1/3 \times (1 - 0.2) = 0.4667$, according to (10.2-2). Because the observed number of correct responses (35) is smaller than the critical value (37), we can conclude that the two brands of the product are similar in sweetness. In other words, we can claim that there is no detectable difference between the two brands on sweetness at a significance level $\alpha = 0.05$ in terms of $p_{do} = 0.2$.

```
> simforce0(100,3,0.2,0.05)
[1] 37
```

---

## Similarity Tests Using the Paired Comparison Method

The paired comparison method (2-AFC) is one of the forced-choice methods. This method involves comparing a proportion with a specified value, 0.5. Unlike the other forced-choice methods, the paired comparison method can be used for both one-sided and two-sided tests and for both intensity and preference tests. This section discusses two-sided similarity testing using the paired comparison method.

Let $p$ denote the probability of selecting a specified product in a pair of products and $D = p - 0.5$. Let $\Delta$, which defines similarity, be the distance from 0.5. The hypotheses for similarity testing are

$$H_0: D \geq +\Delta \quad \text{or} \quad D \leq -\Delta \quad \text{versus} \quad H_a: -\Delta < D < +\Delta \quad (10.2\text{-}3)$$

This is the one-sample version of the interval hypotheses. The hypotheses in (10.2-3) can be decomposed into two sets of one-sided hypotheses:

$$H_{01}: D \geq +\Delta \quad \text{versus} \quad H_{a1}: D < +\Delta, \qquad (10.2\text{-}4)$$

$$H_{02}: D \leq -\Delta \quad \text{versus} \quad H_{a2}: D > -\Delta. \qquad (10.2\text{-}5)$$

The "two one-sided tests" (TOST) was proposed by Westlake (1981) and Schuirmann (1981, 1987) and has become the standard test for similarity and bioequivalence in, for example, the clinical and pharmaceutical fields. The TOST is a simple version of the intersection-union test (IUT). IUTs are the tests with the null hypothesis expressed as a union,

$$H_0: \quad \cup (H_{01}, H_{02}), \qquad (10.2\text{-}6)$$

and the alternative hypothesis expressed as an intersection,

$$H_a: \quad \cap (H_{a1}, H_{a2}). \qquad (10.2\text{-}7)$$

In contrast with similarity testing, the two-sided difference test is another type of test with $H_0: D = 0$ versus $H_a: D \neq 0$. This test is equivalent to the simultaneous tests of the pair of one-sided hypotheses: $H_{01}: D \geq 0$ versus $H_{a1}: D < 0$ and $H_{02}: D \leq 0$ versus $H_{a2}: D > 0$. Hence, the two-sided difference tests are union-intersection tests (UITs) with

$$H_0: \quad \cap (H_{01}, H_{02}) \qquad (10.2\text{-}8)$$

and

$$H_a: \quad \cup (H_{a1}, H_{a2}). \qquad (10.2\text{-}9)$$

See Hochberg and Tamhane (1987, 28) and Dunnett and Gent (1996).

In the UIT problem, the adjustment for multiple testing is to use $\alpha_1 + \alpha_2$ for $\alpha$, whereas in the IUT problem, the adjustment is to use max $(\alpha_1, \alpha_2)$ for $\alpha$, where $\alpha_1$ for $H_{01}$ and $\alpha_2$ for $H_{02}$. If we use the same $\alpha$ level in all the tests, we need to make the adjustment of $2\alpha$ in the UIT; however, no adjustment is needed in the IUT. For the theory of IUT, see Berger (1982), Casella and Berger (1990), and Berger and Hsu (1996). The IUT has a prominent position in similarity testing. Based on the principle of the IUT, $H_0$ can be rejected at the $\alpha$ level if $H_{01}$ and $H_{02}$ are rejected simultaneously at the $\alpha$ level in similarity hypothesis testing. In similarity testing, the significance level, i.e., the Type I error $\alpha$, is the probability of declaring similarity of the two products for comparison when they are not, in fact, similar.

The number of observations selecting a product in a pair of products in $n$ responses is a random variable, $X$, which follows a binomial distribution. Let $c_u$ be the $\alpha$ quantile of the binomial distribution with parameters $n$ and $p_u = 0.5 + \Delta$ and $c_l$ be the $1 - \alpha$ quantile of the binomial distribution with parameters $n$ and $p_l = 0.5 - \Delta$. The number $c_u$ should be the maximum integer satisfying equation (10.2-10), and the number $c_l$ should be the minimum integer satisfying equation (10.2-11):

$$\Pr(X \le c_u | n, p_u) = \sum_{x=0}^{c_u} \binom{n}{x} p_u{}^x (1 - p_u)^{n-x} < \alpha, \qquad (10.2\text{-}10)$$

$$\Pr(X \ge c_l | n, p_l) = \sum_{x=c_l}^{n} \binom{n}{x} p_l{}^x (1 - p_l)^{n-x} < \alpha. \qquad (10.2\text{-}11)$$

Table A.29 provides the critical values $c_l$ and $c_u$ for similarity testing for $\Delta = 0.1$, $0.15$, and $0.2$ and for test levels $\alpha = 0.05$ and $\alpha = 0.1$, respectively. Note that for the situation of $c_l > c_u$, we cannot conclude similarity for any observation for the specified $n$, $\Delta$, and $\alpha$.

If the observed number selecting a product in a pair of products for $n$ responses is smaller than $c_u$, we can reject $H_{01}$ and accept $H_{11}$ at the $\alpha$ level. If the observed number is larger than $c_l$, we can reject $H_{02}$ and accept $H_{12}$ at the $\alpha$ level at most. If the observed number ($c$) falls into the rejection region, $[c_l, c_u]$, i.e., $c_l \le c \le c_u$, the similarity in terms of $\Delta$ can be concluded at the $\alpha$ (rather than $2\alpha$) level.

Using a computer program, for example, the S-PLUS code *paireq2*, we can get the critical values $c_l$ and $c_u$ quickly for any given $n$, $\Delta$, and $\alpha$. For example, for $n = 300$, $\Delta = 0.05$, and $\alpha = 0.1$, then $c_l = 147$ and $c_u = 153$.

```
> paireq2(300,0.05,0.1)
[1] 147 153
```

---

**Example 10.2-2**

In a similarity test for comparison of a sensory intensity between a current product and a new product, 200 consumer panelists were used. A similarity in terms of $\Delta = 0.1$ and a significance level $\alpha = 0.1$ were selected. The observed number, i.e., the number selecting the new product as that with higher sensory intensity, was 108. From Table A.29, the critical values are [90, 110]. Because $90 < 108 < 110$, the conclusion that the two products have similar sensory intensity in terms of $\Delta = 0.1$ can be drawn at $\alpha = 0.1$ level.

## Similarity Tests Using A–Not A and Same-Different Methods

In this section, we discuss similarity testing using the monadic A–Not A and same-different methods, which are the discrimination methods with response bias. The methods in this design involve comparison of two independent proportions.

Dunnett and Gent (1977) suggested a chi-squared statistic and an approximate normal distributed $Z$ statistic for similarity testing for the data in a $2 \times 2$ table. Although the $Z$ statistic is not exactly equivalent to the chi-square statistic in similarity testing, the test results using the two statistics show good agreement. For simplification, only the $Z$ statistic is discussed here.

Let $p_A$ and $p_N$ denote the probabilities of response "A" for sample A and for sample Not A, respectively. The null and alternative hypotheses are

$$H_0: \ p_A - p_N \geq \Delta_0$$

$$H_a: \ 0 \leq p_A - p_N < \Delta_0, \tag{10.2-12}$$

where $\Delta_0$ is an allowable nonzero value defining equivalence or similarity. The $Z$ statistic is as

$$Z = \frac{\widehat{p}_A - \widehat{p}_N - \Delta_0}{\sqrt{\widehat{V}(\widehat{p}_A - \widehat{p}_N)}}, \tag{10.2-13}$$

where $\widehat{V}(\widehat{p}_A - \widehat{p}_N)$ is estimated variance of $\widehat{p}_A - \widehat{p}_N$ under the null hypothesis. There are different methods for estimation of the variance. A method is to use expected proportions of response "A" for samples A and Not A, $\widehat{\pi}_A$ and $\widehat{\pi}_N$ in (10.2-14 and 10.2-15), rather than the observed proportions, $\widehat{p}_A$ and $\widehat{p}_N$, for estimation of the variance:

$$\widehat{\pi}_A = \frac{x + y + n_N \Delta_0}{n_A + n_N}, \tag{10.2-14}$$

$$\widehat{\pi}_N = \frac{x + y - n_A \Delta_0}{n_A + n_N}, \tag{10.2-15}$$

where $x$ and $y$ are observed numbers of response "A" for samples A and Not A, respectively; $n_A$ and $n_N$ are sample sizes for samples A and Not A. The estimated variance using the expected proportions is

$$\widehat{V}(\widehat{p}_A - \widehat{p}_N) = \frac{\widehat{\pi}_A(1 - \widehat{\pi}_A)}{n_A} + \frac{\widehat{\pi}_N(1 - \widehat{\pi}_N)}{n_N}. \tag{10.2-16}$$

Using the $Z$ statistic in (10.2-13), we can reject the null hypothesis and accept the alternative hypothesis in (10.2-12) at the $\alpha$ significance level if the $Z < z_\alpha$, where $z_\alpha$ is the $\alpha$ quantile of the standard normal distribution. The $P$-value is the probability of $Z < z_\alpha$.

With continuity correction, (10.2-13) becomes

$$Z = \frac{\hat{p}_A - \hat{p}_N - \Delta_0 + n'}{\sqrt{\hat{V}(\hat{p}_A - \hat{p}_N)}}, \qquad (10.2\text{-}17)$$

where $n' = (1/n_A + 1/n_N)/2$.

---

**Example 10.2-3**

In order to make sure a product (sample Not A) with substituted ingredients has a similar sensory characteristic with the current product (sample A), researchers conducted a similarity test for two products using a monadic A–Not A method. Two hundred panelists received the A sample, and 200 received the Not A sample; i.e., $n_A = n_N = 200$. The specified allowable limit defining similarity was selected as $\Delta_0 = 0.1$. This means that we regard the two products as similar if the difference of the proportions of response "A" for sample A and sample Not A is not larger than 0.1.

The observed numbers of response "A" for sample A and sample Not A are $x = 45$ and $y = 39$, respectively. According to (10.2-14) and (10.2-15), the expected proportions of response "A" for sample A and response "A" for sample Not A under the null hypothesis are

$$\hat{\pi}_A = \frac{x + y + n_A \Delta_0}{n_A + n_N} = \frac{45 + 39 + 200 \times 0.1}{200 + 200} = 0.26,$$

$$\hat{\pi}_N = \frac{x + y - n_N \Delta_0}{n_A + n_N} = \frac{45 + 39 - 200 \times 0.1}{200 + 200} = 0.16.$$

According to (10.2-16), the variance of $\hat{p}_A - \hat{p}_N$ under the null hypothesis is then

$$\hat{V}(\hat{p}_A - \hat{p}_N) = \frac{0.26 \times (1 - 0.26)}{200} + \frac{0.16 \times (1 - 0.16)}{200} = 0.00163.$$

The value of the test statistic (10.2-17) is

$$Z = \frac{45/200 - 39/200 - 0.1}{\sqrt{0.00163}} = -1.73$$

with the associated $P$-value $= 0.042$. This value suggests that the product with substituted ingredients and the current product have similar sensory characteristics at a significance level $\alpha = 0.05$ and in terms of similarity criteria $\Delta_0 = 0.1$. We also can use the S-PLUS code *simana* to get these results:

```
> simana(45,39,200,200,0.1)
[1] -1.73169744 0.04166372
```

## Similarity Tests for Continuous Data

### Anderson and Hauck's Noncentral t Test

It is widely accepted in the sensory and consumer fields that the hedonic or intensity ratings data using the 9-point scale can be approximately regarded as continuous data. Similarity testing for two hedonic or intensity rating means is often needed. The null hypothesis and alternative hypotheses for the test are

$$H_0: \ |\mu_1 - \mu_2| \geq \Delta_0$$

$$H_a: \ -\Delta_0 < \mu_1 - \mu_2 < \Delta_0. \qquad (10.2\text{-}18)$$

Anderson and Hauck (1983) and Hauck and Anderson (1984) proposed a test statistic that can be used to evaluate the null hypothesis in (10.2-18) directly. The test statistic is

$$T_{AH} = \frac{\overline{X}_1 - \overline{X}_2}{\sqrt{s_1^2/n_1 + s_2^2/n_2}}, \qquad (10.2\text{-}19)$$

where $\overline{X}_1$ and $\overline{X}_2$ are estimated means, $s_1^2$ and $s_2^2$ are estimated variances, and $n_1$ and $n_2$ are sample sizes of $X_1$ and $X_2$. The test statistic (10.2-19) follows a noncentral $t$ distribution with the noncentrality parameter

$$\delta = \frac{\mu_1 - \mu_2}{\sqrt{\sigma_1^2/n_1 + \sigma_2^2/n_2}}, \qquad (10.2\text{-}20)$$

and $v = (n_1 + n_2 - 2)$ DF under the null hypothesis for a completely random design. The noncentrality parameter $\delta$ can be estimated by

$$\widehat{\delta} = \frac{\Delta_0}{\sqrt{s_1^2/n_1 + s_2^2/n_2}}, \qquad (10.2\text{-}21)$$

where $\Delta_0$ is the specified maximum allowable difference between two true rating means for similarity. The P-value is

$$p = F_v(|t_{AH}| - \widehat{\delta}) - F_v(-|t_{AH}| - \widehat{\delta}), \qquad (10.2\text{-}22)$$

where $F_v$ () denotes the distribution function of the central $t$ distribution with degrees of freedom $v$, and $t_{AH}$ denotes the observed value of test statistic $T_{AH}$. The null hypothesis would be rejected in favor of similarity if the observed P-value is less than the significance level $\alpha$.

**Example 10.2-4**

In order to determine whether consumers in two cities (A and B) have similar overall likings for a product, 100 panelists were selected in each of the two cities, and 9-point liking scale was used with 9 = Like extremely and 1 = Dislike extremely. The similarity limit $\Delta_0 = 0.5$ and significance level $\alpha = 0.1$ were selected. The observed overall liking means and their variances for the two cities are $\overline{X}_A = 7.1$, $s_A^2 = 2.0$; $\overline{X}_B = 6.9$, and $s_B^2 = 2.2$. The observed value of the test statistic is

$$T_{AH} = \frac{7.1 - 6.9}{\sqrt{(2 + 2.2)/100}} = 0.976.$$

The estimated noncentrality parameter is

$$\hat{\delta} = \frac{0.5}{\sqrt{(2 + 2.2)/100}} = 2.44.$$

The calculated $P$-value is then

$$p = F_v(0.976 - 2.44) - F_v(-0.976 - 2.44) = F_v(-1.464) - F_v(-3.416)$$
$$= 0.072,$$

where $F_v(-1.464) = 0.0724$ is the probability of the central $t$ distribution with $v = 2 \times (100 - 1) = 198$ DF from $-\infty$ to $-1.464$, and $F_v(-3.416) = 0.0004$ is the probability of the central $t$ distribution with 198 DF from $-\infty$ to $-3.416$. Because the $P$-value (0.072) is smaller than $\alpha = 0.1$, we can conclude at 0.1 significance level that the overall liking for the product is similar between the two cities in terms of the similarity limit $\Delta_0 = 0.5$. We can use the S-PLUS code *ahsimtest* to get these results:

```
> ahsimtest2(7.1,6.9,2,2.2,0.5,100,100)
[1] 0.97590007 2.43975018 0.07202436
```

### Two One-Sided Tests (TOST)

The objective of this example is to test if the difference of the means for product A and B, $|\mu_a - \mu_b|$, is smaller than a specified allowed value, $\Delta_0$. This test involves two sets of one-sided hypotheses:

$$H_{01}: \mu_A - \mu_B \le -\Delta_0 \quad \text{versus} \quad H_{a1}: \mu_A - \mu_B > -\Delta_0 \qquad (10.2\text{-}23)$$

and

$$H_{02}: \mu_A - \mu_B \ge \Delta_0 \quad \text{versus} \quad H_{a2}: \mu_A - \mu_B < \Delta_0. \qquad (10.2\text{-}24)$$

The first set of hypotheses in (10.2-23) is to test for noninferiority of product A to product B. The second set of hypotheses in (10.2-24) is to test for nonsuperiority of product A to product B. We can declare the two products are similar in preference if and only if both $H_{01}$ and $H_{02}$ are rejected at a significance level $\alpha$. The two one-sided tests (TOST) have become a standard test for similarity and bioequivalence in, for example, the clinical and pharmaceutical fields (FDA, 1992; EC-GCP, 1993), as mentioned previously.

In order to do the tests, we first estimate the rating means for the two products, $\overline{X}_A$, $\overline{X}_B$, and the variances, $s_A^2$ and $s_B^2$. For large sample sizes, i.e., both sample sizes for the two products are 30 or more, the test statistics are (10.2-25) and (10.2-26), which follow approximately a standard normal distribution:

$$Z_1 = \frac{(\overline{X}_A - \overline{X}_B) + \Delta_0}{\widehat{\sigma}}, \tag{10.2-25}$$

$$Z_2 = \frac{(\overline{X}_A - \overline{X}_B) - \Delta_0}{\widehat{\sigma}}, \tag{10.2-26}$$

where

$$\widehat{\sigma} = \sqrt{\frac{s_A^2}{n_A} + \frac{s_B^2}{n_B}}.$$

If $Z_1 > z_{1-\alpha}$, we can conclude at $\alpha$ level that product A is noninferior to B in terms of $\Delta_0$. If $Z_2 < z_\alpha$, we can conclude at $\alpha$ level that product A is nonsuperior to B. If both $Z_1 > z_{1-\alpha}$ and $Z_2 < z_\alpha$ hold, we can say the two products A and B are equivalent. The $P$—value of the similarity test is max $(P_1, P_2)$, where $P_1$ and $P_2$ are the $P$—values of the two one-sided tests.

---

**Example 10.2-5**

For this example, we use the data in Example 10.2-4: $\overline{X}_A = 7.1$, $s_A^2 = 2.0$, $\overline{X}_B = 6.9$, $s_B^2 = 2.2$, $\Delta_0 = 0.5$, and $\alpha = 0.1$. Hence,

$$\widehat{\sigma} = \sqrt{\frac{2}{100} + \frac{2.2}{100}} = 0.2049,$$

$$Z_1 = \frac{(7.1 - 6.9) + 0.5}{0.2049} = 3.42,$$

$$Z_2 = \frac{(7.1 - 6.9) - 0.5}{0.2049} = -1.46.$$

Because $Z_1 > z_{0.9} = 1.28$, and $Z_2 < z_{0.1} = -1.28$, we can conclude that the two products are similar in preference at $\alpha = 0.1$. The $P$-values for the two tests are 0.0003 and 0.072, respectively. The $P$—value for the similarity test is the maximum of 0.0003 and 0.072. It is 0.072.

```
> schtest2(7.1,6.9,2,2.2,100,100,0.5,0.1)
[1] 3.416 -1.464 0.000 0.072
```

## Similarity Tests for Correlated Data

Matched pair design is often used in sensory and consumer research. In this design, each panelist evaluates two or multiple samples for comparison. The responses are correlated data.

### Paired t-Test for Similarity

Wellek (2003) proposed a procedure for the tests for continuous data. With symmetric equivalence margins, the null hypothesis and alternative hypothesis are

$$H_0: \ \delta/\sigma_D \leq -\varepsilon \text{ or } \delta/\sigma_D \geq \varepsilon \quad \text{versus} \quad H_a: \ -\varepsilon < \delta/\sigma_D < \varepsilon, \quad (10.2\text{-}27)$$

where $\delta = E(D_i)$, $\sigma_D^2 = Var(D_i)$, $D_i$ is the difference between product A and B for the $i$th panelist, and $i = 1, 2, ..., n$; $\varepsilon$ is an allowed standard difference. The test statistic is

$$T = \sqrt{n}\overline{D}/S_D, \quad (10.2\text{-}28)$$

where $\overline{D}$ and $S_D$ denote the mean and standard deviation of $D_i$. If $|T| < C_{\alpha;n-1}(\varepsilon)$, then the null hypothesis is rejected and the alternative hypothesis is accepted and similarity is concluded. The critical value is

$$C_{\alpha;n-1}(\varepsilon) = \sqrt{F_{1,n-1;\alpha}(\tilde{\varepsilon}^2)}, \quad (10.2\text{-}29)$$

where $F_{1,n-1;\alpha}(\tilde{\varepsilon}^2)$ denotes the upper $100\alpha$ percentile of the noncentral F distribution with 1 and $n-1$ degrees of freedom and a noncentrality parameter of $\tilde{\varepsilon}^2 = n\varepsilon^2$. Table A.30 shows the critical values for $\alpha = 0.05$, $\varepsilon = 0.25$ (0.25) 1.0, and $n = 30$ (1) 50 and 50 (5) 100.

---

**Example 10.2-6**

Some ingredients have been changed in a food product. The manufacturer wants to demonstrate that consumers have similar overall liking for the new product (A) and the current product (B) for a specified allowed difference $\varepsilon = 0.5$ and Type I error level $\alpha = 0.05$. A test with 100 consumer panelists was conducted. Each of the panelists evaluated both products A and B. The data are displayed in Table 10.18. We calculated $\overline{M}_A = 7.98$, $\overline{M}_B = 8.23$, $\overline{D} = -0.21$, and $S_D^2 = 1.319$. Hence,

$$|T| = |\sqrt{100}(-0.21)/\sqrt{1.319}| = 1.829.$$

The corresponding critical value for $\alpha = 0.05$, $\varepsilon = 0.5$, and $n = 100$ is 3.318 from Table A.30. Because $1.829 < 3.318$, we can claim that consumers have similar overall liking for the new product (A) and the current product (B) in the sense that the specified allowed standard difference of overall liking for the two products is 0.5.

| Table 10.18 | Overall Liking for New and Current Products (Example 10.2-6) | | | | | | | | | | | | | | |
|---|---|---|---|---|---|---|---|---|---|---|---|---|---|---|---|
| No. | A | B | Dif. | No. | A | B | Dif. | No. | A | B | Dif. | No. | A | B | Dif. |
| 1 | 6 | 7 | −1 | 26 | 9 | 9 | 0 | 51 | 9 | 9 | 0 | 76 | 9 | 8 | 1 |
| 2 | 8 | 9 | 0 | 27 | 9 | 9 | 0 | 52 | 8 | 8 | 0 | 77 | 8 | 8 | 0 |
| 3 | 7 | 8 | 0 | 28 | 9 | 8 | 1 | 53 | 9 | 9 | 0 | 78 | 6 | 8 | −2 |
| 4 | 9 | 9 | 0 | 29 | 8 | 7 | 1 | 54 | 9 | 9 | 0 | 79 | 5 | 8 | −3 |
| 5 | 9 | 9 | 0 | 30 | 9 | 7 | 2 | 55 | 8 | 8 | 0 | 80 | 6 | 8 | −2 |
| 6 | 6 | 9 | −3 | 31 | 8 | 8 | 0 | 56 | 9 | 9 | 0 | 81 | 9 | 8 | 1 |
| 7 | 7 | 9 | 0 | 32 | 8 | 8 | 0 | 57 | 9 | 8 | 1 | 82 | 8 | 8 | 0 |
| 8 | 9 | 8 | 1 | 33 | 9 | 9 | 0 | 58 | 8 | 8 | 0 | 83 | 8 | 6 | 2 |
| 9 | 4 | 7 | −3 | 34 | 9 | 9 | 0 | 59 | 7 | 7 | 0 | 84 | 9 | 8 | 1 |
| 10 | 8 | 9 | −1 | 35 | 7 | 7 | 0 | 60 | 8 | 7 | 1 | 85 | 9 | 8 | 1 |
| 11 | 8 | 9 | −1 | 36 | 8 | 9 | −1 | 61 | 9 | 9 | 0 | 86 | 7 | 8 | −1 |
| 12 | 9 | 9 | 0 | 37 | 9 | 9 | 0 | 62 | 9 | 9 | 0 | 87 | 7 | 8 | −1 |
| 13 | 6 | 6 | 0 | 38 | 8 | 9 | −1 | 63 | 9 | 8 | 1 | 88 | 8 | 9 | −1 |
| 14 | 6 | 9 | −3 | 39 | 8 | 9 | −1 | 64 | 9 | 9 | 0 | 89 | 8 | 9 | −1 |
| 15 | 6 | 9 | −3 | 40 | 9 | 9 | 0 | 65 | 6 | 7 | −1 | 90 | 9 | 8 | 1 |
| 16 | 9 | 8 | 1 | 41 | 9 | 9 | 0 | 66 | 9 | 8 | 1 | 91 | 8 | 9 | −1 |
| 17 | 8 | 9 | −1 | 42 | 8 | 8 | 0 | 67 | 9 | 8 | 1 | 92 | 8 | 8 | 0 |
| 18 | 9 | 9 | 0 | 43 | 8 | 9 | −1 | 68 | 6 | 8 | −2 | 93 | 7 | 9 | −2 |
| 19 | 9 | 9 | 0 | 44 | 9 | 9 | 0 | 69 | 8 | 9 | −1 | 94 | 5 | 9 | −4 |
| 20 | 9 | 9 | 0 | 45 | 8 | 8 | 0 | 70 | 9 | 9 | 0 | 95 | 6 | 7 | −1 |
| 21 | 9 | 9 | 0 | 46 | 9 | 9 | 0 | 71 | 9 | 9 | 0 | 96 | 7 | 7 | 0 |
| 22 | 7 | 7 | 0 | 47 | 9 | 9 | 0 | 72 | 7 | 7 | 0 | 97 | 7 | 6 | 1 |
| 23 | 6 | 8 | −2 | 48 | 9 | 9 | 0 | 73 | 8 | 7 | 1 | 98 | 9 | 9 | 0 |
| 24 | 8 | 6 | 2 | 49 | 9 | 7 | 2 | 74 | 7 | 7 | 0 | 99 | 8 | 9 | −1 |
| 25 | 9 | 9 | 0 | 50 | 8 | 6 | 2 | 75 | 9 | 9 | 0 | 100 | 6 | 7 | −1 |

### Generalization of McNemar's Test for Similarity

Similarity evaluation for two correlated proportions in matched-pair design is often needed in sensory and consumer research. Statistical procedures for the evaluations have been proposed and developed recently, mainly in the clinical and medical fields. The procedures include Wald-type (sample-based) test statistics (Lu and Bean, 1995); restricted maximum likelihood estimation, i.e., score-type test statistics (see, e.g., Nam, 1997; Tango, 1998; Liu et al., 2002; Tang et al., 2002; Tang et al., 2003); exact and approximate exact uncondi-tional test statistics (Chan et al., 2003); and noncentral chi-square statistic (Wellek, 2003). Some empirical results (see, e.g., Nam, 1997; Tang et al., 2002) showed that the score-type test is superior to the Wald-type test and some other tests in terms of actual Type I error rate. In this section, we discuss the score-type similarity test in matched-pair studies with binary outcomes and with a risk difference.

Let $\theta = p_t - p_s = p_{10} - p_{01}$ denote the difference of response probabil-ities for test and standard samples. The sample estimates of $p_{10}$ and $p_{01}$ are $\widehat{p}_{10} = x_{10}/n$ and $\widehat{p}_{01} = x_{01}/n$, where $x_{10}$ is the number of response "yes" for the test sample and response "no" for the standard sample; $x_{01}$ is the number of response "no" for the test sample and "yes" for the standard sample. Let $\delta$ be a predetermined practically meaningful similarity or equivalence limit.

The null and alternative hypotheses in a similarity test are

$$H_0: p_t - p_s \geq \delta \text{ or } p_t - p_s \leq -\delta \quad \text{versus} \quad H_a: -\delta < p_t - p_s < \delta. \quad (10.2\text{-}30)$$

The interval hypotheses in (10.2-30) can be decomposed into two sets of one-sided hypotheses:

$$H_{0l}: \ p_t - p_s \leq -\delta \quad \text{versus} \quad H_{1l}: \ p_t - p_s > -\delta \qquad (10.2\text{-}31)$$

and

$$H_{0u}: \ p_t - p_s \geq \delta \quad \text{versus} \quad H_{1u}: \ p_t - p_s < \delta. \qquad (10.2\text{-}32)$$

The one-sided hypothesis (10.2-31) has been referred as to the *nonin-feriority* hypothesis. The one-sided hypothesis (10.2-32) has been referred as to the *nonsuperiority* hypothesis. The test statistics are

$$Z_l = \frac{\widehat{\theta} + \delta}{\widehat{\sigma}_l} \geq z_{1-\alpha} \quad \text{and} \qquad (10.2\text{-}33)$$

$$Z_u = \frac{\widehat{\theta} - \delta}{\widehat{\sigma}_u} \leq z_\alpha, \qquad (10.2\text{-}34)$$

where

$$\hat{\sigma}_l^2 = \left[(\tilde{p}_{l01} + \tilde{p}_{l10}) - \delta^2\right]\big/n,$$

$$\tilde{p}_{l01} = \left(-\tilde{a}_l + \sqrt{\tilde{a}_l^2 - 8\tilde{b}_l}\right)\big/4,$$

$$\tilde{p}_{l10} = \tilde{p}_{l01} - \delta,$$

$$\tilde{a}_l = -\widehat{\theta}(1 - \delta) - 2(\widehat{p}_{01} + \delta) \quad \text{and}$$

$$\tilde{b}_l = \delta(1 + \delta)\widehat{p}_{01}.$$

Similarly,

$$\hat{\sigma}_u^2 = \left[(\tilde{p}_{u01} + \tilde{p}_{u10}) - \delta^2\right]/n,$$

$$\tilde{p}_{u01} = \left(-\tilde{a}_u + \sqrt{\tilde{a}_u^2 - 8\tilde{b}_u}\right)\big/4,$$

$$\tilde{p}_{u10} = \tilde{p}_{u01} + \delta,$$

$$\tilde{a}_u = -\widehat{\theta}(1 + \delta) - 2(\widehat{p}_{01} - \delta) \quad \text{and}$$

$$\tilde{b}_u = -\delta(1 - \delta)\widehat{p}_{01}.$$

The $100(1 - 2\alpha)$ percent confidence interval $(\tilde{\theta}_l, \tilde{\theta}_u)$, which corresponds to the $\alpha$ (not $2\alpha$) level hypothesis test, can be obtained numerically by getting the roots of the functions

$$f_l(-\delta) = Z_l(-\delta) - z_{1-\alpha} \quad \text{and} \quad f_u(\delta) = Z_u(\delta) - z_\alpha. \qquad (10.2\text{-}35)$$

It can be shown that when $\delta = 0$, the test statistic becomes

$$\frac{x_{10} - x_{01}}{\sqrt{x_{10} + x_{01}}}, \qquad (10.2\text{-}36)$$

which coincides with the conventional McNemar's test. One should note that, unlike the conventional McNemar's test, the generalization of McNemar's test takes into consideration not only the number of discordant results but also the total number of samples. A computer program such as the S-PLUS *simmcnemar5* should be used for calculating the test.

---

A manufacturer wanted to investigate whether consumers had similar purchase intent for a new product with reduced ingredients versus the current product. A similarity test was conducted with 100 consumer panelists. Each panelist evaluated both the new and current products and gave responses using a 5-point purchase intent scale. The data for the top two boxes are shown in Table 10.19.

**Example 10.2-7**

**Table 10.19** Purchase Intent for Products (Example 10.2-7)

| | | Current Product | | |
| --- | --- | --- | --- | --- |
| | | **"Buy"** | **"Not Buy"** | **Total** |
| New Product | "Buy" | 55 | 8 | 63 |
| | "Not Buy" | 10 | 27 | 37 |
| | Total | 65 | 35 | 100 |

The specified allowed similarity limit is $\delta = 0.1$ and Type I error is $\alpha = 0.05$. We find $\hat{p}_{01} = 0.1$, $\hat{\theta} = 0.63\text{-}0.65 = -0.02$. Hence, $\tilde{a}_l = 0.02(1 - 0.1) - 2(0.1 + 0.1) = -0.382$, $\tilde{b}_l = 0.1(1 + 0.1)0.1 = 0.011$, $\tilde{p}_{l01} = (0.382 + \sqrt{0.382^2 - 8 \times 0.011})/4 = 0.15567$, $\tilde{p}_{l10} = 0.15567 - 0.1 = 0.05567$, and $\hat{\sigma}_l^2 = [(0.15567 + 0.05567) - 0.1^2]/100 = 0.002013$. Similarly, $\tilde{a}_u = 0.022$, $\tilde{b}_u = -0.009$, $\tilde{p}_{u01} = 0.0618$, $\tilde{p}_{u10} = 0.1618$, and $\hat{\sigma}_u^2 = 0.002136$. The values of the test statistics are

$$Z_l = \frac{-0.02 + 0.1}{\sqrt{0.002013}} = 1.7829 > 1.645 = z_{1-0.05} \text{ with } P\text{-value} = 0.0373,$$

$$Z_u = \frac{-0.02 - 0.1}{\sqrt{0.002136}} = -2.5964 < -1.645 = z_{0.05} \text{ with } P\text{-value} = 0.0047.$$

We can conclude at $\alpha = 0.05$ and $\delta = 0.1$ that the purchase intent for the new product is similar with that for the current product in terms of the proportions of the top two boxes. The $P$-value for the test is 0.0373. The $100(1 - 2\alpha)$ (i.e., 90) percent two-sided confidence interval $(\tilde{\theta}_l, \tilde{\theta}_u)$ is $(-0.0933, 0.0523)$, which is calculated by using the S-PLUS built-in function *uniroot*. Because the confidence interval is within $(-\delta, \delta)$, i.e., $(-0.1, 0.1)$, the same conclusion can be obtained. Using S-PLUS code, we can obtain these results quickly:

```
> simmcnemar5(cbind(c(55,10),c(8,27)),0.1)
[1] 1.782905  0.037301  -2.596367  0.004711  0.037301
-0.093304  0.052264
```

## Hypothesis Test and Confidence Interval for Similarity Evaluation

The confidence interval is of interest both in difference testing and similarity testing. Even in situations in which the main interest is to test hypotheses, it is also desirable to provide a confidence interval. There is substantial literature about confidence intervals for bioequivalence (see Westlake, 1972, 1976; Bofinger 1992;

Lui, 1990; Gacula Jr., 1993; Hsu et al., 1994; Berger and Hsu, 1996). There is no argument that the confidence interval can be used to evaluate similarity. But there are arguments about how to construct an appropriate confidence interval associated with a similarity hypothesis test and how to determine sample size for similarity evaluation.

In difference testing, the confidence interval approach is exactly equivalent to a significance testing approach. The test level is 1 minus the confidence level of the interval. However, the two approaches may not be equivalent in a similarity evaluation. A discrepancy between confidence level and test level may occur (see, e.g., Chow and Lui, 1992, 74; Berger and Hsu, 1996). A $100(1 - \alpha)\%$ confidence interval will have, in the long run, at least a $1 - \alpha$ chance to cover the true value of the parameter $p$ under the normality assumption; i.e.,

$$P\{p \in (\widehat{p}_l, \widehat{p}_u)\} = 1 - \alpha,$$

where $\widehat{p}_l = \widehat{p} - z_{1-\alpha/2} s_{\widehat{p}}$ and $\widehat{p}_u = \widehat{p} + z_{1-\alpha/2} s_{\widehat{p}}$ are limits of conventional $100(1 - \alpha)\%$ confidence interval, $s_{\widehat{p}} = \sqrt{\widehat{p}(1 - \widehat{p})/n}$. However, it is not guaranteed that, in the long run, the chance of the $100(1 - \alpha)\%$ confidence interval being within the range of the similarity limits is $1 - \alpha$; namely, the probability $P\{(\widehat{p}_l, \widehat{p}_u) \in (p_l, p_u)\}$ is not necessarily equal to $1 - \alpha$, where $p_l = 0.5 - \Delta$, $p_u = 0.5 + \Delta$, are specified allowed difference limits for similarity.

Another criticism of the practice for similarity evaluation using the conventional confidence interval is aimed at determination of sample size. It is not appropriate to determine sample size on the basis of difference testing power, which is totally irrelevant to similarity testing. There are different statistical logics, models, and equations for determinations of sample sizes for difference testing and similarity testing.

### The Conventional 100 (1−α)% Confidence Interval and the α-Level TOST

The early practice for bioequivalence evaluation and the widely used practice for similarity testing in sensory and consumer research employ the conventional confidence interval approach (see, e.g., Westlake, 1972, 1979, 1981). In this approach, a conventional $100(1 - \alpha)\%$ two-sided confidence interval of a parameter is estimated. If and only if the confidence interval is completely contained within the allowed range, we can conclude similarity.

The conventional $100(1 - \alpha)\%$ two-sided confidence interval is calculated from

$$CI = [\hat{p} - z_{1-\alpha/2}s_{\hat{p}}, \hat{p} + z_{1-\alpha/2}s_{\hat{p}}]. \qquad (10.2\text{-}37)$$

The main problem using the conventional $100(1 - \alpha)\%$ two-sided confidence interval to conduct a similarity test is that it will lead to a too-conservative test. It is uniformly less powerful than the TOST (Hsu et al., 1994).

---

**Example 10.2-8**

For a preference similarity test using a paired comparison method, $n = 100$, $\Delta = 0.15$, and $\alpha = 0.05$, the observed number for preference is 56. From Table A.29, we can find that 56 is in the rejection region [44, 56]. Hence, we can reach a similarity conclusion. If we use the conventional two-sided $100(1 - \alpha)\%$ confidence interval, from (10.2-37), it is

$$CI = [.56 - 1.96 \times \sqrt{.56 \times .44/100}, 0.56 + 1.96 \times \sqrt{(.56 \times .44)/100}], \text{ i.e,}$$

$$CI = [0.463, 0.657].$$

Because [0.463,0.657] is not completely contained within the allowed range [0.35,0.65] (i.e., [0.5 − 0.15, 0.5 + 0.15]), we cannot conclude similarity.

---

### The Conventional 100(1−2α)% Confidence Interval and the α-Level TOST

Another practice is to use a $100(1 - 2\alpha)\%$ rather than a $100(1 - \alpha)\%$ two-sided conventional confidence interval to do $\alpha$-level similarity testing. The conventional two-sided $100(1 - 2\alpha)\%$ confidence interval is

$$CI = [\hat{p} - z_{1-\alpha}s_{\hat{p}}, \hat{p} + z_{1-\alpha}s_{\hat{p}}]. \qquad (10.2\text{-}38)$$

It can be demonstrated that under the normal distribution assumption, the similarity testing in terms of the $100(1 - 2\alpha)\%$ confidence interval is operationally equivalent to the TOST with the $\alpha$ level.

According to the decision rules of the confidence interval approach, $\hat{p}_l = \hat{p} - z_{1-\alpha}s_{\hat{p}} \geq p_l = 0.5 - \Delta$ and $\hat{p}_u = \hat{p} + z_{1-\alpha}s_{\hat{p}} \leq p_u = 0.5 + \Delta$, which can be expressed as

$$\frac{\hat{p} - 0.5 - \Delta}{s_{\hat{p}}} \leq z_{\alpha}, \qquad (10.2\text{-}39)$$

$$\frac{\hat{p} - 0.5 + \Delta}{s_{\hat{p}}} \geq z_{1-\alpha}. \qquad (10.2\text{-}40)$$

Equations (10.2-39) and (10.2-40) represent a TOST. Because the TOST is an $\alpha$-level test according to the principle of the IUT, it means that the similarity testing in terms of the conventional $100(1 - 2\alpha)\%$ confidence interval is also an $\alpha$-level (not $2\alpha$) test.

As pointed out by Berger and Hsu (1996), the $100(1 - 2\alpha)\%$ confidence interval is an $\alpha$-level test only when the interval is "equal-tailed." It is misleading to imply that one may always base an $\alpha$-level test on a $100(1 - 2\alpha)\%$ confidence interval. Examples in Berger and Hsu (1996) showed that the $100(1 - 2\alpha)\%$ confidence intervals can result in both liberal and conservative tests. The mixture of $100(1 - 2\alpha)\%$ confidence intervals and $\alpha$-level tests is confusing and logically discontinuous.

Because of these potential difficulties, Berger and Hsu (1996) believed "it is unwise to attempt to define a size-$\alpha$ test in terms of a $100(1 - 2\alpha)\%$ confidence set." Their conclusion was that this practice "should be abandoned".

---

**Example 10.2-9**

For the example in 10.2-8, the $100(1 - 2\alpha)\%$ confidence interval is calculated from (10.2-38). That is,

$$CI = [.56 - 1.65 \times \sqrt{.56 \times .44/100}, 0.56 + 1.65$$
$$\times \sqrt{(.56 \times .44)/100}], \text{ i.e., } [0.478, 0.642].$$

Because the confidence interval is completely contained within the allowed range [0.35, 0.65], we can conclude similarity.

---

## The Similarity Confidence Interval Corresponding to the $\alpha$-Level TOST

Several procedures have been proposed in the literature to construct a $100(1 - \alpha)\%$ confidence interval associated with an $\alpha$-level test. Westlake (1976) constructed a symmetric $100(1 - \alpha)\%$ confidence interval. Liu (1990) proposed a sharper $100(1 - \alpha)\%$ confidence interval, which is always included in Westlake's symmetric confidence interval. Hsu (1984) and Hsu et al. (1994) derived a $100(1 - \alpha)\%$ confidence interval corresponding exactly to an $\alpha$-level TOST as (10.2-41) in the context of this chapter. The $100(1 - \alpha)\%$ confidence interval associated exactly with an $\alpha$-level test is referred to as a similarity confidence interval:

$$CI = [0.5 + \min(0, \widehat{p}_l - 0.5), 0.5 + \max(0, \widehat{p}_u - 0.5)], \qquad (10.2-41)$$

where $\widehat{p}_l = \widehat{p} - z_{1-\alpha}s_{\widehat{p}}$ and $\widehat{p}_u = \widehat{p} + z_{1-\alpha}s_{\widehat{p}}$ denote the conventional $100(1 - 2\alpha)\%$ lower and upper confidence limits. This type of confidence interval in (10.2-41) was derived by Hsu (1984), Bofinger (1985), and Stefansson et al. (1988) in the multiple comparisons setting, and by Muller-Cohrs (1991), Bofinger (1992), and Hsu et al. (1994) in the bioequivalence setting. The $100(1 - \alpha)\%$ similarity confidence interval in (10.2-41) is equal to the $100(1 - 2\alpha)\%$ confidence interval in (10.2-38) when the interval (10.2-38) contains 0.5. But, when the interval (10.2-38) lies above (or below) 0.5, the interval (10.2-41) is different from the interval (10.2-38). The general form of $(1 - \alpha)\%$ confidence interval corresponding to a size-$\alpha$ TOST is $CI = [\min(0, L), \max(0, U)$, where $L$ and $U$ denote the lower and upper limits for equal-tailed $(1-2\alpha)\%$ confidence interval.

| | |
|---|---|
| **Example 10.2-10** | For the example in 10.2-8, the $100(1 - \alpha)\%$ similarity confidence interval in (10.2-41), which connects exactly to the TOST, is |

$$CI = [0.5 + \min(0, 0.478 - 0.5),\ 0.5 + \max(0, 0.642 - 0.5)],$$

i.e.,

$$CI = [0.478, 0.642],$$

which is the same as that in Example 10.2-9. Note that $\widehat{p}_l = 0.56 - 1.64\sqrt{0.56 \times 0.44/100} = 0.478$ and $\widehat{p}_u = 0.56 + 1.64\sqrt{0.56 \times 0.44/100} = 0.642$. If, for example, the observed number for preference is 60 (rather than 56) in the previous example, then $\widehat{p}_l = 0.6 - 1.64\sqrt{0.6 \times 0.4/100} = 0.519$ and $\widehat{p}_u = 0.6 + 1.64\sqrt{0.6 \times 0.4/100} = 0.681$. The $100(1 - \alpha)\%$ similarity confidence interval in (10.2-41) should be

$$CI = [0.5 + \min(0, 0.519 - 0.5), 0.5 + \max(0, 0.681 - 0.5)],$$

i.e.,

$$CI = [0.5, 0.681].$$

It is slightly wider than the conventional two-sided $100(1 - 2\alpha)\%$ confidence interval in (10.2-38), $CI = [0.519, 0.681]$.

## 10.3   REPLICATED DIFFERENCE AND SIMILARITY TESTS

### Introduction

The motivation behind replicated tests, including replicated difference and similarity tests, is based on the belief that both between and within consumers variations exist. Only replicated tests can account for both sources of variability. In addition, replicated tests may increase dramatically the estimation precision and testing power.

Various models and approaches for analyses of replicated discrimination test data have been suggested in the statistical and sensory literature; e.g., Ferris (1958) for the *k*-visit model of consumer preference testing, Horsnell (1969) for a family of composite models of which the Ferris method is a particular case, Wierenga (1974) for a total stochastic model in paired comparison product testing, Morrison (1978) for a probability model for forced binary choices, Harries and Smith (1982) for the beta-binomial model for the triangular test, Ennis and Bi (1998, 1999) for the beta-binomial model and the Dirichlet-multinomial model for difference and preference tests, Brockhoff and Schlich (1998) for an adjusted overdispersion approach, Kunert and Meyners (1999) for a binomial mixture model, Hunter et al. (2000) for the generalized linear model, and Brockhoff (2003) for corrected versions of the beta-binomial model and the generalized linear model.

There are two main types of models. One is the composite model, and the other is the stochastic model. In the composite model, a consumer population is assumed to be composed of *A*-preferrers and *B*-preferrers, or discriminators and nondiscriminators, and the panelists of a laboratory panel are assumed to have the same discrimination ability. In the stochastic model, however, the personal preference or discrimination ability is regarded as a random variable. The probabilistic interpretation of personal preference and discrimination ability might be more reasonable.

In this section we discuss three stochastic models and their applications in replicated difference and similarity tests. They are the beta-binomial model, the corrected beta-binomial model, and the Dirichlet-multinomial model. We can regard the corrected beta-binomial model and the Dirichlet-multinomial model as different extensions of the beta-binomial model.

## The Beta-Binomial (BB) Model

### Beta-Binomial Distribution

The beta-binomial is a compound distribution of the beta and binomial distributions. It is obtained when the parameter $p$ in the binomial distribution is assumed to follow a beta distribution. Suppose $X$ is the number of a panelist preferring a specified product in a replicated preference test with $n$ replications, and further suppose it follows a binomial distribution conditionally with probability of preference $p$. If $p$ is assumed to have a beta distribution with parameters $a$ and $b$, then the unconditional distribution of $X$ is beta-binomial with its probability distribution function

$$\Pr\left(X = x | n, a, b\right) = \binom{n}{x} \frac{\Gamma(a+b)\Gamma(a+x)\Gamma(n-x+b)}{\Gamma(a)\Gamma(b)\Gamma(a+b+n)}, \quad (10.3\text{-}1)$$

where $\Gamma(.)$ denotes the gamma function. It is useful to define $\mu = \dfrac{a}{a+b}$ and $\theta = \dfrac{1}{a+b}$. Hence, $a = \dfrac{\mu}{\theta}$ and $b = \dfrac{1-\mu}{\theta}$. Sometimes, parameter $\gamma$ is used instead of $\theta$. Parameter $\gamma$ is defined as $\gamma = \dfrac{1}{a+b+1}$; hence, $a = \mu\left(\dfrac{1}{\gamma} - 1\right)$ and $b = (1-\mu)\left(\dfrac{1}{\gamma} - 1\right)$. The relationship of $\gamma$ and $\theta$ is $\theta = \dfrac{\gamma}{1-\gamma}$ or $\gamma = \dfrac{\theta}{1+\theta}$. Usually, $\gamma$ varies between 0 and 1.

The mean and variance of the beta-binomial distribution in (10.3-1) are

$$E(X) = n\mu, \quad (10.3\text{-}2)$$

$$Var(X) = n\mu(1-\mu)[1 + \gamma(n-1)]. \quad (10.3\text{-}3)$$

If $X$ has a binomial distribution with parameters $n$ and $p$, then its mean and variance are $n\mu$ and $n\mu(1-\mu)$, respectively. Thus, the term $1 + (n-1)\gamma$ in the variance of the beta-binomial distribution, multiplier of the binomial variance, is at least 1 and models the overdispersion due to the variance of $p$. The case of pure binomial variation in which parameter $p$ is a constant corresponds to degenerate prior beta distribution of $p$. The maximum variance of the beta-binomial corresponds to $\gamma = 1$ when the maximum inflation factor $1 + (n-1)\gamma$ is $n$. Notice that the variance of the beta-binomial distribution is always greater than the corresponding variance of a binomial distribution. This property is referred to as overdispersion.

Underdispersion (i.e., $\gamma < 0$) is also possible in an extended beta-binomial model (see, e.g., Prentice, 1986). In this book, however, we will discuss only overdispersion because it has more practical meanings. The main reason for using the beta-binomial model in replicated tests is overdispersion ($\gamma > 0$). The possibility of underdispersion ($\gamma < 0$) is of theoretical importance. It means that the binomial distribution (i.e., $\gamma = 0$) does not occur at an endpoint of the allowable range of $\gamma$. Consequently, it is possible to perform a valid likelihood ratio test of the null hypothesis that $\gamma = 0$.

The parameters in the beta-binomial model can be estimated either by the method of maximum likelihood or the method of moments.

### Maximum Likelihood Estimation

The log likelihood of the beta-binomial distribution is as

$$L = \sum_{i=1}^{k} \log[\Pr(x_i|n_i, \mu, \theta)], \tag{10.3-4}$$

where $k$ is the number of panelists, and $x_i$ and $n_i$ are the numbers of preference and replications for the $i$th panelist. The log likelihood can be expressed as (see, e.g., Smith, 1983; Morgan 1992, 242).

$$L = \sum_{i=1}^{k} \left\{ \sum_{j=0}^{x_i-1} \log(\mu + j\theta) + \sum_{j=0}^{n_i-x_i-1} \log(1 - \mu + j\theta) - \sum_{j=0}^{n_i-1} \log(1 + j\theta) \right\} + c, \tag{10.3-5}$$

where $c$ is a constant. The maximum likelihood estimators $\widehat{\mu}$ and $\widehat{\theta}$ are the values of $\mu$ and $\theta$ in (10.3-5), which maximize $L$. The covariance matrix of the estimators $\widehat{\mu}$ and $\widehat{\theta}$ can be obtained from the negative expected values of the second partial derivatives of the parameters. The maximum likelihood estimators $\widehat{\mu}$ and $\widehat{\gamma}$ can be obtained by replacing $\theta$ with $\frac{\gamma}{1-\gamma}$. Using advanced statistical software, we can easily obtain maximum likelihood estimation for the parameters $\mu$ and $\theta$ (or $\gamma$), for example, by using the built-in function *nlminb* in S-PLUS. We can also obtain the covariance matrix for the estimators of the parameters easily by using the built-in function *vcov.nlminb* in the S-PLUS library.

### Moment Estimation

For equal replications, the moment estimates of the parameters $\mu$ and $\gamma$ are (Kleinman, 1973)

$$\widehat{\mu} = \frac{\sum_{i=1}^{k} x_i}{nk}, \tag{10.3-6}$$

$$\widehat{\gamma} = \frac{nS}{k(n-1)(1-\widehat{\mu})\widehat{\mu}} - \frac{1}{(n-1)}, \tag{10.3-7}$$

where $S = \sum_{i=1}^{k} (\widehat{p}_i - \widehat{\mu})^2$ and $\widehat{p}_i = \frac{x_i}{n}$ is the proportion of selections for the $i$th panelist $(i = 1, 2, \ldots, k)$ with $n$ replications. The variance of $\widehat{\mu}$ is (10.3-8). It can be used to estimate the variance of the estimator $\widehat{\mu}$ at a specific value of $\mu$,

$$V(\widehat{\mu}) = \mu(1 - \mu)[(n - 1)\gamma + 1]/nk. \qquad (10.3\text{-}8)$$

Moment estimation in the situation of unequal numbers of replications or observations in the trials is more complicated than the equal replication case because of the weighting requirement. The estimates of $\mu$ and $\gamma$ can be obtained from

$$\widehat{\mu} = \frac{\sum_{i=1}^{k} w_i \widehat{p}_i}{w}, \qquad (10.3\text{-}9)$$

$$\widehat{\gamma} = \frac{S - \widehat{\mu}(1 - \widehat{\mu})\left[\sum_{i=1}^{k}\left(\frac{w_i}{n_i} - \frac{w_i^2}{n_i w}\right)\right]}{\widehat{\mu}(1 - \widehat{\mu})\left[\sum_{i=1}^{k}\left(w_i - \frac{w_i^2}{w}\right) - \sum_{i=1}^{k}\left(\frac{w_i}{n_i} - \frac{w_i^2}{n_i w}\right)\right]}, \qquad (10.3\text{-}10)$$

where $w = \sum_{i=1}^{k} w_i$, $w_i = \dfrac{n_i}{1 + \widehat{\gamma}(n_i - 1)}$, and $S = \sum_{i=1}^{k} w_i(\widehat{p}_i - \widehat{\mu})^2$.

The difficulty is that the weights, $w_i$, contain $\widehat{\gamma}$. We can iteratively use the equations (10.3-9) and (10.3-10) as follows. Start either with $w_i = n_i$ or $w_i = 1$ and obtain estimates of $\mu$ and $\gamma$. From this value of $\gamma$, say $\widehat{\gamma}_0$, compute $w_i = n_i/(1 + \widehat{\gamma}_0(n_i - 1))$ and use these "empirical" weights to form new estimates of $\mu$ and $\gamma$. We can repeat this process until successive estimates are almost unchanged. If $\widehat{\gamma} < 0$ is obtained in the iterative procedure, $\widehat{\gamma} = 0$ is assigned.

The variance of $\widehat{\mu}$ can be estimated from

$$\widehat{V}(\widehat{\mu}) = \frac{\sum_{i=1}^{k} w_i^2(\widehat{p}_i - \widehat{\mu})^2}{w^2 - \sum_{i=1}^{k} w_i^2}. \qquad (10.3\text{-}11)$$

---

**Example 10.3-1**

In a replicated preference test for products A and B with 100 consumer panelists and three replications, the numbers of product A selected for each panelist are displayed in Table 10.20. The maximum likelihood estimates for $\mu$ and $\gamma$ and the covariance matrix for the estimators are $\widehat{\mu} = 0.687$, $\widehat{\gamma} = 0.123$, and

$$\text{Cov}(\widehat{\mu}, \widehat{\gamma}) = \begin{pmatrix} 0.000850 & -0.000041 \\ -0.000041 & 0.004471 \end{pmatrix}.$$

The moment estimates of $\mu$ and $\gamma$ are $\widehat{\mu} = 0.687$, $\widehat{\gamma} = 0.117$, $\widehat{V}(\widehat{\mu}) = 0.0009$, and $\widehat{V}(\widehat{\gamma}) = 0.0874$. The following S-PLUS function calls can be used to generate these estimates:

**Table 10.20** Data for Replicated Preference Tests with Three Replicates (Example 10.3-1)

| No. | For "A" | No. | For "A" | No. | For "A" | No. | For "A" |
|-----|---------|-----|---------|-----|---------|-----|---------|
| 1 | 3 | 26 | 1 | 51 | 2 | 76 | 3 |
| 2 | 2 | 27 | 1 | 52 | 3 | 77 | 2 |
| 3 | 1 | 28 | 3 | 53 | 3 | 78 | 1 |
| 4 | 2 | 29 | 2 | 54 | 3 | 79 | 3 |
| 5 | 3 | 30 | 1 | 55 | 3 | 80 | 2 |
| 6 | 2 | 31 | 3 | 56 | 3 | 81 | 1 |
| 7 | 3 | 32 | 3 | 57 | 2 | 82 | 3 |
| 8 | 2 | 33 | 2 | 58 | 3 | 83 | 2 |
| 9 | 1 | 34 | 3 | 59 | 2 | 84 | 2 |
| 10 | 3 | 35 | 2 | 60 | 0 | 85 | 3 |
| 11 | 2 | 36 | 1 | 61 | 3 | 86 | 2 |
| 12 | 1 | 37 | 2 | 62 | 2 | 87 | 1 |
| 13 | 3 | 38 | 2 | 63 | 0 | 88 | 3 |
| 14 | 1 | 39 | 2 | 64 | 3 | 89 | 2 |
| 15 | 2 | 40 | 3 | 65 | 1 | 90 | 1 |
| 16 | 3 | 41 | 1 | 66 | 1 | 91 | 3 |
| 17 | 2 | 42 | 1 | 67 | 2 | 92 | 2 |
| 18 | 1 | 43 | 3 | 68 | 1 | 93 | 2 |
| 19 | 3 | 44 | 3 | 69 | 1 | 94 | 3 |
| 20 | 2 | 45 | 3 | 70 | 3 | 95 | 2 |
| 21 | 1 | 46 | 3 | 71 | 1 | 96 | 1 |
| 22 | 3 | 47 | 3 | 72 | 0 | 97 | 3 |
| 23 | 1 | 48 | 0 | 73 | 3 | 98 | 1 |
| 24 | 2 | 49 | 3 | 74 | 2 | 99 | 1 |
| 25 | 3 | 50 | 2 | 75 | 2 | 100 | 3 |

```
> x<-nlminb(start = c(0.1,0.2),x = bbest4dat,objective =
  bbmaxg)
> x$par
[1] 0.6872332 0.1230780
> vcov.nlminb(x)
    [,1]    [,2]
[1,] 0.00085038992 -0.00004108528
[2,] -0.00004108528 0.00447065739
> bbmoest2(bbest4dat)
[1] 0.6867 0.1169 0.0009 0.0874
> bbest4dat
  times of selection replications
1  3  3
2  2  3
3  1  3
..........
```

### Chi-square Goodness of Fit Test

We can evaluate the fit of the beta-binomial model to the data in Table 10.20. The observed and expected frequencies are shown in Table 10.21. The predicted frequencies by the beta-binomial model are based on the probability distribution function of the beta-binomial distribution (10.3-1) with parameters $\hat{\mu} = 0.687$, $\hat{\gamma} = 0.123$, and $n = 3$; i.e., $a = 4.898$ and $b = 2.232$. The predicted frequencies by the binomial model are based on the probability distribution function of the binomial distribution with parameters $\hat{p} = 0.687$ and $n = 3$. The values of the $X^2$ goodness of fit statistic in (9.3-2) are 2.43 and 6.16, respectively, for the two models. The statistic follows approximately a chi-square distribution with 2 and 3 DF,

**Table 10.21** Observed and Predicted Frequencies for the Data in Table 10.20

| Number of Preferring A in Three Replications | Observed Frequencies | Expected Frequencies by BB | Expected Frequencies by Binomial |
|---|---|---|---|
| 0 | 4 | 5.8 | 3.1 |
| 1 | 25 | 20.0 | 20.2 |
| 2 | 32 | 36.6 | 44.3 |
| 3 | 39 | 37.7 | 32.4 |
| $X^2$ | | 2.43 (P-value = 0.30) | 6.16 (P-value = 0.10) |

respectively. The associated *P*-values are 0.30 and 0.10, respectively. These values indicate that the beta-binomial model is better than the binomial model to fit the observed data.

```
> bb<-seq(1,4)
> for(i in 1:4){bb[i]<-bbdis(i-1,3,c(4.898,2.232))*100}
> bb
[1] 5.768466 20.028791 36.550064 37.652679
> b<-seq(1,4)
> for(i in 1:4){b[i]<-dbinom(i-1,3,0.687)*100}
> b
[1] 3.06643 20.19141 44.31789 32.42427
> sum((ob-round(bb,1))^2/round(bb,1))
[1] 2.43159
> sum((ob-round(b,1))^2/round(b,1))
[1] 6.161453
> ob
[1] 4 25 32 39
```

### Tarone's Statistic for Goodness of Fit Test

Tarone's $Z$ statistic in (10.3-13) (Tarone, 1979; Paul et al., 1989) can also be used to test the goodness of fit of the binomial distribution against the beta-binomial distribution. The null hypothesis is that the underling distribution is a binomial distribution, whereas the alternative hypothesis is that the underlying distribution is a beta-binomial distribution. In other words,

$$H_0: \gamma = 0 \quad \text{versus} \quad H_a: \gamma > 0. \qquad (10.3\text{-}12)$$

The test statistic is

$$Z = \frac{E - nk}{\sqrt{2kn(n-1)}}, \qquad (10.3\text{-}13)$$

where $E = \dfrac{\sum_{i=1}^{k}(x_i - n\widehat{\mu})^2}{\widehat{\mu}(1 - \widehat{\mu})}$ and $\widehat{\mu} = \dfrac{\sum_{i=1}^{k} x_i}{nk}$.

The statistic $Z$ in (10.3-13) has an asymptotic standard normal distribution under the null hypothesis of a binomial distribution. If $Z > z_{1-\alpha}$, we can conclude at the significance level $\alpha$ that the underlying distribution is beta-binomially distributed. Because the parameter $\gamma$ cannot take negative values, the test is always one-sided. However, if the null hypothesis cannot be rejected, we cannot conclude that the distribution is a binomial. In this case, treating the replicated discrimination testing data as beta-binomial data is more robust than treating them as binomial data. Values of the parameter $\gamma > 0$ reflect different response patterns and suggest existence of latent preference groups.

---

<table>
<tr><td>**Example 10.3-2**</td><td>

For the data in Example 10.3-1, $\hat{\mu} = 0.6867$, $E = 370.15$; hence,

$$Z = \frac{370.15 - 3 \times 100}{\sqrt{2 \times 100 \times 3 \times 2}} = 2.025,$$

with associated $P$-value $= 0.0214$. This value indicates that the data are beta-binomial distributed rather than binomial distributed.

```
> bbdist(bbest4dat[,1],3,100)
[1] 2.0251 0.0214
```

</td></tr>
</table>

---

## Tests Based on the BB Model

### *Replicated Difference and Preference Tests Using the Paired Comparison Method*

#### Test for Parameters in a Single Experiment

The Wald-type test statistics, which are based on the maximum-likelihood estimations of the parameters in the beta-binomial model, can be used for replicated difference and preference tests to test if one or both of the parameters $\mu$ and $\gamma$ are equal to some fixed value(s).

For the test: $H_0$: $\mu = \mu_0$ and $\gamma = \gamma_0$ versus $H_1$: $\mu \neq \mu_0$ and/or $\gamma \neq \gamma_0$, the test statistics is

$$X^2 = (\hat{\mu} - \mu_0, \hat{\gamma} - \gamma_0)[S]^{-1}(\hat{\mu} - \mu_0, \hat{\gamma} - \gamma_0)', \tag{10.3-14}$$

where $[S]^{-1}$ denotes the inverse of the covariance matrix of the estimator of the parameter vector $(\hat{\mu}, \hat{\gamma})$.

$$S = \begin{pmatrix} \hat{\sigma}_1^2 & r\hat{\sigma}_1\hat{\sigma}_2 \\ r\hat{\sigma}_1\hat{\sigma}_2 & \hat{\sigma}_2^2 \end{pmatrix},$$

where $\hat{\sigma}_1$ and $\hat{\sigma}_2$ are estimated variances of $\hat{\mu}$ and $\hat{\gamma}$, respectively, and $r$ is the estimated correlation coefficient between $\hat{\mu}$ and $\hat{\gamma}$. The test statistic $X^2$ follows asymptotically a chi-square distribution with 2 DF. Equation (10.3-14) is algebraically equivalent to (10.3-15):

$$X^2 = \left[\frac{A + B}{\sqrt{2(1 + r)}}\right]^2 + \left[\frac{A - B}{\sqrt{2(1 - r)}}\right]^2, \tag{10.3-15}$$

where $A = \dfrac{\hat{\mu} - \mu_0}{\hat{\sigma}_1}$ and $B = \dfrac{\hat{\gamma} - \gamma_0}{\hat{\sigma}_2}$.

For the test: $H_0: \mu = \mu_0$ versus $H_a: \mu \neq \mu_0$, the test statistics is

$$Z = |\widehat{\mu} - \mu_0| / \sqrt{\widehat{V}(\widehat{\mu})}, \qquad (10.3\text{-}16)$$

where $\widehat{V}(\widehat{\mu})$ denotes the estimated variance of the estimator $\widehat{\mu}$.

For the test: $H_0: \gamma = \gamma_0$ versus $H_a: \gamma \neq \gamma_0$, the test statistics is

$$Z = |\widehat{\gamma} - \gamma_0| / \sqrt{\widehat{V}(\widehat{\gamma})}, \qquad (10.3\text{-}17)$$

where $\widehat{V}(\widehat{\gamma})$ denotes the estimated variance of the estimator $\widehat{\gamma}$.

---

**Example 10.3-3**

Using the results in Example 10.3-1 for the data in Table 10.20, we test $H_0 : \mu = 0.5, \gamma = 0$ versus $H_a: \mu \neq 0.5, \gamma \neq 0$. According to (10.3-15),

$$A = \frac{0.687 - 0.5}{\sqrt{0.00085}} = 6.421, \quad B = \frac{0.123 - 0}{\sqrt{0.0047}} = 1.841, \quad r = \frac{-0.000041}{\sqrt{0.00085 \times 0.00447}}$$

$= -0.021$. Hence, the value of the test statistic is

$$X^2 = \left[ \frac{6.421 + 1.841}{\sqrt{2 \times (1 - 0.021)}} \right]^2 + \left[ \frac{6.421 - 1.841}{\sqrt{2 \times (1 + 0.021)}} \right]^2 = 45.13,$$

with the associated $P$-value $< 0.0001$ from a chi-square distribution with 2 DF. Hence, we can reject the null hypothesis that $H_0: \mu = 0.5, \gamma = 0$ at a 0.05 significance level. The calculations in S-PLUS are as follows:

```
> t(bb4m-a)%*%solve(bb4v)%*%(bb4m-a)
    [,1]
[1,] 45.13022
> a
[1] 0.5 0.0
> bb4m
[1] 0.6872332 0.1230780
> bb4v
    [,1]      [,2]
[1,] 0.00085038992 -0.00004108528
[2,] -0.00004108528 0.00447065739
```

### Test for Parameters in Different Experiments

Asymptotic likelihood ratio tests can be used for comparisons of parameters in $t$ different experiments. For example, consumer preference tests are conducted in different cities for two products. The investigators want to know if consumers in different cities show different preferences and patterns of preference for the two products. The test hypotheses are

$$H_0: \mu_1 = \mu_2 = \ldots = \mu_t, \gamma_1 = \gamma_2 = \ldots = \gamma_t;$$

$$H_a: \mu_i \neq \mu_j; \gamma_i \neq \gamma_j, \text{ which holds at least in one situation, } i, j = 1, 2, \ldots, t.$$

(10.3-18)

The likelihood ratio test statistic is

$$X^2 = 2(L_1 - L_0),$$
(10.3-19)

where $L_1$ is the value of the log-likelihood function maximized subject to no constraints on the values of $\mu_i$, $\gamma_i$, $i = 1, 2, \ldots, t$ (this is obtained by maximizing separately the log-likelihood for each of $t$ experiments, $L_1 = \sum_i^t L_{1i}$) and $L_0$ is the log-likelihood value obtained from pooling the data for all experiments. The likelihood ratio test statistic in (10.3-19) follows asymptotically a chi-square distribution with $2(t - 1)$ DF.

---

**Example 10.3-4**

Replicated preference tests were conducted in three cities. The data are displayed in Table 10.22 as fractions $x/n$. We wanted to test if the consumers in different cities have similar preference responses. In this case, we used the likelihood ratio test statistic. The maximum likelihood estimates and log likelihood for the data from the three cities are, respectively, $\hat{\mu}_1 = 0.4704$, $\hat{\gamma}_1 = 0.1816$, $-L_{11} = 65.516$; $\hat{\mu}_2 = 0.5879$, $\hat{\gamma}_2 = 0.3505$, $-L_{12} = 63.923$; $\hat{\mu}_3 = 0.7013$, $\hat{\gamma}_3 = 0.5019$, $-L_{13} = 54.282$. Hence, $L_1 = -(65.516 + 63.923 + 54.282) = -183.721$. Using the pooled data of the three cities, we get $\hat{\mu}_1 = \hat{\mu}_2 = \hat{\mu}_3 = 0.5834$, $\hat{\gamma}_1 = \hat{\gamma}_2 = \hat{\gamma}_3 = 0.3551$, $-L_0 = 191.386$. The value of the likelihood ratio test statistic is then

$$X^2 = 2(-183.721 + 191.386) = 15.33,$$

with the associated $P$-value $= 0.004$ from a chi-square distribution with 4 DF. The test result suggests that consumers in the different cities have difference preferences for the two products. Pooling the preference data

**Table 10.22** Data for Replicated Preference Tests in Three Cities (Example 10.3-4)

| City 1 | | City 2 | | City 3 | |
|---|---|---|---|---|---|
| 0/2 | 2/3 | 1/3 | 2/3 | 3/3 | 2/3 |
| 2/3 | 2/3 | 2/3 | 3/3 | 2/3 | 1/3 |
| 1/2 | 2/3 | 3/3 | 0/3 | 3/3 | 3/3 |
| 0/3 | 1/3 | 3/3 | 0/3 | 3/3 | 3/3 |
| 1/3 | 0/3 | 1/3 | 3/3 | 3/3 | 1/3 |
| 0/3 | 0/3 | 1/3 | 2/3 | 0/3 | 3/3 |
| 0/3 | 2/3 | 2/3 | 2/3 | 3/3 | 3/3 |
| 2/3 | 2/3 | 1/1 | 3/3 | 2/2 | 3/3 |
| 1/3 | 0/3 | 3/3 | 2/3 | 1/3 | 3/3 |
| 1/2 | 1/3 | 0/3 | 1/3 | 3/3 | 2/3 |
| 2/3 | 2/3 | 3/3 | 1/3 | 1/3 | 0/2 |
| 2/3 | 2/3 | 3/3 | 0/3 | 1/3 | 3/3 |
| 0/3 | 1/3 | 1/3 | 0/3 | 3/3 | 0/2 |
| 3/3 | 2/3 | 0/3 | 3/3 | 2/3 | 0/3 |
| 1/3 | 2/3 | 3/3 | 2/3 | 0/3 | 3/3 |
| 0/3 | 2/3 | 1/3 | 0/3 | 3/3 | 1/3 |
| 2/3 | 1/3 | 2/3 | 3/3 | 3/3 | 0/3 |
| 3/3 | 0/3 | 3/3 | 2/3 | 3/3 | 3/3 |
| 0/3 | 3/3 | 2/3 | 0/3 | 3/3 | 2/3 |
| 3/3 | 0/1 | 1/3 | 2/3 | 2/3 | 3/3 |
| 2/2 | 1/3 | 3/3 | 2/3 | 2/3 | 3/3 |
| 2/2 | 3/3 | 2/2 | 1/2 | 3/3 | |
| 3/3 | 1/3 | 1/3 | 3/3 | 0/3 | |
| 1/3 | 1/2 | 3/3 | | 1/3 | |
| 0/3 | 2/3 | 0/3 | | 3/3 | |

of the different cities does not seem appropriate. The results using the S-PLUS program are as follows:

```
> x<-nlminb(start = c(0.1,0.2),x = city1,objective =
  bbmaxg)
> x$par
[1] 0.4704141 0.1816396
> x$objective
[1] 65.51603
> x<-nlminb(start = c(0.1,0.2),x = city2,objective =
  bbmaxg)
> x$par
[1] 0.5878899 0.3504648
> x$objective
[1] 63.92261
> x<-nlminb(start = c(0.1,0.2),x = city3,objective =
  bbmaxg)
> x$par
[1] 0.7012685 0.5018679
> x$objective
[1] 54.28199
> a123<-rbind(a1,a2,a3)
> x<-nlminb(start = c(0.1,0.2),x = city123,objective =
  bbmaxg)
> x$par
[1] 0.5833675 0.3551310
> x$objective
[1] 191.3859
```

### Replicated Similarity Test Using Paired Comparison Method

Let $p$ denote the probability of selecting a specified product in a pair of products and $D = \mu - 0.5$. Let $\Delta$, which defines similarity, be the distance from 0.5. The hypotheses for similarity testing are

$$H_0: \; D \geq +\Delta \quad \text{or} \quad D \leq -\Delta \quad \text{versus} \quad H_a: \; -\Delta < D < +\Delta. \quad (10.3\text{-}20)$$

This is the one-sample version of the interval hypotheses for similarity testing. The test in (10.3-20) can be broken down into two one-sided tests (TOST):

$$H_{01}: \; D \geq +\Delta \quad \text{versus} \quad H_{a1}: \; D < +\Delta; \quad (10.3\text{-}21)$$

$$H_{02}: \; D \leq -\Delta \quad \text{versus} \quad H_{a2}: \; D > -\Delta. \quad (10.3\text{-}22)$$

The test statistics used to test the hypotheses are

$$\frac{\widehat{\mu} - \mu_{01}}{s_u} \leq z_\alpha, \tag{10.3-23}$$

$$\frac{\widehat{\mu} - \mu_{02}}{s_l} \geq z_{1-\alpha}, \tag{10.3-24}$$

where $\widehat{\mu} = \sum_{i=1}^{k} x_k/nk$, $\mu_{01} = 0.5 + \Delta$, $\mu_{02} = 0.5 - \Delta$, $s_u = \sqrt{V(\widehat{\mu})_{01}}$, $s_l = \sqrt{V(\widehat{\mu})_{02}}$, $V(\widehat{\mu})_{01} = \mu_{01}(1 - \mu_{01})[(n-1)\gamma + 1]/nk$, and $V(\widehat{\mu})_{02} = \mu_{02}(1 - \mu_{02})[(n-1)\gamma + 1]/nk$. The critical values, $\mu_l$ and $\mu_u$, are defined as

$$\mu_u = z_\alpha s_u + 0.5 + \Delta, \tag{10.3-25}$$

$$\mu_l = z_{1-\alpha} s_l + 0.5 - \Delta, \tag{10.3-26}$$

where $z_\alpha$ and $z_{1-\alpha}$ are the 100 $\alpha$ and 100(1 − $\alpha$) percentiles of the standard normal distribution. One should note that in the two-sided similarity testing, no multiplicity adjustment is needed for the significance level $\alpha$ based on the intersection-union test (IUT) principle (Berger, 1982; Berger and Hsu, 1996). Table A.31 shows the critical values for $\gamma = 0.1$, 0.2, 0.3, and 0.4; $k = 20$ to 100; $n = 3$ and 4. The null hypothesis $H_0$ in (10.3-20) will be rejected and the alternative hypothesis $H_a$ in (10.3-20) will be accepted—i.e., similarity will be concluded—if the value of the statistic $\widehat{\mu}$ is in the range of $\mu_l$ and $\mu_u$.

---

**Example 10.3-5**

Some ingredients were changed in a product. It was expected that the change of the ingredients did not seriously affect, if any, the consumer's preference for the product. A similarity preference test was conducted using the paired comparison method for the new product (B) and the current product (A). There were 100 panelists with four replications in the test, and the allowed difference of preference was $\Delta = 0.1$; i.e., the difference of proportion preferring product A and 0.5 was no more than 0.1. The Type I error $\alpha = 0.05$ was selected. The responses for the 100 panelists with four replications appear in Table 10.23.

The maximum likelihood estimates of the parameters are $\widehat{\mu} = 0.5165$ and $\widehat{\gamma} = 0.408$ based on (10.3-1) and (10.3-4). The moment estimates of the parameters are $\widehat{\mu} = 0.5200$ and $\widehat{\gamma} = 0.406$ based on (10.3-6) and (10.3-7). The critical values of the TOST are $\mu_l = 0.4601$ and $\mu_u = 0.5399$ based on (10.3-25) and (10.3-26) for $\Delta = 0.1$, $\alpha = 0.05$, $n = 4$, $k = 100$, and $\gamma = 0.408$. We can also find approximate critical values $\mu_l = 0.4598$ and $\mu_u = 0.5402$ from Table A.31 for $\Delta = 0.1$, $\alpha = 0.05$, $n = 4$, $k = 100$, and $\gamma = 0.4$.

**Table 10.23** Numbers Preferring Sample A in Replicated Similarity Test Using Paired Comparison Method ($k = 100$, $n = 4$) (Example 10.3-5)

| ID | r | ID | r | ID | r | ID | r |
|----|---|----|---|----|---|----|---|
| 1 | 3 | 26 | 0 | 51 | 1 | 76 | 4 |
| 2 | 2 | 27 | 2 | 52 | 0 | 77 | 2 |
| 3 | 3 | 28 | 0 | 53 | 1 | 78 | 3 |
| 4 | 0 | 29 | 2 | 54 | 2 | 79 | 2 |
| 5 | 3 | 30 | 2 | 55 | 4 | 80 | 1 |
| 6 | 1 | 31 | 0 | 56 | 2 | 81 | 1 |
| 7 | 4 | 32 | 4 | 57 | 4 | 82 | 4 |
| 8 | 4 | 33 | 2 | 58 | 2 | 83 | 3 |
| 9 | 0 | 34 | 1 | 59 | 0 | 84 | 4 |
| 10 | 1 | 35 | 2 | 60 | 4 | 85 | 0 |
| 11 | 0 | 36 | 4 | 61 | 0 | 86 | 2 |
| 12 | 2 | 37 | 4 | 62 | 4 | 87 | 2 |
| 13 | 2 | 38 | 0 | 63 | 3 | 88 | 1 |
| 14 | 4 | 39 | 4 | 64 | 4 | 89 | 0 |
| 15 | 1 | 40 | 3 | 65 | 3 | 90 | 3 |
| 16 | 1 | 41 | 3 | 66 | 3 | 91 | 0 |
| 17 | 0 | 42 | 2 | 67 | 4 | 92 | 0 |
| 18 | 4 | 43 | 0 | 68 | 4 | 93 | 4 |
| 19 | 0 | 44 | 3 | 69 | 4 | 94 | 4 |
| 20 | 4 | 45 | 3 | 70 | 3 | 95 | 3 |
| 21 | 3 | 46 | 3 | 71 | 0 | 96 | 3 |
| 22 | 2 | 47 | 0 | 72 | 1 | 97 | 4 |
| 23 | 0 | 48 | 1 | 73 | 0 | 98 | 1 |
| 24 | 3 | 49 | 1 | 74 | 2 | 99 | 3 |
| 25 | 0 | 50 | 0 | 75 | 2 | 100 | 4 |

Because the estimated value of $\mu$ is $\widehat{\mu} = 0.5165$, which is in the range of the critical values 0.4601 and 0.5399, we can conclude that the consumer preferences for the new product (B) and the current product (A) are similar in terms of the allowed difference $\Delta = 0.1$. However, we should also note that the large $\gamma$ value ($\widehat{\gamma} = 0.408$) indicates the higher overdispersion and suggests possible consumer segmentation for preference of the products.

### Replicated Difference Tests Using A–Not A and Same-Different Methods

In this section, we discuss replicated difference tests using the methods with response bias. The same procedure can be used for both the A–Not A and same-different methods.

Bi and Ennis (2001) discussed application of the beta-binomial model in the replicated monadic A–Not A method. In the method, each panelist receives more than one sample of A or Not A but not both. The total numbers of sample A and Not A are fixed in advance. This is a replicated homogeneity test. In this situation, we can use an adjusted Pearson's chi-square statistic

$$\tilde{X}_P^2 = \sum_{i=1}^{2}\sum_{j=1}^{2} \frac{(\tilde{n}_{ij} - \widehat{\tilde{E}}_{ij})^2}{\widehat{\tilde{E}}}, \tag{10.3-27}$$

where $\tilde{n}_{ij} = \dfrac{n_{ij}}{\widehat{C}_j}$, $\widehat{\tilde{E}}_{ij} = \tilde{N}_j \left( \dfrac{\tilde{n}_{i1} + \tilde{n}_{i2}}{\tilde{N}_1 + \tilde{N}_2} \right)$, $\tilde{N}_j = \dfrac{N_j}{\widehat{C}_j}$, $\widehat{C}_j = 1 + (n_j - 1)\widehat{\gamma}_j$, $n_{ij}$ is the number of responses in the $2 \times 2$ table for the A–Not A test; $N_j$ is the sample size for A or Not A; $n_j$ is the number of replications, and $\widehat{\gamma}_j$ is an estimator of the beta-binomial parameter $\gamma_j$ in the $j$ th population, $j = 1$ for the population for sample A, and $j = 2$ for sample Not A. The adjusted Pearson's chi-square statistic follows asymptotically a chi-square distribution with 1DF. Using adjusted observations $\tilde{n}_{ij}$ and the S-PLUS built-in function *chisq.test*, we can get the result easily.

---

A replicated monadic A–Not A test was conducted with 100 panelists and three replications. Fifty panelists received sample A, and 50 panelists received sample Not A. The number of "A" responses for each panelist in the three replications is listed in Table 10.24. For the population for sample A, the maximum likelihood estimation of $\gamma$ is $\widehat{\gamma}_1 = 0.1776$. For the population for sample Not A, $\widehat{\gamma}_2 = 0.0639$. Hence,

| **Example 10.3-6** |

$$\widehat{C}_1 = 1 + (3 - 1) \times 0.1776 = 1.355,$$

$$\widehat{C}_2 = 1 + (3 - 1) \times 0.0639 = 1.128.$$

**Table 10.24** Data for a Replicated Monadic A–Not A Test with Three Replications (Example 10.3-6)

| Sample A | | | | Sample Not A | | | |
|---|---|---|---|---|---|---|---|
| No. | Response "A" | No. | Response "A" | No. | Response "A" | No. | Response "A" |
| 1 | 2 | 26 | 3 | 1 | 3 | 26 | 2 |
| 2 | 1 | 27 | 1 | 2 | 2 | 27 | 3 |
| 3 | 2 | 28 | 3 | 3 | 3 | 28 | 1 |
| 4 | 3 | 29 | 2 | 4 | 1 | 29 | 3 |
| 5 | 1 | 30 | 0 | 5 | 3 | 30 | 3 |
| 6 | 2 | 31 | 2 | 6 | 2 | 31 | 1 |
| 7 | 3 | 32 | 2 | 7 | 3 | 32 | 2 |
| 8 | 3 | 33 | 2 | 8 | 3 | 33 | 2 |
| 9 | 0 | 34 | 2 | 9 | 2 | 34 | 2 |
| 10 | 1 | 35 | 3 | 10 | 1 | 35 | 1 |
| 11 | 3 | 36 | 3 | 11 | 2 | 36 | 3 |
| 12 | 1 | 37 | 2 | 12 | 2 | 37 | 1 |
| 13 | 2 | 38 | 2 | 13 | 1 | 38 | 2 |
| 14 | 3 | 39 | 3 | 14 | 3 | 39 | 2 |
| 15 | 3 | 40 | 2 | 15 | 3 | 40 | 2 |
| 16 | 3 | 41 | 3 | 16 | 1 | 41 | 3 |
| 17 | 3 | 42 | 3 | 17 | 3 | 42 | 3 |
| 18 | 3 | 43 | 1 | 18 | 3 | 43 | 3 |
| 19 | 3 | 44 | 2 | 19 | 1 | 44 | 3 |
| 20 | 2 | 45 | 3 | 20 | 2 | 45 | 2 |
| 21 | 3 | 46 | 3 | 21 | 3 | 46 | 2 |
| 22 | 3 | 47 | 3 | 22 | 0 | 47 | 3 |
| 23 | 1 | 48 | 1 | 23 | 1 | 48 | 2 |
| 24 | 2 | 49 | 3 | 24 | 2 | 49 | 3 |
| 25 | 3 | 50 | 3 | 25 | 2 | 50 | 3 |
| | | $n_{11} = 113$ | | | | $n_{12} = 103$ | |

Then, we can get

$$\tilde{N}_1 = 150/1.355 = 110.701,$$

$$\tilde{N}_2 = 150/1.128 = 132.979,$$

$$\tilde{n}_{11} = 113/1.355 = 83.395,$$

$$\tilde{n}_{12} = 103/1.128 = 91.312,$$

$$\tilde{n}_{21} = (150 - 113)/1.355 = 27.306,$$

$$\tilde{n}_{22} = (150 - 103)/1.128 = 41.667,$$

$$\widehat{\tilde{E}}_{11} = 110.701 \times (83.395 + 91.312)/(110.701 + 132.979) = 79.367,$$

$$\widehat{\tilde{E}}_{12} = 132.979 \times (83.395 + 91.312)/(110.701 + 132.979) = 95.340,$$

$$\widehat{\tilde{E}}_{21} = 110.701 \times (27.306 + 41.667)/(110.701 + 132.979) = 31.334,$$

$$\widehat{\tilde{E}}_{22} = 132.979 \times (27.306 + 41.667)/(110.701 + 132.979) = 37.639.$$

The value of the adjusted Pearson's statistic is

$$\tilde{X}_P^2 = \frac{(83.395 - 79.367)^2}{79.367} + \frac{(91.312 - 95.340)^2}{95.340} + \frac{(27.306 - 31.334)^2}{31.334}$$

$$+ \frac{(41.667 - 37.639)^2}{37.639}$$

$$= 1.32,$$

with associated *P*-value $= 0.25$ for a chi-square distribution with 1 DF. We cannot reject the null hypothesis based on the results. Using adjusted observations, $\tilde{n}_{ij}$, and the S-PLUS built-in function *chisq.test*, we can get the same result:

```
> x<-nlminb(start = c(0.1,0.2),x = repana1,objective =
  bbmaxg)
> x$par
[1] 0.7536001 0.1776106
>x<-nlminb(start = c(0.1,0.2),x = repana2,objective =
  bbmaxg)
> x$par
[1] 0.72678674 0.06394936
> chisq.test(cbind(c(113,37)/1.355,c(103,47)/1.128),
correct = F)
     Pearson's chi-square test without Yates' continuity
      correction
data: cbind(c(113, 37)/1.355, c(103, 47)/1.128)
X-square = 1.3231, df = 1, p-value = 0.25
```

## The Corrected Beta-Binomial (CBB) Model

Morrison (1978) and Brockhoff (2003) pointed out that the beta-binomial model should be adapted for replicated forced-choice methods because the probability of correct responses for each of the panelists for the methods is distributed on the range of $[C, 1]$ rather than $[0, 1]$, where $C$ denotes a guessing probability. In the corrected beta-binomial model, the probability of correct responses $P_c$ is

$$P_c = C + (1 - C)P, \qquad (10.3\text{-}28)$$

where $P$ denotes true discrimination ability. We can assume reasonably that $P$ follows a beta distribution and $P_c$ follows a corrected beta distribution.

### *The Corrected Beta-Binomial Distribution*

### The Probability Distribution Function (pdf)

Morrison (1978) and Brockhoff (2003) derived the corrected beta-binomial distribution. Bi (2006, 2007c) provided a closed-form expression for the probability distribution function

$$P(x|n, a, b, C) = \frac{(1 - C)^n}{B(a, b)} \binom{n}{x} \sum_{i=0}^{x} \binom{x}{i} \left(\frac{C}{1 - C}\right)^{x-i} B(a + i, n + b - x),$$

$$(10.3\text{-}29)$$

where

$B(.,.)$ denotes the beta function, $B(a, b) = \dfrac{\Gamma(a)\Gamma(b)}{\Gamma(a + b)}$ and

$B(a + i, n + b - x) = \dfrac{\Gamma(a + i)\Gamma(n + b - x)}{\Gamma(a + i + n + b - x)}; \ a = \mu\left(\dfrac{1}{\gamma} - 1\right)$ and

$b = (1 - \mu)\left(\dfrac{1}{\gamma} - 1\right).$

For the 2-AFC and duo–trio methods, $C = 1/2$,

$$P(x|n, a, b) = \frac{1}{2^n B(a, b)} \binom{n}{x} \sum_{i=0}^{x} \binom{x}{i} B(a + i, n + b - x). \qquad (10.3\text{-}30)$$

For the 3-AFC and triangle methods, $C = 1/3$,

$$P(x|n, a, b) = \frac{2^{n-x}}{3^n B(a, b)} \binom{n}{x} \sum_{i=0}^{x} \binom{x}{i} 2^i \, B(a + i, n + b - x). \qquad (10.3\text{-}31)$$

When $C = 0$, the corrected beta-binomial distribution becomes the conventional beta-binomial distribution.

### Parameter Estimation

### Maximum Likelihood Estimation

Suppose $X_i$, the number of correct responses in $n$ replications for the $i$th panelist, follows a modified beta-binomial distribution with a probability distribution function in (10.3-30) or (10.3-31). The log likelihood of the corrected beta-binomial distribution is

$$L = \sum_{i=1}^{k} \log[\Pr(x_i|n_i, \mu, \gamma)]. \tag{10.3-32}$$

Using the S-PLUS built-in program *nlminb*, we can obtain the estimated parameters $\hat{\mu}$ and $\hat{\gamma}$ and covariance matrix for the estimators $\hat{\mu}$ and $\hat{\gamma}$. We do not need to have the same number of replications for each trial for the maximum likelihood estimation for parameters.

### Moment Estimation

Let

$$\hat{\pi} = \frac{\sum_{i=1}^{k} X_i}{nk}, \tag{10.3-33}$$

$$S = \sum_{i=1}^{k} (\hat{P}_{ci} - \hat{\pi})^2, \tag{10.3-34}$$

where $\hat{P}_{ci} = \dfrac{X_i}{n}$ is the proportion of correct responses for the $i$th panelist $(i = 1, 2, ..., k)$ with $n$ replications. The moment estimates of $\mu$ and $\gamma$ are

$$\hat{\mu} = \frac{\hat{\pi} - C}{1 - C}, \tag{10.3-35}$$

$$\hat{\gamma} = \frac{nS}{k(n-1)(1-\hat{\pi})(\hat{\pi} - C)} - \frac{\hat{\pi}}{(n-1)(\hat{\pi} - C)}. \tag{10.3-36}$$

It can be demonstrated on the basis of probability rules for conditional means and variances (see, e.g., Rao, 1973, 97; Meyners and Brockhoff, 2003) that the estimated variance of $\hat{\pi}$ is

$$\hat{V}(\hat{\pi}) = (1 - C)(1 - \hat{\mu})[(1 - C)(n - 1)\hat{\mu}\hat{\gamma} + C + (1 - C)\hat{\mu}]/nk. \tag{10.3-37}$$

The estimated variances of $\hat{\mu}$ are

$$\hat{V}(\hat{\mu}) = \hat{V}(\hat{\pi})/(1 - C)^2, \tag{10.3-38}$$

and

$$\hat{V}(\hat{\pi}) = (1 - C)^2 \hat{V}(\hat{\mu}). \tag{10.3-39}$$

**Example 10.3-7**

For a replicated 2-AFC test with 30 panelists and two replications, the correct responses for the panelists are 2, 2, 2, 2, 1, 2, 1, 2, 2, 2, 2, 2, 2, 2, 2, 0, 2, 2, 0, 2, 1, 2, 2, 2, 1, 2, 2, 2, 2, 1, 1. We can calculate

$$\hat{\pi} = \frac{\sum_{i=1}^{k} X_i}{nk} = \frac{2 + 2 + \ldots 1 + 1}{30 \times 2} = 0.833,$$

$$S = \sum_{i=1}^{k} (\hat{p}_{ci} - \hat{\pi}_c)^2 = \sum_{i=1}^{30} (\hat{p}_{ci} - 0.833)^2 = 2.667.$$

Hence, according to (10.3-35) and (10.3-36), the moment estimates of $\mu$ and $\gamma$ are

$$\hat{\mu} = 2 \times 0.833 - 1 = 0.667,$$

$$\hat{\gamma} = \frac{2 \times 2.667 \times 2}{(2 \times 0.833 - 1)(1 - 0.833)(2 - 1) \times 30} - \frac{2 \times 0.833}{(2 \times 0.833 - 1)(2 - 1)}$$

$$= 0.700.$$

According to (10.3-37) and (10.3-38), the estimated variances of $\hat{\pi}$ and $\hat{\mu}$ are

$$\hat{V}(\hat{\pi}) = (1 - 0.5)(1 - 0.667)[(1 - 0.5)(2 - 1)0.667 \times 0.7 + 0.5$$
$$+ (1 - 0.5)0.667]/(2 \times 30)$$

$$= 0.00296,$$

$$\hat{V}(\hat{\mu}) = 0.00296/(1 - 0.5)^2 = 0.01184.$$

```
> forcebbm2(cbbdat1[,1],2,2)
[1]  0.666700000  0.700000000  0.833333333  0.002962745
0.011850981
```

The maximum likelihood estimates of the parameters are $\hat{\mu} = 0.667$, $\hat{\gamma} = 0.700$, and the covariance matrix for $\hat{\mu}$ and $\hat{\gamma}$ is

$$Cov(\hat{\mu}, \hat{\gamma}) = \begin{pmatrix} 0.01016 & -0.02091 \\ -0.02091 & 0.33566 \end{pmatrix},$$

and, according to (10.3-35) and (10.3-39), $\hat{\pi}$ and the variance of $\hat{\pi}$ are

$$\hat{\pi} = 0.5 + (1 - 0.5)0.667 = 0.8335,$$

$$\hat{V}(\hat{\pi}) = 0.01016 \times (1 - 0.5)^2 = 0.00254.$$

```
> x<-nlminb(start = c(0.3,0.4),objective = cbbmaxg,x =
  cbbdat1,cc = 1/2)
> x$par
[1] 0.6666667 0.7000000
```

```
> vcov.nlminb(x)
   [,1] [,2]
[1,] 0.01015895 -0.02091277
[2,] -0.02091277 0.33565613
> forcebbm2(cbbdat1[,1],2)
[1] 0.6667 0.7000
> cbbdat1
integer matrix: 30 rows, 2 columns.
   [,1] [,2]
[1,] 2 2
[2,] 2 2
[3,] 2 2
. . . . . . . . . . . . . .
[30,] 1 2
```

For a replicated 3-AFC test with 30 panelists and four replications for comparisons of two products, the correct responses for the panelists are 4, 0, 4, 4, 4, 4, 4, 4, 4, 4, 3, 3, 3, 4, 2, 4, 4, 4, 4, 3, 4, 4, 2, 4, 2, 4, 3, 4, 3, 2. According to (10.3-33) and (10.3-34), we can calculate

**Example 10.3-8**

$$\widehat{\pi}_c = \frac{\sum\limits_{i=1}^{k} x_i}{nk} = \frac{4 + 0 + \dots 3 + 2}{30 \times 4} = \frac{102}{120} = 0.85,$$

$$S = \sum\limits_{i=1}^{k} (\widehat{p}_{ci} - \widehat{\pi}_c)^2 = \sum\limits_{i=1}^{30} (\widehat{p}_{ci} - 0.85)^2 = 1.7.$$

Hence, according to (10.3-35) and (10.3-36), the moment estimates of $\mu$ and $\gamma$ are

$$\widehat{\mu} = \frac{3 \times 0.85 - 1}{2} = 0.775,$$

$$\widehat{\gamma} = \frac{3 \times 1.7 \times 4}{(3 \times 0.85 - 1)(1 - 0.85)(4 - 1) \times 30} - \frac{3 \times 0.85}{(3 \times 0.85 - 1)(4 - 1)}$$

$$= 0.4265.$$

According to (10.3-37) and (10.3-38), the estimated variances of $\widehat{\pi}$ and $\widehat{\mu}$ are

$$\widehat{V}(\widehat{\pi}) = (1 - 0.33)(1 - 0.775)[(1 - 0.33)(4 - 1)0.775 \times 0.4265 + 0.33$$
$$+ (1 - 0.33)0.775]/(4 \times 30)$$
$$= 0.00189,$$

$$\widehat{V}(\widehat{\mu}) = 0.00189/(1 - 0.33)^2 = 0.00425.$$

```
> forcebbm2(cbbdat2[,1],4,3)
[1]  0.775000000  0.426500000  0.850000000  0.001888844
   0.004249898
```

The maximum likelihood estimates of the parameters are $\hat{\mu} = 0.776$, $\hat{\gamma} = 0.381$, and the covariance matrix for $\hat{\mu}$ and $\hat{\gamma}$ is

$$Cov(\hat{\mu}, \hat{\gamma}) = \begin{pmatrix} 0.00363 & -0.00324 \\ -0.00324 & 0.03464 \end{pmatrix},$$

and, according to (10.3-35) and (10.3-39), $\hat{\pi}$ and the variance of $\hat{\pi}$ are then

$$\hat{\pi} = 1/3 + (1 - 1/3)0.776 = 0.8507,$$

$$\hat{V}(\hat{\pi}) = 0.00363 \times (1 - 1/3)^2 = 0.00161.$$

```
> x<-nlminb(start = c(0.3,0.4),objective = cbbmaxg,x =
   cbbdat2,cc = 1/3)
> x$par
[1] 0.7759800 0.3805537
> vcov.nlminb(x)
   [,1] [,2]
[1,] 0.003632559 -0.00324478
[2,] -0.003244780 0.03464266
> forcebbm3(cbbdat2[,1],4)
[1] 0.7750 0.4265
> cbbdat2
integer matrix: 30 rows, 2 columns.
   [,1] [,2]
[1,] 4 4
[2,] 0 4
[3,] 4 4
.......
```

## Tests Based on the CBB Model

### *Replicated Difference Tests Using Forced-Choice Methods*

We can use a Wald-type statistic to test simultaneously or separately the parameters $\mu$ and $\gamma$ in a single experiment or two or multiple experiments using $\hat{\mu}$ and $\hat{\gamma}$ and variances of $\hat{\mu}$ and $\hat{\gamma}$ estimated from the CBB model. We should be aware that $\mu = 0$ is an endpoint of the range of the parameter in the CBB model. It would be difficult to estimate the parameter using the maximum likelihood estimation if the true value of $\mu$ is near zero. It is invalid for inference involving $\mu = 0$ in the CBB model.

There is a viewpoint that the conventional binomial model can also be used to test $\mu = 0$, i.e., $\pi = C$, in replicated difference tests using forced-choice methods. The key argument is that if there is no difference in the null hypothesis, the number of replications and number of panelists can be interchanged. The argument seems reasonable. However, the naïve binomial model cannot adequately model the replicated testing data in a general situation. Both estimation and test for the parameters $\mu$ and $\gamma$ in the CBB model are necessary. Both the null hypothesis and alternative hypothesis about the parameters should be considered in determination of sample size and testing power.

It is valid to test $H_0: \gamma = 0$ versus $H_a: \gamma > 0$ because zero is not an endpoint of the range of $\gamma$. For a replicated difference test using forced methods, $\gamma \neq 0$ also implies $\mu \neq 0$ in practice. The test statistic is

$$Z = \frac{\widehat{\gamma}}{\sqrt{\widehat{V}(\widehat{\gamma})}}, \qquad (10.3\text{-}40)$$

which follows approximately the standard normal distribution. However, $\gamma = 0$ does not mean $\mu = 0$. Hence, if the null hypothesis of $\gamma = 0$ is not rejected, a further test for $\mu$ is needed.

It is also valid to use a Wald-type statistic to test $H_0: \mu \leq \mu_0$ versus $H_a: \mu > \mu_0$ where $\mu_0 > 0$ is a specified value. The test statistic is

$$Z = \frac{\widehat{\mu} - \mu_0}{\sqrt{\widehat{V}(\widehat{\mu})}}, \qquad (10.3\text{-}41)$$

which also follows approximately the standard normal distribution.

A simultaneous test for the two parameters is that $H_0: \mu = \mu_0$ and $\gamma = \gamma_0$ versus $H_a: \mu \neq \mu_0$ and/or $\gamma \neq \gamma_0$, where $\mu_0$ and $\gamma_0$ are specified values. The test statistic is

$$X^2 = (\widehat{\mu} - \mu_0, \widehat{\gamma} - \gamma_0)[S]^{-1}(\widehat{\mu} - \mu_0, \widehat{\gamma} - \gamma_0)', \qquad (10.3\text{-}42)$$

where $[S]^{-1}$ denotes the inverse of the covariance matrix of the estimator of the parameter vector $(\widehat{\mu}, \widehat{\gamma})$. The test statistic $X^2$ follows asymptotically a chi-square distribution with 2 DF.

**Example 10.3-9**

For the data in Example 10.3-8, we want to know whether panelists can detect the difference. We test $H_0: \gamma = 0$ versus $H_a: \gamma > 0$. Because $\hat{\gamma} = 0.381$, $\hat{V}(\hat{\gamma}) = 0.03464$, the value of the test statistic (10.3-40) is

$$Z = \frac{\hat{\gamma} - 0}{\sqrt{\hat{V}(\hat{\gamma})}} = \frac{0.381}{\sqrt{0.03464}} = 2.05,$$

with P-value = 0.02. The null hypothesis is rejected, and the alternative hypothesis $\gamma > 0$ is accepted. The heterogeneity implies the products for comparison are significantly different.

For the data in this example, suppose that the objective of the experiment is to test if more than 60% of consumers can detect the difference, or if the consumers have more than 60% above chance to detect the difference. The test is $H_0: \mu \leq \mu_0 = 0.6$ versus $H_a: \mu > \mu_0 = 0.6$. Because $\hat{\mu} = 0.776$, $\hat{V}(\hat{\mu}) = 0.00363$, from (10.3-41), the test statistic is

$$Z = \frac{\hat{\mu} - 0.6}{\sqrt{\hat{V}(\hat{\mu})}} = \frac{0.776 - 0.6}{\sqrt{0.00363}} = 2.92,$$

with P-value = 0.002. Hence, we can conclude that more than 60% of consumers can detect the difference between the two products.

For a simultaneous test for the parameters, e.g., $H_0: \mu = 0.7$ and $\gamma = 0.3$ versus $H_a: \mu \neq 0.7$ and/or $\gamma \neq 0.3$, the value of the test statistic (10.3-42) is

$$X^2 = (0.776 - 0.7, 0.381 - 0.3) \begin{pmatrix} 0.00363 & -0.00324 \\ -0.00324 & 0.03464 \end{pmatrix}^{-1} \begin{pmatrix} 0.776 - 0.7 \\ 0.381 - 0.3 \end{pmatrix} = 2.288.$$

The associated P-value = 0.32 from a chi-square distribution with 2 DF. We cannot reject the null hypothesis at a reasonable significance level.

```
> t(a)%*%solve(s1)%*%a
     [,1]
[1,] 2.288373
> 1-pchisq(2.288,2)
[1] 0.3185423
> a
[1] 0.075 0.081
> s1
          [,1]         [,2]
[1,] 0.003632559 -0.00324478
[2,] -0.003244780 0.03464266
```

## Replicated Similarity Tests Using Forced-Choice Methods

Similarity or equivalence testing can be conducted using replicated forced-choice methods. For a given similarity limit, $\mu_0$, the test is to demonstrate that the difference is smaller than the specified limit. This is a one-sided similarity test with the hypotheses $H_0 : \mu \geq \mu_0$ versus $H_a : \mu < \mu_0$. The test statistic is

$$Z = \frac{\hat{\mu} - \mu_0}{\sqrt{\hat{V}(\mu_0)}} < z_\alpha, \qquad (10.3\text{-}43)$$

where $z_\alpha$ denotes the $\alpha$ 100th percentile of the standard normal distribution and

$$\hat{\mu} = \frac{\frac{x}{nk} - C}{1 - C}, \qquad (10.3\text{-}44)$$

$$\hat{V}(\mu_0) = (1 - \mu_0)[(1 - C)(n - 1)\mu_0\hat{\gamma} + C + (1 - C)\mu_0]/[nk(1 - C)]. \qquad (10.3\text{-}45)$$

Table A.32 gives the critical values, $x_0$, for $\mu_0 = 0.1$, $\alpha = 0.05$ for the 2-AFC, duo–trio, 3-AFC, and triangle methods. If the observed numbers of correct responses in a replicated test using a forced-choice method are smaller than the critical value, similarity or equivalence can be claimed in the sense of $\mu_0 = 0.1$ and $\alpha \leq 0.05$.

---

**Example 10.3-10**

For a replicated similarity test using the 2-AFC method with 100 consumer panelists and three replications, the number of correct responses is 148. The similarity limit $\mu_0 = 0.1$ and Type I error $\alpha = 0.05$ are specified. The estimated parameter values are $\hat{\gamma} = 0.4$ and

$$\hat{\mu} = \left(\frac{148}{3 \times 100} - 0.5\right)\Big/(1 - 0.5) = -0.0133.$$

From (10.3-45), the estimated variance of $\hat{\mu}$ at $\mu_0 = 0.1$ is

$$(1 - 0.1)[(1 - 0.5)(3 - 1)0.1 \times 0.4 + 0.5$$
$$+ (1 - 0.5)0.1]/[3 \times 100(1 - 0.5)]$$
$$= 0.00354.$$

The statistic value is

$$Z = \frac{-0.0133 - 0.1}{\sqrt{0.00354}} = -2.24.$$

Because $-2.24 < -1.64 = z_{0.05}$, we can conclude with a 0.05 significance level that the true difference is smaller than the specified limit; i.e., less than 10% of consumers can detect the difference. From Table A.32, the corresponding critical value is 150. Because the observed number of correct responses, 148, is smaller than the critical value, 150, the same conclusion can be drawn.

## The Dirichlet-Multinomial (DM) Model

The Dirichlet-multinomial (DM) model is a natural extension of the beta-binomial model. It can be regarded as a multivariate version of the beta-binomial model. One of the earliest discussions and applications of the DM model appears to be by Morsimann (1962). Ennis and Bi (1999) discussed the application of the DM model in sensory and consumer fields.

### The Dirichlet-Multinomial Distribution

Dirichlet-multinomial distribution is a compound distribution of the Dirichlet distribution and the multinomial distribution. For example, in a replicated consumer preference test, each panelist conducts $n$ replicated tests. If the "no preference" option is allowed, then the $n$ responses should fall into the three categories: "preferring A," "no preference," and "preferring B." The vector of preference counts for a panelist, $\mathbf{X} = (X_1, X_2, \ldots, X_m)$, $\sum_{i=1}^{m} X_i = n$ (here, $m = 3$), follows a conditional multinomial distribution with parameters $n$ and $\mathbf{p} = (p_1, p_2, \ldots p_m)$, where $p_m = 1 - \sum_{i=1}^{m-1} p_i$. Assume that the parameter vector $\mathbf{p} = (p_1, p_2, \ldots p_m)$ is a variable rather than an unknown constant vector over the panelists, and it follows a multivariate beta distribution, i.e., the Dirichlet distribution. Then $\mathbf{X} = (X_1, X_2, \ldots, X_m)$ for any panelist follows a Dirichlet-multinomial distribution with parameters $n$, $\boldsymbol{\pi}$, and $g$, denoted as $DM_m(n, \boldsymbol{\pi}\ g)$, where $\boldsymbol{\pi} = (\pi_1, \pi_2, \ldots, \pi_m)$, $\pi_m = 1 - \sum_{i=1}^{m-1} \pi_i$.

The probability distribution function of the DM distribution at $\mathbf{X} = \mathbf{x}$, $\mathbf{x} = (x_1, x_2, \ldots, x_m)$ is

$$P(\mathbf{x}|n, \boldsymbol{\pi}, g) = \frac{n!}{\Pi_{i=1}^{m} x_i!} \frac{\Gamma(g)\Pi_{i=1}^{m}\Gamma(x_i + \pi_i g)}{\Gamma(n+g)\Pi_{i=1}^{m}\Gamma(\pi_i g)}, \qquad (10.3\text{-}46)$$

where parameter vector $\boldsymbol{\pi} = (\pi_1, \pi_2, \ldots \pi_m)$ is the mean of multinomial parameter vector $\mathbf{p}$ and $g$ is a scale parameter, which measures the variation of $\mathbf{p}$. Parameter $g$ can be reparameterized to $\theta = 1/g$ or $\gamma = \dfrac{1}{1+g}$. Parameter $\gamma$ varies between 0 and 1. There are $m + 1$ parameters: $\boldsymbol{\pi} = (\pi_1, \pi_2, \ldots \pi_m)$ and $g$ in the DM model. However, there are only $m$ independent parameters because $\pi_m = 1 - \sum\limits_{i=1}^{m-1} \pi_i$. When $m = 2$, (10.3-46) becomes the probability distribution function of a beta-binomial distribution with parameters $n$, $\boldsymbol{\pi}$, and $\theta$, where $\theta = 1/g$, $\pi_1 = \pi$ and $\pi_2 = 1 - \pi_1 = 1 - \pi$. We can also use $\gamma$ instead of $\theta$, where

$$\gamma = \frac{1}{1+g} = \frac{\theta}{1+\theta}.$$

The mean of $DM_m(n, \boldsymbol{\pi}, g)$ is $n\boldsymbol{\pi}$ and the covariance matrix is $nC(\Delta_\pi - \boldsymbol{\pi\pi}')$, where $\Delta_\pi$ is a diagonal matrix with entries $(\pi_1, \pi_2, \ldots \pi_m)$ and $C = \dfrac{n+g}{1+g} = 1 + \gamma(n-1)$. It is noted that the covariance matrix of a multinomial distribution is $n(\Delta_\pi - \boldsymbol{\pi\pi}')$; hence, the covariance matrix of a DM distribution is just a constant, $C$, times the corresponding multinomial covariance matrix based on $\boldsymbol{\pi}$. The $C$ value, a heterogeneity factor, which varies between 1 and $n$, is a measure of overdispersion or clustering effect, and it links the DM distribution and the multinomial distribution. When $C = 1$, i.e., $g \to \infty$ or $\gamma = 0$, the DM distribution becomes a conventional multinomial distribution.

### Estimation of Parameters

#### Moment Estimation

Assume there are $k$ panelists in replicated ratings with $m$ categories. The number of replications is $n$ for each panelist. The rating counts are $\mathbf{x}^{(i)} = (x_{i1}, x_{i2}, \ldots x_{im})$, $i = 1, 2, \ldots k$. The moment estimates of $\boldsymbol{\pi} = (\pi_1, \pi_2, \ldots \pi_m)$ and $g$ are

$$\widehat{\boldsymbol{\pi}} = \frac{\sum\limits_{i=1}^{k} \mathbf{x}^{(i)}}{nk}, \tag{10.3-47}$$

$$\widehat{g} = \frac{n - \widehat{C}}{\widehat{C} - 1}, \tag{10.3-48}$$

where $n = n^{(i)} = \sum\limits_{j=1}^{m} x_{ij}$,

$$\widehat{C} = \frac{n}{(k-1)(m-1)} \sum\limits_{j=1}^{m} \frac{1}{\widehat{\pi}_j} \sum\limits_{i=1}^{k} \left(\frac{x_{ij}}{n} - \widehat{\pi}_j\right)^2, \tag{10.3-49}$$

and $\widehat{\pi}_j = \dfrac{\sum_{i=1}^{k} x_{ij}}{nk}$. It is noted that, in analogy with the estimation of variance components, $\widehat{C}$ may fall outside its allowable range, $1 \leq C \leq n$. In these cases $\widehat{C}$ should be truncated. $\widehat{C}$ in (10.3-49) can also be expressed in the form

$$\widehat{C} = \frac{1}{(k-1)(m-1)} \sum_{i=1}^{k} (x^{(i)} - n\widehat{\pi})'(n\Delta_{\widehat{\pi}})^{-1}(x^{(i)} - n\widehat{\pi}), \qquad (10.3\text{-}50)$$

where $\Delta_{\widehat{\pi}}$ denotes a diagonal matrix with diagonal values $\widehat{\pi}$.

---

**Example 10.3-11**

There are 50 panelists ($k = 50$) with three replications ($n = 3$) in a replicated preference test with a "no preference" option ($m = 3$) for products A and B. The data are shown in Table 10.25. Because

$$\widehat{\pi}_1 = \frac{\sum_{i=1}^{50} x_{i1}}{nk} = \frac{61}{3 \times 50} = 0.407,$$

$$\widehat{\pi}_2 = \frac{\sum_{i=1}^{50} x_{i2}}{nk} = \frac{64}{3 \times 50} = 0.427,$$

$$\widehat{\pi}_3 = \frac{\sum_{i=1}^{50} x_{i3}}{nk} = \frac{25}{3 \times 50} = 0.167; \text{hence,}$$

$$\widehat{\pi} = (\widehat{\pi}_1, \widehat{\pi}_2, \widehat{\pi}_3) = (0.407, 0.427, 0.167), \text{and}$$

$$\widehat{C} = \frac{3}{(50-1)(3-1)} \sum_{j=1}^{3} \frac{1}{\widehat{\pi}_j} \sum_{i=1}^{50} \left(\frac{x_{ij}}{3} - \widehat{\pi}_j\right)^2 = 1.223.$$

Using an S-Plus code shown below, we get

$$\widehat{g} = \frac{n - \widehat{C}}{\widehat{C} - 1} = \frac{3 - 1.223}{1.223 - 1} = 7.969,$$

$$\widehat{\theta} = \frac{1}{7.969} = 0.125, \text{ or}$$

$$\widehat{\gamma} = \frac{1}{1 + 7.969} = 0.11.$$

```
> cbval(dmdat2)
[1] 1.223
```

**Table 10.25** Data for a Replicated Preference Test with a "No Preference" Option (Example 10.3-11) ($k = 50$, $n = 3$, $m = 3$)

| No. | $x_{i1}$ | $x_{i2}$ | $x_{i3}$ | No. | $x_{i1}$ | $x_{i2}$ | $x_{i3}$ |
|-----|----------|----------|----------|-----|----------|----------|----------|
| 1 | 3 | 0 | 0 | 26 | 2 | 0 | 1 |
| 2 | 2 | 1 | 0 | 27 | 0 | 3 | 0 |
| 3 | 1 | 2 | 0 | 28 | 2 | 0 | 1 |
| 4 | 1 | 1 | 1 | 29 | 1 | 1 | 1 |
| 5 | 1 | 1 | 1 | 30 | 1 | 0 | 2 |
| 6 | 1 | 2 | 0 | 31 | 1 | 1 | 1 |
| 7 | 0 | 3 | 0 | 32 | 2 | 1 | 0 |
| 8 | 2 | 0 | 1 | 33 | 0 | 3 | 0 |
| 9 | 1 | 2 | 0 | 34 | 0 | 3 | 0 |
| 10 | 1 | 1 | 1 | 35 | 1 | 2 | 0 |
| 11 | 3 | 0 | 0 | 36 | 0 | 2 | 1 |
| 12 | 2 | 0 | 1 | 37 | 0 | 2 | 1 |
| 13 | 1 | 1 | 1 | 38 | 2 | 1 | 0 |
| 14 | 2 | 1 | 0 | 39 | 2 | 1 | 0 |
| 15 | 0 | 2 | 1 | 40 | 0 | 0 | 3 |
| 16 | 3 | 0 | 0 | 41 | 1 | 2 | 0 |
| 17 | 2 | 0 | 1 | 42 | 1 | 0 | 2 |
| 18 | 2 | 1 | 0 | 43 | 2 | 1 | 0 |
| 19 | 0 | 3 | 0 | 44 | 0 | 3 | 0 |
| 20 | 2 | 1 | 0 | 45 | 1 | 2 | 0 |
| 21 | 1 | 2 | 0 | 46 | 0 | 2 | 1 |
| 22 | 1 | 2 | 0 | 47 | 0 | 1 | 2 |
| 23 | 3 | 0 | 0 | 48 | 1 | 2 | 0 |
| 24 | 2 | 0 | 1 | 49 | 2 | 1 | 0 |
| 25 | 1 | 2 | 0 | 50 | 1 | 2 | 0 |

Note: $x_{i1}$ = number of preferring A; $x_{i2}$ = number of preferring B; $x_{i3}$ = number of no preference.

## Maximum Likelihood Estimation

Maximum likelihood estimation for parameters is more accurate than moment estimation. For replicated testing data of $k$ panelists with $n$ replications, for example, we can obtain the log likelihood function for the $k$ independent samples as

$$L = \sum_{j=1}^{k} \log P(\mathbf{x}_j; n, \boldsymbol{\pi}, g), \tag{10.3-51}$$

where $P(\mathbf{x}_j|n, \boldsymbol{\pi}, g)$ is the probability distribution function for the $j$th sample in (10.3-46). Using the built-in function in some statistical software, for example, *nlminb* in S-PLUS, we can get the maximum likelihood estimates $\hat{g}$ and $\hat{\boldsymbol{\pi}} = (\hat{\pi}_1, \hat{\pi}_2, \dots \hat{\pi}_m)$, which are the values of $g$ and $\pi = (\pi_1, \pi_2, \dots \pi_m)$ making $-L$ minimum.

---

**Example 10.3-12**

Using the data in Table 10.25 and an S-PLUS program, we calculate the maximum likelihood estimates. They are $\hat{\boldsymbol{\pi}} = (\hat{\pi}_1, \hat{\pi}_2, \hat{\pi}_3) = (0.408, 0.424, 0.168)$ and $\hat{g} = 9.14$. Hence, we can get $\hat{C} = \dfrac{3 + 9.14}{1 + 9.14} = 1.197$ and $\hat{\gamma} = \dfrac{1}{1 + 9.14} = 0.1$. The results of the maximum likelihood estimates are close to the results of the moment estimated in Example 10.3-11. We also can obtain the covariance matrix for $\hat{\pi}_1(\hat{\pi}_A)$, $\hat{\pi}_2(\hat{\pi}_B)$, and $\hat{g}$.

```
> x<-nlminb(start = c(0.1,0.1,0.1),x = dmdat2,objective
= dmff, low = c(0.1,0.1,0.1))
> x$par
[1] 0.4084775 0.4235756 9.1406427
> vcov.nlminb(x,tol = 0.01)
      [,1] [,2] [,3]
[1,] 0.002057456 -0.001540373 -0.003831526
[2,] -0.001540373 0.002093506 0.023226212
[3,] -0.003831526 0.023226212 56.055145762
```

---

## Goodness of Fit Test

Paul et al. (1989) developed a $Z$ statistic in (10.3-52) for testing multinomial assumption against the Dirichlet-multinomial alternatives. The null hypothesis is that the underlying distribution is a multinomial distribution, and the alternative hypothesis is that the distribution is a Dirichlet-multinomial. The test is equivalent to $H_0: \gamma = 0$:

$$Z = \frac{N \sum_{j=1}^{m} \frac{1}{x_{0j}} \sum_{i=1}^{k} x_{ij}(x_{ij} - 1) - \sum_{i=1}^{k} n_i(n_i - 1)}{\sqrt{2(m-1) \sum_{i=1}^{k} n_i(n_i - 1)}}, \qquad (10.3\text{-}52)$$

where $x_{0j} = \sum_{i=1}^{k} x_{ij}$, $N = \sum_{i=1}^{k} n_i$, and $n_i$ is the number of replications for the $i$th trial (panelist).

The statistic (10.3-52) is a generalization of Tarone's (1979) statistic in (10.3-13). If the null hypothesis is rejected, we can conclude the underlying distribution is a Dirichlet-multinomial. However, if the null hypothesis cannot be rejected, we cannot conclude that the distribution is a multinomial. In this case, treating the replicated ratings as Dirichlet-multinomial data is more robust than treating them as multinomial data. Because the parameter $\gamma$ is assumed to have only non-negative values, the test is always one-sided.

---

For the data in Table 10.25, $k = 50$, $m = 3$, $x_{01} = 61$, $x_{02} = 64$, $x_{03} = 25$, and $N = 3 \times 50 = 150$. Hence,

| **Example 10.3-13** |
|---|

$$Z = \frac{150 \times \sum_{j=1}^{3} \frac{1}{x_{0j}} \sum_{i=1}^{50} x_{ij}(x_{ij} - 1) - 50 \times 3 \times (3 - 1)}{\sqrt{2 \times (3 - 1) \times 50 \times 3 \times (3 - 1)}} = 1.7169,$$

with associated *P*-value = 0.043. This value suggests that we can reject the multinomial distribution and accept the Dirichlet-multinomial model. In other words, we can reject the null hypothesis $\gamma = 0$ at the 0.05 significance level.

We should mention that there is no serious risk to treating the replicated testing data as Dirichlet-multinomial distributed data even in a situation in which the null hypothesis of multinomial distribution is not rejected.

```
> fitdm2(dmdat2)
[1] 1.7169 0.0430
```

---

## Tests Based on the DM Model

### *Replicated Preference Tests with "No Preference" Option*

There are some discussions in sensory and statistical literature about whether a tied response should be allowed in difference and preference tests. The consensus in preference testing seems to be that the "No preference" (or tied) option should be admitted because it adds information. For preference testing with the "No preference" option, the ties transform a binomial system into a two-parameter trinomial one. A multinomial test should be used instead of a binomial test. In the replicated situation, the DM model can be used to deal with both between and within panelist variations. For a test with the hypotheses $H_0$: $\boldsymbol{\pi} = \boldsymbol{\pi}_0$ versus $H_a$: $\boldsymbol{\pi} \neq \boldsymbol{\pi}_0$, the test statistic is

$$X^2 = \frac{nk}{\widehat{C}} \sum_{i=1}^{3} \frac{(\widehat{\pi}_i - \pi_{i0})^2}{\pi_{i0}}, \tag{10.3-53}$$

where $\boldsymbol{\pi} = (\pi_1, \pi_2, \pi_3)$, $\widehat{\boldsymbol{\pi}} = (\widehat{\pi}_1, \widehat{\pi}_2, \widehat{\pi}_3)$, and $\boldsymbol{\pi}_0 = (\pi_{10}, \pi_{20}, \pi_{30})$. These are the theoretical, observed, and specified proportions, respectively, for A-preference, B-preference, and No-preference. The test statistic follows asymptotically a chi-square distribution with 2 DF.

Suppose the interest is only in testing preference proportions of two products, A and B, i.e., $H_0$: $\pi_A = \pi_B$ versus $H_a$: $\pi_A \neq \pi_B$. The results obtained from a maximum likelihood estimation can be used for the test. The test statistic is

$$Z = \frac{\widehat{\pi}_A - \widehat{\pi}_B}{\sqrt{V(\widehat{\pi}_A) + V(\widehat{\pi}_B) - 2Cov(\widehat{\pi}_A, \widehat{\pi}_B)}}, \tag{10.3-54}$$

which follows asymptotically the standard normal distribution.

---

**Example 10.3-14**

For the data in Table 10.25 and Example 10.3-12, we get $\widehat{\boldsymbol{\pi}} = (\widehat{\pi}_1, \widehat{\pi}_2, \widehat{\pi}_3) = (0.408, 0.424, 0.168)$, $\widehat{C} = 1.2$, and $Cov(\widehat{\pi}_1, \widehat{\pi}_2) = \begin{pmatrix} 0.00206 & -0.00154 \\ -0.00154 & 0.00209 \end{pmatrix}$.

We want to test whether there is a difference in preference for the two products compared. Assume the true probability of "No preference" is 0.2. Then the null hypothesis is $H_0$: $\boldsymbol{\pi} = \boldsymbol{\pi}_0$ with the null proportion vector $\boldsymbol{\pi}_0 = (0.4, 0.4, 0.2)$. The value of the test statistic is

$$X^2 = \frac{50 \times 3}{1.2} \left[ \frac{(0.408 - 0.4)^2}{0.4} + \frac{(0.424 - 0.4)^2}{0.4} + \frac{(0.168 - 0.2)^2}{0.2} \right] = 0.84.$$

The associated P-value is 0.34 from a chi-square distribution with 2 DF. This value suggests that we have no evidence to reject the null hypothesis.

For testing $H_0$: $\pi_A = \pi_B$ versus $H_a$: $\pi_A \neq \pi_B$, according to (10.3-54), we get

$$Z = \frac{0.408 - 0.424}{\sqrt{0.00206 + 0.00209 - 2 \times (-0.00154)}} = -0.188.$$

The associated $P$-value is 0.85 from the standard normal distribution.

### Testing Independence of Contingency Tables under Cluster Sampling

For two-way contingency table data, if the total number of rows and columns is not fixed in advance, the interest is testing whether the two variables presented in the rows and columns are independent. For cluster sampling data, e.g., for $k$ clusters with $n$ sample size for each cluster, Brier (1980) pointed out that the data vector $\{n_{ij}\}$ with $I \times J$ categories for each cluster follows a DM distribution with parameters $\boldsymbol{\pi} = (\pi_1, \pi_2, \ldots, \pi_{I \times J})$ and $g$. And a test statistic is given to test independence of rows and columns in a contingency table for the pooled data $\{U_{ij}\}$ with the null hypothesis $H_0$: $\pi_{ij} = \pi_{i+}\pi_{+j}$, $i = 1, 2, \ldots, I$; $j = 1, 2, \ldots, J$.

$$X^2 = \frac{1}{\widehat{C}} \sum_{j=1}^{J} \sum_{i=1}^{I} \frac{(U_{ij} - E_{ij})^2}{E_{ij}}, \tag{10.3-55}$$

where $U_{ij}$ are polled data in an $I \times J$ table, $E_{ij} = U_{i+}U_{+j}/N$, $N = nk$ is the total number of observations for $k$ clusters with $n$ replications, and $\widehat{C}$ is an estimated heterogeneity factor in the DM model. The test statistic (10.3-55) follows asymptotically a chi-square distribution with $(I-1)(J-1)$ DF.

---

**Example 10.3-15**

For a replicated mixed designed A–Not A test with 50 panelists and four replications, the number of replications is fixed for each panelist, whereas the number of sample A and Not A is random; i.e., it is not fixed in advance. The data are presented in Table 10.26 and the pooled data are shown in Table 10.27, where $n_{11}$ is the number of response "A" for sample A; $n_{21}$ is the number of response "Not A" for sample A; $n_{12}$ is the number of response "A" for sample Not A; and $n_{22}$ is the number of response "Not A" for sample Not A. For the data, the maximum likelihood

**Table 10.26** Data for a Replicated Mixed Designed A-Not A Test (Example 10.3-15)

| No. | $n_{11}$ | $n_{21}$ | $n_{12}$ | $n_{22}$ | No. | $n_{11}$ | $n_{21}$ | $n_{12}$ | $n_{22}$ |
|---|---|---|---|---|---|---|---|---|---|
| 1 | 1 | 1 | 1 | 1 | 26 | 1 | 1 | 2 | 0 |
| 2 | 2 | 1 | 1 | 0 | 27 | 1 | 0 | 3 | 0 |
| 3 | 3 | 0 | 0 | 1 | 28 | 1 | 0 | 3 | 0 |
| 4 | 3 | 1 | 0 | 0 | 29 | 2 | 0 | 2 | 0 |
| 5 | 2 | 0 | 0 | 2 | 30 | 1 | 1 | 2 | 0 |
| 6 | 0 | 0 | 2 | 2 | 31 | 0 | 1 | 1 | 2 |
| 7 | 0 | 0 | 3 | 1 | 32 | 3 | 1 | 0 | 0 |
| 8 | 3 | 0 | 0 | 1 | 33 | 1 | 0 | 2 | 1 |
| 9 | 2 | 0 | 1 | 1 | 34 | 2 | 0 | 2 | 0 |
| 10 | 1 | 1 | 2 | 0 | 35 | 0 | 0 | 3 | 1 |
| 11 | 1 | 2 | 0 | 1 | 36 | 2 | 1 | 0 | 1 |
| 12 | 1 | 0 | 3 | 0 | 37 | 2 | 0 | 2 | 0 |
| 13 | 2 | 0 | 1 | 1 | 38 | 1 | 0 | 3 | 0 |
| 14 | 1 | 0 | 3 | 0 | 39 | 2 | 0 | 1 | 1 |
| 15 | 0 | 0 | 2 | 2 | 40 | 3 | 0 | 1 | 0 |
| 16 | 1 | 0 | 3 | 0 | 41 | 3 | 0 | 1 | 0 |
| 17 | 1 | 1 | 1 | 1 | 42 | 1 | 1 | 0 | 2 |
| 18 | 0 | 2 | 1 | 1 | 43 | 1 | 0 | 0 | 3 |
| 19 | 2 | 0 | 2 | 0 | 44 | 3 | 0 | 1 | 0 |
| 20 | 1 | 1 | 2 | 0 | 45 | 2 | 2 | 0 | 0 |
| 21 | 3 | 0 | 1 | 0 | 46 | 0 | 0 | 2 | 2 |
| 22 | 1 | 0 | 2 | 1 | 47 | 2 | 0 | 2 | 0 |
| 23 | 0 | 0 | 1 | 3 | 48 | 1 | 1 | 0 | 2 |
| 24 | 1 | 0 | 3 | 0 | 49 | 0 | 1 | 1 | 2 |
| 25 | 1 | 0 | 1 | 2 | 50 | 2 | 0 | 1 | 1 |
| | | | | | $U_{ij} =$ | 70 | 20 | 71 | 39 |

Note: $n_{11}$: A-"A"; $n_{21}$: A-"Not A"; $n_{12}$: Not A-"A"; $n_{22}$: Not A-"Not A."

**Table 10.27** Pooled Data for a Replicated Mixed A–Not A Test (Example 10.3-15)

| | | Sample | | |
|---|---|---|---|---|
| | | **A** | **Not A** | |
| | "A" | 70 | 71 | 141 |
| Response | "Not A" | 20 | 39 | 59 |
| | | 90 | 110 | $N = 200$ |

estimates are $\hat{\pi} = (0.351, 0.101, 0.355, 0.195)$ and $\hat{g} = 26.22$; hence, $\hat{C} = \dfrac{4 + 26.22}{1 + 26.22} = 1.11$. The value of the test statistic is

$$X^2 = \frac{1}{1.11} \times \left[ \frac{(70 - 141 \times 90/200)^2}{141 \times 90/200} + \frac{(71 - 141 \times 110/200)^2}{141 \times 110/200} \right.$$

$$\left. + \frac{(20 - 59 \times 90/200)^2}{59 \times 90/200} + \frac{(39 - 59 \times 110/200)^2}{59 \times 110/200} \right] = 3.754,$$

with associated $P$-value $= 0.053$. One should note that for the polled data without adjustment, the conventional Pearson's chi-square statistic value is 4.167 with associated $P$-value $= 0.04$. The output of the test using S-PLUS programs is as follows:

```
>x<-nlminb(start  =  c(0.1,0.1,0.1,0.1),x  =  dmdat3,
  objective = dmff,low = c(0.1,0.1,0.1,0.1))
> x$par
[1] 0.3510223 0.1012155 0.3544970 26.2219362
> chisq.test(cbind(c(70,20),c(71,39)),correct = F)
    Pearson's chi-square test without Yates' continuity
    correction
data: cbind(c(70, 20), c(71, 39))
X-square = 4.1674, df = 1, p-value = 0.0412
> 4.167/1.11
[1] 3.754054
> 1-pchisq(3.754,1)
[1] 0.0526813
```

### Testing for Several Independent Overdispersed Proportion Vectors

Brier (1980) and Koehler and Wilson (1986) developed a test statistic based on the DM model to test homogeneity of several independent categorical proportion vectors for overdispersed multinomial data from several populations. Assume that there are $J$ vectors of proportions produced from $J$ experiments with replicated ratings, $\widehat{\boldsymbol{\pi}}_j = (\widehat{\pi}_{1j}, \widehat{\pi}_{2j}, \ldots, \widehat{\pi}_{mj})', j = 1, 2, \ldots, J$. The test is

$$H_0: \boldsymbol{\pi}_1 = \boldsymbol{\pi}_2 = \ldots \boldsymbol{\pi}_J; \quad H_a: \boldsymbol{\pi}_i \neq \boldsymbol{\pi}_j, \tag{10.3-56}$$

where $i, j = 1, 2, \ldots, J; i \neq j$. The test statistic is

$$X^2 = \sum_{j=1}^{J} \frac{N_j}{\widehat{C}_j} \sum_{i=1}^{m} \frac{(\widehat{\pi}_{ij} - \widehat{\pi}_i)^2}{\widehat{\pi}_i}, \tag{10.3-57}$$

where $\widehat{\boldsymbol{\pi}} = (\widehat{\pi}_1, \widehat{\pi}_2, \ldots \widehat{\pi}_m)'$ is the weighted mean of $\widehat{\boldsymbol{\pi}}_j = (\widehat{\boldsymbol{\pi}}_{1j}, \widehat{\boldsymbol{\pi}}_{2j}, \ldots, \widehat{\pi}_{mj})'$. $\widehat{\boldsymbol{\pi}} = \sum_{j=1}^{J} \alpha_j \widehat{\boldsymbol{\pi}}_j$ and $\alpha_j = \frac{N_j}{\widehat{C}_j} \Big/ \left(\sum_{r=1}^{J} \frac{N_r}{\widehat{C}_r}\right)$. The test statistic (10.3-57) follows a chi-square distribution with $(J-1)(m-1)$ DF. The test statistic (10.3-57) reduces to the Pearson chi-square statistic when $\widehat{C}_j = 1$ for each population.

---

**Example 10.3-16**

In the study of housing satisfaction performed by H. S. Stoeckler and M. G. Gage for the U.S. Department of Agriculture (see Brier, 1980), households around Montevideo, Minnesota, were stratified into two populations: those in the metropolitan area and those outside the metropolitan area. There are 17 neighborhoods from the nonmetropolitan area and 18 neighborhoods from the metropolitan area. Five households were sampled from each of the neighborhoods. One response was obtained from the residents of each household concerning their satisfaction with their home. The possible responses were unsatisfied (US), satisfied (S), and very satisfied (VS). The data appeared in Koehler and Wilson (1986) and Wilson (1989) and are reproduced in Tables 10.28 and 10.29. The interest is to test whether there is a significant difference between the residents of the metropolitan area and the nonmetropolitan area with respect to degree of satisfaction with their homes.

The estimated proportions of responses are $\widehat{\boldsymbol{\pi}}_1 = (0.522, 0.422, 0.056)$ for the nonmetropolitan area and $\widehat{\boldsymbol{\pi}}_2 = (0.353, 0.506, 0.141)$ for the metropolitan area. Estimates of the $C$ values for the two populations are

**Table 10.28** Housing Satisfaction Data for Montevideo, Minnesota (Wilson, 1989) (Example 10.3-16)

| Nonmetropolitan | | | | Metropolitan | | | |
|---|---|---|---|---|---|---|---|
| Neighborhood | US | S | VS | Neighborhood | US | S | VS |
| 1 | 3 | 2 | 0 | 1 | 0 | 4 | 1 |
| 2 | 3 | 2 | 0 | 2 | 0 | 5 | 0 |
| 3 | 0 | 5 | 0 | 3 | 0 | 3 | 2 |
| 4 | 3 | 2 | 0 | 4 | 3 | 2 | 0 |
| 5 | 0 | 5 | 0 | 5 | 2 | 3 | 0 |
| 6 | 4 | 1 | 0 | 6 | 1 | 3 | 1 |
| 7 | 3 | 2 | 0 | 7 | 4 | 1 | 0 |
| 8 | 2 | 3 | 0 | 8 | 4 | 0 | 1 |
| 9 | 4 | 0 | 1 | 9 | 0 | 3 | 2 |
| 10 | 0 | 4 | 1 | 10 | 1 | 2 | 2 |
| 11 | 2 | 3 | 0 | 11 | 0 | 5 | 0 |
| 12 | 4 | 1 | 0 | 12 | 3 | 2 | 0 |
| 13 | 4 | 1 | 0 | 13 | 2 | 3 | 0 |
| 14 | 1 | 2 | 2 | 14 | 2 | 2 | 1 |
| 15 | 4 | 1 | 0 | 15 | 4 | 0 | 1 |
| 16 | 1 | 3 | 1 | 16 | 0 | 4 | 1 |
| 17 | 4 | 1 | 0 | 17 | 4 | 1 | 0 |
| 18 | 5 | 0 | 0 | | | | |

*Note: US, unsatisfied; S, satisfied; VS, very satisfied. See Wilson (1989).*

**Table 10.29** Summary of Data in Table 10.28 (Example 10.3-16)

| | US | S | VS | Total |
|---|---|---|---|---|
| Nonmetropolitan | 47 (0.522) | 38 (0.422) | 5 (0.056) | 90 (1.000) |
| Metropolitan | 30 (0.353) | 43 (0.506) | 12 (0.141) | 85 (1.000) |

*Note: Proportions in parentheses.*

$\hat{C}_1 = 1.619$ and $\hat{C}_2 = 1.632$, respectively, from (10.3-49), and $N_1 = 90$ and $N_2 = 85$. Using the DM test statistic (10.3-57), we get $\alpha_1 = 0.516$, $\alpha_2 = 0.484$, and $\hat{\pi} = (0.44, 0.46, 0.097)$; hence, $X^2 = 4.19$ with associated $P = 0.123$ from a chi-square distribution with 2 DF. There was no evidence to support a difference in housing satisfaction between residents of the two areas at a 0.05 significance level. However, if the cluster effects are ignored and the conventional chi-square test is used, the test result is $X^2 = 6.81$ with associated $P = 0.033$ from a chi-square distribution with 2 DF. Overdispersion should be considered in this case.

```
> cbval(t(dmdat11))
[1] 1.619
> cbval(t(dmdat12))
[1] 1.632
>dmtest2(c(90,85),c(1.619,1.632),t(cbind(c(47,38,5)/
90,c(30,43,12)/85)))
[1] 4.190 0.123
> chisq.test(cbind(c(47,38,5),c(30,43,12)))
     Pearson's chi-square test without Yates' continuity
correction
data: cbind(c(47, 38, 5), c(30, 43, 12))
X-square = 6.8069, df = 2, p-value = 0.0333
```

## EXERCISES

**10.1.** Using the duo–trio test, Stone and Bosley (1964, 620) obtained the following results:

|  | Correct Choice | Incorrect Choice |
|---|---|---|
| Control Odd | 39 | 21 |
| Treated Odd | 37 | 23 |

**1.** Test whether the panelists can differentiate between the control and the treated samples.

**2.** Estimate the probability of correct selection by means of a 95% confidence interval.

**10.2.** The preferences expressed by 12 panelists for one of two citrus juice solutions are shown in the following table (Gregson, 1960, 249). A tied preference was scored ½. For our purposes, the no preference vote is discarded.

|  | Preferred A | Preferred B |
|---|---|---|
| **AAB-Type Triangle Test** | | |
| Correct | 10 | 4 |
| Incorrect | 7 | 8 |
| **ABB-Type Triangle Test** | | |
| Correct | 9 | 13 |
| Incorrect | 0.5 | 6.5 |

1. Was the panel able to discriminate between samples $A$ and $B$?
2. Is there a bias in the data in the stage 2 triangle test?
3. Analyze the data using the Gridgeman model.

**10.3.** What is the probability of making a correct selection simply by chance in the following tests? Rationalize your answers.

1. Triangle test
2. Pair test
3. Duo–trio test

**10.4.** Let $\theta$(or $d'$) denote the standardized difference in means between the odd and two identical samples in the triangle or duo–trio test. Suppose you want to test $H_0$: $\theta = 0$ against $H_a$: $\theta > 0$ at $\alpha = 0.05$.

1. Determine the number of tests required so that the probability of rejecting $H_0$ when $\theta = 2.0$ is 0.80. Carry out your calculations for both the triangle and the duo–trio tests.
2. Compare the two tests with respect to the number of tests required.

**10.5.** Repeat Exercise 10.4 when $\theta = 4$ and $\theta = 6$. How does the comparison between the triangle and the duo–trio tests change as $\theta$ increases?

**10.6.** A triangle test was conducted to compare two food products. The results show that 13 out of 30 panelists correctly picked the odd sample. Let $P_T$ denote the probability of correct selection.

1. Test the null hypothesis $H_0$: $P_T = \frac{1}{3}$ at the 5% level of significance.

**2.** Find the 95% confidence interval for $P_T$. How could we use this interval to test the null hypothesis in Part 1?

**10.7.** Repeat Exercise 10.6 assuming the experimental outcomes of a duo–trio test.

**10.8.** Bradley (1964) applied the modified triangle test to a set of experimental data for beverages provided through the courtesy of Mr. Charles C. Beazley, Post Division, General Foods Corporation. A total of 54 panelists participated in a triangle test with 25 correct responses and 29 incorrect responses. The scores given by the panelists are presented in the following table, where $0 =$ no difference; $2 =$ very slight difference; $4 =$ slight difference; $6 =$ moderate difference; $8 =$ large difference; $10 =$ very large difference. Analyze the data using the Bradley–Harmon model for a modified triangle test.

R   2, 2, 2, 4, 6, 2, 4, 2, 6, 6, 2, 2, 2, 2, 4, 6, 4, 4, 2, 4, 2, 8, 8, 4, 0;

W   4, 2, 4, 2, 2, 4, 6, 2, 2, 2, 4, 4, 0, 2, 4, 4, 2, 4, 2, 6, 4, 6, 6, 4, 2, 2, 4, 0, 2;

**10.9.** Calculate the sample sizes, $n$ (with and without continuity correction), needed for the triangle and 3-AFC tests with the power of 0.8 to detect a difference in terms of $d' = 1.5$ (i.e., $p_1 = 0.51$ for the triangle and $p_1 = 0.77$ for the 3-AFC) at the 5% of significance level $\alpha$. See the following table.

|  | Power | $p_1$ ($d' = 1.5$) | $\alpha$ | $n$ | $n$ (With Continuity Correction) |
|---|---|---|---|---|---|
| Triangle | 0.8 | 0.51 | 0.05 | | |
| 3-AFC | 0.8 | 0.77 | 0.05 | | |

**10.10.** Calculate the sample sizes, $n$ (with and without continuity correction), needed for the duo–trio and 2-AFC tests with the power of 0.8 to detect a difference in terms of $d' = 1.5$ (i.e., $p_1 = 0.66$ for the duo–trio and $p_1 = 0.86$ for the 2-AFC) at the 5% of significance level $\alpha$.

|  | Power | $p_1$ ($d' = 1.5$) | $\alpha$ | $n$ | $n$ (With Continuity Correction) |
|---|---|---|---|---|---|
| Duo–trio | 0.8 | 0.66 | 0.05 | | |
| 2-AFC | 0.8 | 0.86 | 0.05 | | |

**10.11.** Calculate the sample sizes, $n$ (with and without continuity correction, $h = 1$), for the same-different and A–Not A tests with the power of 0.8 to detect a difference in terms of $d' = 1.5$ (i.e., $p_S = 0.339$ for the same-different and $p_A = 0.745$ for the A–Not A under the assumption that $p_D = 0.2$ and $p_N = 0.2$, respectively for the two tests) at the 5% of significance level $\alpha$.

| | Power | $p_D$ or $p_N$ | $p_S$ or $p_A$ (for $d' = 1.5$) | $\alpha$ | $n$ | $n$ (With Continuity Correction) |
|---|---|---|---|---|---|---|
| Same-different | 0.8 | 0.2 | 0.339 | 0.05 | | |
| A–Not A | 0.8 | 0.2 | 0.745 | 0.05 | | |

**10.12.** Calculate or find from corresponding tables the critical values $x_0$ for the similarity tests using the forced-choice methods for given sample sizes $n$, similarity limits $p_{d0}$, and Type I errors $\alpha$, in the following table.

| Methods | $n$ | $p_{d0}$ | $\alpha$ | $x_0$ |
|---|---|---|---|---|
| 2-AFC or Duo–trio | 100 | 0.2 | 0.05 | |
| | 200 | 0.2 | 0.05 | |
| 3-AFC or Triangle | 100 | 0.2 | 0.05 | |
| | 200 | 0.2 | 0.05 | |

**10.13.** Calculate or find from corresponding tables the critical values $c_l$ and $c_u$ for paired preference tests for given sample sizes $n$, similarity limits $\Delta$, and Type I errors $\alpha$ in the following table.

| $n$ | $\Delta$ | $\alpha$ | $c_l$ | $c_u$ |
|---|---|---|---|---|
| 100 | 0.1 | 0.05 | | |
| 200 | 0.1 | 0.05 | | |
| 300 | 0.05 | 0.1 | | |
| 400 | 0.05 | 0.1 | | |

**10.14.** Liu et. al (2002) gave a numerical example for equivalence test for paired binary data by using a generalized McNemar's test. The data are presented in the following table. This is an example of a clinical trial to compare the noninvasive MRI alternative procedure and invasive CTAP reference procedure. Let $\Theta = p_t - p_r = X_{10}/N = X_{01}/N$ denote the difference of response probability between the alternative and the reference procedures and $\widehat{\Theta} = \widehat{p}_t - \widehat{p}_r = X_{10}/n - X_{01}/n$ denote the sample estimates. The equivalence test is $H_0: p_t - p_r \geq \delta$ or $p_t - p_r \leq \delta$ versus $H_1: -\delta < p_t - p_r < \delta$, where $\delta = 0.15$ is a given equivalence limit. Carry out the calculations for the equivalence test.

|  |  | CTAP | |
|---|---|---|---|
|  |  | 1 | 0 |
| MRI | 1 | $X_{11} = 39$ | $X_{10} = 5$ |
|  | 0 | $X_{01} = 4$ | $X_{00} = 2$ |
|  |  |  | $n = 40$ |

**10.15.** William (1975) gave a numerical example for the beta-binomial model. The data are from Weil (1970) for an experiment comprising two treatments. One group of 16 pregnant female rats was fed a control diet during pregnancy, and the second group of 16 pregnant females was treated with a chemical. The number $n$ of pups alive at 4 days and the number $x$ of pups that survived were recorded as $x/n$ in the following table. Use the likelihood ratio statistic to test $H_0: \mu_1 = \mu_2, \theta_1 = \theta_2$ against $H_a: \mu_1 \neq \mu_2, \theta_1 \neq \theta_2$.

| Control | 13/13 | 12/12 | 9/9 | 9/9 | 8/8 | 8/8 | 12/13 | 11/12 |
|---|---|---|---|---|---|---|---|---|
|  | 9/10 | 9/10 | 8/9 | 11/13 | 4/5 | 5/7 | 7/10 | 7/10 |
| Treatment | 12/12 | 11/11 | 10/10 | 9/9 | 10/11 | 9/10 | 9/10 | 8/9 |
|  | 8/9 | 4/5 | 7/9 | 4/7 | 5/10 | 3/6 | 3/10 | 0/7 |

**10.16.** Calculate the probabilities of the numbers of correct responses 0, 1, 2, 3 in replicated 2-AFC with three replications from the distribution function of a corrected beta-binomial distribution with parameters $\mu = 1/2$ and $\gamma = 1/3$.

**10.17.** Ferris (1958) developed a $k$-visit method and model for replicated consumer preference testing and provided the data of 900 panelists for a 2-visit test. The data are represented in the following table. Analyze the data by using the DM model. Be aware that the DM model and Ferris' model are philosophically different. (Note: Transfer the data in the following table into a $900 \times 3$ matrix by using a program such as the S-PLUS code *ferrisdatf*—i.e., $y <- ferrisdatf()$)

| Category No. | Category Type* | Observed No. |
|:---:|:---:|:---:|
| 1 | AA | 457 |
| 2 | BB | 343 |
| 3 | AB | 8 |
| 4 | BA | 14 |
| 5 | AO | 14 |
| 6 | OA | 12 |
| 7 | BO | 17 |
| 8 | OB | 11 |
| 9 | OO | 24 |

*Note: AA-response "A" for both visits; BB-response "B" for both visits; AB-response "A" for first visit and "B" for the second visit; BA-response "B" for the first visit and "A" for the second visit; AO-response "A" for the first visit and "no preference" for the second visit; OA-response "no preference" for the first visit and "A" for the second visit; BO-response "B" for the first visit and "no preference" for the second visit; OB-response "no preference" for the first visit and "B" for the second visit; OO-response "no preference" for both visits.*

**10.18.** Brier (1980) analyzed the data from a housing satisfaction study using the DM model to test independence in a two-way table. Twenty neighborhoods were selected, and five homes were selected in each neighborhood. The data for the exercise are only for 18 neighborhoods with five homes (two neighborhoods with three homes were omitted). In each home the family was questioned in two areas: satisfaction with the housing in the neighborhood as a whole and satisfaction with their own home. The data are reproduced and summarized in the following tables. Analyze the data using the DM model.

| US-US | US-S | US-VS | S-US | S-S | S-VS | VS-US | VS-S | VS-VS |
|:---:|:---:|:---:|:---:|:---:|:---:|:---:|:---:|:---:|
| 1 | 0 | 0 | 2 | 2 | 0 | 0 | 0 | 0 |
| 1 | 0 | 0 | 2 | 2 | 0 | 0 | 0 | 0 |
| 0 | 2 | 0 | 0 | 2 | 0 | 0 | 1 | 0 |
| 0 | 1 | 0 | 2 | 1 | 0 | 1 | 0 | 0 |
| 0 | 0 | 0 | 0 | 4 | 0 | 0 | 1 | 0 |
| 1 | 0 | 0 | 3 | 1 | 0 | 0 | 0 | 0 |
| 3 | 0 | 0 | 0 | 1 | 0 | 0 | 1 | 0 |
| 1 | 0 | 0 | 1 | 3 | 0 | 0 | 0 | 0 |
| 3 | 0 | 0 | 0 | 0 | 0 | 1 | 0 | 1 |
| 0 | 1 | 0 | 0 | 3 | 1 | 0 | 0 | 0 |

*(continued)*

| US-US | US-S | US-VS | S-US | S-S | S-VS | VS-US | VS-S | VS-VS |
|-------|------|-------|------|-----|------|-------|------|-------|
| 1 | 1 | 0 | 0 | 2 | 0 | 1 | 0 | 0 |
| 0 | 1 | 0 | 4 | 0 | 0 | 0 | 0 | 0 |
| 0 | 0 | 0 | 4 | 1 | 0 | 0 | 0 | 0 |
| 0 | 0 | 0 | 1 | 2 | 0 | 0 | 0 | 2 |
| 2 | 0 | 0 | 2 | 1 | 0 | 0 | 0 | 0 |
| 0 | 0 | 0 | 1 | 1 | 1 | 0 | 2 | 0 |
| 2 | 0 | 0 | 2 | 1 | 0 | 0 | 0 | 0 |
| 2 | 0 | 0 | 2 | 0 | 0 | 1 | 0 | 0 |
| 17 | 6 | 0 | 26 | 27 | 2 | 4 | 5 | 3 |

| | US (community) | S (community) | VS (community) |
|---|---|---|---|
| US (personal) | 17 | 26 | 4 |
| S (personal) | 6 | 27 | 5 |
| VS (personal) | 0 | 2 | 3 |

# The Method of Paired Comparisons in Sensory Tests and Thurstonian Scaling

The theory and practice of paired comparisons have advanced considerably since Thurstone formulated his law of comparative judgment in 1927. The reader is referred to Thurstone (1959), David (1963b), Bock and Jones (1968), and Bradley (1976) for the historical and mathematical development of the method of paired comparison. Davidson and Farquhar (1976) published a bibliography on the method of paired comparison. In this chapter we discuss some paired comparison (PC) designs and models used in sensory tests for statistical analyses. We also discuss some measures of sensory difference.

## 11.1  PAIRED COMPARISON DESIGNS

There are experimental situations for which PC designs are extremely suitable. For instance, in taste testing experiments involving several objects, it is desirable to compare objects two at a time to account for taste fatigue and also to control lingering aftertaste effects. An attractive aspect of PC designs is that they are easy to administer and analyze. Various PC designs used in subjective and objective assessments of foods and other consumer products are classified in Figure 11.1. These designs are discussed briefly in the following two sections.

### Completely Balanced Paired Comparison Designs

First, we consider completely balanced designs for comparing $t$ objects in pairwise comparisons. There are $m = t(t - 1)/2$ possible different pairs. In a completely balanced design, all the $m$ pairs are compared equally often. We consider $t = 3$ objects and denote the three objects by using $A$, $B$, and $C$. The total number of pairs is $m = 3(3 - 1)/2 = 3$. Let $AB$ denote a paired comparison of $A$ with $B$ when $A$ is presented first and then $B$. Similarly, when $A$ follows $B$, the comparison

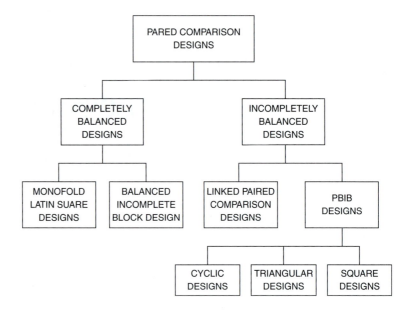

■ **FIGURE 11.1** Classification of paired comparison designs.

is denoted by *BA*. Thus, there are six possible paired comparisons, as shown in the following:

| Pairs | Comparisons |
|---|---|
| 1 | *AB* |
| 2 | *AC* |
| 3 | *BC* |
| 4 | *BA* |
| 5 | *CA* |
| 6 | *CB* |

Every object is compared with every other object in pairs 1−3. To eliminate possible order effects, one presents the objects of each pair in reverse order so that all the pairwise comparisons are repeated, as shown by pair comparisons 4−6.

### *Monofold Latin Square*

As *t* increases, the number of pairs increases rapidly. A systematic scheme for the enumeration and allocation of pairs for comparison among panelists was developed by Gridgeman (1970b). The scheme, based on Latin square designs (see Chapter 4), was called *monofold*

*Latin squares* by Gridgeman. A monofold Latin square consists of identical items (zeros) on the main diagonal and is symmetric; that is, the triangle above the diagonal is an exact mirror image of the triangle below the diagonal. Both the triangles of $m = t(t - 1)/2$ elements comprise $(t - 1)$ different elements; each element is replicated $t/2$ times. A $4 \times 4$ monofold Latin square with its elements denoted by 0, 1, 2, and 3 is shown as follows, the restriction being that none of the numerals 0, 1, 2, and 3 occurs twice in a row or in a column:

$$
\begin{array}{cccc}
0 & 1 & 2 & 3 \\
1 & 0 & 3 & 2 \\
2 & 3 & 0 & 1 \\
3 & 2 & 1 & 0
\end{array}
$$

Let us use this monofold Latin square to design a PC experiment with $t = 4$ items: $A$, $B$, $C$, and $D$. There are $m = 6$ pairs. To the triangle below the diagonal, we assign letters $A$, $B$, $C$, and $D$, as in the following design. For each numerical element in the triangle, there is a corresponding pair. Thus, all six pairs—$AB$, $AC$, $AD$, $BC$, $BD$, and $CD$—are enumerated:

$$
\begin{array}{c|ccc}
B & 1 & & \\
C & 2 & 3 & \\
D & 3 & 2 & 1 \\
\hline
& A & B & C
\end{array}
$$

Suppose we recruit a panel of $(t - 1) = 4 - 1 = 3$ judges and assign them numerical codes (1), (2), and (3). Then we can associate the three numerical codes (judges) with the six pairs to obtain six pairwise comparisons, as shown:

$$
\begin{array}{c|ccc}
B & AB(1) & & \\
C & AC(2) & BC(3) & \\
D & AD(3) & BD(2) & CD(1) \\
\hline
& A & B & C
\end{array}
$$

Each judge compares $t/2 = 2$ pairs. Note that the two pairs compared by each judge involve every one of the four items. Thus, the experimental design is balanced. We may repeat the design by presenting the items of each pair in reverse order if the order effect is suspect. Furthermore, we may also repeat the experiment by permuting the panel members to pairs.

In general, consider $t$ items to be compared pairwise. If $t$ is even, we use a $t \times t$ monofold Latin square with zeros on the diagonal for the enumeration of all pairs and $t - 1$ judges, designated by the numerical codes $(1), (2), \ldots, (t - 1)$. Each judge makes $t/2$ distinct paired comparisons. When $t$ is odd, pair enumeration and allocation of judges are identical, except that we start with a $(t + 1) \times (t + 1)$ monofold Latin square. When this is done, the first column and first row are chopped off, reducing the square to size $t \times t$. Consider $t = 3$. Starting with $t + 1 = 4$, we have the following $4 \times 4$ monofold Latin square:

$$
\begin{array}{cccc}
0 & 1 & 2 & 3 \\
1 & 0 & 3 & 2 \\
2 & 3 & 0 & 1 \\
3 & 2 & 1 & 0
\end{array}
$$

Now chopping off column 1 and row 1 results in a $3 \times 3$ square:

$$
\begin{array}{ccc}
0 & 3 & 2 \\
3 & 0 & 1 \\
2 & 1 & 0
\end{array}
$$

As before, let us take the triangle below the main diagonal and assign treatment pairs to the numerals and judges, as indicated:

$$
\begin{array}{c|cc}
B & AB(3) & \\
C & AC(2) & BC(1) \\
\hline
 & A & B
\end{array}
$$

When $t$ is odd, the number of judges equals $t$. Thus, we have a basic design for enumerating all treatment pairs and their allocation to judges. The basic design can be repeated to account for order and judge effects.

Following the monofold Latin square designs, Table 11.1 lists treatment pairs and their assignment to panel members when $t = 5$ and $t = 6$.

**Table 11.1** Assignment of Treatment Pairs to Panelists Following Monofold Latin Square Designs

| Panelists | Treatment Pairs ($t = 5$) | | Treatment Pairs ($t = 6$) | | |
|---|---|---|---|---|---|
| 1 | BC | DE | AB | CD | EF |
| 2 | CD | AE | AC | DF | BE |
| 3 | AD | BE | AD | BE | CF |
| 4 | AB | CE | BC | AE | DF |
| 5 | AC | BD | BD | CE | AF |

### Balanced Incomplete Block Paired Design

The balanced incomplete block (BIB) design and its parameters were defined in Chapter 5. For PC experiments we need the design with blocks of size $k = 2$. Thus, the parameters of a BIB design for paired comparisons are $b = t(t - 1)/2$, $k = 2$, $r = (t - 1)$, and $\lambda = r(k - 1)/t - 1 = 1$. Note that for a BIB design, $\lambda = 1$ whenever $k = 2$. Obviously, the number of panelists $b$ and the number of replications per treatment $r$ are fixed by the number of treatments $t$. When $t = 4$, we obtain $b = 6$, $r = 3$, and $\lambda = 1$. Hence, the basic design consists of $b = 6$ panelists, $r = 3$ replications per treatment, and $\lambda = 1$. A layout of the BIB design for paired comparison would lead to the following assignment of panelists to pairs:

| Panelists | Treatment Pairs |
|-----------|-----------------|
| 1 | AB |
| 2 | CD |
| 3 | AC |
| 4 | BC |
| 5 | AD |
| 6 | BD |

The panelists are assigned to pairs at random in each replication. When the design is repeated $p$ times, the design parameters become $pb$, $pr$, and $p\lambda$. Table 11.2 contains some BIB designs for paired comparisons when $t = 4$, 5, and 6.

**Table 11.2** Designs for Paired Comparisons[a]

| Parameters | Panelists | Pairs |
|------------|-----------|-------|
| $t = 4$, $k = 2$, $r = 3$, $b = 6$, $\lambda = 1$ | 1 | AB |
| | 2 | CD |
| | 3 | AC |
| | 4 | BD |
| | 5 | AD |
| | 6 | BC |
| $t = 5$, $k = 2$, $r = 4$, $b = 10$, $\lambda = 1$ | 1 | AB |
| | 2 | CD |

*(continued)*

**Table 11.2** Designs for Paired Comparisons[a]—cont...

| Parameters | Panelists | Pairs |
|---|---|---|
| | 3 | BE |
| | 4 | AC |
| | 5 | DE |
| | 6 | AD |
| | 7 | BC |
| | 8 | BE |
| | 9 | AE |
| | 10 | BD |
| $t = 6, k = 2,$ | 1 | AB |
| $r = 5, b = 15,$ | 2 | CD |
| $\lambda = 1$ | 3 | EF |
| | 4 | AC |
| | 5 | BE |
| | 6 | DF |
| | 7 | AD |
| | 8 | BF |
| | 9 | CE |
| | 10 | AE |
| | 11 | BD |
| | 12 | CE |
| | 13 | AE |
| | 14 | BC |
| | 15 | DE |

[a]*Source: Cochran and Cox (1957).*

## Incomplete Balanced Paired Comparison Designs

A practical difficulty with paired comparisons is that a large number of paired comparisons is required if $t$ is large. When only a fraction of comparisons is to be made, we may then use an incomplete PC design. Kendall (1955) suggested that these designs be balanced as to the number of comparisons per judge. We discuss some of these designs in this section.

### Linked Paired Comparison Designs

If an aim of the PC experiment is to measure agreement between judges, then the design used must be symmetric with respect to judges and objects. Bose (1956) defined the so-called linked paired comparison designs, in which each pair of judges has certain comparisons in common. Suppose $t$ objects are to be compared in pairs by $v$ judges. There are $m = t(t-1)/2$ pairs. Suppose each judge makes $r > 1$ distinct paired comparisons. To ensure symmetry between objects and judges, Bose defined PC designs satisfying the following three restrictions:

1. Among the $r$ pairs compared by each judge, each object appears equally often, say $\alpha$ times.
2. Each pair is compared by $k$ judges, $k > 1$.
3. For any two judges there are exactly $\lambda$ pairs that are compared by both judges.

Let us illustrate these requirements for $t = 4$ objects and $v = 3$ judges. The three requirements can be satisfied by having a design consisting of three squares of size $2 \times 2$ as follows:

|   | Square 1 | |   |   | Square 2 | |   |   | Square 3 | |
|---|---|---|---|---|---|---|---|---|---|---|
|   | 1 | 2 |   |   | 1 | 3 |   |   | 1 | 4 |
| 3 | 13 | 23 | 2 |   | 12 | 23 | 2 |   | 12 | 24 |
| 4 | 14 | 24 | 4 |   | 14 | 34 | 3 |   | 13 | 34 |
|   | Judge A | |   |   | Judge B | |   |   | Judge C | |

The four cells of a square represent $r = 4$ pairs to be evaluated by each judge. In each square, every treatment appears twice ($\alpha = 2$) satisfying the first requirement. Of the $m = 6$ pairs, each pair is compared by $k = 2$ judges. Note that the pair (1, 3) appears in squares 1 (judge A) and 3 (judge C); and the pair (2, 3) in squares 1 and 2 (judges A and B). This satisfies the second requirement. The third requirement is fulfilled because any two judges have $\lambda = 2$ pairs of treatment in common. For instance, judges A and B have pairs (1, 4) and (2, 3) in common. The preceding illustration is of a design given by Bose with parameters $t = 4$, $m = 6$, $k = 2$, $\alpha = 2$, $v = 3$, $r = 4$, and $\lambda = 2$. This design is balanced in the sense that each judge evaluates every treatment the same number of times. In addition to such balance between treatments and judges, each judge compares only four pairs instead of all six pairs when $t = 4$; a reduction of $\frac{1}{3}$. This saving in comparisons is an important feature of linked PC designs.

Table 11.3 lists several linked PC designs for $t = 4$ through $t = 8$. The reader can refer to Bose (1956) and David (1963a) for the construction of various linked PC designs. These authors also discussed the correspondence between the linked and BIB PC designs. In applications, it is customary to use more judges than would be needed for a basic design. Hence, the basic design is repeated to have more judges.

### Partially Balanced Incomplete Block Designs

A BIB design is severely limited by its parameters. Bose and associates developed the partially balanced incomplete block (PBIB) design aimed at lessening this limitation. For PC designs, one would be

**Table 11.3** Linked Designs[a]

| No. | Parameters | Design Judge | Design Pairs Assigned to a Judge |
|---|---|---|---|
| 1 | $t = 4, v = 3,$ $b = 6, r = 4,$ $k = 2, \lambda = 2,$ $\alpha = 2$ | A | (1, 4), (1, 3), (2, 4), (2, 3) |
| | | B | (1, 3), (2, 4), (1, 2), (3, 4) |
| | | C | (1, 4), (1, 2), (2, 3), (3, 4) |
| 2 | $t = 5, v = 6,$ $b = 10, r = 5,$ $k = 3, \lambda = 2,$ $\alpha = 2$ | A | (3, 5), (2, 4), (1, 3), (1, 4), (2, 5) |
| | | B | (2, 3), (3, 4), (1, 4), (1, 5), (2, 5) |
| | | C | (2, 3), (3, 5), (1, 2), (4, 5), (1, 4) |
| | | D | (3, 5), (1, 2), (3, 4), (2, 4), (1, 5) |
| | | E | (1, 2), (3, 4), (4, 5), (1, 3), (2, 5) |
| | | F | (2, 3), (4, 5), (2, 4), (1, 3), (1, 5) |
| 3 | $t = 6, v = 10,$ $b = 15, r = 6,$ $k = 4, \lambda = 2,$ $\alpha = 2$ | A | (1, 2), (1, 3), (2, 3), (4, 5), (4, 6), (5, 6) |
| | | B | (2, 3), (2, 4), (3, 4), (1, 5), (1, 6), (5, 6) |
| | | C | (1, 3), (1, 4), (3, 4), (2, 5), (2, 6), (5, 6) |
| | | D | (1, 4), (2, 4), (1, 2), (3, 5), (3, 6), (5, 6) |
| | | E | (1, 4), (1, 6), (4, 6), (2, 3), (2, 5), (3, 5) |
| | | F | (1, 3), (1, 5), (3, 5), (2, 4), (2, 6), (4, 6) |
| | | G | (1, 2), (1, 5), (2, 5), (3, 4), (3, 6), (4, 6) |
| | | H | (1, 4), (1, 5), (4, 5), (2, 3), (2, 6), (3, 6) |
| | | I | (1, 3), (1, 6), (3, 6), (2, 4), (2, 5), (4, 5) |
| | | J | (1, 2), (1, 6), (2, 6), (3, 4), (3, 5), (4, 5) |

*(continued)*

**Table 11.3** Linked Designs[a]—cont...

| No. | Parameters | Sets of Pairs | Design Judge | Design Sets of Pairs Assigned to a Judge |
|-----|-----------|---------------|-------|--------------------------------|
| 4 | $t = 6, \quad v = 5,$ $b = 15, r = 12,$ $k = 4, \quad \lambda = 9,$ $\alpha = 4$ | I 14 23 56 | A | II, III, IV, V |
| | | III 25 34 16 | B | I, III, IV, V |
| | | III 31 45 26 | C | I, II, IV, V |
| | | IV 42 51 36 | D | I, II, III, V |
| | | V 53 12 46 | E | I, II, III, IV |
| 5 | $t = 7, \quad v = 3,$ $b = 21, r = 14,$ $k = 2, \quad \lambda = 7,$ $\alpha = 4$ | I 12 23 34 | A | II, III |
| | | 45 56 67 71 | B | III, I |
| | | II 13 24 35 | C | I, II |
| | | 46 57 61 72 | | |
| | | III 14 25 36 | | |
| | | 47 51 62 73 | | |
| 6 | $t = 8, \quad v = 7,$ $b = 28, r = 12,$ $k = 3, \quad \lambda = 4,$ $\alpha = 3$ | I 16 25 34 78 | A | I, V, VII |
| | | II 27 36 45 18 | B | II, VI, I |
| | | III 31 47 56 28 | C | III, VII, II |
| | | IV 42 51 67 38 | D | IV, I, III |
| | | V 53 62 71 48 | E | V, II, IV |
| | | VI 64 73 12 58 | F | VI, III, V |
| | | VII 75 14 23 68 | G | VII, IV, VI |
| 7 | $t = 8, \quad v = 7,$ $b = 28, r = 16,$ $k = 4, \quad \lambda = 8,$ $\alpha = 4$ | I 16 25 34 78 | A | III, V, VI, VII |
| | | II 27 36 45 18 | B | IV, VI, VII, I |
| | | III 31 47 56 28 | C | V, VII, I, II |
| | | IV 42 51 67 38 | D | VI, I, II, III |

[a]Source: Bose (1956).

interested in PBIB designs having only two "associate" classes and blocks of size 2. The concept of the associate class is rather involved; hence, we do not discuss it any further. The interested reader may refer to Bose and Shimamoto (1952). Three simple PBIB designs useful in PC experimentation are discussed in the following subsections.

### Triangular Designs

Suppose that the number of treatments $t$ can be expressed as $t = n(n-1)/2$, where $n$ is some integer. Then a design can be obtained by arranging treatments in an $n \times n$ array, where elements of the array are the $t$ treatments in numerical order and the principal diagonal of the array consists of zeros. Thus, for $t = 6$, $n$ is 4, and a $4 \times 4$ array with zeros on the diagonal is

$$
\begin{matrix}
0 & 1 & 2 & 3 \\
1 & 0 & 4 & 5 \\
2 & 4 & 0 & 6 \\
3 & 5 & 6 & 0
\end{matrix}
$$

The array is symmetric. A triangular PC design is constructed by forming pairs of treatments from each row or each column and then having a block for each pair. Corresponding to our $4 \times 4$ array, a triangular design is as follows:

| Rows | Blocks | Treatment Pairs |
|------|--------|-----------------|
| 1 | 1 | 1,2 |
|   | 2 | 1,3 |
|   | 3 | 2,3 |
| 2 | 4 | 1,4 |
|   | 5 | 1,5 |
|   | 6 | 4,5 |
| 3 | 7 | 2,4 |
|   | 8 | 2,6 |
|   | 9 | 4,6 |
| 4 | 10 | 3,5 |
|   | 11 | 3,6 |
|   | 12 | 5,6 |

In this triangular PC design, each treatment is replicated $r = 4$ times. For instance, treatment 4 appears in blocks 4, 6, 7, and 9. Table 11.4 lists some triangular PC designs of interest.

### Square Designs

If there exists an integer $s > 1$ such that the number of treatments $t = s^2$, then treatments can be arranged in an $s \times s$ array. For example, when $t = 9$, we have $s = 3$; a $3 \times 3$ square array is

$$
\begin{matrix}
1 & 2 & 3 \\
4 & 5 & 6 \\
7 & 8 & 9
\end{matrix}
$$

**Table 11.4** Some Triangular Designs[a]

| Design | Treatments $t$ | Blocks $b$ | Replications $r$ | Size of $n$ |
|--------|----------------|------------|------------------|-------------|
| 1 | 6 | 12 | 4 | 4 |
| 2 | 10 | 30 | 6 | 5 |
| 3 | 15 | 60 | 8 | 6 |
| 4 | 21 | 105 | 10 | 7 |
| 5 | 10 | 15 | 3 | 5 |
| 6 | 15 | 45 | 6 | 6 |
| 7 | 21 | 105 | 10 | 7 |

[a]Source: Clatworthy (1955); Dykstra (1960).

The pairs of treatments are formed by pairing treatments in the same row and same column of the array. From our $3 \times 3$ array, we get the following design:

| Rows | Blocks | Treatment Pairs |
|------|--------|-----------------|
| 1 | 1 | 1,2 |
|   | 2 | 1,3 |
|   | 3 | 2,3 |
| 2 | 4 | 4,5 |
|   | 5 | 4,6 |
|   | 6 | 5,6 |
| 3 | 7 | 7,8 |
|   | 8 | 7,9 |
|   | 9 | 8,9 |
| Columns | | |
| 1 | 10 | 1,4 |
|   | 11 | 1,7 |
|   | 12 | 4,7 |
| 2 | 13 | 2,5 |
|   | 14 | 2,8 |
|   | 15 | 5,8 |
| 3 | 16 | 3,6 |
|   | 17 | 3,9 |
|   | 18 | 6,9 |

| **Table 11.5** Some Square Designs[a] | | | | |
|---|---|---|---|---|
| **Design** | **Treatments $t$** | **Blocks $b$** | **Replications $r$** | **Size of $s$** |
| 1 | 4 | 4 | 2 | 2 |
| 2 | 9 | 18 | 4 | 3 |
| 3 | 16 | 48 | 6 | 4 |
| 4 | 25 | 100 | 8 | 5 |
| 5 | 36 | 180 | 10 | 6 |
| 6 | 16 | 72 | 9 | 4 |
| [a]Source: Clatworthy (1955); Dykstra (1960). | | | | |

The design compares only 18 pairs instead of the 36 pairs for a complete design. Some square designs are listed in Table 11.5.

### Cyclic Designs

Cyclic designs (CDs) are constructed by the cyclic arrangement of an initial block. The construction of CDs is more complex than that of either triangular or square designs. The interested reader may refer to Clatworthy (1955), David (1963a, b), and John et al. (1972) for detailed treatments of CDs. For our purposes, let us consider an example with $t = 5$ treatments. A complete design for five treatments would have $m = 10$ pairs. Suppose we desire an incomplete PC design satisfying the following two requirements formulated by Kendall (1955): (1) every treatment should occur equally often; (2) treatments cannot be partitioned into two sets; thus, no paired comparison is made between any treatment of one set and any treatment of the other set. Such a design can be obtained by two cyclic arrangements. The first arrangement is simply to write down $t$ treatments in numerical order. The second cyclic arrangement may start with treatment 2. Combining the corresponding elements of cycles leads to pairs and blocks as shown in the following:

| Sets | Cycle I | Cycle II | Pairs of Treatments | Blocks |
|---|---|---|---|---|
| I | 1 | 2 | 1,2 | 1 |
| | 2 | 3 | 2,3 | 2 |
| | 3 | 4 | 3,4 | 3 |
| | 4 | 5 | 4,5 | 4 |
| | 5 | 1 | 5,1 | 5 |

*(continued)*

| Sets | Cycle I | Cycle II | Pairs of Treatments | Blocks |
|------|---------|----------|---------------------|--------|
| II | 1 | 3 | 1,3 | 1 |
| | 2 | 4 | 2,4 | 2 |
| | 3 | 5 | 3,5 | 3 |
| | 4 | 1 | 4,1 | 4 |
| | 5 | 2 | 5,2 | 5 |

| **Table 11.6** Some Cyclic Designs[a] | | | |
|---------|--------------|-----------|----------------|
| **Design** | **Treatments *t*** | **Blocks *b*** | **Replications *r*** |
| 1 | 5 | 5 | 2 |
| 2 | 13 | 39 | 6 |
| 3 | 17 | 68 | 8 |
| [a]*Source: Clatworthy (1955); Dykstra (1960).* | | | |

If the second cycle starts with 3, we get Set II. Either set can be used for an incomplete design. A design that accounts for the order of item presentation is made by repeating the design with the order of each pair reversed. Table 11.6 lists three PC CDs.

## 11.2 PAIRED COMPARISON MODELS

In this section we consider three PC models for the analysis of data from PC designs. The three models are commonly known as the Scheffé, Bradley–Terry, and Thurstone–Mosteller models. In addition, we also consider a least squares analysis formulation by Morrissey (1955) for scaling stimuli.

### The Scheffé Model

Scheffé (1952) developed a model for general consumer preference studies. The model is also applicable to other studies, such as the testing of attitudes or the evaluation of responses to physical or psychological stimuli through PC experiments. The Scheffé model uses a scoring system, such as a 7-point rating scale, and allows for effects due to the order of presentation of items of a pair. For instance, a 7-point scale may consist of the following statements with their corresponding scores indicated when item $i$ is compared with item $j$:

| Statement | Score |
|---|---|
| I prefer *i* to *j* strongly | 3 |
| I prefer *i* to *j* moderately | 2 |
| I prefer *i* to *j* slightly | 1 |
| No preference | 0 |
| I prefer *j* to *i* slightly | −1 |
| I prefer *j* to *i* moderately | −2 |
| I prefer *j* to *i* strongly | −3 |

Letters *i* and *j* may be replaced by codes given to treatments *i* and *j*. Note that in the sample scoring scale sheet (Figure 11.2), treatments are assigned the two-digit codes 52 and 86.

Order effects can be estimated by reversing the order of presentation of *i* and *j* in their paired comparisons. For instance, if the order of presentation is $(i, j)$, that is, in the first comparison treatment *i* is presented first and then treatment *j*, the reverse order $(j, i)$ is used in the second comparison, and so on. To balance judges and orders, we may assign *n* judges to order $(i, j)$ and another *n* judges to the reverse order $(j, i)$. With *t* treatments, a total of $nt(t - 1)$ judges is required, although fewer judges need to be used if the experiment is adequately balanced.

■ **FIGURE 11.2** Sample scoring scale sheet for the Scheffé model.

Date_____

Name_____Product_____

PREFERENCE TEST

Please indicate your overall preference between two items coded 52 and 86 by checking one of the statements below.

Check Here              Statement

_____ I strongly prefer ____52____ to __86__ .

_____ I moderately prefer ____52____ to __86__ .

_____ I slightly prefer ____52____ to __86__ .

_____ No preference.

_____ I slightly prefer ____86____ to __52__ .

_____ I moderately prefer ____86____ to __52__ .

_____ I strongly prefer ____86____ to __52__ .

Comments:_____
_____
_____

For instance, a judge may be asked to compare several pairs so that each judge evaluates the same number of pairs for a balanced design.

The mathematical model of Scheffé is as follows. Let $X_{ijk}$ denote the preference score for $i$ over $j$ of the $k$th judge when $i$ and $j$ are presented in order $(i, j)$. On a 7-point scale $X_{ijk}$ may be any one of the seven possible values: 3, 2, 1, 0, $-1$, $-2$, $-3$. It is assumed that equal but opposite preferences ($i$ over $j$, and $j$ over $i$) correspond to equal but opposite scores. Thus, the preference score of the $k$th judge for $i$ over $j$ when presented in order $(j, i)$ is $- X_{jik}$. It is assumed that the $X_{ijk}$ are independently and normally distributed with mean $\mu_{ij}$ and constant variance $\sigma^2$. The assumptions of normality and constant variance (homogeneity) are not strictly valid. The normality assumption is usually not serious in statistical analyses. The homogeneity of variance assumption is that all judges are equally efficient in their comparisons.

Let the mean preference score of $i$ over $j$, when presented in order $(i, j)$, be $\mu_{ij}$ and, when presented in order $(j, i)$, $-\mu_{ji}$. Hence, the average preference score $\pi_{ij}$ of $i$ over $j$ is the average of $\mu_{ij}$ and $-\mu_{ji}$. That is,

$$\pi_{ij} = \left(\mu_{ij} - \mu_{ji}\right)/2.$$

Let $\delta_{ij}$ measure the difference due to the order of presentation in the mean preferences for $i$ over $j$ as defined by

$$\delta_{ij} = \left(\mu_{ij} + \mu_{ji}\right)/2.$$

Scheffé assumed "subtractivity"; that is, there exist parameters $\alpha_1, \alpha_2, \ldots,$ $\alpha_t$, which for all pairs $i$ and $j$ satisfy $\pi_{ij} = \alpha_i - \alpha_j$ and $\Sigma \alpha_i = 0$. Departures from subtractivity are measured by parameters $\gamma_{ij}$ so that

$$\pi_{ij} = \alpha_i - \alpha_j + \gamma_{ij},$$

where the $\gamma_{ij}$ satisfy conditions

$$\gamma_{ij} = -\gamma_{ji} \quad \text{and} \quad \Sigma \gamma_{ij} = 0.$$

With these parameters, the Scheffé model is

$$X_{ijk} = \mu_{ij} + \delta_{ij} + \varepsilon_{ijk} = \alpha_i - \alpha_j + \gamma_{ij} + \delta_{ij} + \varepsilon_{ijk} \qquad (11.2\text{-}1)$$

where

$$i \neq j, \quad i = 1, \ldots, t, \quad j = 1, \ldots, t, \quad k = 1, \ldots, n,$$

the error term $\varepsilon_{ijk}$ is a random variable, and all other terms on the right side are unknown parameters to be estimated. If desired, one may introduce an average-order-effect parameter $\delta$ defined by

$$\delta = \sum_{i<j} \delta_{ij}/m,$$

where the summation is over all $m = t(t - 1)/2$ pairs. We can also rewrite (11.2-1) as follows:

$$X_{ijk} = \alpha_i - \alpha_j + \gamma_{ij} + \delta + \left(\delta_{ij} - \delta\right) + \varepsilon_{ijk}.$$

Usually, a null hypothesis of interest is

$$H_0: \ \alpha_1 = \alpha_2 = \cdots = \alpha_t$$

against an alternative,

$$H_a: \ \alpha_i \neq \alpha_j, \quad \text{for at least one pair} \quad i \text{ and } j.$$

Also, one may wish to test the presence of order effects by formulating the null and alternative hypotheses as follows:

$$H_0: \ \delta_{ij} = 0, \quad \text{for all pairs} \quad i \text{ and } j.$$

versus

$$H_a: \ \delta_{ij} \neq 0, \quad \text{for at least one pair} \quad i \text{ and } j.$$

A null hypothesis for testing the assumption of subtractivity can be stated as follows:

$$H_0: \ \gamma_{ij} = 0, \quad \text{for all pairs} \quad i \text{ and } j.$$

versus

$$H_a: \ \gamma_{ij} \neq 0, \quad \text{for at least one pair} \quad i \text{ and } j.$$

Now let us proceed to calculations of various sums of squares for the analysis of variance and for testing the preceding hypotheses. First, we may estimate parameters of the model as follows: Estimate $\mu_{ij}$ by

$$\widehat{\mu}_{ij} = \sum_{k}^{n} X_{ijk}/n,$$

where $n$ denotes the number of judges. Then estimate $\pi_{ij}$ and $\delta_{ij}$ by

$$\widehat{\pi}_{ij} = \left(\widehat{\mu}_{ij} - \widehat{\mu}_{ji}\right)/2 \quad \text{and} \quad \widehat{\delta}_{ij} = \left(\widehat{\mu}_{ij} + \widehat{\mu}_{ji}\right)/2,$$

respectively. The average order effect is estimated by

$$\widehat{\delta} = \sum_{i<j} \widehat{\delta}_{ij}/m.$$

Finally, estimate the main effect $\alpha_i$ by

$$\widehat{\alpha}_i = \sum_{j}^{t} \widehat{\pi}_{ij}/t,$$

where $\hat{\pi}_{ii}$ in the summation is defined to be zero. If interest is only in a direct comparison between $i$ and $j$, then $\hat{\pi}_{ij}$ is an appropriate statistic. But usually in practice the interest lies in an overall comparison of treatments relative to each other.

Using estimates of the parameters, the sums of squares due to various sources can be computed as follows:

Due to main effects:

$$SS\alpha = 2nt \sum_i^t \hat{a}_i^2.$$

Due to average preferences:

$$SS\pi = 2n \sum_{i<j} \hat{\pi}_{ij}^2,$$

where $\Sigma_{i<j}$ denotes the sum over $i$ and $j$ with $i < j$.

Due to means:

$$SS\mu = n \sum_i^t \sum_j^t \hat{\mu}_{ij}^2,$$

where $\hat{\pi}_{ii} = 0$ for any $i = 1, \ldots, t$.

Due to subtractivity:

$$SS\gamma = SS\pi - SS\alpha.$$

Due to order effects:

$$SS\delta = SS\mu - SS\pi.$$

Total variation:

$$SSto = \sum_i^t \sum_j^t \sum_k^n X_{ijk}^2,$$

where $X_{iik}$ is defined to be zero.

Due to error:

$$SSe = SSto - SS\mu.$$

These sums of squares are summarized in Table 11.7.

An unbiased estimate of the error variance $\sigma^2$, denoted by $\hat{\sigma}^2$, is the mean square due to errors in the analysis of variance table; that is,

$$\hat{\sigma}^2 = SSe/t(t-1)(n-1).$$

**Table 11.7** Analysis of Variance Table for the Scheffé Model

| Source of Variation | DF | SS | MS[a] |
|---|---|---|---|
| Main effects | $t - 1$ | $SS\alpha$ | $MS\alpha$ |
| Subtractivity | $\dfrac{t(t-1)}{2} - t + 1$ | $SS\gamma$ | $MS\gamma$ |
| Average preferences | $\dfrac{t(t-1)}{2}$ | $SS\pi$ | $MS\pi$ |
| Order effects | $\dfrac{t(t-1)}{2}$ | $SS\delta$ | $MS\delta$ |
| Means | $t(t-1)$ | $SS\mu$ | $MS\mu$ |
| Error | $t(t-1)(n-1)$ | $SSe$ | $MSe$ |
| Total | $nt(t-1)$ | $Ssto$ | |

[a]$MS = SS/DF.$

The standard errors (SEs) of $\widehat{\alpha}_i$ and $\left(\widehat{\alpha}_i - \widehat{\alpha}_j\right)$ are estimated, respectively, by

$$SE(\widehat{\alpha}_i) = \widehat{\sigma}\sqrt{(t-1)/2nt^2}, \quad \text{and} \quad SE(\widehat{\alpha}_i - \widehat{\alpha}_j) = \widehat{\sigma}\sqrt{1/nt}.$$

**Example 11.2-1**

An experiment was carried out to study preferences for frankfurters with varying levels of citric acid. Frankfurters with citric acid at two levels (treatments 2 and 3) and at the control level (treatment 1) were presented in pairs to $2n = 16$ panelists for their preferences. There are three treatments and hence $t = 3$. For $t = 3$, there are three pairs (1,2), (1,3), and (2,3), and for each pair, there are two orders of presentation, as shown here:

| Pairs | Order of Presentation | |
|---|---|---|
| 1 | (1,2) | (2,1) |
| 2 | (1,3) | (3,1) |
| 3 | (2,3) | (3,2) |

Each panelist compared all three pairs one at a time in a single session. The first eight panelists compared the three pairs in one order, and the remaining eight panelists compared the pairs in reverse order. The scoring was done using the questionnaire shown in Figure 11.2. The data are summarized in Table 11.8, along with estimates of $\mu_{ij}$ and $\pi_{ij}$. The numbers in column (2) denote the number of panelists (frequencies) assigning the $X_{ijk}$ scores.

**Table 11.8** Paired Comparison Data for Example 11.2-1

| (1) Orders (i,j) | (2) Frequency of Scores, $X_{ijk}$ | | | | | | | (3) Total Score, $X_{ij\cdot}$ | (4) $\widehat{\mu}_{ij}$ | (5) $\widehat{\pi}_{ij}$ |
|---|---|---|---|---|---|---|---|---|---|---|
| | **−3** | **−2** | **−1** | **0** | **1** | **2** | **3** | | | |
| (1, 2) | | 3 | 2 | 1 | | 2 | | −4 | −0.50 | 0.125 |
| (2, 1) | | 3 | 2 | 2 | | 1 | | −6 | −0.75 | −0.125 |
| (1, 3) | | 1 | | 3 | | 2 | 2 | 8 | 1.00 | 1.00 |
| (3, 1) | 1 | 3 | 1 | 2 | | 1 | | −8 | −1.00 | −1.000 |
| (2, 3) | 1 | 1 | | 4 | 1 | 1 | | −2 | −0.25 | 0.625 |
| (3, 2) | 3 | 1 | 1 | 3 | | | | −12 | −1.50 | −0.625 |
| | 5 | 12 | 6 | 15 | 1 | 7 | 2 | | | |

*Note, in Table 11.8:*

$X_{12\cdot} = 3(-2) + 2(-1) + 1(0) + 2(2) = -4, X_{13\cdot} = 8, X_{23\cdot} = -2,$

$X_{21\cdot} = 3(-2) + 2(-1) + 2(0) + 1(2) = -6, X_{31\cdot} = -8, X_{32\cdot} = -12.$

$\widehat{\mu}_{12} = -\dfrac{4}{8} = -0.50, \widehat{\mu}_{13} = 1.00, \widehat{\mu}_{23} = -0.25,$

$\widehat{\mu}_{21} = -\dfrac{6}{8} = -0.75, \widehat{\mu}_{31} = -1.00, \mu_{32} = -1.50.$

$\widehat{\pi}_{12} = (-0.50 + 0.75)/2 = 0.125, \widehat{\pi}_{13} = -1.00,$

$\widehat{\pi}_{21} = -0.125, \widehat{\pi}_{31} = -1.00,$

$\widehat{\pi}_{23} = (-.25 + 1.50)/2 = 0.625,$

$\widehat{\pi}_{32} = -0.625.$

*Estimates of treatment main effects $\alpha_1$, $\alpha_2$, and $\alpha_3$ are*

$\widehat{\alpha}_1 = (\widehat{\pi}_{12} + \widehat{\pi}_{13})/3 = (0.1250 + 1.0000)/3 = 0.3750,$

$\widehat{\alpha}_2 = (\pi_{21} + \widehat{\pi}_{23})/3 = 0.1667,$

$\widehat{\alpha}_3 = (\widehat{\pi}_{31} + \widehat{\pi}_{32})/3 = -0.5417.$

*Check: $\widehat{\alpha}_1 + \widehat{\alpha}_2 + \widehat{\alpha}_3 = 0$. Various sums of squares are*

$SS\alpha = 2(8)(3)\left[(0.3750)^2 + (0.1667)^2 + (-0.5417)^2\right] = 22.1688,$

$SS\pi = 2(8)\left[(0.1250)^2 + (1.0000)^2 + (-0.6250)^2\right] = 22.5000,$

$SS\mu = 8\left[(-0.5000)^2 + (-0.7500)^2 + \cdots + (-1.5000)^2\right] = 41.0000,$

$SSto = 3^2(5 + 2) + 2^2(12 + 7) + 1^2(6 + 1) = 146.$

$SS\gamma = SS\pi - SS\alpha = 22.5000 - 22.1688 = 0.3312,$

$SS\delta = SS\mu - SS\pi = 41.0000 - 22.5000 = 18.500,$

$SSe = SSto - SS\mu = 146.0000 - 41.0000 = 105.0000.$

For instance, for pair comparison (1,2), three panelists moderately preferred treatment 2 over treatment 1, two panelists slightly preferred treatment 2 over treatment 1, one panelist rated treatments 1 and 2 equal, and two panelists moderately preferred treatment 1 over 2. Columns (3) through (5) are obtained from column (2), as shown in the footnote to the table.

The sums of squares, their DF, and the mean squares are given in Table 11.9 following the analysis of variance table, Table 11.7. Table 11.9 includes the $F$ ratios for testing the null hypotheses mentioned earlier. To test the null hypothesis

$$H_0: \alpha_1 = \alpha_2 = \cdots = \alpha_t,$$

we need to find

$$F_{t-1,t(t-1)(n-1)} = MS\alpha/MSe,$$

which, for our example, is observed to be

$$F_{2,42} = 11.0789/2.5 = 4.43$$

To test at the $\alpha$ significance level, we compare the observed $F$ ratio with the $(1 - \alpha)$ 100th percentile of the $F$ distribution corresponding to 2 and 42 DF of the numerator and the denominator, respectively. The 99th percentile of the $F$ distribution from Table A.4 is nearly 5.18 and is greater than the observed $F = 4.47$. Hence, the null hypothesis cannot be rejected at the 1%

**Table 11.9** Analysis of Variance for Example 11.2-1

| Source of Variance | DF | SS | MS | F Ratio |
|---|---|---|---|---|
| Main Effects $\alpha$ | 2 | 22.1577 | 11.0789 | 4.43[a] |
| Subtractivity $\gamma$ | 1 | 0.3312 | 0.3312 | |
| Average Preferences $\pi$ | 3 | 22.5000 | | |
| Order Effects $\delta$ | 3 | 18.5000 | 6.1667 | |
| Mean | 6 | 41.0000 | | |
| Error | 42 | 105.0000 | 2.5 | |
| | 48 | 146 | | |

[a]Significant at the 5% level (P-value = 0.018).

significance level. But at the 5% level of significance, the null hypothesis would be rejected, since the 95th percentile of the $F$ distribution is 3.40, which is less than the observed $F$ ratio. We can use the S-PLUS code *sfmodel* for the preceding calculations. The results are as follows:

```
> sfmodel(pr,sfdat)
4.433 0.018
              DF  SS  MS
   Main effect   2 22.1667 11.0833
  Subtractivity 1 0.3333 0.3333
Average preference 3 22.5000 7.5000
   Order effect 3 18.5000 6.1667
        Mean    6 41.0000 6.8333
        Error 42 105.0000 2.5000
> pr
  [,1] [,2]
[1,]  1  2
[2,]  2  1
[3,]  1  3
[4,]  3  1
[5,]  2  3
[6,]  3  2
> sfdat
  -3 -2 -1 0 1 2 3
12 0  3  2 1 0 2 0
21 0  3  2 2 0 1 0
13 0  1  0 3 0 2 2
31 1  3  1 2 0 1 0
23 1  1  0 4 1 1 0
32 3  1  1 3 0 0 0
```

Suppose we decide that at the 5% significance level the three treatments are different. Following this decision, we can carry out a multiple-comparison test to judge the significance of the difference between any two treatments. Following Tukey's multiple-comparison procedure, we may find

$$Y_\alpha = Q_{\alpha,v,t}\left(\hat{\alpha}\sqrt{1/2nt}\right),$$

where $Q_{\alpha,*v,t}$ is the critical value at the $\alpha$ level from Table A.6 for Tukey's Studentized range corresponding to $v$ DF for SSe in the analysis of variance and for the number of treatments equal to $t$. For our example, $\alpha = 0.05$,

$v = 42$, and $t = 3$. For these parameters, $Q_{0.05,42,3} = 3.44$. The error mean square, MSe, estimates $\hat{\sigma}^2$. Hence, $\hat{\sigma} = \sqrt{\text{MSe}} = \sqrt{2.4762}$ from Table 11.9, and

$$Y_{0.05} = 3.44 \left[ \sqrt{2.4762/2(8)(3)} \right] = 0.7812.$$

Two treatments are judged to be significantly different with an overall $\alpha$ significance level if the difference between their main effects is greater than $Y_\alpha$. The observed differences between the three main effects are

$$\hat{\alpha}_1 - \hat{\alpha}_2 = 0.3750 - 0.1667 = 0.2083,$$
$$\hat{\alpha}_1 - \hat{\alpha}_3 = 0.3750 - 0.5417 = 0.9167,$$
$$\hat{\alpha}_2 - \hat{\alpha}_3 = 0.1667 + 0.5417 = 0.7084.$$

Of these three differences, only the observed difference between $\hat{\alpha}_1$ and $\hat{\alpha}_3$ exceeds the critical value $Y_{0.05}$. Therefore, $\hat{\alpha}_1$ and $\hat{\alpha}_3$ are significantly different at the overall 5% significance level. The positive values of the treatment's main-effect parameters indicate a preferred treatment; and negative values, a treatment that is not preferred.

One may estimate all the differences $\alpha_i - \alpha_j$ simultaneously by the confidence intervals with a specified overall confidence coefficient. For the 95% overall confidence coefficient, three simultaneous confidence intervals are given by

$$\hat{\alpha}_i - \hat{\alpha}_j - Y_{0.05} \leq \alpha_i - \alpha_j \leq \hat{\alpha}_j - \hat{\alpha}_j + Y_{0.05}$$

for each and every pair. For our example, we estimate differences $(\alpha_1 - \alpha_2)$, $(\alpha_1 - \alpha_3)$, and $(\alpha_2 - \alpha_3)$ with an overall 95% degree of confidence as follows:

$$-0.4412 \leq \alpha_1 - \alpha_2 \leq 0.8578,$$
$$0.2672 \leq \alpha_1 - \alpha_3 \leq 1.5662,$$
$$0.0589 \leq \alpha_2 - \alpha_3 \leq 1.3579.$$

The order of item presentation is an important consideration in the sensory testing of food and other consumer products. The Scheffé model takes into account the order effects and hence permits their estimation, as well as the sum of squares due to the order effects. The sum of squares due to the order effects, $SS\delta$, can be partitioned into the sum of squares due to the average-order-effect parameter $\delta = \Sigma_{i<j} \, \delta_{ij}/m$. We recall that

$$\hat{\delta} = \sum_{i<j} \hat{\delta}_{ij}/m$$

$$= \sum_{i<j} \left( \hat{\mu}_{ij} + \hat{\mu}_{ji} \right)/2m$$

$$= \sum_{i<j} \sum_{k} (X_{ijk} + X_{jik})/2nm;$$

therefore, $\widehat{\delta}$ is really the mean of $2nm$ preference scores $X_{ijk}$ and $X_{jik}$. Hence, the contribution of $SS\delta$ due to the average order effect denoted by the correction factor (CF) is

$$CF = 2nm\widehat{\delta}^2 \quad \text{with} \quad 1 \text{ DF}.$$

The sum of squares $SS\delta'$ due to order effects corrected for the average order effect is

$$SS\delta' = SS\delta - CF$$

with $[t(t-1)/2] - 1$ DF.

For illustration we continue with Example 11.2-1, where $SS\delta = 18.50$ with 3 DF. To find the CF, we need $\widehat{\delta} = \Sigma_{i<j}(\widehat{\mu}_{ij} + \widehat{\mu}_{ji})/2m$. From calculations carried out in Table 11.8, $\widehat{\delta} = -0.500$. Hence,

$$CF = 2(8)(3)(-0.500)^2 = 12.00 \quad \text{with} \quad 1 \text{ DF}$$

and

$$SS\delta' = 18.50 - 12.00 = 6.50 \quad \text{with} \quad 2 \text{ DF}.$$

The $F$ ratio $MS\delta/MSe = 2.49$ for testing the significance of the order effects is not significant even at the 5% significance level because the 95th percentile of the $F$ distribution with 3 and 42 DF for the numerator and denominator, respectively, is nearly 2.84. However, when we partition $SS\delta$ into $SS\delta'$ and CF, the $F$ ratio $CF/MSe = 4.84$, corresponding to the average order effect, is highly significant, whereas the $F$ ratio $MS\delta'/MSe = 1.31$ is not significant. These calculations are summarized in Table 11.10.

**Table 11.10** Analysis of Variance of Order Effects for the Scheffé Model (Example 11.2-1)

| Source of Variance | DF | SS | MS | F Ratio |
|---|---|---|---|---|
| Average Order Effects (CF) | 1 | 12.00 | 12.00 | 4.84[a] |
| Differences among Order Effects ($\delta$) | 2 | 6.50 | 3.25 | 1.31 |
| Order Effects ($\delta$) | 3 | 18.50 | 6.17 | 2.49 |
| Error | 42 | 105.00 | 2.48 | |

[a]*Significant at the 5% level.*

We may notice that the individual order effects from $\widehat{\delta}_{ij} = \left(\widehat{\mu}_{ij} + \widehat{\mu}_{ji}\right)/2$ are $\widehat{\delta}_{12} = -0.62450, \widehat{\delta}_{13} = 0,$ and $\widehat{\delta}_{23} = -0.8750$. Note that there is no order effect associated with the comparison of the pair of treatments 1 and 3. Apparently, the order effects for the other two pairs are responsible for the significance of the $F$ ratio for the average order effect in Table 11.10.

## The Bradley–Terry Model

Bradley and Terry (1952) introduced a simple and appealing model for the analysis of PC designs. Consider $t$ treatments or objects to be compared in pairs with treatment rating parameters $\pi_1, \ldots, \pi_t$. The Bradley–Terry model postulates that treatments have true ratings on a particular subjective continuum such that $\pi_i \geq 0$ and $\Sigma \pi_i = 1$. When treatment $i$ is compared with treatment $j$, the probability $P_{ij}$ that treatment $i$ is ranked over treatment $j$ (or receives a rank of 1) is given by

$$P_{ij} = \pi_i / \left(\pi_i + \pi_j\right) \tag{11.2-2}$$

for each and every pair of treatments.

The Bradley–Terry model (11.2-2) has been found useful in many experiments, especially with regard to consumer testing and sensory evaluation. The use of the Bradley–Terry model for sensory evaluation has been considered by Bradley (1954b), Hopkins (1954), Gridgeman (1955b), Larmond et al. (1969), and many others.

Define $a_{ijk} = 1$ if, in the $k$th paired comparison, treatment $i$ is selected over treatment $j$ and $a_{ijk} = 0$ otherwise. Let treatments $i$ and $j$ be compared $n_{ij}$ independent times. Then $a_{ij} = \Sigma_k a_{ijk}$ is the number of comparisons out of $n_{ij}$ comparisons when the $i$th treatment is selected or preferred over the $j$th treatment. If the rating parameters remain the same from comparison to comparison, then $a_{ij}$ follows the binomial distribution, and the likelihood function of the total experiment is the product of $t(t - 1)/2$ binomial functions. Maximum likelihood estimates $p_i$ of $\pi_i$ are obtained by solving the likelihood equation iteratively. The likelihood equation is

$$p_i = a_i \Big/ \sum_{j \neq i} n_{ij} \left(p_i + p_j\right)^{-1}, \quad i = 1, \ldots, t, \tag{11.2-3}$$

where $a_i = \Sigma_{j \neq i} a_{ij}$. Dykstra (1956) suggested an iterative procedure and initial estimates $p_i^0$ so that the process would converge rapidly.

If the $p_i$ are not too different from each other, Dykstra suggested using initial estimates given by

$$p_i^0 = a_i/[(t-1)n_i - (t-2)a_i], \quad i = 1, \ldots, t, \qquad (11.2\text{-}4)$$

where $n_i = \Sigma_{j \neq i}\, n_{ij}$. Note that $n_i$ is the total number of comparisons involving the $i$th treatment and $a_i$ is the total number of preferences for treatment $i$. For the $k$th iteration, we write (11.2-3) as

$$p_i^k = a_i \left/ \left[ \sum_{j \neq i} n_{ij} \left( p_i^{k-1} + p_j^{k-1} \right)^{-1} \right] \right., \quad i = 1, \ldots, t.$$

The iteration begins with $k = 1$, with initial estimates $p_i^0$ given by (11.2-4). If the assumption that the $p_i$ are not too different from each other is inaccurate, then Dykstra recommended $(p_i^0 - k_i)$, $i = 1, \ldots, t$, for initial estimates, rather than $p_i^0$, where

$$k_i = \frac{[(t-1)R - S^2]/t + a_i n_i - a_i^2}{(t-2)R - S^2 + \sum a_i n_i} \sum \left[ \sum p_i^0 - 1 \right], \quad i = 1, \ldots, t, \quad (11.2\text{-}5)$$

and

$$R = \Sigma a_i^2, \quad S = \Sigma a_i = N.$$

According to Dykstra, in most cases the $(p_i^0 - k_i)$ are sufficiently close to the maximum likelihood estimates $p_i$ and can be used, therefore, at least as a first approximation without going through the iterative process.

A hypothesis of interest in sensory testing is that all treatments are equally preferred or are of equal ratings. In terms of the Bradley–Terry model parameters, the hypothesis that treatments have equal ratings can be expressed as

$$H_0: \quad \pi_i = 1/t \quad \text{for all} \quad i;$$
$$H_a: \quad \pi_i \neq 1/t \quad \text{for some} \quad i.$$

For testing the null hypothesis of equal ratings, the likelihood ratio test turns out to depend on statistic

$$B = \sum_{i<j} n_{ij} \ln (p_i + p_j) - \sum a_i \ln p_i, \qquad (11.2\text{-}6)$$

where ln is the natural logarithmic function to the base $e$. Bradley and Terry (1952) provided tables for the cumulative distribution of statistic $B$ in (11.2-6), under the null hypothesis of equal ratings, and also estimates $p_1, \ldots, p_t$. Their tables were given for $t = 3$,

$n = 1,2, \ldots, 10$; $t = 4$, $n = 1,2, \ldots, 6$. Later, Bradley (1954a) extended the tables to $t = 4$, $n = 7$, 8 and $t = 5$, $n = 1,2, \ldots, 5$. Whenever a paired comparison experiment falls within the range of these tables, experimenters may use them. We have not reproduced the tables here, but they are available in the references cited. Frequently, however, paired comparison experiments are of sizes beyond the scope of the tables. For such large-size experiments, the likelihood ratio test statistic is

$$X^2 = (1.3863)N - 2B, \tag{11.2-7}$$

where $B$ is defined by (11.2-6) and $N$ is the total number of all paired comparisons in the experiment. Note that $2N = \Sigma n_i$, where $n_i$ is defined in (11.2-4). The likelihood ratio test statistic approximates the $\chi^2$ distribution with $(t - 1)$ DF. Bradley (1976) pointed out that the use of the likelihood ratio statistic is satisfactory, even for modest values of $n_{ij}$.

Bradley (1955), Dykstra (1960), and Davidson and Bradley (1970) established the large-sample joint distribution of $p_i$. It is established that the $(p_i - \pi_i)$ have jointly an approximate multivariate normal distribution with variances and covariances obtained as follows. Let $\sigma_{ij}$ denote the covariance between $p_i$ and $p_j$. Note that in this notation, $\sigma_{ij}$ denotes the variance of $p_i$. To estimate $\sigma_{ij}$, we need to consider a matrix $V$ of $t$ rows and $t$ columns defined as follows. The $i$th diagonal element of $V$ is

$$V_{ii} = \frac{1}{p_i} \sum_{j \neq i} \frac{n_{ij}}{N} p_j (p_i + p_j)^{-2}, \quad i = 1, \ldots, t, \tag{11.2-8}$$

and the $(i, j)$th off-diagonal element of $V$ is

$$V_{ii} = -\frac{n_{ij}}{N} (p_i + p_j)^{-2}, \quad i \neq j = 1, \ldots, t. \tag{11.2-9}$$

Note that $V_{ij} = V_{ji}$. Now to obtain the $D$ matrix, augment the $V$ matrix with the $(t + 1)$th row and $(t + 1)$th column in the following way:

$$D = \begin{bmatrix} V_{11} & V_{1t} & 1 \\ V_{t1} & V_{tt} & 1 \\ 1 & 1 & 0 \end{bmatrix}.$$

Let $|D|$ denote the determinant of the augmented matrix $D$, and $|D_{ij}|$, the determinant of the matrix obtained from $D$ by deleting the $i$th row and the $j$th column. Then estimate the covariance $\sigma_{ij}$ by

$$\hat{\sigma}_{ij} = (-1)^{i+j} |D_{ij}| / N |D| \tag{11.2-10}$$

When all $n_{ij}$ are equal to $n$, say, and the null hypothesis of $\pi_i = 1/t$ is true, it follows that

$$\hat{\sigma}_{ii} = 4(t-1)/nt^4, \quad i = 1, \ldots, t;$$

$$\hat{\sigma}_{ij} = -4/nt^4, \quad\quad i \neq j = 1, \ldots, t.$$

There are formulations of the Bradley–Terry model that result in a symmetric distribution function; for example, see David (1963b). In these formulations rating parameters $\pi_i$ are converted to $\ln \pi_i$. On the logarithmic scale $\ln \pi_i$ is called the merit of treatments. Should such a formulation be appealing, we may estimate merit parameters $\ln \pi_i$ by $\ln p_i$. It has been shown that $\ln p_i$ is also approximately normally distributed with variance and covariance estimated by

$$Cov(\ln p_i, \ln p_j) = \hat{\sigma}_{ij}/p_i p_j, \quad i = 1, \ldots, t, \quad\quad (11.2\text{-}11)$$

where $\hat{\sigma}_{ij}$ is given by (11.2-10)

Before considering a numerical example, we wish to point out that paired comparison outcomes for the Bradley–Terry model may also be scored as follows. Let $r_{ijk} = 1$ if the $i$th treatment is preferred over the $j$th treatment in the $k$th comparison, and 0 otherwise. We note that

$$a_{ijk} = 2 - r_{ijk}. \quad\quad (11.2\text{-}12)$$

If treatments $i$ and $j$ are compared $n_{ij}$ times, then from (11.2-12) we get

$$a_{ij} = 2n_{ij} - \Sigma r_{ijk} \quad \text{and} \quad a_i = 2n_i(t-1) - \sum_{j \neq 1} \sum_k r_{ijk}.$$

In their original paper Bradley and Terry (1952) used the $r_{ijk}$ scores and denoted $\Sigma_j \Sigma_k r_{ijk}$ by $\Sigma r_i$ to express their statistic $B$, defined in (11.2-6), in terms of $\Sigma r_i$ quantities.

We now consider an example using the Bradley–Terry model.

---

**Example 11.2-2**

A preference study was carried out using the Bradley–Terry model for establishing a preference ranking of frankfurters with respect to three levels of citric acid, as in Example 11.2-1. Thus, $t = 3$ and there are three pairs (1, 2), (1, 3), and (2, 3). The scoring was done using the questionnaire in Figure 11.3. The scores are recorded in Table 11.11. Each pair was compared by 16 judges; thus, $n_{ij} = n = 16$ for each pair $(i, j)$. To balance for the order of presentation, each pair was compared by 2 judges in opposite order. Note the order of presentation of the three pairs to judges 1–8 and the reverse order of presentation to judges 9–16.

■ **FIGURE 11.3** Questionnaire form for the
Thurstone–Mosteller and Bradley–Terry models.

Date_____

Name_____Product_____

PREFERENCE TEST

Please indicate your preference between the two samples by assigning a rank
of 1 to the preferred sample and a rank of 2 to the other member of the pair.

Sample Number  _____     _____

Rank                     _____     _____

Comments _____

_____

_____

**Table 11.11** Paired Comparison Data for the Bradley–Terry Model Using the Questionnaire in
Figure 11.3

| (1) Judge $k$ | (2) Pairs | (3) Rank $r_{ijk}$ | | | (4) Rank Sums by Judges | | | (5) B |
|---|---|---|---|---|---|---|---|---|
| | | 1 | 2 | 3 | $\Sigma r_1$ | $\Sigma r_2$ | $\Sigma r_3$ | |
| 1 | (1, 2) | 1 | 2 | — | | | | |
| | (1, 3) | 1 | — | 2 | | | | |
| | (2, 3) | — | 1 | 2 | 2 | 3 | 4 | 0 |
| 2 | (1, 2) | 2 | 1 | — | | | | |
| | (1, 3) | 1 | — | 2 | | | | |
| | (2, 3) | — | 1 | 2 | 3 | 2 | 4 | 0 |
| 3 | (1, 2) | 2 | 1 | — | | | | |
| | (1, 3) | 1 | — | 2 | | | | |
| | (2, 3) | — | 1 | 2 | 3 | 2 | 4 | 0 |
| 4 | (1, 2) | 1 | 2 | — | | | | |
| | (1, 3) | 1 | — | 2 | | | | |
| | (2, 3) | — | 2 | 1 | 2 | 4 | 3 | 0 |
| 5 | (1, 2) | 2 | 1 | — | | | | |
| | (1, 3) | 2 | — | 1 | | | | |
| | (2, 3) | — | 1 | 2 | 4 | 2 | 3 | 0 |

*(continued)*

**Table 11.11** Paired Comparison Data for the Bradley–Terry Model Using the Questionnaire in Figure 11.3—*cont...*

| (1) Judge $k$ | (2) Pairs | (3) Rank $r_{ijk}$ | | | (4) Rank Sums by Judges | | | (5) B |
|---|---|---|---|---|---|---|---|---|
| | | 1 | 2 | 3 | $\Sigma r_1$ | $\Sigma r_2$ | $\Sigma r_3$ | |
| 6 | (1, 2) | 1 | 2 | — | | | | |
| | (1, 3) | 2 | — | 1 | | | | |
| | (2, 3) | — | 2 | 1 | 3 | 4 | 2 | 0 |
| 7 | (1, 2) | 1 | 2 | — | | | | |
| | (1, 3) | 1 | — | 2 | | | | |
| | (2, 3) | — | 1 | 2 | 2 | 3 | 4 | 0 |
| 8 | (1, 2) | 1 | 2 | — | | | | |
| | (1, 3) | 1 | — | 2 | | | | |
| | (2, 3) | — | 1 | 2 | 2 | 3 | 4 | 0 |
| 9 | (2, 1) | 1 | 2 | — | | | | |
| | (3, 1) | 1 | — | 2 | | | | |
| | (3, 2) | — | 1 | 2 | 2 | 3 | 4 | 0 |
| 10 | (2, 1) | 1 | 2 | — | | | | |
| | (3, 1) | 1 | — | 2 | | | | |
| | (3, 2) | — | 2 | 1 | 2 | 4 | 3 | 0 |
| 11 | (2, 1) | 1 | 2 | — | | | | |
| | (3, 1) | 1 | — | 2 | | | | |
| | (3, 2) | — | 1 | 2 | 2 | 3 | 4 | 0 |
| 12 | (2, 1) | 1 | 2 | — | | | | |
| | (3, 1) | 2 | — | 1 | | | | |
| | (3, 2) | — | 1 | 2 | 3 | 3 | 3 | 0.903 |
| 13 | (2, 1) | 2 | 1 | — | | | | |
| | (3, 1) | 1 | — | 2 | | | | |
| | (3, 2) | — | 2 | 1 | 3 | 3 | 3 | 0.903 |

*(continued)*

**Table 11.11** Paired Comparison Data for the Bradley–Terry Model Using the Questionnaire in Figure 11.3—cont...

| (1) Judge $k$ | (2) Pairs | (3) Rank $r_{ijk}$ | | | (4) Rank Sums by Judges | | | (5) $B$ |
|---|---|---|---|---|---|---|---|---|
| | | 1 | 2 | 3 | $\Sigma r_1$ | $\Sigma r_2$ | $\Sigma r_3$ | |
| 14 | (2, 1) | 1 | 2 | — | | | | |
| | (3, 1) | 2 | — | 1 | | | | |
| | (3, 2) | — | 1 | 2 | 3 | 3 | 3 | 0.903 |
| 15 | (2, 1) | 1 | 2 | — | | | | |
| | (3, 1) | 1 | — | 2 | | | | |
| | (3, 2) | — | 1 | 2 | 2 | 3 | 4 | 0 |
| 16 | (2, 1) | 2 | 1 | — | | | | |
| | (3, 1) | 1 | — | 2 | | | | |
| | (3, 2) | — | 1 | 2 | 3 | 2 | 4 | 0 |
| | | | | Total $\Sigma r_i. =$ | 41 | 47 | 56 | |
| $n = 16$ | | | $a_i = 2n(t-1) - \Sigma r_i. =$ | | 23 | 17 | 8 | $B^c = 2.709$ |
| | | | | $\hat{p}_i =$ | 0.5864 | 0.3379 | 0.1458 | |

Since each pair was compared 16 times, we have $n_1 = n_2 = n_3 = 32$. To find the maximum likelihood estimates, start with initial estimates given by (11.2-4). They are

$$p_1^0 = \frac{23}{2(32) - 23} = 0.5610, \quad p_2^0 = \frac{17}{2(32) - 17} = 0.3617,$$

and

$$p_3^0 = \frac{8}{2(32) - 8} = 0.1429.$$

Starting with these initial estimates, by iterating (11.2-3) we can find $p_1^1, p_2^1$, and $p_3^1$ corresponding to $k = 1$. Then we iterate again for $k = 2$ until the process converges. Note that, since in our example $n_{ij} = 16$ and $t = 3$, the computations in the iterative process are not too involved. Otherwise, for unequal $n_{ij}$ and $t$ larger than 3, we must use computers to carry out the iterative process. Our calculations lead us to $p_1 = 0.5454$, $p_2 = 0.3197$, and $p_3 = 0.1359$. Now we can compute $B$ in (11.2-6).

The assumption that the $p_i$ are not too different from each other is not really valid; hence, the initial estimates $p_i^0$ may not be optimal choices for starting the iterative process. Instead, let us compute $(p_i^0 - k_i)$ for the initial estimates. To find $k_i$, we need $R$ and $S$. They are

$$R = \sum a_i^2 = (23)^2 + (17)^2 + (8)^2 = 882,$$
$$S = \sum a_i = 23 + 17 + 8 = 48.$$

Hence, from (11.2-5),

$$k_1 = \frac{\left[2(882) - (48)^2\right]\big/3 + (23)(32) - 529}{882 - (48)^2 + 1536}(1.0656 - 1)$$

$$= \frac{-180 + 207}{116}(0.0656)$$

$$= (0.2328)(0.0656) = 0.0153,$$

$$k_2 = \frac{-180.255}{116}(0.0656) = 0.0424,$$

$$k_3 = \frac{-180 + 192}{116}(0.0656) = 0.0068.$$

Then

$$p_1^0 - k_1 = 0.5610 - 0.0153 = 0.5457,$$
$$p_2^0 - k_2 = 0.3617 - 0.0424 = 0.3193,$$
$$p_3^0 - k_3 = 0.1429 - 0.0068 = 0.1361,$$

which are extremely close to the maximum likelihood estimates listed. Now we can compute statistic $B$ as given in (11.2-6):

$$B = 16\left[\ln(0.8650) + \ln(0.6818) + \ln(0.4554)\right] - 23\ln(0.5457)$$
$$\quad - 17\ln(0.3193) - 8\ln(0.1361)$$
$$= -16(0.1450 + 0.3827 + 0.7875) + 23(0.6052) + 17(1.1426)$$
$$\quad + 8(2.0715)$$
$$= -21.0432 + 49.9158 = 28.8726.$$

The $X^2$ statistic in (11.2-7) for testing the equality of ratings is

$$X^2 = 1.3863(48) - 2(28.8726) = 8.7972 \quad \text{with} \quad 2 \text{ DF.}$$

The calculated $X^2$ value 8.7972 is more than the 95th percentile 5.99 of the $\chi^2$ distribution with 2 DF. Hence, the null hypothesis of equal ratings is rejected at any significance level $\alpha \geq 0.05$. However, since the 99th percentile of the $\chi^2$ distribution with 2 DF is 9.21, the null hypothesis cannot be rejected at the 0.01 significance level.

To estimate variances and covariances of $p_i$ and $p_j$, we need matrix $D$ with its elements $V_{ij}$, as defined in (11.2-8) and (11.2-9). From (11.2-8),

$$V_{11} = \frac{1}{3p_1}[p_2(p_1 + p_2)^{-2} + p_3(p_1 + p_3)^{-2}]$$

$$= \frac{1}{1.6371}\left[\frac{0.3193}{(0.8650)^2} + \frac{0.1361}{(0.6818)^2}\right]$$

$$= 0.4396,$$

$$V_{22} = \frac{1}{3p_2}[p_1(p_2 + p_1)^{-2} + p_3(p_2 + p_3)^{-2}]$$

$$= \frac{1}{0.9579}\left[\frac{0.5457}{(0.8650)^2} + \frac{0.1361}{(0.4554)^2}\right]$$

$$= 1.4464,$$

$$V_{33} = \frac{1}{3p_3}[p_1(p_3 + p_1)^{-2} + p_2(p_3 + p_2)^{-2}]$$

$$= \frac{1}{0.4083}\left[\frac{0.5457}{(0.6818)^2} + \frac{0.3193}{(0.4554)^2}\right]$$

$$= 6.6461,$$

$$V_{12} = -\frac{1}{3}(p_1 + p_2)^{-2} = -0.4455,$$

$$V_{13} = -\frac{1}{3}(p_1 + p_3)^{-2} = -0.7172,$$

$$V_{23} = -\frac{1}{3}(p_2 + p_3)^{-2} = -1.6072.$$

Thus,

$$D = \begin{bmatrix} 0.4396 & -0.4455 & -0.7172 & 1 \\ -0.4455 & 1.4464 & -1.6072 & 1 \\ -0.7172 & -1.6072 & 6.6461 & 1 \\ 1 & 1 & 1 & 1 \end{bmatrix}.$$

Although the calculation is not difficult for the present example, in general, we need a computer program to calculate $|D|$ and $|D_{ij}|$. We find

$$|D| = -23,6603, \quad |D_{12}| = 8.5250,$$
$$|D_{11}| = -11.3069, \quad |D_{13}| = 2.7819,$$
$$|D_{22}| = -8.5201, \quad |D_{23}| = -0.0049,$$
$$|D_{33}| = -2.7770.$$

We note that $|D_{12}| = |D_{21}|$, $|D_{13}| = |D_{31}|$, and $|D_{23}| = |D_{32}|$.

Now we can use (11.2-10) to estimate the variances and covariances of $p_i$ and $p_j$. That is,

$$\hat{\sigma}_{11} = 0.0010, \quad \hat{\sigma}_{22} = 0.0075, \quad \hat{\sigma}_{33} = 0.0024,$$
$$\hat{\sigma}_{12} = 0.0075, \quad \hat{\sigma}_{13} = 0.0024, \quad \hat{\sigma}_{23} = 0.0000.$$

Note that covariance $\hat{\sigma}_{23} = 0$ suggests that $p_2$ is independent of $p_3$. These variances and covariances can be used to estimate the variances of $(p_i - p_j)$, denoted by $var(p_i - p_j)$, to test pairwise difference of treatment ratings. For example,

$$var(p_1 - p_2) = \hat{\sigma}_{11} + \hat{\sigma}_{22} - 2\hat{\sigma}_{12}$$
$$= 0.0010 + 0.0075 + 0.0150$$
$$= 0.0235,$$

and a $(1 - \alpha)100\%$ degree of confidence interval for $(\pi_1 - \pi_2)$ is

$$(p_1 - p_2) \pm z_{1-\alpha/2}\sqrt{var(p_1 - p_2)}.$$

Thus, the 95% degree of confidence interval for $(\pi_1 - \pi_2)$ is

$$(0.5454 - 0.3197) \pm (1.96)(0.1533),$$

which simplifies to $(-0.0747, 0.5261)$. Since the 95% confidence interval contains zero, we cannot reject the equality of ratings $\pi_1$ and $\pi_2$.

Similarly, the 95% degree of confidence interval for $(\pi_1 - \pi_3)$ is estimated to be

$$(0.5454 - 0.1359) \pm (1.96)0.0906,$$

which simplifies to $(0.2320, 0.5870)$. Note that this interval does not contain zero, and therefore, one may reject the equality of ratings $\pi_1$ and $\pi_3$ at the 0.05 level of significance. A confidence interval for $(\pi_2 - \pi_3)$ can similarly be found to test whether $\pi_2$ and $\pi_3$ are equal. In formulations where rating parameters $\pi_i$ are converted to $\ln \pi_i$, we use $\ln p_i$ to estimate them. Then we may use (11.2-11) to estimate the variances and the covariances of $p_i$. For our example, var(ln $p_1$) = 0.0034, var(ln $p_2$) = 0.0735, var(ln $p_3$) = 0.0130; cov(ln $p_1$, ln $p_2$) = −0.0430, cov(ln $p_1$, ln $p_3$) = −0.0324; cov(ln $p_2$, ln $p_3$) = 0.0000.

Using the S-PLUS codes *bradley* and *bradleyv*, with the input $a_i$ and $n$, we can obtain the $p_i$, covariances of $p_i$ , and $B$ statistic value:

```
> bradley(c(23,17,8),16)
0.546 0.318 0.137
B1: 28.26
> bradleyv(c(0.546,0.318,0.137),16)
     [,1] [,2] [,3]
[1,] 0.00994 -0.00747 -0.00247
[2,] -0.00747 0.00746 0.00001
[3,] -0.00247 0.00001 0.00246
```

In some situations, particularly in certain types of sensory testing experiments in which a number of judges perform comparisons for each pair, the model with the assumption of homogenous comparisons is not valid. If so, the Bradley–Terry model may be modified by postulating the existence of rating parameters $\pi_{1u}, \ldots, \pi_{tu} \geq 0$ and $\Sigma_i \pi_{iu} = 1$ for the $u$th group of $g$ groups of comparisons in the PC experiment. This means that the PC experiment can be designed to have $g$ groups. Each pair is compared within a group. The simplest design is that in which each pair is compared within a group an equal number of times. Let $n_u$ denote the number of comparisons of each pair in the $u$th group, $\Sigma n_u = n$. Paired comparisons within a group are taken to be homogeneous, but group differences in treatment ratings are permitted. For the modified model, the null and alternative hypotheses are

$$H_0: \pi_{iu} = 1/t \quad \text{for all} \quad i \text{ and } u;$$

$$H_a: \pi_{iu} \neq 1/t \quad \text{for some} \quad i \text{ and } u. \tag{11.2-13}$$

Let $B_u$ be defined to correspond to statistic $B$ in (11.2-6) for $n_u$ homogeneous comparisons of each pair in the $u$th group. A test statistic for testing hypotheses (11.2-13) is defined through

$$B^c = \sum_{u=1}^{g} B_u. \tag{11.2-14}$$

The likelihood ratio test statistic in terms of $B^c$ is

$$X_c^2 = (1.3863)N - 2B^c, \tag{11.2-15}$$

which has, for large values of $n_1, n_2, \ldots, n_g$, a $\chi^2$ distribution with $g(t - 1)$ DF.

If we want to test the homogeneity of groups, then we formulate the following hypotheses:

$$H_0: \pi_{iu} = \pi_i \quad \text{for all} \quad i \text{ and } u;$$

$$H_a: \pi_{iu} \neq \pi_i \quad \text{for some} \quad i \text{ and } u. \tag{11.2-16}$$

The likelihood ratio test statistic for testing the hypotheses in (11.2-16) is

$$X_h^2 = 2(B - B^c), \tag{11.2-17}$$

where $B$ and $B^c$ are as defined earlier in (11.2-6) and (11.2-14), respectively. The distribution of $X_h^2$ for large values of $n_1, n_2, \ldots, n_g$ is the $\chi^2$ distribution with $(g - 1)(t - 1)$ DF.

To calculate statistics $B$ and $B_1, B_2, \ldots, B_g$, we need to have the maximum likelihood estimates $p_i$ and $p_{iu}$ of the respective rating parameters $\pi_i$ and $\pi_{iu}$, $i = 1, \ldots, t$. We can obtain these estimates quite accurately for each group as follows:

$$p_{iu} = p_{iu}^0 - k_{iu}, \qquad \begin{aligned} & i = 1, \ldots, t, \\ & u = 1, \ldots, g, \end{aligned} \tag{11.2-18}$$

where $p_{iu}^0$ and $k_{iu}$ are defined as in (11.2-4) and (11.2-5), respectively, for each group.

---

**Example 11.2-3**

This example is taken from Bradley (1954a) to illustrate the analysis when the PC experiment is carried out in groups, with each group homogeneous, but when the assumption of homogeneity may not hold for the entire experiment. Three rations (treatments) were fed to hogs selected for similar characteristics and from a limited number of litters. $A$ was a corn ration, $B$ was a corn ration supplemented by a small peanut ration, and $C$ was a corn ration supplemented by a larger peanut ration. The experiment was conducted so that the tasters recorded a rank of 1 for the sample in a pair that they thought came from the animal fed the larger amount of peanuts and 2 otherwise. Members of the panel were obtained after preliminary training and selection, and it was believed that they would be able to recognize the effect of a peanut supplemented diet on fresh roast pork. The rank total $a_i$ and $\Sigma r_i = 2n_i - a_i$ are given in Table 11.12 for each judge. Each judge made $n_u = 10$ comparisons of each pair $(A, B)$, $(A, C)$, and $(B, C)$. Thus, each pair was compared $n = \Sigma_{u=1}^5 n_u = 50$ times, and in all there were $N = 3(50) = 150$ paired comparisons.

**Table 11.12** Paired Comparison Experiment with Five Groups (Judges) (Example 11.2-3)

| Judge u | Rations | $a_{ij}$ A | B | C | $\Sigma r_i$ | $p_{1u}$ | $p_{2u}$ | $p_{3u}$ | $B_u^a$ |
|---------|---------|-----|---|---|------|------|------|------|---------|
| 1 | A | | 7 | 5 | 28 | 0.43 | 0.28 | 0.28 | 20.4010 |
|   | B | 3 | | 6 | 31 | | | | |
|   | C | 5 | 4 | | 31 | | | | |
| 2 | A | | 3 | 4 | 33 | 0.21 | 0.42 | 0.37 | 19.8484 |
|   | B | 7 | | 5 | 28 | | | | |
|   | C | 6 | 5 | | 29 | | | | |
| 3 | A | | 5 | 3 | 32 | 0.24 | 0.32 | 0.43 | 20.2629 |
|   | B | 5 | | 5 | 30 | | | | |
|   | C | 7 | 5 | | 28 | | | | |
| 4 | A | | 3 | 4 | 33 | 0.21 | 0.42 | 0.37 | 19.8484 |
|   | B | 7 | | 5 | 28 | | | | |
|   | C | 6 | 5 | | 29 | | | | |
| 5 | A | | 4 | 4 | 32 | 0.23 | 0.54 | 0.23 | 19.1576 |
|   | B | 6 | | 8 | 26 | | | | |
|   | C | 6 | 2 | | 32 | | | | |

$^a$From (11.2-6).

We need to find $p_{iu}^0$ in (11.2-4) and $k_{iu}$ in (11.2-5) for each group in order to find $p_{iu}$ using (11.2-18). We note that regardless of the group, $n_1 = n_2 = n_3 = 20$, $S = \Sigma a_i = 30$, and, of course $t = 3$. Also we have the following information:

| Group u | $a_1$ | $a_2$ | $a_3$ | $R = \Sigma a_i^2$ |
|---------|-------|-------|-------|--------------------|
| 1 | 12 | 9 | 9 | 306 |
| 2 | 7 | 12 | 11 | 314 |
| 3 | 8 | 10 | 12 | 308 |
| 4 | 7 | 12 | 11 | 314 |
| 5 | 8 | 14 | 8 | 324 |
| For all groups | 42 | 57 | 51 | 7614 |

Now we can calculate $p_{iu}^0$ and $k_{iu}$ for each group and then $p_{iu} = p_{iu}^0 - k_{iu}$, $i = 1, 2, 3$. We leave these calculations for the reader (see Exercise 11.4). One will find that the $p_{iu}$ so calculated are very close to the maximum likelihood estimates given in Table 11.12. The estimates $p_{iu}$ given in Table 11.12 were obtained from the Bradley–Terry tables. When all five groups are combined into one under homogeneity, we have $n_{ij} = 50$ for each pair, and $a_1 = 42$, $a_2 = 57$, $a_3 = 51$, $n_1 = n_2 = n_3 = 100$. To find the approximate maximum likelihood estimates $p_1$, $p_2$, and $p_3$ for the combined experiment, we find initial estimates $p_i^0$ from (11.2-4). Then we find $p_i = p_i^0 - k_i$, where $k_i$ is given by (11.2-5), $i = 1, 2, 3$. Thus,

$$p_1^0 = 42/[2(100) - 42] = 0.2658,$$

$$p_2^0 = 57/[2(100) - 57] = 0.3986,$$

$$p_3^0 = 51/[2(100) - 51] = 0.3423.$$

From (11.2-5),

$$k_i = \frac{2(7614) - (150)^2/3 + a_i n_i - a_i^2}{7614 - (150)^2 + 15,000}(0.0067)$$

$$= \frac{-2424 - a_i n_i - a_i^2}{114}(0.0067).$$

Now

$$k_1 = \frac{-2424 + 2436}{114}(0.0067) = 0.0007,$$

$$k_2 = \frac{-3424 + 2451}{114}(0.0067) = 0.0016,$$

$$k_3 = \frac{-2424 + 2499}{114}(0.0067) = 0.0044.$$

Hence, the maximum likelihood estimates are

$$p_1 = p_1^0 - k_1 = 0.2658 - 0.0007 = 0.2651,$$

$$p_2 = p_2^0 - k_2 = 0.3986 - 0.0016 = 0.3970,$$

$$p_3 = p_3^0 - k_3 = 0.3423 - 0.0044 = 0.3379.$$

Note that in these calculations the sum of $p_1 + p_2 + p_3$ is precisely equal to 1, as it should be.

To test the null hypothesis (11.2-13), calculate the $X_c^2$ statistic in (11.2-15). That is,

$$X_c^2 = (1.3863)(150) - 2B^c$$

where from (11.2-44), $B^c = 99.5183$. Hence,

$$X_c^2 = 207.9450 - 199.0366$$
$$= 8.9084 \quad \text{with} \quad 10 \text{ DF.}$$

Since the 95th percentile of the $\chi^2$ distribution with 10 DF is 18.307, we cannot reject the null hypothesis (11.2-13). To test the hypothesis (11.2-16) of the homogeneity of groups, we calculate $X_h^2$ in (11.2-17). To calculate $X_h^2$, we need $B$ from (11.2-6):

$$B = 50[\ln(0.6621) + \ln(0.6030) + \ln(0.7349)]$$
$$-42 \ln(0.2651) - 57 \ln(0.3970) - 51 \ln(0.3379)$$
$$= -50(0.4125 + 0.5059 + 0.3079) + 42(1.3280)$$
$$+57(0.9238) + 51(1.0847)$$
$$= -61.3150 + 163.7523 = 102.4373.$$

Hence,

$$X_h^2 = 2(102.4373 - 99.5183) = 5.838 \quad \text{with} \quad 8 \text{ DF.}$$

Since $\chi_{0.95,8}^2 = 15.507$, the null hypothesis cannot be rejected. Thus, the judges were consistent in their ratings of the three rations. Using the S-PLUS code *bradley*, we can get $p_1, p_2, p_3,$ and $B$ as follows:

```
> bradley(c(42,57,51),50)
0.265 0.397 0.338
B1: 102.44
```

Bradley (1954b) provided test statistics for testing the validity of the Bradley–Terry model. The reader who wishes to test whether, indeed, the Bradley–Terry model is appropriate to use in his or her studies may refer to Bradley (1954b).

## The Thurstone–Mosteller Model

Thurstone (1927a, b) formulated a PC model through a subjective scale on which sensations can be ordered for stimuli of varying and unknown physical amounts. An individual is assumed to receive a sensation in response to a stimulus. The amount of sensation varies for stimuli of the same strength and is governed by a probability model. Specifically, the amount of sensation $X_i$ received as a result of the $i$th stimulus is governed by the normal probability model having mean sensations $S_i$. There are several cases of Thurstone's formulations that pertain to distributional assumptions. The important and widely used distributional model is called Thurstone's Case V. Thurstone's Case V assumes that two different stimuli generate sensation distributions with different mean sensations but with equal standard deviations. Also, the correlation coefficient between sensations of any two stimuli is the same, though unknown. For this case the sensation continuum can be scaled so that when a stimulus $i$ is compared with stimulus $j$, what is in effect perceived is the order of sensations $X_i$ and $X_j$, and

$$P(X_i > X_j) = \Phi(S_i - S_j), \quad i \neq j = 1, \ldots, t, \tag{11.2-19}$$

where $\Phi(X)$ denotes the standard normal cumulative distribution function (see Chapter 1). Figure 11.3 shows the questionnaire form for the Thurstone–Mosteller model. Mosteller (1951a, b) equated observed proportions of sensation $X_i$ greater than sensation $X_j$ in paired comparisons of stimuli $i$ and $j$ to $\Phi(S_i - S_j)$ and developed a method of estimating $S_i$, $i = 1, \ldots, t$.

To carry out Mosteller's estimation procedure, let $a_{ij}$ denote the number of comparisons in which sensation $X_i$ was recorded as stronger than sensation $X_j$ in $n_{ij}$ paired comparisons of stimuli $i$ and $j$. Prepare a matrix of observed proportions $p_{ij} = a_{ij}/n_{ij}$ for each pair $(i, j)$. To follow Mosteller's procedure, we write

$$p_{ij} = \Phi(S_i - S_j) \tag{11.2-20}$$

and use the normal distribution Table A.12 to find $Z_{ij}$ so that

$$p_{ij} = \Phi(Z_{ij}). \tag{11.2-21}$$

Note that $Z_{ij}$ is the $(p_{ij})$th percentile of the standard normal distribution. Hence, from (11.2-20) and (11.2-21) we can write

$$Z_{ij} = S_i - S_j, \quad i \neq j = 1, \ldots, t. \tag{11.2-22}$$

Notice now that the Thurstone–Mosteller model as expressed by (11.2-21) simply equates differences between mean sensations to normal deviates $Z_{ij}$. Thus, for the Thurstone–Mosteller model, there are $t(t-1)/2$ equations equating $(S_i - S_j)$ with $Z_{ij}$. But these equations involve only $t$ unknown mean sensations $S_1, S_2, \ldots, S_t$. Obviously, there are more constraints on the $S_i$ than are necessary to solve for them. All the constraints may not be entirely consistent with each other. The reason is that, after all, $Z_{ij}$ are percentiles corresponding to observed proportions $p_{ij}$, which are subject to sampling errors.

Without any loss we can redefine $S_i$ so that $\Sigma S_i = 0$. Then, from (11.2-22), it follows that estimates of $S_i$ are

$$\widehat{S}_i = \sum_j Z_{ij}/t = \overline{Z}_{i\cdot},$$

where in the summation $Z_{ii} = 0$. Hence, one may estimate $S_i$ by $\overline{Z}_{i\cdot}$. Suppose that instead of defining the $S_i$ so that $\Sigma S_i = 0$, we define them so that $S_t = 0$; then all other sensation parameters need to be estimated in relation to $S_t = 0$. Mosteller argued that a reasonable (least squares) set of estimates $\widehat{S}_i$ of $S_i$ with $S_t = 0$ is

$$\widehat{S}_j = \frac{\sum_{i=1}^{t} Z_{it}}{t} - \frac{\sum_{i=1}^{t} Z_{ij}}{t} = \overline{Z}_{\cdot t} - \overline{Z}_{\cdot j}, \quad j = 1, \dots, t-1. \qquad (11.2\text{-}23)$$

A goodness of fit test of the model with the observed proportion $p_{ij}$ was also developed by Mosteller (1951c) using the $\chi^2$ test. Numerical examples of goodness of fit tests are given in Mosteller (1951c), Guilford (1954), and Torgerson (1958).

| | |
|---|---|
| **Example 11.2-4** | We will use the data from Example 11.2-2 in Table 11.11 to illustrate the use of the Thurstone–Mosteller model. There are $t = 3$ treatments, and each pair was compared by 16 judges. Note that for 11 times, item 1, when compared with item 2, received a rank of 1 out of 16 comparisons. Hence, $p_{12} = 11/16 = 0.6875$. Similarly, we can compute $p_{13}$ and $p_{23}$ to complete matrix $P$ given in Table 11.13. We find the $(p_{ij})$th percentile corresponding to each $p_{ij}$, that is, $Z_{ij}$ of the standard normal distribution. The matrix $Z$ of $Z_{ij}$ is also given in Table 11.13. For instance, if $p_{12} = 0.6875$, the corresponding normal deviate is $Z_{12} = 0.4902$; if $p_{21} = 1 - p_{12} = 0.3125$, $Z_{21} = -0.4902$. It is often desirable to summarize estimates of $S_i$ with a plot such as that in Figure 11.4. |

Mosteller (1951a) demonstrated that the problem of scaling a set of stimuli using the Thurstone–Mosteller model can be viewed in a least squares formulation. Indeed, Morrissey (1955), following Mosteller (1951a), provided a least squares formulation for setting a scale of $t$ stimuli when the underlying PC design is incomplete. In an incomplete PC design, only $r$ pairs are compared, with $r$ a number between $t$ and $t(t - 1)/2$. Let $p_{ij}$ continue to denote the proportion of times treatment $i$ is ranked higher than treatment $j$ in the experiment. Equation (11.2-22) is basic for the least squares formulation. Remember that there will be $r$ equations in the

**Table 11.13** Paired Comparison Data of Table 11.11 Summarized by Matrix P and the Corresponding Matrix Z of Normal Deviates

| *I* | *j* 1 | 2 | 3 | | |
|-----|-----|-----|-----|-----|-----|
| $P = (p_{ij})$, $i > j$, $t = 3$, $n = 16$ | | | | | |
| 1 | 0 | 0.6875 | 0.7500 | | |
| 2 | 0.3125 | 0 | 0.7500 | | |
| 3 | 0.2500 | 0.2500 | 0 | | |

| *i* | *j* 1 | 2 | 3 | $\sum_j Z_{ij}$ | $\bar{Z}_{i\cdot}$ |
|-----|-----|-----|-----|-----|-----|
| $Z_{ij}$ | | | | | |
| 1 | 0 | 0.4902 | 0.6745 | 1.1647 | 0.3882 |
| 2 | −0.4902 | 0 | 0.6745 | 0.1843 | 0.0614 |
| 3 | −0.6745 | −0.6745 | 0 | −1.3490 | −0.4497 |
| $\sum_j Z_{ij}$ | −1.1647 | −0.1843 | 1.3490 | | |
| $\bar{Z}_{\cdot j}$ | −0.3882 | −0.0614 | 0.4497 | | |

Note: We set S3 = 0; then from (11.2-23), we have
$\hat{S}1 = 0.4497 - (-0.3882) = 0.8379$,
$\hat{S}2 = 0.4497 - (-0.0614) = 0.5111$,
$\hat{S}3 = 0$.

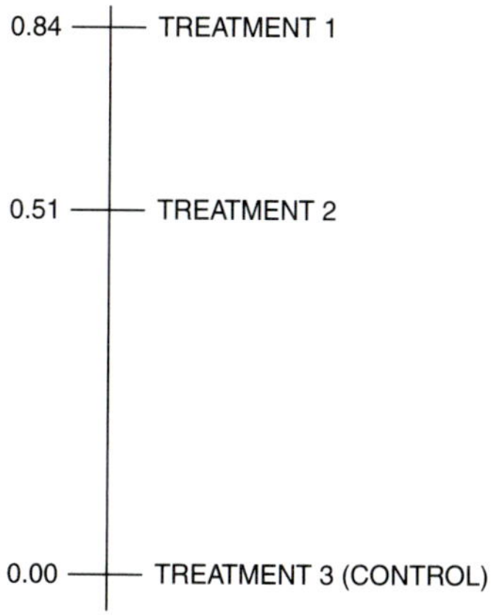

■ **FIGURE 11.4** Plot of scale values on a continuum with the scale origin at zero $(\hat{S}_3 = 0)$.

model, since only r pairs have been compared. Let Z be a column vector of observed differences $Z_{ij}$; S a column vector of $S_1, S_2, \ldots, S_t$, which are true mean sensations or true psychological scale values; and X a matrix of the coefficients of Eq. (11.2-22). In this notation, r equations resulting from comparisons of r pairs can be written in a matrix form as follows:

$$XS = Z.$$

For the existence of the least squares solution we must augment the X matrix by having an additional $(r + 1)$th row consisting entirely of 1s and vector Z by having the $(r + 1)$th element, which is zero. Thus, augmented, the system of equations $XS = Z$ is such that X is a matrix of $(r + 1)$ rows and t columns and Z is a column vector of $(r + 1)$ elements. The least squares solution for S is given by

$$\hat{S} = (X'X)^{-1}X'Z, \tag{11.2-24}$$

where $X'$, called the transpose of $X$, is obtained from $X$ as follows: the first column of $X'$ is the first row of $X$, the second column of $X'$ is the second row of $X$, and so on.

The standard error of $\widehat{S}_i$ is estimated by

$$SE\left(\widehat{S}_i\right) = \sqrt{a_{ii}MSe}, \tag{11.2-25}$$

where

$$MSe = \left(Z - X\widehat{S}\right)'\left(Z - X\widehat{S}\right)\big/(r - t + 1), \tag{11.2-26}$$

$a_{ii} =$ the $i$th diagonal element of $(X'X)^{-1}$.

Suppose we used a CD with $t = 5$ (Design 1, Table 11.6) for a PC experiment. As we mentioned in Section 11.2, a cyclic arrangement using the pair of treatments 1 and 2 in the initial block determines all the other pairs in the design. In the design matrix $X$, the position of coefficients 1 and $-1$ indicates the pair that is being compared. Consider the following design:

| Blocks | 1 | 2 | 3 | 4 | 5 |
|--------|-----|-----|-----|-----|-----|
| Pairs | 1,2 | 2,3 | 3,4 | 4,5 | 1,5 |

Since only five pairs are being compared in the experiment, $r = 5$. Note here that $r$ does not stand for the number of replications. The design matrix $X$ of the five paired comparisons augmented by the last row, the $Z$ vector augmented by the last element 0, and the vector $S$ are as follows:

$$S = \begin{bmatrix} S_1 \\ S_2 \\ S_3 \\ S_4 \\ S_5 \end{bmatrix}, \quad X = \begin{bmatrix} 1 & -1 & 0 & 0 & 0 \\ 0 & 1 & -1 & 0 & 0 \\ 0 & 0 & 1 & -1 & 0 \\ 0 & 0 & 0 & 1 & -1 \\ 1 & 0 & 0 & 0 & -1 \\ 1 & 1 & 1 & 1 & 1 \end{bmatrix}, \quad Z = \begin{bmatrix} Z_{12} \\ Z_{23} \\ Z_{34} \\ Z_{45} \\ Z_{15} \\ 0 \end{bmatrix}.$$

Note that

$$X' = \begin{bmatrix} 1 & 0 & 0 & 0 & 1 & 1 \\ -1 & 1 & 0 & 0 & 0 & 1 \\ 0 & -1 & 1 & 0 & 0 & 1 \\ 0 & 0 & -1 & 1 & 0 & 1 \\ 0 & 0 & 0 & -1 & -1 & 1 \end{bmatrix}.$$

For a complete PC design, $r = t(t - 1)/2$, and the least squares formulation leads to Mosteller's solution.

The data for this example are taken from Guilford (1954, 160) and pertain to a preference study involving nine vegetables. With $t = 9$, a complete design would require $t(t - 1)/2 = 36$ comparisons. In this example, we use only $r = 18$ pairs out of 36. Starting with vegetable pairs (1,2) and (1,3) in the initial blocks, a cyclic enumeration produces the following pairs to be compared:

**Example 11.2-5**

| Blocks | 1 | 2 | 3 | 4 | 5 | 6 | 7 | 8 | 9 |
|---|---|---|---|---|---|---|---|---|---|
| Pairs: Set I | 1,2 | 2,3 | 3,4 | 4,5 | 5,6 | 6,7 | 7,8 | 8,9 | 1,9 |
| Set II | 1,3 | 2,4 | 3,5 | 4,6 | 5,7 | 6,8 | 7,9 | 1,8 | 2,9 |

The design matrix $X$ and the vector $Z$ (augmented) are shown in Table 11.14. As before, the transpose $X'$ is found by interchanging the rows and columns of $X$. We would need $(X'X)^{-1}$. To find elements of $Z$, first calculate the proportion $p_{ij}$ of the time vegetable $i$ was preferred over vegetable $j$. Then find the corresponding normal deviates $Z_{ij}$. These quantities are given in Table 11.15.

**Table 11.14** Design Matrix for $t = 9$ and $r = 18$ of Example 11.2-5 and Representation of $Z$

| Pairs | Items | | | | | | | | | Z |
|---|---|---|---|---|---|---|---|---|---|---|
| | 1 | 2 | 3 | 4 | 5 | 6 | 7 | 8 | 9 | |
| 1, 2 | 1 | −1 | 0 | 0 | 0 | 0 | 0 | 0 | 0 | $Z_{12}$ |
| 1, 3 | 1 | 0 | −1 | 0 | 0 | 0 | 0 | 0 | 0 | $Z_{13}$ |
| 2, 3 | 0 | 1 | −1 | 0 | 0 | 0 | 0 | 0 | 0 | $Z_{23}$ |
| 2, 4 | 0 | 1 | 0 | −1 | 0 | 0 | 0 | 0 | 0 | $Z_{24}$ |
| 3, 4 | 0 | 0 | 1 | −1 | 0 | 0 | 0 | 0 | 0 | $Z_{34}$ |
| 3, 5 | 0 | 0 | 1 | 0 | −1 | 0 | 0 | 0 | 0 | $Z_{35}$ |
| 4, 5 | 0 | 0 | 0 | 1 | −1 | 0 | 0 | 0 | 0 | $Z_{45}$ |
| 4, 6 | 0 | 0 | 0 | 1 | 0 | −1 | 0 | 0 | 0 | $Z_{46}$ |
| 5, 6 | 0 | 0 | 0 | 0 | 1 | −1 | 0 | 0 | 0 | $Z_{56}$ |
| 5, 7 | 0 | 0 | 0 | 0 | 1 | 0 | −1 | 0 | 0 | $Z_{57}$ |
| 6, 7 | 0 | 0 | 0 | 0 | 0 | 1 | −1 | 0 | 0 | $Z_{67}$ |
| 6, 8 | 0 | 0 | 0 | 0 | 0 | 1 | 0 | −1 | 0 | $Z_{68}$ |
| 7, 8 | 0 | 0 | 0 | 0 | 0 | 0 | 1 | −1 | 0 | $Z_{78}$ |
| 7, 9 | 0 | 0 | 0 | 0 | 0 | 0 | 1 | 0 | −1 | $Z_{79}$ |
| 8, 9 | 0 | 0 | 0 | 0 | 0 | 0 | 0 | 1 | −1 | $Z_{89}$ |

*(continued)*

**Table 11.14** Design Matrix for $t = 9$ and $r = 18$ of Example 11.2-5 and Representation of $Z$—*cont...*

| Pairs | \multicolumn{10}{c}{Items} |
|---|---|---|---|---|---|---|---|---|---|---|

| Pairs | 1 | 2 | 3 | 4 | 5 | 6 | 7 | 8 | 9 | Z |
|---|---|---|---|---|---|---|---|---|---|---|
| 1, 8 | 1 | 0 | 0 | 0 | 0 | 0 | 0 | −1 | 0 | $Z_{18}$ |
| 1, 9 | 1 | 0 | 0 | 0 | 0 | 0 | 0 | 0 | −1 | $Z_{19}$ |
| 2, 9 | 0 | 1 | 0 | 0 | 0 | 0 | 0 | 0 | −1 | $Z_{29}$ |
| 0, 0 | 1 | 1 | 1 | 1 | 1 | 1 | 1 | 1 | 1 | 0 |

**Table 11.15** Matrices of Observed Preferred Proportions and the Corresponding Normal Deviates

| I | \multicolumn{9}{c}{J} |
|---|---|---|---|---|---|---|---|---|---|

| I | 1 | 2 | 3 | 4 | 5 | 6 | 7 | 8 | 9 |
|---|---|---|---|---|---|---|---|---|---|
| \multicolumn{10}{c}{Observed Proportions, $i > j$} |
| 1 | 0 | 0.182 | 0.230 | | | | | 0.108 | 0.074 |
| 2 | | 0 | 0.399 | 0.277 | | | | | 0.142 |
| 3 | | | 0 | 0.439 | 0.264 | | | | |
| 4 | | | | 0 | 0.439 | 0.412 | | | |
| 5 | | | | | 0 | 0.507 | 0.426 | | |
| 6 | | | | | | 0 | 0.372 | 0.318 | |
| 7 | | | | | | | 0 | 0.473 | 0.358 |
| 8 | | | | | | | | 0 | 0.372 |
| 9 | | | | | | | | | 0 |
| \multicolumn{10}{c}{Corresponding Normal Deviates $Z_{ij}$} |
| 1 | 0 | −0.908 | −0.739 | | | | | −1.237 | −1.447 |
| 2 | | 0 | −0.256 | −0.592 | | | | | −1.071 |
| 3 | | | 0 | −0.154 | −0.631 | | | | |
| 4 | | | | 0 | −0.154 | −0.222 | | | |
| 5 | | | | | 0 | 0.018 | −0.187 | | |
| 6 | | | | | | 0 | −0.327 | −0.473 | |
| 7 | | | | | | | 0 | −0.068 | −0.364 |
| 8 | | | | | | | | 0 | −0.327 |
| 9 | | | | | | | | | 0 |

For instance, for pair (1,2), $p_{12} = 0.182$ and the corresponding normal deviate $Z_{12}$ is $-0.908$. Thus, the $Z$ column of observed differences, augmented by the last element 0, is

$$Z = \begin{bmatrix} -0.908 \\ -0.739 \\ \vdots \\ -1.447 \\ -1.071 \\ 0.000 \end{bmatrix}.$$

Substitution of matrix $X$ and vector $Z$ into Eq. (11.2-24) leads to estimates of $S_i$. Calculations frequently require computer assistance, and a computer program can be easily written for this task.

Table 11.16 shows estimates $\widehat{S}_i$, along with the estimates obtained by Guilford on the basis of a complete PC design. The two sets of estimates agree closely in rating beets, cabbage, and turnips as the least preferred vegetables, and string beans, peas, and corn as the most preferred. In this example, we see that the incomplete PC design is as effective as the complete PC design in ordering the nine vegetables on the preference continuum.

**Table 11.16** Estimates of Scale Values $\widehat{S}_i$ from Incomplete and Complete Paired Comparison Designs

| Vegetables | Incomplete Design, 18 Pairs, Morrissey Solution | Complete Design, 36 Pairs, Guilford Estimate |
|---|---|---|
| 1. Turnips | −0.991 | −0.988 |
| 2. Cabbage | −0.403 | −0.465 |
| 3. Beets | −0.246 | −0.333 |
| 4. Asparagus | 0.0004 | −0.008 |
| 5. Carrots | 0.199 | 0.129 |
| 6. Spinach | 0.082 | 0.156 |
| 7. String beans | 0.345 | 0.412 |
| 8. Peas | 0.380 | 0.456 |
| 9. Corn | 0.635 | 0.642 |

We can use the S-PLUS code *tmmodel* for the calculations with the input of indicators of the pairs and $p_{ij}$. The output of $\widehat{S}$ and $SE(\widehat{S_i})$ is as follows:

```
> tmmodel(tmdat1,tmdat2)
0.032
[1] -0.991 -0.404 -0.246 0.000 0.199 0.082 0.345 0.380 0.635
    0.089 0.089 0.089 0.089 0.089 0.089 0.089 0.089 0.089
> tmdat1
    [,1] [,2]
[1,] 1 2
[2,] 1 3
[3,] 2 3
...
[18,] 2 9
> tmdat2
[1] 0.182 0.230 0.399 0.277 0.439 0.264 0.439 0.412 0.507
    0.426 0.372 0.318 0.473 0.358 0.372 0.108 0.074 0.142
```

Let us return to Example 11.2-4 and redo the calculations using the least squares formulation. Notice that the data now are from a complete PC design with three treatments. Hence, $r = t(t - 1)/2 = 3$. The matrix $X$ augmented by the last row and the vector $Z$ augmented by the last element corresponding to the data of the example appear in Table 11.17. All the other calculations following the least squares formulation are also

**Table 11.17** Calculation of Variances of Scale Values for the Thurstone–Mosteller Model

| Pair (i, j), i > j | X | Z |
|---|---|---|
| 1, 2 | $\begin{bmatrix} 1 & -1 & 0 \\ 1 & 0 & -1 \\ 0 & 1 & -1 \\ 1 & 1 & 1 \end{bmatrix}$ | $\begin{bmatrix} 0.4902 \\ 0.6745 \\ 0.6745 \\ 0.0000 \end{bmatrix}$ |
| 1, 3 | | |
| 2, 3 | | |
| $\mu$ | | |

$$(X'X)^{-1} = \begin{bmatrix} 0.3333 & 0.0000 & 0.0000 \\ 0.0000 & 0.3333 & 0.0000 \\ 0.0000 & 0.0000 & 0.3333 \end{bmatrix} \qquad X'Z = \begin{bmatrix} 1.1647 \\ 0.1843 \\ -1.3490 \end{bmatrix}$$

$$\widehat{S} = (X'X)^{-1}X'Z = \begin{bmatrix} 0.3882 \\ 0.0614 \\ -0.4497 \end{bmatrix}, \text{ same result as in Example 11.2-3; see}$$

Table 11.13 for $\overline{Z}_{j}$.

*Note:*
*MSe = 0.0801 from formula (11.2-26).*
*SE(Si) = $\sqrt{0.0267} = 0.163$ from formula (11.2-25), i = 1, 2, 3.*

given in the table. Note that the least squares estimates for $\widehat{S}$ are the ones in Example 11.2-4, as they should be. To estimate the standard error of $\widehat{S}_i$, we use (11.2-25) with $r = 3$. One can see from Table 11.17 that

$$(Z - X\widehat{S})'(Z - X\widehat{S}) = 0.0801,$$

and that each $a_{ii} = \frac{1}{3}$. Since $r - t + 1 = 1$, from (11.2-25),

$$SE(S_i) = \sqrt{(1/3)(0.0801)} = \sqrt{0.0267}, \quad i = 1, 2, 3.$$

Since the off-diagonal elements of $(X'X)^{-1}$ are zeros, the $\widehat{S}_i$ are uncorrelated. Therefore, the standard error of the difference is

$$SE(\widehat{S}_i - \widehat{S}_j) = \sqrt{2(0.0267)} = 0.2311.$$

The 99% confidence intervals of the differences between treatments are

$$\widehat{S}_1 - \widehat{S}_2 = 0.3268 \pm 2.576(0.2311) = (-0.2685, 0.9221),$$
$$\widehat{S}_1 - \widehat{S}_3 = 0.8379 \pm 2.576(0.2311) = (0.2426, 1.4332),$$
$$\widehat{S}_2 - \widehat{S}_3 = 0.5111 \pm 2.576(0.2311) = (-0.0842, 1.1064),$$

where 2.576 is the $(1 - \alpha/2)$th percentile of the standard normal distribution. The confidence interval for $S_1 - S_3$ does not include zero in the interval; therefore, treatments 1 and 3 differ significantly at the 1% level.

By using the S-PLUS code *tmmodel*, we can obtain the output of $\widehat{S}$ and $SE(\widehat{S}_i)$ as follows:

```
> tmmodel(tmdat21,tmdat22)
0.08
[1] 0.388 0.061 -0.450 0.163 0.163 0.163
> tmdat21
    [,1] [,2]
[1,] 1 2
[2,] 1 3
[3,] 2 3
> tmdat22
[1] 0.6880038 0.7500033 0.7500033
```

## Choice of Models

We have used the Scheffé, Thurstone–Mosteller, and Bradley–Terry models to analyze the outcomes of the same PC experiment. Estimates of the scale values resulting from the three models are plotted

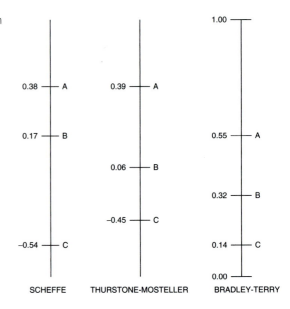

■ **FIGURE 11.5** Plot of the estimated scale values on a continuum obtained from three models.

in Figure 11.5, where $A$ = treatment 1 (control), $B$ = treatment 2 (0.15% citric acid), and $C$ = treatment 3 (0.50% citric acid). The three models were in agreement with respect to the order of stimuli on the preference continuum but were different in the degrees of separation between pairs of stimuli. The result corroborates a large study on color photography by Jackson and Fleckenstein (1957). Notice also that the models were similarly able to establish significance at the 5% level between treatments $A$ and $C$, as shown here:

| Contrasts | Scheffé | Bradley–Terry | Thurstone–Mosteller |
|-----------|---------|---------------|---------------------|
| A−B | 0.2083 | 0.2257 | 0.3268 |
| A−C | 0.9167[a] | 0.4095[a] | 0.8379[a] |
| B−C | 0.7084 | 0.1838 | 0.5111 |

[a]Significance.

This simple illustration shows the relative effectiveness of the three models with regard to scale distances between stimuli. Researchers must take into consideration other factors in the choice of a model, such as purpose and study size. If graded responses are desired, rather than simply "yes" or "no" responses, then the Scheffé model is appropriate. This model also permits the estimation of the order effect due to the order in which the items are presented.

The Thurstone–Mosteller model was originally developed to scale psychological responses on a continuum. This model is preferred for scaling a large number of stimuli. It is computationally simple. The reader may refer to a book by Guilford (1954) for a detailed treatment of this model. The Bradley–Terry model is also computationally simple when the size of the experiment is within the existing tables provided by Bradley and Terry (1952). The model itself is easy to understand and is used to analyze "yes" and "no" types of responses only. The model lends itself to several hypotheses relevant to sensory evaluation. There are other studies dealing with quantitative sensory evaluation, and we may refer the reader to McKeon (1960) and Bock and Jones (1968).

## 11.3 **THURSTONIAN DISCRIMINAL DISTANCE *d'***

### Introduction

Discrimination testing described in Chapter 10 can tell us if there is a significant difference between products for comparison. However, the testing does not tell us the degree or the extent of the difference. Measurement of sensory difference using a suitable index or scale is highly desirable.

### *Proportion of Correct Responses in Forced-Choice Methods*

Proportion of correct responses in a discrimination test using a forced-choice method is an important test statistic. However, it is not a good index to measure sensory difference or discriminability because the index is not independent of methods used. Obviously, for a same pair of stimuli, the proportions of correct responses using the 2-AFC and 3-AFC are different because the two methods contain a different guessing probability. Even for the methods with the same guessing probability—e.g., the 2-AFC and duo-trio methods; the 3-AFC and triangle methods—a same probability of correct responses using difference methods reflects different sensory differences or discriminabilities. The famous so-called paradox of discriminatory nondiscriminators (Gridgeman, 1970a) revealed this fact. In the paradox, for the same stimuli, judges gave the higher proportion of correct responses for the 3-AFC than that for the triangle test. Byer and Abrams (1953) first noted the fact from their experimental data. Many studies have confirmed this fact. The paper by Frijters (1979a) first explained and solved the paradox in theory.

### Difference between Two Proportions in the A–Not A Method and the Same-Different Method

For a given pair of stimuli, A and B, if the A–Not A method is used, we can get two proportions, $p_A$ and $p_N$, where $p_A$ is the proportion of response "A" for sample A and $p_N$ is the proportion of responses "A" for the Not A sample (sample B). If the same-different method is used, the two proportions are $p_{ss}$ and $p_{sd}$, where $p_{ss}$ is the proportion of the "same" responses for the concordant sample pairs, and $p_{sd}$ is the proportion of the "same" responses for the discordant sample pairs. The expected difference between $p_A$ and $p_N$ is not the same as the expected difference between $p_{ss}$ and $p_{sd}$ for a given sensory difference. Hence, the difference between the two proportions cannot be treated as a measure of sensory difference.

### Thurstonian Scaling

Louis Leon Thurstone (1887–1955) is the originator of psychological scaling. He stands as one of the great synthesizers of sensory scaling in the 20th century, and his synthesis remains alive and is increasingly being refined (Luce, 1994). The distance of the two means of the distributions for comparison in terms of a common standard variance is defined as the *Thurstonian discriminal distance*, $\delta = \dfrac{\mu_2 - \mu_1}{\sigma}$, and is used as an index and scale to measure sensory discriminability or sensory difference.

Sensory difference in terms of $\delta$ can be estimated from observed proportion of correct responses or other proportions in different discrimination methods using different psychometric functions based on different decision rules. The psychometric functions for the forced-choice methods describe the relationship between $\delta$ and probability of correct responses, $P_c$. The psychometric functions for the methods with response bias describe the relationship between $\delta$ and probabilities of hit and false alarm. The probability of hit is the probability of response "A" for sample A in the A–Not A method or the probability of response "same" for the concordant sample pair in the same-different method. The probability of false alarm is the probability of response "A" for sample Not A in the A–Not A method or the probability of response "same" for the discordant sample pair in the same-different method.

Thurstonian scaling is related closely to the signal detection theory (Green and Swets, 1966; Macmillan and Creelman, 2005). Signal detection theory (SDT) was established originally in electrical engineering in the early 1950s in the context of visual and auditory detection and has been applied to a wide range of perceptual, cognitive, and other psychological tasks. Thurstonian scaling is, in fact, the basis of SDT. SDT can be viewed as a natural outgrowth of classical psychophysics and Thurstonian scaling (Baird and Noma, 1978, 126). In SDT, one of the most widely used sensitivity measures is $d'$, which is defined as

$$d' = z(H) - z(F), \tag{11.3-1}$$

where $z$ denotes the inverse of the normal distribution function; $H$ and $F$ denote the proportions of "hit" and "false alarm" in a yes-no experiment. Only in the yes-no (or rating) experiment, under the assumption of normality with equal variance,

$$d' = \frac{\mu_2 - \mu_1}{\sigma}. \tag{11.3-2}$$

Although $d'$ can be calculated from different experiments, they are rooted in the yes-no experiment. In sensory fields, $d'$ is often used as an estimate of Thurstonian $\delta$. In this book, $\delta$ and $d'$ are interchangeable for convenience.

## Estimation of $d'$

### Decision Rules and Psychometric Functions for Forced-Choice Methods

Bradley (1957) first derived the psychometric functions based on different decision rules for the 2-AFC (duo), duo-trio, and triangular methods in a memorandum prepared for the General Foods Corporation in the United States. The results were announced in abstracts in 1958 (Bradley, 1958b) and were published in detail in 1963 (Bradley, 1963). Ura (1960) independently derived the psychometric functions for the three methods, and David and Trivedi (1962) gave further details of the results. The psychometric functions for the 3-AFC and ($m$-AFC) were given by Birdsall and Peterson (1954) and Green and Birdsall (1964). Later, Frijters (1979b) gave the logistic variants of the psychometric functions for the 3-AFC and triangular methods.

The decision rules and psychometric functions for the four forced-choice methods are described in the following subsections.

### The 2-AFC Method

Assume $x$, $y$ are sensations evoked by samples A and B, respectively. Sample B has stronger sensory intensity than sample A. A correct response will be given when $y > x$. Based on this decision rule, the probability of correct responses in this method is

$$P_c = \Phi\left(d'/\sqrt{2}\right), \tag{11.3-3}$$

where $\Phi(.)$ is the cumulative distribution function of the standard normal distribution.

### The 3-AFC Method

Assume that $x_1$ and $x_2$ are sensations evoked by two samples of A, and $y$ is a sensation evoked by sample B. Sample B has stronger sensory intensity than sample A. A correct response will be given when $y > x_1$ and $y > x_2$. Based on this decision rule, the probability of correct responses in this method is

$$P_c = \int_{-\infty}^{\infty} \Phi^2(u)\phi(u - d')du, \tag{11.3-4}$$

where $\phi(.)$ is the standard normal density function.

### The Duo-Trio Method

Similarly, assume $x_1$ and $x_2$ are sensations evoked by two samples of A, and $y$ is a sensation evoked by sample B. Sample A is selected as the standard sample. A correct response will be given when $|x_2 - x_1| < |y - x_1|$. The probability of correct responses in this method is

$$P_c = 1 - \Phi\left(d'/\sqrt{2}\right) - \Phi\left(d'/\sqrt{6}\right) + 2\Phi\left(d'/\sqrt{2}\right)\Phi\left(d'/\sqrt{6}\right). \tag{11.3-5}$$

### The Triangular Method

Assume that $x_1$ and $x_2$ are sensations evoked by two samples of A, and $y$ is a sensation evoked by sample B. A correct response will be given when $|x_2 - x_1| < |y - x_1|$ and $|x_2 - x_1| < |y - x_2|$. Based on this decision rule, the probability of correct responses in this method is

$$P_c = 2\int_0^{\infty} \left[\Phi\left(-u\sqrt{3} + d'\sqrt{2/3}\right) + \Phi\left(-u\sqrt{3} - d'\sqrt{2/3}\right)\right]\exp(-u^2/2)/\sqrt{2\pi}du. \tag{11.3-6}$$

The authors who developed the psychometrical functions provided tables in their papers for $P_c$ and $d'$ values for forced-choice methods. The tables were revised, expanded, and reproduced by many authors, e.g., Elliott (1964); Hacker and Ratcliff (1979); Frijters, Kooistra, and Vereijken (1980); Frijters (1982); Craven (1992); Ennis (1993); Versfeld et al. (1996); and ASTM (2003). In this book, $d'$ as a function of $P_c$ are recalculated and given in Tables 11.19 through 11.22, respectively, for the four forced-choice methods.

For a specified $P_c$ value, we can easily find the corresponding $d'$ values in different forced-choice methods from Tables 11.19 through 11.22. We can find that, for a specified proportion of correct responses, for example, $P_c = 0.76$, $d' = 0.9989$ in 2-AFC, $d' = 1.4754$ in 3-AFC, $d' = 2.0839$ in duo-trio, and $d' = 2.8601$ in the triangular method. This means that a specified probability of correct responses represents different sensory differences in terms of $d'$. In other words, for a specified sensory difference in terms of $d'$, it evokes a different probability of correct responses in different methods. Obviously, the proportion of correct responses cannot be used as a pure index to measure sensory difference or discriminability because it is dependent on methods used.

### Decision Rules and Psychometric Functions for the A–Not A and Same-Different Methods

The decision rules and psychometric functions for the A–Not A and same-different methods are based on a monadic design under the assumption that the responses in an experiment are independent of each other.

### The A–Not A Method

Assume that $x$ and $y$ are sensations evoked by samples A and Not A, respectively. A hit is made when $x > c$, and a false alarm is made when $y > c$, where $c$ is a criterion. On this basis, the psychometric function for the A–Not A method is

$$d' = z(P_A) - z(P_N), \quad \text{or} \quad d' = z_N - z_A, \qquad (11.3\text{-}7)$$

where $P_A$ is the probability of response "A" for A sample and $P_N$ is the probability of response "A" for not A sample. $z(P_A)$, $z(P_N)$, $z_N$, and $z_A$ are the quantiles of $P_A$, $P_N$, $1 - P_N$, and $1 - P_A$ for the standard normal distribution. Though $P_N$ and $P_A$ are affected by the criteria that the subjects adopted, $d'$ is not affected by the criteria. Equation (11.3-7) has been discussed adequately in the signal detection theory for the yes/no task (Green and Swets, 1966). Elliott (1964) produced tables of $d'$ values for this method.

### The Same-Different Method

Likewise, assume $x$ and $y$ are sensations evoked by samples A and B, respectively. A hit is made when $|x_1 - x_2| < k$ or $|y_1 - y_2| < k$, where $k$ is a criterion. A false alarm is made when $|x - y| < k$. Macmillan et al. (1977) derived the psychometric function for the same-different method shown here:

$$P_{ss} = 2\Phi\left(\frac{k}{\sqrt{2}}\right) - 1, \tag{11.3-8}$$

$$P_{sd} = \Phi\left(\frac{k - d'}{\sqrt{2}}\right) - \Phi\left(\frac{-k - d'}{\sqrt{2}}\right), \tag{11.3-9}$$

where $P_{ss}$ is the proportion of the "same" responses for the concordant sample pairs: <AA> or <BB>; $P_{sd}$ is the proportion of the "same" responses for the discordant sample pairs <AB> or <BA>; and $k$ is a criterion. For given proportions $P_{ss}$ and $P_{sd}$, $d'$ and $k$ can be estimated numerically. Kaplan et al. (1978) provided tables of $d'$ values for this method.

### *Psychometric Function for the Rating Method*

The rating method is an extension of the A–Not A (yes-no) method. In the rating method, for a given sample A or Not A, the response is a rating of a $k$-point ($k > 2$) rather than a 2-point scale. The rating values indicate the degree of confidence for responses "Not A" and "A"—for example, for comparison between two products for purchase intent using a 5-point scale, from 1 = definitely would not buy it to 5 = definitely would buy it. Thurstonian $d'$ can be estimated from two sets of frequencies of ratings for the two products. The psychometric function for the rating method in the normal-normal equal variance model can be expressed by

$$P_{1j} = \Phi(Z_j) - \Phi(Z_{j-1}), \tag{11.3-10}$$

$$P_{2j} = \Phi(Z_j - d') - \Phi(Z_{j-1} - d'), \tag{11.3-11}$$

where $j = 1, 2, \ldots, k$; $P_{1j}$ and $P_{2j}$ are the probabilities of a rating in the $j$th category of a $k$-point scale for product 1 and 2, respectively; $Z_j$ are cutoffs of the categories of a $k$-point scale on a sensory unidimensional continuum. $Z_0 = -\infty$ and $Z_k = \infty$; hence, $\Phi(Z_0) = 0$ and $\Phi(Z_k) = 1$. There are $k$ independent parameters, i.e., $Z_1, Z_2, \ldots, Z_{k-1}$ and $d'$ in the equations (11.3-10) and (11.3-11).

Ogilvie and Creelman (1968) developed maximum likelihood estimates for the parameters of the Thurstonian model for the rating method by using the logistic distribution. Then Dorfman and Alf (1969) developed similar estimating procedures by using the normal distribution. Later, Grey and Morgan (1972) discussed the estimating procedures for both normal and logistic models.

The log of the likelihood function is

$$\log L = \sum_{j=1}^{k} \left( r_{1j} \log P_{1j} + r_{2j} \log P_{2j} \right), \qquad (11.3\text{-}12)$$

where $r_{1j}$ and $r_{2j}$ are frequency counts of ratings for two products.

Using the built-in function *nlminb* in S-PLUS, we can easily maximize the log likelihood (11.3-12) and obtain estimates of the parameters $z_1, z_2, \ldots, z_{k-1}$ and $d'$, which make the log likelihood (11.3-12) maximum. The covariance matrix of the parameter estimates, $cov(z_1, z_2, \ldots, z_{k-1}, d')$, can also be obtained by using the built-in function *vcov.nlminb*. The starting points of the parameters can be selected by

$$z_j^{(0)} = \Phi^{-1}\left( \varphi_{1j} \right), \qquad (11.3\text{-}13)$$

where $\varphi_{1j} = \sum_{m=1}^{j} \frac{r_{1m}}{n_1}, j = 1, 2, \ldots, k-1; n_1$ is the total number of rating frequencies for the first product; and $d_0' = 0.50$. If the estimated $d'$ value is negative, it means that the intensity of the second product is smaller than that of the first product. Changing the order of the two products can change the sign of $d'$ but not the magnitude of $d'$ and its variance. Inverting the input data, i.e., $r_{1j} \leftrightarrow r_{2,k+1-j}$, $j = 1, 2, \ldots, k$, ought to provide the same estimate value of $d'$ and the variance of $d'$.

---

Ratings data of purchase intent for two products are given in Table 11.18. The values $-0.8778963$, $-0.2275450$, $0.1763742$, $0.9944579$, and $0.50$ are used for the starting values of the parameters according to (11.3-13). The estimated values of the parameters are $d' = 0.5487$, and $z_1, z_2, z_3, z_4$ are $-0.8591$, $-0.2079$, $0.1366$, and $0.9971$, respectively, using an S-PLUS program. The covariance matrix of the estimated parameters can also be obtained. From the covariance matrix, we find that the variance of $d'$ is $0.0234$. Because the sign of the $d'$ is positive, it means that purchase intent for product B is higher than that for product A. Using the estimated parameter values, we can get the expected probabilities of the rating categories and the expected frequency counts shown in Table 11.18 according to (11.3-10) and (11.3-11). We can see that the expected frequencies fit the observed frequencies well in this example.

**Example 11.3-1**

**Table 11.18** Data for Purchase Intent Ratings (Example 11.3-1)

| Rating | Observed Frequencies | | Expected Frequencies | |
|---|---|---|---|---|
| | Product A $r_{1j}$ | Product B $r_{2j}$ | Product A $r_{1j}$ | Product B $r_{2j}$ |
| 1 = Definitely would not buy | 19 | 9 | 19.5 | 8.5 |
| 2 = Probably would not buy | 22 | 16 | 22.3 | 15.5 |
| 3 = Might/ Might not buy | 16 | 10 | 13.7 | 12.4 |
| 4 = Probably would buy | 27 | 37 | 28.6 | 35.6 |
| 5 = Definitely would buy | 16 | 35 | 15.9 | 35.0 |
| Total number of ratings | 100 | 107 | 100 | 107 |

*x<-ratdel(c(19,22,16,27,16),c(9,16,10,37,35))*
*x$parameters:*
*[1] −0.8591414 −0.2079039 0.1365949 0.9970515 0.5487463*
*> vcov.nlminb(x)*
*[,1] [,2] [,3] [,4] [,5]*
*[1,] 0.016947917 0.01150714 0.01026413 0.008641442 0.01072496*
*[2,] 0.011507138 0.01442421 0.01258372 0.010108696 0.01161076*
*[3,] 0.010264127 0.01258372 0.01477983 0.011449246 0.01229549*
*[4,] 0.008641442 0.01010870 0.01144925 0.017597172 0.01368619*
*[5,] 0.010724965 0.01161076 0.01229549 0.013686187 0.02340774*

## Variance of $d'$

Thurstonian $d'$ provides a measure of sensory difference or discriminability. It is theoretically unaffected by the criterion that subjects adopt or the method used. However, the true $d'$ cannot be observed. It can only be estimated from data. The precision of the estimate $d'$ can be expressed by its variance, $Var(d')$.

Variance of $d'$ is important in Thurstonian model. It describes how close the estimated $d'$ value is to a true $d'$ value. Moreover, it provides a basis of statistical inference for $d's$. The magnitude of variance of $d'$ depends not only on sample size but also the method used. Gourevitch and Galanter (1967) gave an estimate of the variance of $d'$ for the yes-no task (i.e., the A−Not A method). Later, Bi et al. (1997) provided estimates and tables for the variance estimates of $d'$ for four forced-choice methods: 2-AFC, 3-AFC, triangular, and duo-trio. Then Bi (2002) provided variance estimates of $d'$, tables, and a computer program for the same-different method (differencing model only).

There are different approaches to estimate variance of $d'$. One is the so-called delta method (Bishop et al., 1975, Section 14.6), which is based on Taylor-series expansion with one and/or two variables. According to the delta method, if $Y$ is a function of a variable $X$, i.e., $Y = f(X)$, then

$$f(X) = f(x) + (X - x)f'_x + o(|X - x|) \text{ as } X \rightarrow x, \qquad (11.3\text{-}14)$$

where $f'_x$ denotes the derivative with respect to $X$, taking the corresponding $x$ value. Hence, the variances of $Y$ and $X$ have the relationship

$$\sigma_Y^2 \approx f'^2_x \sigma_X^2, \qquad (11.3\text{-}15)$$

Another approach is to use the inverse of the second derivative of the maximum likelihood function with respect to $d'$ as that for the rating method. The delta method is used for variance of $d'$ in this section. The advantage of the delta method is that the variance of $d'$ can often be expressed in a precise equation.

### Variance of d′ for Forced-Choice Methods

For the forced-choice methods, the proportion of correct responses $P_c$ is a function of $d'$, i.e., $P_c = f(d')$, where $f(d')$ denotes a psychometrical function for a forced-choice method. We can use the delta method to calculate approximate variances of $d'$ by

$$Var(d') = \frac{Var(P_c)}{f'^2(d'_0)}. \qquad (11.3\text{-}16)$$

Variance of $d'$ for the forced-choice methods is composed of two components. One is sample size $N$, and the other is called $B$ value. $B$ value is determined solely by the method used and is independent of sample size $N$. A general form of variance of $d'$ for the forced-choice methods is

$$Var(d') = \frac{B}{N}. \qquad (11.3\text{-}17)$$

### For the 2-AFC Method

$$B = \frac{2p_c(1 - p_c)}{\phi^2\left(\dfrac{d'}{\sqrt{2}}\right)}, \qquad (11.3\text{-}18)$$

where $p_c$ is the observed proportion of correct responses in the method and $\phi\left(\dfrac{d'}{\sqrt{2}}\right)$ denotes density function of the standard normal distribution evaluated at $\dfrac{d'}{\sqrt{2}}$. The $B$ values can be found in Table 11.19.

### For the 3-AFC Method

$$B = \frac{p_c(1 - p_c)}{P_c'^2(d')}, \qquad (11.3\text{-}19)$$

where $P_c'$ denotes the derivative of $P_c$, the psychometric function of the 3-AFC method.

### For the Duo-Trio Method

$$B = \frac{p_c(1 - p_c)}{P_D'^2(d')}, \qquad (11.3\text{-}20)$$

**Table 11.19** $d'$ and $B$ Values for Variance of $d'$ for the 2-AFC Method

| $p_c$ | 0.00 | 0.01 | 0.02 | 0.03 | 0.04 | 0.05 | 0.06 | 0.07 | 0.08 | 0.09 |
|---|---|---|---|---|---|---|---|---|---|---|
| 0.5 |  | 0.0355 | 0.0709 | 0.1065 | 0.1421 | 0.1777 | 0.2135 | 0.2494 | 0.2855 | 0.3218 |
|  |  | 3.1423 | 3.1445 | 3.1481 | 3.1531 | 3.1597 | 3.1677 | 3.1773 | 3.1885 | 3.2013 |
| 0.6 | 0.3583 | 0.3950 | 0.4320 | 0.4693 | 0.5069 | 0.5449 | 0.5833 | 0.6221 | 0.6614 | 0.7012 |
|  | 3.2159 | 3.2321 | 3.2502 | 3.2702 | 3.2923 | 3.3164 | 3.3428 | 3.3716 | 3.4030 | 3.4371 |
| 0.7 | 0.7416 | 0.7826 | 0.8243 | 0.8666 | 0.9098 | 0.9539 | 0.9989 | 1.0449 | 1.0920 | 1.1405 |
|  | 3.4742 | 3.5145 | 3.5583 | 3.6057 | 3.6573 | 3.7136 | 3.7748 | 3.8416 | 3.9145 | 3.9948 |
| 0.8 | 1.1902 | 1.2415 | 1.2945 | 1.3494 | 1.4064 | 1.4657 | 1.5278 | 1.5930 | 1.6617 | 1.7346 |
|  | 4.0827 | 4.1798 | 4.2873 | 4.4069 | 4.5406 | 4.6906 | 4.8607 | 5.0547 | 5.2779 | 5.5378 |
| 0.9 | 1.8124 | 1.8961 | 1.9871 | 2.0871 | 2.1988 | 2.3262 | 2.4758 | 2.6599 | 2.9044 | 3.2900 |
|  | 5.8443 | 6.2114 | 6.6603 | 7.2222 | 7.9492 | 8.9313 | 10.3417 | 12.5718 | 16.7205 | 27.876 |

*Note: There are two values in a cell corresponding to a $p_c$ value. The first one is the $d'$ value, and the second is the $B$ value. For example, for $p_c = 0.61$, $d' = 0.3950$ and $B = 3.2321$. The variance of $d'$ at 0.3950 is $B/N$, where $N$ is sample size.*

**Table 11.20** *d′* and *B* Values for Variance of *d′* for the 3-AFC Method

| $p_c$ | 0.00 | 0.01 | 0.02 | 0.03 | 0.04 | 0.05 | 0.06 | 0.07 | 0.08 | 0.09 |
|-------|------|------|------|------|------|------|------|------|------|------|
| 0.3 | | | | | 0.0235 | 0.0585 | 0.0932 | 0.1275 | 0.1615 | 0.1953 |
| | | | | | 2.7779 | 2.7576 | 2.7391 | 2.7224 | 2.7073 | 2.6938 |
| 0.4 | 0.2289 | 0.2622 | 0.2953 | 0.3283 | 0.3611 | 0.3939 | 0.4265 | 0.4591 | 0.4916 | 0.5241 |
| | 2.6818 | 2.6713 | 2.6622 | 2.6544 | 2.6479 | 2.6427 | 2.6388 | 2.6361 | 2.6346 | 2.6344 |
| 0.5 | 0.5565 | 0.5890 | 0.6215 | 0.6541 | 0.6867 | 0.7194 | 0.7522 | 0.7852 | 0.8183 | 0.8517 |
| | 2.6353 | 2.6375 | 2.6409 | 2.6456 | 2.6514 | 2.6586 | 2.6670 | 2.6767 | 2.6878 | 2.7004 |
| 0.6 | 0.8852 | 0.9189 | 0.9529 | 0.9872 | 1.0218 | 1.0568 | 1.0921 | 1.1279 | 1.1641 | 1.2007 |
| | 2.7144 | 2.7298 | 2.7469 | 2.7657 | 2.7862 | 2.8086 | 2.8329 | 2.8594 | 2.8881 | 2.9191 |
| 0.7 | 1.2380 | 1.2758 | 1.3142 | 1.3533 | 1.3931 | 1.4338 | 1.4754 | 1.5179 | 1.5615 | 1.6063 |
| | 2.9529 | 2.9894 | 3.0289 | 3.0717 | 3.1182 | 3.1687 | 3.2238 | 3.2836 | 3.3491 | 3.4208 |
| 0.8 | 1.6524 | 1.6999 | 1.7490 | 1.7999 | 1.8527 | 1.9078 | 1.9654 | 2.0260 | 2.0899 | 2.1577 |
| | 3.4996 | 3.5864 | 3.6824 | 3.7893 | 3.9084 | 4.0425 | 4.1940 | 4.3672 | 4.5664 | 4.7982 |
| 0.9 | 2.2302 | 2.3082 | 2.3931 | 2.4865 | 2.5909 | 2.7101 | 2.8504 | 3.0231 | 3.2533 | 3.6179 |
| | 5.0719 | 5.3997 | 5.8008 | 6.3034 | 6.9539 | 7.8335 | 9.0998 | 11.1029 | 14.8452 | 24.9982 |

*Note: There are two values in a cell corresponding to a $p_c$ value. The first one is the d′ value, and the second is the B value. For example, for $p_c = 0.61$, d′=0.9189, and B = 2.7298. The variance of **d′** at 0.9189 is B/N, where N is sample size.*

where $P'_D = \dfrac{-1}{\sqrt{2}}\phi\left(\dfrac{d'}{\sqrt{2}}\right) - \dfrac{1}{\sqrt{6}}\phi\left(\dfrac{d'}{\sqrt{6}}\right) + \sqrt{2}\phi\left(\dfrac{d'}{\sqrt{2}}\right)\Phi\left(\dfrac{d'}{\sqrt{6}}\right) + \sqrt{\dfrac{2}{3}}\,\Phi\left(\dfrac{d'}{\sqrt{2}}\right)\phi\left(\dfrac{d'}{\sqrt{6}}\right)$.

The *B* values can be found in Table 11.21.

### For the Triangular Method

$$B = \frac{p_c(1 - p_c)}{P'^2_\Delta(d')},\qquad (11.3\text{-}21)$$

where $P'_\Delta = \sqrt{\dfrac{2}{3}}\phi\left(\dfrac{d'}{\sqrt{6}}\right)\left[\Phi\left(\dfrac{d'}{\sqrt{2}}\right) - \Phi\left(\dfrac{-d'}{\sqrt{2}}\right)\right]$. The *B* values can be found in Table 11.22.

## Variance of *d′* for the A—Not A and Same-Different Methods

### For the A—Not A Method

$$Var(d') = \frac{Var(P_A)}{\phi^2(z_A)} + \frac{Var(P_N)}{\phi^2(z_N)}.\qquad (11.3\text{-}22)$$

**Table 11.21** $d'$ and $B$ Values for Variance of $d'$ for the Duo-Trio Method

| $p_c$ | 0.00 | 0.01 | 0.02 | 0.03 | 0.04 | 0.05 | 0.06 | 0.07 | 0.08 | 0.09 |
|---|---|---|---|---|---|---|---|---|---|---|
| 0.5 | | 0.3319 | 0.4723 | 0.5821 | 0.6766 | 0.7614 | 0.8397 | 0.9132 | 0.9831 | 1.0503 |
| | | 70.5347 | 36.5723 | 25.2868 | 19.6655 | 16.3216 | 14.1112 | 12.552 | 11.4003 | 10.5206 |
| 0.6 | 1.1152 | 1.1784 | 1.2403 | 1.3011 | 1.3611 | 1.4206 | 1.4796 | 1.5385 | 1.5973 | 1.6561 |
| | 9.8347 | 9.2896 | 8.8510 | 8.4964 | 8.2088 | 7.9758 | 7.7892 | 7.6417 | 7.5288 | 7.4468 |
| 0.7 | 1.7153 | 1.7749 | 1.8350 | 1.8957 | 1.9574 | 2.0200 | 2.0839 | 2.1491 | 2.2158 | 2.2843 |
| | 7.3924 | 7.3640 | 7.3602 | 7.3802 | 7.4235 | 7.4904 | 7.5815 | 7.6976 | 7.8403 | 8.0118 |
| 0.8 | 2.3549 | 2.4277 | 2.5032 | 2.5817 | 2.6635 | 2.7493 | 2.8396 | 2.9352 | 3.0367 | 3.1456 |
| | 8.2149 | 8.4525 | 8.7296 | 9.0515 | 9.4246 | 9.8584 | 10.3637 | 10.9554 | 11.6507 | 12.4774 |
| 0.9 | 3.2631 | 3.3910 | 3.5317 | 3.6886 | 3.8664 | 4.0724 | 4.3183 | 4.6253 | 5.0396 | 5.7009 |
| | 13.4685 | 14.6719 | 16.1565 | 18.0273 | 20.4493 | 23.7071 | 28.3374 | 35.5203 | 48.5851 | 82.7794 |

*Note: There are two values in a cell corresponding to a $p_c$ value. The first one is the $d'$ value, and the second is the $B$ value. For example, for $p_c = 0.61$, $d' = 1.1784$, and $B = 9.2896$. The variance of $d'$ at 1.1784 is $B/N$, where N is sample size.*

**Table 11.22** $d'$ and $B$ Values for Variance of $d'$ for the Triangular Method

| $p_c$ | 0.00 | 0.01 | 0.02 | 0.03 | 0.04 | 0.05 | 0.06 | 0.07 | 0.08 | 0.09 |
|---|---|---|---|---|---|---|---|---|---|---|
| 0.3 | | | | | 0.2702 | 0.4292 | 0.5454 | 0.6425 | 0.7284 | 0.8065 |
| | | | | | 93.246 | 38.8750 | 25.3118 | 19.1675 | 15.6672 | 13.4201 |
| 0.4 | 0.8791 | 0.9475 | 1.0125 | 1.0748 | 1.1349 | 1.1932 | 1.2500 | 1.3055 | 1.3599 | 1.4135 |
| | 11.8579 | 10.7139 | 9.8452 | 9.1660 | 8.6235 | 8.1827 | 7.8200 | 7.5189 | 7.2672 | 7.0554 |
| 0.5 | 1.4663 | 1.5184 | 1.5701 | 1.6213 | 1.6722 | 1.7229 | 1.7735 | 1.8239 | 1.8744 | 1.9249 |
| | 6.8772 | 6.7273 | 6.6013 | 6.4962 | 6.4094 | 6.3387 | 6.2824 | 6.2394 | 6.2084 | 6.1886 |
| 0.6 | 1.9756 | 2.0265 | 2.0776 | 2.1290 | 2.1808 | 2.2331 | 2.2859 | 2.3393 | 2.3933 | 2.4481 |
| | 6.1793 | 6.1800 | 6.1902 | 6.2097 | 6.2381 | 6.2756 | 6.3219 | 6.3773 | 6.4417 | 6.5157 |
| 0.7 | 2.5037 | 2.5602 | 2.6178 | 2.6764 | 2.7363 | 2.7975 | 2.8601 | 2.9244 | 2.9904 | 3.0584 |
| | 6.5993 | 6.6932 | 6.7980 | 6.9140 | 7.0425 | 7.1842 | 7.3400 | 7.5118 | 7.7009 | 7.9093 |
| 0.8 | 3.1286 | 3.2012 | 3.2764 | 3.3546 | 3.4361 | 3.5215 | 3.6112 | 3.7057 | 3.8060 | 3.9129 |
| | 8.1396 | 8.3945 | 8.6770 | 8.9918 | 9.3437 | 9.7401 | 10.1889 | 10.6999 | 11.2887 | 11.9732 |
| 0.9 | 4.0276 | 4.1518 | 4.2875 | 4.4377 | 4.6067 | 4.8007 | 5.0305 | 5.3156 | 5.6983 | 6.3095 |
| | 12.7793 | 13.7448 | 14.9231 | 16.3979 | 18.3058 | 20.8808 | 24.5837 | 30.4445 | 41.3906 | 71.0272 |

*Note: There are two values in a cell corresponding to a $p_c$ value. The first one is the $d'$ value, and the second is the $B$ value. For example, for $p_c = 0.61$, $d' = 2.0265$, and $B = 6.1800$. The variance of $d'$ at 2.0265 is $B/N$, where N is sample size.*

It can be expressed as (11.3-23),

$$Var(d') = \frac{B_N}{N_N} + \frac{B_A}{N_A},\qquad(11.3\text{-}23)$$

where $N_N$ and $N_A$ are sample sizes for sample A and Not A, respectively, and $B_N = \frac{p_N(1-p_N)}{\phi^2(z_N)}$; $B_A = \frac{p_A(1-p_A)}{\phi^2(z_A)}$. We can see that the variance of $d'$ for the A−Not A method depends on $p_N$, $p_A$, total sample size $N_N + N_A$, and sample allocation such as the ratio $N_N/N_A$.

### For the Same-Different Method

$$Var(d') = \frac{B_d}{N_d} + \frac{B_s}{N_s},\qquad(11.3\text{-}24)$$

where $N_d$ and $N_s$ are sample sizes for the concordant sample pairs and discordant sample pairs, respectively, and

$$B_d = \frac{p_{sd}(1-p_{sd})}{w^2},$$

$$B_s = \frac{v^2 p_{ss}(1-p_{ss})}{w^2 u^2},$$

$$w = -\phi\left(\frac{k-d'}{\sqrt{2}}\right)\frac{1}{\sqrt{2}} + \phi\left(\frac{-k-d'}{\sqrt{2}}\right)\frac{1}{\sqrt{2}},$$

$$v = \phi\left(\frac{k-d'}{\sqrt{2}}\right)\frac{1}{\sqrt{2}} + \phi\left(\frac{-k-d'}{\sqrt{2}}\right)\frac{1}{\sqrt{2}},$$

$$u = \sqrt{2}\phi\left(\frac{k}{\sqrt{2}}\right),$$

$$k = \sqrt{2}\Phi^{-1}\left(\frac{p_{ss}+1}{2}\right),$$

where $\Phi^{-1}(.)$ denotes the quantile of the standard normal distribution. The variance of $d'$ for the same-different method depends on $p_{ss}$ and $p_{sd}$, total sample size, $N_d + N_s$, and sample size allocation, i.e., the ratio $N_d/N_s$. Generally speaking, the variance of $d'$ in the same-different method is mainly determined by performance on the discordant sample pairs. Hence, in order to reduce variance of $d'$ in the test, sample size for the discordant sample pairs should be generally larger than that for the concordant sample pairs.

## Tables and S-PLUS Codes for $d'$ and Variance of $d'$

### Tables for Forced-Choice Methods

Tables 11.19 through 11.22 give $d'$ and $B$ values for the four forced-choice methods. In the tables, $d'$ values were calculated as a function of $P_c$. The $B$ value, which is a component of variance of $d'$, can be found in the same tables. For a given $P_c$ or observed $p_c$, there are two values in the tables. The first one is the $d'$ value, and the second one is the $B$ value. The $P_c$ values range from $P_c = P_{c0} + 0.01$ to 0.99; $P_{c0} = 0.5$ for the 2-AFC and duo-trio methods, and $P_{c0} = 0.33$ for the 3-AFC and triangular methods.

| **Example 11.3-2** | One hundred panelists participated in a 3-AFC test. Sixty-three panelists gave correct responses. Hence, $p_c = 63/100 = 0.63$. We therefore can find that $d' = 0.9872$ and $B = 2.7657$ from Table 11.20, corresponding to $p_c = 0.63$. Hence, the variance of the $d'$ at 0.9872 should be $2.7657/100 = 0.027657$. |

### S-PLUS Codes for the A–Not A and Same-Different Methods

The S-PLUS codes *anadvn* and *sddvn* can be used to calculate $d'$ and variance of $d'$ for the A–Not A and same-different methods.

| **Example 11.3-3** | Assume that there are 300 panelists in a monadic A–Not A test. In this test, 55 panelists of the total 100 panelists who received the A sample gave "A" responses, and 30 panelists of the total 200 panelists who received the Not A sample gave "A" responses. We can use the S-PLUS code *anadvn* to get $d'E = E1.1621$ and $Var(d') = 0.028$: |

```
> anadvn(55,100,30,200)
[1] 1.16209474 0.02752512
```

| **Example 11.3-4** | In a same-different test with $N_s = 100$ concordant sample pairs (50 AA and 50 BB) and $N_d = 200$ discordant sample pairs (100 AB and 100 BA), $x_{ss} = 15$ and $x_{sd} = 20$ were observed. We can use the S-PLUS code *sddvn* to get $d' = 1.2811$ and $Var(d') = 0.257$: |

```
> sddvn(15,100,20,200)
[1] 1.281 0.257
```

## Confidence Intervals and Tests for *d'*

Statistical inference can be conducted based on the estimated *d'* and variance of *d'*, particularly for comparisons of two or multiple *d'*s obtained from different experiments using different methods. See Section 11.3.5 for the discussion.

### Confidence Interval of δ

A $(1 - \alpha)100\%$ two-sided confidence interval for δ can be obtained by using

$$d' - z_{1-\alpha/2}\sqrt{Var(d')} < \delta < d' + z_{1-\alpha/2}\sqrt{Var(d')}. \qquad (11.3\text{-}25)$$

---

For Example 11.3-2, $d' = 0.987$, $Var(d') = 0.027657$, and according to (11.3-25), the 90% confidence interval for δ is

**Example 11.3-5**

$$0.987 - 1.645 \times \sqrt{0.027657} < \delta < 0.987 + 1.645 \times \sqrt{0.027657}.$$

It is $0.71 < \delta < 1.26$.

---

### Comparison of a *d'* with a Specified Value

Sometimes, the interest is to test if the true difference in terms of *d'* differs from a specified value $d_0'$, i.e., to test $H_0$: $d' = d_0'$ versus $H_a$: $d' \neq d_0'$. In the situation that *d'* is obtained from a forced-choice method, testing $d' = d_0'$ is equivalent to testing $p_c = p_0$, i.e., testing if the probability of correct responses is equal to a specified probability $p_0$, which corresponds to $d_0'$. In this case, the conventional statistical test for proportion should be used. However, if the *d'* is obtained using the A−Not A method or same-different method or other non-forced-choice method, the test statistic $T$ in (11.3.26) should be used:

$$T = \frac{d' - d_0'}{\sqrt{V(d')}}, \qquad (11.3\text{-}26)$$

where $V(d')$ denotes the variance of $d'$. The test statistic $T$ follows approximately the standard normal distribution. For a two-sided test, if $|T| > z_{1-\alpha/2}$, we can conclude at an $\alpha$ significance level that the true difference in terms of $d'$ is different from $d'_0$.

The test can also be one-sided if the alternative hypothesis is $H_a: d' < d'_0$ or $H_a: d' > d'_0$. For the one-sided test, the null hypothesis is rejected and the alternative hypothesis is accepted if $|T| > z_{1-\alpha}$.

---

**Example 11.3-6**

In Example 11.3-3, we obtained $d' = 1.16$ with variance $V(d') = 0.0275$ for a monadic A–Not A test. We want to conduct a one-sided test for $H_0: d' = 1$ versus $H_a: d' > 1$ at a significance level $\alpha = 0.05$. According to (11.3-26), the test statistic value is

$$T = \frac{1.16 - 1}{\sqrt{0.0275}} = 0.965.$$

Because $T = 0.965 < 1.645$, we cannot reject the null hypothesis $H_0: d' = 1$ at $\alpha = 0.05$.

---

### Comparison of Two d's

Sometimes, the interest is to test if there is a significant difference between two $d's$, i.e., to test $H_0: d'_1 = d'_2$ versus $H_a: d'_1 \neq d'_2$. If the two $d's$ are estimated using a same forced-choice method, the comparison of two $d's$ is equivalent to comparison of two proportions of correct responses. In this situation, we should use the conventional statistical test for two proportions. If the two $d's$ are estimated from different discrimination methods or from a same but non-forced-choice method—e.g., the A–Not A method or the same-different method—the test statistic $T$ in (11.3-27) can be used for the comparison:

$$T = \frac{d'_1 - d'_2}{\sqrt{V(d'_1) + V(d'_2)}}. \tag{11.3-27}$$

The statistic $T$ follows approximately a standard normal distribution. If $|T| > z_{1-\alpha/2}$, the conclusion is that the two $d's$ are significantly different at an $\alpha$ significance level.

**Example 11.3-7**

For this example, two $d'$s are obtained from two studies using different methods. One is from a 2-AFC test, and the other is from a triangular test. The results for the first study are $p_{1c} = 0.83$, $N_1 = 100$; hence, $d'_1 = 1.3494$,

$$V(d'_1) = \frac{B}{N_1} = \frac{4.4069}{100} = 0.044069 \text{ from Table 11.19. The results for the}$$

second study are $p_{2c} = 0.61$, $N_2 = 100$; hence, $d'_2 = 2.0265$,

$$V(d'_2) = \frac{B}{N_2} = \frac{6.18}{100} = 0.0618 \text{ from Table 11.22. The value of the test}$$

statistic $T$ in (11.3-27) is then

$$T = \frac{1.3494 - 2.0265}{\sqrt{0.044069 + 0.0618}} = -2.08.$$

Because $|T| = 2.08 > 1.96$, we can conclude at $\alpha = 0.05$ that the two $d'$s are significantly different.

## Comparison of Multiple d's

Instead of considering whether two $d'$s are significantly different, sometimes we need to compare $k$ ($k > 2$) $d'$s, i.e., to test $H_0$: $d'_1 = d'_2 = \ldots = d'_k$ versus $H_a$: at least $d'_i \neq d'_j$, $i, j = 1, 2, \ldots, k; i \neq j$. If the $k$ $d'$s are obtained from the same forced-choice method, the comparison of the $k$ $d'$s is equivalent to comparison of $k$ proportions of correct responses; the conventional statistical test for proportions can be used for this purpose. If the $k$ $d'$s are estimated from different discrimination methods or from the same but non-forced-choice method, e.g., the A−Not A method or same-different method, the test statistic $T$ in (11.3-28) can be used for the comparison (Marascuilo, 1966, 1970):

$$T = \sum_{i=1}^{k} \widehat{W}_i (d'_i - d'_0)^2, \tag{11.3-28}$$

where $\widehat{W}_i = \frac{1}{V(d'_i)}$, $i = 1, 2, \ldots, k$; $d'_0 = \frac{\sum_{i=1}^{k} \widehat{W}_i d'_i}{\sum_{i=1}^{k} \widehat{W}_i}$. The test statistic $T$ in (11.3-28) follows asymptotically a chi-square distribution with $k - 1$ DF.

**Example 11.3-8**

There are four $d'$s obtained from four different forced-choice methods. They are $d'_1 = 2.1$, $d'_2 = 1.27$, $d'_3 = 1.43$, and $d'_4 = 1.63$. The corresponding variances are $V(d'_1) = 0.155$, $V(d'_2) = 0.217$, $V(d'_3) = 0.079$, and $V(d'_4) = 0.129$. We want to know if the four $d'$s are significantly different. The $d'_0$ value in the statistic (11.3-28) is

$$d_0' = \frac{(2.1/0.155) + (1.27/0.217) + (1.43/0.079) + (1.63/0.129)}{(1/0.155) + (1/0.217) + (1/0.079) + (1/0.129)} = 1.59.$$

The value of the $T$ statistic in (11.3-28) is then

$$T = \frac{(2.1 - 1.59)^2}{0.155} + \frac{(1.27 - 1.59)^2}{0.217} + \frac{(1.43 - 1.59)^2}{0.079} + \frac{(1.63 - 1.59)^2}{0.129}$$

$$= 2.486.$$

The associated $P$-value for a chi-square distribution with $4 - 1 = 3$ DF is 0.478. Hence, we cannot conclude that the four $d$'s are significantly different. The S-PLUS code is as follows:

```
> dstest(dsdat[,1],dsdat[,2])
[1] 1.593 2.486 0.478
> dsdat
  [,1] [,2]
[1,] 2.10 0.155
[2,] 1.27 0.217
[3,] 1.43 0.079
[4,] 1.63 0.129
```

## 11.4   AREA UNDER ROC CURVE AND $R$-INDEX

### Introduction

$R$-index analysis is one of the important methodologies used widely in sensory and consumer research for testing and measuring product effects. This methodology was independently developed by Brown (1974) and strongly advocated by O'Mahony (O'Mahony 1979, 1992; O'Mahony et al., 1978). The $R$-index is an estimate of the area under an empirical receiver operating characteristic (ROC) curve in signal detection theory (Green and Swets, 1966). The validity and merits of the $R$-index are also due to the fact that the area under the ROC curve, when calculated by the trapezoidal rule, is closely related to the Mann-Whitney $U$ statistic, a version of the Wilcoxon statistic (Bamber, 1975; Hanley and McNeil, 1982).

$R$-index is one of the most powerful test statistics. Lehmann (1975, 81) and Bickel and Doksum (1977, 350–353), among others, investigated the relative efficiency of the Wilcoxon test to the $t$-test on the basis of rigorous demonstrations. The conclusions are exactly applicable to the comparison of the $R$-index test and $t$-test. If the underlying distribution is exactly normal, the R-index test is only slightly (about 5 percent) less powerful than the *t*-test. For distribution close to normal, there will typically be little difference in the power of the two tests. When nothing is known about the shape of the distribution, particularly when the normal distribution does not hold, the R-index test is frequently more powerful than the *t*-test. It is generally known that hedonic rating data are more likely to be skewed or bimodal than normal. When one uses the R-index test rather than the *t*-test for the rating data, the largest possible loss of efficiency is slight, while the largest possible gain of efficiency is unbounded.

R-index is a distribution-free and robust test statistic. It is based on an ordinal scale rather than an interval scale. The R-index produced from ratings data is in fact based on the ordinal information of the data. Since the R-index links easily with Thurstonian $d'$, it provides an alternative to the maximum-likelihood estimate of $d'$ from rating data (Dorfman and Alf, 1969).

R-index is a meaningful and intuitive measure of separation of two distributions. It is the estimate of the probability $P(X < Y)$, which is the probability of product $Y$ larger than product $X$ in preference, purchase intent, or sensory intensity. It conforms more closely to the researcher's questions. Both Thurstonian $d'$, which is a distance measure, and the R-index, which is an area measure, are justified as measures of sensory difference. The two measures can be transformed from each other under some assumptions.

## Estimating *R*-Index and Its Variance

Let $X_1, \ldots, X_m$ and $Y_1, \ldots, Y_n$ be two independent samples with sample sizes $m$ and $n$ from distributions $G$ and $H$. The Mann–Whitney $U$ statistic is

$$U = \sum_{i=1}^{m} \sum_{j=1}^{n} \varphi(X_i, Y_j), \qquad (11.4\text{-}1)$$

where $\varphi(X_i, Y_j) = 1$ if $X_i < Y_j$, $\varphi(X_i, Y_j) = 1/2$ if $X_i = Y_j$ and $\varphi(X_i, Y_j) = 0$ otherwise. When $G$ and $H$ are continuous, $P(X_i = Y_j) = 0$. The R-index and the Mann–Whitney $U$ statistic are connected by

$$R\text{-index} = U/mn. \qquad (11.4\text{-}2)$$

For the summarized rating frequency data in Table 11.23, the Mann–Whitney $U$ statistic can be calculated according to (9.2-18) in Chapter 9.

**Table 11.23** Frequency Counts of Ratings for Two Products

| | Rating Scale | | | | | |
|---|---|---|---|---|---|---|
| | $k$ | $k-1$ | $\dots$ | 2 | 1 | Total |
| Control (A) | $a_1$ | $a_2$ | $\dots$ | $a_{k-1}$ | $a_k$ | m |
| Treatment (B) | $b_1$ | $b_2$ | $\dots$ | $b_{k-1}$ | $b_k$ | n |

Note: $a_1$ denotes the frequency counts for rating k in a k-point rating scale, and $a_k$ denotes the frequency counts of 1 in the rating scale for sample A.

The variance of the $R$-index can be estimated by

$$\widehat{V}(R) = \frac{1}{mn}\left[R(1-R) + (n-1)\left(q_1 - R^2\right) + (m-1)\left(q_2 - R^2\right)\right], \quad (11.4\text{-}3)$$

where $q_1 = q_2 = \mathbf{F}(\mathbf{x}_0)$, $\mathbf{x}_0 = \left(d_s/\sqrt{2}, d_s/\sqrt{2}\right)'$, $d_s = \sqrt{2}\Phi^{-1}(R)$, $\Phi^{-1}(.)$ denotes an inverse of the cumulative distribution function of the standard normal distribution, and F(.) denotes a cumulative distribution function of a bivariate normal distribution with mean vector $\boldsymbol{\mu} = (\mu_1, \mu_2) = (0,0)$ and covariance matrix $\mathbf{V} = \begin{pmatrix} 1 & \frac{1}{2} \\ \frac{1}{2} & 1 \end{pmatrix}$; see Bickel and Doksum (1977, 349–351). Hanley and McNeil (1982) suggested using an approximation based on the exponential distribution to estimate $q_1$ and $q_2$ as $\widehat{q}_1 = R/(2-R)$ and $\widehat{q}_2 = 2R^2/(1+R)$.

Van Dantzig (1951), Birnbaum (1956), and Bamber (1975) gave an estimate of the maximum variance of $R = U/mn$, which is

$$\widehat{V}(R)_{\max} = R(1-R)/\min(m, n). \quad (11.4\text{-}4)$$

**Example 11.4-1**

A study was conducted to investigate whether male and female consumers have different evaluations for a food product. Consumers consisting of 200 males and 200 females were selected. The 5-point just-about-right (JAR) scale was used for some sensory attributes. The responses for amount of overall flavor are presented in Table 11.24 with 5 = Much too strong to 1 = Much too weak. The estimate of the $R$-index is 0.576 from (11.4-2) and (9.2-18).

$$
\begin{aligned}
R\text{-index} = [&84(55 + 48 + 20 + 9) + 70(48 + 20 + 9) + 28(20 + 9) \\
&+ 14(9) + 0.5(68 \times 84 + 55 \times 70 + 48 \times 28 \\
&+ 20 \times 14 + 9 \times 4)]/(200 \times 200) \\
= \ &0.5757.
\end{aligned}
$$

**Table 11.24** Rating Frequencies (Example 11.4-1)

|  | 5 | 4 | 3 | 2 | 1 | Total |
|---|---|---|---|---|---|---|
| Male | 68 | 55 | 48 | 20 | 9 | 200 |
| Female | 84 | 70 | 28 | 14 | 4 | 200 |

*Note: 5 = Much too strong; 4 = Somewhat too strong; 3 = Just about right; 2 = Somewhat too weak; 1 = Much too weak.*

For calculation of variance of the $R$-index according to (11.4-3), the key is to calculate $q_1$ and $q_2$. Since the $R$-index $= 0.5757$, $d_s = \sqrt{2}\,\Phi^{-1}(0.5757) = 0.27$, $\mathbf{x}_0 = \left(0.27/\sqrt{2}, 0.27/\sqrt{2}\right)'$. Using the S-PLUS built-in program *pmvnorm*, we get $q_1 = q_2 = \mathbf{F}(\mathbf{x}_0) = 0.4123$. The S-PLUS code is as follows:

```
> pmvnorm(c(0.27/sqrt(2), 0.27/sqrt(2)), c(0, 0), matrix
  (c(1, 0.5, 0.5, 1), 2, 2))
[1] 0.4123544
```

The variance of the $R$-index is

$$\widehat{V}(R) = \Big[0.5757(1 - 0.5757) + (200 - 1)(0.4123 - 0.5757^2)$$
$$+ (200 - 1)(0.4123 - 0.5757^2)\Big]/(200 \times 200)$$
$$= 0.00081.$$

The output of the corresponding S-PLUS code is as follows:

```
> riest(ridat)
R-index: 0.575675 ; Var(R): 0.000811
> ridat
5 4 3 2 1
Control 68 55 48 20 9
Treatment 84 70 28 14 4
```

Using $\widehat{q}_1 = 0.5757/(2 - 0.5757) = 0.4042$; $\widehat{q}_2 = 2 \times 0.5757^2/(1 + 0.5757) = 0.4207$, we also get $\widehat{V}(R) = 0.00081$ from (11.4-3). According to (11.4-4), the maximum variance of the $R$-index is

$$\widehat{V}(R)_{max} = 0.5757(1 - 0.5757)/200 = 0.00122.$$

## Difference Testing Using *R*-Index

In difference testing, the null hypothesis is $H_0$: $P(X < Y) = 1/2$, and the alternative hypothesis is $H_a$: $P(X < Y) \neq 1/2$ for two-sided tests and $H_a$: $P(X < Y) > 1/2$ or $P(X < Y) < 1/2$ for one-sided tests. Here, $P(X < Y)$ denotes the probability that an observation from $Y$

population is greater than an arbitrary observation from $X$ population. A two-sided test is used when there is no information about which treatment is better. The test statistic is

$$Z = \frac{R - 0.5}{\sqrt{V(R)_0}}. \tag{11.4-5}$$

which approximately follows the standard normal distribution. The variance of the $R$-index in the null hypothesis is

$$V(R)_0 = (m + n + 1)/12mn - \sum_{i=1}^{k} (t_i^3 - t_i)/12mn(m + n)(m + n - 1),$$

$$\tag{11.4-6}$$

where $t_i, i = 1, 2, \ldots, k$ is the number of ties for the $i$th category of rating with a $k$-point scale.

---

**Example 11.4-2**

For a difference test using the data in Table 11.24 from Example 11.4-1, we can calculate $\widehat{V}(R)_0$ from (11.4-6). Note that there are 152, 125, 76, 34, and 13 ties, respectively, for rating categories 5, 4, 3, 2, and 1. Therefore,

$$\widehat{V}(R)_0 = [401/12 - (152^3 - 152 + 125^3 - 125 + 76^3 - 76 + 34^3 - 4 + 13^3 - 13)/(12 \times 400 \times 399)]/(200 \times 200) = 0.00076.$$

The value of the $Z$ statistic is $(0.576\text{-}0.5)/\sqrt{0.00076} = 2.75$ with the associated $P$-value $= 0.006$ for the two-sided test and the associated $P$-value $= 0.003$ for the one-sided test. We can conclude at a reasonable significance level that female consumers are more likely than male consumers to feel that the amount of overall flavor of the product is strong. The S-PLUS code for this example is as follows

```
> rdtest(ridat)
R-index: 0.576 ; z: 2.749 ; p-value (2-sided): 0.006 ; p-
value (1-sided): 0.003
```

An alternative practice for difference testing using $R$-index is to use the nonconditional, i.e., non-null variance, $\widehat{V}(R)$, based on (11.4-3). The Wald-type test statistic is

$$Z = \frac{R - 0.5}{\sqrt{\widehat{V}(R)}}. \tag{11.4-7}$$

The critical values in Table A.33, which are the same as those from Bi and O'Mahony (2007), can be used for the tests. The critical values in the table are the $R$-index values satisfying the equation

$$\frac{R - 0.5}{\sqrt{\widehat{V}(R)}} = z_{1-\alpha} \quad \text{or} \quad \frac{R - 0.5}{\sqrt{\widehat{V}(R)}} = z_{1-\alpha/2}, \qquad (11.4\text{-}8)$$

where $z_{1-\alpha}$ and $z_{1-\alpha/2}$ denote the $100(1 - \alpha)$th or $100(1 - \alpha/2)$th percentile of the standard normal distribution. For any observed $R$-index value, if it is equal to or larger than the corresponding value in the table, the null hypothesis is rejected and the alternative hypothesis is accepted.

---

**Example 11.4-3**

For Example 11.4-1 and the data in Table 11.24, $N = m = n = 200$, $R = 0.576$, the critical value for a two-sided difference test with a significance level $\alpha = 0.05$ is $50 + 5.62 = 55.62\%$ or $0.5562$ from Table A.33. Since $0.576 > 0.556$, we can conclude that there is a significant difference between male and female consumers for evaluations of amount of overall flavor in the food product.

---

## Similarity Testing Using R-Index

The $R$-index can also be used for similarity testing. Let $\Delta$ define a similarity limit, which is an allowed difference between the $R$-index and 0.50, i.e., $\Delta = |R - 0.50|$ and let $D = R - 0.50$. In the two-sided similarity testing, the null and alternative hypotheses are

$$H_0: \ D \geq \Delta \quad \text{or} \quad D \leq -\Delta \quad \text{versus} \quad H_a: \ -\Delta < D < \Delta. \qquad (11.4\text{-}9)$$

The hypotheses can be decomposed into two sets of one-sided hypotheses tests (TOSTs):

$$H_{01}: \ D \geq \Delta \quad \text{versus} \quad H_{a1}: \ D < \Delta,$$
$$H_{02}: \ D \leq -\Delta \quad \text{versus} \quad H_{a2}: \ D > -\Delta.$$

The test statistics are

$$\frac{R - 0.50 - \Delta}{\sqrt{V(R)_u}} \leq z_\alpha, \qquad (11.4\text{-}10)$$

$$\frac{R - 0.50 + \Delta}{\sqrt{V(R)_l}} \geq z_{1-\alpha}, \tag{11.4-11}$$

where $V(R)_l$ and $V(R)_u$ denote the variances of the $R$-index at $0.50 - \Delta$ and $0.50 + \Delta$, which can be calculated from (11.4-3). In this case, $V(R)_u = V(R)_l$. The critical values for a given $\Delta$ and $\alpha$ can be calculated from

$$R_u(\Delta, \alpha) = 0.50 + \Delta + z_\alpha \sqrt{V(R)_u}. \tag{11.4-12}$$

$$R_l(\Delta, \alpha) = 0.50 - \Delta + z_{1-\alpha} \sqrt{V(R)_l}. \tag{11.4-13}$$

The null hypothesis, $H_0$, will be rejected, and the alternative hypothesis of "similarity", $H_a$, will be accepted at an $\alpha$ level, if $H_{01}$ and $H_{02}$ are rejected simultaneously at an $\alpha$ level. In this situation the observed $R$-index should be in the rejection region of $R_l(\Delta, \alpha) \leq R \leq R_u(\Delta, \alpha)$. Table A.34 gives the critical values for similarity tests using the $R$-index with $m = n = 30$ to 200; $\Delta = 0.05$ and 0.1; $\alpha = 0.05$ and 0.1. One should note that the critical values can also be used for one-sided similarity testing. No adjustment for $\alpha$ level is needed for one-sided and two-sided similarity tests. If $R_u(\Delta, \alpha) \leq R_l(\Delta, \alpha)$, it means that "similarity" cannot be concluded in the situation even if the observed $R$-index is 0.50.

| **Example 11.4-4** | For the data in Table 11.24 with $m = n = 200$, if the allowed difference in terms of $R$-index is $\Delta = 0.10$ and $\alpha = 0.05$, the critical values for a two-sided similarity test are $cl = 0.446$ and $cu = 0.554$ from Table A.34. Because 0.576 is not in the range of [0.446, 0.554], no similarity conclusion can be drawn. The same conclusion can be obtained using a one-sided similarity test. |

## Linking *R*-Index with *d'*

Green (1964) proved a psychophysical area theorem. Green's theorem states that the area under the ROC curve in a yes-no or ratings method is equal to the proportion of correct responses in a 2-AFC task (Egan, 1975, 46–48 for a proof). Because $R$-index is an estimator of the area, the relationship in (11.4-14) or (11.4-15) approximately holds, according to the psychometric function for the 2-AFC. One should note that for a small number of categories (less than 5) in a rating scale, the true area under the ROC curve and $d'$ may be underestimated by $R$-index due to the trapezoidal rule (Hanley and McNeil, 1982). However, the differences of the areas estimated by using the parametrical maximum likelihood estimation technique and nonparametric R-index method generally are small (Centor and Schwartz, 1985).

$$R = \Phi\left(\frac{d'}{\sqrt{2}}\right), \qquad (11.4\text{-}14)$$

$$d' = \sqrt{2}\,\Phi^{-1}(R), \qquad (11.4\text{-}15)$$

where $\Phi(.)$ denotes the cumulative distribution function of the standard normal distribution and $\Phi^{-1}(.)$ denotes the quantile of the standard normal distribution. Table 11.27 gives $d'$s corresponding to R-index values.

The approximate variance of $d'$, which corresponds to the R-index, can be calculated from (11.4-16) based on Taylor-series expansion:

$$\widehat{V}(d') \approx f \times \widehat{V}(R), \qquad (11.4\text{-}16)$$

where $f = 2/\phi^2\left(d'/\sqrt{2}\right)$, $\phi$ denotes the standard normal density function, $\widehat{V}(R)$ can be calculated from (11.4-3), and $f$ is a coefficient, which can be found in Table 11.26.

---

For Example 11.4-1 and the data in Table 11.264, recall that $R = 0.58$ and $\widehat{V}(R) = 0.00081$. Using these values, we get $d' = 0.286$ from (11.4-15) or Table 11.25 and $f = 13.089$ from (11.4-16) or Table 11.26. Hence, $\widehat{V}(d') = 13.089 \times 0.00081 = 0.0106$. We should mention that using the maximum likelihood estimation technique, we can also estimate $d'$ and its variance. They are $d' = 0.3005$ and $\widehat{V}(d') = 0.0124$. The difference of the results produced by the two methods (parametric and nonparametric) is slight. The S-PLUS code for this example is as follows:

**Example 11.4-5**

| **Table 11.25** Linking R-index with $d'$ | | | | | | | | | |
|---|---|---|---|---|---|---|---|---|---|
| **R** | **0.00** | **0.01** | **0.02** | **0.03** | **0.04** | **0.05** | **0.06** | **0.07** | **0.08** | **0.09** |
| 0.5 | 0.000 | 0.035 | 0.071 | 0.106 | 0.142 | 0.178 | 0.214 | 0.249 | 0.286 | 0.322 |
| 0.6 | 0.358 | 0.395 | 0.432 | 0.469 | 0.507 | 0.545 | 0.583 | 0.622 | 0.661 | 0.701 |
| 0.7 | 0.742 | 0.783 | 0.824 | 0.867 | 0.910 | 0.954 | 0.999 | 1.045 | 1.092 | 1.140 |
| 0.8 | 1.190 | 1.242 | 1.295 | 1.349 | 1.406 | 1.466 | 1.528 | 1.593 | 1.662 | 1.735 |
| 0.9 | 1.812 | 1.896 | 1.987 | 2.087 | 2.199 | 2.326 | 2.476 | 2.660 | 2.904 | 3.290 |

**Table 11.26** Coefficient ($f$) Values for Estimation of Variance of $d'$ from the $R$-index

| R | 0.00 | 0.01 | 0.02 | 0.03 | 0.04 | 0.05 | 0.06 | 0.07 | 0.08 | 0.09 |
|---|---|---|---|---|---|---|---|---|---|---|
| 0.5 | 12.566 | 12.574 | 12.598 | 12.638 | 12.694 | 12.766 | 12.856 | 12.963 | 13.089 | 13.234 |
| 0.6 | 13.399 | 13.586 | 13.796 | 14.029 | 14.289 | 14.578 | 14.897 | 15.250 | 15.639 | 16.069 |
| 0.7 | 16.544 | 17.069 | 17.650 | 18.294 | 19.009 | 19.806 | 20.695 | 21.691 | 22.812 | 24.079 |
| 0.8 | 25.517 | 27.159 | 29.047 | 31.232 | 33.783 | 36.790 | 40.371 | 44.691 | 49.980 | 56.565 |
| 0.9 | 64.936 | 75.842 | 90.491 | 110.939 | 140.942 | 188.024 | 269.327 | 431.993 | 853.127 | 2815.566 |

```
> sqrt(2)*qnorm(0.58)
[1] 0.2855205
> 2/dnorm(0.2855/sqrt(2))^2
[1] 13.08909
x<-ratdel(c(9,20,48,55,68),c(4,14,28,70,84))
> x$par
[1] -1.7174814 -1.0526848 -0.3577325 0.4610797 0.3005212
> vcov.nlminb(x)
      [,1]        [,2]        [,3]        [,4]
[1,] 0.016894915 0.008118922 0.005162067 0.004179991
  0.005563217
[2,]  0.008118922  0.009879261  0.006023602  0.004706897
0.005881604
[3,]  0.005162067  0.006023602  0.007505028  0.005538557
0.006163327
[4,]  0.004179991  0.004706897  0.005538557  0.007712065
0.006596470
[5,] 0.005563217 0.005881604 0.006163327 0.006596470
  0.012380459
```

## Same-Different Area Theorem

Besides the classical Green's area theorem, there is another psycho-physical area theorem: the same-different area theorem (Macmillan et al., 1977; Noreen, 1981; Irwin et al., 1999). In this theorem, the area under the same-different ROC curve equals the proportion of correct decisions of an unbiased observer in a four-interval same-different experiment (i.e., the dual pair protocol). The area under the same-different ROC curve can be estimated from the degree of difference method. The degree of difference method is an extension of the same-different method when the $k$-point scale ($k > 2$) instead of a 2-point scale is used for responses. Two products (A and B) are involved. Panelists are presented with one of the possible sample pairs: A/A, B/B, A/B, and B/A. The panelists' task is to rate the degree of difference for a given sample pair on a scale, where 1 denotes identical and $k$ denotes extremely different.

Let $R_{sd}$ denote an estimate of the area under the ROC curve in the same-different or the degree of difference method. The estimate and statistical tests for $R_{sd}$ are the same as those for the R-index. It is analogous to the fact that the same estimate and significance test can be used for the 3-AFC and triangle data, 2-AFC and duo-trio data, and for the A–Not A and same-different data.

However, the relationship between $R_{sd}$ and $d'$ in the same-different area theorem is dramatically different from the relationship between R-index and $d'$ in Green's area theorem. $R_{sd}$ is not a direct measure of sensory difference of treatments. At most, it is a measure of the difference between the concordant pair and discordant pair. In order to measure effect-size of treatments, we should obtain the corresponding R-index or $d'$ from $R_{sd}$ when the same-different or degree of difference method is used.

### Linking $R_{sd}$ with $d'$

The relationship between $R_{sd}$ and $d'$ is as shown in (11.4-17), according to the same-different area theorem. The relationship between the variance of $R_{sd}$ and the variance of the corresponding $d'$ is as shown in (11.4-18), based on Taylor-series expansion.

$$R_{sd} = [\Phi(d'/2)]^2 + [\Phi(-d'/2)]^2, \qquad (11.4\text{-}17)$$

$$\widehat{V}(d') \approx f_0 \times \widehat{V}(R_{sd}), \qquad (11.4\text{-}18)$$

where

$$f_0 = \frac{1}{\phi^2(d'/2)[2\Phi(d'/2) - 1]^2}.$$

### Linking $R_{sd}$ with the R-Index

The area under the same-different ROC curve estimated by $R_{sd}$ can be linked with the area under the yes-no ROC curve estimated by the R-index in (11.4-19). The variance of the R-index can be estimated from (11.4-20):

$$R = \Phi\left(\sqrt{2}z\left(0.5 + 0.5\sqrt{2R_{sd} - 1}\right)\right), \qquad (11.4\text{-}19)$$

$$\widehat{V}(R) \approx f_1 \times \widehat{V}(R_{sd}), \tag{11.4-20}$$

where

$$f_1 = \frac{\phi^2\left(d'/\sqrt{2}\right)}{2\phi^2(d'/2)[2\Phi(d'/2) - 1]^2}.$$

**Example 11.4-6**

One hundred panelists participated in a degree of difference test for comparison of two products A and B. Half of the panelists (50) evaluated the concordant product pairs A/A or B/B, and the other half of the panelists evaluated the discordant product pairs A/B or B/A. A 5-point scale was used with 1 = identical to 5 = extremely different. The frequencies of the responses are listed in Table 11.27. We can obtain $R_{sd}$ and its variance $\widehat{V}(R_{sd})$ according to (11.4-2) and (11.4-3) as follows:

$$\begin{aligned}
R_{sd} &= [16(9 + 12 + 15 + 11) + 10(12 + 15 + 11) + 6(15 + 11) + 10(11) \\
&\quad + 0.5(3 \times 16 + 9 \times 10 + 12 \times 6 + 15 \times 10 + 11 \times 8)]/(50 \times 50) \\
&= 0.6488.
\end{aligned}$$

Since $R_{sd}$-index = 0.6488, $d_s = \sqrt{2}\,\Phi^{-1}(0.6488) = 0.5403$, $\mathbf{x}_0 = (Z_1 = 0.5403/\sqrt{2}, Z_2 = 0.5403/\sqrt{2})'$, and $q_1 = q_2 = F(\mathbf{x}_0) = 0.4949$, which we can obtain from the S-PLUS built-in program *pmvnorm* as follows:

```
> sqrt(2)*qnorm(0.6488)
[1] 0.5403466
> pmvnorm(c(0.54/sqrt(2), 0.54/sqrt(2)), c(0, 0),
    matrix(c(1, 0.5, 0.5, 1), 2, 2))
[1] 0.4949116
```

**Table 11.27** Data of a Degree of Difference Test (Example 11.4-6)

|                    | 5  | 4  | 3  | 2  | 1  |
|--------------------|----|----|----|----|----|
| Concordant Pairs   | 3  | 9  | 12 | 15 | 11 |
| Discordant Pairs   | 16 | 10 | 6  | 10 | 8  |

Continuing,

$$\widehat{V}(R_{sd}) = (0.6488 \times (1 - 0.6488) + 49 \times (0.4949 - 0.64882)$$
$$+ 49 \times (0.4949 - 0.64882))/(50 \times 50)$$
$$= 0.00299.$$

The output of the S-PLUS code, *riest*, is as follows:

```
> riest(sdrindexdat)
[1] 0.648800 0.002995
> sdrindexdat
5 4 3 2 1
concordent pais 3 9 12 15 11
Discordant pais 16 10 6 10 8
```

We can conduct a difference test based on $R_{sd}$ by using Table A.33. For $N = 50$, $\alpha = 0.05$, the critical value for a two-sided test is $50 + 11.02 = 61.02\%$ or 0.6102. Because $R_{sd} = 0.6488 > 0.6102$, we can conclude that the difference between products A and B is significant.

---

**Example 11.4-7**

Using the results $R_{sd} = 0.6488$ and $\widehat{V}(R_{sd}) = 0.00299$ in Example 11.4-6, we can obtain the corresponding $d' = 1.4972$, $\widehat{V}(d') = 0.11033$; R-index $= 0.8551$, and $\widehat{V}(R) = 0.00286$ from (11.4-17–20) or by using the S-PLUS code *rsddr*. We can claim that the difference between products A and B is 1.4972 in terms of the distance measure $d'$ with a precision of 0.11033 or 0.855 in terms of the area measure R-index with a precision of 0.00286.

```
> rsddr(0.649,0.00299)
[1] 1.497173776 0.855123069 0.110409323 0.002864534
```

---

## EXERCISES

**11.1.** The following table data came from the Biostatistics department files at the Armour Research Center. Four brands of food products were compared by 96 respondents. Each respondent scored on a 7-point scale for the Scheffé model of paired comparisons.

Frequencies of Scores on a Seven-Point Scale

| Brand Pair | Score 3 | 2 | 1 | 0 | −1 | −2 | −3 |
|---|---|---|---|---|---|---|---|
| (1, 2) | 6 | 1 | — | — | 1 | — | — |
| (2, 1) | — | 2 | — | — | 1 | 1 | 4 |
| (1, 3) | 8 | — | — | — | — | — | — |
| (3, 1) | — | 3 | — | — | — | 3 | 2 |
| (1, 4) | 2 | — | 3 | — | — | 2 | 1 |
| (4, 1) | 4 | 2 | — | — | — | — | 2 |
| (2, 3) | 2 | 3 | 1 | — | 1 | — | 1 |
| (3, 2) | 2 | 1 | — | — | 2 | — | 3 |
| (2, 4) | 1 | — | 1 | — | 1 | 1 | 4 |
| (4, 2) | 4 | 1 | — | — | — | 2 | 1 |
| (3, 4) | 1 | — | — | — | 1 | 1 | 5 |
| (4, 3) | 5 | 1 | — | — | — | — | 2 |

1. Estimate the average preference $\pi$ for brand 1 over brand 2.
2. Estimate the main-effect rating parameters $\alpha_i$, $i = 1, 2, 3, 4$.
3. Prepare an analysis of variance table, such as Table 11.7, and test the hypothesis $H_0$: $\alpha_1 = \alpha_2 = \alpha_3 = \alpha_4$.
4. Test for the presence of order effects.
5. Conduct a multiple-comparison test of the preference ratings.

**11.2.** Assume that we have $t = 15$ brands to be compared.

1. Construct an incomplete PC design by cyclic enumeration using brand pairs $(1, 2)$, $(1, 3)$, and $(1, 4)$ in initial blocks.
2. For the incomplete PC design found in part 1, write the design matrix for the least squares solution.

**11.3.** A computer-simulated frequency matrix using the model $Z_{ij} = (S_i - S_j) + e_{ij}$ is shown in the following table. Values of $S_1 = 0.15$, $S_2 = 0.10$, $S_3 = 0.0$, $S_4 = -0.10$, and $S_5 = -0.15$ served as the input parameters in the simulation process.

| Pair (i, j) | Frequency i | Frequency j | Ties (i = j) |
|---|---|---|---|
| (1, 2) | 23 | 27 | 0 |
| (1, 3) | 31 | 19 | 0 |
| (1, 4) | 31 | 19 | 0 |
| (1, 5) | 30 | 20 | 0 |

*(continued)*

| Pair (*i, j*) | Frequency *i* | Frequency *j* | Ties (*i = j*) |
|---|---|---|---|
| (2, 3) | 20 | 29 | 1 |
| (2, 4) | 31 | 18 | 1 |
| (2, 5) | 27 | 23 | 0 |
| (3, 4) | 29 | 20 | 1 |
| (3, 5) | 25 | 25 | 0 |
| (4, 5) | 23 | 27 | 0 |

1. Set up the proportion matrix, that is, the proportion of times *i* was rated over *j*, and obtain the corresponding normal deviates.
2. Estimate the scale values *Si* and its standard error SE(*Si*) using the Thurstone–Mosteller model.
3. Set up the design matrix X for the least squares formulation and find $X' X$ and $(X' X)^{-1}$.
4. Estimate the scale values using the least squares equation $S = (X' X)^{-1} X'Z$, where Z is a column vector of the normal deviates corresponding to the proportion matrix.

**11.4.** Refer to Example 11.2-3.

1. For each group, calculate initial estimates $p_i^0$ and adjustment constants $k_i$ using (11.2-4) and (11.2-5), respectively. Then for each group, find $p_i = p_i^0 - k_i$, $i = 1, 2, 3$. Compare the $p_i$ obtained with the maximum likelihood estimates in Table 11.12 and note that the two are essentially the same.
2. Check the $p_i$ and B values in Table 11.12 for each group.

**11.5.** Find the corresponding $d'$ and variance of $d'$ for each of the forced-choice methods in the following table. Discuss the results.

| | $p_c$ | $n$ | $d'$ | $V(d')$ |
|---|---|---|---|---|
| 2-AFC | 0.78 | 100 | | |
| 3-AFC | 0.83 | 100 | | |
| Duo-Trio | 0.65 | 100 | | |
| Triangle | 0.70 | 100 | | |

**11.6.** Find the corresponding $d'$ and variance of $d'$ for the A–Not A method in the following table. Discuss the results.

| $p_A$ | $p_N$ | $n_A$ | $n_N$ | $d'$ | $V(d')$ |
|-------|-------|-------|-------|------|---------|
| 0.6   | 0.1   | 100   | 100   |      |         |
| 0.75  | 0.2   | 100   | 100   |      |         |
| 0.85  | 0.3   | 100   | 200   |      |         |

**11.7.** Find the corresponding $d'$ and variance of $d'$ for the same-different method in the following table. Discuss the results.

| $p_{ss}$ | $p_{sd}$ | $n_s$ | $n_d$ | $d'$ | $V(d')$ |
|----------|----------|-------|-------|------|---------|
| 0.4      | 0.1      | 100   | 100   |      |         |
| 0.6      | 0.2      | 100   | 100   |      |         |
| 0.8      | 0.3      | 100   | 200   |      |         |

**11.8.** Purchase intent of consumers for a product was investigated. There are 150 consumers in each of two cities. The frequency counts of the responses in a 5-point scale are given in the following table. Calculate $d'$ and variance of $d'$ for the ratings data under the assumption of normal distribution with equal variance. The $d'$ can be used to measure the difference of purchase intent for the consumers in the two cities.

|         | 5  | 4  | 3  | 2  | 1  | Total |
|---------|----|----|----|----|----|-------|
| City A  | 52 | 36 | 29 | 19 | 14 | 150   |
| City B  | 83 | 41 | 13 | 7  | 6  | 150   |

Note: 5 = Definitely would buy to 1 = Definitely would not buy.

**11.9.** For the data in Exercise 11.8, estimate the $R$-index and its variance.

**11.10.** Using Table A.33, test whether the $R$-index estimated in Exercise 11.9 is significantly larger than 0.50 for $\alpha = 0.05$ and a two-sided test.

**11.11.** Using Table A.34, conduct a two-sided similarity test for the $R$-index estimated in Exercise 11.9 with $\alpha = 0.05$ and a similarity limit $\Delta = 0.1$.

**11.12.** The data of a test using the degree of difference method are shown in the following table. Discuss the results.

|  | 9 | 8 | 7 | 6 | 5 | 4 | 3 | 2 | 1 | 0 | Total |
|---|---|---|---|---|---|---|---|---|---|---|---|
| Concordant Pairs | 2 | 8 | 7 | 4 | 9 | 8 | 6 | 10 | 18 | 28 | 100 |
| Discordant Pairs | 3 | 17 | 16 | 8 | 4 | 18 | 6 | 14 | 9 | 5 | 100 |

Note: "0" denotes " Identical" and "9" denotes "Extremely different."

    **1.** Estimate $R_{sd}$ and its variance from the data.
    **2.** Estimate $d'$ and its variance from $R_{sd}$ and its variance.
    **3.** Estimate the $R$-index and its variance from $R_{sd}$ and its variance.

# Descriptive Analysis and Perceptual Mapping

In this chapter we discuss statistical evaluation of descriptive analysis data and how these data relate to consumer test information. Descriptive analysis performed by a trained panel and data from consumer tests have been shown to be useful in research and development (R&D), marketing research, and management for making decisions.

## 12.1 DESCRIPTIVE ANALYSIS

*Descriptive analysis* (DA) has a long history, and its application has spread to various consumer products from food, where it started, to the textile and cosmetic industries. Descriptive analysis defines what is in the product in terms of sensory impressions using the basic human senses of taste, smell, hearing, touch, and sight. The senses that are perceived are quantified and related to product acceptance and preference, which results in a large data matrix where multivariate procedures come into play. Relevant multivariate procedures are discussed in Section 12.4. Descriptive analysis is discussed in detail by Stone and Sidel (2004), Lawless and Heymann (1998), and Meilgaard et al. (1999).

The sensory score for descriptive attributes is quantified by a statistical function ($f$) of the explanatory attributes

$$\text{Score} = f(\text{appearance}, \text{texture}, \text{flavor}) + \text{error} \qquad (12.1\text{-}1)$$

where the trained panel evaluates the explanatory attributes in the following order: appearance, texture, and flavor. Depending on the types of products, the number of sensory characteristics has ranged from 20 to 50, in which most of them are in texture and flavor. A successful descriptive analysis depends on four factors:

1. *Training and experience of the judges.* Training is product dependent because the sensory attributes vary among products; i.e., attributes

**641**

for lotion products are different from those of wines. Some products require longer training than others. Experienced judges, by virtue of product exposure and product usage, should not be considered trained judges because they were not taught in scaling procedures, attribute definition, and other aspects of product-related training. The ideal situation is the existence of experienced and trained judges in an organization.

2. *The sensory execution.* Steps in sensory execution include the choice of reference standards, choice of test design, and conduct of the test, among others.

3. *The panel leader.* The panel leader or program administrator has a critical role in the establishment and maintenance of a descriptive panel, particularly in maintaining motivation of panel members.

4. *Long-term commitment by company management.* Management commitment is a prime mover for a successful sensory program in both academia and industry. Development and maintenance of a descriptive analysis program require time and a special physical facility that needs capital investment.

Items 1 and 2 are critical because they also contribute to a large degree on the product variability in the data.

## 12.2 **CONSUMER TESTS**

Measurement of consumer responses on products is a necessity in product development. Because of the cost in conducting consumer tests, various forms of gathering consumer information have been used in practice and presented in several books (Stone and Sidel, 2004; Lawless and Heymann, 1998; Resurreccion, 1998; Moskowitz et al., 2003). The book by Sudman and Wansink (2002) contains detailed coverage of consumer panels.

### In-House Consumer Test

In the in-house consumer tests, consumers come to the research and development laboratory to evaluate products. These consumers come from areas near the location of the company's technical center but may not be a good representation of the consumer population. The main purpose of in-house tests is to obtain preliminary hedonic

and attribute intensity information on product formulations and formulations in a product optimization setting, i.e., studying the effects
of various combinations of ingredients following a design of experiment. These in-house consumer tests are limited to the evaluation
of products that do not require use of the products at home, such
as personal care products like bar soaps, shampoos, and cosmetics.
Thus, in-house tests are limited to the evaluation of foods, beverages,
fabrics, and other similar products in which the consumer responses
are for taste, flavor, appearance, and hand-feel.

The statistical model to describe an observation for in-house consumer tests is a fixed-effects model given by

$$X_{ij} = \mu + T_i + P_j + \varepsilon_{ij} \qquad (12.1\text{-}2)$$

where

$X_{ij}$ = observed score for the $i$th product given by the $j$th panelist for a
particular sensory attribute;

$\mu$ = grand mean;

$T_i$ = the effect of the $i$th product or treatment, $i = 1, 2 \dots, t$ products;

$P_j$ = the effect of the $j$th panelist, $j = 1, 2, \dots, p$ panelists;

$\varepsilon_{ij}$ = random errors.

See Chapter 3 for statistical properties of fixed-effects model. Unless
consumer tests are replicated, product and panelist interaction is
estimable.

## Home-Use Consumer Test

To solve the problem of in-house tests, such as testing shampoos and
lotions, researchers employ home-use tests. For home-use tests, the
panelists pick up the products either in R&D facilities and/or in a cen-
tralized location such as in a shopping center. It is suggested that the
incomplete block design discussed in Chapter 5 (Example 5.1-1)
should be used where the respondents evaluate only two samples.
Such design can minimize confusion, mistakes, and other logistic
problems that can occur when products are used at home. Moreover,
home-use tests provide normal home conditions in product usage;
thus, they are more realistic. Uncontrolled factors during product
use are included in the evaluation process. The results of home-use
tests can be considered a result from one consumer location. The statistical model is the same as that given by (12.1-2).

## Central Location Consumer Test

When product prototypes have been developed, they should be tested together with a target brand within their product category using the central location test (CLT). It is critical that two to three locations should be used to provide the desired demographics in obtaining reliable perception and acceptance of the tested products. In addition to sensory questions relating to appearance, texture, and flavor, marketing questions are asked, such as price, purchase intent, and demographic information, i.e., gender, education, household income, frequency of product usage. The preparations of the products and product exposure time, among others, are factors controlled in CLT. These factors should be considered in interpreting the CLT results. Several books have addressed these issues and should be consulted (see Stone and Sidel, 1993; Lawless and Heymann, 1998; Resurreccion, 1998), including published articles on these issues (Giffin et al., 1990; King et al., 2004; King et al., 2007). The CLT statistical model is that given by (12.1-2).

Useful information obtained in CLT is consumer segmentation effect, an effect due to demographics. An example of this effect is given here using data from two cities comparing six products. Consumer responses for all products showed lower overall liking mean scores in Orlando than in Philadelphia; products A, C, and E were significantly more well liked ($p < 0.05$).

| CLT City* | A | B | C | D | E | F |
|---|---|---|---|---|---|---|
| Orlando | 5.73a | 5.86 | 6.04a | 6.10 | 4.98a | 5.38 |
| Philadelphia | 6.15b | 6.11 | 6.47b | 6.30 | 5.68b | 5.69 |
| Difference | −0.42 | −0.25 | −0.43 | −0.20 | −0.70 | −0.31 |

*Courtesy Rich Products Corporation.*

## 12.3 HEDONIC AND INTENSITY RATING SCALES

The hedonic scale was thoroughly discussed in Chapter 2, and after more than 50 years (Peryam and Girardot, 1952), it is still the rating scale used to gather like/dislike and acceptance information of products and other perceived stimuli from consumers. In parallel with the development of the hedonic scale is the Flavor Profile (Cairncross and Sjostrom, 1950), which describes the sensory properties of a product in terms of attribute intensity that leads to the development of a sensory technique known as descriptive analysis, as described in Section 12.1. To maximize the sensory information obtained in consumer tests and descriptive analysis, researchers use perceptual mapping, which became an important technique in data analysis. Sensory mapping is discussed in Section 12.4.

## Just-About-Right Rating Scale

The intensity scale was discussed in Chapter 2. A modification of this scale anchors the middle category of the scale by "just about right" (JAR), as shown here using color as an attribute.

| Too light | | Just about right | | Too dark |
|---|---|---|---|---|
| 1 | 2 | 3 | 4 | 5 |

To use this type of scale, we must know the meaning of "just about right" as used by naïve consumers for product evaluation; this knowledge is immensely critical. The meaning of JAR in this scale was first reported by Gacula et al. (2007) using consumer and in-house respondents to refer to "prefer product," "very good." "I like the product," "like it very much," "okay," "best for the situation," among others. Thus, we can assume the middle category of 3 as the target point in the scale as perceived by the consumer.

The JAR rating scale is an intensity discrimination task with only category 3 defined as both intensity and acceptability. As an example for evaluating the level of moistness in a product, the rating scale is given in Table 12.1.

| **Table 12.1** Just-About-Right Rating Scale ||
|---|---|
| **Anchor Description** | **Scale Category** |
| Much too moist | 1 |
| Little too moist | 2 |
| Just about right | 3 |
| Little too dry | 4 |
| Much too dry | 5 |

Since the ideal category is at the middle of the scale, the analysis of variance for evaluating the hedonic and intensity data as discussed in Chapter 3 can no longer be applied. A common method for analyzing the JAR ratings is to use only the %JAR (% of category 3) in the data. We present a signal-to-noise ratio method for the analysis of the just-about-right intensity score that uses all the data.

## Signal-to-Noise Ratio

Taguchi (1986) introduced a concept in achieving robustness of products and processes that uses a target value instead of process specifications in quality engineering. Taguchi contended that it is difficult to generate a real measure of process variability because it changes markedly due to uncontrolled environmental factors. But one can estimate this variability as a deviation from the target value, the resulting statistic being called signal-to-noise ratio (SNR). In this ratio, the estimate of treatment effect or product effect is maximized, and the estimate of variability (noise effect) is minimized.

The SNR statistic can be applied nicely to the just-about-right rating scale where the target value is the middle category of the scale, with a value of 3 on a 5-point scale. For a data set with $n$ panelists, $i = 1, 2, \ldots, n$, the average SNR for a particular treatment or product is computed by the formula

$$\text{SNR} = \sum_{i=1}^{n} \left[ -10^* \log((X_i - 3)^2 + k) \right] / n \qquad (12.3\text{-}1)$$

In the formula, $X_i$ is the response of the $i$th panelist on the just-about-right rating scale. The constant 3 in the formula is the just-about-right scale category value, and the constant $k$ is used to avoid taking the logarithm of zero, which is undefined. It can range from 0.1 to 0.5.

Table 12.2 shows the basic calculations needed to convert the just-about-right rating scale to the SNR statistic.

The plot of the intensity scale and SNR forms an inverted letter $V$, indicating that a desirable SNR should be large; this suggests a robust product or production process (Figure 12.1). When the responses are close to the target value of category 3 on the 5-point scale, SNR increases as the squared deviations from the target value of 3 decrease, lowering the variance of the deviation estimate. The SNR calculations can be easily implemented using SAS® (SAS Institute Inc., 1999).

**Table 12.2** Calculations of SNR for Just-About-Right Rating Scale

| Scale Category, X | $(X - 3)^2 + k$ | Signal-to-Noise Ratio |
|---|---|---|
| 1: much too strong | 4.0 + 0.25 | −6.284 |
| 2: too strong | 1.0 + 0.25 | −0.969 |
| 3: just about right | 0.0 + 0.25 | 6.021 |
| 4: too weak | 1.0 + 0.25 | −0.969 |
| 5: much too weak | 4.0 + 0.25 | −6.284 |
| Calculations: | | |
| $(1–3)^2 + 0.25 = 4.25$ | $−10*\log(4.25)/1 = −6.284$ | |
| $(2–3)^2 + 0.25 = 1.25$ | $−10*\log(1.25)/1 = −0.969$ | |
| $(3–3)^2 + 0.25 = 0.25$ | $−10*\log(0.25)/1 = 6.021$ | |
| $(4–3)^2 + 0.25 = 1.25$ | $−10*\log(1.25)/1 = −0.969$ | |
| $(5–3)^2 + 0.25 = 4.25$ | $−10*\log(4.25)/1 = −6.284$ | |

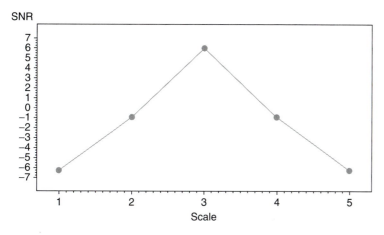

SNR

Scale

■**FIGURE 12.1** The relationship between signal-to-noise ratio and the 5-point JAR rating scale with scale category of 3 anchored by just about right.

The plot in Figure 12.1 is similar to the frequency distribution of the JAR 5-point scale; the highest point corresponds to the most frequent category on the scale. For the SNR, which is a deviation from the center point of the scale, the highest point is always, by definition, category 3 on the scale.

**Example 12.3-1**

Table 12.3 shows a typical format for consumer test data typed in an Excel spreadsheet. Using copy and paste, we can insert the data into a SAS program. This example illustrates the analysis of just-about-right ratings for "inside color" and relate it to overall liking. The calculations following Table 12.3 show conversion of the intensity score $X_{ij}$ used in the analysis of variance. Using the model in Section 3.1 of Chapter 3, we use a one-way analysis of variance to analyze the SNR data:

$$X_{ij} = \mu + T_i + \varepsilon_{ij} \tag{12.3-2}$$

For a larger data set, the panelist effect should also be included in the model as a source of variation.

The analysis of variance is shown in Table 12.4, indicating significant difference among products ($F_{3,16} = 3.42$, $p = 0.0428$). Table 12.5 shows a portion of SAS output for the multiple comparisons of the four products; calculations in this table for Duncan's Multiple Range Test are illustrated in Chapter 3. The separation of the products following the multiple range test at the 5% level is as follows:

| Product ID | 128 | 435 | 617 | 340 |
|---|---|---|---|---|
| Means | 6.021a | 1.099ab | 0.764ab | −3.095b |

Products 128 and 340 are significantly different from each other at the 5% level. The other grouping is indicated by the letter attached to the mean, as shown in Table 12.5. This format of presenting the results is used in practice and in scientific publications.

The plot of the overall liking product means against the SNR is shown in Figure 12.2. As we emphasized earlier, a high value of SNR indicates that the product scored close to just about right. In this example all the panelists ($n = 5$) scored product 128 with a 3, resulting in an average SNR of 6.021 for this product; see also Figure 12.1. As a result, the %JAR is 100% as shown. Graphically, this result suggests that "inside color" intensity of the products is positively correlated with overall liking.

**Table 12.3** Raw Data

| Panelist | City | Product | Inside Color Intensity Score, 5-Point Scale | Overall Liking Score, 9-Point Scale |
|---|---|---|---|---|
| 1 | 1 | 617 | 3 | 7 |
| 1 | 1 | 435 | 1 | 5 |
| 1 | 1 | 128 | 3 | 7 |
| 1 | 1 | 340 | 2 | 4 |
| 2 | 1 | 128 | 3 | 6 |
| 2 | 1 | 617 | 4 | 4 |
| 2 | 1 | 340 | 2 | 3 |
| 2 | 1 | 435 | 3 | 6 |
| 3 | 1 | 340 | 2 | 8 |
| 3 | 1 | 128 | 3 | 7 |
| 3 | 1 | 435 | 3 | 5 |
| 3 | 1 | 617 | 3 | 5 |
| 4 | 1 | 435 | 3 | 7 |
| 4 | 1 | 340 | 1 | 7 |
| 4 | 1 | 617 | 4 | 8 |
| 4 | 1 | 128 | 3 | 9 |
| 5 | 1 | 617 | 1 | 3 |
| 5 | 1 | 435 | 1 | 4 |
| 5 | 1 | 128 | 3 | 6 |
| 5 | 1 | 340 | 1 | 6 |
| Calculation of SNR for intensity score: | | | | |
| Panelist 1: | $-10 * \log(3-3)^2 + 0.25 = 6.0206$ | | | |
| Panelist 2: | $-10 * \log(1-3)^2 + 0.25 = -6.2839$ | | | |
| . | . | . | | |
| . | . | . | | |
| Panelist 5: | $-10 * \log(1-3)^2 + 0.25 = -6.2839$ | | | |

**Table 12.4** Analysis of Variance SAS Output

The GLM Procedure

Dependent Variable: SNR

| Source | DF | Sum of Squares | Mean Square | F Value | Pr > F |
|---|---|---|---|---|---|
| Model | 3 | 209.4290063 | 69.8096688 | 3.42 | 0.0428 |
| Error | 16 | 326.5206273 | 20.4075392 | | |
| Corrected Total | 19 | 535.9496337 | | | |

| R-Square | Coeff Var | Root MSE | SNR Mean |
|---|---|---|---|
| 0.390762 | 377.3828 | 4.517470 | 1.197053 |

| Source | DF | Type I SS | Mean Square | F Value | Pr > F |
|---|---|---|---|---|---|
| Product | 3 | 209.4290063 | 69.8096688 | 3.42 | 0.0428 |

---

**Table 12.5** Multiple Comparison SAS Output

The GLM Procedure

Duncan's Multiple Range Test for SNR

NOTE: This test controls the Type I comparisonwise error rate, not the experimentwise error rate.

| Alpha | 0.05 |
|---|---|
| Error Degrees of Freedom | 16 |
| Error Mean Square | 20.40754 |

| Number of Means | 2 | 3 | 4 |
|---|---|---|---|
| Critical Range | 6.057 | 6.351 | 6.535 |

Means with the same letter are not significantly different.

| Duncan Grouping | Mean | N | Product |
|---|---|---|---|
| A | 6.021 | 5 | 128 |
| B A | 1.099 | 5 | 435 |
| B A | 0.764 | 5 | 617 |
| B | −3.095 | 5 | 340 |

Overall Liking

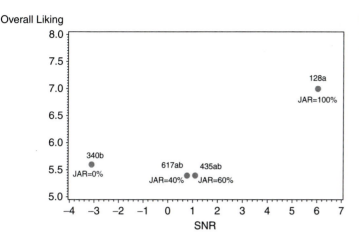

■ **FIGURE 12.2** Plot of products in the space of overall liking and inside color SNR. Products with one letter in common are not significantly different from each other (alpha or $p = 0.05$) for the SNR variable.

---

This example illustrates the effect of skewness and kurtosis on SNR estimates and how a product is penalized due to the shape of its frequency distribution. Consumer segmentation can cause negative skewness, indicating a long tail to the left, or positive skewness, indicating a long tail to the right. Kurtosis indicates the extent to which scores are concentrated in the center and the tails. A kurtosis with concentration of score in the center of the scale on a 5-point just-about-right rating scale is desirable (see Section 2.5 of Chapter 2).

In our example using color intensity, some panelists might judge the product as "too light" falling in the 1–2 categories and "too dark" falling in the 4–5 categories on the 5-point intensity scale with category 3 as "just about right." In some instances, panelists use all the scale categories in their judgments for a particular product to form flatness in the frequency distribution. This example illustrates consumer/panelist segmentation and how the SNR statistic takes into account instances stated previously in separating the products. It uses artificially generated data from 10 panelists. The calculations of SNR scores are illustrated in Example 12.3-1.

Figure 12.3 shows the frequency distribution for product A, portraying the desired distribution with high value of SNR since the observations are close to category 3. For product B, the panelists perceived it to have a light inside color (Figure 12.4), product C has a flat distribution (Figure 12.5), and finally panelists perceived product D to have darker

**Example 12.3-2**

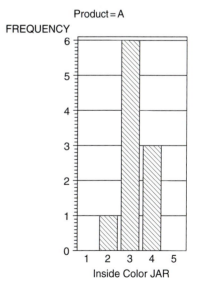

■ **FIGURE 12.3** Desired frequency distribution (unimodal kurtosis) for "just-about-right" inside color.

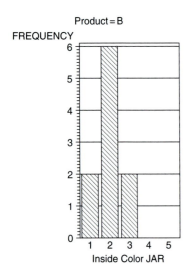

**■ FIGURE 12.4** A nearly positive skewness frequency distribution for "too light" inside color.

inside color (Figure 12.6). Products B, C, and D will be penalized due to the shape of their frequency distributions. The penalty is confirmed in Figure 12.7, where product A has the highest SNR compared to the other products with larger penalties. In this figure, the %JAR is used as the vertical axis.

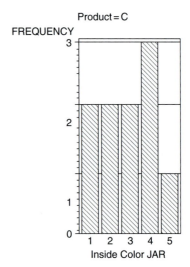

**■ FIGURE 12.5** Approaches a flatness frequency distribution where panelists disagree in their judgments for inside color.

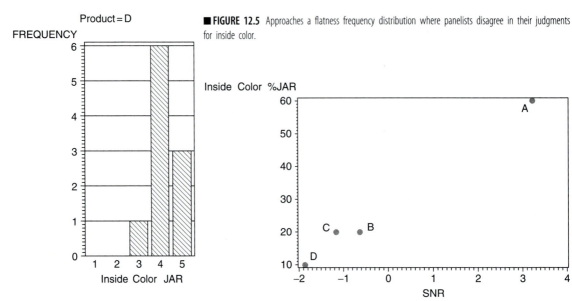

**■ FIGURE 12.6** Frequency distribution with negative skewness for "too dark" inside color.

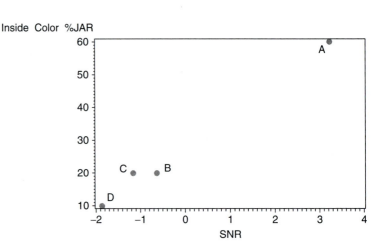

**■ FIGURE 12.7** Plot of average signal-to-noise ratio and %JAR for four products.

**Table 12.6** Modified Frequency Distribution to Illustrate Penalty of Products

| Product | 1 + 2 'Too Strong' | 3 'Just About Right' | 4 + 5 'Too Weak' | Total |
|---------|--------------------|----------------------|------------------|-------|
| A | 1 | 6 | 3 | 10 |
|   | 10% | 60% | 30% | 100% |
| B | 8 | 2 | 0 | 10 |
|   | 80% | 20% | 0% | 100% |
| C | 4 | 2 | 4 | 10 |
|   | 40% | 20% | 40% | 100% |
| D | 0 | 1 | 9 | 10 |
|   | 0% | 10% | 90% | 100% |

Using Figures 12.3 through 12.6, we can quantify the penalty by combining categories 1 and 2 and categories 4 and 5 of the rating scale, as shown in Table 12.6. Assuming that product A is the "control," the frequency distribution is converted into percentages as a measure of penalty expressed as %below and %above "just about right." For example, product B is 80% penalized as "too strong"; product C as 40% "too strong" and 40% "too weak"; finally, product D is 90% penalized for being "too weak." This type of information is useful to R&D and marketing research.

## Questionnaire Design

The focus in this section is on questionnaire design for obtaining consumer responses. Four important issues must be addressed: (1) form of the questionnaire, which is tied to the experimental design of the proposed study; (2) length of the questionnaire, which is dependent on the sensory properties of the products; (3) types of rating scales; and (4) the order of attribute evaluation dealing with hedonics and intensity ratings.

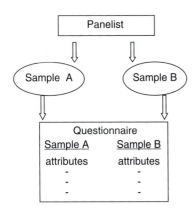

■ **FIGURE 12.8** Questionnaire for a paired comparison design.

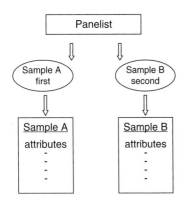

■ **FIGURE 12.9** Questionnaire for a sequential monadic design. Separate questionnaire sheet for each product.

If only two samples are to be evaluated, such as in paired comparison, the resulting questionnaire appears as shown in Figure 12.8. Here, the products are evaluated side by side in random order or in a balanced order such as AB, BA. The questionnaire sheet is one page that includes the sensory score for both samples. For more than two samples for evaluation, a sequential monadic questionnaire is used (Figure 12.9). Each sample is evaluated singly with the assumption that panelists can provide absolute judgment; the frame of reference during evaluation is internal to the panelists. The order of sample evaluation is randomized or balanced following the experimental designs discussed in Chapters 4 and 5.

There is no established location of hedonic and intensity attributes in the questionnaire. In practice, the location of overall liking and intensity attributes varies: some prefer overall liking as the first question followed by intensity attributes, whereas others place overall liking as the last question. To satisfy all types of products from foods to personal care that require extended use before being evaluated, such as bar soaps and cosmetic products (lip gloss, mascara, anti-aging products), the overall liking is generally the last question given at the end of the period of product evaluation; this statement is supported by a research finding by Gacula et al. (2008).

Considering overall liking as a sensory description that includes hedonic and intensity attributes, one can define overall liking as a function ($f$) of hedonic attributes and intensity attributes that are a diagnostic expression of what is in the product. In particular,

$$\text{Overall liking} = f(\text{appearance, texture, flavor}) + \text{error} \qquad (12.3\text{-}3)$$

where $f$ refers to the functional form of the relation between overall liking and diagnostic attributes—appearance, texture, and flavor. The diagnostic attributes can include both hedonic and intensity. Depending on the product, appearance can have two to three descriptors (inside and outside color, color uniformity, etc.), and most of the time, texture (firmness, cohesiveness, crispness, etc.) and flavor (sweetness, chocolate flavor, etc.) have the largest number of sensory descriptors. With the large number of attributes indicated in (12.3-4), the important role of multivariate methods in the analysis of sensory data is apparent.

**Example 12.3-3**

This example illustrates a questionnaire design in a consumer study with $N = 105$ consumers, its analysis, and interpretation of results. Table 12.7 shows a brief description of the 22 sensory descriptors. The hedonic scale is 9-point: $1 =$ dislike extremely, $5 =$ neither like nor dislike, $9 =$ like extremely. The JAR intensity scale is the same as described earlier in Section 12.3. In this example, JAR flavor of crumbs is used for illustration, and for hedonics, overall acceptance and preference ranking are used. The regression analysis method used in this example is described in Section 8.3 (Chapter 8). The regression model is

$$\text{Overall liking} = \beta_0 + \beta_1 X + \varepsilon \qquad (12.3\text{-}4)$$

where $X$ is a JAR intensity attribute expressed as a percentage, and the remaining regression parameters are defined in Section 8.3. A SAS program was written to obtain the regression analysis (Table 12.8). Under CARDS, a statement contains the product means for overall acceptance and preference ranking, and the JAR intensities are expressed as percent just about right based on $N = 105$ respondents.

Figure 12.10 shows a useful plot in consumer studies. This figure shows the regression of overall acceptance on %JAR for flavor of crumb, which contains useful information for sensory analysts and product developers. Products with low %JAR emerged as a result of flavor crumb intensity ratio—below and above JAR of 3 (the middle category of the scale). For example, low FlavorCrumbJAR of 45% (horizontal axis) is a result of the

**Table 12.7** Order of Evaluation of 22 Sensory Descriptors

| Attributes in Order of Evaluation | Sensory Descriptors |
| --- | --- |
| Appearance | 5 hedonics |
| Texture | 5 hedonics |
| Flavor | 5 hedonics |
| Overall liking/acceptance | 1 hedonics |
| Just about right | 5 JAR intensities |
| Preference ranking | Rank |

**Table 12.8** SAS Program for Producing Figures 12.8 and 12.9

```
*prog Product Z Reg.sas;
options nodate;
data a;
input Code Prod$ y1 y3-y5 y7;
label
y1='OverallAcceptance'
y3='FlavorCrumbJAR'
y4='MoistnessCrumbJAR'
y5='SoftnessCrumbJAR'
y7='PreferenceRank'
;
cards;
1 125 5.51 45 49 60 3.62
2 150 6.55 61 70 73 2.99
3 ALM 6.56 64 66 78 2.70
4 PIY 6.47 60 64 72 2.71
5 RTX 6.62 59 75 73 2.98
;
goptions reset=global gunit=pct
     ftext=swissb htitle=4 htext=4;

legend1 position=(bottom left)
     across=1 cborder=red offset=(0,0)
     shape=symbol(3,1) label=none
     value=(height=3);

symbol1 c=red v=dot h=4;
symbol2 c=red;
symbol3 c=black;
symbol4 c=black;
symbol5 c=black;
proc reg data=a;
     model y1=y3 / r cli clm;
     plot y1*y3 / mse
     caxis=black ctext=black cframe=white
     legend=legend1;
     run;
proc reg data=a;
     model y1=y4 / r cli clm;
     plot y1*y4 / mse
     caxis=black ctext=black cframe=white
     legend=legend1;
     run;
```

*(continued)*

**Table 12.8** SAS Program for Producing Figures 12.8 and 12.9—
*cont...*

```
proc reg data=a;
    model y1=y5 / r cli clm;
    plot y1*y5 / mse
    caxis=black ctext=black cframe=white
    legend=legend1;
    run;
proc reg data=a;
    model y1=y7 / r cli clm;
    plot y1*y7 / mse
    caxis=black ctext=black cframe=white
    legend=legend1;
    run;
```

$y1 = 2.8333 + 0.0607\ y3$

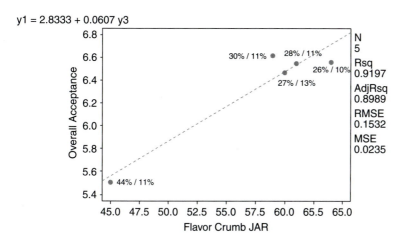

■ **FIGURE 12.10** Regression of overall acceptance (y1) on percent JAR (y3) for flavor of crumb. Percentage values in the plot refer to "too low/too high."

product being 44% "too low" in crumb flavor and 11% "too high," which is indicated in the plot as 44%/11%. Note that $45 + 44 + 11 = 100\%$. These % JAR values can be obtained from frequency distribution, as illustrated in Table 12.6.

Several statistics in Figure 12.10 are the results of various options, such as *r*, and *mse*, in the SAS program. They are part of the regression outputs in Table 12.9. A useful statistic is the R-square, which is 0.9197. That is, 92% of the total variability in overall acceptance is accounted for by flavor of

**Table 12.9** SAS Partial Output of the Regression of Overall Acceptance on %JAR

The REG Procedure
Model: MODEL1
Dependent Variable: y1 OverallAcceptance

| | |
|---|---|
| Number of Observations Read | 5 |
| Number of Observations Used | 5 |

Analysis of Variance

| Source | DF | Sum of Squares | Mean Square | F Value | Pr > F |
|---|---|---|---|---|---|
| Model | 1 | 0.80627 | 0.80627 | 34.35 | 0.0099 |
| Error | 3 | 0.07041 | 0.02347 | | |
| Corrected Total | 4 | 0.87668 | | | |

| | | | |
|---|---|---|---|
| Root MSE | 0.15320 | R-Square | 0.9197 |
| Dependent Mean | 6.34200 | Adj R-Sq | 0.8929 |
| Coeff Var | 2.41566 | | |

Parameter Estimates

| Variable | Label | Parameter DF | Standard Estimate | Error | t Value | Pr > \|t\| |
|---|---|---|---|---|---|---|
| Intercept | Intercept | 1 | 2.83332 | 0.60255 | 4.70 | 0.0182 |
| y3 | FlavorCrumbJAR | 1 | 0.06070 | 0.01036 | 5.86 | 0.0099 |

crumb. In the case of only one independent variable in the regression, the square root of R-square with appropriate sign is the correlation coefficient. In this example, $\sqrt{0.9197} = 0.96$ is the amount of positive correlation coefficient between overall acceptance and flavor of crumb. It is possible to reproduce Table 12.9 manually when regression analysis has only one independent variable by using formulas discussed in Section 8.3 (Chapter 8).

An important point in this example is the location of overall acceptance/liking in the questionnaire design. The JAR intensity attributes were located between overall acceptance and preference ranking, both of which are hedonic attributes. If there is no effect of the intermediate JAR question, the correlation should be high between overall acceptance rating and preference ranking. Figure 12.11 shows that this is true, which implies that JAR intensity questions did not affect the hedonic; that is, the

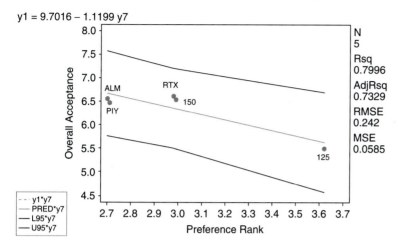

y1 = 9.7016 − 1.1199 y7

N 5
Rsq 0.7996
AdjRsq 0.7329
RMSE 0.242
MSE 0.0585

■ **FIGURE 12.11** Regression of overall acceptance (y1) on preference ranking (y7), where the highest preference is indicated by rank 1 and the least preference by rank 5.

JAR assessment was likely performed independently. Thus, the JAR intensity attributes were evaluated based solely on "correct" intensity, where the word *correct* is a semantic property of the meaning of JAR (Gacula et al., 2008). These observations support the hypothesis that overall acceptance/liking can be the last question in questionnaire design to satisfy all types of food and personal care products that consumers evaluate.

## 12.4 PERCEPTUAL MAPPING

With the advancing computer technology, software has been written to simplify the use of advanced statistical methods for which their theory has been known for over 50 years. *Perceptual mapping* is one such method for creating a pictorial representation of the relationships among sensory stimuli such as perceived responses of consumers and responses from a trained panel using multivariate statistical methods. As the name implies, multivariate analysis involves several variables with a large number of observations involved. Applied and theoretical books dealing with multivariate methods include Harman (1967), Affifi and Azen (1979), Seber (1984), Jobson (1992), Tacq (1997), and Dijksterhuis (1997).

In practice, the overall liking or overall acceptance from consumer tests is used to correlate product acceptance to descriptive analysis attributes as identified in the perceptual map. Through regression analysis, the predictors of overall liking/acceptance are obtained. This concept is illustrated in Example 12.4-1.

## Factor Analysis

For a product with $p$ attributes, a typical data matrix resulting from $n$ panelists consists of $n$ rows and $p$ columns, such as represented here:

$$\begin{bmatrix} X_{11} & X_{12} & . & . & X_{1p} \\ X_{21} & X_{22} & . & . & X_{2p} \\ . & & . & . & . \\ . & & . & . & . \\ X_{n1} & X_{n2} & . & . & X_{np} \end{bmatrix} \qquad (12.4\text{-}1)$$

We modify the matrix (12.4-1) by adding product and panelist columns as indicated in (12.4-2) and use it as a format for typing the data in Excel:

$$\begin{array}{ccc} \text{Product} & \text{Panelist} & \text{Attributes} \end{array}$$

$$\begin{bmatrix} 1 & 1 & X_{111} & X_{112} & . & . & X_{11p} \\ 1 & 2 & X_{121} & X_{122} & . & . & X_{12p} \\ . & . & . & & . & . & . \\ . & . & . & & . & . & . \\ 1 & n & X_{1n1} & X_{1n2} & . & . & X_{1np} \end{bmatrix} \qquad (12.4\text{-}2)$$

This matrix is repeated for the other products in the test and then "copied and pasted" into an SAS program. There will be $k$ products and $p$ sensory attributes that $n$ panelists evaluate for each product in the test, and $X_{1ij}$ is the corresponding sensory score for product 1. In practice, the number of attributes in consumer tests has ranged from 10 to 20; and for descriptive analysis, from 15 to 50. For this large number of attributes (variables), multivariate methods are applied to reduce the number of variables into components that put together the attributes into independent (*orthogonal*) factors/components for describing the sensory properties of the product. Because they are independent components, a reliable regression analysis can now be applied, such as regression of overall liking or acceptability on the components.

One multivariate method for reducing the size of matrix (12.4-1) is factor analysis. Factor analysis reduces the number of attributes into smaller numbers, called *common factors*. It is assumed that the observed variables, i.e., sensory attributes, are functions of the *unobserved common factors.* Now the sensory attributes are linear combinations of the common factors as indicated in the factor analysis model that follows:

$$\begin{bmatrix} Y_1 \\ Y_2 \\ . \\ . \\ Y_p \end{bmatrix} = \begin{bmatrix} B_{11} & B_{12} & . & . & B_{1m} \\ B_{21} & B_{22} & . & . & B_{2m} \\ . & . & . & . & . \\ . & . & . & . & . \\ B_{p1} & B_{p2} & . & . & B_{pm} \end{bmatrix} \begin{bmatrix} X_1 \\ X_2 \\ . \\ . \\ X_m \end{bmatrix} + \begin{bmatrix} E_1 \\ E_2 \\ . \\ . \\ E_p \end{bmatrix} \qquad (12.4\text{-}3)$$

In matrix notation the model is

$$Y = XB + E \qquad (12.4\text{-}4)$$

where

$X = p \times 1$ matrix of observed ratings for $p$ attributes;

$Y = m \times 1$ matrix of *common factors* extracted from the original number of sensory attributes; common factors are also known in various terms, such as *principal factor, factor scores, latent factor, unobservable construct;*

$B = p \times m$ matrix of coefficients or *factor pattern loadings* that reflect the importance of the attribute, given as *rotated factor pattern* in an SAS output;

$E = p \times 1$ matrix of error or noise component plus those unaccounted for by the common factors, this error is generally known as *unique variation.*

Given the vector $X$ of ratings, factor analysis seeks a small number of common factors, that is, a small $m$, so that $Y$ is a vector of a small number of common factors to adequately represent the dependency structure in the observed data. The analysis requires estimating the matrix $B$ of factor loadings and the specific variances to eventually decide the number $m$ of common factors to adequately account for the dependency structure in the observed data matrix. Then the estimated matrix of factor loadings, that is, $B$ of $p$ rows and $m$ columns (number of common factors), is used to estimate the vector $Y$ of $m$ common factors, called *factor score vector.* For vector $X_j$ of ratings from the $j$th panel, we estimate a vector $Y_j$ of factor scores, $j = 1, 2, \ldots, n$. Factor scores for Factor 1 and Factor 2 are obtained using (12.4-5) as follows:

$$\begin{aligned} \text{Factor1} \quad Y_1 &= B_{11}X_1 + B_{12}X_2 + \cdots + B_{1p}X_p \\ \text{Factor2} \quad Y_2 &= B_{21}X_1 + B_{22}X_2 + \cdots + B_{2p}X_p \end{aligned} \qquad (12.4\text{-}5)$$

where the values of the right side of the equations are SAS outputs and likewise for the remaining factors. The factor score corresponding to each attribute is obtained by substituting the "standardized sensory ratings" with mean = 0 and standard deviation = 1 into the resultant common factor equation given as "standardized scoring coefficient" in the SAS output. The calculation is illustrated in Example 12.4-1.

Since factor analysis uses either the covariance matrix or the correlation matrix among the attributes as input in the analysis, an important consideration in its application is the length of the scale. A rating scale having a reasonably large number of categories would be desirable in obtaining a realistic correlation measure between attributes. Again, as assumed in practice, the category intervals must be nearly equal. Thus, choice of the length of the scale and category intervals in factor analysis are important considerations for their usefulness in the analysis of sensory and consumer data. Similar requirements for the rating scales are also desirable in the *principal component analysis* (PCA) to be considered in the next section.

---

**Example 12.4-1**

This example uses only a portion of the descriptive data for analysis of three products with respect to 12 sensory attributes; each product is evaluated by seven trained panelists. In the analysis, the input is the score of each panelist. To provide an idea of the data matrix, Table 12.10 shows the matrix along with the SAS program. In this format, the sensory data in Excel can be copied and pasted into an SAS program after the CARDS statement. It is convenient for the sensory analyst and researcher to run the program. The INPUT statement shows the order of the data, starting with Product ID, Panelist ID, and the 25 attributes. In this example only the first 12 attributes and three products are included for illustration. The descriptive analysis scale is a 15-point rating scale (0–15) with 0.5 increments, a desirable scale length for factor analysis.

The first output to examine is the SCREE plot that indicates the number of factors to be included in the analysis. A popular criterion is the use of eigenvalue equal to 1 (Kaiser, 1960), as indicated in Figure 12.12, which corresponds to 3 and is given in the SAS program as NFACTOR=3. With this criterion, the total variance (eigenvalues) accounted for by the three factors is 77.1% partitioned as follows: Factor1 = 49.7%, Factor2 = 19.9%, and Factor3 = 7.6% (Table 12.11). The three factors comprise 77.2% of the variability in the data and may be sufficient to define the sensory characteristics among products.

> **Table 12.10** Descriptive Analysis Data and SAS Program That Uses
> the Principal Component Solution under METHOD = prin
>
> ```
> _
> *prog. DA.sas;
> options nodate;
> data a;
>
> input Product Panelist y1-y25;
>
> if product=982 then delete;
>
> if product=421 then delete;
>
> label
> y1='Cocoa'
> y2='ChocoComplex'
> y3='DarkRoast'
> y4='Buttery'
> y5='Caramelized'
> y6='Fruity'
> y7='Cherry'
> y8='Eggy'
> y9='Oily'
> y10='Wheat'
> y11='Leavening'
> y12='Vanilla'
> y13='Vanillin'
>
> y14='Alcohol-like'
> y15='Almond-like'
> y16='Lemon-like'
> y17='Cardboard'
> y18='Waxy'
> y19='OverallSweet'
> y20='Sweet'
> y21='Salty'
> y22='Bitter'
> y23='Sour'
> y24='Astringent'
> y25='Metallic'
> ;
> cards;
>
> 878     8    8.0  9.5  4.0  2.0  1.5  0.0  0.0  2.0  3.0  2.0
>              2.0  2.0  1.0  0.0  0.0  0.0  0.0  0.0  6.5  4.5  2.5
>              3.5  1.5  0.0  0.0
> ```

*(continued)*

**Table 12.10** Descriptive Analysis Data and SAS Program That Uses the Principal Component Solution under METHOD = prin—*cont...*

```
878    34   7.5  9.0   3.5  1.5  3.0  1.0  0.0  2.0  4.0  3.0
            3.0  1.5  2.0   0.0  0.0  0.0  0.0  0.0  7.0  5.5  3.5
            3.0  1.5  1.0   0.0
878    38   9.0 10.0   4.5  3.0  2.5  0.0  0.0  2.0  4.0  2.0
            3.0  2.5  1.0   0.0  1.0  0.0  0.0  0.0  5.5  3.5  3.5
            4.0  2.0  1.0   0.0
878    45   8.0  9.0   3.5  3.0  2.5  0.0  0.0  1.5  4.0  2.0
            2.5  2.0  1.5   1.0  0.0  0.0  0.0  0.0  6.5  5.0  4.0
            4.0  2.0  2.0   0.0
878    47   7.0  9.0   3.0  3.0  2.0  1.0  0.0  2.0  4.0  2.0
            3.0  1.0  2.0   0.0  0.5  0.0  0.0  0.0  5.5  5.0  3.0
            3.5  2.0  1.0   0.0
878    54   7.0  8.5   3.5  3.0  2.0  0.0  0.0  1.5  3.0  2.5
            3.0  2.0  1.5   0.0  0.0  0.0  0.0  0.0  5.5  3.5  3.0
            3.0  2.0  1.0   1.5
878    57   8.0  8.0   4.0  3.0  3.0  0.0  0.0  1.5  4.0  2.0
            3.0  2.0  1.0   1.0  0.0  0.0  0.0  0.0  5.0  3.0  2.0
            3.0  2.0  2.0   0.0
.          .    .    .     .    .    .    .    .    .    .    .


(Same format for the remaining products)
;
proc sort data=a;
    by product;
    run;

proc standard data=a
    mean=0
    std=1
    out=normal;
    var y1-y12;
    run;

proc print data=normal;
    var y1-y2 y11-y12;
    by product;
    title 'Standardized Raw Data';
    run;
```

*(continued)*

**Table 12.10** Descriptive Analysis Data and SAS Program That Uses the Principal Component Solution under METHOD = prin—*cont...*

```
proc means mean n std min max maxdec=2 data=a;
     var y1-y12;
     by product;
     title 'Descriptive Analysis';
     run;

proc factor data=a
     scree
     nfactor=3
     outstat=factor
     method=prin
     rotate=varimax
     reorder
     score;
     var y1-y12;
     title 'Descriptive Analysis';
     run;

proc score data=a
     score=factor
     out=fscore;
     var y1-y12;
     run;

proc print data=fscore;
     by product;
     run;

proc sort data=fscore;
     by product;
     run;

proc means mean data=fscore maxdec=3;
     var factor1-factor3;
     by product;
     output out=fscore1 mean=;
     title 'Descriptive Analysis';
     run;

goptions reset=global hsize=12 in vsize=12 in;
%plotit (data=fscore1, labelvar=product,
     plotvars=factor1 factor2, color=black, vref=0, href=0);
     run;
```

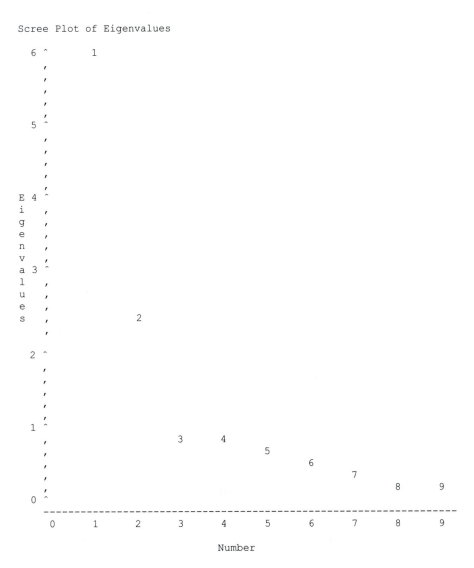

■**FIGURE 12.12** SCREE plot that shows the recommended number of factors is 3.

The second output to look at is the "rotated factor pattern," which shows the attributes that define the Factor1 Dimension (Table 12.12). The values in this table are the correlations of each component's attribute to its respective factors. Factors with both positive and negative correlations are known as *bipolar factor*. The higher the correlations, the more influence of

**Table 12.11** SAS Output: Proportion of the Variance Accounted For by the Factors

The FACTOR Procedure
Initial Factor Method: Principal Components
Prior Communality Estimates: ONE
Eigenvalues of the Correlation Matrix: Total = 12 Average = 1

|    | Eigenvalue | Difference | Proportion | Cumulative |
|----|------------|------------|------------|------------|
| 1  | 5.96233763 | 3.58043229 | 0.4969     | 0.4969     |
| 2  | 2.38190533 | 1.47072135 | 0.1985     | 0.6954     |
| 3  | 0.91118398 | 0.14111591 | 0.0759     | 0.7713     |
| 4  | 0.77006807 | 0.15569558 | 0.0642     | 0.8355     |
| 5  | 0.61437249 | 0.17349029 | 0.0512     | 0.8867     |
| 6  | 0.44088220 | 0.02753480 | 0.0367     | 0.9234     |
| 7  | 0.41334739 | 0.22760551 | 0.0344     | 0.9578     |
| 8  | 0.18574188 | 0.04755777 | 0.0155     | 0.9733     |
| 9  | 0.13818411 | 0.04570117 | 0.0115     | 0.9848     |
| 10 | 0.09248294 | 0.02175629 | 0.0077     | 0.9925     |
| 11 | 0.07072665 | 0.05195931 | 0.0059     | 0.9984     |
| 12 | 0.01876734 |            | 0.0016     | 1.0000     |

**Table 12.12** SAS Output: Correlation of the Sensory Attributes with the Factor Dimension

Rotated Factor Pattern

|     |              | Factor1   | Factor2   | Factor3   |
|-----|--------------|-----------|-----------|-----------|
| y2  | ChocoComplex | 0.93675   | 0.07825   | −0.03308  |
| y1  | Cocoa        | 0.93611   | 0.04371   | −0.10300  |
| y9  | Oily         | 0.87899   | 0.18815   | −0.05566  |
| y3  | DarkRoast    | 0.87597   | −0.33458  | −0.04935  |
| y5  | Caramelized  | 0.77969   | 0.14083   | 0.34188   |
| y4  | Buttery      | 0.77424   | 0.04314   | 0.15754   |
| y12 | Vanilla      | 0.64123   | −0.59786  | −0.07777  |
| y10 | Wheat        | −0.77390  | −0.03221  | 0.40402   |
| y7  | Cherry       | −0.02779  | 0.81997   | 0.12299   |
| y6  | Fruity       | 0.34129   | 0.76273   | 0.25647   |
| y8  | Eggy         | 0.16230   | 0.28917   | 0.81063   |
| y11 | Leavening    | −0.53579  | 0.13111   | 0.59738   |

the component attribute to the Factor Dimension. For example, "chocolate complex (0.94)," "cocoa (0.94)," "oily (0.88)," and "dark roast (0.88)" define the Factor1 Dimension, and these attributes contribute the most in the factor score of Factor1 as will be illustrated later. The negative correlation has a contrast effect that reduces a factor score. Because of the presence of negative correlation, Factor1 is known as a bipolar factor. Factor2 consists of "cherry (0.82)" and "fruity(0.76)," and Factor3 consists of "eggy (0.81)" and "leavening (0.60)." Thus, we have reduced the 12 sensory attributes into three factors that differentiate the products.

The sensory analyst must develop an integrated sensory name for each factor with the help of the standard sensory lexicon; for example, Factor1 may be called "chocolate aromatics"; and Factor2, "fruity aromatics."

Table 12.13 shows the "standardized scoring coefficients," a regression coefficient for obtaining the factor score. The Factor1 ($Y_1$) average score for product 349 is computed using (12.4-6).

$$Y_1 = (B_1X_1 + B_2X_2 + \cdots + B_6X_6 + B_7X_7)/n$$

$$Y_1 = [(0.15448)(-1.16795) + (0.15708)(-0.92144) + \cdots$$

$$+ (-0.06766)(-0.57456) + (0.11458)(1.13790)]/7$$

$$= -0.5753$$

where the values in the equation are highlighted in Table 12.13 and Table 12.14, and the divisor $n = 7$ is the number of judges or panelists. Table 12.15 shows the factor score for each product where the value

**Table 12.13** SAS Output: Regression Coefficients for Calculating the Factor Score

| | | Standardized Scoring Coefficients | | |
|---|---|---|---|---|
| | | Factor1 | Factor2 | Factor3 |
| y2 | ChocoComplex | 0.15708 | 0.04763 | −0.01367 |
| y1 | Cocoa | 0.15448 | 0.04765 | −0.06336 |
| y9 | Oily | 0.14469 | 0.12674 | −0.07085 |
| y3 | DarkRoast | 0.15199 | −0.21322 | 0.10151 |
| y5 | Caramelized | 0.14544 | −0.02944 | 0.28509 |
| y4 | Buttery | 0.13805 | −0.03362 | 0.15606 |
| y12 | Vanilla | 0.11458 | −0.37098 | 0.15092 |
| y10 | Wheat | −0.11414 | −0.13874 | 0.32791 |
| y7 | Cherry | −0.01132 | 0.49129 | −0.15642 |
| y6 | Fruity | 0.05810 | 0.40630 | −0.00649 |
| y8 | Eggy | 0.05813 | −0.07438 | 0.61820 |
| y11 | Leavening | −0.06766 | −0.09835 | 0.45375 |

**Table 12.14** SAS Output: Standardized Sensory Data with Mean = 0 and Standard Deviation = 1

| Prod | Obs | $y1$ | $y2$ | . . | $y11$ | $y12$ |
|------|-----|------|------|-----|-------|-------|
| 349 | 1 | −1.16795 | −0.92144 | . . | 0.63201 | −1.03445 |
| | 2 | −0.81249 | −1.22859 | . . | 1.83858 | 0.05172 |
| | 3 | 0.60936 | 0.30715 | . . | 1.23530 | −1.03445 |
| | 4 | −1.16795 | −0.92144 | . . | 1.83858 | −1.03445 |
| | 5 | −0.81249 | −0.61430 | . . | 0.02873 | −1.03445 |
| | 6 | −0.81249 | −0.61430 | . . | 1.23530 | 0.05172 |
| | 7 | −0.81249 | −1.22859 | . . | −0.57456 | 1.13790 |
| 593 | 8 | −0.81249 | −0.61430 | . . | 0.02873 | 0.05172 |
| | 9 | −0.81249 | −0.61430 | . . | 0.63201 | −1.03445 |
| | 10 | 0.60936 | 0.30715 | . . | −0.57456 | 0.05172 |
| | 11 | −0.45702 | −0.30715 | . . | 0.63201 | 0.05172 |
| | 12 | −0.10156 | −0.30715 | . . | −1.17784 | −1.03445 |
| | 13 | −0.81249 | −0.61430 | . . | 0.63201 | 0.05172 |
| | 14 | −0.81249 | −1.22859 | . . | −0.57456 | −1.03445 |
| 878 | 15 | 1.32029 | 1.53574 | . . | −1.78112 | 1.13790 |
| | 16 | 0.96483 | 1.22859 | . . | −0.57456 | 0.05172 |
| | 17 | 2.03122 | 1.84289 | . . | −0.57456 | 2.22407 |
| | 18 | 1.32029 | 1.22859 | . . | −1.17784 | 1.13790 |
| | 19 | 0.60936 | 1.22859 | . . | −0.57456 | −1.03445 |
| | 20 | 0.60936 | 0.92144 | . . | −0.57456 | 1.13790 |
| | 21 | 1.32029 | 0.61430 | . . | −0.57456 | 1.13790 |

**Table 12.15** SAS Output: Factor Mean Scores by Product

```
                The MEANS Procedure
Variable                    Product 349 Mean
Factor1                         −0.5753
Factor2                          0.3939
Factor3                          0.1448

Variable                    Product 593 Mean
Factor1                         −0.7010
Factor2                         −0.2633
Factor3                          0.0647

Variable                    Product 878 Mean
Factor1                          1.2763
Factor2                         −0.1307
Factor3                         −0.2095
```

−0.5753 is indicated. The same procedure is followed for obtaining the other product average factor scores.

The key point in this illustration is that the calculation of factor score includes all the 12 attributes with Factor1 attributes contributing the most to factor score, followed by Factor2, and so forth. In practice, the map of Factor1 and Factor2 would be sufficient to separate the products in the perceptual map, especially when the variance accounted for by both factors is experimentally reasonable. Furthermore, this example illustrates the complex nature of multivariate analysis as compared to the earlier chapters and the need to include the SAS program in the learning process.

In this example, Factor1 and Factor2 accounted for 70% of the total variance (Table 12.11). The next step in the interpretation of the analysis result is to produce a cluster tree that separates the products into clusters where each cluster has similar sensory properties. The cluster tree is shown in Figure 12.13 where Products 349 and 593 belong to cluster 1, and Product 878 in cluster 2 with an R-square of over 0.75. Finally, the perceptual map is shown in Figure 12.14 using the Factor1 and Factor2 factor scores. The two product clusters are indicated in the map. Product 878 is characterized by a "high" amount of Factor1, but "medium" in the Factor2 dimension. Products 593 and 349 have a "low" amount of

■ **FIGURE 12.13** Cluster Tree for Factor1 and Factor2 Dimensions.

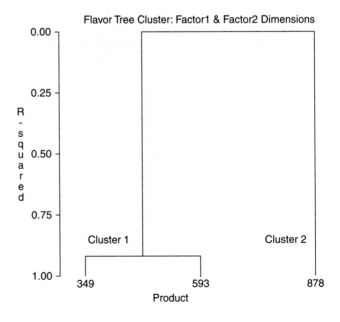

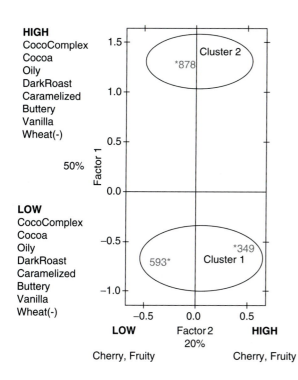

■ **FIGURE** 12.14 Perceptual map of Factor1 and Factor2 Dimensions.

Factor1, with 593 "low" in Factor2 and 349 "high" in Factor2. It should be noted that "high" does not mean the higher scale category on the 15-point descriptive analysis scale; it is the high point based on the standardized ratings, and this also applies to the "medium" and "low." Based on the perceptual map, we were able to characterize the sensory properties of the three products, a result useful in R&D direction. This result becomes fully actionable when the descriptive analysis information is correlated with hedonic data.

At the end of the SAS program given in Table 12.10, we can add the following statements to produce the plot shown in Figure 12.15. Three SAS procedures are involved: *proc aceclus, proc cluster,* and *proc tree.*

```
proc means mean data=fscore1 maxdec=4;
  var factor1-factor3;
  by product;
  output out=result mean=;
  run;
proc aceclus data=result out=ace p=.03;
  var factor1-factor2;
  run;
```

```
proc cluster data=ace outtree=Tree method=ward
 ccc pseudo print=15;
 var can1 can2;
 id product;
 run;
goptions vsize=8 in hsize=8 in htext=3pct htitle=4pct;
 axis1 order=(0 to 1 by 0.2);
proc tree data=Tree out=New nclusters=2
 graphics haxis=axis1;
 height _rsq_;
 copy can1 can2;
 id product;
 title 'Flavor Tree Cluster: Factor1 & Factor2 Dimensions';
 run;
```

■ **FIGURE 12.15** Perceptual map of all the intensity attributes with product location superimposed.

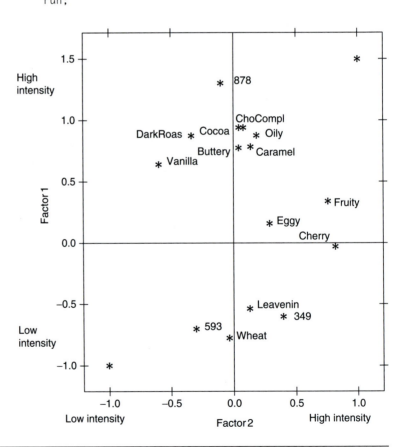

This example is a continuation of the preceding example. The output data in Table 12.12 are used to create another perceptual map where all the 12 sensory attributes are mapped together with the factor means scores of the three products given in Table 12.15. This perceptual map provides an overall view of all the sensory attributes that were used in obtaining the product factor scores, including the location of the products in the map. A SAS program is needed (Table 12.16) using the rotated factor information given in Table 12.12 and the factor mean scores in Table 12.15. Again, the main data under the CARDS statement were obtained by copying and pasting from the main SAS program output. The *x1*, *x2*, and *p1–p3* data were typed in.

**Example 12.4-2**

The perceptual map for Factor1 and Factor2 dimensions is shown in Figure 12.15. Note that the unit of attribute rotated factor ranges from $-1$ to $+1$, being the correlation of the sensory attributes to factor dimension, whereas the product factor scores have no limit but the limit is mostly closer

---

**Table 12.16** SAS Program for Mapping Intensity Attributes and Products into the Factor1 and Factor2 Dimensions

```
*prog. rotated factor Example 12.4-1.sas;
options nodate;
data a;
input Attribute Name$ Factor1-Factor3;
cards;
```

| | | | | |
|---|---|---|---|---|
| y2 | ChoComplex | 0.93675 | 0.07825 | −0.03308 |
| y1 | Cocoa | 0.93611 | 0.04371 | −0.10300 |
| y9 | Oily | 0.87899 | 0.18815 | −0.05566 |
| y3 | DarkRoast | 0.87597 | −0.33458 | −0.04935 |
| y5 | Caramel | 0.77969 | 0.14083 | 0.34188 |
| y4 | Buttery | 0.77424 | 0.04314 | 0.15754 |
| y12 | Vanilla | 0.64123 | −0.59786 | −0.07777 |
| y10 | Wheat | −0.77390 | −0.03221 | 0.40402 |
| y7 | Cherry | −0.02779 | 0.81997 | 0.12299 |
| y6 | Fruity | 0.34129 | 0.76273 | 0.25647 |
| y8 | Eggy | 0.16230 | 0.28917 | 0.81063 |
| y11 | Leavening | −0.53579 | 0.13111 | 0.59738 |
| p1 | 349 | −0.5753 | 0.3939 | 0.1448 |
| p2 | 593 | −0.7010 | −0.2633 | 0.0647 |
| p3 | 878 | 1.2763 | −0.1307 | −0.2095 |
| x1 | . | 1.5 | 1.0 | 1.0 |
| x2 | . | −1.0 | −1.0 | −1.0 |

*(continued)*

**Table 12.16** SAS Program for Mapping Intensity Attributes and Products into the Factor1 and Factor2 Dimensions—*cont...*

```
;
goptions reset=global hsize=12 in vsize=12 in;
%plotit (data=a, labelvar=Name,
      plotvars=factor1 factor2, color=black, vref=0, href=0);
      run;

goptions reset=global hsize=12 in vsize=12 in;
%plotit (data=a, labelvar=Name,
      plotvars=factor1 factor3, color=black, vref=0, href=0);
      run;
```

*Note: x1 and x2 rows are for plotting purposes.*

to $\pm 1$. As shown in Figure 12.15, product 878 is associated with dark roast, chocolate complex, cocoa, oily, vanilla, buttery, and caramelized. Similarly, products 593 and 349 are associated with leavening and wheaty aromatics. In practice, the sensory and research analysts must find a sensory description associated with the factors, especially for Factor1 and Factor2.

## Principal Component Analysis

The model for principal component analysis is

$$\begin{bmatrix} Y_1 \\ Y_2 \\ . \\ . \\ Y_p \end{bmatrix} = \begin{bmatrix} B_{11} & B_{12} & . & . & B_{1p} \\ B_{21} & B_{22} & . & . & B_{2p} \\ . & . & . & . & . \\ . & . & . & . & . \\ B_{p1} & B_{p2} & . & . & B_{ij} \end{bmatrix} \begin{bmatrix} X_1 \\ X_2 \\ . \\ . \\ X_i \end{bmatrix} \qquad (12.4\text{-}6)$$

where

$Y_i =$ linear combinations of the original attributes, $i = 1, 2, \ldots, p$; $Y_1$ is the first principal component score denoted by Prin1 in SAS output; $Y_2$ is the second principal component score denoted by Prin2, and so on;

$X_i =$ matrix of observed sensory ratings of the $p$th attributes;

$B_{ij} =$ regression coefficient of $i$th component and $j$th attribute, mathematically derived from the eigenvector of the covariance matrix of the original attribute scores.

In matrix form, the model is simply

$$Y = BX \qquad (12.4\text{-}7)$$

where the matrix for $Y$ and $X$ is of size $p \times 1$ and $B$ is a $p \times p$ matrix. Notice that there is no error term $E$ as compared to the factor analysis model because in principal component analysis the component score is directly constructed from the observed variables, i.e., sensory ratings. In (12.4-6), the model has no intercept since the observed data were standardized in the analysis with mean 0 and variance 1.

As one can see, the principal component $Y_i$ is a function of the observed variable $X_i$. Thus, the equation for Prin1 is

$$Y_1 = X_1 B_{11} + X_2 B_{12} + \cdots + X_i B_{1p} \qquad (12.4\text{-}8)$$

and for Prin2,

$$Y_2 = X_1 B_{21} + X_2 B_{22} + \cdots + X_i B_{2p}$$

and so on for remaining components. Substitution of standardized ratings into the preceding equation produces the principal component scores. There is no SCREE plot in the principal component method; instead, the contribution of each sensory attribute in terms of variance or eigenvalue is produced as output, and from there, one can determine the number of reasonable principal components using each contribution to the total variance. This output is indicated as "Eigenvalues of the Covariance Matrix."

**Example 12.4-2**

In this example the same sensory data in factor analysis are used to illustrate the principal component analysis. The PROC PRINCOMP procedure is run using the program in Table 12.17 to obtain the output that gives the total number of principal components, among others. In this program, the variance-covariance matrix was used in the analysis as indicated by COV in the PROC PRINCOMP statement. When the correlation matrix is used, the standardized data is the information needed in the PROC PRINCOMP statement, which is written as PROC PRINCOMP DATA = NORMAL and the same statements follow without COV.

Using the variance-covariance method, the result is shown in Table 12.18, where 75% of the variance is accounted for by Prin1, 7% by Prin2, and 6% by Prin3, resulting in a cumulative value of 88%. We see that cocoa, chococomplex, darkroast, oily, and wheat aromatics are basic components of Prin1 (Table 12.19).

Using equation 12.4-9, we can compute the principal component score using the standardized data in Table 12.14 and the eigenvectors in Table 12.19,

**Table 12.17** SAS Program for Principal Component Analysis

```
*prog principal component.sas;
options nodate;
        .
        . (Same statements as in Table 12.10)
        .

proc sort data=a;
     by product;
     run;

proc standard data=a
     mean=0
     std=1
     out=normal;
     var y1-y12;
     run;

proc princomp data=a
     cov
     n=3 (In the first run, n is not specified)
     out=prin;
     var y1-y12;
     run;

proc sort data=normal;
     by product;

proc print data=normal;
     var y1-y12;
     by product;
     title 'Standardized Raw data';
     run;

proc sort data=prin;
     by product;
     run;

proc means mean data=prin maxdec=3;
     var prin1-prin3;
     by product;
     output out=pscore mean=;
     title 'Example 12.4-2';
     run;

goptions reset=global hsize=12 in vsize=12 in;
%plotit (data=pscore, labelvar=product,
     plotvars=prin1 prin2, color=black, vref=0, href=0);

goptions reset=global hsize=12 in vsize=12 in;
%plotit (data=pscore, labelvar=product,
     plotvars=prin1 prin3, color=black, vref=0, href=0);
               run;
```

**Table 12.18** SAS Output for Determining the Number of Components for Inclusion in the Analysis

Eigenvalues of the Covariance Matrix

| | Eigenvalue | Difference | Proportion | Cumulative |
|---|---|---|---|---|
| 1 | 6.98888 | 6.31295 | 0.7493 | 0.7493 |
| 2 | 0.67593 | 0.09590 | 0.0725 | 0.8218 |
| 3 | 0.58002 | 0.25793 | 0.0622 | 0.8839 |
| 4 | 0.32209 | 0.10721 | 0.0345 | 0.9185 |
| 5 | 0.21488 | 0.02156 | 0.0230 | 0.9415 |
| 6 | 0.19331 | 0.04289 | 0.0207 | 0.9622 |
| 7 | 0.15042 | 0.07459 | 0.0161 | 0.9784 |
| 8 | 0.07582 | 0.02253 | 0.0081 | 0.9865 |
| 9 | 0.05329 | 0.01524 | 0.0057 | 0.9922 |
| 10 | 0.03804 | 0.01453 | 0.0041 | 0.9963 |

**Table 12.19** The Eigenvectors (Regression Coefficient) for Each Principal Component

| | | Prin1 | Prin2 | Prin3 |
|---|---|---|---|---|
| y1 | Cocoa | 0.5141 | 0.0770 | −.2937 |
| y2 | ChocoComplex | 0.5976 | 0.2059 | −.3223 |
| y3 | DarkRoast | 0.3897 | −.3759 | 0.5486 |
| y4 | Buttery | 0.1991 | 0.2676 | 0.3019 |
| y5 | Caramelized | 0.1676 | 0.3582 | 0.0658 |
| y6 | Fruity | 0.0418 | 0.2555 | −.0831 |
| y7 | Cherry | −.0046 | 0.1624 | −.0596 |
| y8 | Eggy | 0.0138 | 0.2356 | −.0240 |
| y9 | Oily | 0.2352 | 0.1694 | 0.2819 |
| y10 | Wheat | −.2253 | 0.1856 | −.3029 |
| y11 | Leavening | −.1956 | 0.5912 | 0.4563 |
| y12 | Vanilla | 0.1116 | −.2245 | 0.1509 |

$$Y_1 = [(-1.16975)(0.5141) + (-0.92144)(0.5976) + \cdots$$
$$+ (-0.57456)(-0.1956) + (1.13790)(0.1116)]/7$$
$$= -1.827$$

for Prin1 Product 349 (Table 12.20). The same procedure is followed for the calculation of the remaining principal components. One can see the similarity of result between factor analysis and principal component analysis in the perceptual map (Figure 12.16); however, the estimates of factor/component scores are lower for the factor analysis.

**Table 12.20** SAS Output: Principal Component Mean Scores by Product Averaged across Judges

| Variable | Product 349 Mean |
|---|---|
| Prin1 | −1.827 |
| Prin2 | 0.297 |
| Prin3 | 0.442 |

| Variable | Product 593 Mean |
|---|---|
| Prin1 | −1.586 |
| Prin2 | −0.164 |
| Prin3 | −0.595 |

| Variable | Product 878 Mean |
|---|---|
| Prin1 | 3.413 |
| Prin2 | −0.133 |
| Prin3 | 0.153 |

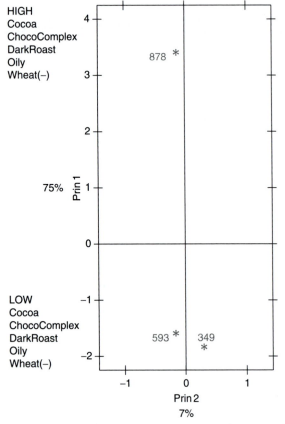

■ **FIGURE 12.16** Perceptual map for Prin1 and Prin2 components.

## 12.5 PREFERENCE MAPPING

In Example 12.4-1, the descriptive sensory data were evaluated using factor analysis in obtaining the sensory characteristics of the products in terms of independent factors where each factor consists of a number of sensory attributes, as shown in Figure 12.14. A common practice in sensory evaluation is to run a consumer test of the products after obtaining factor analysis or principal component analysis information on the descriptive analysis data. This analysis provides information on the relationship between consumer response and trained panel evaluation. This type of analysis is also called *external preference mapping*, and if only consumer data are evaluated, this analysis is called *internal preference mapping* in the literature. A detailed discussion of external and internal mapping is reported by Van Kleef et al. (2006). In practice, the external preference mapping approach is well received in R&D and marketing research.

**Example 12.5-1**

Consider the information in Table 12.21 where F1–F4 are average factor scores for each product; this table provides output from factor analysis of the descriptive data. The average overall liking score obtained from a consumer test is included in this table. Preliminary evaluation of the data shows that a linear regression model to study the relationship between overall liking and factor components is appropriate. For Factor1,

$$Y = \beta_0 + \beta_1 F1 + e \qquad (12.5\text{-}1)$$

where

$Y$ = overall liking score,
$\beta_0$ = intercept,
$\beta_1$ = slope of the regression equation,
$F1$ = factor average score for Factor1,
$e$ = error term.

and similarly for the other factor components. Table 12.22 shows the SAS program to plot the information in Table 12.21, which can be easily implemented by research and sensory analysts.

**Table 12.21** Average Overall Liking and Factor Scores for Each Product

| Product ID | Overall Liking | F1 Ave Score | F2 Ave Score | F3 Ave Score | F4 Ave Score |
|---|---|---|---|---|---|
| 349 | 5.69 | −0.1800 | 0.5195 | 0.8811 | −0.3205 |
| 593 | 5.68 | −0.9240 | 0.3114 | −0.2219 | 0.2286 |
| 878 | 6.47 | 1.1040 | −0.2082 | −0.6591 | 0.0919 |
| 982 | 6.10 | 1.0001 | 0.0010 | −0.3465 | −0.1262 |
| 421 | 5.90 | −0.5621 | −0.1000 | −0.1652 | 0.3001 |

**Table 12.22** Linear Regression Program to Evaluate the Relationship between Overall Liking and Product Factor Scores

```
*prog factor mapping.sas;
options nodate;
data a;
input prod y F1-F4;

label
F1='F1 AveScore'
F2='F2 AveScore'
F3='F3 AveScore'
F4='F4 AveScore'
y='OverallLiking'
;

cards;
349 5.69 −0.1800 0.5195 0.8811 −0.3205 (etc.)
;

goptions reset=global gunit=pct
ftext=swissb htitle=4 htext=4;
legend1 position=(bottom left outside)
across=1 cborder=red offset=(0,0)
shape=symbol(3,1) label=none
value=(height=3);

symbol1 c=red v=dot h=4;
symbol2 c=red;
symbol3 c=black;
symbol4 c=black;
symbol5 c=black;

proc reg data=a;
model y=F1 /r;
plot y*F1 / mse
caxis=black ctext=black cframe=white
legend=legend1;
run;

proc reg data=a;
model y=F2 /r;
plot y*F2 / mse
caxis=black ctext=black cframe=white
legend=legend1;
run;
```

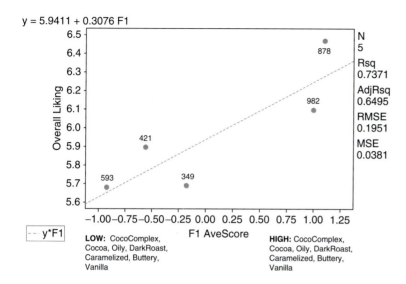

y = 5.9411 + 0.3076 F1

Overall Liking (y-axis): 6.5, 6.4, 6.3, 6.2, 6.1, 6.0, 5.9, 5.8, 5.7, 5.6

F1 AveScore (x-axis): −1.00 −0.75 −0.50 −0.25 0.00 0.25 0.50 0.75 1.00 1.25

N 5
Rsq 0.7371
AdjRsq 0.6495
RMSE 0.1951
MSE 0.0381

-- y*F1

LOW: CocoComplex, Cocoa, Oily, DarkRoast, Caramelized, Buttery, Vanilla

HIGH: CocoComplex, Cocoa, Oily, DarkRoast, Caramelized, Buttery, Vanilla

■ **FIGURE 12.17** Linear regression of overall liking on Factor1 product average score.

Since mapping is our interest, the regression analysis output is not given in this example, nor are the confidence limits of the regression line. Figure 12.17 shows the plot of overall liking and product factor score average for Factor1. One can see that products 878 and 982 have a "high" factor score for Factor1 components, and products 593 and 421 have a "low" factor score. This result indicates that a higher perceived amount of chococomplex, cocoa, oily, darkroast, caramelized, buttery, and vanilla increases overall liking of the product as indicated by the regression line with an R-square (Rsq) of 74%, which is a good prediction value considering the small number of data points in the linear regression analysis. In this plot the regression equation is indicated in the upper right corner of the graph. As a review, refer to Chapter 8 on regression analysis (Section 8.3).

Figure 12.18 shows the results for the Factor2 product average score, which indicates that the desirable level is from a "lower" degree to the "medium" level (0.0) of cherry and fruity as evaluated by the descriptive analysis panel. The information in Figures 12.17 and 12.18 is extremely useful for R&D and marketing research as a guide in the product improvement process.

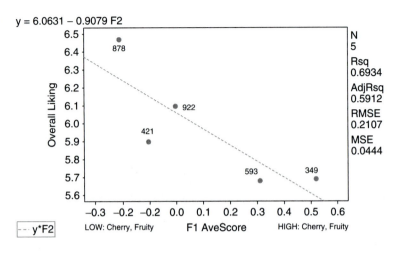

$$y = 6.0631 - 0.9079\ F2$$

| | N | 5 |
| Rsq | 0.6934 |
| AdjRsq | 0.5912 |
| RMSE | 0.2107 |
| MSE | 0.0444 |

LOW: Cherry, Fruity    F1 AveScore    HIGH: Cherry, Fruity

--- y*F2

**Example 12.5-2**

In this example internal preference mapping is illustrated. The data came from a two-product in-house consumer test consisting of 13 hedonic attributes and 6 sensory intensity attributes evaluated in one questionnaire by 65 panelists. The hedonic attributes were evaluated using a 9-point scale (1 = dislike extremely, 5 = neither like nor dislike, 9 = like extremely) and the intensity attributes by a 5-point scale such as 1 = too light to 5 = too dark for color. The data are in the following format for data entry:

| Product | Panelist | Y1 | Y2 | . | . | Y18 | Y19 |
|---|---|---|---|---|---|---|---|
| 1 | 1 | $X_{111}$ | $X_{112}$ | . | . | $X_{1118}$ | $X_{1119}$ |
| 1 | 2 | $X_{121}$ | $X_{122}$ | . | . | $X_{1218}$ | $X_{1219}$ |
| . | . | | | . | . | | |
| 2 | 1 | $X_{211}$ | $X_{212}$ | . | . | $X_{2118}$ | $X_{2119}$ |
| 2 | 2 | $X_{221}$ | $X_{222}$ | . | . | $X_{2218}$ | $X_{2219}$ |
| . | . | | | . | . | | |

(12.5-2)

where $Y1$ = overall acceptance, $Y2$ = color liking, ... , $Y19$ = moistness intensity, $X_{111}$ = rating for product 1 given by panelist 1, and so on. Once the data are typed into an Excel file, they are directly copied and pasted into a SAS program. Using the factor analysis SAS program given in Table 12.10 with appropriate changes for products, attributes, and data, the "rotated factor pattern" output was saved as a .DAT file and later copied and pasted into the SAS program (Table 12.23) for plotting purposes.

**Table 12.23** SAS Program along with the Data for Plotting Internal Preference Map*

```
*prog. rotated factor.sas;
options nodate;
data a;
input Attribute Name$ Factor1-Factor4;
cards;
```

| | | | | | |
|---|---|---|---|---|---|
| y1 | ACCEPTANCE | 0.84084 | 0.35576 | −0.01951 | 0.09410 |
| y2 | Color | 0.74526 | 0.42522 | 0.14642 | −0.16755 |
| y3 | OverallFlavor | 0.72516 | 0.28158 | 0.35601 | 0.29224 |
| y16 | MilFlaIN | 0.71498 | −0.06053 | 0.04199 | 0.04086 |
| y6 | Bitterness | 0.70914 | 0.24754 | 0.44162 | 0.07323 |
| y5 | MilkyFlavor | 0.65875 | 0.18961 | 0.41199 | 0.12851 |
| y4 | ChocoFlavor. | 0.58588 | 0.31546 | 0.34944 | 0.45736 |
| y7 | Sweetness | 0.56226 | 0.34701 | 0.55000 | 0.14459 |
| y9 | Moistness | 0.11223 | 0.90550 | 0.07979 | 0.23053 |
| y8 | OverallTexture | 0.33870 | 0.87168 | 0.15903 | 0.10571 |
| y11 | Denseness | 0.23600 | 0.83762 | 0.15311 | −0.07957 |
| y10 | Fluffiness | 0.26933 | 0.82389 | 0.16048 | 0.09892 |
| y19 | MoistIN | −0.41930 | 0.46210 | 0.09613 | 0.17742 |
| y13 | Aftertaste | 0.50073 | 0.07643 | 0.70367 | 0.05356 |
| y17 | BitterIN | −0.01693 | −0.14609 | −0.84643 | −0.04753 |
| y15 | ChoFlaIN | 0.21317 | 0.27425 | 0.06345 | 0.78558 |
| y14 | ColorIN | −0.29242 | −0.14671 | −0.04865 | 0.54218 |
| y18 | SweetIN | 0.22589 | 0.16126 | 0.32850 | 0.43186 |
| x1 | . | −1.0 | −1.0 | −1.0 | . |

```
;
goptions reset=global hsize=12 in vsize=12 in;
%plotit (data=a, labelvar=Name,
plotvars=factor1 factor2, color=black, vref=0, href=0);
run;

goptions reset=global hsize=12 in vsize=12 in;
%plotit (data=a, labelvar=Name,
plotvars=factor2 factor3, color=black, vref=0, href=0);
run;
```

*The x1 row is for plot scaling purposes.*

In this study, the two products are the same but manufactured at different times and were found not significantly different from each other in all the sensory and intensity attributes. Thus, the preference map represents the two products. Figure 12.19 shows the preference map for Factor1 and Factor2. Since the overall acceptance—the first question asked—was

■ **FIGURE 12.19** Internal preference map for Factor1 and Factor2.

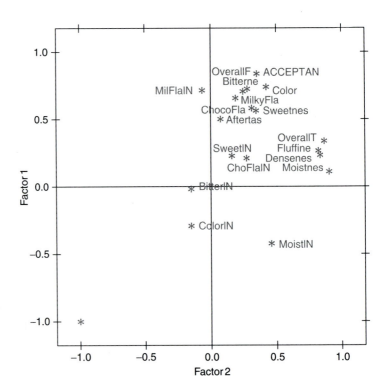

included in the factor analysis, it is expected that sensory attributes associated with overall acceptance should map close to overall acceptance. The analysis shows that overall acceptance, color, overall flavor, milky flavor intensity, bitterness, milky flavor, chocolate flavor, and sweetness form an integrated definition of Factor1 dimension. Factor1 can be called a "flavor-acceptability" dimension. Likewise, Factor2 can be called a "texture" dimension, which is an integration of moistness, overall texture, denseness, and fluffiness. This is clearly indicated in the preference map (Figure 12.19), which also shows a higher amount of flavor-acceptability dimension and a higher amount of texture dimension in the product as perceived by the consumer.

Factor3 is bipolar, consisting of aftertaste and bitter intensity, which is negatively correlated to the Factor3 dimension (Figure 12.20). In terms of sensory characteristics, Factor3 can be called a "bitter-aftertaste" dimension.

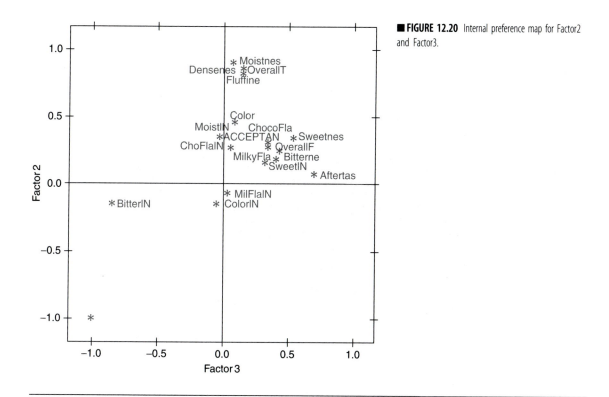

■ **FIGURE 12.20** Internal preference map for Factor2 and Factor3.

## EXERCISES

**12.1.** Create an experimental situation in which one would conduct the following tests and provide the reasons why. In addition, provide the statistical model of each test.

1. Consumer test
2. In-house consumer test
3. Central location test
4. Home-use consumer test

**12.2.** Explain sensory descriptive analysis and discuss its uses in research and development.

**12.3.** An experiment was conducted to determine the robustness of three products in terms of signal-to-noise ratio (SNR). The intensity data are tabulated here based on a 5-point intensity scale: 1 = not enough moistness, 3 = just about right, 5 = too much moistness. For overall liking, the rating is based on a 9-point hedonic scale: $1 =$ dislike extremely, $5 =$ neither like/nor dislike, $9 =$ like extremely.

| Panelist | Product | Moistness | Overall Liking |
|---|---|---|---|
| 1 | A | 4 | 5 |
| 1 | B | 3 | 6 |
| 1 | C | 2 | 5 |
| 2 | B | 3 | 7 |
| 2 | A | 4 | 6 |
| 2 | C | 2 | 5 |
| 3 | C | 2 | 6 |
| 3 | B | 3 | 7 |
| 3 | A | 4 | 5 |
| 4 | A | 3 | 5 |
| 4 | B | 2 | 6 |
| 4 | C | 2 | 6 |
| 5 | B | 4 | 5 |
| 5 | A | 4 | 6 |
| 5 | C | 3 | 5 |
| 6 | C | 2 | 5 |
| 6 | B | 3 | 7 |
| 6 | A | 4 | 5 |

1. Conduct an analysis of variance on the moistness data and compute the SNR of each product as well as the product multiple comparison test.
2. Graph the overall liking as the dependent variable with SNR and discuss the relationship.

**12.4.** What is the main purpose of multivariate statistical methods such as factor analysis in sensory evaluation and consumer tests? Define the following terms:

1. Factor dimension
2. Rotated factor pattern
3. Factor score
4. Bipolar factor
5. External and internal preference map

# Sensory Evaluation in Cosmetic Studies

In wineries, experts evaluate the effect of age on wine quality and acceptance; in cosmetic evaluations, similar procedures are used. Cosmetic evaluations have not been fully covered in sensory evaluation books to this point. In this chapter, except for different terminologies used, cosmetic procedures are the same as used in the sensory evaluation of foods, beverages, textiles, and personal care products. In addition, experimental designs are discussed.

In cosmetic studies, three important types of research data are gathered: (1) instrumental measurements, (2) expert evaluation, and (3) consumer tests. In most cases, these types of data are collected in every cosmetic research study. For a useful book on cosmetic testing, see Moskowitz (1984), in which sensory descriptive analysis is discussed; see also the paper by Aust et al. (1987). A book edited by Aust (1998) deals with claims substantiation on cosmetics and toiletries products.

## 13.1 EXPERIMENTAL DESIGNS

The principles of experimental design discussed in Chapter 1 still apply. Like in foods and personal care products, the initial study of cosmetics is performed under laboratory conditions to determine the efficacy of formulation ingredients, similar to Phase I in medical clinical trials. For example, in anti-aging studies, the instrumental data dealing with skin conditions are critical, followed by expert and sensory evaluation by panelists under home-use conditions. Standardized clinical photography to assess skin appearance (lines, wrinkles, skin clarity, age spots, etc.) and skin texture (firmness, moisturization) provides critical information, whereas expert and subject evaluations at times cannot clearly see the microanatomical changes on the surface of the skin before and after treatment applications.

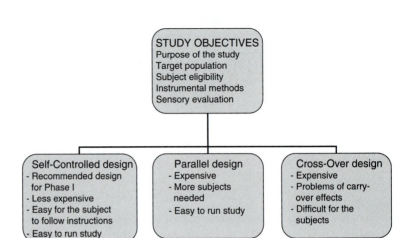

■ **FIGURE 13.1** Commonly used experimental designs in cosmetic studies.

Of interest to sensory analysts are the papers by Griffiths et al. (1992), Larnier et al. (1994), Barel (2001), Rona and Berardesca (2001), and Kappes and Elsner (2003). For strong claims substantiation, the desire is that treatment differences by instrumental methods and expert and subject evaluations can be physically observed on the treated skin. The procedure for conducting a cosmetic trial is summarized in Figure 13.1.

When the objectives of the study have been determined, the next steps are to (1) define the target population, such as gender and age; (2) define subject screening for eligibility, such as chronic skin allergies, medications that affect/interfere with treatments; (3) determine the instrumental measurement needed, such as transepidermal water loss (TEWL) readings to measure skin hydration; and (4) define the sensory effects of the products such as skin clarity, moistness, firmness, and overall appearance. These steps are contained in the Study Protocol, which includes an Informed Consent Form to be filled in by the subjects; such forms are standard in all types of sensory and clinical studies.

## Clinical Study Data

As indicated earlier, instrumental, expert, and subject responses are the basic experimental data in cosmetic clinical studies. Examples of instrumental data are TEWL, which measures transepidermal water loss on the skin; cutometer, which essentially measures skin texture by creating a negative pressure on the skin; and corneometer, which measures dielectric constants occurring in the stratum corneum to be able to evaluate moisturization, radiance, and clarity on the skin (Pinnagoda et al., 1990; Berndt and Elsner, 2001; Wickett, 2001). An advantage of these instrumental measures is that the condition of the skin after treatment can be evaluated by the experts and subjects, and their evaluations correlated to the instrumental measures. The key point here is the performance of the treatment on the skin. Moskowitz (1984) reported the relative importance of characteristics of a facial moisturizing lotion based on a 100-point scale:

| | |
|---|---|
| Color/appearance | 10 |
| Product texture | 21 |
| Fragrance | 14 |
| Performance on skin | 55 |

This table shows that color/appearance (10) is less important, and performance on the skin is the most important (55).

In practice, questionnaires given to experts and subjects entail sensory attributes related to performance of the product on the skin. Examples are the following questions in which the subjects respond using a 7-point scale (1 = strongly disagree, 4 = neither agree nor disagree, 7 = strongly agree):

■ Product is easy to apply
■ Is the most unique facial cleanser I have used
■ My skin feels smoother and more refined
■ My skin looks radiant
■ Gently eliminates make-up buildup
■ Long lasting in skin smoothness
■ Product nonirritating
■ Fragrance is just about right

In many studies, 15 to 20 similar questions are generally asked.

In clinical studies, the baseline measurement is critical in evaluating effectiveness or efficacy of a product. Post-treatment evaluations are compared to the baseline to provide data to measure intervention effect or product effectiveness. Both subjects and experts evaluate texture, pores, skin tone, fine lines, wrinkles, age spots, and other similar skin conditions using a 6-point intensity scale (1 = none, 2 = very slight, 3 = slight, 4 = moderate, 5 = moderately severe, 6 = severe) known in cosmetic clinical studies as *visual-analog*, a terminology not used in sensory evaluations. Depending on the cosmetic products, post-treatment evaluations have intervals of 7 days, 14 days, 21 days, and even up to 82 days of product use. At times, two to four product regimens are investigated. In the analysis, post-treatment data are subtracted from the baseline measurement. Table 13.1 shows some examples of rating scales used in cosmetic studies. The need for a descriptive analysis lexicon is apparent, including the reference standard. Instruments such as photographic imaging are used to obtain the sensory descriptions.

**Table 13.1** Examples of Cosmetic Rating Scales

| Attribute/Questions | Rating Scale |
|---|---|
| Facial pores, etc. | 1 = very small, barely visible (0 cm)<br>2 = small but visible (2.5 cm)<br>3 = medium (5.0 cm)<br>4 = large (7.5 cm)<br>5 = very large (over 7.5 cm) |
| Skin texture, etc. | 1 = poor, leathery/flaky skin (0 cm)<br>2 = slight, coarse/rough skin (2 cm)<br>3 = moderately firm, not too rough/not too smooth (4 cm),<br>4 = moderately smooth (6 cm)<br>5 = excellent texture, fine, silky smooth (8 cm) |
| Skin tone/radiance, etc. | 1 = poor, dull (0 cm)<br>2 = slight, somewhat dull, skin tone uneven (2.5 cm)<br>3 = moderate, skin radiant with small areas of dullness (5.0 cm)<br>4 = moderately strong, very radiant/luminescent/even tone (7.5 cm)<br>5 = excellent skin tone (over 7.5 cm) |
| Product is long lasting.<br>Product not clumping.<br>Product texture smooth.<br>Product is sticky.<br>Product is firm, etc. | 1 = strongly disagree<br>2 = slightly disagree<br>3 = moderately disagree<br>4 = neither disagree nor agree<br>5 = slightly agree<br>6 = moderately agree<br>7 = strongly agree |

*(continued)*

| **Table 13.1** Examples of Cosmetic Rating Scales—*cont…* | |
|---|---|
| **Attribute/Questions** | **Rating Scale** |
| Tanning cleanser | 1 = no visual tanning |
| | 2 = slight tanning, some skin discoloration still present |
| | 3 = moderate self-tanning evident |
| | 4 = strong self-tanning, skin discoloration clearly evident |
| | 5 = very strong skin discoloration |
| Tanning Makeup foundation, etc. | 1 = much too dark |
| | 2 = slightly dark |
| | 3 = just about right |
| | 4 = slightly light |
| | 5 = much too light |
| Coarse wrinkling (eye sites) | 1 = none |
| Fine lines (eye sites) | 2 = very slight |
| Mottled hyperpigmentation (entire face) | 3 = slight |
| Yellowing (eye face), etc. | 4 = moderate |
| | 5 = moderately severe |
| | 6 = very severe |
| Appearance of the skin | 1 = absent, not visible |
| | 2 = slightly marked/visible |
| | 3 = moderately marked/visible |
| | 4 = highly visible |
| | 5 = very visible |
| | 1 = not visible/absent |
| | 2 = slightly visible |
| | 3 = just about right (JAR) |
| | 4 = high visibility |
| | 5 = very visible/highly marked |

Note: Some rating scales courtesy of Bioscreen Testing Services Inc. with category scale numbers modified.

## Self-Controlled Design

In a self-controlled design, as the name implies, each subject serves as his or her own control, as described by Louis et al. (1984). This design nicely fits in cosmetic studies, where pre- and post-treatment observations are collected for each subject. The self-controlled design has been in use for over 30 years in medical clinical trials. Louis et al. (1984) discussed the problems involved in the design, which fortunately do not apply to cosmetic clinical studies. The reason is that in cosmetic studies, the subjects are observed at the same time increments from the beginning to the end of the study. The resulting data are naturally paired as being from the same subject. Regardless of the types of data—whether instrumental, expert, or sensory—the basic data for statistical analysis are different from baseline or time periods of interest. Thus, the self-controlled design is basically the same as a paired comparison in other fields of applications, as discussed in Chapter 4. Figure 13.2 shows the layout of the self-controlled design.

The statistical model for self-controlled design is

$$Y_i = \mu + (X_{2i} - X_{1i}) + \varepsilon_i, \quad i = 1, 2, .., n \qquad (13.1\text{-}1)$$

where

$Y_i$ = response differences observed on the $i$th subject for a particular variable;

$\mu$ = grand mean;

$X_{2i}$ = post-treatment measurements on the $i$th subject;

$X_{1i}$ = pre-treatment or baseline measurements on the $i$th subject;

$\varepsilon_i$ = random errors.

Since the expected value of $Y$ is zero, that is, $E(Y) = 0$, the null hypothesis is $H_0$: $Y = 0$ against the alternative $H_a$: $Y \neq 0$, since the average $Y$ could be either positive or negative. It is apparent that the paired $t$-test (Chapter 4) and the sign test (Chapter 9) can be used to test the null hypothesis of no difference.

■ **FIGURE 13.2** Layout of self-controlled design with two post-treatment measurements.

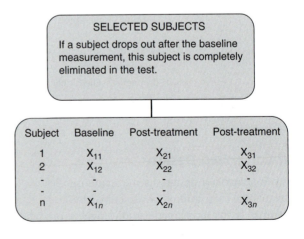

In this example, the well-known TEWL (transepidermal water loss), which measures the alteration of normal barrier function of the skin (stratum corneum), was used in personal care products as an indicator of product mildness. Sample data are tabulated in Table 13.2 to illustrate the statistical analysis using 10 subjects.

**Example 13.1-1**

Following is the calculation of differences from baseline measurements:

$$\text{Week } 3:$$
$$Y1 = 1.9 - 5.7 = -3.8$$
$$Y2 = 2.8 - 8.6 = -5.8$$
$$\cdot \quad \cdot \quad \cdot$$
$$Y10 = 5.2 - 7.8 = -2.6$$
$$\text{Week } 6:$$
$$Y1 = 4.0 - 5.7 = -1.7$$
$$Y2 = 2.5 - 8.6 = -6.1$$
$$\cdot \quad \cdot \quad \cdot \quad \cdot$$
$$Y10 = 6.3 - 7.8 = -1.5$$

**Table 13.2** TEWL Instrumental Measurements for a Cosmetic Product

| Subject | Baseline | Week 3 | Week 6 |
|---------|----------|--------|--------|
| 1 | 5.7 | 1.9 | 4.0 |
| 2 | 8.6 | 2.8 | 2.5 |
| 3 | 3.4 | 4.2 | 3.9 |
| 4 | 5.2 | 4.6 | 7.7 |
| 5 | 8.3 | 3.7 | 7.2 |
| 6 | 12.6 | 7.6 | 10.8 |
| 7 | 7.6 | 7.4 | 3.2 |
| 8 | 9.7 | 8.1 | 4.3 |
| 9 | 6.2 | 4.6 | 5.3 |
| 10 | 7.8 | 5.2 | 6.3 |

The statistical analysis is the same as that of paired comparison designs discussed in Chapter 4. This example uses an SAS program that can be easily run by the research analyst (Table 13.3). The key point in this calculation being from the same subject is that the degree of difference is most likely due to treatment effects. The results of the analysis are shown in Table 13.4. One can see that the treatment was

---

**Table 13.3** SAS Program for a Self-Controlled Design

```
*prog Example 13.1—1 Tew1.sas;
options nodate;
data    a;
input Subject B1 Wk3 Wk6;

D3 = Wk3 — B1;
D6 = Wk6 — B1;
/*PercentD3 = ((B1 — Wk3)/B1)*100;
PercentD6 = ((B1 — Wk6)/B1)*100;*/

label
B1='Baseline'
D3='Week3 Differences'
D6='Week6 Differences'
;
cards;
  1    5.7   1.9    4.0
  2    8.6   2.8    2.5
  3    3.4   4.2    3.9
  4    5.2   4.6    7.7
  5    8.3   3.7    7.2
  6   12.6   7.6   10.8
  7    7.6   7.4    3.2
  8    9.7   8.1    4.3
  9    6.2   4.6    5.3
 10    7.8   5.2    6.3
;
proc print data=a;
    var Subject D3 D6;
    title 'Example 13.1';
    run;

proc means mean n stderr prt clm maxdec=2 data=a;
    var D3 D6;
    title 'Example 13.1';
run;
```

**Table 13.4** SAS Output of Results of Statistical Analysis for Example 13.1. Positive Difference Indicates No Improvement in Transepidermal Water Loss (TEWL).

| Subject | D3 | D6 |
|---|---|---|
| 1 | −3.8 | −1.7 |
| 2 | −5.8 | −6.1 |
| 3 | 0.8 | 0.5 |
| 4 | −0.6 | 2.5 |
| 5 | −4.6 | −1.1 |
| 6 | −5.0 | −1.8 |
| 7 | −0.2 | −4.4 |
| 8 | −1.6 | −5.4 |
| 9 | −1.6 | −0.9 |
| 10 | −2.6 | −1.5 |

| Variable | Label | Mean | N | Std Error | Pr > \|t\| |
|---|---|---|---|---|---|
| D3 | Week3 Differences | −2.50 | 10 | 0.70 | 0.0062 |
| D6 | Week6 Differences | −1.99 | 10 | 0.84 | 0.0414 |

*Note: For simplicity, the 95% confidence intervals of the mean are not reported in this table. This output is generated by the clm option in statement proc means.*

effective in reducing transepidermal water loss on the treated skin after 3 weeks, with a mean difference from baseline of −2.50 ($p = 0.0062$) and at 6 weeks with a mean difference of −1.99 ($p = 0.0414$) of product use.

The results also can be expressed as a percentage to facilitate interpretation by computing the %D3 and %D6 in this example. When we use the SAS program symbols given in Table 13.3, the formula changes that are in the SAS code to obtain positive values are as follows:

$$\text{PercentD3} = ((\text{B1} - \text{Wk3})/\text{B1}) * 100$$

$$\text{PercentD6} = ((\text{B1} - \text{Wk6})/\text{B1}) * 100$$

The analyses show that at Week 3 the mean transepidermal water loss from baseline is 29.55% ($p < 0.0107$), and at Week 6 it is 21.28% ($p < 0.0903$), an information format highly acceptable in practice for reporting and interpreting the results. In other words, 70.45% (100 − 29.55) of water was retained at 3 weeks, and 78.72% was retained at 6 weeks of product use. We can confirm these findings using the SAS code in Table 13.3 with appropriate symbol changes for D3 and D6. The probability level differs slightly from those based on difference due to the wider range in the raw data, especially at Week 6, which ranges from 2.5 to 10.8.

---

| **Example 13.1-2** | In the previous example, only one product was evaluated for efficacy—a common type of study in cosmetics. In some studies more than two products are evaluated at the same time, and the analysis uses the completely randomized design model discussed in Section 4.2 of Chapter 4. The statistical model is |

$$Y_{ij} = \mu + P_i + \varepsilon_{ij}, \quad i = 1, 2, .., t$$

$$j = 1, 2, ..., n$$

where $Y_{ij}$ is the response difference observed for the $i$th product on the $j$th subject, $\mu$ is the grand mean, and $\varepsilon_{ij}$ are the random errors. For this example, we use a modification of the SAS code shown in Table 13.3. First, we add the product ID denoted by ProdA and ProdB as another column in the CARDS statement as follows:

```
cards;
ProdA  1     5.7     1.9     4.0
ProdA  2     8.6     2.8     2.5
ProdA  3     3.4     4.2     3.9
ProdA  4     5.2     4.6     7.7
ProdA  5     8.3     3.7     7.2

ProdB  6    12.6     7.6    10.8
ProdB  7     7.6     7.4     3.2
ProdB  8     9.7     8.1     4.3
ProdB  9     6.2     4.6     5.3
ProdB 10     7.8     5.2     6.3
;
proc sort data=a;
by product;
run;

proc means mean n stderr PRT clm maxdec=2 data=a;
var D3 D6 ;
by product;
title 'Example 13.1-2';
run;

proc glm data=a;
class product;
model D3 D6 = product / ss1;
means product / bon;
title 'Example 13.1-2';
run;
```

For this example, the original data were divided into two groups representing products A and B. We added the BY PRODUCT statement and PROC GLM to the SAS code to determine significant differences between products. The results are shown in Table 13.5. The MEANS

---

**Table 13.5** Simplified SAS Output for Example 13.2

| Obs | Product | Subject | D3 | D6 |
|-----|---------|---------|------|------|
| 1 | ProdA | 1 | −3.8 | −1.7 |
| 2 | ProdA | 2 | −5.8 | −6.1 |
| 3 | ProdA | 3 | 0.8 | 0.5 |
| 4 | ProdA | 4 | −0.6 | 2.5 |
| 5 | ProdA | 5 | −4.6 | −1.1 |
| 6 | ProdB | 6 | −5.0 | −1.8 |
| 7 | ProdB | 7 | −0.2 | −4.4 |
| 8 | ProdB | 8 | −1.6 | −5.4 |
| 9 | ProdB | 9 | −1.6 | −0.9 |
| 10 | ProdB | 10 | −2.6 | −1.5 |

The MEANS Procedure
Product = ProdA

| Variable | Label | Mean | N | Std Error | Pr > \|t\| |
|----------|-------|------|---|-----------|----------|
| D3 | Week3 Differences | −2.80 | 5 | 1.25 | 0.0879 |
| D6 | Week6 Differences | −1.18 | 5 | 1.43 | 0.4554 |

Product = ProdB

| Variable | Label | Mean | N | Std Error | Pr > \|t\| |
|----------|-------|------|---|-----------|----------|
| D3 | Week3 Differences | −2.20 | 5 | 0.80 | 0.0509 |
| D6 | Week6 Differences | −2.80 | 5 | 0.88 | 0.0339 |

The GLM Procedure
Bonferroni (Dunn) t Tests for D3

Means with the same letter are not significantly different (5% level)

| Bon Grouping | Mean | N | Product |
|--------------|--------|---|---------|
| A | −2.200 | 5 | ProdB |
| A | −2.800 | 5 | ProdA |

The GLM Procedure
Bonferroni (Dunn) t Tests for D6

Means with the same letter are not significantly different (5% level)

| Bon Grouping | Mean | N | Product |
|--------------|--------|---|---------|
| A | −1.180 | 5 | ProdA |
| A | −2.800 | 5 | ProdB |

procedure provides results for effectiveness of each product at weeks 3 and 6 as the differences from the baseline. The GLM procedure provides differences between products at weeks 3 and 6 using the Bonferroni $t$-test. As in Example 13.1-1, the difference from baseline can be expressed as a percentage for statistical analysis.

## Parallel Design

A parallel design is a completely randomized design with two treatments being evaluated; this is called a group-comparison design in Chapter 4. Chow and Liu (1998) named it *parallel design* in clinical trials. Since human subjects are the experimental unit, attention should be given in subject selection as well as in the entire course of the experimental process. Figure 13.3 displays the layout of the parallel design. The null and alternative hypotheses as well as the statistical analysis in the parallel design are the same with the group-comparison design discussed in Chapter 4.

The statistical model given in (4.3-1) is given as

$$X_{ij} = \mu + T_i + S_j + \varepsilon_{ij}, \quad i = 1, ..., t,$$

$$j = 1, ..., n,$$

(13.1-2)

where

$X_{ij} =$ observed measurement for the $i$th treatment condition (Control or Treatment);

$\mu =$ grand mean;

$T_i =$ effect due to the $i$th treatment condition;

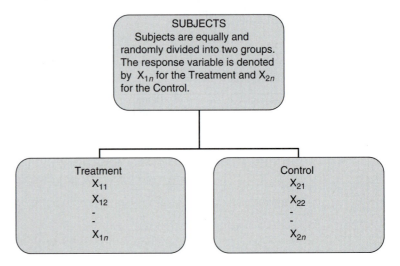

■ FIGURE 13.3 Layout of a parallel design.

$S_j$ = effect of jth subject;

$\varepsilon_{ij}$ = random errors.

Inclusion of the subject in the model assumes that the 10 subjects used as controls provide a good representation of the other 10 subjects used in the treatment group. In the model, the effect of the subject is removed as a source of variance from the random error component, which is used as the denominator of the $F$-test.

**Example 13.1-3**

In this example, the data for baseline and week 6 in Example 13.1-1 are used for illustration purposes. The baseline will be denoted as Control and week 6 as Treatment. Ten subjects were used for Control and another 10 subjects for Treatment, resulting in a total of 20 subjects. The use of different subjects for Control and Treatment occurs in some experimental situations to avoid carryover effects. Thus, the parallel or group-comparison design fits in with this situation. The statistical analysis shown in Chapter 4 can be used. For this example, the generalized linear model is used, and the SAS program that includes the data under the CARDS statement is shown in Table 13.6. Notice the "subject" column

**Table 13.6** SAS Program for Parallel Design Using the Generalized Linear Model (GLM)

```
*prog Example 13.1-2 Tew1.sas;
options nodate;
data a;
input Product$ Subject X;

*C=Control;
*T=Treatment;

cards;
C       1        5.7
C       2        8.6
C       3        3.4
C       4        5.2
C       5        8.3
C       6       12.6
C       7        7.6
C       8        9.7
C       9        6.2
C      10        7.8
```

*(continued)*

**Table 13.6** SAS Program for Parallel Design Using the Generalized Linear Model (GLM)—*cont...*

```
T      1      4.0
T      2      2.5
T      3      3.9
T      4      7.7
T      5      7.2
T      6      10.8
T      7      3.2
T      8      4.3
T      9      5.3
T      10     6.3
;
proc print data=a;
      var Subject Product X;
      title 'Example 13.1-2';
      run;
proc sort data = a;
      by product subject;
      run;
proc means mean n std clm maxdec=2 data=a;
      var X;
      by product;
      title 'Example 13.1-2';
      run;
proc glm data=a;
      class product subject;
      model X = product subject/ss1;
      title 'Example 13.1-2';
      run;
```

where the subject identification is from 1–10 for modeling purposes, as discussed earlier. For large data sets, the data are typed in an Excel format and then copied and pasted into the SAS program. If more than two treatments are involved, the SAS code can still be used.

Table 13.7 shows the output of the analysis, which indicates a significant difference between control and treatment ($p < 0.0414$) and subject effect is close to significance ($p < 0.0769$), suggesting that subject variability should be part of the model, as was done. If subject effect is excluded from the model, the difference between control and treatment is not significant ($p < 0.0978$) but not reported here.

**Table 13.7** GLM Output for Testing Difference between Control and Treatment

The MEANS Procedure
Analysis Variable $X$

| Product | Mean | N | Std Dev |
|---------|------|-----|---------|
| C | 7.51 | 10 | 2.58 |
| T | 5.52 | 10 | 2.52 |

The GLM Procedure
Class Level Information

Dependent Variable: $X$

| Source | DF | Sum of Squares | Mean Square | F Value | Pr > F |
|--------|-----|---------------|-------------|---------|--------|
| Model | 10 | 105.1510000 | 10.5151000 | 3.00 | 0.0565 |
| Error | 9 | 31.5145000 | 3.5016111 | | |
| Corrected Total | 19 | 136.6655000 | | | |

| R-Square | Coeff Var | Root MSE | $X$ Mean |
|----------|-----------|----------|----------|
| 0.769404 | 28.72232 | 1.871259 | 6.515000 |

| Source | DF | Type I SS | Mean Square | F Value | Pr > F |
|--------|-----|-----------|-------------|---------|--------|
| Product | 1 | 19.80050000 | 19.80050000 | 5.65 | 0.0414 |
| Subject | 9 | 85.35050000 | 9.48338889 | 2.71 | 0.0769 |

## Cross-Over Design

The cross-over design was discussed in Chapter 4, and the same method of analysis applies with the addition of the SAS program in this section. A book by Gacula (1993) discussed the application of cross-over design in sensory evaluation; an extensive statistical discussion of this design was done by Jones and Kenward (1989). In this chapter on cosmetic studies, we deal only with a lower order of the cross-over designs, consisting of two products evaluated in two periods because of the high carryover effects in cosmetic clinical studies that bias estimation of product efficacy. Most cosmetic studies deal with two products at a time; each "product" could be a regimen of various products considered as a treatment. Figure 13.4 shows the layout of the cross-over design.

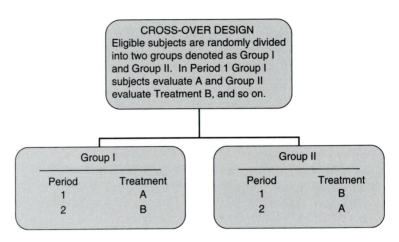

■ **FIGURE 13.4** Cross–over design layout.

**Example 13.1-4**

The data in this example consist of subject evaluation of facial skin texture after 3 weeks of customary product use. The rating scale is $1-2 =$ flaky/leathery skin, $3-4 =$ firm/not too rough/not too smooth, $5-6 =$ smooth, $7-8 =$ fine, silky smooth. Table 13.8 shows the SAS program to obtain the statistics for calculating the treatment and carryover effects not given in Chapter 4 for a two-product test. It is important to examine the INPUT

---

**Table 13.8** SAS Program for Cross-Over Design to Obtain Basic Statistics

```
*prog example 13.1-3 crossover.sas;
options nodate;
data a;
input Group$ period subject y1 y2;

*y1=treatment A;
*y2=treatment B;

if period=1 then do;
d1=y1-y2;                            *A - B;
end;

if period=2 then do;
d2=y2-y1;                            *A - B;
end;
```

*(continued)*

**Table 13.8** SAS Program for Cross-Over Design to Obtain Basic Statistics—*cont...*

```
cards;
I    1    1   6   6
I    1    2   5   3
I    1    3   3   2
I    1    4   5   6
I    1    5   6   3
I    1    6   4   3
I    1    7   6   5
I    1    8   5   3
I    1    9   6   4
I    1   10   6   5

II   2    1   6   4
II   2    2   4   4
II   2    3   6   5
II   2    4   5   4
II   2    5   3   4
II   2    6   4   4
II   2    7   6   5
II   2    8   6   5
II   2    9   5   4
II   2   10   4   5
;
proc print data=a ;
var subject y1 y2 d1 y2 y1 d2 ;
title 'Example 13.1-3 Cosmetic';
run;

proc sort data=a;
by period;
run;

proc means mean n std maxdec=2;
var y1 y2;
by period;
title 'Example 13.1-3 Cosmetic';
run;

proc means mean n std var t prt maxdec=2 data=a;
var d1 d2 ;
title 'Example 13.1-3 Cosmetic';
run;
```

$y1$ and $y2$ and the statements that follow, where $d1$ and $d2$ are the differences between products A and B.

The result is shown in Table 13.9, familiar basic statistics for describing the data. Notice the note below the table that defines $y1$ and $y2$. From the mean estimates in Periods 1 and 2, the difference between treatments for Groups I and II, where the order of product use was reversed, is as follows; also shown is the paired $t$-test probability level.

$$\text{Group I}: d1 = y1 - y2 = 5.20 - 4.00 = 1.20, p = 0.0086$$

$$\text{Group II}: d2 = y2 - y1 = 4.40 - 4.90 = \text{-}0.50, p = 0.1382$$

Using $d1$ and $d2$ estimates, which are the differences between A and B, we can calculate the estimate of treatment effects across groups by using the formula

$$T = (d1 + d2)/2$$

---

**Table 13.9** SAS Output for Obtaining Basic Statistics for Example 13.1-4

Group=I
The MEANS Procedure

| Variable | Mean | N | Std Dev |
|---|---|---|---|
| y1 | 5.20 | 10 | 1.03 |
| y2 | 4.00 | 10 | 1.41 |

Group=II
The MEANS Procedure

| Variable | Mean | N | Std Dev |
|---|---|---|---|
| y1 | 4.90 | 10 | 1.10 |
| y2 | 4.40 | 10 | 0.52 |

The MEANS Procedure

| Variable | Mean | N | Std Dev | Variance | t Value | Pr > \|t\| |
|---|---|---|---|---|---|---|
| d1 | 1.20 | 10 | 1.14 | 1.29 | 3.34 | 0.0086 |
| d2 | −0.50 | 10 | 0.97 | 0.94 | −1.63 | 0.1382 |

Note:
Group I: y1 = Treatment A,  y2 = Treatment B
Group II: y1 = Treatment B,  y2 = Treatment A

Group I: d1 = y1 − y2 = A − B
Group II: d2 = y2 − y1 = A − B

and substituting $d1$ and $d2$, we can obtain the estimate of treatment effects

$$T = (1.20 + (-0.50))/2 = 0.35.$$

For the estimate of carryover effects, the formula is

$$C = (d1 - d2)/2;$$

therefore, the estimate of carryover is

$$C = (1.20 - (-0.50))/2 = 0.85.$$

To test the significance of $T$ and $C$, we obtain the estimate of standard error SE from the variance in Table 13.9. It is obtained by pooling the variance of $d1$ denoted by $S_1^2$ and $d2$ denoted by $S_2^2$ , where $n$ is the number of subjects that can vary between groups:

$$SE = \left(\frac{1}{2}\right)\sqrt{(S_1^2/n) + (S_2^2/n)}$$

When we substitute the variances and the number of subjects, $n = 10$, that completed the test,

$$SE = \left(\frac{1}{2}\right)\sqrt{(1.29/10) + (0.94/10)}$$

$$= 0.24$$

Using this SE estimate, we can obtain the Student's $t$ statistic for treatment effects,

$$t_T = 0.35/0.24 = 1.46, \text{ with 18 DF}$$

and for carryover effects,

$$t_C = 0.85/0.24 = 3.54, \text{ also with 18 DF.}$$

Referring to Table A.2 in the appendix, the treatment effect is not significant ($p < 0.10$), but the carryover effect is significant ($p < 0.01$). We can see the carryover effect by plotting the mean score estimates by period, as shown in Figure 13.5. The crossing of the lines for A and B indicates the significant cross-over effects. The difference between treatments was larger in Period 1 than in Period 2, which was opposite the direction given earlier by $d1$ and $d2$. When this situation exists, the treatment A data in Group I should be compared to treatment B in Group II. As a preliminary result, this comparison is used, pending further confirmation data. Of course, the ideal result at the bottom part of Table 13.9 is that both $d1$ and $d2$ should be in the same direction and statistically significant by the $t$-test, suggesting that a product is favored in both orders of product use.

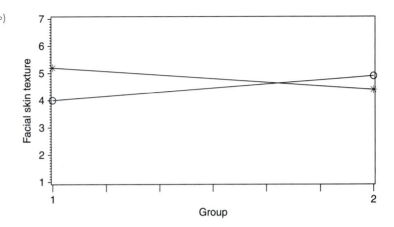

■ **FIGURE 13.5** Plot of treatments A(*) and B(○) facial mean scores by group to illustrate the presence of carryover effects.

## 13.2  **REGRESSION ANALYSIS**

### **Polynomial Regression**

In Chapter 8, the regression analysis was discussed as applied to shelf life testing, and various computational formulas were presented. Similar principles are applied in this section, with the addition of nonlinear regression, where the parameters are specified in advance. In this section the emphasis is graphical to meet the needs of most sensory analysts and research scientists. We will write computer programs using SAS to analyze the sensory data. The simplicity and effectiveness of the SAS statement INTERPOL, which provides linear, quadratic, and cubic regressions, are illustrated. As a review, the linear and quadratic regression models and the definition of the parameters were given in Chapter 8 as follows:

$$\text{Linear (8.3-1):}\quad Y = \beta_0 + \beta_1 X + \varepsilon \qquad (13.2\text{-}1)$$

$$\text{Quadratic (8.3-23):}\quad Y = \beta_0 + \beta_1 X + \beta_2 X^2 + \varepsilon \qquad (13.2\text{-}2)$$

Like the quadratic, the cubic is simply an extension of the linear model; that is,

$$\text{Cubic:}\quad Y = \beta_0 + \beta_1 X + \beta_2 X^2 + \beta_3 X^3 + \varepsilon \qquad (13.2\text{-}3)$$

**Example 13.2-1**

This example illustrates the use of the INTERPOL statement in SAS for linear and polynomial regression analyses. The statement requires three changes for each run as follows:

Linear:     interpol = *rLclm95*
Quadratic: interpol = *rQclm95*
Cubic:     interpol = *rCclm95*

where *r* denotes a regression, *L* denotes a linear, *Q* denotes a quadratic, *C* denotes a cubic, and *clm95* refers to the 95% confidence limits of the mean predicted values. For the 95% confidence limits of the individual predicted values, the symbol is *cli95* and is expected to be wider in width. Table 13.10 contains data on cosmetic changes on lip gloss appearance over time, collected initially, at 4 hours, and at 8 hours. The appearance intensity scale used is as follows:

0 = absent   1 = slightly marked   2 = moderately marked   3 = marked
4 = very marked

As expected, a higher score will be observed initially and lower scores over time of product use if the product is not effective.

The result for the linear regression analysis is shown in Figure 13.6, along with the 95% confidence limits produced by the SAS statement

$$\text{symbol interpol} = \text{rLclm95}$$

**Table 13.10** Sensory Data on Cosmetic Longevity Represented by Three Subjects

| Subject | Time, hr X | Appearance Score, Y |
|---------|------------|---------------------|
| 1 | 0 | 4 |
| 2 | 0 | 3 |
| 3 | 0 | 4 |
| 1 | 4 | 3 |
| 2 | 4 | 3 |
| 3 | 4 | 2 |
| 1 | 8 | 0 |
| 2 | 8 | 1 |
| 3 | 8 | 0 |

■ **FIGURE 13.6** Linear regression plot between appearance (*Y*) and time (*X*) where subject rating is indicated (∗). The horizontal dashed line denotes the upper limit of the intensity scale, and values above this limit should be ignored. The upper and lower 95% confidence limits are denoted by (- - - -).

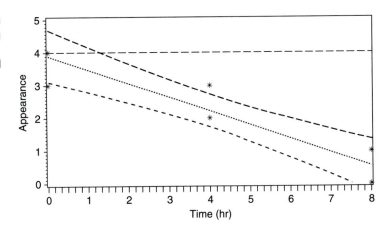

which is found in the lower part of the SAS program code in Table 13.11. The resulting linear regression is Appearance = 3.889 − 0.417(Time); since the regression coefficient is negative, the lip gloss appearance decreases over time. At 4 hours, the predicted longevity based on the linear model was 2.2 or "moderately marked":

$$Y = 3.889 - 0.417(4) = 2.2$$

**Table 13.11** SAS Program Code for the Lip Gloss Longevity Study

```
*prog example 13.2-1.sas;
options nodate;
%let title=Example 13.2-1;
data a;
input Subj Time Appearance;

cards;
1  0  4
2  0  3
3  0  4
1  4  3
2  4  3
3  4  2
1  8  0
2  8  1
3  8  0
;
proc sort data=a;
by time;
run;
```

*(continued)*

---

**TABLE 13.11** SAS Program Code for the Lip Gloss Longevity Study—*cont...*

```
proc means mean n std maxdec=2;
var appearance;
by time;
title"&title";
run;

goptions reset=all;
symbol1 color=black interpol=join height=2 width=2 line=2;
symbol2 color=black interpol=join height=2 width=2 line=2;
symbol3 color=black interpol=join height=2 width=2 line=2;

axis1 order=(0 to 8 by 1)
major=(height=1)
minor=(height=1 number=5)
value=(font=zapf height=2)
label=(font=zapf height=2  justify=center 'Time(hr)');

axis2 order=(0 to 5 by 1)
major=(height=1)
value=(font=zapf height=2)
label=(font=zapf height=2 angle=90 justify=center
        'Appearance');

proc sort data=a;
by time;
run;

title"&title";
interpol=rQclm95
value=star
height=3
width=2;

proc gplot data=a;
plot Appearance*Time /
haxis=axis1
vaxis=axis2
vref=4
lvref=2;
run;
```

---

Figure 13.7 shows the results for the quadratic regression obtained by

$$\text{symbol interpol} = \text{rQclm95}$$

Notice that the letter *L* was changed to *Q*, as stated earlier, for each program run. The quadratic regression equation is Appearance = 3.667 − 0.083(Time) −

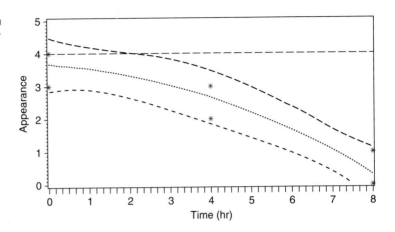

■ **FIGURE 13.7** Quadratic regression plot between appearance (Y) and time (X). The upper and lower 95% confidence limits are denoted by (- - - -).

0.0417(Time$^2$). At 4 hours, the quadratic estimate is 2.7 or "marked," indicating that the quadratic model gives a longer longevity estimate than linear—providing information on the importance of model building:

$$Y = 3.667 - 0.083(4) - 0.0417(16) = 2.7$$

## Nonlinear Regression

The nonlinear regression in this section refers to nonlinear in at least one of the parameters, i.e., $\alpha$, $\beta$, $\gamma$, etc., of the model. For an excellent book on nonlinear regression models, see Ratkowsky (1989); this book is recommended for scientists because it contains both theory and applied statistics. Books by Neter et al. (1990) and Seber and Wild (1989) are also recommended. For a researcher, the initial step is to plot the data with Y, the observation/response variable, in the vertical axis and the independent variable X in the horizontal axis. This plot provides a view of the relationship between Y and X. Is the plot linear or nonlinear? In differential equations, we are looking for a model in the form

$$\frac{dy}{dx} = f(x; \theta) \tag{13.2-4}$$

where $y$ is a function of $x$ and parameters $\theta$. In a regression form, (13.2-4) reduces to a model

$$Y = f(X; \theta) + \varepsilon \tag{13.2-5}$$

where $\varepsilon$ denotes an error term. Fortunately, through the use of the preceding differential equation, several models have already been

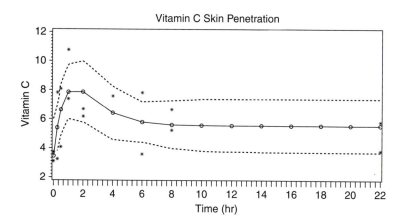

**■ FIGURE 13.8** Nonlinear regression plot between vitamin C (*Y*) and time (*X*).

solved over the years in various scientific fields, and plots are available as a guide in model building of one's experimental data.

An example of the application of a four-parameter $(a, b, c, d)$ nonlinear regression model was reported by Gacula (2003) on skin penetration of vitamin C over time (hours). The equation in the form of (13.2-5) is

$$Y = (a + b^*X)\exp(-c^*X) + d$$

$$Y = (-1.2886 + 2.6068^*X)\exp(-0.3198^*X) + 4.8979$$

The graph of the regression model is shown in Figure 13.8. The 95% confidence limits (lower curve line and upper curve line) of the mean predicted values are appropriately indicated. The nonlinear regression line is indicated by a circle (o) and the actual observation by a star (∗). Several nonlinear forms useful in various sciences also are described in Ratkowsky's book. The use of the SAS software in nonlinear regression analysis is illustrated in this section.

---

In this example, flavor intensity of the product was evaluated based on a 7-point intensity scale (1 = low, 7 = high). Table 13.12 displays the data for three judges. Using the SAS program given in Table 13.14, a nonlinear regression plot was produced (Figure 13.9) using the extended Langmuir model, as described by Ratkowsky (1989),

**Example 13.2-2**

$$Y = ab(X^{1-c})/(1 + bX^{1-c})$$

**Table 13.12** Flavor Intensity Scores

| Panelist | Day, X | Score, Y |
|---|---|---|
| 1 | 1 | 7 |
| 2 | 1 | 6 |
| 3 | 1 | 6 |
| 1 | 7 | 5 |
| 2 | 7 | 6 |
| 3 | 7 | 5 |
| 1 | 14 | 3 |
| 2 | 14 | 2 |
| 3 | 14 | 2 |
| 1 | 21 | 1 |
| 2 | 21 | 1 |
| 3 | 21 | 2 |

■ **FIGURE 13.9** Nonlinear regression·plot of changes in flavor intensity over 21 days.

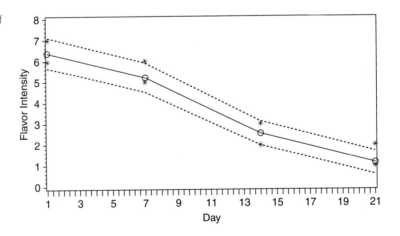

with parameters *a*, *b*, and *c*. The 95% confidence limits of the predicted mean values are indicated. Based on the SAS output, the estimates of the parameters are substituted into the equation as

$$Y = 6.39(946.10)(X^{1-3.76})/(1 + 946.10(X^{1-3.76}))$$

Using the preceding equation, we can substitute $X$ values of interest to obtain the estimated $Y$.

A commonly observed output is that shown in Table 13.13, which contains the actual data, the predicted output, and its corresponding lower and upper limits. The headings in this table are the words used in the SAS program, as indicated in the PROC GPLOT statement (Table 13.14). For educational purposes, one can use the SAS program for other data sets by changing the model in the PROC NLIN statement and the appropriate labels as well as the size of the vertical and horizontal axes. Also, in the PROC NLIN, the PARMS statement can include negative values; the choice of the magnitude of the parameters is based on experience and from published information. The practical approach is to try several values and check that the iterative process went through. When one examines the output, it is important to look for *"Note: Convergence criterion met"* given in the Iterative Phase of the output. If it did not converge, the model used for the data is not appropriate, or one may need to make changes in the parameter values by trial and error.

**Table 13.13** Lower and Upper Limits of the Mean Predicted Values at the 95% Confidence Limits

| Obs | X | Y | Predict | Lower | Upper |
|-----|-----|-----|---------|---------|---------|
| 1 | 1 | 7 | 6.38689 | 5.64894 | 7.12484 |
| 2 | 1 | 6 | 6.38689 | 5.64894 | 7.12484 |
| 3 | 1 | 6 | 6.38689 | 5.64894 | 7.12484 |
| 4 | 7 | 5 | 5.21599 | 4.52170 | 5.91028 |
| 5 | 7 | 6 | 5.21599 | 4.52170 | 5.91028 |
| 6 | 7 | 5 | 5.21599 | 4.52170 | 5.91028 |
| 7 | 14 | 3 | 2.53132 | 1.95151 | 3.11112 |
| 8 | 14 | 2 | 2.53132 | 1.95151 | 3.11112 |
| 9 | 14 | 2 | 2.53132 | 1.95151 | 3.11112 |
| 10 | 21 | 1 | 1.12848 | 0.56253 | 1.69443 |
| 11 | 21 | 1 | 1.12848 | 0.56253 | 1.69443 |
| 12 | 21 | 2 | 1.12848 | 0.56253 | 1.69443 |

*Note: Obs 1–3 have upper limits outside the scale length of 7 and should be ignored.*

**Table 13.14** SAS Program Code for Flavor Intensity

```
*prog example 13.2-2.sas;
options nodate;
%let title=Flavor Intensity;
data a;
input Subj x y;

cards;
1    1    7
2    1    6
3    1    6

1    7    5
2    7    6
3    7    5

1    14   3
2    14   2
3    14   2

1    21   1
2    21   1
3    21   2
;
goptions reset=all;
symbol1 color=black value=star height=2;
symbol2 color=black interpol=join value=circle height=2;
symbol3 color=black interpol=join value=none height=2
        width=2 line=2;
symbol4 color=black interpol=join value=none height=2
        width=2 line=2;

axis1 order=(1 to 21 by 2)
major=(height=1)
minor=(height=1 number=5)
value=(font=zapf height=2)
label=(font=zapf height=2 justify=center 'Day');

axis2 order=(0 to 8 by 1)
major=(height=1)
value=(font=zapf height=2)
label=(font=zapf height=2 angle=90 justify=center 'Flavor
        Intensity');

data b;
do time=1 to 21 by 2;
output;
end;
run;
```

*(continued)*

**Table 13.14** SAS Program Code for Flavor Intensity—*cont...*

```
data a;
set a b;
run;

proc nlin data=a best=4;
parms
a=-2 to 8 by .5
b=-2 to 8 by .5
c=-2 to 8 by .5;
temp=x**(1-c);
model y=a*b*(temp) / (1+b*temp);
output out=result p=predict l95m=lower u95m=upper;
title"&title";
run;

proc sort data=result;
by time;
run;

proc print data=result;
var x y predict lower upper;
run;

proc gplot data=result;
plot y*x predict*x lower*x upper*x / overlay
haxis=axis1
vaxis=axis2;
run;
```

In this example, another form of nonlinear regression plot observed in cosmetic studies is the sensory evaluation of "fine lines" around the eyes. Table 13.15 contains the data as the difference from baseline at various time periods (weeks). The rating scale is as follows:

**Example 13.2-3**

0–1 = none
2–3 = mild
4–5 = moderate
6–7 = large
8–9 = very large

**Table 13.15** Average Difference from Baseline by Weeks

| Week $X$ | Difference from Baseline, $Y$ |
|:---:|:---:|
| 2 | 0.00 |
| 4 | −0.20 |
| 6 | −0.25 |
| 8 | −0.40 |
| 10 | −0.42 |
| 12 | −0.41 |

At baseline, the fine lines can range from moderate to very large, which is undesirable. Photographs of the eye sites were evaluated by the clinical expert using the preceding scale. An increment of 0.5 on the scale is permitted, such as 0.5, 4.5, and so on. In this example the average of $N = 24$ subjects was used. The average difference from baseline is given in Table 13.15. As expected, if the product is effective in reducing fine lines, a lower score will be given, resulting in a negative value of $Y$, as shown by the formula

$$Y = X - Baseline.$$

The SAS program given in the previous example (Table 13.14) is used with the following modifications to fit the title, data range, and model:

```
%let title = Fine Lines;
. . .
input week Y;
. . .
axis1 order = (2 to 12 by 2)
. . .
axis2 order = (.20 to -.60 by -.10)
. . .
do week = 0 to 12 by 2;
```

The model is defined by the PROC NLIN statements:

```
proc nlin data = a best = 4;
parms
a = −10 to 8 by 1
b = −10 to 8 by 1
c = −10 to 8 by 1;
temp = exp(b/(week + c));
```

```
model y = a*temp;
output out = result p = predict l95m = lower u95m = upper;
run;
```

The initial step in finding a nonlinear model is to plot the $Y$ and $X$ and then consult the book by Ratkowsky (1989) or other books and look for a graph that is close to the data plot. Applications of the various models to the fine lines data show that the Gunary model, as discussed by Ratkowsky's equation (4.3.33), did converge and this model is selected. Since we do not know the range values of the parameters, it is advisable to include negative values, i.e., $a = -10$ to 8 by 1, in the iteration process for fitting the selected nonlinear model.

In this example the model is found to be

$$Y = a^* \exp(b/(\text{week} + c))$$

resulting in parameter estimates of

$$Y = -0.5753^* \exp(-3.0882/(\text{week} - 1.3054))$$

A useful SAS output is shown in Table 13.16. After 13 iterations, the estimate of the parameters was obtained as indicated by the note on convergence. Figure 13.10 shows the plot of fine lines $(Y)$ and week $(X)$ along with the 95% confidence limits of the predicted values. The plot indicates effectiveness of the product for reducing fine lines over time.

**Table 13.16** SAS Output: Number of Iterations to Obtain the Three Parameters of the Model Using the Gauss-Newton Method

The NLIN Procedure
Dependent Variable Y
Method: Gauss-Newton
Iterative Phase

| Iter | a | b | c | Sum of Squares |
|---|---|---|---|---|
| 0 | −1.0000 | −10.0000 | 1.0000 | 0.01370 |
| 1 | −0.5316 | −2.9173 | −1.0459 | 0.00653 |
| 2 | −0.5742 | −3.1548 | −1.2451 | 0.00479 |
| 3 | −0.5740 | −3.0616 | −1.3262 | 0.00475 |
| 4 | −0.5760 | −3.1021 | −1.2944 | 0.00475 |
| 5 | −0.5750 | −3.0821 | −1.3102 | 0.00475 |
| 6 | −0.5754 | −3.0913 | −1.3030 | 0.00475 |
| 7 | −0.5752 | −3.0869 | −1.3065 | 0.00475 |

*(continued)*

**Table 13.16** SAS Output: Number of Iterations to Obtain the Three Parameters of the Model Using the Gauss-Newton Method—*cont*...

| | | | | |
|---|---|---|---|---|
| 8 | −0.5753 | −3.0889 | −1.3048 | 0.00475 |
| 9 | −0.5753 | −3.0879 | −1.3056 | 0.00475 |
| 10 | −0.5753 | −3.0884 | −1.3052 | 0.00475 |
| 11 | −0.5753 | −3.0882 | −1.3054 | 0.00475 |
| 12 | −0.5753 | −3.0883 | −1.3053 | 0.00475 |
| 13 | −0.5753 | −3.0882 | −1.3054 | 0.00475 |

NOTE: Convergence criterion met.

| Parameter | Estimate | Std Error | Approximate 95% | Confidence Limits |
|---|---|---|---|---|
| a | −0.5753 | 0.1092 | −0.9230 | −0.2276 |
| b | −3.0882 | 1.8175 | −8.8724 | 2.6959 |
| c | −1.3054 | 1.1645 | −5.0113 | 2.4005 |

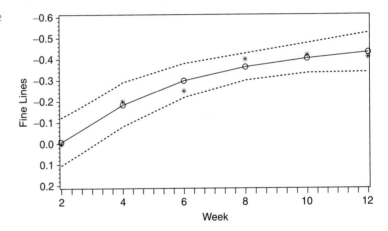

■ **FIGURE 13.10** Nonlinear regression plot for fine lines as a difference from baseline over 12 weeks.

## Fit of Nonlinear Regression Model

In the linear model, the fit of the data to the model is evaluated by the multiple correlation coefficient $R^2$ or $r^2$, as discussed in Chapter 8, defined as the proportion of the variance accounted for by the regression model. The higher the proportion, the better the fit of the model to the data. However, $R^2$ does not apply to the nonlinear regression model for assessing the fit of the model to the data, an issue that has been addressed and reviewed (Ratkowsky,

1989). Instead, the residual sums of squares are used—that is, the smaller the residual variance, the better the fit; this usually occurs when the iterative process of estimating the parameters has converged. In some data sets, despite convergence, the confidence limits are still wide and the possibility exists that outliers may be present in the data. In sensory studies, it is practical to examine the 95% confidence limits of the predicted values whether the lower and upper limits are reasonable in terms of sensory evaluations.

## EXERCISES

**13.1.** Define the following experimental designs with an illustration using a cosmetic or food product:
  **1.** Self-controlled design.
  **2.** Parallel design.
  **3.** Cross-over design.

**13.2.** Discuss the role of baseline measurements in cosmetic clinical studies.

**13.3.** The following data came from an anti-aging cream study. Analyze the data to determine the effectiveness of the cream in reducing skin wrinkles. A 1–7 rating scale was used: $0 =$ none, $1 =$ very slight, $2 =$ slight, $3 =$ moderate, $4 =$ moderately visible, $5 =$ highly visible, $6 =$ very visible.

| Subject | Baseline (Pre-treatment | Week 4 (Post-treatment) |
|---------|-------------------------|-------------------------|
| 1       | 6                       | 2                       |
| 2       | 4                       | 3                       |
| 3       | 6                       | 5                       |
| 4       | 6                       | 4                       |
| 5       | 5                       | 0                       |
| 6       | 5                       | 4                       |
| 7       | 6                       | 4                       |
| 8       | 5                       | 3                       |
| 9       | 6                       | 3                       |
| 10      | 6                       | 5                       |

1. Estimate the significant difference between baseline and Week 4.
2. Express the difference as a percentage from baseline and determine whether the difference is significant.
3. Are probability values similar between item 1 and 2 results?

**13.4.** Estimate the parameters in the nonlinear model given by

$$Y = X/(a + bX + c\sqrt{X})$$

using the SAS program shown in Table 13.13. The data are tabulated as follows:

| Time, X | Assay, Y |
|---------|----------|
| 0 | 99.6 |
| 1 | 102.5 |
| 2 | 98.1 |
| 3 | 96.2 |
| 6 | 98.6 |
| 9 | 97.5 |

1. Discuss the results of the model building.

# Statistical Tables

**TABLES**

**Table A.1** Normal Curve Areas[a]

| Z | Decimal Fraction of Z | | | | | | | | | |
|---|---|---|---|---|---|---|---|---|---|---|
|  | **.00** | **.01** | **.02** | **.03** | **.04** | **.05** | **.06** | **.07** | **.08** | **.09** |
| 0.0 | .0000 | .0040 | .0080 | .0120 | .0160 | .0199 | .0239 | .0279 | .0319 | .0359 |
| 0.1 | .0398 | .0438 | .0478 | .0517 | .0557 | .0596 | .0636 | .0675 | .0714 | .0753 |
| 0.2 | .0793 | .0832 | .0871 | .0910 | .0948 | .0987 | .1026 | .1064 | .1103 | .1141 |
| 0.3 | .1179 | .1217 | .1255 | .1293 | .1331 | .1368 | .1406 | .1443 | .1480 | .1517 |
| 0.4 | .1554 | .1591 | .1628 | .1664 | .1700 | .1736 | .1772 | .1808 | .1844 | .1879 |
| 0.5 | .1915 | .1950 | .1985 | .2019 | .2054 | .2088 | .2123 | .2157 | .2190 | .2224 |
| 0.6 | .2257 | .2291 | .2324 | .2357 | .2389 | .2422 | .2454 | .2486 | .2517 | .2549 |

*(continued)*

**Table A.1** Normal Curve Areas[a]—*cont...*

| Z | .00 | .01 | .02 | .03 | .04 | .05 | .06 | .07 | .08 | .09 |
|---|-----|-----|-----|-----|-----|-----|-----|-----|-----|-----|
| | | | | | **Decimal Fraction of Z** | | | | | |
| 0.7 | .2580 | .2611 | .2642 | .2673 | .2704 | .2734 | .2764 | .2794 | .2823 | .2852 |
| 0.8 | .2881 | .2910 | .2939 | .2967 | .2995 | .3023 | .3051 | .3078 | .3106 | .3133 |
| 0.9 | .3159 | .3186 | .3212 | .3238 | .3264 | .3289 | .3315 | .3340 | .3365 | .3389 |
| 1.0 | .3413 | .3438 | .3461 | .3485 | .3508 | .3531 | .3554 | .3577 | .3599 | .3621 |
| 1.1 | .3643 | .3665 | .3686 | .3708 | .3729 | .3749 | .3770 | .3790 | .3810 | .3830 |
| 1.2 | .3849 | .3869 | .3888 | .3907 | .3925 | .3944 | .3962 | .3980 | .3997 | .4015 |
| 1.3 | .4032 | .4049 | .4066 | .4082 | .4099 | .4115 | .4131 | .4147 | .4162 | .4177 |
| 1.4 | .4192 | .4207 | .4222 | .4236 | .4251 | .4265 | .4279 | .4292 | .4306 | .4319 |
| 1.5 | .4332 | .4345 | .4357 | .4370 | .4382 | .4394 | .4406 | .4418 | .4429 | .4441 |
| 1.6 | .4452 | .4463 | .4474 | .4484 | .4495 | .4505 | .4515 | .4525 | .4535 | .4545 |
| 1.7 | .4554 | .4564 | .4573 | .4582 | .4591 | .4599 | .4608 | .4616 | .4625 | .4633 |
| 1.8 | .4641 | .4649 | .4656 | .4664 | .4671 | .4678 | .4686 | .4693 | .4699 | .4706 |
| 1.9 | .4713 | .4719 | .4726 | .4732 | .4738 | .4744 | .4750 | .4756 | .4761 | .4767 |
| 2.0 | .4772 | .4778 | .4783 | .4788 | .4793 | .4798 | .4803 | .4808 | .4812 | .4817 |
| 2.1 | .4821 | .4826 | .4830 | .4834 | .4838 | .4842 | .4846 | .4850 | .4854 | .4857 |
| 2.2 | .4861 | .4864 | .4868 | .4871 | .4875 | .4878 | .4881 | .4884 | .4887 | .4890 |
| 2.3 | .4893 | .4896 | .4898 | .4901 | .4904 | .4906 | .4909 | .4911 | .4913 | .4916 |
| 2.4 | .4918 | .4920 | .4922 | .4925 | .4927 | .4929 | .4931 | .4932 | .4934 | .4936 |
| 2.5 | .4938 | .4940 | .4941 | .4943 | .4945 | .4946 | .4948 | .4949 | .4951 | .4952 |
| 2.6 | .4953 | .4955 | .4956 | .4957 | .4959 | .4960 | .4961 | .4962 | .4963 | .4964 |
| 2.7 | .4965 | .4966 | .4967 | .4968 | .4969 | .4970 | .4971 | .4972 | .4973 | .4974 |
| 2.8 | .4974 | .4975 | .4976 | .4977 | .4977 | .4978 | .4979 | .4979 | .4980 | .4981 |
| 2.9 | .4981 | .4982 | .4982 | .4983 | .4984 | .4984 | .4985 | .4985 | .4986 | .4986 |
| 3.0 | .4987 | .4987 | .4987 | .4988 | .4988 | .4989 | .4989 | .4989 | .4990 | .4990 |

[a]*Generated using an SAS program written by R. W. Washam II, Armour Research Center, Scottsdale, Arizona.*

**Table A.2** The *t* Distribution[a]

| | | | | | $\alpha$ | | | | |
|---|---|---|---|---|---|---|---|---|---|
| DF | .50 | .40 | .30 | .20 | .10 | .05 | .02 | .01 | .001 |
| 1 | 1.000 | 1.376 | 1.963 | 3.078 | 6.314 | 12.706 | 31.821 | 63.657 | 636.619 |
| 2 | .816 | 1.061 | 1.386 | 1.886 | 2.920 | 4.303 | 6.965 | 9.925 | 31.598 |
| 3 | .765 | .978 | 1.250 | 1.638 | 2.353 | 3.182 | 4.541 | 5.841 | 12.924 |
| 4 | .741 | .941 | 1.190 | 1.533 | 2.132 | 2.776 | 3.747 | 4.604 | 8.610 |
| 5 | .727 | .920 | 1.156 | 1.476 | 2.015 | 2.571 | 3.365 | 4.032 | 6.869 |
| 6 | .718 | .906 | 1.134 | 1.440 | 1.943 | 2.447 | 3.143 | 3.707 | 5.959 |
| 7 | .711 | .896 | 1.119 | 1.415 | 1.895 | 2.365 | 2.998 | 3.499 | 5.408 |
| 8 | .706 | .889 | 1.108 | 1.397 | 1.860 | 2.306 | 2.896 | 3.355 | 5.041 |
| 9 | .703 | .883 | 1.100 | 1.383 | 1.833 | 2.262 | 2.821 | 3.250 | 4.781 |
| 10 | .700 | .879 | 1.093 | 1.372 | 1.812 | 2.228 | 2.764 | 3.169 | 4.587 |
| 11 | .697 | .876 | 1.088 | 1.363 | 1.796 | 2.201 | 2.718 | 3.106 | 4.437 |
| 12 | .695 | .873 | 1.083 | 1.356 | 1.782 | 2.179 | 2.681 | 3.055 | 4.318 |
| 13 | .694 | .870 | 1.079 | 1.350 | 1.771 | 2.160 | 2.650 | 3.012 | 4.221 |
| 14 | .692 | .868 | 1.076 | 1.345 | 1.761 | 2.145 | 2.624 | 2.977 | 4.140 |
| 15 | .691 | .866 | 1.074 | 1.341 | 1.753 | 2.131 | 2.602 | 2.947 | 4.073 |
| 16 | .690 | .865 | 1.071 | 1.337 | 1.746 | 2.120 | 2.583 | 2.921 | 4.015 |
| 17 | .689 | .863 | 1.069 | 1.333 | 1.740 | 2.110 | 2.567 | 2.898 | 3.965 |
| 18 | .688 | .862 | 1.067 | 1.330 | 1.734 | 2.101 | 2.552 | 2.878 | 3.922 |
| 19 | .688 | .861 | 1.066 | 1.328 | 1.729 | 2.093 | 2.539 | 2.861 | 3.883 |
| 20 | .687 | .860 | 1.064 | 1.325 | 1.725 | 2.086 | 2.528 | 2.845 | 3.850 |
| 21 | .686 | .859 | 1.063 | 1.323 | 1.721 | 2.080 | 2.518 | 2.831 | 3.819 |
| 22 | .686 | .858 | 1.061 | 1.321 | 1.717 | 2.074 | 2.508 | 2.819 | 3.792 |
| 23 | .685 | .858 | 1.060 | 1.319 | 1.714 | 2.069 | 2.500 | 2.807 | 3.767 |
| 24 | .685 | .857 | 1.059 | 1.318 | 1.711 | 2.064 | 2.492 | 2.797 | 3.745 |
| 25 | .684 | .856 | 1.058 | 1.316 | 1.708 | 2.060 | 2.485 | 2.787 | 3.725 |
| 26 | .684 | .856 | 1.058 | 1.315 | 1.706 | 2.056 | 2.479 | 2.779 | 3.707 |

*(continued)*

**Table A.2** The *t* Distribution[a]—*cont...*

| DF | .50 | .40 | .30 | .20 | .10 | .05 | .02 | .01 | .001 |
|---|---|---|---|---|---|---|---|---|---|
| 27 | .684 | .855 | 1.057 | 1.314 | 1.703 | 2.052 | 2.473 | 2.771 | 3.690 |
| 28 | .683 | .855 | 1.056 | 1.313 | 1.701 | 2.048 | 2.467 | 2.763 | 3.674 |
| 29 | .683 | .854 | 1.055 | 1.311 | 1.699 | 2.045 | 2.462 | 2.756 | 3.659 |
| 30 | .683 | .854 | 1.055 | 1.310 | 1.697 | 2.042 | 2.457 | 2.750 | 3.646 |
| 40 | .681 | .851 | 1.050 | 1.303 | 1.684 | 2.021 | 2.423 | 2.704 | 3.551 |
| 60 | .679 | .848 | 1.046 | 1.296 | 1.671 | 2.000 | 2.390 | 2.660 | 3.460 |
| 120 | .677 | .845 | 1.041 | 1.289 | 1.658 | 1.980 | 2.358 | 2.617 | 3.373 |
| ∞ | .674 | .842 | 1.036 | 1.282 | 1.645 | 1.960 | 2.326 | 2.576 | 3.291 |

The column header spanning the $\alpha$ values is $\alpha$.

[a]*Generated using an SAS program written by R. W. Washam II, Armour Research Center, Scottsdale, Arizona.*

**Table A.3** The $\chi^2$ Distribution[a]

| DF | .50 | .30 | .20 | .10 | .05 | .02 | .01 | .001 |
|---|---|---|---|---|---|---|---|---|
| 1 | .455 | 1.074 | 1.642 | 2.706 | 3.841 | 5.412 | 6.635 | 10.827 |
| 2 | 1.386 | 2.408 | 3.219 | 4.605 | 5.991 | 7.824 | 9.210 | 13.815 |
| 3 | 2.366 | 3.665 | 4.642 | 6.251 | 7.815 | 9.837 | 11.345 | 16.266 |
| 4 | 3.357 | 4.878 | 5.989 | 7.779 | 9.488 | 11.668 | 13.277 | 18.266 |
| 5 | 4.351 | 6.064 | 7.289 | 9.236 | 11.070 | 13.388 | 15.086 | 20.515 |
| 6 | 5.348 | 7.231 | 8.558 | 10.645 | 12.592 | 15.033 | 16.812 | 22.457 |
| 7 | 6.346 | 8.383 | 9.803 | 12.017 | 14.067 | 16.622 | 18.475 | 24.322 |
| 8 | 7.344 | 9.524 | 11.030 | 13.362 | 15.507 | 18.168 | 20.090 | 26.125 |
| 9 | 8.343 | 10.656 | 12.242 | 14.684 | 16.919 | 19.679 | 21.666 | 27.877 |
| 10 | 9.342 | 11.781 | 13.442 | 15.987 | 18.307 | 21.161 | 23.209 | 29.588 |
| 11 | 10.341 | 12.899 | 14.631 | 17.275 | 19.675 | 22.618 | 24.725 | 31.264 |
| 12 | 11.340 | 14.011 | 15.812 | 18.549 | 21.026 | 24.054 | 26.217 | 32.909 |

The column header spanning the values is $\alpha$.

*(continued)*

**Table A.3** The $\chi^2$ Distribution[a]—*cont...*

| DF | .50 | .30 | .20 | .10 | .05 | .02 | .01 | .001 |
|----|-----|-----|-----|-----|-----|-----|-----|------|
| 13 | 12.340 | 15.119 | 16.985 | 19.812 | 22.362 | 25.472 | 27.688 | 34.528 |
| 14 | 13.339 | 16.222 | 18.151 | 21.064 | 23.685 | 26.873 | 29.141 | 36.123 |
| 15 | 14.339 | 17.322 | 19.311 | 22.307 | 24.996 | 28.259 | 30.578 | 37.697 |
| 16 | 15.338 | 18.418 | 20.465 | 23.542 | 26.296 | 29.633 | 32.000 | 39.252 |
| 17 | 16.338 | 19.511 | 21.615 | 24.769 | 27.587 | 30.995 | 33.409 | 40.790 |
| 18 | 17.338 | 20.601 | 22.760 | 25.989 | 28.869 | 32.346 | 34.805 | 42.312 |
| 19 | 18.338 | 21.689 | 23.900 | 27.204 | 30.144 | 33.687 | 36.191 | 43.820 |
| 20 | 19.337 | 22.775 | 25.038 | 28.412 | 31.410 | 35.020 | 37.566 | 45.315 |
| 21 | 20.337 | 23.858 | 26.171 | 29.615 | 32.671 | 36.343 | 38.932 | 46.797 |
| 22 | 21.337 | 24.939 | 27.301 | 30.813 | 33.924 | 37.659 | 40.289 | 48.268 |
| 23 | 22.337 | 26.018 | 28.429 | 32.007 | 35.172 | 38.968 | 41.638 | 49.728 |
| 24 | 23.337 | 27.096 | 29.553 | 33.196 | 36.415 | 40.270 | 42.980 | 51.179 |
| 25 | 24.337 | 28.172 | 30.675 | 34.382 | 37.652 | 41.566 | 44.314 | 52.620 |
| 26 | 25.336 | 29.246 | 31.795 | 35.563 | 38.885 | 42.856 | 45.642 | 54.052 |
| 27 | 26.336 | 30.319 | 32.912 | 36.741 | 40.113 | 44.140 | 46.963 | 55.476 |
| 28 | 27.336 | 31.391 | 34.027 | 37.916 | 41.337 | 45.419 | 43.278 | 56.893 |
| 29 | 28.336 | 32.461 | 35.139 | 39.087 | 42.557 | 46.693 | 49.588 | 58.302 |
| 30 | 29.336 | 33.530 | 36.250 | 40.256 | 43.773 | 47.962 | 50.892 | 59.703 |

[a]*Generated using an SAS program written by R. W. Washam II, Armour Research Center, Scottsdale, Arizona.*

**Table A.4** The *F* Distribution[a]

| DF for Denominator | 5% Level | | | | | | | | | |
|---|---|---|---|---|---|---|---|---|---|---|
| | DF for Numerator | | | | | | | | | |
| | 1 | 2 | 3 | 4 | 5 | 6 | 8 | 12 | 24 | $\infty$ |
| 1 | 161.4 | 199.5 | 215.7 | 224.6 | 230.2 | 234.0 | 238.9 | 243.9 | 249.0 | 254.3 |
| 2 | 18.51 | 19.00 | 19.16 | 19.25 | 19.30 | 19.33 | 19.37 | 19.41 | 19.45 | 19.50 |
| 3 | 10.13 | 9.55 | 9.28 | 9.12 | 9.01 | 8.94 | 8.84 | 8.74 | 8.64 | 8.53 |
| 4 | 7.71 | 6.94 | 6.59 | 6.39 | 6.26 | 6.16 | 6.04 | 5.91 | 5.77 | 5.63 |
| 5 | 6.61 | 5.79 | 5.41 | 5.19 | 5.05 | 4.95 | 4.82 | 4.68 | 4.53 | 4.36 |
| 6 | 5.99 | 5.14 | 4.76 | 4.53 | 4.39 | 4.28 | 4.15 | 4.00 | 3.84 | 3.67 |
| 7 | 5.59 | 4.74 | 4.35 | 4.12 | 3.97 | 3.87 | 3.73 | 3.57 | 3.41 | 3.23 |
| 8 | 5.32 | 4.46 | 4.07 | 3.84 | 3.69 | 3.58 | 3.44 | 3.28 | 3.12 | 2.93 |
| 9 | 5.12 | 4.26 | 3.86 | 3.63 | 3.48 | 3.37 | 3.23 | 3.07 | 2.90 | 2.71 |
| 10 | 4.96 | 4.10 | 3.71 | 3.48 | 3.33 | 3.22 | 3.07 | 2.91 | 2.74 | 2.54 |
| 11 | 4.84 | 3.98 | 3.59 | 3.36 | 3.20 | 3.09 | 2.95 | 2.79 | 2.61 | 2.40 |
| 12 | 4.75 | 3.88 | 3.49 | 3.26 | 3.11 | 3.00 | 2.85 | 2.69 | 2.50 | 2.30 |
| 13 | 4.67 | 3.80 | 3.41 | 3.18 | 3.02 | 2.92 | 2.77 | 2.60 | 2.42 | 2.21 |
| 14 | 4.60 | 3.74 | 3.34 | 3.11 | 2.96 | 2.85 | 2.70 | 2.53 | 2.35 | 2.13 |
| 15 | 4.54 | 3.68 | 3.29 | 3.06 | 2.90 | 2.79 | 2.64 | 2.48 | 2.29 | 2.07 |
| 16 | 4.49 | 3.63 | 3.24 | 3.01 | 2.85 | 2.74 | 2.59 | 2.42 | 2.24 | 2.01 |
| 17 | 4.45 | 3.59 | 3.20 | 2.96 | 2.81 | 2.70 | 2.55 | 2.38 | 2.19 | 1.96 |
| 18 | 4.41 | 3.55 | 3.16 | 2.93 | 2.77 | 2.66 | 2.51 | 2.34 | 2.15 | 1.92 |
| 19 | 4.38 | 3.52 | 3.13 | 2.90 | 2.74 | 2.63 | 2.48 | 2.31 | 2.11 | 1.88 |
| 20 | 4.35 | 3.49 | 3.10 | 2.87 | 2.71 | 2.60 | 2.45 | 2.28 | 2.08 | 1.84 |
| 21 | 4.32 | 3.47 | 3.07 | 2.84 | 2.68 | 2.57 | 2.42 | 2.25 | 2.05 | 1.81 |
| 22 | 4.30 | 3.44 | 3.05 | 2.82 | 2.66 | 2.55 | 2.40 | 2.23 | 2.03 | 1.78 |
| 23 | 4.28 | 3.42 | 3.03 | 2.80 | 2.64 | 2.53 | 2.38 | 2.20 | 2.00 | 1.76 |
| 24 | 4.26 | 3.40 | 3.01 | 2.78 | 2.62 | 2.51 | 2.36 | 2.18 | 1.98 | 1.73 |
| 25 | 4.24 | 3.38 | 2.99 | 2.76 | 2.60 | 2.49 | 2.34 | 2.16 | 1.96 | 1.71 |

*(continued)*

**Table A.4**  The *F* Distribution[a]—*cont...*

| | 5% Level | | | | | | | | | |
|---|---|---|---|---|---|---|---|---|---|---|
| | DF for Numerator | | | | | | | | | |
| DF for Denominator | 1 | 2 | 3 | 4 | 5 | 6 | 8 | 12 | 24 | ∞ |
| 26 | 4.22 | 3.37 | 2.98 | 2.74 | 2.59 | 2.47 | 2.32 | 2.15 | 1.95 | 1.69 |
| 27 | 4.21 | 3.35 | 2.96 | 2.73 | 2.57 | 2.46 | 2.30 | 2.13 | 1.93 | 1.67 |
| 28 | 4.20 | 3.34 | 2.95 | 2.71 | 2.56 | 2.44 | 2.29 | 2.12 | 1.91 | 1.65 |
| 29 | 4.18 | 3.33 | 2.93 | 2.70 | 2.54 | 2.43 | 2.28 | 2.10 | 1.90 | 1.64 |
| 30 | 4.17 | 3.32 | 2.92 | 2.69 | 2.53 | 2.42 | 2.27 | 2.09 | 1.89 | 1.62 |
| 40 | 4.08 | 3.23 | 2.84 | 2.61 | 2.45 | 2.34 | 2.18 | 2.00 | 1.79 | 1.51 |
| 60 | 4.00 | 3.15 | 2.76 | 2.52 | 2.37 | 2.25 | 2.10 | 1.92 | 1.70 | 1.39 |
| 120 | 3.92 | 3.07 | 2.68 | 2.45 | 2.29 | 2.17 | 2.02 | 1.83 | 1.61 | 1.25 |
| ∞ | 3.84 | 2.99 | 2.60 | 2.37 | 2.21 | 2.10 | 1.94 | 1.75 | 1.52 | 1.00 |

| | 1% Level | | | | | | | | | |
|---|---|---|---|---|---|---|---|---|---|---|
| | DF for Numerator | | | | | | | | | |
| DF for Denominator | 1 | 2 | 3 | 4 | 5 | 6 | 8 | 12 | 24 | ∞ |
| 1 | 4052 | 4999 | 5403 | 5625 | 5764 | 5859 | 5982 | 6106 | 6234 | 6366 |
| 2 | 98.50 | 99.00 | 99.17 | 99.25 | 99.30 | 99.33 | 99.37 | 99.42 | 99.46 | 99.50 |
| 3 | 34.12 | 30.82 | 29.46 | 28.71 | 28.24 | 27.91 | 27.49 | 27.05 | 26.60 | 26.12 |
| 4 | 21.20 | 18.00 | 16.69 | 15.98 | 15.52 | 15.21 | 14.80 | 14.37 | 13.93 | 13.46 |
| 5 | 16.26 | 13.27 | 12.06 | 11.39 | 10.97 | 10.67 | 10.29 | 9.89 | 9.47 | 9.02 |
| 6 | 13.74 | 10.92 | 9.78 | 9.15 | 8.75 | 8.47 | 8.10 | 7.72 | 7.31 | 6.88 |
| 7 | 12.25 | 9.55 | 8.45 | 7.85 | 7.46 | 7.19 | 6.84 | 6.47 | 6.07 | 5.65 |
| 8 | 11.26 | 8.65 | 7.59 | 7.01 | 6.63 | 6.37 | 6.03 | 5.67 | 5.28 | 4.86 |
| 9 | 10.56 | 8.02 | 6.99 | 6.42 | 6.06 | 5.80 | 5.47 | 5.11 | 4.73 | 4.31 |
| 10 | 10.04 | 7.56 | 6.55 | 5.99 | 5.64 | 5.39 | 5.06 | 4.71 | 4.33 | 3.91 |
| 11 | 9.65 | 7.20 | 6.22 | 5.67 | 5.32 | 5.07 | 4.74 | 4.40 | 4.02 | 3.60 |
| 12 | 9.33 | 6.93 | 5.95 | 5.41 | 5.06 | 4.82 | 4.50 | 4.16 | 3.78 | 3.36 |
| 13 | 9.07 | 6.70 | 5.74 | 5.20 | 4.86 | 4.62 | 4.30 | 3.96 | 3.59 | 3.16 |
| 14 | 8.86 | 6.51 | 5.56 | 5.03 | 4.69 | 4.46 | 4.14 | 3.80 | 3.43 | 3.00 |

*(continued)*

**Table A.4** The *F* Distribution[a]—*cont...*

| DF for Denominator | 1% Level | | | | | | | | | |
|---|---|---|---|---|---|---|---|---|---|---|
| | DF for Numerator | | | | | | | | | |
| | **1** | **2** | **3** | **4** | **5** | **6** | **8** | **12** | **24** | **∞** |
| 15 | 8.68 | 6.36 | 5.42 | 4.89 | 4.56 | 4.32 | 4.00 | 3.67 | 3.29 | 2.87 |
| 16 | 8.53 | 6.23 | 5.29 | 4.77 | 4.44 | 4.20 | 3.89 | 3.55 | 3.18 | 2.75 |
| 17 | 8.40 | 6.11 | 5.18 | 4.67 | 4.34 | 4.10 | 3.79 | 3.45 | 3.08 | 2.65 |
| 18 | 8.28 | 6.01 | 5.09 | 4.58 | 4.25 | 4.01 | 3.71 | 3.37 | 3.00 | 2.57 |
| 19 | 8.18 | 5.93 | 5.01 | 4.50 | 4.17 | 3.94 | 3.63 | 3.30 | 2.92 | 2.49 |
| 20 | 8.10 | 5.85 | 4.94 | 4.43 | 4.10 | 3.87 | 3.56 | 3.23 | 2.86 | 2.42 |
| 21 | 8.02 | 5.78 | 4.87 | 4.37 | 4.04 | 3.81 | 3.51 | 3.17 | 2.80 | 2.36 |
| 22 | 7.94 | 5.72 | 4.82 | 4.31 | 3.99 | 3.76 | 3.45 | 3.12 | 2.75 | 2.31 |
| 23 | 7.88 | 5.66 | 4.76 | 4.26 | 3.94 | 3.71 | 3.41 | 3.07 | 2.70 | 2.26 |
| 24 | 7.82 | 5.61 | 4.72 | 4.22 | 3.90 | 3.67 | 3.36 | 3.03 | 2.66 | 2.21 |
| 25 | 7.77 | 5.57 | 4.68 | 4.18 | 3.86 | 3.63 | 3.32 | 2.99 | 2.62 | 2.17 |
| 26 | 7.72 | 5.53 | 4.46 | 4.14 | 3.82 | 3.59 | 3.29 | 2.96 | 2.58 | 2.13 |
| 27 | 7.68 | 5.49 | 4.60 | 4.11 | 3.78 | 3.56 | 3.26 | 2.93 | 2.55 | 2.10 |
| 28 | 7.64 | 5.45 | 4.57 | 4.07 | 3.75 | 3.53 | 3.23 | 2.90 | 2.52 | 2.06 |
| 29 | 7.60 | 5.42 | 4.54 | 4.04 | 3.73 | 3.50 | 3.20 | 2.87 | 2.49 | 2.03 |
| 30 | 7.56 | 5.39 | 4.51 | 4.02 | 3.70 | 3.47 | 3.17 | 2.84 | 2.47 | 2.01 |
| 40 | 7.31 | 5.18 | 4.31 | 3.83 | 3.51 | 3.29 | 2.99 | 2.66 | 2.29 | 1.80 |
| 60 | 7.08 | 4.98 | 4.13 | 3.65 | 3.34 | 3.12 | 2.82 | 2.50 | 2.12 | 1.60 |
| 120 | 6.85 | 4.79 | 3.95 | 3.48 | 3.17 | 2.96 | 2.66 | 2.34 | 1.95 | 1.38 |
| ∞ | 6.64 | 4.60 | 3.78 | 3.32 | 3.02 | 2.80 | 2.51 | 2.18 | 1.79 | 1.00 |

[a]Source: M. Merrington and C. M. Thompson. (1943). Tables of percentage points of the inverted beta (F) distribution. *Biometrika*, **33**, 73–99. Reproduced with permission of the Biometrika Trustees.

**Table A.5** Percentage Points of Dunnett's Multiple-Comparison Test[a]

| | | | | | | | | | | | | | | | | | | |
|---|---|---|---|---|---|---|---|---|---|---|---|---|---|---|---|---|---|---|
| | | | | | | | | $D_{\alpha,k,v}$ **(One-Tailed)** | | | | | | | | | | |
| | | | | $\alpha = .05$ | | | | | | | | | $\alpha = .01$ | | | | | |
| $v$ \ $k$ | 1 | 2 | 3 | 4 | 5 | 6 | 7 | 8 | 9 | 1 | 2 | 3 | 4 | 5 | 6 | 7 | 8 | 9 |
| 5 | 2.02 | 2.44 | 2.68 | 2.85 | 2.98 | 3.08 | 3.16 | 3.24 | 3.30 | 3.37 | 3.90 | 4.21 | 4.43 | 4.60 | 4.73 | 4.85 | 4.94 | 5.03 |
| 6 | 1.94 | 2.34 | 2.56 | 2.71 | 2.83 | 2.92 | 3.00 | 3.07 | 3.12 | 3.14 | 3.61 | 3.88 | 4.07 | 4.21 | 4.33 | 4.43 | 4.51 | 4.59 |
| 7 | 1.89 | 2.27 | 2.48 | 2.62 | 2.73 | 2.82 | 2.89 | 2.95 | 3.01 | 3.00 | 3.42 | 3.66 | 3.83 | 3.96 | 4.07 | 4.15 | 4.23 | 4.30 |
| 8 | 1.86 | 2.22 | 2.42 | 2.55 | 2.66 | 2.74 | 2.81 | 2.87 | 2.92 | 2.90 | 3.29 | 3.51 | 3.67 | 3.79 | 3.88 | 3.96 | 4.03 | 4.09 |
| 9 | 1.83 | 2.18 | 2.37 | 2.50 | 2.60 | 2.68 | 2.75 | 2.81 | 2.86 | 2.82 | 3.19 | 3.40 | 3.55 | 3.66 | 3.75 | 3.82 | 3.89 | 3.94 |
| 10 | 1.81 | 2.15 | 2.34 | 2.47 | 2.56 | 2.64 | 2.70 | 2.76 | 2.81 | 2.76 | 3.11 | 3.31 | 3.45 | 3.56 | 3.64 | 3.71 | 3.78 | 3.83 |
| 11 | 1.80 | 2.13 | 2.31 | 2.44 | 2.53 | 2.60 | 2.67 | 2.72 | 2.77 | 2.72 | 3.06 | 3.25 | 3.38 | 3.48 | 3.56 | 3.63 | 3.69 | 3.74 |
| 12 | 1.78 | 2.11 | 2.29 | 2.41 | 2.50 | 2.58 | 2.64 | 2.69 | 2.74 | 2.68 | 3.01 | 3.19 | 3.32 | 3.42 | 3.50 | 3.56 | 3.62 | 3.67 |
| 13 | 1.77 | 2.09 | 2.27 | 2.39 | 2.48 | 2.55 | 2.61 | 2.66 | 2.71 | 2.65 | 2.97 | 3.15 | 3.27 | 3.37 | 3.44 | 3.51 | 3.56 | 3.61 |
| 14 | 1.76 | 2.08 | 2.25 | 2.37 | 2.46 | 2.53 | 2.59 | 2.64 | 2.69 | 2.62 | 2.94 | 3.11 | 3.23 | 3.32 | 3.40 | 3.46 | 3.51 | 3.56 |
| 15 | 1.75 | 2.07 | 2.24 | 2.36 | 2.44 | 2.51 | 2.57 | 2.62 | 2.67 | 2.60 | 2.91 | 3.08 | 3.20 | 3.29 | 3.36 | 3.42 | 3.47 | 3.52 |
| 16 | 1.75 | 2.06 | 2.23 | 2.34 | 2.43 | 2.50 | 2.56 | 2.61 | 2.65 | 2.58 | 2.88 | 3.05 | 3.17 | 3.26 | 3.33 | 3.39 | 3.44 | 3.48 |
| 17 | 1.74 | 2.05 | 2.22 | 2.33 | 2.42 | 2.49 | 2.54 | 2.59 | 2.64 | 2.57 | 2.86 | 3.03 | 3.14 | 3.23 | 3.30 | 3.36 | 3.41 | 3.45 |
| 18 | 1.73 | 2.04 | 2.21 | 2.32 | 2.41 | 2.48 | 2.53 | 2.58 | 2.62 | 2.55 | 2.84 | 3.01 | 3.12 | 3.21 | 3.27 | 3.33 | 3.38 | 3.42 |
| 19 | 1.73 | 2.03 | 2.20 | 2.31 | 2.40 | 2.47 | 2.52 | 2.57 | 2.61 | 2.54 | 2.83 | 2.99 | 3.10 | 3.18 | 3.25 | 3.31 | 3.36 | 3.40 |
| 20 | 1.72 | 2.03 | 2.19 | 2.30 | 2.39 | 2.46 | 2.51 | 2.56 | 2.60 | 2.53 | 2.81 | 2.97 | 3.08 | 3.17 | 3.23 | 3.29 | 3.34 | 3.38 |
| 24 | 1.71 | 2.01 | 2.17 | 2.28 | 2.36 | 2.43 | 2.48 | 2.53 | 2.57 | 2.49 | 2.77 | 2.92 | 3.03 | 3.11 | 3.17 | 3.22 | 3.27 | 3.31 |
| 30 | 1.70 | 1.99 | 2.15 | 2.25 | 2.33 | 2.40 | 2.45 | 2.50 | 2.54 | 2.46 | 2.72 | 2.87 | 2.97 | 3.05 | 3.11 | 3.16 | 3.21 | 3.24 |
| 40 | 1.68 | 1.97 | 2.13 | 2.23 | 2.31 | 2.37 | 2.42 | 2.47 | 2.51 | 2.42 | 2.68 | 2.82 | 2.92 | 2.99 | 3.05 | 3.10 | 3.14 | 3.18 |
| 60 | 1.67 | 1.95 | 2.10 | 2.21 | 2.28 | 2.35 | 2.39 | 2.44 | 2.48 | 2.39 | 2.64 | 2.78 | 2.87 | 2.94 | 3.00 | 3.04 | 3.08 | 3.12 |
| 120 | 1.66 | 1.93 | 2.08 | 2.18 | 2.26 | 2.32 | 2.37 | 2.41 | 2.45 | 2.36 | 2.60 | 2.73 | 2.82 | 2.89 | 2.94 | 2.99 | 3.03 | 3.06 |
| $\infty$ | 1.64 | 1.92 | 2.06 | 2.16 | 2.23 | 2.29 | 2.34 | 2.38 | 2.42 | 2.33 | 2.56 | 2.68 | 2.77 | 2.84 | 2.89 | 2.93 | 2.97 | 3.00 |

*(continued)*

**Table A.5** Percentage Points of Dunnett's Multiple-Comparison Test[a]—*cont...*

| | $|D|_{\alpha, k, v}$ (Two-Tailed) | | | | | | | | | | | | | |
|---|---|---|---|---|---|---|---|---|---|---|---|---|---|---|
| | $\alpha = .05$ | | | | | | | | | | | | | |
| $v$ \ $k$ | 1 | 2 | 3 | 4 | 5 | 6 | 7 | 8 | 9 | 10 | 11 | 12 | 15 | 20 |
| 5 | 2.57 | 3.03 | 3.29 | 3.48 | 3.62 | 3.73 | 3.82 | 3.90 | 3.97 | 4.03 | 4.09 | 4.14 | 4.26 | 4.42 |
| 6 | 2.45 | 2.86 | 3.10 | 3.26 | 3.39 | 3.49 | 3.57 | 3.64 | 3.71 | 3.76 | 3.81 | 3.86 | 3.97 | 4.11 |
| 7 | 2.36 | 2.75 | 2.97 | 3.12 | 3.24 | 3.33 | 3.41 | 3.47 | 3.53 | 3.58 | 3.63 | 3.67 | 3.78 | 3.91 |
| 8 | 2.31 | 2.67 | 2.88 | 3.02 | 3.13 | 3.22 | 3.29 | 3.35 | 3.41 | 3.46 | 3.50 | 3.54 | 3.64 | 3.76 |
| 9 | 2.26 | 2.61 | 2.81 | 2.95 | 3.05 | 3.14 | 3.20 | 3.26 | 3.32 | 3.36 | 3.40 | 3.44 | 3.53 | 3.65 |
| 10 | 2.23 | 2.57 | 2.76 | 2.89 | 2.99 | 3.07 | 3.14 | 3.19 | 3.24 | 3.29 | 3.33 | 3.36 | 3.45 | 3.57 |
| 11 | 2.20 | 2.53 | 2.72 | 2.84 | 2.94 | 3.02 | 3.08 | 3.14 | 3.19 | 3.23 | 3.27 | 3.30 | 3.39 | 3.50 |
| 12 | 2.18 | 2.50 | 2.68 | 2.81 | 2.90 | 2.98 | 3.04 | 3.09 | 3.14 | 3.18 | 3.22 | 3.25 | 3.34 | 3.45 |
| 13 | 2.16 | 2.48 | 2.65 | 2.78 | 2.87 | 2.94 | 3.00 | 3.06 | 3.10 | 3.14 | 3.18 | 3.21 | 3.29 | 3.40 |
| 14 | 2.14 | 2.46 | 2.63 | 2.75 | 2.84 | 2.91 | 2.97 | 3.02 | 3.07 | 3.11 | 3.14 | 3.18 | 3.26 | 3.36 |
| 15 | 2.13 | 2.44 | 2.61 | 2.73 | 2.82 | 2.89 | 2.95 | 3.00 | 3.04 | 3.08 | 3.12 | 3.15 | 3.23 | 3.33 |
| 16 | 2.12 | 2.42 | 2.59 | 2.71 | 2.80 | 2.87 | 2.92 | 2.97 | 3.02 | 3.06 | 3.09 | 3.12 | 3.20 | 3.30 |
| 17 | 2.11 | 2.41 | 2.58 | 2.69 | 2.78 | 2.85 | 2.90 | 2.95 | 3.00 | 3.03 | 3.07 | 3.10 | 3.18 | 3.27 |
| 18 | 2.10 | 2.40 | 2.56 | 2.68 | 2.76 | 2.83 | 2.89 | 2.94 | 2.98 | 3.01 | 3.05 | 3.08 | 3.16 | 3.25 |
| 19 | 2.09 | 2.39 | 2.55 | 2.66 | 2.75 | 2.81 | 2.87 | 2.92 | 2.96 | 3.00 | 3.03 | 3.06 | 3.14 | 3.23 |
| 20 | 2.09 | 2.38 | 2.54 | 2.65 | 2.73 | 2.80 | 2.86 | 2.90 | 2.95 | 2.98 | 3.02 | 3.05 | 3.12 | 3.22 |
| 24 | 2.06 | 2.35 | 2.51 | 2.61 | 2.70 | 2.76 | 2.81 | 2.86 | 2.90 | 2.94 | 2.97 | 3.00 | 3.07 | 3.16 |
| 30 | 2.04 | 2.32 | 2.47 | 2.58 | 2.66 | 2.72 | 2.77 | 2.82 | 2.86 | 2.89 | 2.92 | 2.95 | 3.02 | 3.11 |
| 40 | 2.02 | 2.29 | 2.44 | 2.54 | 2.62 | 2.68 | 2.73 | 2.77 | 2.81 | 2.85 | 2.87 | 2.90 | 2.97 | 3.06 |
| 60 | 2.00 | 2.27 | 2.41 | 2.51 | 2.58 | 2.64 | 2.69 | 2.73 | 2.77 | 2.80 | 2.83 | 2.86 | 2.92 | 3.00 |
| 120 | 1.98 | 2.24 | 2.38 | 2.47 | 2.55 | 2.60 | 2.65 | 2.69 | 2.73 | 2.76 | 2.79 | 2.81 | 2.87 | 2.95 |
| $\infty$ | 1.96 | 2.21 | 2.35 | 2.44 | 2.51 | 2.57 | 2.61 | 2.65 | 2.69 | 2.72 | 2.74 | 2.77 | 2.83 | 2.91 |

*(continued)*

**Table A.5** Percentage Points of Dunnett's Multiple-Comparison Test[a]—*cont...*

| | | | | | | $\alpha = .01$ | | | | | | | | |
|---|---|---|---|---|---|---|---|---|---|---|---|---|---|---|
| v \ k | 1 | 2 | 3 | 4 | 5 | 6 | 7 | 8 | 9 | 10 | 11 | 12 | 15 | 20 |
| 5 | 4.03 | 4.63 | 4.98 | 5.22 | 5.41 | 5.56 | 5.69 | 5.80 | 5.89 | 5.98 | 6.05 | 6.12 | 6.30 | 6.52 |
| 6 | 3.71 | 4.21 | 4.51 | 4.71 | 4.87 | 5.00 | 5.10 | 5.20 | 5.28 | 5.35 | 5.41 | 5.47 | 5.62 | 5.81 |
| 7 | 3.50 | 3.95 | 4.21 | 4.39 | 4.53 | 4.64 | 4.74 | 4.82 | 4.89 | 4.95 | 5.01 | 5.06 | 5.19 | 5.36 |
| 8 | 3.36 | 3.77 | 4.00 | 4.17 | 4.29 | 4.40 | 4.48 | 4.56 | 4.62 | 4.68 | 4.73 | 4.78 | 4.90 | 5.05 |
| 9 | 3.25 | 3.63 | 3.85 | 4.01 | 4.12 | 4.22 | 4.30 | 4.37 | 4.43 | 4.48 | 4.53 | 4.57 | 4.68 | 4.82 |
| 10 | 3.17 | 3.53 | 3.74 | 3.88 | 3.99 | 4.08 | 4.16 | 4.22 | 4.28 | 4.33 | 4.37 | 4.42 | 4.52 | 4.65 |
| 11 | 3.11 | 3.45 | 3.65 | 3.79 | 3.89 | 3.98 | 4.05 | 4.11 | 4.16 | 4.21 | 4.25 | 4.29 | 4.39 | 4.52 |
| 12 | 3.05 | 3.39 | 3.58 | 3.71 | 3.81 | 3.89 | 3.96 | 4.02 | 4.07 | 4.12 | 4.16 | 4.19 | 4.29 | 4.41 |
| 13 | 3.01 | 3.33 | 3.52 | 3.65 | 3.74 | 3.82 | 3.89 | 3.94 | 3.99 | 4.04 | 4.08 | 4.11 | 4.20 | 4.32 |
| 14 | 2.98 | 3.29 | 3.47 | 3.59 | 3.69 | 3.76 | 3.83 | 3.88 | 3.93 | 3.97 | 4.01 | 4.05 | 4.13 | 4.24 |
| 15 | 2.95 | 3.25 | 3.43 | 3.55 | 3.64 | 3.71 | 3.78 | 3.83 | 3.88 | 3.92 | 3.95 | 3.99 | 4.07 | 4.18 |
| 16 | 2.92 | 3.22 | 3.39 | 3.51 | 3.60 | 3.67 | 3.73 | 3.78 | 3.83 | 3.87 | 3.91 | 3.94 | 4.02 | 4.13 |
| 17 | 2.90 | 3.19 | 3.36 | 3.47 | 3.56 | 3.63 | 3.69 | 3.74 | 3.79 | 3.83 | 3.86 | 3.90 | 3.98 | 4.08 |
| 18 | 2.88 | 3.17 | 3.33 | 3.44 | 3.53 | 3.60 | 3.66 | 3.71 | 3.75 | 3.79 | 3.83 | 3.86 | 3.94 | 4.04 |
| 19 | 2.86 | 3.15 | 3.31 | 3.42 | 3.50 | 3.57 | 3.63 | 3.68 | 3.72 | 3.76 | 3.79 | 3.83 | 3.90 | 4.00 |
| 20 | 2.85 | 3.13 | 3.29 | 3.40 | 3.48 | 3.55 | 3.60 | 3.65 | 3.69 | 3.73 | 3.77 | 3.80 | 3.87 | 3.97 |
| 24 | 2.80 | 3.07 | 3.22 | 3.32 | 3.40 | 3.47 | 3.52 | 3.57 | 3.61 | 3.64 | 3.68 | 3.70 | 3.78 | 3.87 |
| 30 | 2.75 | 3.01 | 3.15 | 3.25 | 3.33 | 3.39 | 3.44 | 3.49 | 3.52 | 3.56 | 3.59 | 3.62 | 3.69 | 3.78 |
| 40 | 2.70 | 2.95 | 3.09 | 3.19 | 3.26 | 3.32 | 3.37 | 3.41 | 3.44 | 3.48 | 3.51 | 3.53 | 3.60 | 3.68 |
| 60 | 2.66 | 2.90 | 3.03 | 3.12 | 3.19 | 3.25 | 3.29 | 3.33 | 3.37 | 3.40 | 3.42 | 3.45 | 3.51 | 3.59 |
| 120 | 2.62 | 2.85 | 2.97 | 3.06 | 3.12 | 3.18 | 3.22 | 3.26 | 3.29 | 3.32 | 3.35 | 3.37 | 3.43 | 3.51 |
| $\infty$ | 2.58 | 2.79 | 2.92 | 3.00 | 3.06 | 3.11 | 3.15 | 3.19 | 3.22 | 3.25 | 3.27 | 3.29 | 3.35 | 3.42 |

[a]*Source: R. G. Miller, Jr. (1981). Simultaneous Statistical Inference. Springer-Verlag, New York. Reproduced with permission of the publisher and The Biometric Society.*

**Table A.6** Critical Values $Q_{\alpha,k,\nu}$ of Tukey's Studentized Range[a]

| DF | $\alpha$ | 2 | 3 | 4 | 5 | 6 | 7 | 8 | 9 | 10 | 11 | 12 | 13 | 14 | 15 | 16 | 17 | 18 | 19 | 20 | $\alpha$ | DF |
|----|----|----|----|----|----|----|----|----|----|----|----|----|----|----|----|----|----|----|----|----|----|----|
| | | | | | | | | | | $k$ = Number of Means | | | | | | | | | | | | |
| 5 | .05 | 3.64 | 4.60 | 5.22 | 5.67 | 6.03 | 6.33 | 6.58 | 6.80 | 6.99 | 7.17 | 7.32 | 7.47 | 7.60 | 7.72 | 7.83 | 7.93 | 8.03 | 8.12 | 8.21 | .05 | 5 |
| | .01 | 5.70 | 6.98 | 7.80 | 8.42 | 8.91 | 9.32 | 9.67 | 9.97 | 10.24 | 10.48 | 10.70 | 10.89 | 11.08 | 11.24 | 11.40 | 11.55 | 11.68 | 11.81 | 11.93 | .01 | |
| 6 | .05 | 3.46 | 4.34 | 4.90 | 5.30 | 5.63 | 5.90 | 6.12 | 6.32 | 6.49 | 6.65 | 6.79 | 6.92 | 7.03 | 7.14 | 7.24 | 7.34 | 7.43 | 7.51 | 7.59 | .05 | 6 |
| | .01 | 5.24 | 6.33 | 7.03 | 7.56 | 7.97 | 8.32 | 8.61 | 8.87 | 9.10 | 9.30 | 9.48 | 9.65 | 9.81 | 9.95 | 10.08 | 10.21 | 10.32 | 10.43 | 10.54 | .01 | |
| 7 | .05 | 3.34 | 4.16 | 4.68 | 5.06 | 5.36 | 5.61 | 5.82 | 6.00 | 6.16 | 6.30 | 6.43 | 6.55 | 6.66 | 6.76 | 6.85 | 6.94 | 7.02 | 7.10 | 7.17 | .05 | 7 |
| | .01 | 4.95 | 5.92 | 6.54 | 7.01 | 7.37 | 7.68 | 7.94 | 8.17 | 8.37 | 8.55 | 8.71 | 8.86 | 9.00 | 9.12 | 9.24 | 9.35 | 9.46 | 9.55 | 9.65 | .01 | |
| 8 | .05 | 3.26 | 4.04 | 4.53 | 4.89 | 5.17 | 5.40 | 5.60 | 5.77 | 5.92 | 6.05 | 6.18 | 6.29 | 6.39 | 6.48 | 6.57 | 6.65 | 6.73 | 6.80 | 6.87 | .05 | 8 |
| | .01 | 4.75 | 5.64 | 6.20 | 6.62 | 6.96 | 7.24 | 7.47 | 7.68 | 7.86 | 8.03 | 8.18 | 8.31 | 8.44 | 8.55 | 8.66 | 8.76 | 8.85 | 8.94 | 9.03 | .01 | |
| 9 | .05 | 3.20 | 3.95 | 4.41 | 4.76 | 5.02 | 5.24 | 5.43 | 5.59 | 5.74 | 5.87 | 5.98 | 6.09 | 6.19 | 6.28 | 6.36 | 6.44 | 6.51 | 6.58 | 6.64 | .05 | 9 |
| | .01 | 4.60 | 5.43 | 5.96 | 6.35 | 6.66 | 6.91 | 7.13 | 7.33 | 7.49 | 7.65 | 7.78 | 7.91 | 8.03 | 8.13 | 8.23 | 8.33 | 8.41 | 8.49 | 8.57 | .01 | |
| 10 | .05 | 3.15 | 3.88 | 4.33 | 4.65 | 4.91 | 5.12 | 5.30 | 5.46 | 5.60 | 5.72 | 5.83 | 5.93 | 6.03 | 6.11 | 6.19 | 6.27 | 6.34 | 6.40 | 6.47 | .05 | 10 |
| | .01 | 4.48 | 5.27 | 5.77 | 6.14 | 6.43 | 6.67 | 6.87 | 7.05 | 7.21 | 7.36 | 7.49 | 7.60 | 7.71 | 7.81 | 7.91 | 7.99 | 8.08 | 8.15 | 8.23 | .01 | |
| 11 | .05 | 3.11 | 3.82 | 4.26 | 4.57 | 4.82 | 5.03 | 5.20 | 5.35 | 5.49 | 5.61 | 5.71 | 5.81 | 5.90 | 5.98 | 6.06 | 6.13 | 6.20 | 6.27 | 6.33 | .05 | 11 |
| | .01 | 4.39 | 5.15 | 5.62 | 5.97 | 6.25 | 6.48 | 6.67 | 6.84 | 6.99 | 7.13 | 7.25 | 7.36 | 7.46 | 7.56 | 7.65 | 7.73 | 7.81 | 7.88 | 7.95 | .01 | |
| 12 | .05 | 3.08 | 3.77 | 4.20 | 4.51 | 4.75 | 4.95 | 5.12 | 5.27 | 5.39 | 5.51 | 5.61 | 5.71 | 5.80 | 5.88 | 5.95 | 6.02 | 6.09 | 6.15 | 6.21 | .05 | 12 |
| | .01 | 4.32 | 5.05 | 5.50 | 5.84 | 6.10 | 6.32 | 6.51 | 6.67 | 6.81 | 6.94 | 7.06 | 7.17 | 7.26 | 7.36 | 7.44 | 7.52 | 7.59 | 7.66 | 7.73 | .01 | |
| 13 | .05 | 3.06 | 3.73 | 4.15 | 4.45 | 4.69 | 4.88 | 5.05 | 5.19 | 5.32 | 5.43 | 5.53 | 5.63 | 5.71 | 5.79 | 5.86 | 5.93 | 5.99 | 6.05 | 6.11 | .05 | 13 |
| | .01 | 4.26 | 4.96 | 5.40 | 5.73 | 5.98 | 6.19 | 6.37 | 6.53 | 6.67 | 6.79 | 6.90 | 7.01 | 7.10 | 7.19 | 7.27 | 7.35 | 7.42 | 7.48 | 7.55 | .01 | |
| 14 | .05 | 3.03 | 3.70 | 4.11 | 4.41 | 4.64 | 4.83 | 4.99 | 5.13 | 5.25 | 5.36 | 5.46 | 5.55 | 5.64 | 5.71 | 5.79 | 5.85 | 5.91 | 5.97 | 6.03 | .05 | 14 |
| | .01 | 4.21 | 4.89 | 5.32 | 5.63 | 5.88 | 6.08 | 6.26 | 6.41 | 6.54 | 6.66 | 6.77 | 6.87 | 6.96 | 7.05 | 7.13 | 7.20 | 7.27 | 7.33 | 7.39 | .01 | |
| 15 | .05 | 3.01 | 3.67 | 4.08 | 4.37 | 4.59 | 4.78 | 4.94 | 5.08 | 5.20 | 5.31 | 5.40 | 5.49 | 5.57 | 5.65 | 5.72 | 5.78 | 5.85 | 5.90 | 5.96 | .05 | 15 |
| | .01 | 4.17 | 4.84 | 5.25 | 5.56 | 5.80 | 5.99 | 6.16 | 6.31 | 6.44 | 6.55 | 6.66 | 6.76 | 6.84 | 6.93 | 7.00 | 7.07 | 7.14 | 7.20 | 7.26 | .01 | |
| 16 | .05 | 3.00 | 3.65 | 4.05 | 4.33 | 4.56 | 4.74 | 4.90 | 5.03 | 5.15 | 5.26 | 5.35 | 5.44 | 5.52 | 5.59 | 5.66 | 5.73 | 5.79 | 5.84 | 5.90 | .05 | 16 |
| | .01 | 4.13 | 4.79 | 5.19 | 5.49 | 5.72 | 5.92 | 6.08 | 6.22 | 6.35 | 6.46 | 6.56 | 6.66 | 6.74 | 6.82 | 6.90 | 6.97 | 7.03 | 7.09 | 7.15 | .01 | |
| 17 | .05 | 2.98 | 3.63 | 4.02 | 4.30 | 4.52 | 4.70 | 4.86 | 4.99 | 5.11 | 5.21 | 5.31 | 5.39 | 5.47 | 5.54 | 5.61 | 5.67 | 5.73 | 5.79 | 5.84 | .05 | 17 |
| | .01 | 4.10 | 4.74 | 5.14 | 5.43 | 5.66 | 5.85 | 6.01 | 6.15 | 6.27 | 6.38 | 6.48 | 6.57 | 6.66 | 6.73 | 6.81 | 6.87 | 6.94 | 7.00 | 7.05 | .01 | |

(continued)

**Table A.6** Critical Values $Q_{\alpha,k,\nu}$ of Tukey's Studentized Range[a] —cont...

| DF | α | 2 | 3 | 4 | 5 | 6 | 7 | 8 | 9 | 10 | 11 | 12 | 13 | 14 | 15 | 16 | 17 | 18 | 19 | 20 | α | DF |
|----|---|---|---|---|---|---|---|---|---|----|----|----|----|----|----|----|----|----|----|----|---|----|
| 18 | .05 | 2.97 | 3.61 | 4.00 | 4.28 | 4.49 | 4.67 | 4.82 | 4.96 | 5.07 | 5.17 | 5.27 | 5.35 | 5.43 | 5.50 | 5.57 | 5.63 | 5.69 | 5.74 | 5.79 | .05 | 18 |
|    | .01 | 4.07 | 4.70 | 5.09 | 5.38 | 5.60 | 5.79 | 5.94 | 6.08 | 6.20 | 6.31 | 6.41 | 6.50 | 6.58 | 6.65 | 6.73 | 6.79 | 6.85 | 6.91 | 6.97 | .01 |    |
| 19 | .05 | 2.96 | 3.59 | 3.98 | 4.25 | 4.47 | 4.65 | 4.79 | 4.92 | 5.04 | 5.14 | 5.23 | 5.31 | 5.39 | 5.46 | 5.53 | 5.59 | 5.65 | 5.70 | 5.75 | .05 | 19 |
|    | .01 | 4.05 | 4.67 | 5.05 | 5.33 | 5.55 | 5.73 | 5.89 | 6.02 | 6.14 | 6.25 | 6.34 | 6.43 | 6.51 | 6.58 | 6.65 | 6.72 | 6.78 | 6.84 | 6.89 | .01 |    |
| 20 | .05 | 2.95 | 3.58 | 3.96 | 4.23 | 4.45 | 4.62 | 4.77 | 4.90 | 5.01 | 5.11 | 5.20 | 5.28 | 5.36 | 5.43 | 5.49 | 5.55 | 5.61 | 5.66 | 5.71 | .05 | 20 |
|    | .01 | 4.02 | 4.64 | 5.02 | 5.29 | 5.51 | 5.69 | 5.84 | 5.97 | 6.09 | 6.19 | 6.28 | 6.37 | 6.45 | 6.52 | 6.59 | 6.65 | 6.71 | 6.77 | 6.82 | .01 |    |
| 24 | .05 | 2.92 | 3.53 | 3.90 | 4.17 | 4.37 | 4.54 | 4.68 | 4.81 | 4.92 | 5.01 | 5.10 | 5.18 | 5.25 | 5.32 | 5.38 | 5.44 | 5.49 | 5.55 | 5.59 | .05 | 24 |
|    | .01 | 3.96 | 4.55 | 4.91 | 5.17 | 5.37 | 5.54 | 5.69 | 5.81 | 5.92 | 6.02 | 6.11 | 6.19 | 6.26 | 6.33 | 6.39 | 6.45 | 6.51 | 6.56 | 6.61 | .01 |    |
| 30 | .05 | 2.89 | 3.49 | 3.85 | 4.10 | 4.30 | 4.46 | 4.60 | 4.72 | 4.82 | 4.92 | 5.00 | 5.08 | 5.15 | 5.21 | 5.27 | 5.33 | 5.38 | 5.43 | 5.47 | .05 | 30 |
|    | .01 | 3.89 | 4.45 | 4.80 | 5.05 | 5.24 | 5.40 | 5.54 | 5.65 | 5.76 | 5.85 | 5.93 | 6.01 | 6.08 | 6.14 | 6.20 | 6.26 | 6.31 | 6.36 | 6.41 | .01 |    |
| 40 | .05 | 2.86 | 3.44 | 3.79 | 4.04 | 4.23 | 4.39 | 4.52 | 4.63 | 4.73 | 4.82 | 4.90 | 4.98 | 5.04 | 5.11 | 5.16 | 5.22 | 5.27 | 5.31 | 5.36 | .05 | 40 |
|    | .01 | 3.82 | 4.37 | 4.70 | 4.93 | 5.11 | 5.26 | 5.39 | 5.50 | 5.60 | 5.69 | 5.76 | 5.83 | 5.90 | 5.96 | 6.02 | 6.07 | 6.12 | 6.16 | 6.21 | .01 |    |
| 60 | .05 | 2.83 | 3.40 | 3.74 | 3.98 | 4.16 | 4.31 | 4.44 | 4.55 | 4.65 | 4.73 | 4.81 | 4.88 | 4.94 | 5.00 | 5.06 | 5.11 | 5.15 | 5.20 | 5.24 | .05 | 60 |
|    | .01 | 3.76 | 4.28 | 4.59 | 4.82 | 4.99 | 5.13 | 5.25 | 5.36 | 5.45 | 5.53 | 5.60 | 5.67 | 5.73 | 5.78 | 5.84 | 5.89 | 5.93 | 5.97 | 6.01 | .01 |    |
| 120 | .05 | 2.80 | 3.36 | 3.68 | 3.92 | 4.10 | 4.24 | 4.36 | 4.47 | 4.56 | 4.64 | 4.71 | 4.78 | 4.84 | 4.90 | 4.95 | 5.00 | 5.04 | 5.09 | 5.13 | .05 | 120 |
|    | .01 | 3.70 | 4.20 | 4.50 | 4.71 | 4.87 | 5.01 | 5.12 | 5.21 | 5.30 | 5.37 | 5.44 | 5.50 | 5.56 | 5.61 | 5.66 | 5.71 | 5.75 | 5.79 | 5.83 | .01 |    |
| ∞ | .05 | 2.77 | 3.31 | 3.63 | 3.86 | 4.03 | 4.17 | 4.29 | 4.39 | 4.47 | 4.55 | 4.62 | 4.68 | 4.74 | 4.80 | 4.85 | 4.89 | 4.93 | 4.97 | 5.01 | .05 | ∞ |
|    | .01 | 3.64 | 4.12 | 4.40 | 4.60 | 4.76 | 4.88 | 4.99 | 5.08 | 5.16 | 5.23 | 5.29 | 5.35 | 5.40 | 5.45 | 5.49 | 5.54 | 5.57 | 5.61 | 5.65 | .01 |    |

k = Number of Means

[a]Source: E. S. Pearson and H. O. Hartley (Eds). (1970). Biometrika Tables for Statisticians, Vol. I, Cambridge University Press, London, pp. 192–193. Reproduced with permission of the Biometrika Trustees.

**Table A.7** Significant Ranges $r_{\alpha,p,v}$ for Duncan's Multiple-Range Test[a]

| | | $p$ = Number of Means for Range Being Tested | | | | | | | | | | | | | |
|---|---|---|---|---|---|---|---|---|---|---|---|---|---|---|---|
| DF | $\alpha$ | 2 | 3 | 4 | 5 | 6 | 7 | 8 | 9 | 10 | 12 | 14 | 16 | 18 | 20 |
| 1 | .05 | 18.0 | 18.0 | 18.0 | 18.0 | 18.0 | 18.0 | 18.0 | 18.0 | 18.0 | 18.0 | 18.0 | 18.0 | 18.0 | 18.0 |
| | .01 | 90.0 | 90.0 | 90.0 | 90.0 | 90.0 | 90.0 | 90.0 | 90.0 | 90.0 | 90.0 | 90.0 | 90.0 | 90.0 | 90.0 |
| 2 | .05 | 6.09 | 6.09 | 6.09 | 6.09 | 6.09 | 6.09 | 6.09 | 6.09 | 6.09 | 6.09 | 6.09 | 6.09 | 6.09 | 6.09 |
| | .01 | 14.0 | 14.0 | 14.0 | 14.0 | 14.0 | 14.0 | 14.0 | 14.0 | 14.0 | 14.0 | 14.0 | 14.0 | 14.0 | 14.0 |
| 3 | .05 | 4.50 | 4.50 | 4.50 | 4.50 | 4.50 | 4.50 | 4.50 | 4.50 | 4.50 | 4.50 | 4.50 | 4.50 | 4.50 | 4.50 |
| | .01 | 8.26 | 8.5 | 8.6 | 8.7 | 8.8 | 8.9 | 8.9 | 9.0 | 9.0 | 9.0 | 9.1 | 9.2 | 9.3 | 9.3 |
| 4 | .05 | 3.93 | 4.01 | 4.02 | 4.02 | 4.02 | 4.02 | 4.02 | 4.02 | 4.02 | 4.02 | 4.02 | 4.02 | 4.02 | 4.02 |
| | .01 | 6.51 | 6.8 | 6.9 | 7.0 | 7.1 | 7.1 | 7.2 | 7.2 | 7.3 | 7.3 | 7.4 | 7.4 | 7.5 | 7.5 |
| 5 | .05 | 3.64 | 3.74 | 3.79 | 3.83 | 3.83 | 3.83 | 3.83 | 3.83 | 3.83 | 3.83 | 3.83 | 3.83 | 3.83 | 3.83 |
| | .01 | 5.70 | 5.96 | 6.11 | 6.18 | 6.26 | 6.33 | 6.40 | 6.44 | 6.5 | 6.6 | 6.6 | 6.7 | 6.7 | 6.8 |
| 6 | .05 | 3.46 | 3.58 | 3.64 | 3.68 | 3.68 | 3.68 | 3.68 | 3.68 | 3.68 | 3.68 | 3.68 | 3.68 | 3.68 | 3.68 |
| | .01 | 5.24 | 5.51 | 5.65 | 5.73 | 5.81 | 5.88 | 5.95 | 6.00 | 6.0 | 6.1 | 6.2 | 6.2 | 6.3 | 6.3 |
| 7 | .05 | 3.35 | 3.47 | 3.54 | 3.58 | 3.60 | 3.61 | 3.61 | 3.61 | 3.61 | 3.61 | 3.61 | 3.61 | 3.61 | 3.61 |
| | .01 | 4.95 | 5.22 | 5.37 | 5.45 | 5.53 | 5.61 | 5.69 | 5.73 | 5.8 | 5.8 | 5.9 | 5.9 | 6.0 | 6.0 |
| 8 | .05 | 3.26 | 3.39 | 3.47 | 3.52 | 3.55 | 3.56 | 3.56 | 3.56 | 3.56 | 3.56 | 3.56 | 3.56 | 3.56 | 3.56 |
| | .01 | 4.74 | 5.00 | 5.14 | 5.23 | 5.32 | 5.40 | 5.47 | 5.51 | 5.5 | 5.6 | 5.7 | 5.7 | 5.8 | 5.8 |
| 9 | .05 | 3.20 | 3.34 | 3.41 | 3.47 | 3.50 | 3.52 | 3.52 | 3.52 | 3.52 | 3.52 | 3.52 | 3.52 | 3.52 | 3.52 |
| | .01 | 4.60 | 4.86 | 4.99 | 5.08 | 5.17 | 5.25 | 5.32 | 5.36 | 5.4 | 5.5 | 5.5 | 5.6 | 5.7 | 5.7 |
| 10 | .05 | 3.15 | 3.30 | 3.37 | 3.43 | 3.46 | 3.47 | 3.47 | 3.47 | 3.47 | 3.47 | 3.47 | 3.47 | 3.47 | 3.48 |
| | .01 | 4.48 | 4.73 | 4.88 | 4.96 | 5.06 | 5.13 | 5.20 | 5.24 | 5.28 | 5.36 | 5.42 | 5.48 | 5.54 | 5.55 |
| 11 | .05 | 3.11 | 3.27 | 3.35 | 3.39 | 3.43 | 3.44 | 3.45 | 3.46 | 3.46 | 3.46 | 3.46 | 3.46 | 3.47 | 3.48 |
| | .01 | 4.39 | 4.63 | 4.77 | 4.86 | 4.94 | 5.01 | 5.06 | 5.12 | 5.15 | 5.24 | 5.28 | 5.34 | 5.38 | 5.39 |

*(continued)*

**Table A.7** Significant Ranges $r_{\alpha,p,v}$ for Duncan's Multiple-Range Test[a]—cont...

| DF | α | p = Number of Means for Range Being Tested | | | | | | | | | | | | | |
|---|---|---|---|---|---|---|---|---|---|---|---|---|---|---|---|
| | | 2 | 3 | 4 | 5 | 6 | 7 | 8 | 9 | 10 | 12 | 14 | 16 | 18 | 20 |
| 12 | .05 | 3.08 | 3.23 | 3.33 | 3.36 | 3.40 | 3.42 | 3.44 | 3.44 | 3.46 | 3.46 | 3.46 | 3.46 | 3.47 | 3.48 |
| | .01 | 4.32 | 4.55 | 4.68 | 4.76 | 4.84 | 4.92 | 4.96 | 5.02 | 5.07 | 5.13 | 5.17 | 5.22 | 5.24 | 5.26 |
| 13 | .05 | 3.06 | 3.21 | 3.30 | 3.35 | 3.38 | 3.41 | 3.42 | 3.44 | 3.45 | 3.45 | 3.46 | 3.46 | 3.47 | 3.47 |
| | .01 | 4.26 | 4.48 | 4.62 | 4.69 | 4.74 | 4.84 | 4.88 | 4.94 | 4.98 | 5.04 | 5.08 | 5.13 | 5.14 | 5.15 |
| 14 | .05 | 3.03 | 3.18 | 3.27 | 3.33 | 3.37 | 3.39 | 3.41 | 3.42 | 3.44 | 3.45 | 3.46 | 3.46 | 3.47 | 3.47 |
| | .01 | 4.21 | 4.42 | 4.55 | 4.63 | 4.70 | 4.78 | 4.83 | 4.87 | 4.91 | 4.96 | 5.00 | 5.04 | 5.06 | 5.07 |
| 15 | .05 | 3.01 | 3.16 | 3.25 | 3.31 | 3.36 | 3.38 | 3.40 | 3.42 | 3.43 | 3.44 | 3.45 | 3.46 | 3.47 | 3.47 |
| | .01 | 4.17 | 4.37 | 4.50 | 4.58 | 4.64 | 4.72 | 4.77 | 4.81 | 4.84 | 4.90 | 4.94 | 4.97 | 4.99 | 5.00 |
| 16 | .05 | 3.00 | 3.15 | 3.23 | 3.30 | 3.34 | 3.37 | 3.39 | 3.41 | 3.43 | 3.44 | 3.45 | 3.46 | 3.47 | 3.47 |
| | .01 | 4.13 | 4.34 | 4.45 | 4.54 | 4.60 | 4.67 | 4.72 | 4.76 | 4.79 | 4.84 | 4.88 | 4.91 | 4.93 | 4.94 |
| 17 | .05 | 2.98 | 3.13 | 3.22 | 3.28 | 3.33 | 3.36 | 3.38 | 3.40 | 3.42 | 3.44 | 3.45 | 3.46 | 3.47 | 3.47 |
| | .01 | 4.10 | 4.30 | 4.41 | 4.50 | 4.56 | 4.63 | 4.68 | 4.72 | 4.75 | 4.80 | 4.83 | 4.86 | 4.88 | 4.89 |
| 18 | .05 | 2.97 | 3.12 | 3.21 | 3.27 | 3.32 | 3.35 | 3.37 | 3.39 | 3.41 | 3.43 | 3.45 | 3.46 | 3.47 | 3.47 |
| | .01 | 4.07 | 4.27 | 4.38 | 4.46 | 4.53 | 4.59 | 4.64 | 4.68 | 4.71 | 4.76 | 4.79 | 4.82 | 4.84 | 4.85 |
| 19 | .05 | 2.96 | 3.11 | 3.19 | 3.26 | 3.31 | 3.35 | 3.37 | 3.39 | 3.41 | 3.43 | 3.44 | 3.46 | 3.47 | 3.47 |
| | .01 | 4.05 | 4.24 | 4.35 | 4.43 | 4.50 | 4.56 | 4.61 | 4.64 | 4.67 | 4.72 | 4.76 | 4.79 | 4.81 | 4.82 |
| 20 | .05 | 2.95 | 3.10 | 3.18 | 3.25 | 3.30 | 3.34 | 3.36 | 3.38 | 3.40 | 3.43 | 3.44 | 3.46 | 3.46 | 3.47 |
| | .01 | 4.02 | 4.22 | 4.33 | 4.40 | 4.47 | 4.53 | 4.58 | 4.61 | 4.65 | 4.69 | 4.73 | 4.76 | 4.78 | 4.79 |

| | | | | | | | | | | | | | | | |
|---|---|---|---|---|---|---|---|---|---|---|---|---|---|---|---|
| 22 | .05 | 2.93 | 3.08 | 3.17 | 3.24 | 3.29 | 3.32 | 3.35 | 3.37 | 3.39 | 3.42 | 3.44 | 3.45 | 3.46 | 3.47 |
| | .01 | 3.99 | 4.17 | 4.28 | 4.36 | 4.42 | 4.48 | 4.53 | 4.57 | 4.60 | 4.65 | 4.68 | 4.71 | 4.74 | 4.75 |
| 24 | .05 | 2.92 | 3.07 | 3.15 | 3.22 | 3.28 | 3.31 | 3.34 | 3.37 | 3.38 | 3.41 | 3.44 | 3.45 | 3.46 | 3.47 |
| | .01 | 3.96 | 4.14 | 4.24 | 4.33 | 4.39 | 4.44 | 4.49 | 4.53 | 4.57 | 4.62 | 4.64 | 4.67 | 4.70 | 4.72 |
| 26 | .05 | 2.91 | 3.06 | 3.14 | 3.21 | 3.27 | 3.30 | 3.34 | 3.36 | 3.38 | 3.41 | 3.43 | 3.45 | 3.46 | 3.47 |
| | .01 | 3.93 | 4.11 | 4.21 | 4.30 | 4.36 | 4.41 | 4.46 | 4.50 | 4.53 | 4.58 | 4.62 | 4.65 | 4.67 | 4.69 |
| 28 | .05 | 2.90 | 3.04 | 3.13 | 3.20 | 3.26 | 3.30 | 3.33 | 3.35 | 3.37 | 3.40 | 3.43 | 3.45 | 3.46 | 3.47 |
| | .01 | 3.91 | 4.08 | 4.18 | 4.28 | 4.34 | 4.39 | 4.43 | 4.47 | 4.51 | 4.56 | 4.60 | 4.62 | 4.65 | 4.67 |
| 30 | .05 | 2.89 | 3.04 | 3.12 | 3.20 | 3.25 | 3.29 | 3.32 | 3.35 | 3.37 | 3.40 | 3.43 | 3.44 | 3.46 | 3.47 |
| | .01 | 3.89 | 4.06 | 4.16 | 4.22 | 4.32 | 4.36 | 4.41 | 4.45 | 4.48 | 4.54 | 4.58 | 4.61 | 4.63 | 4.65 |
| 40 | .05 | 2.86 | 3.01 | 3.10 | 3.17 | 3.22 | 3.27 | 3.30 | 3.33 | 3.35 | 3.39 | 3.42 | 3.44 | 3.46 | 3.47 |
| | .01 | 3.82 | 3.99 | 4.10 | 4.17 | 4.24 | 4.30 | 4.34 | 4.37 | 4.41 | 4.46 | 4.51 | 4.54 | 4.57 | 4.59 |
| 60 | .05 | 2.83 | 2.98 | 3.08 | 3.14 | 3.20 | 3.24 | 3.28 | 3.31 | 3.33 | 3.37 | 3.40 | 3.43 | 3.45 | 3.47 |
| | .01 | 3.76 | 3.92 | 4.03 | 4.12 | 4.17 | 4.23 | 4.27 | 4.31 | 4.34 | 4.39 | 4.44 | 4.47 | 4.50 | 4.53 |
| 100 | .05 | 2.80 | 2.95 | 3.05 | 3.12 | 3.18 | 3.22 | 3.26 | 3.29 | 3.32 | 3.36 | 3.40 | 3.42 | 3.45 | 3.47 |
| | .01 | 3.71 | 3.86 | 3.93 | 4.06 | 4.11 | 4.17 | 4.21 | 4.25 | 4.29 | 4.35 | 4.38 | 4.42 | 4.45 | 4.48 |
| ∞ | .05 | 2.77 | 2.92 | 3.02 | 3.09 | 3.15 | 3.19 | 3.23 | 3.26 | 3.29 | 3.34 | 3.38 | 3.41 | 3.44 | 3.47 |
| | .01 | 3.64 | 3.80 | 3.90 | 3.98 | 4.04 | 4.09 | 4.14 | 4.17 | 4.20 | 4.26 | 4.31 | 4.34 | 4.38 | 4.41 |

[a]Source: D. B. Duncan. (1955). Multiple range and multiple F tests. Biometrics, **11**, 1–42. Reproduced with permission of The Biometric Society.

**Table A.8** Values of the Maximum Standardized Range $\tau$ for Sample Size Estimation in Completely Random Design[a]

**Number of Treatments, $k = 2$**

| $n$ / $\beta$ | $\alpha = 0.01$ | | | | | | $\alpha = 0.05$ | | | | | |
|---|---|---|---|---|---|---|---|---|---|---|---|---|
| | 0.005 | 0.01 | 0.05 | 0.1 | 0.2 | 0.3 | 0.005 | 0.01 | 0.05 | 0.1 | 0.2 | 0.3 |
| 2 | 23.054 | 21.490 | 17.322 | 15.179 | 12.678 | 10.954 | 10.375 | 9.665 | 7.772 | 6.796 | 5.653 | 4.863 |
| 3 | 7.762 | 7.322 | 6.141 | 5.527 | 4.800 | 4.288 | 5.160 | 4.853 | 4.024 | 3.589 | 3.071 | 2.703 |
| 4 | 5.260 | 4.982 | 4.233 | 3.840 | 3.371 | 3.037 | 3.919 | 3.695 | 3.088 | 2.767 | 2.381 | 2.104 |
| 5 | 4.220 | 4.005 | 3.422 | 3.114 | 2.744 | 2.480 | 3.308 | 3.122 | 2.616 | 2.348 | 2.024 | 1.792 |
| 6 | 3.628 | 3.446 | 2.953 | 2.691 | 2.376 | 2.150 | 2.924 | 2.761 | 2.317 | 2.081 | 1.796 | 1.590 |
| 7 | 3.235 | 3.074 | 2.638 | 2.406 | 2.127 | 1.926 | 2.652 | 2.505 | 2.104 | 1.890 | 1.632 | 1.446 |
| 8 | 2.949 | 2.803 | 2.407 | 2.197 | 1.943 | 1.761 | 2.446 | 2.311 | 1.941 | 1.745 | 1.507 | 1.335 |
| 10 | 2.552 | 2.427 | 2.086 | 1.905 | 1.686 | 1.528 | 2.149 | 2.030 | 1.706 | 1.534 | 1.325 | 1.175 |
| 12 | 2.283 | 2.172 | 1.868 | 1.706 | 1.511 | 1.370 | 1.940 | 1.833 | 1.541 | 1.385 | 1.197 | 1.061 |
| 14 | 2.085 | 1.984 | 1.707 | 1.559 | 1.381 | 1.252 | 1.783 | 1.684 | 1.416 | 1.273 | 1.100 | 0.975 |
| 16 | 1.932 | 1.838 | 1.582 | 1.445 | 1.280 | 1.161 | 1.658 | 1.567 | 1.318 | 1.185 | 1.024 | 0.908 |
| 18 | 1.808 | 1.720 | 1.480 | 1.353 | 1.198 | 1.087 | 1.557 | 1.471 | 1.237 | 1.112 | 0.961 | 0.852 |
| 20 | 1.706 | 1.623 | 1.397 | 1.276 | 1.130 | 1.025 | 1.472 | 1.391 | 1.170 | 1.052 | 0.909 | 0.806 |
| 22 | 1.619 | 1.540 | 1.326 | 1.211 | 1.073 | 0.973 | 1.400 | 1.323 | 1.113 | 1.000 | 0.865 | 0.767 |
| 24 | 1.544 | 1.469 | 1.265 | 1.156 | 1.024 | 0.929 | 1.338 | 1.264 | 1.063 | 0.956 | 0.826 | 0.733 |
| 26 | 1.479 | 1.407 | 1.211 | 1.107 | 0.981 | 0.889 | 1.283 | 1.212 | 1.020 | 0.917 | 0.792 | 0.703 |
| 28 | 1.421 | 1.352 | 1.164 | 1.064 | 0.942 | 0.855 | 1.235 | 1.167 | 0.981 | 0.882 | 0.762 | 0.676 |
| 30 | 1.370 | 1.304 | 1.122 | 1.026 | 0.909 | 0.824 | 1.191 | 1.126 | 0.947 | 0.851 | 0.736 | 0.652 |
| 40 | 1.177 | 1.120 | 0.964 | 0.881 | 0.781 | 0.708 | 1.027 | 0.971 | 0.816 | 0.734 | 0.634 | 0.562 |
| 60 | 0.954 | 0.908 | 0.782 | 0.714 | 0.633 | 0.574 | 0.835 | 0.789 | 0.664 | 0.597 | 0.516 | 0.457 |
| 80 | 0.823 | 0.783 | 0.674 | 0.616 | 0.546 | 0.495 | 0.722 | 0.682 | 0.573 | 0.516 | 0.446 | 0.395 |

| n | \alpha = 0.10 | | | | | | \alpha = 0.20 | | | | | |
|---|---|---|---|---|---|---|---|---|---|---|---|---|
| $\beta$ | 0.005 | 0.01 | 0.05 | 0.1 | 0.2 | 0.3 | 0.005 | 0.01 | 0.05 | 0.1 | 0.2 | 0.3 |
| 100 | 0.735 | 0.699 | 0.602 | 0.550 | 0.487 | 0.442 | 0.645 | 0.609 | 0.512 | 0.461 | 0.398 | 0.353 |
| 200 | 0.517 | 0.492 | 0.424 | 0.387 | 0.343 | 0.311 | 0.455 | 0.430 | 0.361 | 0.325 | 0.281 | 0.249 |
| 500 | 0.326 | 0.311 | 0.267 | 0.244 | 0.216 | 0.196 | 0.287 | 0.271 | 0.228 | 0.205 | 0.177 | 0.157 |
| 1000 | 0.231 | 0.219 | 0.189 | 0.173 | 0.153 | 0.139 | 0.203 | 0.192 | 0.161 | 0.145 | 0.125 | 0.111 |
| 2 | 7.393 | 6.882 | 5.516 | 4.809 | 3.979 | 3.401 | 5.310 | 4.934 | 3.925 | 3.399 | 2.775 | 2.334 |
| 3 | 4.313 | 4.045 | 3.321 | 2.939 | 2.482 | 2.155 | 3.585 | 3.348 | 2.703 | 2.361 | 1.949 | 1.653 |
| 4 | 3.422 | 3.215 | 2.653 | 2.355 | 1.995 | 1.737 | 2.955 | 2.762 | 2.236 | 1.956 | 1.618 | 1.373 |
| 5 | 2.944 | 2.768 | 2.288 | 2.033 | 1.725 | 1.503 | 2.585 | 2.416 | 1.958 | 1.714 | 1.418 | 1.204 |
| 6 | 2.629 | 2.473 | 2.046 | 1.818 | 1.544 | 1.346 | 2.330 | 2.179 | 1.766 | 1.546 | 1.280 | 1.087 |
| 7 | 2.401 | 2.258 | 1.869 | 1.662 | 1.411 | 1.230 | 2.140 | 2.001 | 1.622 | 1.420 | 1.176 | 0.999 |
| 8 | 2.224 | 2.092 | 1.732 | 1.540 | 1.308 | 1.141 | 1.990 | 1.861 | 1.509 | 1.321 | 1.094 | 0.929 |
| 10 | 1.965 | 1.848 | 1.531 | 1.361 | 1.156 | 1.008 | 1.767 | 1.653 | 1.340 | 1.174 | 0.972 | 0.825 |
| 12 | 1.780 | 1.675 | 1.387 | 1.233 | 1.048 | 0.914 | 1.606 | 1.502 | 1.218 | 1.067 | 0.883 | 0.750 |
| 14 | 1.639 | 1.542 | 1.277 | 1.136 | 0.965 | 0.842 | 1.482 | 1.386 | 1.124 | 0.984 | 0.815 | 0.693 |
| 16 | 1.528 | 1.437 | 1.190 | 1.059 | 0.900 | 0.785 | 1.383 | 1.294 | 1.049 | 0.919 | 0.761 | 0.646 |
| 18 | 1.436 | 1.351 | 1.119 | 0.996 | 0.846 | 0.738 | 1.302 | 1.218 | 0.988 | 0.865 | 0.716 | 0.608 |
| 20 | 1.359 | 1.279 | 1.059 | 0.942 | 0.801 | 0.698 | 1.233 | 1.154 | 0.936 | 0.819 | 0.678 | 0.576 |
| 22 | 1.294 | 1.217 | 1.008 | 0.897 | 0.762 | 0.665 | 1.175 | 1.099 | 0.891 | 0.780 | 0.646 | 0.549 |
| 24 | 1.237 | 1.164 | 0.964 | 0.858 | 0.729 | 0.636 | 1.124 | 1.051 | 0.852 | 0.747 | 0.618 | 0.525 |
| 26 | 1.187 | 1.117 | 0.925 | 0.823 | 0.699 | 0.610 | 1.079 | 1.009 | 0.818 | 0.717 | 0.593 | 0.504 |
| 28 | 1.143 | 1.075 | 0.890 | 0.792 | 0.673 | 0.587 | 1.039 | 0.972 | 0.788 | 0.690 | 0.571 | 0.486 |
| 30 | 1.103 | 1.038 | 0.860 | 0.765 | 0.650 | 0.567 | 1.003 | 0.938 | 0.761 | 0.666 | 0.552 | 0.469 |

*(continued)*

**Table A.8** Values of the Maximum Standardized Range $\tau$ for Sample Size Estimation in Completely Random Design[a]—cont...

| $\beta$ \ n | $\alpha = 0.10$ | | | | | | $\alpha = 0.20$ | | | | | |
|---|---|---|---|---|---|---|---|---|---|---|---|---|
| | 0.005 | 0.01 | 0.05 | 0.1 | 0.2 | 0.3 | 0.005 | 0.01 | 0.05 | 0.1 | 0.2 | 0.3 |
| 40 | 0.952 | 0.896 | 0.742 | 0.660 | 0.561 | 0.489 | 0.867 | 0.811 | 0.658 | 0.576 | 0.477 | 0.405 |
| 60 | 0.775 | 0.729 | 0.604 | 0.537 | 0.457 | 0.398 | 0.707 | 0.661 | 0.536 | 0.470 | 0.389 | 0.330 |
| 80 | 0.670 | 0.631 | 0.522 | 0.465 | 0.395 | 0.344 | 0.611 | 0.572 | 0.464 | 0.406 | 0.336 | 0.286 |
| 100 | 0.599 | 0.564 | 0.467 | 0.415 | 0.353 | 0.308 | 0.547 | 0.511 | 0.415 | 0.363 | 0.301 | 0.256 |
| 200 | 0.423 | 0.398 | 0.330 | 0.293 | 0.249 | 0.217 | 0.386 | 0.361 | 0.293 | 0.257 | 0.212 | 0.180 |
| 500 | 0.267 | 0.251 | 0.208 | 0.185 | 0.157 | 0.137 | 0.244 | 0.228 | 0.185 | 0.162 | 0.134 | 0.114 |
| 1000 | 0.189 | 0.178 | 0.147 | 0.131 | 0.111 | 0.097 | 0.173 | 0.161 | 0.131 | 0.115 | 0.095 | 0.081 |

**Number of Treatments, $k = 3$**

| $\beta$ \ n | $\alpha = 0.01$ | | | | | | $\alpha = 0.05$ | | | | | |
|---|---|---|---|---|---|---|---|---|---|---|---|---|
| | 0.005 | 0.01 | 0.05 | 0.1 | 0.2 | 0.3 | 0.005 | 0.01 | 0.05 | 0.1 | 0.2 | 0.3 |
| 2 | 16.539 | 15.536 | 12.857 | 11.473 | 9.846 | 8.712 | 9.560 | 8.966 | 7.375 | 6.548 | 5.570 | 4.883 |
| 3 | 7.281 | 6.903 | 5.886 | 5.352 | 4.716 | 4.264 | 5.371 | 5.073 | 4.265 | 3.838 | 3.325 | 2.957 |
| 4 | 5.257 | 4.999 | 4.300 | 3.931 | 3.487 | 3.169 | 4.168 | 3.944 | 3.334 | 3.010 | 2.618 | 2.335 |
| 5 | 4.322 | 4.115 | 3.553 | 3.255 | 2.895 | 2.636 | 3.541 | 3.354 | 2.841 | 2.568 | 2.236 | 1.997 |
| 6 | 3.761 | 3.583 | 3.099 | 2.842 | 2.530 | 2.306 | 3.139 | 2.973 | 2.521 | 2.280 | 1.987 | 1.775 |
| 7 | 3.376 | 3.217 | 2.786 | 2.556 | 2.277 | 2.076 | 2.851 | 2.701 | 2.292 | 2.073 | 1.808 | 1.615 |
| 8 | 3.090 | 2.946 | 2.552 | 2.343 | 2.088 | 1.905 | 2.631 | 2.493 | 2.116 | 1.915 | 1.670 | 1.492 |
| 10 | 2.688 | 2.563 | 2.222 | 2.040 | 1.820 | 1.660 | 2.313 | 2.192 | 1.861 | 1.684 | 1.469 | 1.313 |
| 12 | 2.412 | 2.300 | 1.995 | 1.832 | 1.634 | 1.491 | 2.088 | 1.979 | 1.681 | 1.521 | 1.327 | 1.186 |
| 14 | 2.207 | 2.105 | 1.826 | 1.677 | 1.496 | 1.365 | 1.919 | 1.819 | 1.545 | 1.398 | 1.220 | 1.090 |
| 16 | 2.046 | 1.952 | 1.693 | 1.555 | 1.388 | 1.267 | 1.785 | 1.692 | 1.437 | 1.301 | 1.135 | 1.015 |
| 18 | 1.917 | 1.829 | 1.587 | 1.457 | 1.300 | 1.187 | 1.676 | 1.589 | 1.350 | 1.222 | 1.066 | 0.953 |
| 20 | 1.809 | 1.726 | 1.498 | 1.376 | 1.227 | 1.120 | 1.585 | 1.502 | 1.276 | 1.155 | 1.008 | 0.901 |

| n | α = 0.10 | | | | | | α = 0.20 | | | | | |
|---|---|---|---|---|---|---|---|---|---|---|---|---|
| | 0.005 | 0.01 | 0.05 | 0.1 | 0.2 | 0.3 | 0.005 | 0.01 | 0.05 | 0.1 | 0.2 | 0.3 |
| 22 | 1.718 | 1.639 | 1.422 | 1.306 | 1.166 | 1.064 | 1.507 | 1.429 | 1.214 | 1.099 | 0.959 | 0.857 |
| 24 | 1.640 | 1.564 | 1.357 | 1.247 | 1.112 | 1.015 | 1.440 | 1.365 | 1.160 | 1.050 | 0.916 | 0.819 |
| 26 | 1.571 | 1.498 | 1.300 | 1.194 | 1.066 | 0.973 | 1.381 | 1.309 | 1.112 | 1.007 | 0.878 | 0.785 |
| 28 | 1.510 | 1.441 | 1.250 | 1.148 | 1.025 | 0.935 | 1.329 | 1.260 | 1.070 | 0.969 | 0.845 | 0.756 |
| 30 | 1.456 | 1.389 | 1.205 | 1.107 | 0.988 | 0.902 | 1.282 | 1.215 | 1.032 | 0.935 | 0.815 | 0.729 |
| 40 | 1.252 | 1.194 | 1.037 | 0.952 | 0.850 | 0.776 | 1.105 | 1.048 | 0.890 | 0.806 | 0.703 | 0.629 |
| 60 | 1.015 | 0.969 | 0.841 | 0.772 | 0.689 | 0.629 | 0.899 | 0.852 | 0.724 | 0.655 | 0.572 | 0.511 |
| 80 | 0.876 | 0.836 | 0.726 | 0.667 | 0.595 | 0.543 | 0.776 | 0.736 | 0.625 | 0.566 | 0.494 | 0.442 |
| 100 | 0.782 | 0.746 | 0.648 | 0.595 | 0.531 | 0.485 | 0.694 | 0.657 | 0.559 | 0.506 | 0.441 | 0.394 |
| 200 | 0.551 | 0.526 | 0.456 | 0.419 | 0.374 | 0.341 | 0.489 | 0.464 | 0.394 | 0.357 | 0.311 | 0.278 |
| 500 | 0.348 | 0.332 | 0.288 | 0.264 | 0.236 | 0.215 | 0.309 | 0.293 | 0.249 | 0.225 | 0.197 | 0.176 |
| 1000 | 0.246 | 0.234 | 0.203 | 0.187 | 0.167 | 0.152 | 0.218 | 0.207 | 0.176 | 0.159 | 0.139 | 0.124 |

| β \ n | α = 0.10 | | | | | | α = 0.20 | | | | | |
|---|---|---|---|---|---|---|---|---|---|---|---|---|
| | 0.005 | 0.01 | 0.05 | 0.1 | 0.2 | 0.3 | 0.005 | 0.01 | 0.05 | 0.1 | 0.2 | 0.3 |
| 2 | 7.500 | 7.023 | 5.740 | 5.071 | 4.274 | 3.709 | 5.826 | 5.439 | 4.393 | 3.841 | 3.177 | 2.700 |
| 3 | 4.655 | 4.384 | 3.646 | 3.254 | 2.780 | 2.438 | 3.975 | 3.725 | 3.044 | 2.680 | 2.236 | 1.912 |
| 4 | 3.718 | 3.506 | 2.927 | 2.618 | 2.243 | 1.971 | 3.262 | 3.060 | 2.507 | 2.210 | 1.847 | 1.581 |
| 5 | 3.202 | 3.021 | 2.526 | 2.261 | 1.939 | 1.705 | 2.844 | 2.669 | 2.188 | 1.930 | 1.614 | 1.383 |
| 6 | 2.859 | 2.698 | 2.257 | 2.022 | 1.734 | 1.526 | 2.558 | 2.401 | 1.969 | 1.737 | 1.453 | 1.245 |
| 7 | 2.609 | 2.463 | 2.061 | 1.846 | 1.584 | 1.394 | 2.345 | 2.201 | 1.806 | 1.593 | 1.333 | 1.142 |
| 8 | 2.416 | 2.281 | 1.909 | 1.710 | 1.468 | 1.292 | 2.179 | 2.045 | 1.678 | 1.480 | 1.239 | 1.062 |
| 10 | 2.133 | 2.013 | 1.686 | 1.510 | 1.297 | 1.141 | 1.931 | 1.813 | 1.488 | 1.313 | 1.098 | 0.941 |
| 12 | 1.931 | 1.823 | 1.527 | 1.368 | 1.174 | 1.034 | 1.753 | 1.646 | 1.350 | 1.192 | 0.997 | 0.855 |
| 14 | 1.778 | 1.678 | 1.406 | 1.259 | 1.081 | 0.952 | 1.617 | 1.518 | 1.245 | 1.099 | 0.920 | 0.788 |

*(continued)*

**Table A.8** Values of the Maximum Standardized Range $\tau$ for Sample Size Estimation in Completely Random Design[a] —cont...

| | $\alpha = 0.10$ | | | | | | $\alpha = 0.20$ | | | | | |
| $n$ \ $\beta$ | 0.005 | 0.01 | 0.05 | 0.1 | 0.2 | 0.3 | 0.005 | 0.01 | 0.05 | 0.1 | 0.2 | 0.3 |
|---|---|---|---|---|---|---|---|---|---|---|---|---|
| 16 | 1.656 | 1.563 | 1.309 | 1.173 | 1.007 | 0.887 | 1.508 | 1.416 | 1.162 | 1.025 | 0.858 | 0.735 |
| 18 | 1.556 | 1.469 | 1.231 | 1.103 | 0.947 | 0.833 | 1.419 | 1.332 | 1.093 | 0.964 | 0.807 | 0.692 |
| 20 | 1.473 | 1.390 | 1.165 | 1.044 | 0.896 | 0.789 | 1.344 | 1.261 | 1.035 | 0.913 | 0.764 | 0.655 |
| 22 | 1.401 | 1.323 | 1.108 | 0.993 | 0.853 | 0.751 | 1.279 | 1.201 | 0.986 | 0.870 | 0.728 | 0.624 |
| 24 | 1.340 | 1.265 | 1.059 | 0.949 | 0.815 | 0.717 | 1.223 | 1.148 | 0.943 | 0.832 | 0.696 | 0.597 |
| 26 | 1.285 | 1.213 | 1.017 | 0.911 | 0.782 | 0.688 | 1.174 | 1.102 | 0.905 | 0.798 | 0.668 | 0.573 |
| 28 | 1.237 | 1.168 | 0.978 | 0.877 | 0.753 | 0.663 | 1.131 | 1.061 | 0.871 | 0.769 | 0.643 | 0.551 |
| 30 | 1.194 | 1.127 | 0.944 | 0.846 | 0.727 | 0.640 | 1.092 | 1.025 | 0.841 | 0.742 | 0.621 | 0.532 |
| 40 | 1.030 | 0.973 | 0.815 | 0.730 | 0.627 | 0.552 | 0.943 | 0.885 | 0.727 | 0.641 | 0.537 | 0.460 |
| 60 | 0.838 | 0.792 | 0.663 | 0.594 | 0.510 | 0.449 | 0.768 | 0.721 | 0.592 | 0.522 | 0.437 | 0.375 |
| 80 | 0.725 | 0.684 | 0.573 | 0.514 | 0.441 | 0.388 | 0.664 | 0.624 | 0.512 | 0.452 | 0.378 | 0.324 |
| 100 | 0.648 | 0.612 | 0.512 | 0.459 | 0.394 | 0.347 | 0.594 | 0.558 | 0.458 | 0.404 | 0.338 | 0.290 |
| 200 | 0.457 | 0.432 | 0.362 | 0.324 | 0.278 | 0.245 | 0.419 | 0.394 | 0.323 | 0.285 | 0.239 | 0.205 |
| 500 | 0.289 | 0.273 | 0.228 | 0.205 | 0.176 | 0.155 | 0.265 | 0.249 | 0.204 | 0.180 | 0.151 | 0.129 |
| 1000 | 0.204 | 0.193 | 0.161 | 0.145 | 0.124 | 0.109 | 0.187 | 0.176 | 0.144 | 0.127 | 0.107 | 0.091 |

**Number of Treatments, $k = 4$**

| | $\alpha = 0.01$ | | | | | | $\alpha = 0.05$ | | | | | |
| $n$ \ $\beta$ | 0.005 | 0.01 | 0.05 | 0.1 | 0.2 | 0.3 | 0.005 | 0.01 | 0.05 | 0.1 | 0.2 | 0.3 |
|---|---|---|---|---|---|---|---|---|---|---|---|---|
| 2 | 13.991 | 13.204 | 11.096 | 10.002 | 8.709 | 7.801 | 9.106 | 8.574 | 7.143 | 6.395 | 5.504 | 4.872 |
| 3 | 7.047 | 6.701 | 5.765 | 5.272 | 4.681 | 4.258 | 5.467 | 5.176 | 4.386 | 3.967 | 3.460 | 3.094 |
| 4 | 5.261 | 5.015 | 4.343 | 3.987 | 3.557 | 3.248 | 4.303 | 4.080 | 3.473 | 3.148 | 2.754 | 2.468 |
| 5 | 4.385 | 4.184 | 3.634 | 3.341 | 2.986 | 2.730 | 3.676 | 3.488 | 2.973 | 2.698 | 2.362 | 2.119 |
| 6 | 3.843 | 3.668 | 3.190 | 2.935 | 2.625 | 2.402 | 3.266 | 3.100 | 2.645 | 2.401 | 2.104 | 1.888 |
| 7 | 3.464 | 3.307 | 2.878 | 2.649 | 2.371 | 2.170 | 2.971 | 2.820 | 2.407 | 2.186 | 1.916 | 1.719 |

| n | α = 0.10 | | | | | | α = 0.20 | | | | | |
|---|---|---|---|---|---|---|---|---|---|---|---|---|
| β | 0.005 | 0.01 | 0.05 | 0.1 | 0.2 | 0.3 | 0.005 | 0.01 | 0.05 | 0.1 | 0.2 | 0.3 |
| 8 | 3.180 | 3.036 | 2.644 | 2.434 | 2.179 | 1.995 | 2.745 | 2.606 | 2.224 | 2.020 | 1.771 | 1.590 |
| 10 | 2.775 | 2.650 | 2.309 | 2.126 | 1.905 | 1.744 | 2.415 | 2.293 | 1.958 | 1.778 | 1.559 | 1.400 |
| 12 | 2.495 | 2.383 | 2.076 | 1.912 | 1.713 | 1.569 | 2.182 | 2.072 | 1.769 | 1.607 | 1.409 | 1.266 |
| 14 | 2.285 | 2.183 | 1.903 | 1.753 | 1.570 | 1.438 | 2.006 | 1.904 | 1.627 | 1.478 | 1.296 | 1.164 |
| 16 | 2.121 | 2.026 | 1.766 | 1.627 | 1.458 | 1.335 | 1.866 | 1.772 | 1.514 | 1.375 | 1.206 | 1.083 |
| 18 | 1.989 | 1.900 | 1.656 | 1.525 | 1.367 | 1.252 | 1.753 | 1.664 | 1.422 | 1.292 | 1.133 | 1.017 |
| 20 | 1.878 | 1.794 | 1.564 | 1.441 | 1.291 | 1.182 | 1.658 | 1.574 | 1.345 | 1.222 | 1.071 | 0.962 |
| 22 | 1.784 | 1.704 | 1.486 | 1.369 | 1.226 | 1.123 | 1.576 | 1.497 | 1.279 | 1.162 | 1.019 | 0.915 |
| 24 | 1.703 | 1.627 | 1.418 | 1.306 | 1.171 | 1.072 | 1.506 | 1.430 | 1.222 | 1.110 | 0.974 | 0.874 |
| 26 | 1.632 | 1.559 | 1.359 | 1.252 | 1.122 | 1.028 | 1.445 | 1.372 | 1.172 | 1.065 | 0.934 | 0.839 |
| 28 | 1.569 | 1.499 | 1.307 | 1.204 | 1.079 | 0.988 | 1.390 | 1.320 | 1.128 | 1.025 | 0.899 | 0.807 |
| 30 | 1.513 | 1.445 | 1.260 | 1.161 | 1.040 | 0.953 | 1.341 | 1.274 | 1.088 | 0.989 | 0.867 | 0.779 |
| 40 | 1.302 | 1.244 | 1.084 | 0.999 | 0.895 | 0.820 | 1.156 | 1.098 | 0.938 | 0.852 | 0.748 | 0.671 |
| 60 | 1.056 | 1.009 | 0.880 | 0.811 | 0.727 | 0.665 | 0.940 | 0.893 | 0.763 | 0.693 | 0.608 | 0.546 |
| 80 | 0.912 | 0.871 | 0.760 | 0.700 | 0.627 | 0.575 | 0.813 | 0.772 | 0.659 | 0.599 | 0.525 | 0.472 |
| 100 | 0.814 | 0.778 | 0.678 | 0.625 | 0.560 | 0.513 | 0.726 | 0.689 | 0.589 | 0.535 | 0.469 | 0.421 |
| 200 | 0.574 | 0.548 | 0.478 | 0.440 | 0.395 | 0.361 | 0.512 | 0.486 | 0.415 | 0.377 | 0.331 | 0.297 |
| 500 | 0.362 | 0.346 | 0.302 | 0.278 | 0.249 | 0.228 | 0.323 | 0.307 | 0.262 | 0.238 | 0.209 | 0.188 |
| 1000 | 0.256 | 0.244 | 0.213 | 0.196 | 0.176 | 0.161 | 0.228 | 0.217 | 0.185 | 0.168 | 0.148 | 0.133 |
| 2 | 7.481 | 7.030 | 5.811 | 5.170 | 4.401 | 3.852 | 6.050 | 5.665 | 4.618 | 4.063 | 3.389 | 2.899 |
| 3 | 4.829 | 4.558 | 3.819 | 3.425 | 2.946 | 2.598 | 4.191 | 3.937 | 3.240 | 2.865 | 2.407 | 2.069 |
| 4 | 3.885 | 3.671 | 3.086 | 2.772 | 2.389 | 2.111 | 3.444 | 3.237 | 2.670 | 2.364 | 1.988 | 1.712 |

*(continued)*

**Table A.8** Values of the Maximum Standardized Range $\tau$ for Sample Size Estimation in Completely Random Design[a]—cont...

| $\beta$ \ $n$ | $\alpha = 0.10$ | | | | | | $\alpha = 0.20$ | | | | | |
|---|---|---|---|---|---|---|---|---|---|---|---|---|
| | 0.005 | 0.01 | 0.05 | 0.1 | 0.2 | 0.3 | 0.005 | 0.01 | 0.05 | 0.1 | 0.2 | 0.3 |
| 5 | 3.353 | 3.170 | 2.667 | 2.398 | 2.068 | 1.828 | 3.002 | 2.823 | 2.330 | 2.064 | 1.737 | 1.496 |
| 6 | 2.997 | 2.834 | 2.386 | 2.146 | 1.852 | 1.637 | 2.699 | 2.539 | 2.096 | 1.857 | 1.563 | 1.346 |
| 7 | 2.736 | 2.588 | 2.180 | 1.960 | 1.692 | 1.496 | 2.474 | 2.327 | 1.922 | 1.703 | 1.433 | 1.235 |
| 8 | 2.535 | 2.397 | 2.020 | 1.817 | 1.568 | 1.387 | 2.298 | 2.161 | 1.785 | 1.582 | 1.332 | 1.147 |
| 10 | 2.238 | 2.117 | 1.784 | 1.604 | 1.385 | 1.225 | 2.036 | 1.915 | 1.582 | 1.402 | 1.180 | 1.017 |
| 12 | 2.026 | 1.917 | 1.615 | 1.453 | 1.255 | 1.110 | 1.847 | 1.738 | 1.436 | 1.272 | 1.071 | 0.923 |
| 14 | 1.865 | 1.765 | 1.487 | 1.338 | 1.155 | 1.022 | 1.703 | 1.602 | 1.324 | 1.173 | 0.988 | 0.851 |
| 16 | 1.738 | 1.644 | 1.385 | 1.246 | 1.076 | 0.952 | 1.588 | 1.494 | 1.235 | 1.094 | 0.921 | 0.794 |
| 18 | 1.633 | 1.545 | 1.302 | 1.172 | 1.012 | 0.895 | 1.494 | 1.405 | 1.161 | 1.029 | 0.867 | 0.747 |
| 20 | 1.546 | 1.462 | 1.232 | 1.109 | 0.957 | 0.847 | 1.415 | 1.331 | 1.100 | 0.975 | 0.821 | 0.707 |
| 22 | 1.471 | 1.391 | 1.173 | 1.055 | 0.911 | 0.806 | 1.347 | 1.267 | 1.047 | 0.928 | 0.781 | 0.673 |
| 24 | 1.406 | 1.330 | 1.121 | 1.008 | 0.871 | 0.770 | 1.288 | 1.212 | 1.001 | 0.887 | 0.747 | 0.644 |
| 26 | 1.349 | 1.276 | 1.076 | 0.968 | 0.836 | 0.739 | 1.236 | 1.163 | 0.961 | 0.852 | 0.717 | 0.618 |
| 28 | 1.298 | 1.228 | 1.035 | 0.931 | 0.804 | 0.711 | 1.190 | 1.120 | 0.925 | 0.820 | 0.691 | 0.595 |
| 30 | 1.253 | 1.185 | 0.999 | 0.899 | 0.776 | 0.687 | 1.149 | 1.081 | 0.893 | 0.792 | 0.667 | 0.574 |
| 40 | 1.081 | 1.023 | 0.862 | 0.776 | 0.670 | 0.593 | 0.993 | 0.934 | 0.772 | 0.684 | 0.576 | 0.496 |
| 60 | 0.880 | 0.832 | 0.702 | 0.631 | 0.545 | 0.482 | 0.808 | 0.760 | 0.628 | 0.557 | 0.469 | 0.404 |
| 80 | 0.761 | 0.720 | 0.607 | 0.546 | 0.471 | 0.417 | 0.699 | 0.658 | 0.543 | 0.482 | 0.406 | 0.349 |
| 100 | 0.680 | 0.643 | 0.542 | 0.488 | 0.421 | 0.373 | 0.625 | 0.588 | 0.486 | 0.431 | 0.363 | 0.312 |
| 200 | 0.480 | 0.454 | 0.382 | 0.344 | 0.297 | 0.263 | 0.441 | 0.415 | 0.343 | 0.304 | 0.256 | 0.221 |
| 500 | 0.303 | 0.287 | 0.242 | 0.217 | 0.188 | 0.166 | 0.279 | 0.262 | 0.217 | 0.192 | 0.162 | 0.139 |
| 1000 | 0.214 | 0.203 | 0.171 | 0.154 | 0.133 | 0.117 | 0.197 | 0.185 | 0.153 | 0.136 | 0.114 | 0.098 |

**Number of Treatments, $k = 5$**

| | $\alpha = 0.01$ | | | | | | $\alpha = 0.05$ | | | | | |
|---|---|---|---|---|---|---|---|---|---|---|---|---|
| $n$ / $\beta$ | 0.005 | 0.01 | 0.05 | 0.1 | 0.2 | 0.3 | 0.005 | 0.01 | 0.05 | 0.1 | 0.2 | 0.3 |
| 2 | 12.732 | 12.053 | 10.232 | 9.283 | 8.155 | 7.359 | 8.871 | 8.375 | 7.036 | 6.333 | 5.490 | 4.889 |
| 3 | 6.938 | 6.610 | 5.721 | 5.251 | 4.685 | 4.278 | 5.544 | 5.259 | 4.480 | 4.065 | 3.562 | 3.197 |
| 4 | 5.286 | 5.046 | 4.391 | 4.042 | 3.620 | 3.315 | 4.405 | 4.183 | 3.576 | 3.251 | 2.856 | 2.568 |
| 5 | 4.444 | 4.245 | 3.702 | 3.412 | 3.060 | 2.805 | 3.777 | 3.589 | 3.072 | 2.795 | 2.457 | 2.211 |
| 6 | 3.911 | 3.738 | 3.263 | 3.009 | 2.701 | 2.477 | 3.363 | 3.196 | 2.738 | 2.492 | 2.191 | 1.973 |
| 7 | 3.536 | 3.379 | 2.952 | 2.723 | 2.445 | 2.243 | 3.062 | 2.911 | 2.494 | 2.271 | 1.997 | 1.798 |
| 8 | 3.252 | 3.108 | 2.716 | 2.506 | 2.251 | 2.066 | 2.831 | 2.691 | 2.307 | 2.100 | 1.848 | 1.664 |
| 10 | 2.844 | 2.719 | 2.377 | 2.194 | 1.971 | 1.809 | 2.493 | 2.370 | 2.032 | 1.850 | 1.628 | 1.466 |
| 12 | 2.560 | 2.448 | 2.140 | 1.976 | 1.775 | 1.629 | 2.254 | 2.142 | 1.837 | 1.673 | 1.472 | 1.326 |
| 14 | 2.347 | 2.245 | 1.963 | 1.812 | 1.628 | 1.495 | 2.072 | 1.970 | 1.690 | 1.539 | 1.354 | 1.220 |
| 16 | 2.180 | 2.085 | 1.823 | 1.683 | 1.513 | 1.389 | 1.929 | 1.834 | 1.573 | 1.432 | 1.261 | 1.135 |
| 18 | 2.045 | 1.955 | 1.710 | 1.579 | 1.419 | 1.302 | 1.812 | 1.722 | 1.477 | 1.345 | 1.184 | 1.066 |
| 20 | 1.932 | 1.847 | 1.616 | 1.491 | 1.340 | 1.231 | 1.714 | 1.629 | 1.397 | 1.273 | 1.120 | 1.009 |
| 22 | 1.836 | 1.755 | 1.535 | 1.417 | 1.274 | 1.169 | 1.630 | 1.550 | 1.329 | 1.210 | 1.065 | 0.960 |
| 24 | 1.752 | 1.676 | 1.466 | 1.353 | 1.216 | 1.117 | 1.557 | 1.481 | 1.270 | 1.157 | 1.018 | 0.917 |
| 26 | 1.680 | 1.606 | 1.405 | 1.297 | 1.166 | 1.070 | 1.494 | 1.420 | 1.218 | 1.109 | 0.976 | 0.879 |
| 28 | 1.615 | 1.545 | 1.351 | 1.247 | 1.121 | 1.029 | 1.437 | 1.367 | 1.172 | 1.068 | 0.940 | 0.846 |
| 30 | 1.558 | 1.490 | 1.303 | 1.203 | 1.081 | 0.993 | 1.387 | 1.319 | 1.131 | 1.030 | 0.907 | 0.817 |
| 40 | 1.341 | 1.283 | 1.122 | 1.036 | 0.931 | 0.855 | 1.196 | 1.137 | 0.975 | 0.888 | 0.782 | 0.704 |
| 60 | 1.089 | 1.041 | 0.911 | 0.841 | 0.756 | 0.694 | 0.973 | 0.925 | 0.793 | 0.722 | 0.636 | 0.573 |
| 80 | 0.940 | 0.899 | 0.786 | 0.726 | 0.653 | 0.599 | 0.841 | 0.799 | 0.685 | 0.624 | 0.550 | 0.495 |
| 100 | 0.839 | 0.803 | 0.702 | 0.648 | 0.583 | 0.535 | 0.751 | 0.714 | 0.612 | 0.558 | 0.491 | 0.442 |

*(continued)*

**Table A.8** Values of the Maximum Standardized Range $\tau$ for Sample Size Estimation in Completely Random Design[a] —cont...

**Number of Treatments, $k = 5$**

| | | $\alpha = 0.01$ | | | | | | | | $\alpha = 0.05$ | | | |
|---|---|---|---|---|---|---|---|---|---|---|---|---|---|
| $\beta$ $\diagdown$ $n$ | 0.005 | 0.01 | 0.05 | 0.1 | 0.2 | 0.3 | 0.005 | 0.01 | 0.05 | 0.1 | 0.2 | 0.3 |
| 200 | 0.592 | 0.566 | 0.495 | 0.457 | 0.411 | 0.377 | 0.530 | 0.504 | 0.432 | 0.393 | 0.346 | 0.312 |
| 500 | 0.373 | 0.357 | 0.312 | 0.288 | 0.259 | 0.238 | 0.334 | 0.318 | 0.273 | 0.248 | 0.218 | 0.197 |
| 1000 | 0.264 | 0.252 | 0.221 | 0.204 | 0.183 | 0.168 | 0.236 | 0.225 | 0.193 | 0.176 | 0.154 | 0.139 |

| | | $\alpha = 0.10$ | | | | | | | | $\alpha = 0.20$ | | | |
|---|---|---|---|---|---|---|---|---|---|---|---|---|---|
| $\beta$ $\diagdown$ $n$ | 0.005 | 0.01 | 0.05 | 0.1 | 0.2 | 0.3 | 0.005 | 0.01 | 0.05 | 0.1 | 0.2 | 0.3 |
| 2 | 7.486 | 7.051 | 5.873 | 5.249 | 4.498 | 3.957 | 6.203 | 5.820 | 4.776 | 4.218 | 3.538 | 3.040 |
| 3 | 4.955 | 4.685 | 3.945 | 3.550 | 3.067 | 2.714 | 4.345 | 4.088 | 3.382 | 3.001 | 2.531 | 2.185 |
| 4 | 4.007 | 3.792 | 3.202 | 2.885 | 2.497 | 2.214 | 3.577 | 3.367 | 2.791 | 2.479 | 2.094 | 1.809 |
| 5 | 3.465 | 3.280 | 2.773 | 2.500 | 2.165 | 1.920 | 3.119 | 2.937 | 2.436 | 2.164 | 1.829 | 1.581 |
| 6 | 3.100 | 2.936 | 2.483 | 2.239 | 1.940 | 1.721 | 2.805 | 2.642 | 2.191 | 1.947 | 1.646 | 1.423 |
| 7 | 2.832 | 2.682 | 2.269 | 2.046 | 1.773 | 1.573 | 2.570 | 2.421 | 2.009 | 1.786 | 1.509 | 1.305 |
| 8 | 2.624 | 2.485 | 2.103 | 1.897 | 1.644 | 1.458 | 2.387 | 2.249 | 1.866 | 1.659 | 1.402 | 1.212 |
| 10 | 2.318 | 2.195 | 1.858 | 1.676 | 1.453 | 1.289 | 2.115 | 1.992 | 1.654 | 1.470 | 1.243 | 1.074 |
| 12 | 2.099 | 1.988 | 1.683 | 1.518 | 1.316 | 1.168 | 1.919 | 1.808 | 1.500 | 1.334 | 1.128 | 0.975 |
| 14 | 1.933 | 1.831 | 1.549 | 1.398 | 1.212 | 1.075 | 1.769 | 1.667 | 1.383 | 1.230 | 1.040 | 0.899 |
| 16 | 1.800 | 1.705 | 1.444 | 1.302 | 1.129 | 1.002 | 1.650 | 1.554 | 1.290 | 1.147 | 0.970 | 0.838 |
| 18 | 1.692 | 1.603 | 1.357 | 1.224 | 1.061 | 0.942 | 1.552 | 1.462 | 1.213 | 1.079 | 0.912 | 0.789 |
| 20 | 1.602 | 1.517 | 1.284 | 1.158 | 1.004 | 0.891 | 1.469 | 1.384 | 1.149 | 1.021 | 0.864 | 0.747 |
| 22 | 1.524 | 1.444 | 1.222 | 1.102 | 0.956 | 0.848 | 1.399 | 1.318 | 1.094 | 0.972 | 0.822 | 0.711 |
| 24 | 1.457 | 1.380 | 1.168 | 1.054 | 0.914 | 0.811 | 1.338 | 1.260 | 1.046 | 0.930 | 0.786 | 0.680 |
| 26 | 1.398 | 1.324 | 1.121 | 1.011 | 0.877 | 0.778 | 1.284 | 1.209 | 1.004 | 0.893 | 0.755 | 0.653 |
| 28 | 1.345 | 1.274 | 1.079 | 0.973 | 0.844 | 0.749 | 1.236 | 1.164 | 0.967 | 0.859 | 0.727 | 0.628 |

| n | 0.005 | 0.01 | 0.05 | 0.1 | 0.2 | 0.3 | 0.005 | 0.01 | 0.05 | 0.1 | 0.2 | 0.3 |
|---|---|---|---|---|---|---|---|---|---|---|---|---|
| 30 | 1.298 | 1.230 | 1.041 | 0.939 | 0.814 | 0.723 | 1.193 | 1.124 | 0.933 | 0.830 | 0.701 | 0.607 |
| 40 | 1.121 | 1.061 | 0.899 | 0.811 | 0.703 | 0.624 | 1.031 | 0.971 | 0.806 | 0.717 | 0.606 | 0.524 |
| 60 | 0.912 | 0.864 | 0.731 | 0.660 | 0.572 | 0.508 | 0.839 | 0.791 | 0.656 | 0.584 | 0.493 | 0.427 |
| 80 | 0.788 | 0.747 | 0.632 | 0.570 | 0.494 | 0.439 | 0.726 | 0.684 | 0.568 | 0.505 | 0.427 | 0.369 |
| 100 | 0.704 | 0.667 | 0.565 | 0.510 | 0.442 | 0.392 | 0.649 | 0.611 | 0.507 | 0.451 | 0.381 | 0.330 |
| 200 | 0.497 | 0.471 | 0.399 | 0.360 | 0.312 | 0.277 | 0.458 | 0.432 | 0.358 | 0.318 | 0.269 | 0.233 |
| 500 | 0.314 | 0.297 | 0.252 | 0.227 | 0.197 | 0.175 | 0.289 | 0.273 | 0.226 | 0.201 | 0.170 | 0.147 |
| 1000 | 0.222 | 0.210 | 0.178 | 0.161 | 0.139 | 0.123 | 0.205 | 0.193 | 0.160 | 0.142 | 0.120 | 0.104 |

**Number of Treatments, $k = 6$**

| | $\alpha = 0.01$ | | | | | | $\alpha = 0.05$ | | | | | |
|---|---|---|---|---|---|---|---|---|---|---|---|---|
| $n$ \ $\beta$ | 0.005 | 0.01 | 0.05 | 0.1 | 0.2 | 0.3 | 0.005 | 0.01 | 0.05 | 0.1 | 0.2 | 0.3 |
| 2 | 12.011 | 11.397 | 9.745 | 8.881 | 7.851 | 7.120 | 8.747 | 8.274 | 6.993 | 6.317 | 5.505 | 4.922 |
| 3 | 6.890 | 6.574 | 5.714 | 5.258 | 4.707 | 4.311 | 5.615 | 5.333 | 4.561 | 4.149 | 3.647 | 3.283 |
| 4 | 5.321 | 5.085 | 4.440 | 4.096 | 3.678 | 3.376 | 4.491 | 4.270 | 3.663 | 3.337 | 2.940 | 2.650 |
| 5 | 4.500 | 4.303 | 3.763 | 3.475 | 3.125 | 2.871 | 3.861 | 3.673 | 3.154 | 2.876 | 2.535 | 2.287 |
| 6 | 3.973 | 3.800 | 3.327 | 3.074 | 2.765 | 2.541 | 3.443 | 3.275 | 2.815 | 2.567 | 2.264 | 2.042 |
| 7 | 3.598 | 3.443 | 3.015 | 2.786 | 2.508 | 2.305 | 3.138 | 2.985 | 2.566 | 2.341 | 2.065 | 1.863 |
| 8 | 3.314 | 3.171 | 2.778 | 2.568 | 2.311 | 2.125 | 2.903 | 2.762 | 2.375 | 2.166 | 1.911 | 1.725 |
| 10 | 2.903 | 2.778 | 2.435 | 2.251 | 2.027 | 1.864 | 2.558 | 2.434 | 2.093 | 1.910 | 1.685 | 1.521 |
| 12 | 2.616 | 2.503 | 2.195 | 2.029 | 1.827 | 1.680 | 2.313 | 2.201 | 1.893 | 1.727 | 1.524 | 1.376 |
| 14 | 2.400 | 2.297 | 2.014 | 1.862 | 1.677 | 1.542 | 2.128 | 2.025 | 1.742 | 1.589 | 1.402 | 1.266 |
| 16 | 2.230 | 2.135 | 1.872 | 1.730 | 1.558 | 1.433 | 1.981 | 1.885 | 1.621 | 1.479 | 1.306 | 1.178 |
| 18 | 2.092 | 2.002 | 1.756 | 1.623 | 1.462 | 1.345 | 1.861 | 1.771 | 1.523 | 1.390 | 1.226 | 1.107 |
| 20 | 1.977 | 1.892 | 1.659 | 1.534 | 1.382 | 1.271 | 1.760 | 1.675 | 1.441 | 1.315 | 1.160 | 1.047 |
| 22 | 1.879 | 1.798 | 1.577 | 1.458 | 1.313 | 1.208 | 1.674 | 1.593 | 1.371 | 1.251 | 1.104 | 0.996 |

*(continued)*

**Table A.8** Values of the Maximum Standardized Range $\tau$ for Sample Size Estimation in Completely Random Design[a]—cont...

**Number of Treatments, $k = 6$**

| $\beta$ \ $n$ | $\alpha = 0.01$ | | | | | | $\alpha = 0.05$ | | | | | |
|---|---|---|---|---|---|---|---|---|---|---|---|---|
| | 0.005 | 0.01 | 0.05 | 0.1 | 0.2 | 0.3 | 0.005 | 0.01 | 0.05 | 0.1 | 0.2 | 0.3 |
| 24 | 1.794 | 1.717 | 1.506 | 1.392 | 1.254 | 1.153 | 1.600 | 1.523 | 1.310 | 1.195 | 1.055 | 0.952 |
| 26 | 1.720 | 1.646 | 1.444 | 1.335 | 1.202 | 1.106 | 1.535 | 1.461 | 1.256 | 1.146 | 1.012 | 0.913 |
| 28 | 1.654 | 1.583 | 1.389 | 1.284 | 1.156 | 1.064 | 1.477 | 1.405 | 1.209 | 1.103 | 0.974 | 0.879 |
| 30 | 1.596 | 1.527 | 1.339 | 1.238 | 1.115 | 1.026 | 1.425 | 1.356 | 1.167 | 1.065 | 0.940 | 0.848 |
| 40 | 1.374 | 1.315 | 1.153 | 1.067 | 0.961 | 0.884 | 1.229 | 1.170 | 1.006 | 0.918 | 0.810 | 0.732 |
| 60 | 1.116 | 1.068 | 0.937 | 0.866 | 0.780 | 0.718 | 1.000 | 0.951 | 0.818 | 0.747 | 0.659 | 0.595 |
| 80 | 0.964 | 0.922 | 0.809 | 0.748 | 0.674 | 0.620 | 0.864 | 0.822 | 0.707 | 0.645 | 0.570 | 0.514 |
| 100 | 0.861 | 0.824 | 0.722 | 0.668 | 0.602 | 0.553 | 0.772 | 0.735 | 0.632 | 0.577 | 0.509 | 0.459 |
| 200 | 0.607 | 0.581 | 0.509 | 0.471 | 0.424 | 0.390 | 0.544 | 0.518 | 0.446 | 0.407 | 0.359 | 0.324 |
| 500 | 0.382 | 0.366 | 0.321 | 0.297 | 0.267 | 0.246 | 0.344 | 0.327 | 0.281 | 0.257 | 0.227 | 0.204 |
| 1000 | 0.270 | 0.259 | 0.227 | 0.210 | 0.189 | 0.174 | 0.243 | 0.231 | 0.199 | 0.181 | 0.160 | 0.145 |

| $\beta$ \ $n$ | $\alpha = 0.10$ | | | | | | $\alpha = 0.20$ | | | | | |
|---|---|---|---|---|---|---|---|---|---|---|---|---|
| | 0.005 | 0.01 | 0.05 | 0.1 | 0.2 | 0.3 | 0.005 | 0.01 | 0.05 | 0.1 | 0.2 | 0.3 |
| 2 | 7.512 | 7.088 | 5.935 | 5.323 | 4.582 | 4.046 | 6.326 | 5.945 | 4.902 | 4.342 | 3.657 | 3.154 |
| 3 | 5.058 | 4.789 | 4.049 | 3.651 | 3.165 | 2.809 | 4.469 | 4.210 | 3.496 | 3.109 | 2.632 | 2.278 |
| 4 | 4.107 | 3.891 | 3.297 | 2.977 | 2.585 | 2.297 | 3.684 | 3.472 | 2.888 | 2.571 | 2.179 | 1.888 |
| 5 | 3.557 | 3.371 | 2.859 | 2.583 | 2.244 | 1.995 | 3.214 | 3.030 | 2.522 | 2.246 | 1.904 | 1.650 |
| 6 | 3.185 | 3.019 | 2.562 | 2.315 | 2.012 | 1.789 | 2.890 | 2.726 | 2.269 | 2.021 | 1.714 | 1.485 |
| 7 | 2.911 | 2.759 | 2.342 | 2.117 | 1.840 | 1.636 | 2.649 | 2.498 | 2.080 | 1.853 | 1.572 | 1.362 |

| | | | | | | | | | | | | |
|---|---|---|---|---|---|---|---|---|---|---|---|---|
| 8 | 2.698 | 2.558 | 2.171 | 1.962 | 1.706 | 1.517 | 2.461 | 2.320 | 1.932 | 1.721 | 1.460 | 1.266 |
| 10 | 2.384 | 2.260 | 1.919 | 1.734 | 1.508 | 1.341 | 2.180 | 2.056 | 1.712 | 1.525 | 1.294 | 1.122 |
| 12 | 2.159 | 2.047 | 1.738 | 1.571 | 1.366 | 1.215 | 1.978 | 1.865 | 1.554 | 1.384 | 1.174 | 1.018 |
| 14 | 1.988 | 1.885 | 1.601 | 1.447 | 1.258 | 1.119 | 1.823 | 1.720 | 1.432 | 1.276 | 1.083 | 0.939 |
| 16 | 1.852 | 1.756 | 1.491 | 1.348 | 1.172 | 1.043 | 1.700 | 1.604 | 1.336 | 1.190 | 1.010 | 0.875 |
| 18 | 1.741 | 1.651 | 1.402 | 1.267 | 1.102 | 0.980 | 1.599 | 1.508 | 1.257 | 1.119 | 0.950 | 0.823 |
| 20 | 1.648 | 1.562 | 1.327 | 1.199 | 1.043 | 0.928 | 1.514 | 1.428 | 1.190 | 1.060 | 0.899 | 0.780 |
| 22 | 1.568 | 1.487 | 1.263 | 1.141 | 0.992 | 0.883 | 1.442 | 1.360 | 1.133 | 1.009 | 0.856 | 0.742 |
| 24 | 1.499 | 1.421 | 1.207 | 1.091 | 0.949 | 0.844 | 1.379 | 1.300 | 1.083 | 0.965 | 0.819 | 0.710 |
| 26 | 1.438 | 1.364 | 1.158 | 1.047 | 0.910 | 0.810 | 1.323 | 1.248 | 1.040 | 0.926 | 0.786 | 0.681 |
| 28 | 1.384 | 1.313 | 1.115 | 1.008 | 0.876 | 0.779 | 1.274 | 1.201 | 1.001 | 0.892 | 0.756 | 0.656 |
| 30 | 1.336 | 1.267 | 1.076 | 0.973 | 0.846 | 0.752 | 1.230 | 1.160 | 0.966 | 0.861 | 0.730 | 0.633 |
| 40 | 1.153 | 1.093 | 0.929 | 0.839 | 0.730 | 0.649 | 1.062 | 1.002 | 0.835 | 0.744 | 0.631 | 0.547 |
| 60 | 0.938 | 0.890 | 0.756 | 0.683 | 0.594 | 0.528 | 0.865 | 0.816 | 0.680 | 0.605 | 0.514 | 0.445 |
| 80 | 0.811 | 0.769 | 0.653 | 0.591 | 0.514 | 0.457 | 0.748 | 0.706 | 0.588 | 0.524 | 0.444 | 0.385 |
| 100 | 0.725 | 0.687 | 0.584 | 0.528 | 0.459 | 0.408 | 0.669 | 0.631 | 0.525 | 0.468 | 0.397 | 0.344 |
| 200 | 0.512 | 0.485 | 0.412 | 0.372 | 0.324 | 0.288 | 0.472 | 0.445 | 0.371 | 0.330 | 0.280 | 0.243 |
| 500 | 0.323 | 0.306 | 0.260 | 0.235 | 0.204 | 0.182 | 0.298 | 0.281 | 0.234 | 0.209 | 0.177 | 0.153 |
| 1000 | 0.228 | 0.217 | 0.184 | 0.166 | 0.145 | 0.129 | 0.211 | 0.199 | 0.166 | 0.148 | 0.125 | 0.109 |

*Source:* M. A. Kastenbaum, D. G. Hoel, and K. O. Bowman. (1970). Sample size requirements: One-way analysis of variance. *Biometrika,* **57,** 421–430. Reproduced with permission of the Biometrika Trustees.

**Table A.9** Values of the Maximum Standardized Range $\tau$ for Sample Size Estimation in a Randomized Block Design[a]

| k | b | n | | α = 0.05 | | | | | | α = 0.01 | | | | | β/n | b | k |
|---|---|---|---|---|---|---|---|---|---|---|---|---|---|---|---|---|---|
| | | β | 0.005 | 0.01 | 0.05 | 0.1 | 0.2 | 0.3 | 0.005 | 0.01 | 0.05 | 0.1 | 0.2 | 0.3 | | | |
| 2 | 2 | 2 | 4.469 | 4.203 | 3.485 | 3.108 | 2.659 | 2.341 | 6.723 | 6.341 | 5.318 | 4.786 | 4.157 | 3.714 | 2 | 2 | 2 |
| | | 3 | 3.020 | 2.850 | 2.388 | 2.143 | 1.848 | 1.636 | 3.853 | 3.656 | 3.124 | 2.843 | 2.505 | 2.264 | 3 | | |
| | | 4 | 2.481 | 2.343 | 1.968 | 1.768 | 1.526 | 1.352 | 3.026 | 2.876 | 2.467 | 2.251 | 1.989 | 1.802 | 4 | | |
| | | 5 | 2.166 | 2.046 | 1.719 | 1.545 | 1.335 | 1.183 | 2.588 | 2.461 | 2.115 | 1.931 | 1.708 | 1.548 | 5 | | |
| | 3 | 1 | 8.471 | 7.891 | 6.346 | 5.549 | 4.616 | 3.970 | 18.823 | 17.547 | 14.144 | 12.394 | 10.352 | 8.944 | 1 | 3 | |
| | | 2 | 3.200 | 3.017 | 2.521 | 2.259 | 1.944 | 1.718 | 4.295 | 4.068 | 3.457 | 3.136 | 2.752 | 2.480 | 2 | | |
| | | 3 | 2.339 | 2.209 | 1.855 | 1.667 | 1.439 | 1.275 | 2.853 | 2.711 | 2.326 | 2.122 | 1.876 | 1.699 | 3 | | |
| | | 4 | 1.962 | 1.853 | 1.558 | 1.400 | 1.209 | 1.072 | 2.329 | 2.215 | 1.904 | 1.739 | 1.539 | 1.395 | 4 | | |
| | | 5 | 1.728 | 1.633 | 1.373 | 1.234 | 1.066 | 0.945 | 2.027 | 1.928 | 1.659 | 1.515 | 1.342 | 1.217 | 5 | | |
| | 4 | 1 | 5.193 | 4.868 | 3.998 | 3.545 | 3.009 | 2.632 | 8.853 | 8.315 | 6.878 | 6.135 | 5.261 | 4.652 | 1 | 4 | |
| | | 2 | 2.615 | 2.468 | 2.068 | 1.856 | 1.600 | 1.417 | 3.337 | 3.166 | 2.705 | 2.462 | 2.170 | 1.961 | 2 | | |
| | | 3 | 1.977 | 1.868 | 1.569 | 1.411 | 1.218 | 1.080 | 2.363 | 2.247 | 1.931 | 1.763 | 1.560 | 1.413 | 3 | | |
| | | 4 | 1.673 | 1.581 | 1.329 | 1.195 | 1.032 | 0.915 | 1.963 | 1.867 | 1.606 | 1.467 | 1.299 | 1.178 | 4 | | |
| | 5 | 1 | 3.997 | 3.759 | 3.117 | 2.780 | 2.379 | 2.094 | 6.013 | 5.671 | 4.757 | 4.281 | 3.718 | 3.322 | 1 | 5 | |
| | | 2 | 2.265 | 2.139 | 1.794 | 1.612 | 1.391 | 1.232 | 2.810 | 2.670 | 2.287 | 2.084 | 1.841 | 1.666 | 2 | | |
| | | 3 | 1.744 | 1.648 | 1.385 | 1.245 | 1.076 | 0.953 | 2.060 | 1.960 | 1.685 | 1.539 | 1.362 | 1.235 | 3 | | |
| | | 4 | 1.483 | 1.402 | 1.178 | 1.060 | 0.916 | 0.812 | 1.728 | 1.644 | 1.415 | 1.292 | 1.145 | 1.038 | 4 | | |
| 3 | 2 | 1 | 14.487 | 13.496 | 10.853 | 9.490 | 7.895 | 6.790 | 32.522 | 30.315 | 24.436 | 21.413 | 17.885 | 15.453 | 1 | 2 | 3 |
| | | 2 | 4.651 | 4.393 | 3.693 | 3.324 | 2.879 | 2.561 | 6.306 | 5.978 | 5.097 | 4.635 | 4.084 | 3.693 | 2 | | |
| | | 3 | 3.233 | 3.061 | 2.593 | 2.344 | 2.041 | 1.823 | 3.946 | 3.757 | 3.244 | 2.971 | 2.642 | 2.406 | 3 | | |
| | | 4 | 2.667 | 2.527 | 2.144 | 1.939 | 1.691 | 1.511 | 3.158 | 3.010 | 2.606 | 2.391 | 2.130 | 1.942 | 4 | | |

| | | | | | | | | | | | | | | | | | |
|---|---|---|---|---|---|---|---|---|---|---|---|---|---|---|---|---|---|
| 3 | | 5 | 2.330 | 2.208 | 1.875 | 1.696 | 1.480 | 1.322 | 2.721 | 2.594 | 2.248 | 2.064 | 1.840 | 1.679 | 5 | | |
| | 3 | 1 | 6.430 | 6.052 | 5.035 | 4.503 | 3.870 | 3.420 | 9.812 | 9.259 | 7.777 | 7.007 | 6.096 | 5.457 | 1 | 3 | |
| | | 2 | 3.403 | 3.220 | 2.722 | 2.458 | 2.137 | 1.906 | 4.292 | 4.082 | 3.511 | 3.210 | 2.847 | 2.588 | 2 | | |
| | | 3 | 2.514 | 2.382 | 2.021 | 1.828 | 1.594 | 1.424 | 2.977 | 2.837 | 2.457 | 2.254 | 2.008 | 1.831 | 3 | | |
| | | 4 | 2.111 | 2.001 | 1.699 | 1.537 | 1.341 | 1.199 | 2.454 | 2.340 | 2.029 | 1.862 | 1.661 | 1.515 | 4 | | |
| | 4 | 1 | 4.651 | 4.393 | 3.693 | 3.324 | 2.879 | 2.561 | 6.306 | 5.978 | 5.097 | 4.635 | 4.084 | 3.693 | 1 | 4 | |
| | | 2 | 2.800 | 2.651 | 2.246 | 2.030 | 1.768 | 1.579 | 3.417 | 3.254 | 2.809 | 2.573 | 2.288 | 2.084 | 2 | | |
| | | 3 | 2.127 | 2.016 | 1.711 | 1.549 | 1.351 | 1.207 | 2.484 | 2.368 | 2.052 | 1.884 | 1.680 | 1.533 | 3 | | |
| | | 4 | 1.801 | 1.707 | 1.450 | 1.312 | 1.145 | 1.023 | 2.075 | 1.979 | 1.717 | 1.576 | 1.406 | 1.283 | 4 | | |
| | 5 | 1 | 3.828 | 3.621 | 3.057 | 2.758 | 2.397 | 2.136 | 4.913 | 4.669 | 4.008 | 3.660 | 3.241 | 2.942 | 1 | 5 | |
| | | 2 | 2.431 | 2.303 | 1.953 | 1.766 | 1.539 | 1.375 | 2.913 | 2.775 | 2.400 | 2.201 | 1.960 | 1.786 | 2 | | |
| | | 3 | 1.877 | 1.779 | 1.511 | 1.367 | 1.192 | 1.066 | 2.174 | 2.073 | 1.798 | 1.651 | 1.472 | 1.343 | 3 | | |
| | | 4 | 1.597 | 1.514 | 1.286 | 1.164 | 1.015 | 0.908 | 1.830 | 1.746 | 1.515 | 1.391 | 1.241 | 1.133 | 4 | | |
| 4 | 2 | 1 | 11.302 | 10.602 | 8.725 | 7.751 | 6.599 | 5.789 | 19.671 | 18.479 | 15.295 | 13.651 | 11.718 | 10.372 | 1 | 2 | 4 |
| | | 2 | 4.734 | 4.483 | 3.799 | 3.435 | 2.996 | 2.680 | 6.103 | 5.803 | 4.993 | 4.566 | 4.054 | 3.688 | 2 | | |
| | | 3 | 3.355 | 3.184 | 2.714 | 2.463 | 2.157 | 1.935 | 4.003 | 3.819 | 3.317 | 3.050 | 2.726 | 2.492 | 3 | | |
| | | 4 | 2.779 | 2.638 | 2.252 | 2.044 | 1.792 | 1.608 | 3.240 | 3.093 | 2.692 | 2.478 | 2.218 | 2.030 | 4 | | |
| | | 5 | 2.432 | 2.309 | 1.972 | 1.791 | 1.570 | 1.410 | 2.805 | 2.679 | 2.333 | 2.148 | 1.924 | 1.762 | 5 | | |
| | 3 | 1 | 6.039 | 5.708 | 4.812 | 4.339 | 3.770 | 3.362 | 8.235 | 7.811 | 6.670 | 6.072 | 5.359 | 4.852 | 1 | 3 | |
| | | 2 | 3.513 | 3.332 | 2.835 | 2.570 | 2.249 | 2.015 | 4.296 | 4.094 | 3.546 | 3.256 | 2.905 | 2.652 | 2 | | |
| | | 3 | 2.620 | 2.487 | 2.123 | 1.928 | 1.689 | 1.516 | 3.055 | 2.916 | 2.538 | 2.336 | 2.091 | 1.914 | 3 | | |
| | | 4 | 2.204 | 2.093 | 1.787 | 1.623 | 1.424 | 1.278 | 2.533 | 2.419 | 2.108 | 1.941 | 1.739 | 1.592 | 4 | | |
| | 4 | 1 | 4.584 | 4.342 | 3.685 | 3.336 | 2.913 | 2.607 | 5.803 | 5.522 | 4.763 | 4.361 | 3.879 | 3.533 | 1 | 4 | |
| | | 2 | 2.906 | 2.757 | 2.350 | 2.133 | 1.868 | 1.675 | 3.467 | 3.307 | 2.873 | 2.641 | 2.361 | 2.158 | 2 | | |

*(continued)*

**Table A.9** Values of the Maximum Standardized Range $\tau$ for Sample Size Estimation in a Randomized Block Design[a] —cont...

| k | b | β \ n | α = 0.05 | | | | | | α = 0.01 | | | | | | β/n | b | k |
|---|---|---|---|---|---|---|---|---|---|---|---|---|---|---|---|---|---|
| | | n | 0.005 | 0.01 | 0.05 | 0.1 | 0.2 | 0.3 | 0.005 | 0.01 | 0.05 | 0.1 | 0.2 | 0.3 | n | b | k |
| 5 | | 3 | 2.220 | 2.108 | 1.800 | 1.635 | 1.433 | 1.287 | 2.561 | 2.445 | 2.130 | 1.961 | 1.756 | 1.608 | 3 | | |
| | | 4 | 1.882 | 1.787 | 1.527 | 1.387 | 1.216 | 1.092 | 2.148 | 2.052 | 1.788 | 1.647 | 1.476 | 1.351 | 4 | | |
| | 5 | 1 | 3.849 | 3.650 | 3.106 | 2.816 | 2.463 | 2.208 | 4.706 | 4.485 | 3.885 | 3.566 | 3.182 | 2.905 | 1 | 5 | |
| | | 2 | 2.530 | 2.401 | 2.049 | 1.860 | 1.629 | 1.462 | 2.976 | 2.841 | 2.471 | 2.273 | 2.034 | 1.861 | 2 | | |
| | | 3 | 1.961 | 1.861 | 1.590 | 1.444 | 1.266 | 1.137 | 2.247 | 2.146 | 1.870 | 1.722 | 1.543 | 1.413 | 3 | | |
| 5 | 2 | 1 | 10.169 | 9.577 | 7.985 | 7.152 | 6.161 | 5.459 | 15.688 | 14.807 | 12.447 | 11.222 | 9.775 | 8.759 | 1 | 2 | 5 |
| | | 2 | 4.801 | 4.554 | 3.880 | 3.520 | 3.084 | 2.769 | 6.008 | 5.724 | 4.954 | 4.547 | 4.057 | 3.705 | 2 | | |
| | | 3 | 3.448 | 3.276 | 2.805 | 2.552 | 2.243 | 2.018 | 4.056 | 3.875 | 3.379 | 3.114 | 2.793 | 2.561 | 3 | | |
| | | 4 | 2.864 | 2.723 | 2.333 | 2.124 | 1.868 | 1.682 | 3.307 | 3.161 | 2.761 | 2.547 | 2.287 | 2.098 | 4 | | |
| | 3 | 1 | 5.921 | 5.610 | 4.765 | 4.316 | 3.773 | 3.381 | 7.658 | 7.285 | 6.276 | 5.745 | 5.108 | 4.653 | 1 | 3 | |
| | | 2 | 3.597 | 3.416 | 2.920 | 2.655 | 2.332 | 2.097 | 4.316 | 4.120 | 3.585 | 3.300 | 2.955 | 2.707 | 2 | | |
| | | 3 | 2.701 | 2.567 | 2.200 | 2.003 | 1.762 | 1.586 | 3.118 | 2.980 | 2.603 | 2.402 | 2.156 | 1.978 | 3 | | |
| | | 4 | 2.276 | 2.164 | 1.855 | 1.689 | 1.486 | 1.338 | 2.596 | 2.482 | 2.170 | 2.003 | 1.799 | 1.651 | 4 | | |
| | 4 | 1 | 4.598 | 4.364 | 3.725 | 3.383 | 2.968 | 2.667 | 5.632 | 5.371 | 4.662 | 4.286 | 3.831 | 3.504 | 1 | 4 | |
| | | 2 | 2.986 | 2.837 | 2.429 | 2.210 | 1.943 | 1.748 | 3.513 | 3.356 | 2.926 | 2.697 | 2.419 | 2.218 | 2 | | |
| | | 3 | 2.291 | 2.178 | 1.867 | 1.700 | 1.496 | 1.347 | 2.622 | 2.507 | 2.191 | 2.022 | 1.816 | 1.667 | 3 | | |
| | | 4 | 1.944 | 1.849 | 1.585 | 1.444 | 1.270 | 1.144 | 2.206 | 2.109 | 1.844 | 1.702 | 1.529 | 1.404 | 4 | | |

| | | | | | | | | | | | | | | | | | |
|---|---|---|---|---|---|---|---|---|---|---|---|---|---|---|---|---|---|
| 5 | | 1 | 3.898 | 3.703 | 3.167 | 2.879 | 2.530 | 2.275 | 4.655 | 4.444 | 3.869 | 3.563 | 3.192 | 2.924 | 1 | 5 |
| | | 2 | 2.605 | 2.475 | 2.121 | 1.930 | 1.697 | 1.528 | 3.030 | 2.895 | 2.527 | 2.331 | 2.092 | 1.919 | 2 | |
| | | 3 | 2.025 | 1.925 | 1.650 | 1.503 | 1.323 | 1.191 | 2.304 | 2.203 | 1.926 | 1.778 | 1.597 | 1.466 | 3 | |
| 6 | 2 | 1 | 9.630 | 9.094 | 7.646 | 6.885 | 5.974 | 5.324 | 13.862 | 13.125 | 11.146 | 10.114 | 8.890 | 8.025 | 1 | 6 |
| | | 2 | 4.863 | 4.618 | 3.950 | 3.593 | 3.159 | 2.843 | 5.967 | 5.693 | 4.949 | 4.554 | 4.077 | 3.733 | 2 | 2 |
| | | 3 | 3.525 | 3.353 | 2.880 | 2.625 | 2.314 | 2.087 | 4.108 | 3.928 | 3.436 | 3.172 | 2.852 | 2.620 | 3 | |
| | | 4 | 2.935 | 2.793 | 2.401 | 2.190 | 1.931 | 1.743 | 3.366 | 3.220 | 2.820 | 2.606 | 2.346 | 2.156 | 4 | |
| | 3 | 1 | 5.891 | 5.591 | 4.772 | 4.335 | 3.805 | 3.421 | 7.387 | 7.041 | 6.102 | 5.605 | 5.007 | 4.577 | 1 | 3 |
| | | 2 | 3.667 | 3.486 | 2.991 | 2.725 | 2.400 | 2.164 | 4.345 | 4.152 | 3.625 | 3.344 | 3.003 | 2.757 | 2 | |
| | | 3 | 2.767 | 2.633 | 2.263 | 2.064 | 1.821 | 1.643 | 3.173 | 3.036 | 2.659 | 2.457 | 2.211 | 2.033 | 3 | |
| | | 4 | 2.335 | 2.222 | 1.911 | 1.743 | 1.538 | 1.388 | 2.650 | 2.536 | 2.223 | 2.055 | 1.850 | 1.701 | 4 | |
| | 4 | 1 | 4.636 | 4.406 | 3.775 | 3.438 | 3.026 | 2.726 | 5.569 | 5.319 | 4.636 | 4.273 | 3.833 | 3.515 | 1 | 4 |
| | | 2 | 3.053 | 2.904 | 2.494 | 2.273 | 2.004 | 1.808 | 3.557 | 3.402 | 2.975 | 2.747 | 2.470 | 2.269 | 2 | |
| | | 3 | 2.350 | 2.236 | 1.923 | 1.754 | 1.548 | 1.397 | 2.674 | 2.559 | 2.243 | 2.073 | 1.866 | 1.716 | 3 | |
| | | 4 | 1.996 | 1.900 | 1.634 | 1.491 | 1.315 | 1.187 | 2.254 | 2.157 | 1.891 | 1.749 | 1.575 | 1.448 | 4 | |
| | 5 | 1 | 3.954 | 3.760 | 3.227 | 2.941 | 2.591 | 2.337 | 4.653 | 4.448 | 3.887 | 3.587 | 3.223 | 2.959 | 1 | 5 |
| | | 2 | 2.667 | 2.537 | 2.180 | 1.988 | 1.753 | 1.582 | 3.078 | 2.944 | 2.577 | 2.381 | 2.142 | 1.969 | 2 | |
| | | 3 | 2.078 | 1.977 | 1.700 | 1.551 | 1.369 | 1.236 | 2.354 | 2.252 | 1.974 | 1.825 | 1.643 | 1.511 | 3 | |

[a]Source: M. A. Kastenbaum and D. G. Hoel. (1970). Sample size requirements: Randomized block designs. Biometrika, **57**, 573–577. Reproduced with permission of the Biometrika Trustees.

**Table A.10** Values of $d/S$ for Sample Size Calculation for a Specified Confidence Interval[a,b]

| DF[c] | Sample Size for New Estimate of Standard Deviation | | | | | | | | | | | | | | | | | |
|---|---|---|---|---|---|---|---|---|---|---|---|---|---|---|---|---|---|---|
| | 2 | 3 | 4 | 5 | 6 | 7 | 8 | 9 | 10 | 11 | 13 | 16 | 21 | 25 | 31 | 41 | 61 | 121 |
| 1 | 114.2 | 35.09 | 23.37 | 18.61 | 15.92 | 14.15 | 12.86 | 11.88 | 11.09 | 10.45 | 9.44 | 8.36 | 7.17 | 6.51 | 5.80 | 5.00 | 4.07 | 2.86 |
| 2 | 38.66 | 10.83 | 6.97 | 5.45 | 4.61 | 4.07 | 3.68 | 3.38 | 3.15 | 2.96 | 2.66 | 2.35 | 2.01 | 1.82 | 1.62 | 1.39 | 1.13 | 0.796 |
| 3 | 28.59 | 7.68 | 4.85 | 3.75 | 3.15 | 2.76 | 2.49 | 2.29 | 2.12 | 1.99 | 1.79 | 1.57 | 1.34 | 1.21 | 1.08 | 0.925 | 0.750 | .526 |
| 4 | 24.95 | 6.55 | 4.08 | 3.14 | 2.62 | 2.30 | 2.06 | 1.89 | 1.75 | 1.64 | 1.47 | 1.29 | 1.10 | 0.984 | 0.879 | 0.755 | 0.611 | .428 |
| 5 | 23.10 | 5.98 | 3.70 | 2.83 | 2.36 | 2.06 | 1.85 | 1.69 | 1.56 | 1.46 | 1.31 | 1.14 | 0.972 | .878 | .778 | .667 | .539 | .378 |
| 6 | 21.98 | 5.63 | 3.47 | 2.64 | 2.20 | 1.91 | 1.72 | 1.56 | 1.45 | 1.35 | 1.21 | 1.06 | .896 | .809 | .716 | .613 | .495 | .346 |
| 7 | 21.25 | 5.41 | 3.32 | 2.52 | 2.09 | 1.82 | 1.63 | 1.48 | 1.37 | 1.28 | 1.14 | 0.998 | .845 | .762 | .674 | .577 | .466 | .325 |
| 8 | 20.72 | 5.25 | 3.21 | 2.43 | 2.02 | 1.75 | 1.56 | 1.42 | 1.32 | 1.23 | 1.10 | .956 | .808 | .729 | .644 | .551 | .444 | .310 |
| 9 | 20.32 | 5.12 | 3.13 | 2.37 | 1.96 | 1.70 | 1.52 | 1.38 | 1.28 | 1.19 | 1.06 | .924 | .780 | .703 | .621 | .531 | .428 | .298 |
| 10 | 20.02 | 5.03 | 3.06 | 2.32 | 1.91 | 1.66 | 1.48 | 1.35 | 1.24 | 1.16 | 1.03 | .899 | .758 | .683 | .603 | .515 | .415 | .289 |
| 11 | 19.78 | 4.96 | 3.01 | 2.28 | 1.88 | 1.63 | 1.45 | 1.32 | 1.22 | 1.14 | 1.01 | .879 | .740 | .667 | .588 | .502 | .404 | .282 |
| 12 | 19.58 | 4.90 | 2.97 | 2.24 | 1.85 | 1.60 | 1.43 | 1.30 | 1.20 | 1.12 | 0.990 | .862 | .726 | .653 | .576 | .492 | .396 | .275 |
| 13 | 19.41 | 4.85 | 2.94 | 2.21 | 1.82 | 1.58 | 1.41 | 1.28 | 1.18 | 1.10 | .975 | .848 | .714 | .642 | .566 | .483 | .388 | .270 |
| 14 | 19.27 | 4.80 | 2.91 | 2.19 | 1.80 | 1.56 | 1.39 | 1.26 | 1.16 | 1.08 | .962 | .836 | .703 | .633 | .557 | .475 | .382 | .266 |
| 15 | 19.15 | 4.77 | 2.88 | 2.17 | 1.79 | 1.54 | 1.38 | 1.25 | 1.15 | 1.07 | .951 | .826 | .694 | .624 | .550 | .469 | .376 | .262 |
| 16 | 19.05 | 4.74 | 2.86 | 2.15 | 1.77 | 1.53 | 1.36 | 1.24 | 1.14 | 1.06 | .941 | .817 | .687 | .617 | .543 | .463 | .372 | .258 |
| 17 | 18.96 | 4.71 | 2.84 | 2.14 | 1.76 | 1.52 | 1.35 | 1.23 | 1.13 | 1.05 | .932 | .809 | .680 | .611 | .538 | .458 | .367 | .255 |

| 18 | 18.88 | 4.68 | 2.83 | 2.12 | 1.75 | 1.51 | 1.34 | 1.22 | 1.12 | 1.04 | .925 | .803 | .674 | .605 | .532 | .453 | .364 | .252 |
|---|---|---|---|---|---|---|---|---|---|---|---|---|---|---|---|---|---|---|
| 19 | 18.81 | 4.66 | 2.81 | 2.11 | 1.74 | 1.50 | 1.33 | 1.21 | 1.11 | 1.04 | .918 | .796 | .668 | .600 | .528 | .449 | .360 | .250 |
| 20 | 18.74 | 4.64 | 2.80 | 2.10 | 1.73 | 1.49 | 1.33 | 1.20 | 1.11 | 1.03 | .912 | .791 | .663 | .596 | .524 | .446 | .357 | .248 |
| 21 | 18.68 | 4.62 | 2.79 | 2.09 | 1.72 | 1.48 | 1.32 | 1.20 | 1.10 | 1.02 | .906 | .786 | .659 | .592 | .520 | .442 | .355 | .246 |
| 22 | 18.63 | 4.61 | 2.78 | 2.08 | 1.71 | 1.48 | 1.31 | 1.19 | 1.10 | 1.02 | .902 | .782 | .655 | .588 | .517 | .439 | .352 | .244 |
| 23 | 18.59 | 4.60 | 2.77 | 2.08 | 1.70 | 1.47 | 1.31 | 1.18 | 1.09 | 1.01 | .897 | .777 | .651 | .584 | .514 | .437 | .350 | .242 |
| 24 | 18.54 | 4.58 | 2.76 | 2.07 | 1.70 | 1.46 | 1.30 | 1.18 | 1.08 | 1.01 | .893 | .774 | .648 | .581 | .511 | .434 | .348 | .241 |
| 25 | 18.50 | 4.57 | 2.75 | 2.06 | 1.69 | 1.46 | 1.30 | 1.18 | 1.08 | 1.00 | .889 | .770 | .645 | .578 | .508 | .432 | .346 | .239 |
| 26 | 18.47 | 4.56 | 2.74 | 2.06 | 1.69 | 1.46 | 1.29 | 1.17 | 1.08 | 1.00 | .886 | .767 | .642 | .576 | .506 | .430 | .344 | .238 |
| 27 | 18.44 | 4.55 | 2.74 | 2.05 | 1.68 | 1.45 | 1.29 | 1.17 | 1.07 | 0.997 | .882 | .764 | .640 | .573 | .504 | .428 | .342 | .237 |
| 28 | 18.40 | 4.54 | 2.73 | 2.05 | 1.68 | 1.45 | 1.28 | 1.16 | 1.07 | .994 | .879 | .761 | .637 | .571 | .501 | .426 | .341 | .236 |
| 29 | 18.38 | 4.53 | 2.73 | 2.04 | 1.67 | 1.44 | 1.28 | 1.16 | 1.07 | .991 | .876 | .759 | .635 | .569 | .500 | .424 | .339 | .235 |
| 30 | 18.35 | 4.52 | 2.72 | 2.04 | 1.67 | 1.44 | 1.28 | 1.16 | 1.06 | .988 | .874 | .756 | .633 | .567 | .498 | .422 | .338 | .234 |
| 40 | 18.16 | 4.47 | 2.68 | 2.00 | 1.64 | 1.41 | 1.25 | 1.14 | 1.04 | .968 | .855 | .739 | .617 | .553 | .484 | .411 | .328 | .226 |
| 60 | 17.97 | 4.41 | 2.64 | 1.97 | 1.62 | 1.39 | 1.23 | 1.11 | 1.02 | .948 | .837 | .722 | .602 | .538 | .471 | .398 | .317 | .218 |
| 120 | 17.79 | 4.35 | 2.60 | 1.94 | 1.59 | 1.36 | 1.21 | 1.09 | 1.00 | .929 | .818 | .705 | .586 | .524 | .457 | .386 | .306 | .209 |
| ∞ | 17.61 | 4.30 | 2.57 | 1.91 | 1.56 | 1.34 | 1.18 | 1.07 | 0.98 | .909 | .800 | .688 | .570 | .508 | .443 | .373 | .294 | .199 |

[a]Source: E. E. Johnson. (1963). Sample size for specified confidence interval. Ind Qual Control. **20**, 40–41. Reproduced with permission. Copyright 1963 American Society for Quality Control, Inc. Reprinted by permission.

[b]d/S Values are for 95% assurance that the specified 95% confidence interval will not be exceeded: d = specified half-confidence interval; S = standard deviation estimated from previous measurements.
[c]Degrees of freedom of previous estimate of standard deviation S.

**Table A.11** The Cumulative Normal Distribution Function: Cumulative Proportions to Unit Normal Deviates[a] (for $\alpha < 0.5$, $Z_{1-\alpha} = -Z_\alpha$)[b]

| 100α | | | | | Decimal Fraction of 100α | | | | | |
|---|---|---|---|---|---|---|---|---|---|---|
|  | .0 | .1 | .2 | .3 | .4 | .5 | .6 | .7 | .8 | .9 |
| 50 | 0.000 | 0.003 | 0.005 | 0.008 | 0.010 | 0.013 | 0.015 | 0.018 | 0.020 | 0.023 |
| 51 | 0.025 | 0.028 | 0.030 | 0.033 | 0.035 | 0.038 | 0.040 | 0.043 | 0.045 | 0.048 |
| 52 | 0.050 | 0.053 | 0.055 | 0.058 | 0.060 | 0.063 | 0.065 | 0.068 | 0.070 | 0.073 |
| 53 | 0.075 | 0.078 | 0.080 | 0.083 | 0.085 | 0.088 | 0.090 | 0.093 | 0.095 | 0.098 |
| 54 | 0.100 | 0.103 | 0.105 | 0.108 | 0.111 | 0.113 | 0.116 | 0.118 | 0.121 | 0.123 |
| 55 | 0.126 | 0.128 | 0.131 | 0.133 | 0.136 | 0.138 | 0.141 | 0.143 | 0.146 | 0.148 |
| 56 | 0.151 | 0.154 | 0.156 | 0.159 | 0.161 | 0.164 | 0.166 | 0.169 | 0.171 | 0.174 |
| 57 | 0.176 | 0.179 | 0.181 | 0.184 | 0.187 | 0.189 | 0.192 | 0.194 | 0.197 | 0.199 |
| 58 | 0.202 | 0.204 | 0.207 | 0.210 | 0.212 | 0.215 | 0.217 | 0.220 | 0.222 | 0.225 |
| 59 | 0.228 | 0.230 | 0.233 | 0.235 | 0.238 | 0.240 | 0.243 | 0.246 | 0.248 | 0.251 |
| 60 | 0.253 | 0.256 | 0.259 | 0.261 | 0.264 | 0.266 | 0.269 | 0.272 | 0.274 | 0.277 |
| 61 | 0.279 | 0.282 | 0.285 | 0.287 | 0.290 | 0.292 | 0.295 | 0.298 | 0.300 | 0.303 |
| 62 | 0.305 | 0.308 | 0.311 | 0.313 | 0.316 | 0.319 | 0.321 | 0.324 | 0.327 | 0.329 |
| 63 | 0.332 | 0.335 | 0.337 | 0.340 | 0.342 | 0.345 | 0.348 | 0.350 | 0.353 | 0.356 |
| 64 | 0.358 | 0.361 | 0.364 | 0.366 | 0.369 | 0.372 | 0.375 | 0.377 | 0.380 | 0.383 |
| 65 | 0.385 | 0.388 | 0.391 | 0.393 | 0.396 | 0.399 | 0.402 | 0.404 | 0.407 | 0.410 |
| 66 | 0.412 | 0.415 | 0.418 | 0.421 | 0.423 | 0.426 | 0.429 | 0.432 | 0.434 | 0.437 |
| 67 | 0.440 | 0.443 | 0.445 | 0.448 | 0.451 | 0.454 | 0.457 | 0.459 | 0.462 | 0.465 |
| 68 | 0.468 | 0.470 | 0.473 | 0.476 | 0.479 | 0.482 | 0.485 | 0.487 | 0.490 | 0.493 |
| 69 | 0.496 | 0.499 | 0.502 | 0.504 | 0.507 | 0.510 | 0.513 | 0.516 | 0.519 | 0.522 |
| 70 | 0.524 | 0.527 | 0.530 | 0.533 | 0.536 | 0.539 | 0.542 | 0.545 | 0.548 | 0.550 |
| 71 | 0.553 | 0.556 | 0.559 | 0.562 | 0.565 | 0.568 | 0.571 | 0.574 | 0.577 | 0.580 |
| 72 | 0.583 | 0.586 | 0.589 | 0.592 | 0.595 | 0.598 | 0.601 | 0.604 | 0.607 | 0.610 |
| 73 | 0.613 | 0.616 | 0.619 | 0.622 | 0.625 | 0.628 | 0.631 | 0.634 | 0.637 | 0.640 |
| 74 | 0.643 | 0.646 | 0.650 | 0.653 | 0.656 | 0.659 | 0.662 | 0.665 | 0.668 | 0.671 |

*(continued)*

**Table A.11** The Cumulative Normal Distribution Function: Cumulative Proportions to Unit Normal Deviates[a] (for $\alpha < 0.5$, $Z_{1-\alpha} = -Z_\alpha$)[b]—cont...

| 100$\alpha$ | \.0 | .1 | .2 | .3 | .4 | .5 | .6 | .7 | .8 | .9 |
|---|---|---|---|---|---|---|---|---|---|---|
| | | | | | **Decimal Fraction of 100$\alpha$** | | | | | |
| 75 | 0.674 | 0.678 | 0.681 | 0.684 | 0.687 | 0.690 | 0.693 | 0.697 | 0.700 | 0.703 |
| 76 | 0.706 | 0.710 | 0.713 | 0.716 | 0.719 | 0.722 | 0.726 | 0.729 | 0.732 | 0.736 |
| 77 | 0.739 | 0.742 | 0.745 | 0.749 | 0.752 | 0.755 | 0.759 | 0.762 | 0.765 | 0.769 |
| 78 | 0.772 | 0.776 | 0.779 | 0.782 | 0.786 | 0.789 | 0.793 | 0.796 | 0.800 | 0.803 |
| 79 | 0.806 | 0.810 | 0.813 | 0.817 | 0.820 | 0.824 | 0.827 | 0.831 | 0.834 | 0.838 |
| 80 | 0.842 | 0.845 | 0.849 | 0.852 | 0.856 | 0.860 | 0.863 | 0.867 | 0.871 | 0.874 |
| 81 | 0.878 | 0.882 | 0.885 | 0.889 | 0.893 | 0.896 | 0.900 | 0.904 | 0.908 | 0.912 |
| 82 | 0.915 | 0.919 | 0.923 | 0.927 | 0.931 | 0.935 | 0.938 | 0.942 | 0.946 | 0.950 |
| 83 | 0.954 | 0.958 | 0.962 | 0.966 | 0.970 | 0.974 | 0.978 | 0.982 | 0.986 | 0.990 |
| 84 | 0.994 | 0.999 | 1.003 | 1.007 | 1.011 | 1.015 | 1.019 | 1.024 | 1.028 | 1.032 |
| 85 | 1.036 | 1.041 | 1.045 | 1.049 | 1.054 | 1.058 | 1.063 | 1.067 | 1.071 | 1.076 |
| 86 | 1.080 | 1.085 | 1.089 | 1.094 | 1.098 | 1.103 | 1.108 | 1.112 | 1.117 | 1.122 |
| 87 | 1.126 | 1.131 | 1.136 | 1.141 | 1.146 | 1.150 | 1.155 | 1.160 | 1.165 | 1.170 |
| 88 | 1.175 | 1.180 | 1.185 | 1.190 | 1.195 | 1.200 | 1.206 | 1.211 | 1.216 | 1.221 |
| 89 | 1.227 | 1.232 | 1.237 | 1.243 | 1.248 | 1.254 | 1.259 | 1.265 | 1.270 | 1.276 |
| 90 | 1.282 | 1.287 | 1.293 | 1.299 | 1.305 | 1.311 | 1.317 | 1.323 | 1.329 | 1.335 |
| 91 | 1.341 | 1.347 | 1.353 | 1.359 | 1.366 | 1.372 | 1.379 | 1.385 | 1.392 | 1.398 |
| 92 | 1.405 | 1.412 | 1.419 | 1.426 | 1.433 | 1.440 | 1.447 | 1.454 | 1.461 | 1.468 |
| 93 | 1.476 | 1.483 | 1.491 | 1.499 | 1.506 | 1.514 | 1.522 | 1.530 | 1.538 | 1.546 |
| 94 | 1.555 | 1.563 | 1.572 | 1.580 | 1.589 | 1.598 | 1.607 | 1.616 | 1.626 | 1.635 |
| 95 | 1.645 | 1.655 | 1.665 | 1.675 | 1.685 | 1.695 | 1.706 | 1.717 | 1.728 | 1.739 |
| 96 | 1.751 | 1.762 | 1.774 | 1.787 | 1.799 | 1.812 | 1.825 | 1.838 | 1.852 | 1.866 |
| 97 | 1.881 | 1.896 | 1.911 | 1.927 | 1.943 | 1.960 | 1.977 | 1.995 | 2.014 | 2.034 |
| 98 | 2.054 | 2.075 | 2.097 | 2.120 | 2.144 | 2.170 | 2.197 | 2.226 | 2.257 | 2.290 |
| 99 | 2.326 | 2.366 | 2.409 | 2.457 | 2.512 | 2.576 | 2.652 | 2.748 | 2.878 | 3.090 |

[a]*Generated by an SAS program written by R. W. Washam II, Armour Research Center, Scottsdale, Arizona.*
[b]*To use this table for the Thurstone–Mosteller model, replace $\alpha$ with $p_{ij}$. See Eq. (11.2-21).*

**Table A.12** The Cumulative Normal Distribution Function: Unit Normal Deviates to Cumulative Proportions[a]

$\alpha' = \Phi(U_\alpha)$ for $-4.99 \leq U_\alpha \leq 0.00$

| $U_\alpha$ | .00 | .01 | .02 | .03 | .04 | .05 | .06 | .07 | .08 | .09 |
|---|---|---|---|---|---|---|---|---|---|---|
| 0.0 | .5000000 | .4960106 | .4920217 | .4880335 | .4840466 | .4800612 | .4760778 | .4720968 | .4681186 | .4641436 |
| −0.1 | .4601722 | .4562047 | .4522416 | .4482832 | .4443300 | .4403823 | .4364405 | .4325051 | .4285763 | .4246546 |
| −0.2 | .4207403 | .4168338 | .4129356 | .4090459 | .4051651 | .4012937 | .3974319 | .3935801 | .3897388 | .3859081 |
| −0.3 | .3820886 | .3782805 | .3744842 | .3707000 | .3669283 | .3631693 | .3594236 | .3556912 | .3519727 | .3482683 |
| −0.4 | .3445783 | .3409030 | .3372427 | .3335978 | .3299686 | .3263552 | .3227581 | .3191775 | .3156137 | .3120669 |
| −0.5 | .3085375 | .3050257 | .3015318 | .2980560 | .2945985 | .2911597 | .2877397 | .2843388 | .2809573 | .2775953 |
| −0.6 | .2742531 | .2709309 | .2676289 | .2643473 | .2610863 | .2578461 | .2546269 | .2514289 | .2482522 | .2450971 |
| −0.7 | .2419637 | .2388521 | .2357625 | .2326951 | .2296500 | .2266274 | .2236273 | .2206499 | .2176954 | .2147639 |
| −0.8 | .2118554 | .2089701 | .2061081 | .2032694 | .2004542 | .1976625 | .1948945 | .1921502 | .1894297 | .1867329 |
| −0.9 | .1840601 | .1814113 | .1787864 | .1761855 | .1736088 | .1710561 | .1685276 | .1660232 | .1635431 | .1610871 |
| −1.0 | .1586553 | .1562476 | .1538642 | .1515050 | .1491700 | .1468591 | .1445723 | .1423097 | .1400711 | .1378566 |
| −1.1 | .1356661 | .1334995 | .1313569 | .1292381 | .1271432 | .1250719 | .1230244 | .1210005 | .1190001 | .1170232 |
| −1.2 | .1150697 | .1131394 | .1112324 | .1093486 | .1074877 | .1056498 | .1038347 | .1020423 | .1002726 | .0985253 |
| −1.3 | .0968005 | .0950979 | .0934175 | .0917591 | .0901227 | .0885080 | .0869150 | .0853435 | .0837933 | .0822644 |
| −1.4 | .0807567 | .0792698 | .0778038 | .0763585 | .0749337 | .0735293 | .0721450 | .0707809 | .0694366 | .0681121 |
| −1.5 | .0668072 | .0655217 | .0642555 | .0630084 | .0617802 | .0605708 | .0593799 | .0582076 | .0570534 | .0559174 |
| −1.6 | .0547993 | .0536989 | .0526161 | .0515507 | .0505026 | .0494715 | .0484572 | .0474597 | .0464787 | .0455140 |
| −1.7 | .0445655 | .0436329 | .0427162 | .0418151 | .0409295 | .0400592 | .0392039 | .0383636 | .0375380 | .0367270 |
| −1.8 | .0359303 | .0351479 | .0343795 | .0336250 | .0328841 | .0321568 | .0314428 | .0307419 | .0300540 | .0293790 |
| −1.9 | .0287166 | .0280666 | .0274289 | .0268034 | .0261898 | .0255881 | .0249979 | .0244192 | .0238518 | .0232955 |
| −2.0 | .0227501 | .0222156 | .0216917 | .0211783 | .0206752 | .0201822 | .0196993 | .0192262 | .0187628 | .0183089 |

| | | | | | | | | | | |
|---|---|---|---|---|---|---|---|---|---|---|
| −2.1 | .0178644 | .0174292 | .0170030 | .0165858 | .0161774 | .0157776 | .0153863 | .0150034 | .0146287 | .0142621 |
| −2.2 | .0139034 | .0135526 | .0132094 | .0128737 | .0125455 | .0122245 | .0119106 | .0116038 | .0113038 | .0110107 |
| −2.3 | .0107241 | .0104441 | .0101704 | .0099031 | .0096419 | .0093867 | .0091375 | .0088940 | .0086563 | .0084242 |
| −2.4 | .0081975 | .0079763 | .0077603 | .0075494 | .0073436 | .0071428 | .0069469 | .0067557 | .0065691 | .0063872 |
| −2.5 | .0062097 | .0060366 | .0058677 | .0057031 | .0055426 | .0053861 | .0052336 | .0050849 | .0049400 | .0047988 |
| −2.6 | .0046612 | .0045271 | .0043965 | .0042692 | .0041453 | .0040246 | .0039070 | .0037926 | .0036811 | .0035726 |
| −2.7 | .0034670 | .0033642 | .0032641 | .0031667 | .0030720 | .0029798 | .0028901 | .0028028 | .0027179 | .0026354 |
| −2.8 | .0025551 | .0024771 | .0024012 | .0023274 | .0022557 | .0021860 | .0021182 | .0020524 | .0019884 | .0019262 |
| −2.9 | .0018658 | .0018071 | .0017502 | .0016948 | .0016411 | .0015889 | .0015382 | .0014890 | .0014412 | .0013949 |
| −3.0 | .0013499 | .0013062 | .0012639 | .0012228 | .0011829 | .0011442 | .0011067 | .0010703 | .0010350 | .0010008 |
| −3.1 | .0009676 | .0009354 | .0009043 | .0008740 | .0008447 | .0008164 | .0007888 | .0007622 | .0007364 | .0007114 |
| −3.2 | .0006871 | .0006637 | .0006410 | .0006190 | .0005976 | .0005770 | .0005571 | .0005377 | .0005190 | .0005009 |
| −3.3 | .0004834 | .0004665 | .0004501 | .0004342 | .0004189 | .0004041 | .0003897 | .0003758 | .0003624 | .0003495 |
| −3.4 | .0003369 | .0003248 | .0003131 | .0003018 | .0002909 | .0002803 | .0002701 | .0002602 | .0002507 | .0002415 |
| −3.5 | .0002326 | .0002241 | .0002158 | .0002078 | .0002001 | .0001926 | .0001854 | .0001785 | .0001718 | .0001653 |
| −3.6 | .0001591 | .0001531 | .0001473 | .0001417 | .0001363 | .0001311 | .0001261 | .0001213 | .0001166 | .0001121 |
| −3.7 | .0001078 | .0001036 | .0000996 | .0000957 | .0000920 | .0000884 | .0000850 | .0000816 | .0000784 | .0000753 |
| −3.8 | .0000723 | .0000695 | .0000667 | .0000641 | .0000615 | .0000591 | .0000567 | .0000544 | .0000522 | .0000501 |
| −3.9 | .0000481 | .0000461 | .0000443 | .0000425 | .0000407 | .0000391 | .0000375 | .0000359 | .0000345 | .0000330 |
| −4.0 | .0000317 | .0000304 | .0000291 | .0000279 | .0000267 | .0000256 | .0000245 | .0000235 | .0000225 | .0000216 |
| −4.1 | .0000207 | .0000198 | .0000189 | .0000181 | .0000174 | .0000166 | .0000159 | .0000152 | .0000146 | .0000139 |
| −4.2 | .0000133 | .0000128 | .0000122 | .0000117 | .0000112 | .0000107 | .0000102 | .0000098 | .0000093 | .0000089 |
| −4.3 | .0000085 | .0000082 | .0000078 | .0000075 | .0000071 | .0000068 | .0000065 | .0000062 | .0000059 | .0000057 |
| −4.4 | .0000054 | .0000052 | .0000049 | .0000047 | .0000045 | .0000043 | .0000041 | .0000039 | .0000037 | .0000036 |

*(continued)*

**Table A.12** The Cumulative Normal Distribution Function: Unit Normal Deviates to Cumulative Proportions[a]—cont...

$\alpha' = \Phi(U_\alpha)$ for $-4.99 \leq U_\alpha \leq 0.00$

| $U_\alpha$ | .00 | .01 | .02 | .03 | .04 | .05 | .06 | .07 | .08 | .09 |
|---|---|---|---|---|---|---|---|---|---|---|
| -4.5 | .0000034 | .0000032 | .0000031 | .0000029 | .0000028 | .0000027 | .0000026 | .0000024 | .0000023 | .0000022 |
| -4.6 | .0000021 | .0000020 | .0000019 | .0000018 | .0000017 | .0000017 | .0000016 | .0000015 | .0000014 | .0000014 |
| -4.7 | .0000013 | .0000012 | .0000012 | .0000011 | .0000011 | .0000010 | .0000010 | .0000009 | .0000009 | .0000008 |
| -4.8 | .0000008 | .0000008 | .0000007 | .0000007 | .0000006 | .0000006 | .0000006 | .0000006 | .0000005 | .0000005 |
| -4.9 | .0000006 | .0000006 | .0000006 | .0000005 | .0000005 | .0000004 | .0000004 | .0000003 | .0000003 | .0000003 |

For $0.00 \leq U_\alpha \leq 4.99$

| $U_\alpha$ | .00 | .01 | .02 | .03 | .04 | .05 | .06 | .07 | .08 | .09 |
|---|---|---|---|---|---|---|---|---|---|---|
| 0.0 | .5000000 | .5039894 | .5079783 | .5119665 | .5159534 | .5199388 | .5239222 | .5279032 | .5318814 | .5358564 |
| 0.1 | .5398278 | .5437953 | .5477584 | .5517168 | .5556700 | .5596177 | .5635595 | .5674949 | .5714237 | .5753454 |
| 0.2 | .5792597 | .5831662 | .5870644 | .5909541 | .5948349 | .5987063 | .6025681 | .6064199 | .6102612 | .6140919 |
| 0.3 | .6179114 | .6217195 | .6255158 | .6293000 | .6330717 | .6368307 | .6405764 | .6443088 | .6480273 | .6517317 |
| 0.4 | .6554217 | .6590970 | .6627573 | .6664022 | .6700314 | .6736448 | .6772419 | .6808225 | .6843863 | .6879331 |
| 0.5 | .6914625 | .6949743 | .6984682 | .7019440 | .7054015 | .7088403 | .7122603 | .7156612 | .7190427 | .7224047 |
| 0.6 | .7257469 | .7290691 | .7323711 | .7356527 | .7389137 | .7421539 | .7453731 | .7485711 | .7517478 | .7549029 |
| 0.7 | .7580363 | .7611479 | .7642375 | .7673049 | .7703500 | .7733726 | .7763727 | .7793501 | .7823046 | .7852361 |
| 0.8 | .7881446 | .7910299 | .7938919 | .7967306 | .7995458 | .8023375 | .8051055 | .8078498 | .8105703 | .8132671 |
| 0.9 | .8159399 | .8185887 | .8212136 | .8238145 | .8263912 | .8289439 | .8314724 | .8339768 | .8364569 | .8389129 |
| 1.0 | .8413447 | .8437524 | .8461358 | .8484950 | .8508300 | .8531409 | .8554277 | .8576903 | .8599289 | .8621434 |
| 1.1 | .8643339 | .8665005 | .8686431 | .8707619 | .8728568 | .8749281 | .8769756 | .8789995 | .8809999 | .8829768 |
| 1.2 | .8849303 | .8868606 | .8887676 | .8906514 | .8925123 | .8943502 | .8961653 | .8979577 | .8997274 | .9014747 |
| 1.3 | .9031995 | .9049021 | .9065825 | .9082409 | .9098773 | .9114920 | .9130850 | .9146565 | .9162067 | .9177356 |
| 1.4 | .9192433 | .9207302 | .9221962 | .9236415 | .9250663 | .9264707 | .9278550 | .9292191 | .9305634 | .9318879 |

| | .00 | .01 | .02 | .03 | .04 | .05 | .06 | .07 | .08 | .09 |
|-----|-----|-----|-----|-----|-----|-----|-----|-----|-----|-----|
| 1.5 | .9331928 | .9344783 | .9357445 | .9369916 | .9382198 | .9394292 | .9406201 | .9417924 | .9429466 | .9440826 |
| 1.6 | .9452007 | .9463011 | .9473839 | .9484493 | .9494974 | .9505285 | .9515428 | .9525403 | .9535213 | .9544860 |
| 1.7 | .9554345 | .9563671 | .9572838 | .9581849 | .9590705 | .9599408 | .9607961 | .9616364 | .9624620 | .9632730 |
| 1.8 | .9640697 | .9648521 | .9656205 | .9663750 | .9671159 | .9678432 | .9685572 | .9692581 | .9699460 | .9706210 |
| 1.9 | .9712834 | .9719334 | .9725711 | .9731966 | .9738102 | .9744119 | .9750021 | .9755808 | .9761482 | .9767045 |
| 2.0 | .9772499 | .9777844 | .9783083 | .9788217 | .9793248 | .9798178 | .9803007 | .9807738 | .9812372 | .9816911 |
| 2.1 | .9821356 | .9825708 | .9829970 | .9834142 | .9838226 | .9842224 | .9846137 | .9849966 | .9853713 | .9857379 |
| 2.2 | .9860966 | .9864474 | .9867906 | .9871263 | .9874545 | .9877755 | .9880894 | .9883962 | .9886962 | .9889893 |
| 2.3 | .9892759 | .9895559 | .9898296 | .9900969 | .9903581 | .9906133 | .9908625 | .9911060 | .9913437 | .9915758 |
| 2.4 | .9918025 | .9920237 | .9922397 | .9924506 | .9926564 | .9928572 | .9930531 | .9932443 | .9934309 | .9936128 |
| 2.5 | .9937903 | .9939634 | .9941323 | .9942969 | .9944574 | .9946139 | .9947664 | .9949151 | .9950600 | .9952012 |
| 2.6 | .9953388 | .9954729 | .9956035 | .9957308 | .9958547 | .9959754 | .9960930 | .9962074 | .9963189 | .9964274 |
| 2.7 | .9965330 | .9966358 | .9967359 | .9968333 | .9969280 | .9970202 | .9971099 | .9971972 | .9972821 | .9973646 |
| 2.8 | .9974449 | .9975229 | .9975988 | .9976726 | .9977443 | .9978140 | .9978818 | .9979476 | .9980116 | .9980738 |
| 2.9 | .9981342 | .9981929 | .9982498 | .9983052 | .9983589 | .9984111 | .9984618 | .9985110 | .9985588 | .9986051 |
| 3.0 | .9986501 | .9986938 | .9987361 | .9987772 | .9988171 | .9988558 | .9988933 | .9989297 | .9989650 | .9989992 |
| 3.1 | .9990324 | .9990646 | .9990957 | .9991260 | .9991553 | .9991836 | .9992112 | .9992378 | .9992636 | .9992886 |
| 3.2 | .9993129 | .9993363 | .9993590 | .9993810 | .9994024 | .9994230 | .9994429 | .9994623 | .9994810 | .9994991 |
| 3.3 | .9995166 | .9995335 | .9995499 | .9995658 | .9995811 | .9995959 | .9996103 | .9996242 | .9996376 | .9996505 |
| 3.4 | .9996631 | .9996752 | .9996869 | .9996982 | .9997091 | .9997197 | .9997299 | .9997398 | .9997493 | .9997585 |
| 3.5 | .9997674 | .9997759 | .9997842 | .9997922 | .9997999 | .9998074 | .9998146 | .9998215 | .9998282 | .9998347 |
| 3.6 | .9998409 | .9998469 | .9998527 | .9998583 | .9998637 | .9998689 | .9998739 | .9998787 | .9998834 | .9998879 |
| 3.7 | .9998922 | .9998964 | .9999004 | .9999043 | .9999080 | .9999116 | .9999150 | .9999184 | .9999216 | .9999247 |

*(continued)*

**Table A.12** The Cumulative Normal Distribution Function: Unit Normal Deviates to Cumulative Proportions[a] —cont...

| $U_\alpha$ | .00 | .01 | .02 | .03 | .04 | .05 | .06 | .07 | .08 | .09 |
|---|---|---|---|---|---|---|---|---|---|---|
| | | | | | For $0.00 \leq U_\alpha \leq 4.99$ | | | | | |
| 3.8 | .9999277 | .9999305 | .9999333 | .9999359 | .9999385 | .9999409 | .9999433 | .9999456 | .9999478 | .9999499 |
| 3.9 | .9999519 | .9999539 | .9999557 | .9999575 | .9999593 | .9999609 | .0000625 | .0000641 | .9999655 | .9999670 |
| 4.0 | .9999683 | .9999696 | .9999709 | .9999721 | .9999733 | .9999744 | .9999755 | .9999765 | .9999775 | .9999784 |
| 4.1 | .9999793 | .9999802 | .9999811 | .9999819 | .9999826 | .9999834 | .9999841 | .9999848 | .9999854 | .9999861 |
| 4.2 | .9999867 | .9999872 | .9999878 | .9999883 | .9999888 | .9999893 | .9999898 | .9999902 | .9999907 | .9999911 |
| 4.3 | .9999915 | .9999918 | .9999922 | .9999925 | .9999929 | .9999932 | .9999935 | .9999938 | .9999941 | .9999943 |
| 4.4 | .9999946 | .9999948 | .9999951 | .9999953 | .9999955 | .9999957 | .9999959 | .9999961 | .9999963 | .9999964 |
| 4.5 | .9999966 | .9999968 | .9999969 | .9999971 | .9999972 | .9999973 | .9999974 | .9999976 | .9999977 | .9999978 |
| 4.6 | .9999979 | .9999980 | .9999981 | .9999982 | .9999983 | .9999983 | .9999984 | .9999985 | .9999986 | .9999986 |
| 4.7 | .9999987 | .9999988 | .9999988 | .9999989 | .9999989 | .9999980 | .9999990 | .9999991 | .9999991 | .9999992 |
| 4.8 | .9999992 | .9999992 | .9999993 | .9999993 | .9999994 | .9999994 | .9999994 | .9999994 | .9999995 | .9999995 |
| 4.9 | .9999995 | .9999995 | .9999996 | .9999996 | .9999996 | .9999996 | .9999996 | .9999997 | .9999997 | .9999997 |

[a]Generated using an SAS program written by R. W. Washam II, Armour Research Center, Scottsdale, Arizona.

**Table A.13** Ordinates of the Normal Distribution[a] $[Y = \phi (Z)]$

| Z | .00 | .01 | .02 | .03 | .04 | .05 | .06 | .07 | .08 | .09 |
|---|---|---|---|---|---|---|---|---|---|---|
| | | | | **Decimal Fraction of Z** | | | | | | |
| 0.0 | .3989 | .3989 | .3989 | .3988 | .3986 | .3984 | .3982 | .3980 | .3977 | .3973 |
| 0.1 | .3970 | .3965 | .3961 | .3956 | .3951 | .3945 | .3939 | .3932 | .3925 | .3918 |
| 0.2 | .3910 | .3902 | .3894 | .3885 | .3876 | .3867 | .3857 | .3847 | .3836 | .3825 |
| 0.3 | .3814 | .3802 | .3790 | .3778 | .3765 | .3752 | .3739 | .3725 | .3712 | .3697 |
| 0.4 | .3683 | .3668 | .3653 | .3637 | .3621 | .3605 | .3589 | .3572 | .3555 | .3538 |
| 0.5 | .3521 | .3503 | .3485 | .3467 | .3448 | .3429 | .3410 | .3391 | .3372 | .3352 |
| 0.6 | .3332 | .3312 | .3292 | .3271 | .3251 | .3230 | .3209 | .3187 | .3166 | .3144 |
| 0.7 | .3123 | .3101 | .3079 | .3056 | .3034 | .3011 | .2989 | .2966 | .2943 | .2920 |
| 0.8 | .2897 | .2874 | .2850 | .2827 | .2803 | .2780 | .2756 | .2732 | .2709 | .2685 |
| 0.9 | .2661 | .2637 | .2613 | .2589 | .2565 | .2541 | .2516 | .2492 | .2468 | .2444 |
| 1.0 | .2420 | .2396 | .2371 | .2347 | .2323 | .2299 | .2275 | .2251 | .2227 | .2203 |
| 1.1 | .2179 | .2155 | .2131 | .2107 | .2083 | .2059 | .2036 | .2012 | .1989 | .1965 |
| 1.2 | .1942 | .1919 | .1895 | .1872 | .1849 | .1826 | .1804 | .1781 | .1758 | .1736 |
| 1.3 | .1714 | .1691 | .1669 | .1647 | .1626 | .1604 | .1582 | .1561 | .1539 | .1518 |
| 1.4 | .1497 | .1456 | .1456 | .1435 | .1415 | .1394 | .1374 | .1354 | .1334 | .1315 |
| 1.5 | .1295 | .1276 | .1257 | .1238 | .1219 | .1200 | .1182 | .1163 | .1145 | .1127 |
| 1.6 | .1109 | .1092 | .1074 | .1057 | .1040 | .1023 | .1006 | .0989 | .0973 | .0957 |
| 1.7 | .0940 | .0925 | .0909 | .0893 | .0878 | .0863 | .0848 | .0833 | .0818 | .0804 |
| 1.8 | .0790 | .0775 | .0761 | .0748 | .0734 | .0721 | .0707 | .0694 | .0681 | .0669 |
| 1.9 | .0656 | .0644 | .0632 | .0620 | .0608 | .0596 | .0584 | .0573 | .0562 | .0551 |
| 2.0 | .0540 | .0529 | .0519 | .0508 | .0498 | .0488 | .0478 | .0468 | .0459 | .0449 |
| 2.1 | .0440 | .0431 | .0422 | .0413 | .0404 | .0396 | .0387 | .0379 | .0371 | .0363 |
| 2.2 | .0355 | .0347 | .0339 | .0332 | .0325 | .0317 | .0310 | .0303 | .0297 | .0290 |
| 2.3 | .0283 | .0277 | .0270 | .0264 | .0258 | .0252 | .0246 | .0241 | .0235 | .0229 |
| 2.4 | .0224 | .0219 | .0213 | .0208 | .0203 | .0198 | .0194 | .0189 | .0184 | .0180 |
| 2.5 | .0175 | .0171 | .0167 | .0163 | .0158 | .0154 | .0151 | .0147 | .0143 | .0139 |
| 2.6 | .0136 | .0132 | .0129 | .0126 | .0122 | .0119 | .0116 | .0113 | .0110 | .0107 |
| 2.7 | .0104 | .0101 | .0099 | .0096 | .0093 | .0091 | .0088 | .0086 | .0084 | .0081 |

*(continued)*

**Table A.13** Ordinates of the Normal Distribution[a] $[Y = \phi\ (Z)]$—cont...

| Z | Decimal Fraction of Z | | | | | | | | | |
|---|---|---|---|---|---|---|---|---|---|---|
| | **.00** | **.01** | **.02** | **.03** | **.04** | **.05** | **.06** | **.07** | **.08** | **.09** |
| 2.8 | .0079 | .0077 | .0075 | .0073 | .0071 | .0069 | .0067 | .0065 | .0063 | .0061 |
| 2.9 | .0060 | .0058 | .0056 | .0055 | .0053 | .0051 | .0050 | .0048 | .0047 | .0046 |
| 3.0 | .0044 | .0043 | .0042 | .0040 | .0039 | .0038 | .0037 | .0036 | .0035 | .0034 |
| 3.1 | .0033 | .0032 | .0031 | .0030 | .0029 | .0028 | .0027 | .0026 | .0025 | .0025 |
| 3.2 | .0024 | .0023 | .0022 | .0022 | .0021 | .0020 | .0020 | .0019 | .0018 | .0018 |
| 3.3 | .0017 | .0017 | .0016 | .0016 | .0015 | .0015 | .0014 | .0014 | .0013 | .0013 |
| 3.4 | .0012 | .0012 | .0012 | .0011 | .0011 | .0010 | .0010 | .0010 | .0009 | .0009 |
| 3.5 | .0009 | .0008 | .0008 | .0008 | .0008 | .0007 | .0007 | .0007 | .0007 | .0006 |
| 3.6 | .0006 | .0006 | .0006 | .0005 | .0005 | .0005 | .0005 | .0005 | .0005 | .0004 |
| 3.7 | .0004 | .0004 | .0004 | .0004 | .0004 | .0004 | .0003 | .0003 | .0003 | .0003 |
| 3.8 | .0003 | .0003 | .0003 | .0003 | .0003 | .0002 | .0002 | .0002 | .0002 | .0002 |
| 3.9 | .0002 | .0002 | .0002 | .0002 | .0002 | .0002 | .0002 | .0002 | .0001 | .0001 |

[a]*Generated using an SAS program written by R. W. Washam II, Armour Research Center, Scottsdale, Arizona.*

**Table A.14** Expected Values of the Normal-Order Statistics[a]

| Rank | Number of Treatments | | | | | | | | |
|---|---|---|---|---|---|---|---|---|---|
| | **2** | **3** | **4** | **5** | **6** | **7** | **8** | **9** | **10** |
| 1 | 0.564 | 0.846 | 1.029 | 1.163 | 1.267 | 1.352 | 1.424 | 1.485 | 1.539 |
| 2 | −0.564 | 0.000 | 0.297 | 0.495 | 0.642 | 0.757 | 0.852 | 0.932 | 1.001 |
| 3 | | −0.846 | −0.297 | 0.000 | 0.202 | 0.353 | 0.473 | 0.572 | 0.656 |
| 4 | | | −1.029 | −0.495 | −0.202 | 0.000 | 0.153 | 0.275 | 0.376 |
| 5 | | | | −1.163 | −0.642 | −0.353 | −0.153 | 0.000 | 0.123 |
| 6 | | | | | −1.267 | −0.757 | −0.473 | −0.275 | −0.123 |
| 7 | | | | | | −1.352 | −0.852 | −0.572 | −0.376 |
| 8 | | | | | | | −1.424 | −0.932 | −0.656 |
| 9 | | | | | | | | −1.485 | −1.001 |
| 10 | | | | | | | | | −1.539 |

[a]*Source: H. L. Harter. (1960). Expected Values of Normal Order Statistics. Technical Report 60–292, Aeronautical Research Laboratories, Wright–Patterson Air Force Base, OH. Reproduced with permission.*

**Table A.15** Critical Values and Probability Levels for the Wilcoxon Rank Sum Test[a]

| n | m = 3 | m = 4 | m = 5 | m = 6 | m = 7 | m = 8 | m = 9 | m = 10 |
|---|---|---|---|---|---|---|---|---|
| *P* = .05 (one-sided), *P* = .10 (two-sided) | | | | | | | | |
| 3 | 5, 16 .0000 | | | | | | | |
|   | 6, 15 .0500 | | | | | | | |
| 4 | 6, 18 .0286 | 11, 25 .0286 | | | | | | |
|   | 7, 17 .0571 | 12, 24 .0571 | | | | | | |
| 5 | 7, 20 .0357 | 12, 28 .0317 | 19, 36 .0476 | | | | | |
|   | 8, 19 .0714 | 13, 27 .0556 | 20, 35 .0754 | | | | | |
| 6 | 8, 22 .0476 | 13, 31 .0333 | 20, 40 .0411 | 28, 50 .0465 | | | | |
|   | 9, 21 .0833 | 14, 30 .0571 | 21, 39 .0628 | 29, 49 .0660 | | | | |
| 7 | 8, 25 0.333 | 14, 34 .0364 | 21, 44 .0366 | 29, 55 .0367 | 39, 66 .0487 | | | |
|   | 9, 24 .0583 | 15, 33 .0545 | 22, 43 .0530 | 30, 54 .0507 | 40, 65 .0641 | | | |
| 8 | 9, 27 .0424 | 15, 37 .0364 | 23, 47 .0466 | 31, 59 .0406 | 41, 71 .0469 | 51, 85 .0415 | | |
|   | 10, 26 .0667 | 16, 36 .0545 | 24, 46 .0637 | 32, 58 .0539 | 42, 70 .0603 | 52, 84 .0524 | | |
| 9 | 9, 30 .0318 | 16, 40 .0378 | 24, 51 .0415 | 33, 63 .0440 | 43, 76 .0454 | 54, 90 .0464 | 66, 105 .0470 | |
|   | 10, 29 .0500 | 17, 39 .0531 | 25, 50 .0559 | 34, 62 .0567 | 44, 75 .0571 | 55, 89 .0570 | 67, 104 .0567 | |
| 10 | 10, 32 .0385 | 17, 43 .0380 | 26, 54 .0496 | 35, 67 .0467 | 45, 81 .0439 | 56, 96 .0416 | 69, 111 .0474 | 82, 128 .0446 |
|   | 11, 31 .0559 | 18, 42 .0529 | 27, 53 .0646 | 36, 66 .0589 | 46, 80 .0544 | 57, 95 .0506 | 70, 110 .0564 | 83, 127 .0526 |
| *P* = .025 (one-sided), *P* = .05 (two-sided) | | | | | | | | |
| 3 | 5, 16 .0000 | | | | | | | |
|   | 6, 15 .0500 | | | | | | | |
| 4 | 5, 19 .0000 | 10, 26 .0143 | | | | | | |
|   | 6, 18 .0286 | 11, 25 .0286 | | | | | | |
| 5 | 6, 21 .0179 | 11, 29 .0159 | 17, 38 .0159 | | | | | |
|   | 7, 20 .0357 | 12, 28 .0317 | 18, 37 .0278 | | | | | |

*(continued)*

**Table A.15** Critical Values and Probability Levels for the Wilcoxon Rank Sum Test[a]—cont...

| n | m = 3 | m = 4 | m = 5 | m = 6 | m = 7 | m = 8 | m = 9 | m = 10 |
|---|---|---|---|---|---|---|---|---|
| 6 | 7, 23 .0238<br>8, 22 .0476 | 12, 32 .0190<br>13, 31 .0333 | 18, 42 .0152<br>19, 41 .0260 | 26, 52 .0206<br>27, 51 .0325 | | | | |
| 7 | 7, 26 .0167<br>8, 25 .0333 | 13, 35 .0212<br>14, 34 .0364 | 20, 45 .0240<br>21, 44 .0366 | 27, 57 .0175<br>28, 56 .0256 | 36, 69 .0189<br>37, 68 .0265 | | | |
| 8 | 8, 28 .0242<br>9, 27 .0424 | 14, 38 .0242<br>15, 37 .0364 | 21, 49 .0225<br>22, 48 .0326 | 29, 61 .0213<br>30, 60 .0296 | 38, 74 .0200<br>39, 73 .0270 | 49, 87 .0249<br>50, 86 .0325 | | |
| 9 | 8, 31 .0182<br>9, 30 .0318 | 14, 42 .0168<br>15, 41 .0252 | 22, 53 .0210<br>23, 52 .0300 | 31, 65 .0248<br>32, 64 .0332 | 40, 79 .0209<br>41, 78 .0274 | 51, 93 .0232<br>52, 92 .0296 | 62, 109 .0200<br>63, 108 .0252 | |
| 10 | 9, 33 .0245<br>10, 32 .0385 | 15, 45 .0180<br>16, 44 .0270 | 23, 57 .0200<br>24, 56 .0276 | 32, 70 .0210<br>33, 69 .0280 | 42, 84 .0215<br>43, 83 .0277 | 53, 99 .0217<br>54, 98 .0273 | 65, 115 .0217<br>66, 114 .0267 | 78, 132 .0216<br>79, 131 .0262 |

$P = .01$ (one-sided), $P = .02$ (two-sided)

| n | m = 3 | m = 4 | m = 5 | m = 6 | m = 7 | m = 8 | m = 9 | m = 10 |
|---|---|---|---|---|---|---|---|---|
| 3 | 5, 16 .0000<br>6, 15 .0500 | | | | | | | |
| 4 | 5, 19 .0000<br>6, 18 .0286 | 9, 27 .0000<br>10, 26 .0143 | | | | | | |
| 5 | 5, 22 .0000<br>6, 21 .0179 | 10, 30 .0079<br>11, 29 .0159 | 16, 39 .0079<br>17, 38 .0159 | | | | | |
| 6 | 5, 25 .0000<br>6, 24 .0119 | 11, 33 .0095<br>12, 32 .0190 | 17, 43 .0087<br>18, 42 .0152 | 24, 54 .0076<br>25, 53 .0130 | | | | |
| 7 | 6, 27 .0083<br>7, 26 .0167 | 11, 37 .0061<br>12, 36 .0121 | 18, 47 .0088<br>19, 46 .0152 | 25, 59 .0070<br>26, 58 .0111 | 34, 71 .0087<br>35, 70 .0131 | | | |
| 8 | 6, 30 .0061<br>7, 29 .0121 | 12, 40 .0081<br>13, 39 .0141 | 19, 51 .0093<br>20, 50 .0148 | 27, 63 .0100<br>28, 62 .0147 | 35, 77 .0070<br>36, 76 .0103 | 45, 91 .0074<br>46, 90 .0103 | | |

| 9 | 7, 32 .0091 | 13, 43 .0098 | 20, 55 .0095 | 28, 68 .0088 | 37, 82 .0082 | 47, 97 .0076 | 59, 112 .0094 | |
|---|---|---|---|---|---|---|---|---|
| | 8, 31 .0182 | 14, 42 .0168 | 21, 54 .0145 | 29, 67 .0128 | 38, 81 .0115 | 48, 96 .0103 | 60, 111 .0122 | |
| 10 | 7, 35 .0070 | 13, 47 .0070 | 21, 59 .0097 | 29, 73 .0080 | 39, 87 .0093 | 49, 103 .0078 | 61, 119 .0086 | 74, 136 .0093 |
| | 8, 34 .0140 | 14, 46 .0120 | 22, 58 .0140 | 30, 72 .0112 | 40, 86 .0125 | 50, 102 .0103 | 62, 118 .0110 | 75, 135 .0116 |

$P = .005$ (one-sided), $P = .01$ (two-sided)

| 3 | 5, 16 .0000 | | | | | | | |
|---|---|---|---|---|---|---|---|---|
| | 6, 15 .0500 | | | | | | | |
| 4 | 5, 19 .0000 | 9, 27 .0000 | | | | | | |
| | 6, 18 .0286 | 10, 26 .0143 | | | | | | |
| 5 | 5, 22 .0000 | 9, 31 .0000 | 15, 40 .0040 | | | | | |
| | 6, 21 .0179 | 10, 30 .0079 | 16, 39 .0079 | | | | | |
| 6 | 5, 25 .0000 | 10, 34 .0048 | 16, 44 .0043 | 23, 55 .0043 | | | | |
| | 6, 24 .0119 | 11, 33 .0095 | 17, 43 .0087 | 24, 54 .0076 | | | | |
| 7 | 5, 28 .0000 | 10, 38 .0030 | 16, 49 .0025 | 24, 60 .0041 | 32, 73 .0035 | | | |
| | 6, 27 .0083 | 11, 37 .0061 | 17, 48 .0051 | 25, 59 .0070 | 33, 72 .0055 | | | |
| 8 | 5, 31 .0000 | 11, 41 .0040 | 17, 53 .0031 | 25, 65 .0040 | 34, 78 .0047 | 43, 93 .0035 | | |
| | 6, 30 .0061 | 12, 40 .0081 | 18, 52 .0054 | 26, 64 .0063 | 35, 77 .0070 | 44, 92 .0052 | | |
| 9 | 6, 33 .0045 | 11, 45 .0028 | 18, 57 .0035 | 26, 70 .0038 | 35, 84 .0039 | 45, 99 .0039 | 56, 115 .0039 | |
| | 7, 32 .0091 | 12, 44 .0056 | 19, 56 .0060 | 27, 69 .0060 | 36, 83 .0058 | 46, 98 .0056 | 57, 114 .0053 | |
| 10 | 6, 36 .0035 | 12, 48 .0040 | 19, 61 .0040 | 27, 75 .0037 | 37, 89 .0048 | 47, 105 .0043 | 58, 122 .0038 | 71, 139 .0045 |
| | 7, 35 .0070 | 13, 47 .0070 | 20, 60 .0063 | 28, 74 .0055 | 38, 88 .0068 | 48, 104 .0058 | 59, 121 .0051 | 72, 138 .0057 |

[a]Source: F. Wilcoxon, S. K. Katti, and R. A. Wilcox. (1970). *Critical values and probability levels for the Wilcoxon rank sum test and the Wilcoxon signed rank test.* In *Selected Tables in Mathematical Statistics* (H. L. Harter and D. B. Owen, eds.) Markham, Chicago, IL: Vol. 1, pp. 171–259. From *Some Rapid Approximate Statistical Procedures.* Copyright 1949, 1964 Lederle Laboratories Division of American Cyanamid Company. All rights reserved and reprinted with permission.

**Table A.16** Critical Values for the Rank Sum Test in Randomized Blocks[a,b]

**Number of Blocks = 1**

| n | 0.002 / 0.001 | 0.01 / 0.005 | 0.02 / 0.01 | 0.05 / 0.025 | 0.10 / 0.05 | 0.20 / 0.10 | 2-Sided / 1-Sided |
|---|---|---|---|---|---|---|---|
| 1 | — | — | — | — | — | — | |
| 2 | — | — | — | — | — | — | |
| 3 | — | — | — | — | 6 | 7 | |
| 4 | — | — | — | 10+ | 11+ | 13 | |
| 5 | — | 15 | 16 | 17+ | 19 | 20+ | |
| 6 | — | 23 | 24 | 26 | 28 | 30 | |
| 7 | 29+ | 32+ | 34 | 36+ | 39 | 41+ | |
| 8 | 40 | 43+ | 45+ | 49 | 51+ | 55 | |
| 9 | 52 | 56+ | 59 | 62+ | 66 | 70 | |
| 10 | 65+ | 71 | 74 | 78+ | 82+ | 87 | |
| 11 | 81 | 87+ | 91 | 96 | 100+ | 106 | |
| 12 | 98 | 105+ | 109+ | 115+ | 120+ | 127 | |
| 13 | 117 | 125+ | 130 | 136+ | 142+ | 149+ | |
| 14 | 137+ | 147+ | 152+ | 160 | 166+ | 174 | |
| 15 | 160 | 171 | 176+ | 184+ | 192 | 200+ | |
| 16 | 184 | 196 | 202 | 211+ | 219+ | 229 | |
| 17 | 210 | 223 | 230 | 240 | 249 | 259+ | |
| 18 | 237 | 252 | 259+ | 270 | 280 | 291+ | |
| 19 | 267 | 283 | 291 | 303 | 313+ | 325 | |
| 20 | 298 | 315+ | 324 | 337 | 348+ | 361+ | |
| 21 | 331 | 349+ | 359 | 373 | 385+ | 399+ | |

**Number of Blocks = 2**

| n | 0.002 / 0.001 | 0.01 / 0.005 | 0.02 / 0.01 | 0.05 / 0.025 | 0.10 / 0.05 | 0.20 / 0.10 | 2-Sided / 1-Sided |
|---|---|---|---|---|---|---|---|
| 1 | — | — | — | — | — | — | |
| 2 | — | — | — | — | 6 | 7 | |
| 3 | — | 12 | 13 | 14 | 15 | 16 | |
| 4 | 21 | 23 | 24 | 25+ | 27 | 29 | |
| 5 | 34+ | 37+ | 39 | 41 | 43 | 45+ | |
| 6 | 51 | 55 | 57 | 60 | 62+ | 66 | |
| 7 | 71+ | 76+ | 79 | 82+ | 86 | 90 | |
| 8 | 95 | 101 | 104+ | 109 | 113 | 118 | |
| 9 | 122 | 129+ | 133+ | 139 | 144 | 149+ | |
| 10 | 153 | 162 | 166 | 172+ | 178+ | 185 | |
| 11 | 187+ | 197+ | 202+ | 210 | 217 | 224+ | |
| 12 | 225 | 237 | 242+ | 251+ | 259 | 267+ | |
| 13 | 266+ | 280 | 286+ | 296+ | 305 | 315 | |
| 14 | 312 | 327 | 334 | 345 | 354+ | 365+ | |
| 15 | 360+ | 377 | 385+ | 397+ | 408 | 420+ | |
| 16 | 413 | 431+ | 440+ | 454 | 465+ | 479 | |
| 17 | 469+ | 489+ | 499 | 514 | 526+ | 541+ | |
| 18 | 529 | 551 | 562 | 578 | 591+ | 608 | |
| 19 | 592+ | 616+ | 628 | 645+ | 660+ | 678 | |
| 20 | 659+ | 685+ | 698 | 717 | 733 | 752 | |
| 21 | 730+ | 758+ | 772 | 792 | 809+ | 830 | |

**(Number of Blocks = 2, continued)**

| n | 0.002 / 0.001 | 0.01 / 0.005 | 0.02 / 0.01 | 0.05 / 0.025 | 0.10 / 0.05 | 0.20 / 0.10 |
|---|---|---|---|---|---|---|
| 22 | 365+ | 386 | 396 | 411 | 424 | 439+ |
| 23 | 402 | 424 | 434+ | 451 | 465 | 481 |
| 24 | 440+ | 464 | 475+ | 492+ | 507+ | 525 |
| 25 | 480+ | 505+ | 517+ | 536 | 552 | 570+ |
| 26 | 522+ | 549 | 562 | 581+ | 598+ | 618 |
| 27 | 566+ | 594+ | 608 | 628+ | 646+ | 667+ |
| 28 | 612 | 641+ | 656 | 678 | 697 | 719 |
| 29 | 659+ | 690+ | 706 | 729 | 749 | 772 |
| 30 | 709 | 741+ | 758 | 782 | 803 | 827 |

**Number of Blocks = 3**

| | 2-Sided | | | | | |
|---|---|---|---|---|---|---|
| n | 0.002 | 0.01 | 0.02 | 0.05 | 0.10 | 0.20 |
| | 1-Sided | | | | | |
| | 0.001 | 0.005 | 0.01 | 0.025 | 0.05 | 0.10 |
| 1 | — | — | — | — | — | — |
| 2 | — | 9 | 9 | 10 | 10+ | 11+ |
| 3 | 19 | 21 | 22 | 23 | 24 | 25+ |
| 4 | 35+ | 38 | 39+ | 41+ | 43+ | 45+ |
| 5 | 57 | 61 | 62+ | 65+ | 68 | 71 |
| 6 | 84 | 89 | 91+ | 95 | 98+ | 102+ |
| 7 | 116 | 122+ | 125+ | 130+ | 134+ | 139+ |
| 8 | 153+ | 161+ | 165 | 171 | 176 | 182 |
| 9 | 196+ | 206 | 210+ | 217+ | 223+ | 230+ |
| 10 | 244+ | 256 | 261+ | 269+ | 276+ | 285 |
| 11 | 298+ | 311+ | 317+ | 327 | 335+ | 345 |
| 12 | 358 | 372+ | 380 | 390+ | 400 | 410+ |
| 13 | 422+ | 439+ | 447+ | 459+ | 470 | 482+ |

**(Number of Blocks = 3, continued)**

| n | 0.002 / 0.001 | 0.01 / 0.005 | 0.02 / 0.01 | 0.05 / 0.025 | 0.10 / 0.05 | 0.20 / 0.10 |
|---|---|---|---|---|---|---|
| 22 | 805 | 835 | 849+ | 871+ | 890 | 912 |
| 23 | 883+ | 915+ | 931 | 954+ | 974+ | 997+ |
| 24 | 965+ | 999+ | 1016+ | 1041 | 1062+ | 1087 |
| 25 | 1051+ | 1087+ | 1105+ | 1131+ | 1154+ | 1180+ |
| 26 | 1141 | 1179+ | 1198 | 1226 | 1250 | 1278 |
| 27 | 1234 | 1275 | 1294+ | 1324 | 1349+ | 1379+ |
| 28 | 1331 | 1374 | 1395 | 1426+ | 1453 | 1484+ |
| 29 | 1431+ | 1477 | 1499 | 1532 | 1560+ | 1593+ |
| 30 | 1536 | 1584 | 1607+ | 1642 | 1672 | 1706+ |

**Number of Blocks = 4**

| | 2-Sided | | | | | |
|---|---|---|---|---|---|---|
| n | 0.002 | 0.01 | 0.02 | 0.05 | 0.10 | 0.20 |
| | 1-Sided | | | | | |
| | 0.001 | 0.005 | 0.01 | 0.025 | 0.05 | 0.10 |
| 1 | — | — | — | — | — | 4 |
| 2 | 12 | 13 | 13+ | 14 | 15 | 16 |
| 3 | 28 | 30 | 31 | 32+ | 33+ | 35+ |
| 4 | 50+ | 53+ | 55+ | 57+ | 60 | 62+ |
| 5 | 80+ | 85 | 87 | 90+ | 93+ | 97 |
| 6 | 117+ | 123+ | 126+ | 131 | 134+ | 139 |
| 7 | 162 | 169+ | 173 | 178+ | 183+ | 189 |
| 8 | 213+ | 222+ | 227 | 234 | 240 | 247 |
| 9 | 272 | 283+ | 289 | 297 | 304 | 312 |
| 10 | 338+ | 351+ | 358 | 367+ | 375+ | 385+ |
| 11 | 412 | 427+ | 434+ | 445+ | 455 | 466 |
| 12 | 493+ | 510+ | 519 | 531+ | 542 | 555 |
| 13 | 582 | 601+ | 611 | 625 | 637 | 651 |

(continued)

**Table A.16** Critical Values for the Rank Sum Test in Randomized Blocks[a,b]—cont...

**Number of Blocks = 3**

| 2-Sided | 0.002 | 0.01 | 0.02 | 0.05 | 0.10 | 0.20 | 2-Sided |
| 1-Sided | 0.001 | 0.005 | 0.01 | 0.025 | 0.05 | 0.10 | 1-Sided |
| n | | | | | | | n |
|---|---|---|---|---|---|---|---|
| 14 | 493 | 512 | 521 | 534+ | 546 | 560 | 14 |
| 15 | 569 | 590 | 600 | 615 | 628 | 643 | 15 |
| 16 | 650+ | 673+ | 685 | 701+ | 715+ | 732 | 16 |
| 17 | 738 | 763 | 775 | 793+ | 809 | 827 | 17 |
| 18 | 830+ | 858 | 871+ | 891 | 908 | 928 | 18 |
| 19 | 929 | 958+ | 973 | 994+ | 1013 | 1034+ | 19 |
| 20 | 1033 | 1065 | 1080+ | 1104 | 1124 | 1147 | 20 |
| 21 | 1142+ | 1177 | 1194 | 1219 | 1240+ | 1265+ | 21 |
| 22 | 1258 | 1295 | 1313 | 1340 | 1363 | 1389+ | 22 |
| 23 | 1379 | 1418+ | 1438 | 1466+ | 1491 | 1519+ | 23 |
| 24 | 1505+ | 1547+ | 1568+ | 1599 | 1625 | 1655+ | 24 |
| 25 | 1637+ | 1682+ | 1704+ | 1737 | 1765 | 1797 | 25 |
| 26 | 1775+ | 1823+ | 1846+ | 1881 | 1910+ | 1945 | 26 |
| 27 | 1919+ | 1969+ | 1994+ | 2030+ | 2062 | 2098+ | 27 |
| 28 | 2068+ | 2122 | 2148 | 2186 | 2219+ | 2257+ | 28 |
| 29 | 2223+ | 2280 | 2307 | 2347+ | 2382+ | 2423 | 29 |
| 30 | 2384 | 2443+ | 2472+ | 2515 | 2551+ | 2594 | 30 |

**Number of Blocks = 4**

| 2-Sided | 0.002 | 0.01 | 0.02 | 0.05 | 0.10 | 0.20 | 2-Sided |
| 1-Sided | 0.001 | 0.005 | 0.01 | 0.025 | 0.05 | 0.10 | 1-Sided |
| n | | | | | | | n |
|---|---|---|---|---|---|---|---|
| 14 | 678 | 699+ | 710+ | 726 | 739+ | 755+ | 14 |
| 15 | 781+ | 805+ | 817+ | 835 | 850 | 867+ | 15 |
| 16 | 892+ | 919 | 932 | 951+ | 968 | 987 | 16 |
| 17 | 1011 | 1040+ | 1054+ | 1075+ | 1093+ | 1114+ | 17 |
| 18 | 1137+ | 1169 | 1184+ | 1207+ | 1227 | 1250 | 18 |
| 19 | 1271 | 1305+ | 1322+ | 1347 | 1368+ | 1393+ | 19 |
| 20 | 1412 | 1449+ | 1467+ | 1494+ | 1517+ | 1554+ | 20 |
| 21 | 1561 | 1601 | 1620+ | 1649+ | 1674+ | 1703+ | 21 |
| 22 | 1717+ | 1760 | 1781+ | 1812+ | 1839 | 1870 | 22 |
| 23 | 1881+ | 1927+ | 1950 | 1983 | 2011+ | 2044+ | 23 |
| 24 | 2053 | 2102 | 2126 | 2161+ | 2191+ | 2227 | 24 |
| 25 | 2232+ | 2284+ | 2310 | 2347+ | 2379+ | 2417 | 25 |
| 26 | 2419 | 2474+ | 2501+ | 2541 | 2575+ | 2615 | 26 |
| 27 | 2613+ | 2672 | 2701 | 2743 | 2779 | 2821 | 27 |
| 28 | 2816 | 2877+ | 2908 | 2952 | 2990+ | 3034+ | 28 |
| 29 | 3025+ | 3091 | 3122+ | 3169+ | 3209+ | 3256+ | 29 |
| 30 | 3243 | 3311+ | 3345 | 3394 | 3436+ | 3485+ | 30 |

**Number of Blocks = 5**

| n | 0.002 / 0.001 | 0.01 / 0.005 | 0.02 / 0.01 | 0.05 / 0.025 | 0.10 / 0.05 | 0.20 / 0.10 |
|---|---|---|---|---|---|---|
| | 2-Sided / 1-Sided | | | | | |
| 1 | — | — | — | — | 5 | 5 |
| 2 | 16 | 17 | 17$^+$ | 18$^+$ | 19$^+$ | 20$^+$ |
| 3 | 36$^+$ | 39 | 40 | 42 | 43$^+$ | 45 |
| 4 | 66 | 69$^+$ | 71$^+$ | 74 | 76$^+$ | 79$^+$ |
| 5 | 104$^+$ | 109$^+$ | 112 | 116 | 119 | 123 |
| 6 | 152 | 158$^+$ | 162 | 167 | 171$^+$ | 176$^+$ |
| 7 | 208$^+$ | 217 | 221 | 227$^+$ | 233 | 239 |
| 8 | 274$^+$ | 285 | 290 | 297$^+$ | 304 | 312 |
| 9 | 349$^+$ | 362 | 368 | 377 | 385 | 394 |
| 10 | 434 | 448$^+$ | 455$^+$ | 466$^+$ | 475$^+$ | 486$^+$ |
| 11 | 527$^+$ | 544$^+$ | 553 | 565 | 575$^+$ | 588 |
| 12 | 630$^+$ | 650 | 659$^+$ | 673$^+$ | 685$^+$ | 699$^+$ |
| 13 | 743 | 765 | 775$^+$ | 791$^+$ | 805 | 821 |
| 14 | 865$^+$ | 889$^+$ | 901$^+$ | 919 | 934 | 952 |
| 15 | 996$^+$ | 1023$^+$ | 1036$^+$ | 1056 | 1073 | 1092$^+$ |
| 16 | 1137 | 1167 | 1181$^+$ | 1203 | 1221$^+$ | 1243 |
| 17 | 1287$^+$ | 1320 | 1336 | 1359$^+$ | 1380 | 1403$^+$ |
| 18 | 1447 | 1483 | 1500 | 1526 | 1548 | 1573$^+$ |
| 19 | 1616$^+$ | 1655 | 1674 | 1701$^+$ | 1725$^+$ | 1753$^+$ |
| 20 | 1795 | 1837 | 1857$^+$ | 1887$^+$ | 1913 | 1943 |
| 21 | 1983$^+$ | 2028$^+$ | 2050$^+$ | 2082$^+$ | 2110$^+$ | 2142$^+$ |
| 22 | 2181 | 2229$^+$ | 2253 | 2287$^+$ | 2317$^+$ | 2352 |

**Number of Blocks = 6**

| n | 0.002 / 0.001 | 0.01 / 0.005 | 0.02 / 0.01 | 0.05 / 0.025 | 0.10 / 0.05 | 0.20 / 0.10 |
|---|---|---|---|---|---|---|
| | 2-Sided / 1-Sided | | | | | |
| 1 | — | — | — | — | — | 6$^+$ |
| 2 | 20 | 21 | 22 | 23 | 24 | 25 |
| 3 | 45$^+$ | 48 | 49$^+$ | 51$^+$ | 53 | 55 |
| 4 | 81$^+$ | 85$^+$ | 87$^+$ | 90$^+$ | 93$^+$ | 96$^+$ |
| 5 | 128$^+$ | 134$^+$ | 137 | 141$^+$ | 145 | 149 |
| 6 | 186$^+$ | 194 | 198 | 203$^+$ | 208 | 213$^+$ |
| 7 | 255$^+$ | 265 | 270 | 276$^+$ | 282$^+$ | 289$^+$ |
| 8 | 336 | 347$^+$ | 353 | 361$^+$ | 369 | 377$^+$ |
| 9 | 427$^+$ | 441 | 448 | 458 | 466$^+$ | 476$^+$ |
| 10 | 530 | 546 | 554 | 566 | 576 | 587$^+$ |
| 11 | 644 | 662$^+$ | 671$^+$ | 685 | 697 | 710$^+$ |
| 12 | 769 | 790$^+$ | 801 | 816 | 829$^+$ | 845 |
| 13 | 905$^+$ | 929$^+$ | 941$^+$ | 958$^+$ | 973$^+$ | 991 |
| 14 | 1053$^+$ | 1080$^+$ | 1093$^+$ | 1113 | 1129$^+$ | 1149 |
| 15 | 1213 | 1242$^+$ | 1257 | 1278$^+$ | 1297 | 1318$^+$ |
| 16 | 1383$^+$ | 1416$^+$ | 1432$^+$ | 1456 | 1476$^+$ | 1500 |
| 17 | 1565$^+$ | 1601$^+$ | 1619 | 1645 | 1667 | 1693 |
| 18 | 1759 | 1798$^+$ | 1817$^+$ | 1845$^+$ | 1870 | 1898 |
| 19 | 1964 | 2006 | 2027$^+$ | 2058 | 2084 | 2114$^+$ |
| 20 | 2180$^+$ | 2226$^+$ | 2249 | 2282 | 2310$^+$ | 2343 |
| 21 | 2408$^+$ | 2458 | 2482 | 2517$^+$ | 2548 | 2583$^+$ |
| 22 | 2648 | 2701 | 2727 | 2765 | 2797$^+$ | 2835$^+$ |

(continued)

**Table A.16** Critical Values for the Rank Sum Test in Randomized Blocks[a,b]—cont...

**Number of Blocks = 5**

| n | 2-Sided 0.002 / 1-Sided 0.001 | 0.01 / 0.005 | 0.02 / 0.01 | 0.05 / 0.025 | 0.10 / 0.05 | 0.20 / 0.10 | 2-Sided / 1-Sided |
|---|---|---|---|---|---|---|---|
| 23 | 2388+ | 2440 | 2465+ | 2502+ | 2534+ | 2571 | |
| 24 | 2605+ | 2660+ | 2687+ | 2727 | 2761 | 2800 | |
| 25 | 2832 | 2890+ | 2919 | 2961 | 2997 | 3039 | |
| 26 | 3068 | 3130 | 3160+ | 3205 | 3243 | 3287+ | |
| 27 | 3314 | 3379+ | 3411+ | 3458+ | 3499 | 3546 | |
| 28 | 3569 | 3638+ | 3672 | 3722 | 3765 | 3814 | |
| 29 | 3834 | 3907 | 3942+ | 3995 | 4040 | 4092+ | |
| 30 | 4108+ | 4185+ | 4223 | 4278 | 4325+ | 4380+ | |

**Number of Blocks = 7**

| n | 2-Sided 0.002 / 1-Sided 0.001 | 0.01 / 0.005 | 0.02 / 0.01 | 0.05 / 0.025 | 0.10 / 0.05 | 0.20 / 0.10 | 2-Sided / 1-Sided |
|---|---|---|---|---|---|---|---|
| 1 | — | — | 7 | 7 | 7+ | 8 | |
| 2 | 24 | 25+ | 26+ | 27+ | 28+ | 30 | |
| 3 | 54+ | 57+ | 59 | 61 | 63 | 65 | |
| 4 | 97+ | 102 | 104 | 107+ | 110 | 113+ | |
| 5 | 153 | 159+ | 162+ | 167 | 171 | 175+ | |
| 6 | 222 | 230 | 234 | 240 | 245 | 251 | |
| 7 | 303+ | 313+ | 319 | 326 | 332+ | 340 | |
| 8 | 398 | 410+ | 417 | 426 | 434 | 443 | |
| 9 | 506 | 521 | 528 | 539 | 548+ | 559+ | |
| 10 | 627 | 644+ | 653 | 665+ | 676+ | 689+ | |
| 11 | 761 | 781+ | 791 | 806 | 818+ | 833 | |

**Number of Blocks = 6**

| n | 2-Sided 0.002 / 1-Sided 0.001 | 0.01 / 0.005 | 0.02 / 0.01 | 0.05 / 0.025 | 0.10 / 0.05 | 0.20 / 0.10 | 2-Sided / 1-Sided |
|---|---|---|---|---|---|---|---|
| 23 | 2899 | 2955+ | 2983 | 3024 | 3059 | 3099+ | |
| 24 | 3161+ | 3222 | 3251 | 3294+ | 3332 | 3375 | |
| 25 | 3435+ | 3499+ | 3531 | 3577 | 3616+ | 3662+ | |
| 26 | 3721 | 3789 | 3822 | 3871 | 3913 | 3961+ | |
| 27 | 4018 | 4090 | 4125 | 4177 | 4221+ | 4272+ | |
| 28 | 4326+ | 4403 | 4440 | 4494+ | 4541+ | 4595+ | |
| 29 | 4647+ | 4727 | 4766 | 4823+ | 4873 | 4930+ | |
| 30 | 4978+ | 5063 | 5104 | 5164+ | 5216+ | 5277 | |

**Number of Blocks = 8**

| n | 2-Sided 0.002 / 1-Sided 0.001 | 0.01 / 0.005 | 0.02 / 0.01 | 0.05 / 0.025 | 0.10 / 0.05 | 0.20 / 0.10 | 2-Sided / 1-Sided |
|---|---|---|---|---|---|---|---|
| 1 | — | 8 | 8 | 8+ | 9 | 9+ | |
| 2 | 28+ | 30 | 31 | 32 | 33 | 34+ | |
| 3 | 63+ | 67 | 68+ | 70+ | 72+ | 75 | |
| 4 | 113+ | 118 | 120+ | 124 | 127 | 130+ | |
| 5 | 178 | 184+ | 188 | 193 | 197 | 202 | |
| 6 | 257 | 266 | 270+ | 276+ | 282 | 288+ | |
| 7 | 351+ | 362+ | 368 | 376 | 383 | 391 | |
| 8 | 460+ | 474 | 481 | 490+ | 499 | 508+ | |
| 9 | 585 | 601 | 609 | 620+ | 630+ | 642 | |
| 10 | 724+ | 743 | 752+ | 766 | 777+ | 791 | |
| 11 | 879 | 900+ | 911 | 927 | 940+ | 956 | |

**Number of Blocks = 9** (continued)

| n | | | | | | |
|---|---|---|---|---|---|---|
| 12 | 908+ | 931+ | 943 | 959+ | 974 | 990+ |
| 13 | 1069 | 1095+ | 1108 | 1126+ | 1143 | 1161+ |
| 14 | 1243 | 1272+ | 1286+ | 1307+ | 1325+ | 1346+ |
| 15 | 1430+ | 1463 | 1478+ | 1502 | 1522 | 1545 |
| 16 | 1631+ | 1667 | 1684 | 1709+ | 1732 | 1757 |
| 17 | 1845+ | 1884+ | 1903+ | 1931+ | 1955+ | 1983 |
| 18 | 2073 | 2115+ | 2136 | 2166+ | 2192+ | 2223 |
| 19 | 2314 | 2360 | 2382 | 2415 | 2443+ | 2476+ |
| 20 | 2568 | 2618 | 2642 | 2677+ | 2708+ | 2744 |
| 21 | 2836 | 2889+ | 2915+ | 2953+ | 2986+ | 3025 |
| 22 | 3117 | 3174+ | 3202+ | 3243+ | 3279 | 3319+ |
| 23 | 3411+ | 3473 | 3503 | 3547 | 3584+ | 3628+ |
| 24 | 3720 | 3785+ | 3817 | 3864 | 3904 | 3950+ |
| 25 | 4041+ | 4111 | 4145 | 4194+ | 4237+ | 4287 |
| 26 | 4376+ | 4450+ | 4486+ | 4539 | 4584+ | 4637 |
| 27 | 4725+ | 4803+ | 4841+ | 4897 | 4945 | 5000+ |
| 28 | 5087+ | 5170 | 5210 | 5269 | 5319+ | 5378 |
| 29 | 5463+ | 5550 | 5592+ | 5654+ | 5708 | 5769+ |
| 30 | 5852+ | 5944 | 5988+ | 6053+ | 6110 | 6175 |

**Number of Blocks = 9**

| n | 0.002 / 0.001 | 0.01 / 0.005 | 0.02 / 0.01 | 0.05 / 0.025 | 0.10 / 0.05 | 0.20 / 0.10 | 2-Sided / 1-Sided |
|---|---|---|---|---|---|---|---|
| 1 | — | 9 | 9 | 10 | 10 | 11 | |
| 2 | 32+ | 34+ | 35+ | 36+ | 38 | 39 | |
| 3 | 73 | 76 | 78 | 80+ | 82+ | 85 | |

**Number of Blocks = 10** (continued)

| n | | | | | | |
|---|---|---|---|---|---|---|
| 12 | 1048+ | 1073+ | 1085+ | 1103+ | 1118+ | 1136+ |
| 13 | 1233+ | 1261+ | 1275 | 1295 | 1312+ | 1332+ |
| 14 | 1434 | 1465 | 1480+ | 1502+ | 1522 | 1544+ |
| 15 | 1649+ | 1684 | 1701 | 1725+ | 1747 | 1772 |
| 16 | 1880 | 1918+ | 1937 | 1964 | 1988 | 2015 |
| 17 | 2126+ | 2168 | 2188+ | 2218+ | 2244 | 2274 |
| 18 | 2388 | 2433+ | 2455+ | 2488 | 2516 | 2548+ |
| 19 | 2665 | 2714 | 2738 | 2773+ | 2804 | 2839 |
| 20 | 2957 | 3010+ | 3036+ | 3074+ | 3107 | 3145 |
| 21 | 3265 | 3322 | 3350 | 3391 | 3426+ | 3467 |
| 22 | 3588 | 3649+ | 3679 | 3723 | 3761 | 3804+ |
| 23 | 3926+ | 3992 | 4024 | 4071 | 4111+ | 4158 |
| 24 | 4280+ | 4350+ | 4384+ | 4434+ | 4477+ | 4527+ |
| 25 | 4650 | 4724+ | 4760+ | 4813+ | 4859+ | 4912+ |
| 26 | 5035 | 5113+ | 5152 | 5208+ | 5257 | 5313 |
| 27 | 5435 | 5518+ | 5559 | 5619 | 5670+ | 5729+ |
| 28 | 5851 | 5939 | 5982 | 6045 | 6099+ | 6162 |
| 29 | 6282+ | 6375+ | 6420+ | 6487 | 6544 | 6610 |
| 30 | 6729+ | 6827 | 6874+ | 6944+ | 7004+ | 7074 |

**Number of Blocks = 10**

| n | 0.002 / 0.001 | 0.01 / 0.005 | 0.02 / 0.01 | 0.05 / 0.025 | 0.10 / 0.05 | 0.20 / 0.10 | 2-Sided / 1-Sided |
|---|---|---|---|---|---|---|---|
| 1 | 10 | 10 | 10+ | 11 | 11+ | 12 | |
| 2 | 37 | 39 | 40 | 41 | 42+ | 44 | |
| 3 | 82 | 86 | 87+ | 90 | 92+ | 95 | |

(continued)

**Table A.16** Critical Values for the Rank Sum Test in Randomized Blocks[a,b]—cont...

Number of Blocks = 10

| n | 2-Sided 0.002 / 1-Sided 0.001 | 0.01 / 0.005 | 0.02 / 0.01 | 0.05 / 0.025 | 0.10 / 0.05 | 0.20 / 0.10 | 2-Sided / 1-Sided |
|---|---|---|---|---|---|---|---|
| 4 | 146 | 151 | 154 | 158 | 161 | 165 | |
| 5 | 228 | 235+ | 239 | 244+ | 249+ | 255 | |
| 6 | 328+ | 338+ | 343+ | 350+ | 357 | 364 | |
| 7 | 448+ | 461 | 467 | 476 | 483+ | 492+ | |
| 8 | 587 | 602 | 609+ | 620+ | 629+ | 640+ | |
| 9 | 744 | 762+ | 771 | 784 | 795+ | 808+ | |
| 10 | 920+ | 942 | 952 | 967+ | 980+ | 995+ | |
| 11 | 1116 | 1140+ | 1152+ | 1170 | 1185 | 1202+ | |
| 12 | 1330+ | 1358+ | 1372 | 1392 | 1409 | 1429 | |
| 13 | 1564+ | 1596 | 1611 | 1633+ | 1653 | 1675 | |
| 14 | 1817+ | 1852+ | 1869+ | 1894+ | 1916 | 1941 | |
| 15 | 2089+ | 2128 | 2147 | 2175 | 2199 | 2226+ | |
| 16 | 2381 | 2423+ | 2444+ | 2475 | 2501 | 2531+ | |
| 17 | 2691+ | 2738 | 2761 | 2794+ | 2823 | 2856+ | |
| 18 | 3021 | 3072 | 3097 | 3133+ | 3165 | 3201 | |
| 19 | 3370+ | 3425+ | 3452+ | 3492 | 3526 | 3565+ | |
| 20 | 3739 | 3798+ | 3827+ | 3870 | 3907 | 3949+ | |
| 21 | 4126+ | 4191 | 4222 | 4268 | 4307+ | 4353 | |
| 22 | 4534 | 4602+ | 4636 | 4685+ | 4727+ | 4776+ | |
| 23 | 4960+ | 5034 | 5069+ | 5122 | 5167+ | 5219+ | |
| 24 | 5406+ | 5484+ | 5522+ | 5578+ | 5627 | 5682+ | |
| 25 | 5871+ | 5955 | 5995+ | 6055 | 6106 | 6165+ | |

Number of Blocks = 9

| n | 2-Sided 0.002 / 1-Sided 0.001 | 0.01 / 0.005 | 0.02 / 0.01 | 0.05 / 0.025 | 0.10 / 0.05 | 0.20 / 0.10 | 2-Sided / 1-Sided |
|---|---|---|---|---|---|---|---|
| 4 | 129+ | 134+ | 137 | 141 | 144 | 148 | |
| 5 | 203 | 210 | 213+ | 218+ | 223 | 228+ | |
| 6 | 293 | 302 | 307 | 313+ | 319+ | 326 | |
| 7 | 400 | 411+ | 417+ | 426 | 433 | 441+ | |
| 8 | 523+ | 538 | 545 | 555+ | 564 | 574+ | |
| 9 | 664+ | 681+ | 690 | 702 | 713 | 725 | |
| 10 | 822+ | 842+ | 852 | 866+ | 879 | 893+ | |
| 11 | 997+ | 1020+ | 1031+ | 1048+ | 1062+ | 1079 | |
| 12 | 1189+ | 1216 | 1228+ | 1247+ | 1264 | 1282+ | |
| 13 | 1398+ | 1428+ | 1443 | 1464 | 1482+ | 1503+ | |
| 14 | 1625+ | 1658+ | 1674+ | 1698+ | 1719 | 1742+ | |
| 15 | 1869 | 1906 | 1923+ | 1950 | 1973 | 1999 | |
| 16 | 2130 | 2170+ | 2190+ | 2219+ | 2244+ | 2273 | |
| 17 | 2408+ | 2453 | 2474+ | 2506 | 2533+ | 2565 | |
| 18 | 2704 | 2752+ | 2776 | 2810+ | 2840+ | 2874+ | |
| 19 | 3017 | 3069+ | 3095 | 3132+ | 3164+ | 3202 | |
| 20 | 3347+ | 3404 | 3431+ | 3472 | 3507 | 3547 | |
| 21 | 3695 | 3756 | 3785+ | 3829 | 3866+ | 3910 | |
| 22 | 4060+ | 4125+ | 4157 | 4204 | 4244 | 4290+ | |
| 23 | 4443 | 4512+ | 4546+ | 4596 | 4639 | 4688+ | |
| 24 | 4842+ | 4917 | 4953 | 5006 | 5052 | 5104+ | |
| 25 | 5260 | 5339 | 5377+ | 5434 | 5482+ | 5538+ | |

## Number of Blocks = 11

| n | 2-Sided 0.002 / 1-Sided 0.001 | 0.01 / 0.005 | 0.02 / 0.01 | 0.05 / 0.025 | 0.10 / 0.05 | 0.20 / 0.10 |
|---|---|---|---|---|---|---|
| 1 | 11 | 11+ | 12 | 12+ | 13 | 13+ |
| 2 | 41+ | 43+ | 44+ | 46 | 47 | 48+ |
| 3 | 91+ | 95+ | 97 | 100 | 102 | 105 |
| 4 | 162 | 168 | 170+ | 175 | 178+ | 182+ |
| 5 | 253 | 261 | 265 | 270+ | 275+ | 281+ |
| 6 | 364+ | 375 | 380 | 387+ | 394 | 401+ |
| 7 | 497 | 510 | 516+ | 526 | 534 | 543+ |
| 8 | 650 | 666 | 674 | 685+ | 695+ | 706+ |
| 9 | 824+ | 843+ | 852+ | 866 | 878 | 891+ |
| 10 | 1019+ | 1041+ | 1052+ | 1068+ | 1082 | 1098 |
| 11 | 1235+ | 1261 | 1273+ | 1292 | 1307+ | 1326 |
| 12 | 1472+ | 1501+ | 1516 | 1536+ | 1554+ | 1575+ |
| 13 | 1730+ | 1763+ | 1779+ | 1803 | 1823+ | 1847 |
| 14 | 2010 | 2046+ | 2064+ | 2091 | 2113+ | 2139+ |
| 15 | 2310+ | 2351 | 2371 | 2400 | 2425 | 2454 |
| 16 | 2632 | 2677 | 2698+ | 2731 | 2758+ | 2790+ |
| 17 | 2975 | 3024 | 3048 | 3083 | 3113+ | 3148+ |
| 26 | 5695 | 5778+ | 5819 | 5879 | 5930+ | 5990 |
| 27 | 6147 | 6235+ | 6278+ | 6342 | 6396+ | 6459+ |
| 28 | 6616+ | 6710 | 6755+ | 6822 | 6880 | 6946+ |
| 29 | 7103+ | 7202+ | 7250+ | 7320+ | 7381+ | 7451+ |
| 30 | 7608+ | 7712 | 7762+ | 7836+ | 7900+ | 7974 |

## Number of Blocks = 12

| n | 2-Sided 0.002 / 1-Sided 0.001 | 0.01 / 0.005 | 0.01 / 0.01 | 0.05 / 0.025 | 0.10 / 0.05 | 0.20 / 0.10 |
|---|---|---|---|---|---|---|
| 1 | 12 | 13 | 13 | 14 | 14+ | 15 |
| 2 | 46 | 48 | 49 | 50+ | 52 | 53+ |
| 3 | 101 | 105 | 107 | 109+ | 112 | 115 |
| 4 | 178+ | 184+ | 187+ | 192 | 195+ | 200 |
| 5 | 278+ | 286+ | 291 | 297 | 302 | 308 |
| 6 | 401 | 411+ | 417 | 425 | 431+ | 439+ |
| 7 | 546 | 559+ | 566+ | 576 | 584+ | 594+ |
| 8 | 714 | 730+ | 738+ | 750+ | 761 | 773 |
| 9 | 904+ | 924+ | 934 | 948+ | 960+ | 975 |
| 10 | 1118 | 1141+ | 1153 | 1169+ | 1184 | 1200+ |
| 11 | 1355 | 1381+ | 1394+ | 1414 | 1430+ | 1449+ |
| 12 | 1614+ | 1645 | 1660 | 1681+ | 1700 | 1722 |
| 13 | 1897 | 1931+ | 1948+ | 1973 | 1994 | 2018+ |
| 14 | 2203 | 2241+ | 2260 | 2287+ | 2311 | 2338+ |
| 15 | 2532 | 2574+ | 2595 | 2625+ | 2652 | 2682 |
| 16 | 2884 | 2931 | 2953+ | 2987 | 3016 | 3049+ |
| 17 | 3259 | 3310+ | 3335+ | 3372 | 3404 | 3440+ |
| 26 | 6356+ | 6444+ | 6487+ | 6550+ | 6605 | 6667+ |
| 27 | 6860+ | 6954 | 6999+ | 7066 | 7123+ | 7190 |
| 28 | 7384 | 7482+ | 7530+ | 7601 | 7662 | 7732 |
| 29 | 7927 | 8031 | 8081+ | 8156 | 8220 | 8293+ |
| 30 | 8489+ | 8599 | 8652 | 8730 | 8797+ | 8875 |

*(continued)*

**Table A.16** Critical Values for the Rank Sum Test in Randomized Blocks[a,b]—cont...

**Number of Blocks = 11**

| n | 0.002 | 0.01 | 0.02 | 0.05 | 0.10 | 0.20 | 2-Sided |
|---|---|---|---|---|---|---|---|
| | 0.001 | 0.005 | 0.01 | 0.025 | 0.05 | 0.10 | 1-Sided |
| 18 | 3339 | 3392+ | 3418+ | 3457 | 3490 | 3528 | |
| 19 | 3724+ | 3782+ | 3810+ | 3852 | 3888 | 3929 | |
| 20 | 4131 | 4194 | 4224 | 4269 | 4307+ | 4352 | |
| 21 | 4559 | 4626+ | 4659+ | 4707+ | 4749 | 4796+ | |
| 22 | 5008+ | 5080+ | 5116 | 5167+ | 5212 | 5263 | |
| 23 | 5479 | 5556+ | 5594 | 5649 | 5696+ | 5751 | |
| 24 | 5971 | 6053+ | 6093+ | 6152 | 6202+ | 6261 | |
| 25 | 6484+ | 6572 | 6614+ | 6677 | 6730+ | 6792+ | |
| 26 | 7019 | 7112 | 7157 | 7223 | 7280 | 7346 | |
| 27 | 7575+ | 7673+ | 7721 | 7791 | 7851+ | 7921 | |
| 28 | 8153 | 8256+ | 8306+ | 8380+ | 8444+ | 8518 | |
| 29 | 8751+ | 8861 | 8914 | 8992 | 9059 | 9136+ | |
| 30 | 9372 | 9487 | 9542+ | 9624+ | 9695 | 9776+ | |

**Number of Blocks = 12**

| n | 0.002 | 0.01 | 0.01 | 0.05 | 0.10 | 0.20 | 2-Sided |
|---|---|---|---|---|---|---|---|
| | 0.001 | 0.005 | 0.01 | 0.025 | 0.05 | 0.10 | 1-Sided |
| 18 | 3657+ | 3713+ | 3741 | 3780+ | 3815 | 3855 | |
| 19 | 4079+ | 4140 | 4169+ | 4213 | 4250 | 4293 | |
| 20 | 4524+ | 4590 | 4621+ | 4668+ | 4708+ | 4755 | |
| 21 | 4992+ | 5063 | 5097 | 5147+ | 5190+ | 5240+ | |
| 22 | 5484 | 5559+ | 5596 | 5650 | 5696+ | 5750 | |
| 23 | 5999 | 6079+ | 6118+ | 6176+ | 6226 | 6283 | |
| 24 | 6537 | 6623 | 6664+ | 6726 | 6779 | 6840 | |
| 25 | 7098+ | 7190 | 7234 | 7299+ | 7355+ | 7420+ | |
| 26 | 7683+ | 7780 | 7827 | 7896+ | 7956 | 8024+ | |
| 27 | 8291+ | 8394 | 8443+ | 8517 | 8580 | 8652+ | |
| 28 | 8923 | 9031 | 9083+ | 9161 | 9227+ | 9304+ | |
| 29 | 9578 | 9692 | 9747+ | 9828+ | 9899 | 9979+ | |
| 30 | 10256 | 10376 | 10434+ | 10520 | 10594 | 10679 | |

[a]Source: L. S. Nelson. (1970). Tables for Wilcoxon's rank sum test in randomized blocks. J Qual Technol, **2**, 207–217. Copyright American Society for Quality Control, Inc. Reprinted by permission.
[b]A superscript +indicates that the next larger integer yields a significance level closer to, but greater than, that shown.

**Table A.17** Percentage Points $R_{a,k,n}$ of Steel's Minimum-Rank Sum[a]

| n = Number in Treatment | $\alpha$ | k = Number of Treatments Being Tested | | | | | | | |
|---|---|---|---|---|---|---|---|---|---|
| | | 3 | 4 | 5 | 6 | 7 | 8 | 9 | 10 |
| 4 | .10 | 10 | 10 | — | — | — | — | — | — |
| | .05 | — | — | — | — | — | — | — | — |
| | .01 | — | — | — | — | — | — | — | — |
| 5 | .10 | 17 | 16 | 15 | 15 | — | — | — | — |
| | .05 | 16 | 15 | — | — | — | — | — | — |
| | .01 | — | — | — | — | — | — | — | — |
| 6 | .10 | 26 | 24 | 23 | 22 | 22 | 21 | 21 | — |
| | .05 | 24 | 23 | 22 | 21 | — | — | — | — |
| | .01 | — | — | — | — | — | — | — | — |
| 7 | .10 | 36 | 34 | 33 | 32 | 31 | 30 | 30 | 29 |
| | .05 | 34 | 32 | 31 | 30 | 29 | 28 | 28 | — |
| | .01 | 29 | 28 | — | — | — | — | — | — |
| 8 | .10 | 48 | 46 | 44 | 43 | 42 | 41 | 40 | 40 |
| | .05 | 45 | 43 | 42 | 40 | 40 | 39 | 38 | 38 |
| | .01 | 40 | 38 | 37 | 36 | — | — | — | — |
| 9 | .10 | 62 | 59 | 57 | 56 | 55 | 54 | 53 | 52 |
| | .05 | 59 | 56 | 54 | 53 | 52 | 51 | 50 | 49 |
| | .01 | 52 | 50 | 48 | 47 | 46 | 45 | — | — |
| 10 | .10 | 77 | 74 | 72 | 70 | 69 | 68 | 67 | 66 |
| | .05 | 74 | 71 | 68 | 67 | 66 | 64 | 63 | 63 |
| | .01 | 66 | 63 | 62 | 60 | 59 | 58 | 57 | 56 |

[a]*Source: R. G. D. Steel. (1961). Some rank sum multiple comparison tests. Biometrics, **17**, 539–552. Reproduced with permission of The Biometric Society.*

**Table A.18** Critical Values $r_0$ and Significance Probabilities $P(r_0)$ for the Sign Test[a,b]

| n | $r_0$ | $P(r_0)$ | n | $r_0$ | $P(r_0)$ | n | $r_0$ | $P(r_0)$ | n | $r_0$ | $P(r_0)$ | n | $r_0$ | $P(r_0)$ |
|---|---|---|---|---|---|---|---|---|---|---|---|---|---|---|
| 10 | 0 | .001953 | | 4 | .049042 | | 7 | .133799 | | 5 | .001514 | | 8 | .010674 |
| | 1 | .021484 | | 5 | .143463 | | 8 | .286274 | | 6 | .005925 | | 9 | .029448 |
| | 2 | .109375 | | 6 | .332304 | 23 | 1 | .000006 | | 7 | .019157 | | 10 | .070753 |
| | 3 | .343749 | 18 | 0 | .000008 | | 2 | .000066 | | 8 | .052238 | | 11 | .149605 |
| 11 | 0 | .000977 | | 1 | .000145 | | 3 | .000488 | | 9 | .122076 | | 12 | .281027 |
| | 1 | .011719 | | 2 | .001312 | | 4 | .002599 | | 10 | .247781 | 32 | 3 | .000003 |
| | 2 | .065430 | | 3 | .007538 | | 5 | .010622 | 28 | 2 | .000003 | | 4 | .000019 |
| | 3 | .226562 | | 4 | .030884 | | 6 | .034689 | | 3 | .000027 | | 5 | .000113 |
| 12 | 0 | .000488 | | 5 | .096252 | | 7 | .093138 | | 4 | .000180 | | 6 | .000535 |
| | 1 | .006348 | | 6 | .237882 | | 8 | .210035 | | 5 | .000912 | | 7 | .002102 |
| | 2 | .038574 | 19 | 0 | .000004 | 24 | 1 | .000003 | | 6 | .003719 | | 8 | .007000 |
| | 3 | .145996 | | 1 | .000076 | | 2 | .000036 | | 7 | .012541 | | 9 | .020061 |
| | 4 | .387694 | | 2 | .000729 | | 3 | .000277 | | 8 | .035698 | | 10 | .050100 |
| 13 | 0 | .000244 | | 3 | .004425 | | 4 | .001544 | | 9 | .087157 | | 11 | .110179 |
| | 1 | .003418 | | 4 | .019211 | | 5 | .006611 | | 10 | .184928 | | 12 | .215317 |
| | 2 | .022461 | | 5 | .063568 | | 6 | .022656 | | 11 | .344916 | 33 | 3 | .000001 |
| | 3 | .092285 | | 6 | .167066 | | 7 | .063914 | 29 | 2 | .000002 | | 4 | .000011 |
| | 4 | .266845 | | 7 | .359276 | | 8 | .151586 | | 3 | .000015 | | 5 | .000066 |
| 14 | 0 | .000123 | 20 | 0 | .000002 | | 9 | .307446 | | 4 | .000104 | | 6 | .000324 |
| | 1 | .001831 | | 1 | .000040 | 25 | 1 | .000002 | | 5 | .000546 | | 7 | .001319 |
| | 2 | .012939 | | 2 | .000402 | | 2 | .000019 | | 6 | .002316 | | 8 | .004551 |
| | 3 | .057373 | | 3 | .002577 | | 3 | .000157 | | 7 | .008130 | | 9 | .013530 |
| | 4 | .179565 | | 4 | .011818 | | 4 | .000911 | | 8 | .024120 | | 10 | .035081 |
| 15 | 0 | .000061 | | 5 | .041389 | | 5 | .004077 | | 9 | .061427 | | 11 | .080140 |
| | 1 | .000977 | | 6 | .115318 | | 6 | .014633 | | 10 | .136042 | | 12 | .162750 |
| | 2 | .007385 | | 7 | .263172 | | 7 | .043285 | | 11 | .264924 | | 13 | .296195 |
| | 3 | .035156 | 21 | 1 | .000021 | | 8 | .107751 | 30 | 3 | .000008 | 34 | 4 | .000006 |
| | 4 | .118469 | | 2 | .000221 | | 9 | .229518 | | 4 | .000059 | | 5 | .000039 |
| | 5 | .301757 | | 3 | .001490 | 26 | 2 | .000011 | | 5 | .000325 | | 6 | .000195 |
| 16 | 0 | .000031 | | 4 | .007197 | | 3 | .000088 | | 6 | .001431 | | 7 | .000821 |
| | 1 | .000519 | | 5 | .026604 | | 4 | .000534 | | 7 | .005223 | | 8 | .002935 |
| | 2 | .004181 | | 6 | .078354 | | 5 | .002494 | | 8 | .016124 | | 9 | .009041 |
| | 3 | .021271 | | 7 | .189245 | | 6 | .009355 | | 9 | .042773 | | 10 | .024305 |
| | 4 | .076813 | | 8 | .383302 | | 7 | .028959 | | 10 | .098732 | | 11 | .057610 |
| | 5 | .210113 | 22 | 1 | .000011 | | 8 | .075518 | | 11 | .200478 | | 12 | .121444 |
| 17 | 0 | .000015 | | 2 | .000121 | | 9 | .168634 | 31 | 3 | .000005 | | 13 | .229473 |
| | 1 | .000275 | | 3 | .000855 | | 10 | .326930 | | 4 | .000034 | 35 | 4 | .000003 |
| | 2 | .002350 | | 4 | .004344 | 27 | 2 | .000006 | | 5 | .000192 | | 5 | .000022 |
| | 3 | .012726 | | 5 | .016901 | | 3 | .000049 | | 6 | .000878 | | 6 | .000117 |
| | | | | 6 | .052479 | | 4 | .000311 | | 7 | .003327 | | 7 | .000508 |

*(continued)*

**Table A.18** Critical Values $r_0$ and Significance Probabilities $P(r_0)$ for the Sign Test[a,b]—*cont...*

| n | $r_0$ | $P(r_0)$ | n | $r_0$ | $P(r_0)$ | n | $r_0$ | $P(r_0)$ | n | $r_0$ | $P(r_0)$ | n | $r_0$ | $P(r_0)$ |
|---|---|---|---|---|---|---|---|---|---|---|---|---|---|---|
| | 8 | .001878 | **39** | 5 | .000002 | | 11 | .002887 | | 17 | .135151 | **49** | 8 | .000002 |
| | 9 | .005988 | | 6 | .000014 | | 12 | .007916 | | 18 | .232685 | | 9 | .000009 |
| | 10 | .016673 | | 7 | .000070 | | 13 | .019520 | **46** | 7 | .000002 | | 10 | .000038 |
| | 11 | .040958 | | 8 | .000294 | | 14 | .043556 | | 8 | .000009 | | 11 | .000142 |
| | 12 | .089528 | | 9 | .001065 | | 15 | .088426 | | 9 | .000041 | | 12 | .000470 |
| | 13 | .175459 | | 10 | .003378 | | 16 | .164142 | | 10 | .000156 | | 13 | .001403 |
| | 14 | .310495 | | 11 | .009475 | | 17 | .279942 | | 11 | .000536 | | 14 | .003802 |
| **36** | 4 | .000002 | | 12 | .023702 | **43** | 6 | .000002 | | 12 | .001641 | | 15 | .009399 |
| | 5 | .000013 | | 13 | .053250 | | 7 | .000009 | | 13 | .004534 | | 16 | .021293 |
| | 6 | .000070 | | 14 | .108124 | | 8 | .000042 | | 14 | .011351 | | 17 | .044382 |
| | 7 | .000313 | | 15 | .199582 | | 9 | .000170 | | 15 | .025895 | | 18 | .085429 |
| | 8 | .001193 | | 16 | .336769 | | 10 | .000606 | | 16 | .054074 | | 19 | .152401 |
| | 9 | .003933 | **40** | 5 | .000001 | | 11 | .001914 | | 17 | .103801 | | 20 | .252859 |
| | 10 | .011330 | | 6 | .000008 | | 12 | .005401 | | 18 | .183917 | **50** | 8 | .000001 |
| | 11 | .028815 | | 7 | .000042 | | 13 | .013718 | | 19 | .301984 | | 9 | .000006 |
| | 12 | .065243 | | 8 | .000182 | | 14 | .031538 | **47** | 7 | .000001 | | 10 | .000024 |
| | 13 | .132493 | | 9 | .000680 | | 15 | .065992 | | 8 | .000006 | | 11 | .000090 |
| | 14 | .242976 | | 10 | .002221 | | 16 | .126285 | | 9 | .000025 | | 12 | .000306 |
| **37** | 4 | .000001 | | 11 | .006426 | | 17 | .222044 | | 10 | .000098 | | 13 | .000936 |
| | 5 | .000007 | | 12 | .016588 | **44** | 7 | .000005 | | 11 | .000346 | | 14 | .002602 |
| | 6 | .000041 | | 13 | .038476 | | 8 | .000025 | | 12 | .001088 | | 15 | .006600 |
| | 7 | .000191 | | 14 | .080687 | | 9 | .000106 | | 13 | .003088 | | 16 | .015346 |
| | 8 | .000753 | | 15 | .153853 | | 10 | .000388 | | 14 | .007942 | | 17 | .032838 |
| | 9 | .002563 | | 16 | .268176 | | 11 | .001260 | | 15 | .018623 | | 18 | .064906 |
| | 10 | .007632 | **41** | 6 | .000005 | | 12 | .003658 | | 16 | .039984 | | 19 | .118915 |
| | 11 | .020073 | | 7 | .000025 | | 13 | .009559 | | 17 | .078938 | | 20 | .202631 |
| | 12 | .047029 | | 8 | .000112 | | 14 | .022628 | | 18 | .143859 | | 21 | .322224 |
| | 13 | .098868 | | 9 | .000431 | | 15 | .048765 | | 19 | .242951 | **51** | 9 | .000003 |
| | 14 | .187736 | | 10 | .001450 | | 16 | .096138 | **48** | 8 | .000003 | | 10 | .000015 |
| | 15 | .324000 | | 11 | .004324 | | 17 | .174164 | | 9 | .000015 | | 11 | .000057 |
| **38** | 5 | .000004 | | 12 | .011507 | | 18 | .291203 | | 10 | .000062 | | 12 | .000198 |
| | 6 | .000024 | | 13 | .027532 | **45** | 7 | .000003 | | 11 | .000222 | | 13 | .000621 |
| | 7 | .000116 | | 14 | .059581 | | 8 | .000015 | | 12 | .000717 | | 14 | .001769 |
| | 8 | .000472 | | 15 | .117270 | | 9 | .000066 | | 13 | .002088 | | 15 | .004601 |
| | 9 | .001658 | | 16 | .211014 | | 10 | .000247 | | 14 | .005515 | | 16 | .010973 |
| | 10 | .005098 | | 17 | .348874 | | 11 | .000824 | | 15 | .013283 | | 17 | .024092 |
| | 11 | .013853 | **42** | 6 | .000003 | | 12 | .002459 | | 16 | .029304 | | 18 | .048872 |
| | 12 | .033551 | | 7 | .000015 | | 13 | .006609 | | 17 | .059461 | | 19 | .091911 |
| | 13 | .072949 | | 8 | .000069 | | 14 | .016094 | | 18 | .111399 | | 20 | .160774 |
| | 14 | .143301 | | 9 | .000272 | | 15 | .035696 | | 19 | .193407 | | 21 | .262430 |
| | 15 | .255867 | | 10 | .000941 | | 16 | .072451 | | 20 | .312318 | **52** | 9 | .000002 |

*(continued)*

**Table A.18** Critical Values $r_0$ and Significance Probabilities $P(r_0)$ for the Sign Test[a,b]—cont...

| n | $r_0$ | $P(r_0)$ | n | $r_0$ | $P(r_0)$ | n | $r_0$ | $P(r_0)$ | n | $r_0$ | $P(r_0)$ | n | $r_0$ | $P(r_0)$ |
|---|---|---|---|---|---|---|---|---|---|---|---|---|---|---|
|  | 10 | .000009 | 55 | 10 | .000002 |  | 24 | .289243 |  | 23 | .092459 |  | 23 | .042956 |
|  | 11 | .000036 |  | 11 | .000009 | 58 | 11 | .000002 |  | 24 | .154999 |  | 24 | .076925 |
|  | 12 | .000128 |  | 12 | .000033 |  | 12 | .000008 |  | 25 | .245057 |  | 25 | .129916 |
|  | 13 | .000410 |  | 13 | .000114 |  | 13 | .000030 | 61 | 13 | .000008 |  | 26 | .207365 |
|  | 14 | .001195 |  | 14 | .000355 |  | 14 | .000100 |  | 14 | .000027 |  | 27 | .313500 |
|  | 15 | .003185 |  | 15 | .001016 |  | 15 | .000307 |  | 15 | .000088 | 64 | 14 | .000007 |
|  | 16 | .007787 |  | 16 | .002667 |  | 16 | .000862 |  | 16 | .000264 |  | 15 | .000024 |
|  | 17 | .017533 |  | 17 | .006456 |  | 17 | .002233 |  | 17 | .000730 |  | 16 | .000077 |
|  | 18 | .036482 |  | 18 | .014454 |  | 18 | .005355 |  | 18 | .001868 |  | 17 | .000227 |
|  | 19 | .070392 |  | 19 | .030028 |  | 19 | .011928 |  | 19 | .004444 |  | 18 | .000617 |
|  | 20 | .126342 |  | 20 | .058063 |  | 20 | .024746 |  | 20 | .009853 |  | 19 | .001563 |
|  | 21 | .211601 |  | 21 | .104787 |  | 21 | .047939 |  | 21 | .020414 |  | 20 | .003690 |
|  | 22 | .331739 |  | 22 | .176999 |  | 22 | .086947 |  | 22 | .039616 |  | 21 | .008147 |
| 53 | 9 | .000001 |  | 23 | .280606 |  | 23 | .148004 |  | 23 | .072176 |  | 22 | .016858 |
|  | 10 | .000006 | 56 | 11 | .000005 |  | 24 | .237044 |  | 24 | .123730 |  | 23 | .032765 |
|  | 11 | .000022 |  | 12 | .000021 |  | 25 | .358139 |  | 25 | .200030 |  | 24 | .059940 |
|  | 12 | .000082 |  | 13 | .000073 | 59 | 12 | .000005 |  | 26 | .305675 |  | 25 | .103420 |
|  | 13 | .000269 |  | 14 | .000234 |  | 13 | .000019 | 62 | 13 | .000005 |  | 26 | .168641 |
|  | 14 | .000802 |  | 15 | .000686 |  | 14 | .000065 |  | 14 | .000017 |  | 27 | .260433 |
|  | 15 | .002190 |  | 16 | .001842 |  | 15 | .000204 |  | 15 | .000058 | 65 | 14 | .000004 |
|  | 16 | .005486 |  | 17 | .004561 |  | 16 | .000584 |  | 16 | .000176 |  | 15 | .000016 |
|  | 17 | .012660 |  | 18 | .010455 |  | 17 | .001547 |  | 17 | .000497 |  | 16 | .000051 |
|  | 18 | .027008 |  | 19 | .022241 |  | 18 | .003794 |  | 18 | .001299 |  | 17 | .000152 |
|  | 19 | .053437 |  | 20 | .044046 |  | 19 | .008641 |  | 19 | .003156 |  | 18 | .000422 |
|  | 20 | .098368 |  | 21 | .081426 |  | 20 | .018337 |  | 20 | .007149 |  | 19 | .001090 |
|  | 21 | .168973 |  | 22 | .140893 |  | 21 | .036343 |  | 21 | .015134 |  | 20 | .002626 |
|  | 22 | .271672 |  | 23 | .228803 |  | 22 | .067443 |  | 22 | .030015 |  | 21 | .005918 |
| 54 | 10 | .000003 |  | 24 | .349679 |  | 23 | .117476 |  | 23 | .055896 |  | 22 | .012502 |
|  | 11 | .000014 | 57 | 11 | .000003 |  | 24 | .192525 |  | 24 | .097953 |  | 23 | .024812 |
|  | 12 | .000052 |  | 12 | .000013 |  | 25 | .297591 |  | 25 | .161879 |  | 24 | .046353 |
|  | 13 | .000175 |  | 13 | .000047 | 60 | 12 | .000003 |  | 26 | .252851 |  | 25 | .081681 |
|  | 14 | .000535 |  | 14 | .000154 |  | 13 | .000012 | 63 | 13 | .000003 |  | 26 | .136030 |
|  | 15 | .001496 |  | 15 | .000460 |  | 14 | .000042 |  | 14 | .000011 |  | 27 | .214537 |
|  | 16 | .003838 |  | 16 | .001264 |  | 15 | .000135 |  | 15 | .000038 |  | 28 | .321081 |
|  | 17 | .009073 |  | 17 | .003201 |  | 16 | .000394 |  | 16 | .000117 | 66 | 14 | .000003 |
|  | 18 | .019834 |  | 18 | .007508 |  | 17 | .001066 |  | 17 | .000337 |  | 15 | .000010 |
|  | 19 | .040223 |  | 19 | .016348 |  | 18 | .002670 |  | 18 | .000898 |  | 16 | .000033 |
|  | 20 | .075903 |  | 20 | .033143 |  | 19 | .006217 |  | 19 | .002228 |  | 17 | .000101 |
|  | 21 | .133672 |  | 21 | .062736 |  | 20 | .013489 |  | 20 | .005152 |  | 18 | .000287 |
|  | 22 | .220325 |  | 22 | .111160 |  | 21 | .027339 |  | 21 | .011141 |  | 19 | .000756 |
|  | 23 | .340886 |  | 23 | .184848 |  | 22 | .051893 |  | 22 | .022575 |  | 20 | .001858 |

*(continued)*

**Table A.18** Critical Values $r_0$ and Significance Probabilities $P(r_0)$ for the Sign Test[a,b]—cont...

| n | $r_0$ | $P(r_0)$ | n | $r_0$ | $P(r_0)$ | n | $r_0$ | $P(r_0)$ | n | $r_0$ | $P(r_0)$ | n | $r_0$ | $P(r_0)$ |
|---|---|---|---|---|---|---|---|---|---|---|---|---|---|---|
| | 21 | .004272 | | 17 | .000029 | | 27 | .056814 | | 20 | .000096 | | 28 | .028626 |
| | 22 | .009210 | | 18 | .000088 | | 28 | .095922 | | 21 | .000256 | | 29 | .050451 |
| | 23 | .018657 | | 19 | .000244 | | 29 | .153911 | | 22 | .000643 | | 30 | .084644 |
| | 24 | .035582 | | 20 | .000636 | | 30 | .235095 | | 23 | .001516 | | 31 | .135382 |
| | 25 | .064017 | | 21 | .001550 | | 31 | .342469 | | 24 | .003372 | | 32 | .206732 |
| | 26 | .108856 | | 22 | .003545 | **72** | 16 | .000002 | | 25 | .007084 | | 33 | .301867 |
| | 27 | .175285 | | 23 | .007621 | | 17 | .000008 | | 26 | .014080 | | | |
| | 28 | .267811 | | 24 | .015432 | | 18 | .000026 | | 27 | .026516 | **77** | 18 | .000003 |
| **67** | 15 | .000006 | | 25 | .029492 | | 19 | .000076 | | 28 | .047392 | | 19 | .000010 |
| | 16 | .000022 | | 26 | .053288 | | 20 | .000208 | | 29 | .080507 | | 20 | .000029 |
| | 17 | .000067 | | 27 | .091185 | | 21 | .000535 | | 30 | .130176 | | 21 | .000082 |
| | 18 | .000194 | | 28 | .148029 | | 22 | .001294 | | 31 | .200677 | | 22 | .000217 |
| | 19 | .000522 | | 29 | .228395 | | 23 | .002943 | | 32 | .295413 | | 23 | .000539 |
| | 20 | .001307 | | 30 | .335551 | | 24 | .006310 | **75** | 17 | .000002 | | 24 | .001263 |
| | 21 | .003065 | **70** | 16 | .000006 | | 25 | .012774 | | 18 | .000007 | | 25 | .002799 |
| | 22 | .006741 | | 17 | .000019 | | 26 | .024460 | | 19 | .000022 | | 26 | .005870 |
| | 23 | .013934 | | 18 | .000058 | | 27 | .044370 | | 20 | .000065 | | 27 | .011672 |
| | 24 | .027120 | | 19 | .000166 | | 28 | .076368 | | 21 | .000176 | | 28 | .022032 |
| | 25 | .049800 | | 20 | .000440 | | 29 | .124916 | | 22 | .000450 | | 29 | .039538 |
| | 26 | .086436 | | 21 | .001093 | | 30 | .194503 | | 23 | .001080 | | 30 | .067546 |
| | 27 | .142070 | | 22 | .002548 | | 31 | .288782 | | 24 | .002444 | | 31 | .110011 |
| | 28 | .221547 | | 23 | .005582 | **73** | 17 | .000005 | | 25 | .005228 | | 32 | .171054 |
| | 29 | .328430 | | 24 | .011526 | | 18 | .000017 | | 26 | .010582 | | 33 | .254295 |
| **68** | 15 | .000004 | | 25 | .022462 | | 19 | .000051 | | 27 | .020298 | | 34 | .362019 |
| | 16 | .000014 | | 26 | .041390 | | 20 | .000142 | | 28 | .036954 | **78** | 19 | .000006 |
| | 17 | .000045 | | 27 | .072236 | | 21 | .000371 | | 29 | .063949 | | 20 | .000020 |
| | 18 | .000131 | | 28 | .119607 | | 22 | .000914 | | 30 | .105340 | | 21 | .000056 |
| | 19 | .000358 | | 29 | .188213 | | 23 | .002118 | | 31 | .165425 | | 22 | .000149 |
| | 20 | .000914 | | 30 | .281974 | | 24 | .004626 | | 32 | .248043 | | 23 | .000378 |
| | 21 | .002186 | **71** | 16 | .000004 | | 25 | .009542 | | 33 | .355696 | | 24 | .000901 |
| | 22 | .004903 | | 17 | .000013 | | 26 | .018617 | **76** | 18 | .000005 | | 25 | .002031 |
| | 23 | .010337 | | 18 | .000039 | | 27 | .034415 | | 19 | .000015 | | 26 | .004335 |
| | 24 | .020526 | | 19 | .000112 | | 28 | .060369 | | 20 | .000044 | | 27 | .008771 |
| | 25 | .038459 | | 20 | .000303 | | 29 | .100641 | | 21 | .000121 | | 28 | .016852 |
| | 26 | .068117 | | 21 | .000767 | | 30 | .159709 | | 22 | .000313 | | 29 | .030785 |
| | 27 | .114252 | | 22 | .001820 | | 31 | .241643 | | 23 | .000765 | | 30 | .053542 |
| | 28 | .181806 | | 23 | .004065 | | 32 | .349180 | | 24 | .001762 | | 31 | .088779 |
| | 29 | .274984 | | 24 | .008554 | **74** | 17 | .000003 | | 25 | .003836 | | 32 | .140534 |
| **69** | 15 | .000003 | | 25 | .016994 | | 18 | .000011 | | 26 | .007905 | | 33 | .212676 |
| | 16 | .000009 | | 26 | .031926 | | 19 | .000034 | | 27 | .015440 | | 34 | .308158 |

*(continued)*

**Table A.18** Critical Values $r_0$ and Significance Probabilities $P(r_0)$ for the Sign Test[a,b]—cont...

| n | $r_0$ | $P(r_0)$ | n | $r_0$ | $P(r_0)$ | n | $r_0$ | $P(r_0)$ | n | $r_0$ | $P(r_0)$ | n | $r_0$ | $P(r_0)$ |
|---|---|---|---|---|---|---|---|---|---|---|---|---|---|---|
| 79 | 19 | .000004 | | 26 | .001689 | | 32 | .047523 | 86 | 21 | .000002 | | 27 | .000374 |
| | 20 | .000013 | | 27 | .003596 | | 33 | .078417 | | 22 | .000007 | | 28 | .000846 |
| | 21 | .000038 | | 28 | .007276 | | 34 | .123849 | | 23 | .000019 | | 29 | .001824 |
| | 22 | .000103 | | 29 | .014000 | | 35 | .187454 | | 24 | .000051 | | 30 | .003746 |
| | 23 | .000264 | | 30 | .025653 | | 36 | .272261 | | 25 | .000130 | | 31 | .007343 |
| | 24 | .000639 | | 31 | .044827 | | 37 | .379988 | | 26 | .000317 | | 32 | .013750 |
| | 25 | .001466 | | 32 | .074786 | 84 | 21 | .000005 | | 27 | .000732 | | 33 | .024622 |
| | 26 | .003183 | | 33 | .119269 | | 22 | .000015 | | 28 | .001606 | | 34 | .042209 |
| | 27 | .006553 | | 34 | .182071 | | 23 | .000041 | | 29 | .003353 | | 35 | .069343 |
| | 28 | .012812 | | 35 | .266404 | | 24 | .000107 | | 30 | .006674 | | 36 | .109290 |
| | 29 | .023819 | | 36 | .374163 | | 25 | .000266 | | 31 | .012672 | | 37 | .165433 |
| | 30 | .042164 | 82 | 20 | .000004 | | 26 | .000628 | | 32 | .022982 | | 38 | .240782 |
| | 31 | .071161 | | 21 | .000011 | | 27 | .001404 | | 33 | .039853 | | 39 | .337384 |
| | 32 | .114657 | | 22 | .000032 | | 28 | .002985 | | 34 | .066151 | 89 | 23 | .000006 |
| | 33 | .176606 | | 23 | .000087 | | 29 | .006038 | | 35 | .105223 | | 24 | .000016 |
| | 34 | .260418 | | 24 | .000222 | | 30 | .011634 | | 36 | .160575 | | 25 | .000043 |
| | 35 | .368177 | | 25 | .000535 | | 31 | .021383 | | 37 | .235374 | | 26 | .000110 |
| 80 | 19 | .000003 | | 26 | .001220 | | 32 | .037529 | | 38 | .331827 | | 27 | .000266 |
| | 20 | .000009 | | 27 | .002642 | | 33 | .062970 | 87 | 22 | .000004 | | 28 | .000610 |
| | 21 | .000025 | | 28 | .005436 | | 34 | .101134 | | 23 | .000013 | | 29 | .001335 |
| | 22 | .000070 | | 29 | .010637 | | 35 | .155652 | | 24 | .000035 | | 30 | .002785 |
| | 23 | .000183 | | 30 | .019826 | | 36 | .229859 | | 25 | .000091 | | 31 | .005545 |
| | 24 | .000452 | | 31 | .035240 | | 37 | .326126 | | 26 | .000224 | | 32 | .010547 |
| | 25 | .001053 | | 32 | .059806 | 85 | 21 | .000003 | | 27 | .000524 | | 33 | .019186 |
| | 26 | .002324 | | 33 | .097027 | | 22 | .000010 | | 28 | .001169 | | 34 | .033415 |
| | 27 | .004868 | | 34 | .150669 | | 23 | .000028 | | 29 | .002479 | | 35 | .055776 |
| | 28 | .009682 | | 35 | .224236 | | 24 | .000074 | | 30 | .005013 | | 36 | .089317 |
| | 29 | .018315 | | 36 | .320283 | | 25 | .000187 | | 31 | .009672 | | 37 | .137362 |
| | 30 | .032991 | 83 | 20 | .000002 | | 26 | .000447 | | 32 | .017827 | | 38 | .203108 |
| | 31 | .056662 | | 21 | .000008 | | 27 | .001016 | | 33 | .031417 | | 39 | .289084 |
| | 32 | .092909 | | 22 | .000022 | | 28 | .002195 | | 34 | .053001 | 90 | 23 | .000004 |
| | 33 | .145631 | | 23 | .000060 | | 29 | .004512 | | 35 | .085685 | | 24 | .000011 |
| | 34 | .218512 | | 24 | .000155 | | 30 | .008836 | | 36 | .132896 | | 25 | .000030 |
| | 35 | .314298 | | 25 | .000378 | | 31 | .016508 | | 37 | .197971 | | 26 | .000077 |
| 81 | 20 | .000006 | | 26 | .000878 | | 32 | .029455 | | 38 | .283595 | | 27 | .000188 |
| | 21 | .000017 | | 27 | .001931 | | 33 | .050249 | 88 | 22 | .000003 | | 28 | .000438 |
| | 22 | .000048 | | 28 | .004039 | | 34 | .082052 | | 23 | .000009 | | 29 | .000973 |
| | 23 | .000127 | | 29 | .008036 | | 35 | .128393 | | 24 | .000024 | | 30 | .002060 |
| | 24 | .000317 | | 30 | .015232 | | 36 | .192775 | | 25 | .000063 | | 31 | .004165 |
| | 25 | .000752 | | 31 | .027533 | | 37 | .277991 | | 26 | .000157 | | 32 | .008046 |

(continued)

**Table A.18** Critical Values $r_0$ and Significance Probabilities $P(r_0)$ for the Sign Test[a,b]—*cont...*

| n | $r_0$ | $P(r_0)$ | n | $r_0$ | $P(r_0)$ | n | $r_0$ | $P(r_0)$ | n | $r_0$ | $P(r_0)$ | n | $r_0$ | $P(r_0)$ | n | $r_0$ | $P(r_0)$ |
|---|---|---|---|---|---|---|---|---|---|---|---|---|---|---|---|---|---|
| | 33 | .014866 | | 39 | .174974 | **95** | 25 | .000004 | | 29 | .000093 | | 32 | .000562 | | | |
| | 34 | .026301 | | 40 | .251309 | | 26 | .000012 | | 30 | .000219 | | 33 | .001185 | | | |
| | 35 | .044596 | | 41 | .348123 | | 27 | .000031 | | 31 | .000490 | | 34 | .002395 | | | |
| | 36 | .072546 | | | | | 28 | .000078 | | 32 | .001050 | | 35 | .004640 | | | |
| | 37 | .113339 | **93** | 24 | .000003 | | 29 | .000186 | | 33 | .002151 | | 36 | .008633 | | | |
| | 38 | .170235 | | 25 | .000009 | | 30 | .000425 | | 34 | .004226 | | 37 | .015432 | | | |
| | 39 | .246096 | | 26 | .000025 | | 31 | .000925 | | 35 | .007959 | | 38 | .026524 | | | |
| **91** | 23 | .000003 | | 27 | .000065 | | 32 | .001924 | | 36 | .014389 | | 39 | .043873 | | | |
| | 24 | .000007 | | 28 | .000158 | | 33 | .003833 | | 37 | .024990 | | 40 | .069897 | | | |
| | 25 | .000020 | | 29 | .000366 | | 34 | .007313 | | 38 | .041727 | | 41 | .107346 | | | |
| | 26 | .000053 | | 30 | .000810 | | 35 | .013378 | | 39 | .067048 | | 42 | .159061 | | | |
| | 27 | .000132 | | 31 | .001713 | | 36 | .023487 | | 40 | .103764 | | 43 | .227614 | | | |
| | 28 | .000313 | | 32 | .003462 | | 37 | .039606 | | 41 | .154807 | | 44 | .314863 | | | |
| | 29 | .000705 | | 33 | .006695 | | 38 | .064209 | | 42 | .222866 | | | | **100** | 27 | .000005 |
| | 30 | .001516 | | 34 | .012400 | | 39 | .100168 | | 43 | .309916 | | | | | 28 | .000013 |
| | 31 | .003113 | | 35 | .022018 | | 40 | .150510 | **98** | 26 | .000004 | | | | | 29 | .000032 |
| | 32 | .006105 | | 36 | .037513 | | 41 | .218042 | | 27 | .000010 | | | | | 30 | .000078 |
| | 33 | .011456 | | 37 | .061383 | | 42 | .304869 | | 28 | .000026 | | | | | 31 | .000183 |
| | 34 | .020583 | | 38 | .096560 | **96** | 25 | .000003 | | 29 | .000066 | | | | | 32 | .000409 |
| | 35 | .035448 | | 39 | .146169 | | 26 | .000008 | | 30 | .000156 | | | | | 33 | .000874 |
| | 36 | .058571 | | 40 | .213142 | | 27 | .000021 | | 31 | .000355 | | | | | 34 | .001790 |
| | 37 | .092942 | | 41 | .299176 | | 28 | .000055 | | 32 | .000770 | | | | | 35 | .003517 |
| | 38 | .141787 | **94** | 24 | .000002 | | 29 | .000132 | | 33 | .001601 | | | | | 36 | .006637 |
| | 39 | .208165 | | 25 | .000006 | | 30 | .000306 | | 34 | .003189 | | | | | 37 | .012033 |
| | 40 | .294456 | | 26 | .000017 | | 31 | .000675 | | 35 | .006092 | | | | | 38 | .020978 |
| **92** | 24 | .000005 | | 27 | .000045 | | 32 | .001424 | | 36 | .011174 | | | | | 39 | .035199 |
| | 25 | .000014 | | 28 | .000111 | | 33 | .002878 | | 37 | .019689 | | | | | 40 | .056885 |
| | 26 | .000037 | | 29 | .000262 | | 34 | .005573 | | 38 | .033359 | | | | | 41 | .088622 |
| | 27 | .000093 | | 30 | .000588 | | 35 | .010345 | | 39 | .054388 | | | | | 42 | .133204 |
| | 28 | .000222 | | 31 | .001261 | | 36 | .018433 | | 40 | .085406 | | | | | 43 | .193338 |
| | 29 | .000509 | | 32 | .002587 | | 37 | .031547 | | 41 | .129286 | | | | | 44 | .271240 |
| | 30 | .001111 | | 33 | .005078 | | 38 | .051908 | | 42 | .188838 | | | | | 45 | .368182 |
| | 31 | .002315 | | 34 | .009548 | | 39 | .082188 | | 43 | .266392 | | | | | | |
| | 32 | .004609 | | 35 | .017209 | | 40 | .125338 | | 44 | .363335 | | | | | | |
| | 33 | .008781 | | 36 | .029765 | | 41 | .184275 | **99** | 26 | .000002 | | | | | | |
| | 34 | .016020 | | 37 | .049448 | | 42 | .261453 | | 27 | .000007 | | | | | | |
| | 35 | .028016 | | 38 | .078972 | | 43 | .358375 | | 28 | .000018 | | | | | | |
| | 36 | .047009 | | 39 | .121365 | **97** | 26 | .000005 | | 29 | .000046 | | | | | | |
| | 37 | .075756 | | 40 | .179656 | | 27 | .000015 | | 30 | .000111 | | | | | | |
| | 38 | .117364 | | 41 | .256429 | | 28 | .000038 | | 31 | .000255 | | | | | | |

[a] Source: M. C. Gacula, Jr., M. J. Moran, and J. B. Reaume. (1971). Use of the sign test in sensory testing. Food Prod Dev., October. Reproduced with permission of the Gorman Publishing Co.
[b] n = Sample size or number of replications; $r_0$ = critical value for the less frequent sign; $P(r_0)$ = probability of obtaining $r_0$ or fewer signs out of n observations.

**Table A.19** Critical Values and Probability Levels for the Wilcoxon Signed Rank Test[a]

| N | α = .05 (One-Sided) α = .10 (Two-Sided) | | α = .025 (One-Sided) α = .05 (Two-Sided) | | α = .01 (One-Sided) α = .02 (Two-Sided) | | α = .005 (One-Sided) α = .01 (Two-Sided) | |
|---|---|---|---|---|---|---|---|---|
| 5 | 0 | .0313 | | | | | | |
|   | 1 | .0625 | | | | | | |
| 6 | 2 | .0469 | 0 | .0156 | | | | |
|   | 3 | .0781 | 1 | .0313 | | | | |
| 7 | 3 | .0391 | 2 | .0234 | 0 | .0078 | | |
|   | 4 | .0547 | 3 | .0391 | 1 | .0156 | | |
| 8 | 5 | .0391 | 3 | .0195 | 1 | .0078 | 0 | .0039 |
|   | 6 | .0547 | 4 | .0273 | 2 | .0117 | 1 | .0078 |
| 9 | 8 | .0488 | 5 | .0195 | 3 | .0098 | 1 | .0039 |
|   | 9 | .0645 | 6 | .0273 | 4 | .0137 | 2 | .0059 |
| 10 | 10 | .0420 | 8 | .0244 | 5 | .0098 | 3 | .0049 |
|   | 11 | .0527 | 9 | .0322 | 6 | .0137 | 4 | .0068 |
| 11 | 13 | .0415 | 10 | .0210 | 7 | .0093 | 5 | .0049 |
|   | 14 | .0508 | 11 | .0269 | 8 | .0122 | 6 | .0068 |
| 12 | 17 | .0461 | 13 | .0212 | 9 | .0081 | 7 | .0046 |
|   | 18 | .0549 | 14 | .0261 | 10 | .0105 | 8 | .0061 |
| 13 | 21 | .0471 | 17 | .0239 | 12 | .0085 | 9 | .0040 |
|   | 22 | .0549 | 18 | .0287 | 13 | .0107 | 10 | .0052 |
| 14 | 25 | .0453 | 21 | .0247 | 15 | .0083 | 12 | .0043 |
|   | 26 | .0520 | 22 | .0290 | 16 | .0101 | 13 | .0054 |
| 15 | 30 | .0473 | 25 | .0240 | 19 | .0090 | 15 | .0042 |
|   | 31 | .0535 | 26 | .0277 | 20 | .0108 | 16 | .0051 |
| 16 | 35 | .0467 | 29 | .0222 | 23 | .0091 | 19 | .0046 |
|   | 36 | .0523 | 30 | .0253 | 24 | .0107 | 20 | .0055 |
| 17 | 41 | .0492 | 34 | .0224 | 27 | .0087 | 23 | .0047 |
|   | 42 | .0544 | 35 | .0253 | 28 | .0101 | 24 | .0055 |
| 18 | 47 | .0494 | 40 | .0241 | 32 | .0091 | 27 | .0045 |
|   | 48 | .0542 | 41 | .0269 | 33 | .0104 | 28 | .0052 |
| 19 | 53 | .0478 | 46 | .0247 | 37 | .0090 | 32 | .0047 |
|   | 54 | .0521 | 47 | .0273 | 38 | .0102 | 33 | .0054 |
| 20 | 60 | .0487 | 52 | .0242 | 43 | .0096 | 37 | .0047 |
|   | 61 | .0527 | 53 | .0266 | 44 | .0107 | 38 | .0053 |

*(continued)*

**Table A.19** Critical Values and Probability Levels for the Wilcoxon Signed Rank Test[a]—*cont...*

| N | $\alpha = .05$ (One-Sided) $\alpha = .10$ (Two-Sided) | | $\alpha = .025$ (One-Sided) $\alpha = .05$ (Two-Sided) | | $\alpha = .01$ (One-Sided) $\alpha = .02$ (Two-Sided) | | $\alpha = .005$ (One-Sided) $\alpha = .01$ (Two-Sided) | |
|---|---|---|---|---|---|---|---|---|
| 21 | 67 | .0479 | 58 | .0230 | 49 | .0097 | 42 | .0045 |
|    | 68 | .0516 | 59 | .0251 | 50 | .0108 | 43 | .0051 |
| 22 | 75 | .0492 | 65 | .0231 | 55 | .0095 | 48 | .0046 |
|    | 76 | .0527 | 66 | .0250 | 56 | .0104 | 49 | .0052 |
| 23 | 83 | .0490 | 73 | .0242 | 62 | .0098 | 54 | .0046 |
|    | 84 | .0523 | 74 | .0261 | 63 | .0107 | 55 | .0051 |
| 24 | 91 | .0475 | 81 | .0245 | 69 | .0097 | 61 | .0048 |
|    | 92 | .0505 | 82 | .0263 | 70 | .0106 | 62 | .0053 |
| 25 | 100 | .0479 | 89 | .0241 | 76 | .0094 | 68 | .0048 |
|    | 101 | .0507 | 90 | .0258 | 77 | .0101 | 69 | .0053 |
| 26 | 110 | .0497 | 98 | .0247 | 84 | .0095 | 75 | .0047 |
|    | 111 | .0524 | 99 | .0263 | 85 | .0102 | 76 | .0051 |
| 27 | 119 | .0477 | 107 | .0246 | 92 | .0093 | 83 | .0048 |
|    | 120 | .0502 | 108 | .0260 | 93 | .0100 | 84 | .0052 |
| 28 | 130 | .0496 | 116 | .0239 | 101 | .0096 | 91 | .0048 |
|    | 131 | .0521 | 117 | .0252 | 102 | .0102 | 92 | .0051 |
| 29 | 140 | .0482 | 126 | .0240 | 110 | .0095 | 100 | .0049 |
|    | 141 | .0504 | 127 | .0253 | 111 | .0101 | 101 | .0053 |
| 30 | 151 | .0481 | 137 | .0249 | 120 | .0098 | 109 | .0050 |
|    | 152 | .0502 | 138 | .0261 | 121 | .0104 | 110 | .0053 |
| 31 | 163 | .0491 | 147 | .0239 | 130 | .0099 | 118 | .0049 |
|    | 164 | .0512 | 148 | .0251 | 131 | .0105 | 119 | .0052 |
| 32 | 175 | .0492 | 159 | .0249 | 140 | .0097 | 128 | .0050 |
|    | 176 | .0512 | 160 | .0260 | 141 | .0103 | 129 | .0053 |
| 33 | 187 | .0485 | 170 | .0242 | 151 | .0099 | 138 | .0049 |
|    | 188 | .0503 | 171 | .0253 | 152 | .0104 | 139 | .0052 |
| 34 | 200 | .0488 | 182 | .0242 | 162 | .0098 | 148 | .0048 |
|    | 201 | .0506 | 183 | .0252 | 163 | .0103 | 149 | .0051 |
| 35 | 213 | .0484 | 195 | .0247 | 173 | .0096 | 159 | .0048 |
|    | 214 | .0501 | 196 | .0257 | 174 | .0100 | 160 | .0051 |
| 36 | 227 | .0489 | 208 | .0248 | 185 | .0096 | 171 | .0050 |
|    | 228 | .0505 | 209 | .0258 | 186 | .0100 | 172 | .0052 |

*(continued)*

**Table A.19** Critical Values and Probability Levels for the Wilcoxon Signed Rank Test[a]—cont...

| N | $\alpha = .05$ (One-Sided) $\alpha = .10$ (Two-Sided) | | $\alpha = .025$ (One-Sided) $\alpha = .05$ (Two-Sided) | | $\alpha = .01$ (One-Sided) $\alpha = .02$ (Two-Sided) | | $\alpha = .005$ (One-Sided) $\alpha = .01$ (Two-Sided) | |
|---|---|---|---|---|---|---|---|---|
| 37 | 241 | .0487 | 221 | .0245 | 198 | .0099 | 182 | .0048 |
|    | 242 | .0503 | 222 | .0254 | 199 | .0103 | 183 | .0050 |
| 38 | 256 | .0493 | 235 | .0247 | 211 | .0099 | 194 | .0048 |
|    | 257 | .0509 | 236 | .0256 | 212 | .0104 | 195 | .0050 |
| 39 | 271 | .0493 | 249 | .0246 | 224 | .0099 | 207 | .0049 |
|    | 272 | .0507 | 250 | .0254 | 225 | .0103 | 208 | .0051 |
| 40 | 286 | .0486 | 264 | .0249 | 238 | .0100 | 220 | .0049 |
|    | 287 | .0500 | 265 | .0257 | 239 | .0104 | 221 | .0051 |
| 41 | 302 | .0488 | 279 | .0248 | 252 | .0100 | 233 | .0048 |
|    | 303 | .0501 | 280 | .0256 | 253 | .0103 | 234 | .0050 |
| 42 | 319 | .0496 | 294 | .0245 | 266 | .0098 | 247 | .0049 |
|    | 320 | .0509 | 295 | .0252 | 267 | .0102 | 248 | .0051 |
| 43 | 336 | .0498 | 310 | .0245 | 281 | .0098 | 261 | .0048 |
|    | 337 | .0511 | 311 | .0252 | 282 | .0102 | 262 | .0050 |
| 44 | 353 | .0495 | 327 | .0250 | 296 | .0097 | 276 | .0049 |
|    | 354 | .0507 | 328 | .0257 | 297 | .0101 | 277 | .0051 |
| 45 | 371 | .0498 | 343 | .0244 | 312 | .0098 | 291 | .0049 |
|    | 372 | .0510 | 344 | .0251 | 313 | .0101 | 292 | .0051 |
| 46 | 389 | .0497 | 361 | .0249 | 328 | .0098 | 307 | .0050 |
|    | 390 | .0508 | 362 | .0256 | 329 | .0101 | 308 | .0052 |
| 47 | 407 | .0490 | 378 | .0245 | 345 | .0099 | 322 | .0048 |
|    | 408 | .0501 | 379 | .0251 | 346 | .0102 | 323 | .0050 |
| 48 | 426 | .0490 | 396 | .0244 | 362 | .0099 | 339 | .0050 |
|    | 427 | .0500 | 397 | .0251 | 363 | .0102 | 340 | .0051 |
| 49 | 446 | .0495 | 415 | .0247 | 379 | .0098 | 355 | .0049 |
|    | 447 | .0505 | 416 | .0253 | 380 | .0100 | 356 | .0050 |
| 50 | 466 | .0495 | 434 | .0247 | 397 | .0098 | 373 | .0050 |
|    | 467 | .0506 | 435 | .0253 | 398 | .0101 | 374 | .0051 |

[a]Source: F. Wilcoxon, S. K. Katti, and R. A. Wilcox. (1970). Critical values and probability levels for the Wilcoxon rank sum test and the Wilcoxon signed rank test. In Selected Tables in Mathematical Statistics (H. L. Harter and D. B. Owen, eds.) Markham, Chicago, IL: Vol. 1, pp. 171–259. From Some Rapid Approximate Statistical Procedures. Copyright 1949, 1964, Lederle Laboratories Division of American Cyanamid Company. All rights reserved and reprinted with permission.

**Table A.20** Distribution of *D* Used in the Rank Correlation Coefficient[a]

| n | D | α | n | D | α |
|---|---|---|---|---|---|
| 4 | 18 | .167 | | 124 | .121 |
| | 20 | .042 | | 128 | .096 |
| 5 | 32 | .175 | | 132 | .074 |
| | 34 | .117 | | 136 | .055 |
| | 36 | .067 | | 140 | .040 |
| | 38 | .042 | | 144 | .028 |
| | 40 | .008 | | 148 | .018 |
| | | | | 152 | .011 |
| 6 | 54 | .149 | | 156 | .006 |
| | 56 | .121 | | 160 | .003 |
| | 58 | .088 | | 164 | .001 |
| | 60 | .068 | | | |
| | 62 | .051 | 9 | 172 | .125 |
| | 64 | .029 | | 176 | .106 |
| | 66 | .017 | | 180 | .088 |
| | 68 | .008 | | 184 | .073 |
| | 70 | .001 | | 188 | .059 |
| | | | | 192 | .047 |
| 7 | 80 | .177 | | 196 | .037 |
| | 82 | .151 | | 200 | .028 |
| | 84 | .133 | | 208 | .015 |
| | 86 | .118 | | 216 | .007 |
| | 88 | .100 | | 224 | .003 |
| | 90 | .083 | | 232 | .001 |
| | 92 | .069 | | | |
| | 94 | .055 | 10 | 228 | .140 |
| | 96 | .044 | | 236 | .109 |
| | 98 | .033 | | 244 | .083 |
| | 100 | .024 | | 252 | .061 |
| | 102 | .017 | | 260 | .043 |
| | 104 | .012 | | 268 | .029 |
| | 106 | .006 | | 276 | .019 |
| | 108 | .003 | | 284 | .011 |
| | 110 | .001 | | 292 | .006 |
| | | | | 300 | .003 |
| 8 | 116 | .181 | | 308 | .001 |
| | 120 | .149 | | | |

[a] *Source: E. G. Olds. (1938). Distributions of sums of squares of rank differences for small numbers of individuals. Annals Math Stat, **9**, 133–148. Reproduced with permission of the Institute of Mathematical Statistics.*

**Table A.21** Probability of m or More Correct Judgments in N Comparisons for the Triangle Test[a,b]

| N \ m | 0 | 1 | 2 | 3 | 4 | 5 | 6 | 7 | 8 | 9 | 10 | 11 | 12 | 13 | 14 | 15 | 16 | 17 | 18 | 19 | 20 | 21 | 22 | 23 | 24 | 25 | 26 | 27 | 28 |
|---|---|---|---|---|---|---|---|---|---|---|---|---|---|---|---|---|---|---|---|---|---|---|---|---|---|---|---|---|---|
| 5 | | 868 | 539 | 210 | 045 | 004 | | | | | | | | | | | | | | | | | | | | | | | |
| 6 | | 912 | 649 | 320 | 100 | 018 | 001 | | | | | | | | | | | | | | | | | | | | | | |
| 7 | | 941 | 737 | 429 | 173 | 045 | 007 | | | | | | | | | | | | | | | | | | | | | | |
| 8 | | 961 | 805 | 532 | 259 | 088 | 020 | 003 | | | | | | | | | | | | | | | | | | | | | |
| 9 | | 974 | 857 | 623 | 350 | 145 | 042 | 008 | 001 | | | | | | | | | | | | | | | | | | | | |
| 10 | | 983 | 896 | 701 | 441 | 213 | 077 | 020 | 003 | | | | | | | | | | | | | | | | | | | | |
| 11 | | 988 | 925 | 766 | 527 | 289 | 122 | 039 | 009 | 001 | | | | | | | | | | | | | | | | | | | |
| 12 | | 992 | 946 | 819 | 607 | 368 | 178 | 066 | 019 | 004 | 001 | | | | | | | | | | | | | | | | | | |
| 13 | | 995 | 961 | 861 | 678 | 448 | 241 | 104 | 035 | 009 | 002 | | | | | | | | | | | | | | | | | | |
| 14 | | 997 | 973 | 895 | 739 | 524 | 310 | 149 | 058 | 017 | 004 | 001 | | | | | | | | | | | | | | | | | |
| 15 | | 998 | 981 | 921 | 791 | 596 | 382 | 203 | 088 | 031 | 008 | 002 | | | | | | | | | | | | | | | | | |
| 16 | | 998 | 986 | 941 | 834 | 661 | 453 | 263 | 126 | 050 | 016 | 004 | 001 | | | | | | | | | | | | | | | | |
| 17 | | 999 | 990 | 956 | 870 | 719 | 522 | 326 | 172 | 075 | 027 | 008 | 002 | | | | | | | | | | | | | | | | |
| 18 | | 999 | 993 | 967 | 898 | 769 | 588 | 391 | 223 | 108 | 043 | 014 | 004 | 001 | | | | | | | | | | | | | | | |
| 19 | | 999 | 995 | 976 | 921 | 812 | 648 | 457 | 279 | 146 | 065 | 024 | 007 | 002 | | | | | | | | | | | | | | | |
| 20 | | | 997 | 982 | 940 | 848 | 703 | 521 | 339 | 191 | 092 | 038 | 013 | 004 | 001 | | | | | | | | | | | | | | |
| 21 | | | 998 | 987 | 954 | 879 | 751 | 581 | 399 | 240 | 125 | 056 | 021 | 007 | 002 | | | | | | | | | | | | | | |
| 22 | | | 998 | 991 | 965 | 904 | 794 | 638 | 460 | 293 | 163 | 079 | 033 | 012 | 003 | 001 | | | | | | | | | | | | | |
| 23 | | | 999 | 993 | 974 | 924 | 831 | 690 | 519 | 349 | 206 | 107 | 048 | 019 | 006 | 002 | | | | | | | | | | | | | |
| 24 | | | 999 | 995 | 980 | 941 | 862 | 737 | 576 | 406 | 254 | 140 | 068 | 028 | 010 | 003 | 001 | | | | | | | | | | | | |
| 25 | | | 999 | 996 | 985 | 954 | 888 | 778 | 630 | 462 | 304 | 178 | 092 | 042 | 016 | 006 | 002 | | | | | | | | | | | | |
| 26 | | | | 997 | 989 | 964 | 910 | 815 | 679 | 518 | 357 | 220 | 121 | 058 | 025 | 009 | 003 | 001 | | | | | | | | | | | |
| 27 | | | | 998 | 992 | 972 | 928 | 847 | 725 | 572 | 411 | 266 | 154 | 079 | 036 | 014 | 005 | 002 | | | | | | | | | | | |
| 28 | | | | 998 | 994 | 979 | 943 | 874 | 765 | 623 | 464 | 314 | 191 | 104 | 050 | 022 | 008 | 003 | 001 | | | | | | | | | | |
| 29 | | | | 999 | 996 | 984 | 955 | 897 | 801 | 670 | 517 | 364 | 232 | 133 | 068 | 031 | 013 | 005 | 001 | | | | | | | | | | |
| 30 | | | | 999 | 997 | 988 | 965 | 916 | 833 | 714 | 568 | 415 | 276 | 166 | 090 | 043 | 019 | 007 | 002 | 001 | | | | | | | | | |

| n | | | | | | | | | | | | | | | | | | | | | | |
|---|---|---|---|---|---|---|---|---|---|---|---|---|---|---|---|---|---|---|---|---|---|---|
| 31 | 998 | 991 | 972 | 932 | 861 | 754 | 617 | 466 | 322 | 203 | 115 | 059 | 027 | 011 | 004 | 001 | | | | | | |
| 32 | 998 | 993 | 978 | 946 | 885 | 789 | 662 | 516 | 370 | 243 | 144 | 078 | 038 | 016 | 006 | 002 | 001 | | | | | |
| 33 | 999 | 995 | 983 | 957 | 905 | 821 | 705 | 565 | 419 | 285 | 177 | 100 | 051 | 023 | 010 | 004 | 001 | | | | | |
| 34 | 999 | 996 | 987 | 965 | 922 | 849 | 744 | 612 | 468 | 330 | 213 | 126 | 067 | 033 | 014 | 006 | 002 | 001 | | | | |
| 35 | 999 | 997 | 990 | 973 | 937 | 873 | 779 | 656 | 516 | 376 | 252 | 155 | 087 | 044 | 020 | 009 | 003 | 001 | | | | |
| 36 | 998 | 992 | 978 | 949 | 895 | 810 | 697 | 562 | 422 | 293 | 187 | 109 | 058 | 028 | 012 | 005 | 002 | 001 | | | | |
| 37 | 998 | 994 | 984 | 959 | 913 | 838 | 735 | 607 | 469 | 336 | 223 | 135 | 075 | 038 | 018 | 007 | 003 | 001 | | | | |
| 38 | 999 | 996 | 987 | 967 | 928 | 863 | 769 | 650 | 515 | 381 | 261 | 164 | 095 | 051 | 025 | 011 | 004 | 002 | 001 | | | |
| 39 | 999 | 997 | 990 | 973 | 941 | 885 | 800 | 689 | 560 | 425 | 301 | 196 | 118 | 066 | 033 | 016 | 007 | 003 | 001 | | | |
| 40 | 999 | 997 | 992 | 979 | 952 | 903 | 829 | 726 | 603 | 470 | 342 | 231 | 144 | 083 | 044 | 021 | 010 | 004 | 001 | | | |
| 41 | 998 | 994 | 983 | 961 | 920 | 854 | 761 | 644 | 515 | 385 | 268 | 173 | 104 | 057 | 029 | 014 | 006 | 002 | 001 | | | |
| 42 | 999 | 995 | 987 | 968 | 933 | 876 | 791 | 683 | 558 | 428 | 307 | 205 | 127 | 073 | 038 | 019 | 008 | 003 | 001 | | | |
| 43 | 999 | 996 | 990 | 974 | 945 | 895 | 820 | 719 | 600 | 471 | 347 | 239 | 153 | 091 | 050 | 025 | 012 | 005 | 002 | 001 | | |
| 44 | 999 | 998 | 994 | 980 | 955 | 912 | 845 | 753 | 639 | 514 | 389 | 275 | 182 | 111 | 063 | 033 | 016 | 007 | 003 | 001 | | |
| 45 | 999 | 998 | 994 | 984 | 963 | 926 | 867 | 783 | 677 | 556 | 430 | 313 | 213 | 135 | 079 | 043 | 022 | 010 | 004 | 002 | 001 | |
| 46 | 998 | 995 | 987 | 970 | 938 | 887 | 811 | 713 | 596 | 472 | 352 | 246 | 161 | 098 | 055 | 029 | 014 | 006 | 003 | 001 | | |
| 47 | 999 | 996 | 990 | 976 | 949 | 904 | 836 | 745 | 635 | 514 | 392 | 282 | 189 | 119 | 070 | 038 | 019 | 009 | 004 | 002 | 001 | |
| 48 | 999 | 997 | 992 | 980 | 958 | 919 | 859 | 776 | 672 | 554 | 433 | 318 | 220 | 142 | 086 | 048 | 025 | 012 | 006 | 002 | 001 | |
| 49 | 999 | 998 | 994 | 984 | 965 | 932 | 879 | 803 | 706 | 593 | 473 | 356 | 253 | 168 | 105 | 061 | 033 | 017 | 008 | 003 | 001 | |
| 50 | 999 | 998 | 995 | 987 | 972 | 943 | 896 | 829 | 739 | 631 | 513 | 395 | 287 | 196 | 126 | 076 | 042 | 022 | 011 | 005 | 002 | 001 |

[a]Source: E. B. Roessler, R. M. Pangborn, J. L. Sidel, and H. Stone. (1978). Expanded statistical tables for estimating significance in paired-preference, paired difference, duo–trio, and triangle tests. J Food Sci. **43**, 940–943. Reproduced with permission of the editor.
[b]Note that the initial decimal point has been omitted.

**Table A.22** Critical Values for the Gridgeman S Statistic[a]

| Score = ≤ S or ≥ (3N − S), Where S Is | Number of Panelists (Judgments) N | | | | | | | | | | | | | | |
|---|---|---|---|---|---|---|---|---|---|---|---|---|---|---|---|
| | 2 | 4 | 6 | 8 | 10 | 12 | 14 | 16 | 18 | 20 | 22 | 24 | 26 | 28 | 30 |
| 0 | 0.0556 | 0.0015 | | | | | | | | | | | | | |
| 1 | 0.2767 | 0.0139 | 0.0006 | | | | | | | | | | | | |
| 2 | | 0.0633 | 0.0036 | 0.0002 | | | | | | | | | | | |
| 3 | | 0.1929 | 0.0159 | 0.0010 | 0.0001 | | | | | | | | | | |
| 4 | | | 0.0519 | 0.0042 | 0.0003 | | | | | | | | | | |
| 5 | | | 0.1347 | 0.0142 | 0.0011 | 0.0001 | | | | | | | | | |
| 6 | | | 0.2890 | 0.0398 | 0.0039 | 0.0003 | | | | | | | | | |
| 7 | | | | 0.0951 | 0.0116 | 0.0011 | 0.0001 | | | | | | | | |
| 8 | | | | 0.1967 | 0.0299 | 0.0034 | 0.0003 | | | | | | | | |
| 9 | | | | | 0.0676 | 0.0092 | 0.0010 | 0.0001 | | | | | | | |
| 10 | | | | | 0.1367 | 0.0222 | 0.0028 | 0.0003 | | | | | | | |
| 11 | | | | | 0.2487 | 0.0485 | 0.0071 | 0.0008 | 0.0001 | | | | | | |
| 12 | | | | | | 0.0963 | 0.0164 | 0.0022 | 0.0002 | | | | | | |
| 13 | | | | | | 0.1749 | 0.0349 | 0.0054 | 0.0007 | 0.0001 | | | | | |
| 14 | | | | | | 0.2925 | 0.0685 | 0.0121 | 0.0017 | 0.0002 | | | | | |
| 15 | | | | | | | 0.1242 | 0.0253 | 0.0041 | 0.0005 | 0.0001 | | | | |
| 16 | | | | | | | 0.2094 | 0.0490 | 0.0090 | 0.0013 | 0.0002 | | | | |
| 17 | | | | | | | | 0.0888 | 0.0184 | 0.0031 | 0.0004 | | | | |
| 18 | | | | | | | | 0.1508 | 0.0353 | 0.0066 | 0.0010 | 0.0001 | | | |

| | | | | | | | |
|---|---|---|---|---|---|---|---|
| 19 | 0.2406 | 0.0639 | 0.0134 | 0.0023 | 0.0003 | | |
| 20 | | 0.1090 | 0.0256 | 0.0049 | 0.0008 | 0.0001 | |
| 21 | | 0.1758 | 0.0462 | 0.0098 | 0.0017 | 0.0003 | |
| 22 | | 0.2686 | 0.0791 | 0.0186 | 0.0036 | 0.0006 | 0.0001 |
| 23 | | | 0.1287 | 0.0335 | 0.0072 | 0.0013 | 0.0002 |
| 24 | | | 0.1992 | 0.0576 | 0.0135 | 0.0027 | 0.0005 | 0.0001 |
| 25 | | | 0.2940 | 0.0943 | 0.0244 | 0.0053 | 0.0010 | 0.0002 |
| 26 | | | | 0.1476 | 0.0420 | 0.0099 | 0.0020 | 0.0003 |
| 27 | | | | 0.2210 | 0.0693 | 0.0178 | 0.0039 | 0.0007 |
| 28 | | | | | 0.1094 | 0.0308 | 0.0072 | 0.0015 |
| 29 | | | | | 0.1658 | 0.0510 | 0.0130 | 0.0028 |
| 30 | | | | | 0.2413 | 0.0811 | 0.0226 | 0.0053 |
| 31 | | | | | | 0.1242 | 0.0375 | 0.0096 |
| 32 | | | | | | 0.1831 | 0.0602 | 0.0166 |
| 33 | | | | | | 0.2603 | 0.0929 | 0.0277 |
| 34 | | | | | | | 0.1386 | 0.0446 |
| 35 | | | | | | | 0.1996 | 0.0695 |
| 36 | | | | | | | 0.2781 | 0.1047 |
| 37 | | | | | | | | 0.1526 |
| 38 | | | | | | | | 0.2154 |
| 39 | | | | | | | | 0.2947 |

[a]Source: N. T. Gridgeman. (1970a). A re-examination of the two-stage triangle test for the perception of sensory differences. *J Food Sci*, **35**, 87–91.

**Table A.23** Probability of $m$ or More Correct Judgments in $N$ Comparisons for the Duo–Trio and Pair Tests[a,b]

One-Sided

| N \ m | 0 | 1 | 2 | 3 | 4 | 5 | 6 | 7 | 8 | 9 | 10 | 11 | 12 | 13 | 14 | 15 | 16 | 17 | 18 | 19 | 20 | 21 | 22 | 23 | 24 | 25 | 26 | 27 | 28 | 29 | 30 | 31 | 32 | 33 | 34 | 35 | 36 |
|---|---|---|---|---|---|---|---|---|---|---|---|---|---|---|---|---|---|---|---|---|---|---|---|---|---|---|---|---|---|---|---|---|---|---|---|---|---|
| 5 | | 969 | 812 | 500 | 188 | 031 | | | | | | | | | | | | | | | | | | | | | | | | | | | | | | | |
| 6 | | 984 | 891 | 656 | 344 | 109 | 016 | | | | | | | | | | | | | | | | | | | | | | | | | | | | | | |
| 7 | | 992 | 938 | 773 | 500 | 227 | 062 | 008 | | | | | | | | | | | | | | | | | | | | | | | | | | | | | |
| 8 | | 996 | 965 | 855 | 637 | 363 | 145 | 035 | 004 | | | | | | | | | | | | | | | | | | | | | | | | | | | | |
| 9 | | 998 | 980 | 910 | 746 | 500 | 254 | 090 | 020 | 002 | | | | | | | | | | | | | | | | | | | | | | | | | | | |
| 10 | | 999 | 989 | 945 | 828 | 623 | 377 | 172 | 055 | 011 | 001 | | | | | | | | | | | | | | | | | | | | | | | | | | |
| 11 | | | 994 | 967 | 887 | 726 | 500 | 274 | 113 | 033 | 006 | | | | | | | | | | | | | | | | | | | | | | | | | | |
| 12 | | | 997 | 981 | 927 | 806 | 613 | 387 | 194 | 073 | 019 | 003 | | | | | | | | | | | | | | | | | | | | | | | | | |
| 13 | | | 998 | 989 | 954 | 867 | 709 | 500 | 291 | 133 | 046 | 011 | 002 | | | | | | | | | | | | | | | | | | | | | | | | |
| 14 | | | 999 | 994 | 971 | 910 | 788 | 605 | 395 | 212 | 090 | 029 | 006 | 001 | | | | | | | | | | | | | | | | | | | | | | | |
| 15 | | | | 996 | 982 | 941 | 849 | 696 | 500 | 304 | 151 | 059 | 018 | 004 | | | | | | | | | | | | | | | | | | | | | | | |
| 16 | | | | 998 | 989 | 962 | 895 | 773 | 598 | 402 | 227 | 105 | 038 | 011 | 002 | | | | | | | | | | | | | | | | | | | | | | |
| 17 | | | | 999 | 994 | 975 | 928 | 834 | 685 | 500 | 315 | 166 | 072 | 025 | 006 | 001 | | | | | | | | | | | | | | | | | | | | | |
| 18 | | | | 999 | 996 | 985 | 952 | 881 | 760 | 593 | 407 | 240 | 119 | 048 | 015 | 004 | 001 | | | | | | | | | | | | | | | | | | | | |
| 19 | | | | | 998 | 990 | 968 | 916 | 820 | 676 | 500 | 324 | 180 | 084 | 032 | 010 | 002 | | | | | | | | | | | | | | | | | | | | |
| 20 | | | | | 999 | 994 | 979 | 942 | 868 | 748 | 588 | 412 | 252 | 132 | 058 | 021 | 006 | 001 | | | | | | | | | | | | | | | | | | | |
| 21 | | | | | 999 | 996 | 987 | 961 | 905 | 808 | 668 | 500 | 332 | 192 | 095 | 039 | 013 | 004 | 001 | | | | | | | | | | | | | | | | | | |
| 22 | | | | | | 998 | 992 | 974 | 933 | 857 | 738 | 584 | 416 | 262 | 143 | 067 | 026 | 008 | 002 | | | | | | | | | | | | | | | | | | |
| 23 | | | | | | 999 | 995 | 983 | 953 | 895 | 798 | 661 | 500 | 339 | 202 | 105 | 047 | 017 | 005 | 001 | | | | | | | | | | | | | | | | | |
| 24 | | | | | | 999 | 997 | 989 | 968 | 924 | 846 | 729 | 581 | 419 | 271 | 154 | 076 | 032 | 011 | 003 | 001 | | | | | | | | | | | | | | | | |
| 25 | | | | | | | 998 | 993 | 978 | 946 | 885 | 788 | 655 | 500 | 345 | 212 | 115 | 054 | 022 | 007 | 002 | 001 | | | | | | | | | | | | | | | |
| 26 | | | | | | | 999 | 995 | 986 | 962 | 916 | 837 | 721 | 577 | 423 | 279 | 163 | 084 | 038 | 014 | 005 | 001 | | | | | | | | | | | | | | | |
| 27 | | | | | | | 999 | 997 | 990 | 974 | 939 | 876 | 779 | 649 | 500 | 351 | 221 | 124 | 061 | 026 | 010 | 003 | 001 | | | | | | | | | | | | | | |
| 28 | | | | | | | | 998 | 994 | 982 | 956 | 908 | 828 | 714 | 575 | 425 | 286 | 172 | 092 | 044 | 018 | 006 | 002 | | | | | | | | | | | | | | |

| | | | | | | | | | | | | | | | | | | | | | | | |
|---|---|---|---|---|---|---|---|---|---|---|---|---|---|---|---|---|---|---|---|---|---|---|---|
| 29 | 999 | 996 | 988 | 969 | 932 | 868 | 771 | 644 | 500 | 356 | 229 | 132 | 068 | 031 | 012 | 004 | 001 | | | | | | |
| 30 | 999 | 997 | 992 | 979 | 951 | 900 | 819 | 708 | 572 | 428 | 292 | 181 | 100 | 049 | 021 | 008 | 003 | 001 | | | | | |
| 31 | 998 | 995 | 985 | 965 | 925 | 859 | 763 | 640 | 500 | 360 | 237 | 141 | 075 | 035 | 015 | 005 | 002 | | | | | | |
| 32 | 999 | 997 | 990 | 975 | 945 | 892 | 811 | 702 | 570 | 430 | 298 | 189 | 108 | 055 | 025 | 010 | 004 | 001 | | | | | |
| 33 | 999 | 998 | 993 | 982 | 960 | 919 | 852 | 757 | 636 | 500 | 364 | 243 | 148 | 081 | 040 | 018 | 007 | 002 | | | | | |
| 34 | 999 | 995 | 988 | 971 | 939 | 885 | 804 | 696 | 568 | 432 | 304 | 196 | 115 | 061 | 029 | 012 | 005 | 002 | | | | | |
| 35 | 999 | 997 | 992 | 980 | 955 | 912 | 845 | 750 | 632 | 500 | 368 | 250 | 155 | 088 | 045 | 020 | 008 | 003 | 001 | | | | |
| 36 | 999 | 998 | 994 | 986 | 967 | 934 | 879 | 797 | 691 | 566 | 434 | 309 | 203 | 121 | 066 | 033 | 014 | 006 | 002 | 001 | | | |
| 37 | 999 | 996 | 990 | 976 | 951 | 906 | 838 | 744 | 629 | 500 | 371 | 256 | 162 | 094 | 049 | 024 | 010 | 004 | 001 | | | | |
| 38 | 999 | 997 | 993 | 983 | 964 | 928 | 872 | 791 | 686 | 564 | 436 | 314 | 209 | 128 | 072 | 036 | 017 | 007 | 003 | 001 | | | |
| 39 | 999 | 998 | 995 | 988 | 973 | 946 | 900 | 832 | 739 | 625 | 500 | 375 | 261 | 168 | 100 | 054 | 027 | 012 | 005 | 002 | 001 | | |
| 40 | 999 | 997 | 992 | 981 | 960 | 923 | 866 | 785 | 682 | 563 | 437 | 318 | 215 | 134 | 077 | 040 | 019 | 008 | 003 | 001 | | | |
| 41 | 999 | 998 | 994 | 986 | 970 | 941 | 894 | 826 | 734 | 622 | 500 | 378 | 266 | 174 | 106 | 059 | 030 | 014 | 006 | 002 | 001 | | |
| 42 | 999 | 996 | 990 | 978 | 956 | 918 | 860 | 780 | 678 | 561 | 439 | 322 | 220 | 140 | 082 | 044 | 022 | 010 | 004 | 001 | | | |
| 43 | 999 | 997 | 993 | 984 | 967 | 937 | 889 | 820 | 729 | 620 | 500 | 380 | 271 | 180 | 111 | 063 | 033 | 016 | 007 | 003 | 001 | | |
| 44 | 999 | 998 | 995 | 989 | 976 | 952 | 913 | 854 | 774 | 674 | 560 | 440 | 326 | 226 | 146 | 087 | 048 | 024 | 011 | 005 | 002 | 001 | |
| 45 | 999 | 997 | 992 | 982 | 964 | 932 | 884 | 814 | 724 | 617 | 500 | 383 | 276 | 186 | 116 | 068 | 036 | 018 | 008 | 003 | 001 | | |
| 46 | 999 | 998 | 994 | 987 | 973 | 948 | 908 | 849 | 769 | 671 | 558 | 442 | 329 | 231 | 151 | 092 | 052 | 027 | 013 | 006 | 002 | 001 | |
| 47 | 999 | 998 | 996 | 991 | 980 | 961 | 928 | 879 | 809 | 720 | 615 | 500 | 385 | 280 | 191 | 121 | 072 | 039 | 020 | 009 | 004 | 002 | 001 |
| 48 | 999 | 997 | 993 | 985 | 970 | 944 | 903 | 844 | 765 | 667 | 557 | 443 | 333 | 235 | 156 | 097 | 056 | 030 | 015 | 007 | 003 | 001 | |
| 49 | 999 | 998 | 995 | 989 | 978 | 957 | 924 | 874 | 804 | 716 | 612 | 500 | 388 | 284 | 196 | 126 | 076 | 043 | 022 | 012 | 005 | 002 | 001 |
| 50 | 999 | 997 | 992 | 984 | 968 | 941 | 899 | 839 | 760 | 664 | 556 | 444 | 336 | 240 | 161 | 101 | 059 | 032 | 016 | 008 | 003 | 001 | |

*(continued)*

## Table A.23 Probability of m or More Correct Judgments in N Comparisons for the Duo–Trio and Pair Tests[a,b]—cont...

Two-Sided

| N \ m | 3 | 4 | 5 | 6 | 7 | 8 | 9 | 10 | 11 | 12 | 13 | 14 | 15 | 16 | 17 | 18 | 19 | 20 | 21 | 22 | 23 | 24 | 25 | 26 | 27 | 28 | 29 | 30 | 31 | 32 | 33 | 34 | 35 | 36 | 37 |
|---|---|---|---|---|---|---|---|---|---|---|---|---|---|---|---|---|---|---|---|---|---|---|---|---|---|---|---|---|---|---|---|---|---|---|---|
| 5 | 625 | 312 | 062 | | | | | | | | | | | | | | | | | | | | | | | | | | | | | | | | |
| 6 | | 688 | 219 | 031 | | | | | | | | | | | | | | | | | | | | | | | | | | | | | | | |
| 7 | | | 453 | 125 | 016 | | | | | | | | | | | | | | | | | | | | | | | | | | | | | | |
| 8 | | | 727 | 289 | 070 | 008 | | | | | | | | | | | | | | | | | | | | | | | | | | | | | |
| 9 | | | | 508 | 180 | 039 | 004 | | | | | | | | | | | | | | | | | | | | | | | | | | | | |
| 10 | | | | 754 | 344 | 109 | 021 | 002 | | | | | | | | | | | | | | | | | | | | | | | | | | | |
| 11 | | | | | 549 | 227 | 065 | 011 | 001 | | | | | | | | | | | | | | | | | | | | | | | | | | |
| 12 | | | | | 774 | 388 | 146 | 039 | 006 | | | | | | | | | | | | | | | | | | | | | | | | | | |
| 13 | | | | | | 581 | 267 | 092 | 022 | 003 | | | | | | | | | | | | | | | | | | | | | | | | | |
| 14 | | | | | | 791 | 424 | 180 | 057 | 013 | 002 | | | | | | | | | | | | | | | | | | | | | | | | |
| 15 | | | | | | | 607 | 302 | 118 | 035 | 007 | 001 | | | | | | | | | | | | | | | | | | | | | | | |
| 16 | | | | | | | 804 | 454 | 210 | 077 | 021 | 004 | 001 | | | | | | | | | | | | | | | | | | | | | | |
| 17 | | | | | | | | 629 | 332 | 143 | 049 | 013 | 002 | | | | | | | | | | | | | | | | | | | | | | |
| 18 | | | | | | | | 815 | 481 | 238 | 096 | 031 | 008 | 001 | | | | | | | | | | | | | | | | | | | | | |
| 19 | | | | | | | | | 648 | 359 | 167 | 064 | 019 | 004 | 001 | | | | | | | | | | | | | | | | | | | | |
| 20 | | | | | | | | | 824 | 503 | 263 | 115 | 041 | 012 | 003 | | | | | | | | | | | | | | | | | | | | |
| 21 | | | | | | | | | | 664 | 383 | 189 | 078 | 027 | 007 | 001 | | | | | | | | | | | | | | | | | | | |
| 22 | | | | | | | | | | 832 | 523 | 286 | 134 | 052 | 017 | 004 | 001 | | | | | | | | | | | | | | | | | | |
| 23 | | | | | | | | | | | 678 | 405 | 210 | 093 | 035 | 011 | 003 | | | | | | | | | | | | | | | | | | |
| 24 | | | | | | | | | | | 839 | 541 | 307 | 152 | 064 | 023 | 007 | 002 | | | | | | | | | | | | | | | | | |
| 25 | | | | | | | | | | | | 690 | 424 | 230 | 108 | 043 | 015 | 004 | 001 | | | | | | | | | | | | | | | | |
| 26 | | | | | | | | | | | | 845 | 557 | 327 | 169 | 076 | 029 | 009 | 002 | 001 | | | | | | | | | | | | | | | |
| 27 | | | | | | | | | | | | | 701 | 442 | 248 | 122 | 052 | 019 | 006 | 002 | | | | | | | | | | | | | | | |
| 28 | | | | | | | | | | | | | 851 | 572 | 345 | 185 | 087 | 036 | 013 | 004 | 001 | | | | | | | | | | | | | | |

| | | | | | | | | | | | | | | | | | | | | | | |
|---|---|---|---|---|---|---|---|---|---|---|---|---|---|---|---|---|---|---|---|---|---|---|
| 29 | 711 | 458 | 265 | 136 | 061 | 024 | 008 | 002 | 001 | | | | | | | | | | | | | |
| 30 | 856 | 585 | 362 | 200 | 099 | 043 | 016 | 005 | 001 | | | | | | | | | | | | | |
| 31 | | 720 | 473 | 281 | 150 | 071 | 030 | 011 | 003 | 001 | | | | | | | | | | | | |
| 32 | | 860 | 597 | 377 | 215 | 100 | 050 | 020 | 007 | 002 | 001 | | | | | | | | | | | |
| 33 | | | 728 | 487 | 296 | 163 | 080 | 035 | 014 | 005 | 001 | | | | | | | | | | | |
| 34 | | | 864 | 608 | 392 | 229 | 121 | 058 | 024 | 009 | 003 | 001 | | | | | | | | | | |
| 35 | | | | 736 | 500 | 310 | 175 | 090 | 041 | 017 | 006 | 002 | | | | | | | | | | |
| 36 | | | | 868 | 681 | 405 | 243 | 132 | 065 | 029 | 011 | 004 | 001 | | | | | | | | | |
| 37 | | | | | 743 | 511 | 324 | 188 | 099 | 047 | 020 | 008 | 003 | 001 | | | | | | | | |
| 38 | | | | | 871 | 627 | 418 | 256 | 143 | 073 | 034 | 014 | 005 | 002 | | | | | | | | |
| 39 | | | | | | 749 | 522 | 337 | 200 | 108 | 053 | 024 | 009 | 003 | 001 | | | | | | | |
| 40 | | | | | | 875 | 636 | 430 | 268 | 154 | 081 | 038 | 017 | 006 | 002 | 001 | | | | | | |
| 41 | | | | | | | 755 | 533 | 349 | 211 | 117 | 060 | 028 | 012 | 004 | 001 | | | | | | |
| 42 | | | | | | | 878 | 644 | 441 | 280 | 164 | 088 | 044 | 020 | 008 | 003 | 001 | | | | | |
| 43 | | | | | | | | 761 | 542 | 360 | 222 | 126 | 066 | 032 | 014 | 005 | 002 | 001 | | | | |
| 44 | | | | | | | | 880 | 652 | 451 | 291 | 174 | 096 | 049 | 023 | 010 | 004 | 001 | | | | |
| 45 | | | | | | | | | 766 | 551 | 371 | 233 | 135 | 072 | 036 | 016 | 007 | 002 | 001 | | | |
| 46 | | | | | | | | | 883 | 659 | 461 | 302 | 184 | 104 | 054 | 026 | 011 | 005 | 002 | 001 | | |
| 47 | | | | | | | | | | 771 | 560 | 382 | 243 | 144 | 079 | 040 | 019 | 008 | 003 | 001 | | |
| 48 | | | | | | | | | | 885 | 665 | 471 | 312 | 193 | 111 | 059 | 029 | 013 | 006 | 003 | 001 | |
| 49 | | | | | | | | | | | 775 | 568 | 392 | 253 | 152 | 085 | 044 | 021 | 009 | 004 | 001 | |
| 50 | | | | | | | | | | | 888 | 672 | 480 | 322 | 203 | 119 | 065 | 033 | 015 | 007 | 003 | 001 |

[a]Source: E. B. Roessler, R. M. Pangborn, J. L. Sidel, and H. Stone. (1978). Expanded statistical tables for estimating significance in paired-preference, paired-difference, duo–trio, and triangle tests. J Food Sci, **43**, 940–943. Reproduced with permission of the editor.
[b]Note that initial decimal point has been omitted.

**Table A.24** Probability of Correct Selection in the Triangle and Duo–Trio Tests as a Function of $\theta$, the Standardized Difference in Means between the Odd and the Identical Samples[a]

| $\theta$ | $P_T$ | $P_D$ | $P_{DH}$ | $P_{DD}$ | $\theta$ | $P_T$ | $P_D$ | $P_{DH}$ | $P_{DD}$ |
|---|---|---|---|---|---|---|---|---|---|
| 0 | .33333 | .50000 | .50000 | .50000 | 3.0 | .78143 | .87646 | .83607 | .87731 |
| .1 | .33425 | .50092 | .50069 | .50103 | 3.1 | .79596 | .88589 | .84697 | .88639 |
| .2 | .33700 | .50366 | .50275 | .50411 | 3.2 | .80984 | .89473 | .85738 | .89493 |
| .3 | .34154 | .50819 | .50616 | .50918 | 3.3 | .82306 | .90298 | .86729 | .90294 |
| .4 | .34784 | .51444 | .51088 | .51614 | 3.4 | .83562 | .91067 | .87671 | .91045 |
| .5 | .35583 | .52235 | .51687 | .52489 | 3.5 | .84752 | .91783 | .88564 | .91747 |
| .6 | .36544 | .53179 | .52408 | .53527 | 3.6 | .85878 | .92449 | .89408 | .92402 |
| .7 | .37658 | .54267 | .53244 | .54712 | 3.7 | .86941 | .93068 | .90206 | .93014 |
| .8 | .38913 | .55484 | .54185 | .56025 | 3.8 | .87942 | .93642 | .90956 | .93584 |
| .9 | .40300 | .56816 | .55225 | .57448 | 3.9 | .88883 | .94174 | .91662 | .94114 |
| 1.0 | .41805 | .58250 | .56354 | .58961 | 4.0 | .89766 | .94666 | .92324 | .94607 |
| 1.1 | .43415 | .59763 | .57561 | .60545 | 4.1 | .90593 | .95122 | .92945 | .95064 |
| 1.2 | .45118 | .61346 | .58839 | .62181 | 4.2 | .91366 | .95543 | .93524 | .95488 |
| 1.3 | .46900 | .62981 | .60175 | .63853 | 4.3 | .92087 | .95932 | .94065 | .95881 |
| 1.4 | .48747 | .64653 | .61560 | .65544 | 4.4 | .92759 | .96291 | .94569 | .96243 |
| 1.5 | .50646 | .66346 | .62985 | .67238 | 4.5 | .93383 | .96622 | .95037 | .96578 |
| 1.6 | .52583 | .68047 | .64437 | .68924 | 4.6 | .93963 | .96927 | .95472 | .96888 |
| 1.7 | .54547 | .69742 | .65910 | .70590 | 4.7 | .94500 | .97207 | .95875 | .97172 |
| 1.8 | .56526 | .71420 | .67394 | .72226 | 4.8 | .94996 | .97465 | .96247 | .97434 |
| 1.9 | .58507 | .73069 | .68880 | .73823 | 4.9 | .95455 | .97702 | .96591 | .97675 |
| 2.0 | .60481 | .74682 | .70361 | .75375 | 5.0 | .95878 | .97919 | .96908 | .97896 |
| 2.1 | .62437 | .76250 | .71828 | .76875 | 5.5 | .97525 | .98758 | .98144 | .98747 |
| 2.2 | .64368 | .77766 | .73276 | .78322 | 6.0 | .98569 | .99283 | .98927 | .99279 |
| 2.3 | .66266 | .79225 | .74699 | .79711 | 6.5 | .99204 | .99602 | .99403 | .99600 |
| 2.4 | .68123 | .80624 | .76092 | .81039 | 7.0 | .99573 | .99786 | .99680 | .99786 |
| 2.5 | .69934 | .81959 | .77451 | .82307 | 8.0 | .99891 | .99945 | .99918 | .99945 |
| 2.6 | .71693 | .83228 | .78770 | .83513 | 9.0 | .99976 | .99988 | .99982 | .99988 |
| 2.7 | .73397 | .84431 | .80048 | .84658 | 10.0 | .99995 | .99998 | .99996 | .99997 |
| 2.8 | .75041 | .85568 | .81281 | .85741 | 15.0 | 1.00000 | 1.00000 | 1.00000 | 1.00000 |
| 2.9 | .76624 | .86639 | .82468 | .86765 | | | | | |

[a] Source: R. A. Bradley. (1963). Some relationships among sensory difference tests. Biometrics, **19**, 385–397. Reproduced with permission of The Biometric Society.

**Table A.25** Minimum Number of Correct Responses for Difference and Preference Tests Using Forced Choice Methods

| n | The 2-AFC $p_0 = 1/2$ (Two-Sided) | | | The 2-AFC and Duo–Trio $p_0 = 1/2$ (One-Sided) | | | The 3-AFC and Triangular $p_0 = 1/3$ (One-Sided) | | |
|---|---|---|---|---|---|---|---|---|---|
| | $\alpha = 0.01$ | $\alpha = 0.05$ | $\alpha = 0.1$ | $\alpha = 0.01$ | $\alpha = 0.05$ | $\alpha = 0.1$ | $\alpha = 0.01$ | $\alpha = 0.05$ | $\alpha = 0.1$ |
| 10 | 10 | 9 | 9 | 10 | 9 | 8 | 8 | 7 | 6 |
| 11 | 11 | 10 | 9 | 10 | 9 | 9 | 8 | 7 | 7 |
| 12 | 11 | 10 | 10 | 11 | 10 | 9 | 9 | 8 | 7 |
| 13 | 12 | 11 | 10 | 12 | 10 | 10 | 9 | 8 | 7 |
| 14 | 13 | 12 | 11 | 12 | 11 | 10 | 10 | 9 | 8 |
| 15 | 13 | 12 | 12 | 13 | 12 | 11 | 10 | 9 | 8 |
| 16 | 14 | 13 | 12 | 14 | 12 | 12 | 11 | 9 | 9 |
| 17 | 15 | 13 | 13 | 14 | 13 | 12 | 11 | 10 | 9 |
| 18 | 15 | 14 | 13 | 15 | 13 | 13 | 12 | 10 | 10 |
| 19 | 16 | 15 | 14 | 15 | 14 | 13 | 12 | 11 | 10 |
| 20 | 17 | 15 | 15 | 16 | 15 | 14 | 13 | 11 | 10 |
| 21 | 17 | 16 | 15 | 17 | 15 | 14 | 13 | 12 | 11 |
| 22 | 18 | 17 | 16 | 17 | 16 | 15 | 14 | 12 | 11 |
| 23 | 19 | 17 | 16 | 18 | 16 | 16 | 14 | 12 | 12 |
| 24 | 19 | 18 | 17 | 19 | 17 | 16 | 14 | 13 | 12 |
| 25 | 20 | 18 | 18 | 19 | 18 | 17 | 15 | 13 | 12 |
| 26 | 20 | 19 | 18 | 20 | 18 | 17 | 15 | 14 | 13 |
| 27 | 21 | 20 | 19 | 20 | 19 | 18 | 16 | 14 | 13 |
| 28 | 22 | 20 | 19 | 21 | 19 | 18 | 16 | 14 | 13 |
| 29 | 22 | 21 | 20 | 22 | 20 | 19 | 17 | 15 | 14 |
| 30 | 23 | 21 | 20 | 22 | 20 | 20 | 17 | 15 | 14 |
| 31 | 24 | 22 | 21 | 23 | 21 | 20 | 17 | 16 | 15 |
| 32 | 24 | 23 | 22 | 24 | 22 | 21 | 18 | 16 | 15 |
| 33 | 25 | 23 | 22 | 24 | 22 | 21 | 18 | 16 | 15 |
| 34 | 25 | 24 | 23 | 25 | 23 | 22 | 19 | 17 | 16 |

*(continued)*

**Table A.25** Minimum Number of Correct Responses for Difference and Preference Tests Using Forced Choice Methods—*cont*...

| n | The 2-AFC $p_0 = 1/2$ (Two-Sided) | | | The 2-AFC and Duo–Trio $p_0 = 1/2$ (One-Sided) | | | The 3-AFC and Triangular $p_0 = 1/3$ (One-Sided) | | |
|---|---|---|---|---|---|---|---|---|---|
| | $\alpha = 0.01$ | $\alpha = 0.05$ | $\alpha = 0.1$ | $\alpha = 0.01$ | $\alpha = 0.05$ | $\alpha = 0.1$ | $\alpha = 0.01$ | $\alpha = 0.05$ | $\alpha = 0.1$ |
| 35 | 26 | 24 | 23 | 25 | 23 | 22 | 19 | 17 | 16 |
| 36 | 27 | 25 | 24 | 26 | 24 | 23 | 20 | 18 | 17 |
| 37 | 27 | 25 | 24 | 27 | 24 | 23 | 20 | 18 | 17 |
| 38 | 28 | 26 | 25 | 27 | 25 | 24 | 20 | 18 | 17 |
| 39 | 28 | 27 | 26 | 28 | 26 | 24 | 21 | 19 | 18 |
| 40 | 29 | 27 | 26 | 28 | 26 | 25 | 21 | 19 | 18 |
| 41 | 30 | 28 | 27 | 29 | 27 | 26 | 22 | 20 | 18 |
| 42 | 30 | 28 | 27 | 29 | 27 | 26 | 22 | 20 | 19 |
| 43 | 31 | 29 | 28 | 30 | 28 | 27 | 23 | 20 | 19 |
| 44 | 31 | 29 | 28 | 31 | 28 | 27 | 23 | 21 | 20 |
| 45 | 32 | 30 | 29 | 31 | 29 | 28 | 23 | 21 | 20 |
| 46 | 33 | 31 | 30 | 32 | 30 | 28 | 24 | 22 | 20 |
| 47 | 33 | 31 | 30 | 32 | 30 | 29 | 24 | 22 | 21 |
| 48 | 34 | 32 | 31 | 33 | 31 | 29 | 25 | 22 | 21 |
| 49 | 34 | 32 | 31 | 34 | 31 | 30 | 25 | 23 | 21 |
| 50 | 35 | 33 | 32 | 34 | 32 | 31 | 25 | 23 | 22 |
| 51 | 36 | 33 | 32 | 35 | 32 | 31 | 26 | 23 | 22 |
| 52 | 36 | 34 | 33 | 35 | 33 | 32 | 26 | 24 | 23 |
| 53 | 37 | 35 | 33 | 36 | 33 | 32 | 27 | 24 | 23 |
| 54 | 37 | 35 | 34 | 36 | 34 | 33 | 27 | 25 | 23 |
| 55 | 38 | 36 | 35 | 37 | 35 | 33 | 27 | 25 | 24 |
| 56 | 39 | 36 | 35 | 38 | 35 | 34 | 28 | 25 | 24 |
| 57 | 39 | 37 | 36 | 38 | 36 | 34 | 28 | 26 | 24 |
| 58 | 40 | 37 | 36 | 39 | 36 | 35 | 29 | 26 | 25 |
| 59 | 40 | 38 | 37 | 39 | 37 | 35 | 29 | 26 | 25 |

*(continued)*

**Table A.25** Minimum Number of Correct Responses for Difference and Preference Tests Using Forced Choice Methods—*cont...*

| n | The 2-AFC $p_0 = 1/2$ (Two-Sided) | | | The 2-AFC and Duo–Trio $p_0 = 1/2$ (One-Sided) | | | The 3-AFC and Triangular $p_0 = 1/3$ (One-Sided) | | |
|---|---|---|---|---|---|---|---|---|---|
| | $\alpha = 0.01$ | $\alpha = 0.05$ | $\alpha = 0.1$ | $\alpha = 0.01$ | $\alpha = 0.05$ | $\alpha = 0.1$ | $\alpha = 0.01$ | $\alpha = 0.05$ | $\alpha = 0.1$ |
| 60 | 41 | 39 | 37 | 40 | 37 | 36 | 29 | 27 | 25 |
| 61 | 41 | 39 | 38 | 41 | 38 | 37 | 30 | 27 | 26 |
| 62 | 42 | 40 | 38 | 41 | 38 | 37 | 30 | 28 | 26 |
| 63 | 43 | 40 | 39 | 42 | 39 | 38 | 31 | 28 | 27 |
| 64 | 43 | 41 | 40 | 42 | 40 | 38 | 31 | 28 | 27 |
| 65 | 44 | 41 | 40 | 43 | 40 | 39 | 31 | 29 | 27 |
| 66 | 44 | 42 | 41 | 43 | 41 | 39 | 32 | 29 | 28 |
| 67 | 45 | 42 | 41 | 44 | 41 | 40 | 32 | 30 | 28 |
| 68 | 46 | 43 | 42 | 45 | 42 | 40 | 33 | 30 | 28 |
| 69 | 46 | 44 | 42 | 45 | 42 | 41 | 33 | 30 | 29 |
| 70 | 47 | 44 | 43 | 46 | 43 | 41 | 33 | 31 | 29 |
| 71 | 47 | 45 | 43 | 46 | 43 | 42 | 34 | 31 | 30 |
| 72 | 48 | 45 | 44 | 47 | 44 | 42 | 34 | 31 | 30 |
| 73 | 48 | 46 | 45 | 47 | 45 | 43 | 35 | 32 | 30 |
| 74 | 49 | 46 | 45 | 48 | 45 | 44 | 35 | 32 | 31 |
| 75 | 50 | 47 | 46 | 49 | 46 | 44 | 35 | 33 | 31 |
| 76 | 50 | 48 | 46 | 49 | 46 | 45 | 36 | 33 | 31 |
| 77 | 51 | 48 | 47 | 50 | 47 | 45 | 36 | 33 | 32 |
| 78 | 51 | 49 | 47 | 50 | 47 | 46 | 37 | 34 | 32 |
| 79 | 52 | 49 | 48 | 51 | 48 | 46 | 37 | 34 | 32 |
| 80 | 52 | 50 | 48 | 51 | 48 | 47 | 37 | 34 | 33 |
| 81 | 53 | 50 | 49 | 52 | 49 | 47 | 38 | 35 | 33 |
| 82 | 54 | 51 | 49 | 52 | 49 | 48 | 38 | 35 | 34 |
| 83 | 54 | 51 | 50 | 53 | 50 | 48 | 39 | 36 | 34 |
| 84 | 55 | 52 | 51 | 54 | 51 | 49 | 39 | 36 | 34 |

*(continued)*

**Table A.25** Minimum Number of Correct Responses for Difference and Preference Tests Using Forced Choice Methods—*cont...*

| n | The 2-AFC $p_0 = 1/2$ (Two-Sided) | | | The 2-AFC and Duo–Trio $p_0 = 1/2$ (One-Sided) | | | The 3-AFC and Triangular $p_0 = 1/3$ (One-Sided) | | |
|---|---|---|---|---|---|---|---|---|---|
| | $\alpha = 0.01$ | $\alpha = 0.05$ | $\alpha = 0.1$ | $\alpha = 0.01$ | $\alpha = 0.05$ | $\alpha = 0.1$ | $\alpha = 0.01$ | $\alpha = 0.05$ | $\alpha = 0.1$ |
| 85 | 55 | 53 | 51 | 54 | 51 | 49 | 39 | 36 | 35 |
| 86 | 56 | 53 | 52 | 55 | 52 | 50 | 40 | 37 | 35 |
| 87 | 56 | 54 | 52 | 55 | 52 | 50 | 40 | 37 | 35 |
| 88 | 57 | 54 | 53 | 56 | 53 | 51 | 41 | 37 | 36 |
| 89 | 58 | 55 | 53 | 56 | 53 | 52 | 41 | 38 | 36 |
| 90 | 58 | 55 | 54 | 57 | 54 | 52 | 41 | 38 | 36 |
| 91 | 59 | 56 | 54 | 58 | 54 | 53 | 42 | 38 | 37 |
| 92 | 59 | 56 | 55 | 58 | 55 | 53 | 42 | 39 | 37 |
| 93 | 60 | 57 | 55 | 59 | 55 | 54 | 42 | 39 | 38 |
| 94 | 60 | 57 | 56 | 59 | 56 | 54 | 43 | 40 | 38 |
| 95 | 61 | 58 | 57 | 60 | 57 | 55 | 43 | 40 | 38 |
| 96 | 62 | 59 | 57 | 60 | 57 | 55 | 44 | 40 | 39 |
| 97 | 62 | 59 | 58 | 61 | 58 | 56 | 44 | 41 | 39 |
| 98 | 63 | 60 | 58 | 61 | 58 | 56 | 44 | 41 | 39 |
| 99 | 63 | 60 | 59 | 62 | 59 | 57 | 45 | 41 | 40 |
| 100 | 64 | 61 | 59 | 63 | 59 | 57 | 45 | 42 | 40 |
| 101 | 64 | 61 | 60 | 63 | 60 | 58 | 46 | 42 | 40 |
| 102 | 65 | 62 | 60 | 64 | 60 | 58 | 46 | 43 | 41 |
| 103 | 66 | 62 | 61 | 64 | 61 | 59 | 46 | 43 | 41 |
| 104 | 66 | 63 | 61 | 65 | 61 | 60 | 47 | 43 | 41 |
| 105 | 67 | 64 | 62 | 65 | 62 | 60 | 47 | 44 | 42 |
| 106 | 67 | 64 | 62 | 66 | 62 | 61 | 47 | 44 | 42 |
| 107 | 68 | 65 | 63 | 66 | 63 | 61 | 48 | 44 | 43 |
| 108 | 68 | 65 | 64 | 67 | 64 | 62 | 48 | 45 | 43 |
| 109 | 69 | 66 | 64 | 68 | 64 | 62 | 49 | 45 | 43 |
| 110 | 69 | 66 | 65 | 68 | 65 | 63 | 49 | 45 | 44 |

**Table A.26** Critical Values ($l_0$) for the Page Tests ($\alpha = 0.05$)

| n | k 3 | 4 | 5 | 6 | 7 | n | k 3 | 4 | 5 | 6 | 7 |
|---|---|---|---|---|---|---|---|---|---|---|---|
| 3 | 41 | 84 | 150 | 243 | 369 | 28 | 349 | 726 | 1304 | 2127 | 3236 |
| 4 | 53 | 110 | 197 | 320 | 486 | 29 | 361 | 751 | 1350 | 2201 | 3350 |
| 5 | 66 | 136 | 244 | 397 | 603 | 30 | 373 | 777 | 1396 | 2276 | 3463 |
| 6 | 78 | 162 | 291 | 473 | 719 | 31 | 385 | 802 | 1441 | 2351 | 3577 |
| 7 | 91 | 188 | 337 | 549 | 834 | 32 | 398 | 827 | 1487 | 2425 | 3691 |
| 8 | 103 | 214 | 384 | 625 | 950 | 33 | 410 | 853 | 1533 | 2500 | 3805 |
| 9 | 115 | 240 | 430 | 701 | 1065 | 34 | 422 | 878 | 1578 | 2575 | 3918 |
| 10 | 128 | 266 | 477 | 776 | 1180 | 35 | 434 | 904 | 1624 | 2649 | 4032 |
| 11 | 140 | 291 | 523 | 852 | 1295 | 36 | 446 | 929 | 1670 | 2724 | 4145 |
| 12 | 153 | 317 | 569 | 927 | 1410 | 37 | 459 | 954 | 1716 | 2798 | 4259 |
| 13 | 165 | 343 | 615 | 1002 | 1524 | 38 | 471 | 980 | 1761 | 2873 | 4372 |
| 14 | 177 | 368 | 661 | 1078 | 1639 | 39 | 483 | 1005 | 1807 | 2947 | 4486 |
| 15 | 190 | 394 | 707 | 1153 | 1753 | 40 | 495 | 1031 | 1853 | 3022 | 4599 |
| 16 | 202 | 419 | 753 | 1228 | 1868 | 41 | 507 | 1056 | 1898 | 3096 | 4713 |
| 17 | 214 | 445 | 799 | 1303 | 1982 | 42 | 520 | 1081 | 1944 | 3171 | 4826 |
| 18 | 226 | 471 | 845 | 1378 | 2096 | 43 | 532 | 1107 | 1989 | 3245 | 4940 |
| 19 | 239 | 496 | 891 | 1453 | 2210 | 44 | 544 | 1132 | 2035 | 3320 | 5053 |
| 20 | 251 | 522 | 937 | 1528 | 2325 | 45 | 556 | 1157 | 2081 | 3394 | 5167 |
| 21 | 263 | 547 | 983 | 1603 | 2439 | 46 | 568 | 1183 | 2126 | 3469 | 5280 |
| 22 | 275 | 573 | 1029 | 1678 | 2553 | 47 | 580 | 1208 | 2172 | 3543 | 5393 |
| 23 | 288 | 598 | 1075 | 1753 | 2667 | 48 | 593 | 1233 | 2217 | 3618 | 5507 |
| 24 | 300 | 624 | 1121 | 1828 | 2781 | 49 | 605 | 1259 | 2263 | 3692 | 5620 |
| 25 | 312 | 649 | 1167 | 1902 | 2895 | 50 | 617 | 1284 | 2309 | 3767 | 5733 |
| 26 | 324 | 675 | 1212 | 1977 | 3008 | 51 | 629 | 1309 | 2354 | 3841 | 5847 |
| 27 | 337 | 700 | 1258 | 2052 | 3122 | 52 | 641 | 1335 | 2400 | 3915 | 5960 |

*(continued)*

**Table A.26** Critical Values ($I_0$) for the Page Tests ($\alpha = 0.05$)—*cont...*

| n | k | | | | | n | k | | | | |
|---|---|---|---|---|---|---|---|---|---|---|---|
| | 3 | 4 | 5 | 6 | 7 | | 3 | 4 | 5 | 6 | 7 |
| 53 | 653 | 1360 | 2445 | 3990 | 6073 | 77 | 945 | 1967 | 3538 | 5773 | 8789 |
| 54 | 666 | 1385 | 2491 | 4064 | 6187 | 78 | 957 | 1992 | 3583 | 5847 | 8903 |
| 55 | 678 | 1411 | 2536 | 4138 | 6300 | 79 | 969 | 2018 | 3629 | 5921 | 9016 |
| 56 | 690 | 1436 | 2582 | 4213 | 6413 | 80 | 981 | 2043 | 3674 | 5996 | 9129 |
| 57 | 702 | 1461 | 2628 | 4287 | 6526 | 81 | 993 | 2068 | 3720 | 6070 | 9242 |
| 58 | 714 | 1487 | 2673 | 4362 | 6640 | 82 | 1006 | 2093 | 3765 | 6144 | 9355 |
| 59 | 726 | 1512 | 2719 | 4436 | 6753 | 83 | 1018 | 2119 | 3810 | 6218 | 9468 |
| 60 | 739 | 1537 | 2764 | 4510 | 6866 | 84 | 1030 | 2144 | 3856 | 6292 | 9581 |
| 61 | 751 | 1563 | 2810 | 4585 | 6979 | 85 | 1042 | 2169 | 3901 | 6367 | 9694 |
| 62 | 763 | 1588 | 2855 | 4659 | 7093 | 86 | 1054 | 2195 | 3947 | 6441 | 9807 |
| 63 | 775 | 1613 | 2901 | 4733 | 7206 | 87 | 1066 | 2220 | 3992 | 6515 | 9920 |
| 64 | 787 | 1638 | 2946 | 4807 | 7319 | 88 | 1078 | 2245 | 4038 | 6589 | 10033 |
| 65 | 799 | 1664 | 2992 | 4882 | 7432 | 89 | 1090 | 2270 | 4083 | 6663 | 10146 |
| 66 | 811 | 1689 | 3037 | 4956 | 7545 | 90 | 1103 | 2296 | 4129 | 6738 | 10259 |
| 67 | 824 | 1714 | 3083 | 5030 | 7658 | 91 | 1115 | 2321 | 4174 | 6812 | 10372 |
| 68 | 836 | 1740 | 3128 | 5105 | 7772 | 92 | 1127 | 2346 | 4219 | 6886 | 10485 |
| 69 | 848 | 1765 | 3174 | 5179 | 7885 | 93 | 1139 | 2371 | 4265 | 6960 | 10598 |
| 70 | 860 | 1790 | 3219 | 5253 | 7998 | 94 | 1151 | 2397 | 4310 | 7034 | 10711 |
| 71 | 872 | 1816 | 3265 | 5327 | 8111 | 95 | 1163 | 2422 | 4356 | 7108 | 10824 |
| 72 | 884 | 1841 | 3310 | 5402 | 8224 | 96 | 1175 | 2447 | 4401 | 7183 | 10937 |
| 73 | 896 | 1866 | 3356 | 5476 | 8337 | 97 | 1187 | 2472 | 4446 | 7257 | 11050 |
| 74 | 909 | 1891 | 3401 | 5550 | 8450 | 98 | 1200 | 2498 | 4492 | 7331 | 11163 |
| 75 | 921 | 1917 | 3447 | 5624 | 8563 | 99 | 1212 | 2523 | 4537 | 7405 | 11276 |
| 76 | 933 | 1942 | 3492 | 5699 | 8676 | 100 | 1224 | 2548 | 4583 | 7479 | 11389 |

*(continued)*

**Table A.26** Critical Values ($l_0$) for the Page Tests ($\alpha = 0.05$)—*cont...*

| n | k | | | | | n | k | | | | |
|---|---|---|---|---|---|---|---|---|---|---|---|
| | **3** | **4** | **5** | **6** | **7** | | **3** | **4** | **5** | **6** | **7** |
| 3 | 40 | 82 | 147 | 238 | 362 | 28 | 346 | 720 | 1294 | 2112 | 3214 |
| 4 | 52 | 108 | 193 | 315 | 478 | 29 | 358 | 745 | 1340 | 2186 | 3327 |
| 5 | 65 | 134 | 240 | 390 | 593 | 30 | 370 | 771 | 1386 | 2260 | 3441 |
| 6 | 77 | 160 | 286 | 466 | 708 | 31 | 383 | 796 | 1431 | 2335 | 3554 |
| 7 | 89 | 185 | 332 | 542 | 823 | 32 | 395 | 821 | 1477 | 2409 | 3667 |
| 8 | 102 | 211 | 379 | 617 | 938 | 33 | 407 | 847 | 1522 | 2484 | 3781 |
| 9 | 114 | 237 | 425 | 692 | 1052 | 34 | 419 | 872 | 1568 | 2558 | 3894 |
| 10 | 126 | 262 | 471 | 767 | 1167 | 35 | 431 | 897 | 1613 | 2632 | 4007 |
| 11 | 139 | 288 | 517 | 842 | 1281 | 36 | 443 | 923 | 1659 | 2707 | 4120 |
| 12 | 151 | 313 | 563 | 917 | 1395 | 37 | 456 | 948 | 1704 | 2781 | 4234 |
| 13 | 163 | 339 | 609 | 992 | 1509 | 38 | 468 | 973 | 1750 | 2855 | 4347 |
| 14 | 175 | 364 | 654 | 1067 | 1623 | 39 | 480 | 999 | 1796 | 2930 | 4460 |
| 15 | 188 | 390 | 700 | 1142 | 1737 | 40 | 492 | 1024 | 1841 | 3004 | 4573 |
| 16 | 200 | 415 | 746 | 1217 | 1851 | 41 | 504 | 1049 | 1887 | 3078 | 4686 |
| 17 | 212 | 441 | 792 | 1291 | 1965 | 42 | 516 | 1074 | 1932 | 3153 | 4799 |
| 18 | 224 | 466 | 838 | 1366 | 2079 | 43 | 528 | 1100 | 1978 | 3227 | 4913 |
| 19 | 236 | 492 | 883 | 1441 | 2192 | 44 | 541 | 1125 | 2023 | 3301 | 5026 |
| 20 | 249 | 517 | 929 | 1515 | 2306 | 45 | 553 | 1150 | 2068 | 3375 | 5139 |
| 21 | 261 | 542 | 975 | 1590 | 2420 | 46 | 565 | 1176 | 2114 | 3450 | 5252 |
| 22 | 273 | 568 | 1021 | 1665 | 2533 | 47 | 577 | 1201 | 2159 | 3524 | 5365 |
| 23 | 285 | 593 | 1066 | 1739 | 2647 | 48 | 589 | 1226 | 2205 | 3598 | 5478 |
| 24 | 297 | 619 | 1112 | 1814 | 2760 | 49 | 601 | 1251 | 2250 | 3672 | 5591 |
| 25 | 310 | 644 | 1158 | 1888 | 2874 | 50 | 613 | 1277 | 2296 | 3746 | 5704 |
| 26 | 322 | 669 | 1203 | 1963 | 2987 | 51 | 625 | 1302 | 2341 | 3821 | 5817 |
| 27 | 334 | 695 | 1249 | 2037 | 3101 | 52 | 638 | 1327 | 2387 | 3895 | 5930 |

*(continued)*

**Table A.26** Critical Values ($l_0$) for the Page Tests ($\alpha = 0.05$)—cont...

| n | k 3 | 4 | 5 | 6 | 7 | n | k 3 | 4 | 5 | 6 | 7 |
|---|---|---|---|---|---|---|---|---|---|---|---|
| 53 | 650 | 1352 | 2432 | 3969 | 6043 | 77 | 940 | 1958 | 3522 | 5748 | 8753 |
| 54 | 662 | 1378 | 2478 | 4043 | 6156 | 78 | 953 | 1983 | 3567 | 5822 | 8866 |
| 55 | 674 | 1403 | 2523 | 4117 | 6269 | 79 | 965 | 2008 | 3612 | 5896 | 8979 |
| 56 | 686 | 1428 | 2568 | 4192 | 6382 | 80 | 977 | 2034 | 3658 | 5970 | 9092 |
| 57 | 698 | 1453 | 2614 | 4266 | 6495 | 81 | 989 | 2059 | 3703 | 6044 | 9204 |
| 58 | 710 | 1479 | 2659 | 4340 | 6608 | 82 | 1001 | 2084 | 3749 | 6118 | 9317 |
| 59 | 722 | 1504 | 2705 | 4414 | 6721 | 83 | 1013 | 2109 | 3794 | 6192 | 9430 |
| 60 | 735 | 1529 | 2750 | 4488 | 6834 | 84 | 1025 | 2134 | 3839 | 6266 | 9543 |
| 61 | 747 | 1554 | 2796 | 4562 | 6947 | 85 | 1037 | 2160 | 3885 | 6340 | 9656 |
| 62 | 759 | 1580 | 2841 | 4636 | 7060 | 86 | 1049 | 2185 | 3930 | 6415 | 9768 |
| 63 | 771 | 1605 | 2886 | 4711 | 7173 | 87 | 1061 | 2210 | 3975 | 6489 | 9881 |
| 64 | 783 | 1630 | 2932 | 4785 | 7286 | 88 | 1074 | 2235 | 4021 | 6563 | 9994 |
| 65 | 795 | 1655 | 2977 | 4859 | 7399 | 89 | 1086 | 2260 | 4066 | 6637 | 10107 |
| 66 | 807 | 1681 | 3023 | 4933 | 7512 | 90 | 1098 | 2286 | 4111 | 6711 | 10219 |
| 67 | 819 | 1706 | 3068 | 5007 | 7624 | 91 | 1110 | 2311 | 4157 | 6785 | 10332 |
| 68 | 831 | 1731 | 3113 | 5081 | 7737 | 92 | 1122 | 2336 | 4202 | 6859 | 10445 |
| 69 | 844 | 1756 | 3159 | 5155 | 7850 | 93 | 1134 | 2361 | 4247 | 6933 | 10558 |
| 70 | 856 | 1781 | 3204 | 5229 | 7963 | 94 | 1146 | 2386 | 4293 | 7007 | 10671 |
| 71 | 868 | 1807 | 3249 | 5304 | 8076 | 95 | 1158 | 2412 | 4338 | 7081 | 10783 |
| 72 | 880 | 1832 | 3295 | 5378 | 8189 | 96 | 1170 | 2437 | 4383 | 7155 | 10896 |
| 73 | 892 | 1857 | 3340 | 5452 | 8302 | 97 | 1182 | 2462 | 4429 | 7229 | 11009 |
| 74 | 904 | 1882 | 3386 | 5526 | 8415 | 98 | 1194 | 2487 | 4474 | 7303 | 11122 |
| 75 | 916 | 1908 | 3431 | 5600 | 8527 | 99 | 1207 | 2512 | 4519 | 7377 | 11234 |
| 76 | 928 | 1933 | 3476 | 5674 | 8640 | 100 | 1219 | 2537 | 4565 | 7451 | 11347 |

**Table A.27** Maximum Number of Correct Responses for Similarity Testing Using the 2-AFC and Duo–Trio Methods

| | $\alpha = 0.05$ | | | | | $\alpha = 0.1$ | | | | |
| | $p_{do}$ | | | | | $p_{do}$ | | | | |
| n | 0.1 | 0.2 | 0.3 | 0.4 | 0.5 | 0.1 | 0.2 | 0.3 | 0.4 | 0.5 |
|---|---|---|---|---|---|---|---|---|---|---|
| 5 | 0 | 0 | 0 | 1 | 1 | 0 | 1 | 1 | 1 | 1 |
| 6 | 0 | 1 | 1 | 1 | 2 | 1 | 1 | 1 | 2 | 2 |
| 7 | 1 | 1 | 1 | 2 | 2 | 1 | 2 | 2 | 2 | 3 |
| 8 | 1 | 2 | 2 | 2 | 3 | 2 | 2 | 2 | 3 | 3 |
| 9 | 2 | 2 | 2 | 3 | 4 | 2 | 3 | 3 | 4 | 4 |
| 10 | 2 | 2 | 3 | 4 | 4 | 2 | 3 | 4 | 4 | 5 |
| 11 | 2 | 3 | 3 | 4 | 5 | 3 | 4 | 4 | 5 | 5 |
| 12 | 3 | 3 | 4 | 5 | 5 | 3 | 4 | 5 | 5 | 6 |
| 13 | 3 | 4 | 5 | 5 | 6 | 4 | 5 | 5 | 6 | 7 |
| 14 | 4 | 4 | 5 | 6 | 7 | 4 | 5 | 6 | 7 | 7 |
| 15 | 4 | 5 | 6 | 6 | 7 | 5 | 6 | 6 | 7 | 8 |
| 16 | 5 | 5 | 6 | 7 | 8 | 5 | 6 | 7 | 8 | 9 |
| 17 | 5 | 6 | 7 | 8 | 9 | 6 | 7 | 8 | 8 | 9 |
| 18 | 5 | 6 | 7 | 8 | 9 | 6 | 7 | 8 | 9 | 10 |
| 19 | 6 | 7 | 8 | 9 | 10 | 7 | 8 | 9 | 10 | 11 |
| 20 | 6 | 7 | 8 | 10 | 11 | 7 | 8 | 9 | 10 | 11 |
| 21 | 7 | 8 | 9 | 10 | 11 | 8 | 9 | 10 | 11 | 12 |
| 22 | 7 | 8 | 10 | 11 | 12 | 8 | 9 | 10 | 12 | 13 |
| 23 | 8 | 9 | 10 | 11 | 13 | 9 | 10 | 11 | 12 | 14 |
| 24 | 8 | 9 | 11 | 12 | 13 | 9 | 10 | 12 | 13 | 14 |
| 25 | 9 | 10 | 11 | 13 | 14 | 10 | 11 | 12 | 14 | 15 |
| 26 | 9 | 10 | 12 | 13 | 15 | 10 | 11 | 13 | 14 | 16 |
| 27 | 10 | 11 | 12 | 14 | 15 | 11 | 12 | 13 | 15 | 16 |
| 28 | 10 | 12 | 13 | 15 | 16 | 11 | 12 | 14 | 15 | 17 |
| 29 | 11 | 12 | 14 | 15 | 17 | 12 | 13 | 15 | 16 | 18 |

*(continued)*

**Table A.27** Maximum Number of Correct Responses for Similarity Testing Using the 2-AFC and Duo–Trio Methods—*cont*...

| | $\alpha = 0.05$ | | | | | $\alpha = 0.1$ | | | | |
| | $p_{do}$ | | | | | $p_{do}$ | | | | |
| **n** | **0.1** | **0.2** | **0.3** | **0.4** | **0.5** | **0.1** | **0.2** | **0.3** | **0.4** | **0.5** |
|---|---|---|---|---|---|---|---|---|---|---|
| 30 | 11 | 13 | 14 | 16 | 17 | 12 | 14 | 15 | 17 | 18 |
| 35 | 13 | 15 | 17 | 19 | 21 | 14 | 16 | 18 | 20 | 22 |
| 40 | 16 | 18 | 20 | 22 | 24 | 17 | 19 | 21 | 23 | 25 |
| 45 | 18 | 21 | 23 | 25 | 28 | 19 | 22 | 24 | 27 | 29 |
| 50 | 21 | 23 | 26 | 29 | 31 | 22 | 25 | 27 | 30 | 33 |
| 60 | 26 | 29 | 32 | 35 | 38 | 27 | 30 | 33 | 36 | 40 |
| 70 | 31 | 34 | 38 | 42 | 45 | 32 | 36 | 39 | 43 | 47 |
| 80 | 36 | 40 | 44 | 48 | 53 | 37 | 41 | 46 | 50 | 54 |
| 90 | 41 | 45 | 50 | 55 | 60 | 42 | 47 | 52 | 56 | 61 |
| 100 | 46 | 51 | 56 | 61 | 67 | 48 | 53 | 58 | 63 | 68 |

**Table A.28** Maximum Number of Correct Responses for Similarity Testing Using the 3-AFC and Triangular Methods

| | $\alpha = 0.05$ | | | | | $\alpha = 0.1$ | | | | |
| | $p_{do}$ | | | | | $p_{do}$ | | | | |
| **n** | **0.1** | **0.2** | **0.3** | **0.4** | **0.5** | **0.1** | **0.2** | **0.3** | **0.4** | **0.5** |
|---|---|---|---|---|---|---|---|---|---|---|
| 5 | 0 | 0 | 0 | 0 | 1 | 0 | 0 | 0 | 1 | 1 |
| 6 | 0 | 0 | 0 | 1 | 1 | 0 | 0 | 1 | 1 | 1 |
| 7 | 0 | 0 | 1 | 1 | 2 | 0 | 1 | 1 | 2 | 2 |
| 8 | 0 | 0 | 1 | 2 | 2 | 0 | 1 | 1 | 2 | 3 |
| 9 | 0 | 1 | 1 | 2 | 3 | 1 | 1 | 2 | 3 | 3 |
| 10 | 1 | 1 | 2 | 2 | 3 | 1 | 2 | 2 | 3 | 4 |
| 11 | 1 | 1 | 2 | 3 | 4 | 1 | 2 | 3 | 4 | 4 |
| 12 | 1 | 2 | 3 | 3 | 4 | 2 | 2 | 3 | 4 | 5 |
| 13 | 1 | 2 | 3 | 4 | 5 | 2 | 3 | 4 | 5 | 5 |

*(continued)*

**Table A.28** Maximum Number of Correct Responses for Similarity Testing Using the 3-AFC and Triangular Methods—*cont...*

| | $\alpha = 0.05$ | | | | | $\alpha = 0.1$ | | | | |
| | $p_{do}$ | | | | | $p_{do}$ | | | | |
| n | 0.1 | 0.2 | 0.3 | 0.4 | 0.5 | 0.1 | 0.2 | 0.3 | 0.4 | 0.5 |
|---|---|---|---|---|---|---|---|---|---|---|
| 14 | 2 | 3 | 3 | 4 | 5 | 2 | 3 | 4 | 5 | 6 |
| 15 | 2 | 3 | 4 | 5 | 6 | 3 | 4 | 5 | 6 | 7 |
| 16 | 2 | 3 | 4 | 5 | 7 | 3 | 4 | 5 | 6 | 7 |
| 17 | 3 | 4 | 5 | 6 | 7 | 3 | 4 | 5 | 7 | 8 |
| 18 | 3 | 4 | 5 | 6 | 8 | 4 | 5 | 6 | 7 | 8 |
| 19 | 3 | 4 | 6 | 7 | 8 | 4 | 5 | 6 | 8 | 9 |
| 20 | 3 | 5 | 6 | 7 | 9 | 4 | 5 | 7 | 8 | 10 |
| 21 | 4 | 5 | 6 | 8 | 9 | 5 | 6 | 7 | 9 | 10 |
| 22 | 4 | 5 | 7 | 8 | 10 | 5 | 6 | 8 | 9 | 11 |
| 23 | 4 | 6 | 7 | 9 | 11 | 5 | 7 | 8 | 10 | 11 |
| 24 | 5 | 6 | 8 | 9 | 11 | 6 | 7 | 9 | 10 | 12 |
| 25 | 5 | 7 | 8 | 10 | 12 | 6 | 7 | 9 | 11 | 13 |
| 26 | 5 | 7 | 9 | 10 | 12 | 6 | 8 | 10 | 11 | 13 |
| 27 | 6 | 7 | 9 | 11 | 13 | 7 | 8 | 10 | 12 | 14 |
| 28 | 6 | 8 | 10 | 12 | 13 | 7 | 9 | 11 | 12 | 14 |
| 29 | 6 | 8 | 10 | 12 | 14 | 7 | 9 | 11 | 13 | 15 |
| 30 | 7 | 9 | 11 | 13 | 15 | 8 | 10 | 11 | 14 | 16 |
| 35 | 8 | 11 | 13 | 15 | 18 | 9 | 12 | 14 | 16 | 19 |
| 40 | 10 | 13 | 15 | 18 | 21 | 11 | 14 | 16 | 19 | 22 |
| 45 | 12 | 15 | 17 | 21 | 24 | 13 | 16 | 19 | 22 | 25 |
| 50 | 13 | 17 | 20 | 23 | 27 | 15 | 18 | 21 | 25 | 28 |
| 60 | 17 | 21 | 25 | 29 | 33 | 18 | 22 | 26 | 30 | 34 |
| 70 | 20 | 25 | 29 | 34 | 39 | 22 | 26 | 31 | 36 | 41 |
| 80 | 24 | 29 | 34 | 40 | 45 | 25 | 31 | 36 | 41 | 47 |
| 90 | 27 | 33 | 39 | 45 | 52 | 29 | 35 | 41 | 47 | 53 |
| 100 | 31 | 37 | 44 | 51 | 58 | 33 | 39 | 46 | 53 | 60 |

**Table A.29** Critical Number of Selecting a Product in Similarity Testing Using the Paired Comparison Method ($\alpha = 0.05$)

| n | $\Delta = 0.1$ | | $\Delta = 0.15$ | | $\Delta = 0.2$ | |
|---|---|---|---|---|---|---|
|  | $c_l$ | $c_u$ | $c_l$ | $c_u$ | $c_l$ | $c_u$ |
| 20 | – | – | – | – | 10 | 10 |
| 21 | – | – | – | – | – | – |
| 22 | – | – | – | – | 11 | 11 |
| 23 | – | – | – | – | – | – |
| 24 | – | – | – | – | 12 | 12 |
| 25 | – | – | – | – | 12 | 13 |
| 26 | – | – | – | – | 13 | 13 |
| 27 | – | – | – | – | 13 | 14 |
| 28 | – | – | – | – | 13 | 15 |
| 29 | – | – | – | – | 14 | 15 |
| 30 | – | – | – | – | 14 | 16 |
| 31 | – | – | – | – | 15 | 16 |
| 32 | – | – | – | – | 15 | 17 |
| 33 | – | – | – | – | 15 | 18 |
| 34 | – | – | – | – | 16 | 18 |
| 35 | – | – | – | – | 16 | 19 |
| 36 | – | – | 18 | 18 | 16 | 20 |
| 37 | – | – | – | – | 17 | 20 |
| 38 | – | – | 19 | 19 | 17 | 21 |
| 39 | – | – | – | – | 17 | 22 |
| 40 | – | – | 20 | 20 | 18 | 22 |
| 41 | – | – | 20 | 21 | 18 | 23 |
| 42 | – | – | 21 | 21 | 19 | 23 |
| 43 | – | – | 21 | 22 | 19 | 24 |
| 44 | – | – | 22 | 22 | 19 | 25 |

(continued)

**Table A.29** Critical Number of Selecting a Product in Similarity Testing Using the Paired Comparison Method ($\alpha = 0.05$)—*cont...*

| n | $\Delta = 0.1$ | | $\Delta = 0.15$ | | $\Delta = 0.2$ | |
|---|---|---|---|---|---|---|
| | $c_l$ | $c_u$ | $c_l$ | $c_u$ | $c_l$ | $c_u$ |
| 45 | – | – | 22 | 23 | 20 | 25 |
| 46 | – | – | 22 | 24 | 20 | 26 |
| 47 | – | – | 23 | 24 | 20 | 27 |
| 48 | – | – | 23 | 25 | 21 | 27 |
| 49 | – | – | 24 | 25 | 21 | 28 |
| 50 | – | – | 24 | 26 | 21 | 29 |
| 55 | – | – | 26 | 29 | 23 | 32 |
| 60 | – | – | 28 | 32 | 25 | 35 |
| 65 | – | – | 30 | 35 | 27 | 38 |
| 70 | – | – | 32 | 38 | 28 | 42 |
| 75 | – | – | 34 | 41 | 30 | 45 |
| 80 | 40 | 40 | 36 | 44 | 32 | 48 |
| 85 | 42 | 43 | 38 | 47 | 34 | 51 |
| 90 | 45 | 45 | 40 | 50 | 35 | 55 |
| 95 | 47 | 48 | 42 | 53 | 37 | 58 |
| 100 | 49 | 51 | 44 | 56 | 39 | 61 |
| 110 | 53 | 57 | 48 | 62 | 42 | 68 |
| 120 | 58 | 62 | 52 | 68 | 45 | 75 |
| 130 | 62 | 68 | 56 | 74 | 49 | 81 |
| 140 | 67 | 73 | 59 | 81 | 52 | 88 |
| 150 | 71 | 79 | 63 | 87 | 55 | 95 |
| 160 | 75 | 85 | 67 | 93 | 59 | 101 |
| 170 | 80 | 90 | 71 | 99 | 62 | 108 |
| 180 | 84 | 96 | 75 | 105 | 65 | 115 |
| 190 | 88 | 102 | 78 | 112 | 68 | 122 |
| 200 | 92 | 108 | 82 | 118 | 72 | 128 |

*(continued)*

**Table A.29** Critical Number of Selecting a Product in Similarity Testing Using the Paired Comparison Method ($\alpha = 0.05$)—*cont...*

| n | $\Delta = 0.1$ | | $\Delta = 0.15$ | | $\Delta = 0.2$ | |
|---|---|---|---|---|---|---|
| | $c_l$ | $c_u$ | $c_l$ | $c_u$ | $c_l$ | $c_u$ |
| 20 | – | – | – | – | 10 | 10 |
| 21 | – | – | – | – | 10 | 11 |
| 22 | – | – | – | – | 10 | 12 |
| 23 | – | – | – | – | 11 | 12 |
| 24 | – | – | 12 | 12 | 11 | 13 |
| 25 | – | – | – | – | 11 | 14 |
| 26 | – | – | 13 | 13 | 12 | 14 |
| 27 | – | – | – | – | 12 | 15 |
| 28 | – | – | 14 | 14 | 13 | 15 |
| 29 | – | – | 14 | 15 | 13 | 16 |
| 30 | – | – | 15 | 15 | 13 | 17 |
| 31 | – | – | 15 | 16 | 14 | 17 |
| 32 | – | – | 16 | 16 | 14 | 18 |
| 33 | – | – | 16 | 17 | 14 | 19 |
| 34 | – | – | 16 | 18 | 15 | 19 |
| 35 | – | – | 17 | 18 | 15 | 20 |
| 36 | – | – | 17 | 19 | 15 | 21 |
| 37 | – | – | 18 | 19 | 16 | 21 |
| 38 | – | – | 18 | 20 | 16 | 22 |
| 39 | – | – | 18 | 21 | 16 | 23 |
| 40 | – | – | 19 | 21 | 17 | 23 |
| 41 | – | – | 19 | 22 | 17 | 24 |
| 42 | – | – | 20 | 22 | 17 | 25 |
| 43 | – | – | 20 | 23 | 18 | 25 |
| 44 | – | – | 20 | 24 | 18 | 26 |
| 45 | – | – | 21 | 24 | 18 | 27 |

*(continued)*

**Table A.29** Critical Number of Selecting a Product in Similarity Testing Using the Paired Comparison Method ($\alpha = 0.05$)—*cont...*

| n | $\Delta = 0.1$ | | $\Delta = 0.15$ | | $\Delta = 0.2$ | |
|---|---|---|---|---|---|---|
| | $c_l$ | $c_u$ | $c_l$ | $c_u$ | $c_l$ | $c_u$ |
| 46 | – | – | 21 | 25 | 19 | 27 |
| 47 | – | – | 22 | 25 | 19 | 28 |
| 48 | – | – | 22 | 26 | 20 | 28 |
| 49 | – | – | 22 | 27 | 20 | 29 |
| 50 | 25 | 25 | 23 | 27 | 20 | 30 |
| 55 | – | – | 25 | 30 | 22 | 33 |
| 60 | 30 | 30 | 27 | 33 | 24 | 36 |
| 65 | 32 | 33 | 29 | 36 | 25 | 40 |
| 70 | 34 | 36 | 31 | 39 | 27 | 43 |
| 75 | 36 | 39 | 33 | 42 | 29 | 46 |
| 80 | 39 | 41 | 34 | 46 | 30 | 50 |
| 85 | 41 | 44 | 36 | 49 | 32 | 53 |
| 90 | 43 | 47 | 38 | 52 | 34 | 56 |
| 95 | 45 | 50 | 40 | 55 | 35 | 60 |
| 100 | 47 | 53 | 42 | 58 | 37 | 63 |
| 110 | 52 | 58 | 46 | 64 | 40 | 70 |
| 120 | 56 | 64 | 50 | 70 | 43 | 77 |
| 130 | 60 | 70 | 53 | 77 | 47 | 83 |
| 140 | 64 | 76 | 57 | 83 | 50 | 90 |
| 150 | 69 | 81 | 61 | 89 | 53 | 97 |
| 160 | 73 | 87 | 65 | 95 | 56 | 104 |
| 170 | 77 | 93 | 68 | 102 | 60 | 110 |
| 180 | 81 | 99 | 72 | 108 | 63 | 117 |
| 190 | 86 | 104 | 76 | 114 | 66 | 124 |
| 200 | 90 | 110 | 80 | 120 | 69 | 131 |

*Note: "–" means "similarity" cannot be concluded for any observation in the situation; $\Delta = |P - 0.5|$ is allowed difference for similarity.*

**Table A.30** Critical Values for Paired $t$-Tests for Similarity ($\alpha = 0.05$)

| n | 0.25 | 0.50 | 0.75 | 1.00 |
|---|---|---|---|---|
| | | | $\varepsilon$ | |
| 20 | 0.119 | 0.614 | 1.672 | 2.709 |
| 21 | 0.122 | 0.662 | 1.752 | 2.816 |
| 22 | 0.126 | 0.711 | 1.830 | 2.919 |
| 23 | 0.130 | 0.759 | 1.907 | 3.021 |
| 24 | 0.134 | 0.808 | 1.982 | 3.120 |
| 25 | 0.138 | 0.856 | 2.055 | 3.218 |
| 26 | 0.142 | 0.903 | 2.127 | 3.314 |
| 27 | 0.147 | 0.950 | 2.198 | 3.408 |
| 28 | 0.151 | 0.997 | 2.268 | 3.500 |
| 29 | 0.156 | 1.042 | 2.336 | 3.591 |
| 30 | 0.161 | 1.087 | 2.403 | 3.680 |
| 31 | 0.166 | 1.132 | 2.469 | 3.768 |
| 32 | 0.171 | 1.175 | 2.535 | 3.854 |
| 33 | 0.176 | 1.218 | 2.599 | 3.940 |
| 34 | 0.182 | 1.261 | 2.662 | 4.024 |
| 35 | 0.187 | 1.302 | 2.724 | 4.107 |
| 36 | 0.193 | 1.344 | 2.786 | 4.188 |
| 37 | 0.199 | 1.384 | 2.846 | 4.269 |
| 38 | 0.205 | 1.425 | 2.906 | 4.348 |
| 39 | 0.211 | 1.464 | 2.965 | 4.427 |
| 40 | 0.218 | 1.503 | 3.024 | 4.505 |
| 41 | 0.224 | 1.542 | 3.081 | 4.581 |
| 42 | 0.231 | 1.580 | 3.138 | 4.657 |
| 43 | 0.238 | 1.618 | 3.195 | 4.732 |
| 44 | 0.245 | 1.655 | 3.250 | 4.806 |
| 45 | 0.252 | 1.692 | 3.305 | 4.880 |
| 46 | 0.260 | 1.729 | 3.360 | 4.952 |

*(continued)*

**Table A.30** Critical Values for Paired *t*-Tests for Similarity
($\alpha = 0.05$)—*cont...*

| n | 0.25 | 0.50 | 0.75 | 1.00 |
|---|---|---|---|---|
| | | | $\varepsilon$ | |
| 47 | 0.267 | 1.765 | 3.414 | 5.024 |
| 48 | 0.275 | 1.801 | 3.467 | 5.095 |
| 49 | 0.283 | 1.836 | 3.520 | 5.165 |
| 50 | 0.292 | 1.871 | 3.572 | 5.235 |
| 55 | 0.336 | 2.041 | 3.826 | 5.573 |
| 60 | 0.384 | 2.204 | 4.069 | 5.897 |
| 65 | 0.437 | 2.360 | 4.303 | 6.208 |
| 70 | 0.493 | 2.510 | 4.528 | 6.507 |
| 75 | 0.552 | 2.655 | 4.745 | 6.797 |
| 80 | 0.613 | 2.796 | 4.955 | 7.077 |
| 85 | 0.674 | 2.932 | 5.159 | 7.348 |
| 90 | 0.736 | 3.064 | 5.357 | 7.612 |
| 95 | 0.797 | 3.193 | 5.550 | 7.869 |
| 100 | 0.858 | 3.318 | 5.737 | 8.119 |

**Table A.31** Critical Proportions for Replicated Similarity Testing Using the Paired Comparison Method ($\Delta = 0.1$, $\alpha = 0.05$)

| | $\gamma = 0.1$ | | | | $\gamma = 0.2$ | | | | $\gamma = 0.3$ | | | | $\gamma = 0.4$ | | | |
| | $n = 3$ | | $n = 4$ | | $n = 3$ | | $n = 4$ | | $n = 3$ | | $n = 4$ | | $n = 3$ | | $n = 4$ | |
| $k$ | $\mu_l$ | $\mu_u$ | $\mu_l$ | $\mu_u$ | $\mu_l$ | $\mu_u$ | $\mu_l$ | $\mu_u$ | $\mu_l$ | $\mu_u$ | $\mu_l$ | $\mu_u$ | $\mu_l$ | $\mu_u$ | $\mu_l$ | $\mu_u$ |
|---|---|---|---|---|---|---|---|---|---|---|---|---|---|---|---|---|
| 20 | – | – | – | – | – | – | – | – | – | – | – | – | – | – | – | – |
| 21 | – | – | – | – | – | – | – | – | – | – | – | – | – | – | – | – |
| 22 | – | – | 0.4979 | 0.5021 | – | – | – | – | – | – | – | – | – | – | – | – |
| 23 | – | – | 0.4958 | 0.5042 | – | – | – | – | – | – | – | – | – | – | – | – |
| 24 | – | – | 0.4938 | 0.5062 | – | – | – | – | – | – | – | – | – | – | – | – |
| 25 | – | – | 0.4919 | 0.5081 | – | – | – | – | – | – | – | – | – | – | – | – |
| 26 | 0.4999 | 0.5001 | 0.4901 | 0.5099 | – | – | 0.4999 | 0.5001 | – | – | – | – | – | – | – | – |
| 27 | 0.4981 | 0.5019 | 0.4884 | 0.5116 | – | – | 0.4981 | 0.5019 | – | – | – | – | – | – | – | – |
| 28 | 0.4963 | 0.5037 | 0.4868 | 0.5132 | – | – | 0.4963 | 0.5037 | – | – | – | – | – | – | – | – |
| 29 | 0.4946 | 0.5054 | 0.4853 | 0.5147 | – | – | 0.4946 | 0.5054 | – | – | – | – | – | – | – | – |
| 30 | 0.4930 | 0.5070 | 0.4839 | 0.5161 | – | 0.5070 | 0.4930 | 0.5070 | – | – | – | – | – | – | – | – |
| 35 | 0.4861 | 0.5139 | 0.4776 | 0.5224 | 0.4930 | 0.5070 | 0.4861 | 0.5139 | 0.4995 | 0.5005 | 0.4939 | 0.5061 | – | – | – | – |

| | | | | | | | | | | | | | | | | |
|---|---|---|---|---|---|---|---|---|---|---|---|---|---|---|---|---|
| 40 | 0.4806 | 0.5194 | 0.4726 | 0.5274 | 0.4870 | 0.5130 | 0.4806 | 0.5194 | 0.4930 | 0.5070 | 0.4878 | 0.5122 | 0.4987 | 0.5013 | 0.4945 | 0.5055 |
| 45 | 0.4760 | 0.5240 | 0.4685 | 0.5315 | 0.4821 | 0.5179 | 0.4760 | 0.5240 | 0.4877 | 0.5123 | 0.4828 | 0.5172 | 0.4930 | 0.5070 | 0.4891 | 0.5109 |
| 50 | 0.4721 | 0.5279 | 0.4650 | 0.5350 | 0.4778 | 0.5222 | 0.4721 | 0.5279 | 0.4832 | 0.5168 | 0.4785 | 0.5215 | 0.4883 | 0.5117 | 0.4845 | 0.5155 |
| 55 | 0.4687 | 0.5313 | 0.4619 | 0.5381 | 0.4742 | 0.5258 | 0.4687 | 0.5313 | 0.4794 | 0.5206 | 0.4749 | 0.5251 | 0.4842 | 0.5158 | 0.4806 | 0.5194 |
| 60 | 0.4658 | 0.5342 | 0.4593 | 0.5407 | 0.4711 | 0.5289 | 0.4658 | 0.5342 | 0.4760 | 0.5240 | 0.4717 | 0.5283 | 0.4806 | 0.5194 | 0.4772 | 0.5228 |
| 65 | 0.4632 | 0.5368 | 0.4570 | 0.5430 | 0.4683 | 0.5317 | 0.4632 | 0.5368 | 0.4730 | 0.5270 | 0.4689 | 0.5311 | 0.4774 | 0.5226 | 0.4741 | 0.5259 |
| 70 | 0.4609 | 0.5391 | 0.4549 | 0.5451 | 0.4658 | 0.5342 | 0.4609 | 0.5391 | 0.4703 | 0.5297 | 0.4664 | 0.5336 | 0.4746 | 0.5254 | 0.4714 | 0.5286 |
| 75 | 0.4588 | 0.5412 | 0.4530 | 0.5470 | 0.4636 | 0.5364 | 0.4588 | 0.5412 | 0.4680 | 0.5320 | 0.4641 | 0.5359 | 0.4721 | 0.5279 | 0.4690 | 0.5310 |
| 80 | 0.4570 | 0.5430 | 0.4514 | 0.5486 | 0.4615 | 0.5385 | 0.4570 | 0.5430 | 0.4658 | 0.5342 | 0.4621 | 0.5379 | 0.4698 | 0.5302 | 0.4668 | 0.5332 |
| 85 | 0.4553 | 0.5447 | 0.4498 | 0.5502 | 0.4597 | 0.5403 | 0.4553 | 0.5447 | 0.4638 | 0.5362 | 0.4602 | 0.5398 | 0.4677 | 0.5323 | 0.4648 | 0.5352 |
| 90 | 0.4537 | 0.5463 | 0.4484 | 0.5516 | 0.4580 | 0.5420 | 0.4537 | 0.5463 | 0.4620 | 0.5380 | 0.4585 | 0.5415 | 0.4658 | 0.5342 | 0.4630 | 0.5370 |
| 95 | 0.4523 | 0.5477 | 0.4471 | 0.5529 | 0.4565 | 0.5435 | 0.4523 | 0.5477 | 0.4604 | 0.5396 | 0.4570 | 0.5430 | 0.4640 | 0.5360 | 0.4613 | 0.5387 |
| 100 | 0.4510 | 0.5490 | 0.4459 | 0.5541 | 0.4550 | 0.5450 | 0.4510 | 0.5490 | 0.4588 | 0.5412 | 0.4555 | 0.5445 | 0.4624 | 0.5376 | 0.4598 | 0.5402 |

*Note: "−" means "similarity" cannot be concluded for any observation in the situation; $\Delta$ is allowed difference for similarity; $\mu_l$ and $\mu_u$ are critical proportions; $\gamma$ is a scale parameter in the beta-binomial model; $n$ is the number of replications, $k$ is the number of trials (panelists); and $\alpha$ is a Type I error.*

**Table A.32** Maximum Allowed Numbers of Correct Responses for Replicated Similarity Testing Using Forced-Choice Methods ($\Delta = \mu_0 = 0.10$, $\alpha = 0.05$)

| | 2-AFC and Duo–Trio Methods | | | | | | | | 3-AFC and Triangle Methods | | | | | | | |
| --- | --- | --- | --- | --- | --- | --- | --- | --- | --- | --- | --- | --- | --- | --- | --- | --- |
| | $\gamma = 0.20$ | | $\gamma = 0.40$ | | $\gamma = 0.60$ | | $\gamma = 0.80$ | | $\gamma = 0.20$ | | $\gamma = 0.40$ | | $\gamma = 0.60$ | | $\gamma = 0.80$ | |
| $k$ | $n=3$ | $n=4$ | $n=3$ | $n=4$ | $n=3$ | $n=4$ | $n=3$ | $n=4$ | $n=3$ | $n=4$ | $n=3$ | $n=4$ | $n=3$ | $n=4$ | $n=3$ | $n=4$ |
| 20 | 26 | 36 | 26 | 36 | 26 | 36 | 26 | 35 | 17 | 24 | 17 | 24 | 17 | 23 | 16 | 23 |
| 21 | 28 | 38 | 27 | 38 | 27 | 38 | 27 | 37 | 18 | 25 | 18 | 25 | 18 | 25 | 18 | 24 |
| 22 | 29 | 40 | 29 | 40 | 29 | 40 | 29 | 39 | 19 | 27 | 19 | 26 | 19 | 26 | 19 | 26 |
| 23 | 31 | 42 | 30 | 42 | 30 | 42 | 30 | 41 | 20 | 28 | 20 | 28 | 20 | 27 | 20 | 27 |
| 24 | 32 | 44 | 32 | 44 | 32 | 44 | 32 | 43 | 21 | 30 | 21 | 29 | 21 | 29 | 21 | 29 |
| 25 | 34 | 46 | 33 | 46 | 33 | 46 | 33 | 45 | 22 | 31 | 22 | 31 | 22 | 30 | 22 | 30 |
| 26 | 35 | 48 | 35 | 48 | 35 | 48 | 35 | 47 | 23 | 32 | 23 | 32 | 23 | 32 | 23 | 31 |
| 27 | 37 | 50 | 36 | 50 | 36 | 50 | 36 | 50 | 24 | 34 | 24 | 34 | 24 | 33 | 24 | 33 |
| 28 | 38 | 52 | 38 | 52 | 38 | 52 | 38 | 52 | 25 | 35 | 25 | 35 | 25 | 35 | 25 | 34 |
| 29 | 40 | 54 | 39 | 54 | 39 | 54 | 39 | 54 | 27 | 37 | 26 | 36 | 26 | 36 | 26 | 36 |
| 30 | 41 | 56 | 41 | 56 | 41 | 56 | 41 | 56 | 28 | 38 | 27 | 38 | 27 | 37 | 27 | 37 |

| | | | | | | | | | | | | | | | | | |
|---|---|---|---|---|---|---|---|---|---|---|---|---|---|---|---|---|---|
| 35 | 49 | 67 | 49 | 66 | 48 | 66 | 48 | 66 | 33 | 46 | 33 | 45 | 32 | 45 | 32 | 44 |
| 40 | 56 | 77 | 56 | 77 | 56 | 76 | 56 | 76 | 38 | 53 | 38 | 52 | 38 | 52 | 38 | 51 |
| 45 | 64 | 87 | 64 | 87 | 64 | 87 | 64 | 86 | 44 | 60 | 44 | 60 | 43 | 59 | 43 | 59 |
| 50 | 72 | 98 | 72 | 97 | 71 | 97 | 71 | 97 | 49 | 68 | 49 | 67 | 49 | 67 | 48 | 66 |
| 55 | 80 | 108 | 79 | 108 | 79 | 107 | 79 | 107 | 55 | 75 | 54 | 74 | 54 | 74 | 54 | 73 |
| 60 | 87 | 118 | 87 | 118 | 87 | 118 | 87 | 118 | 60 | 82 | 60 | 82 | 60 | 81 | 59 | 81 |
| 65 | 95 | 129 | 95 | 129 | 95 | 128 | 95 | 128 | 66 | 90 | 66 | 89 | 65 | 89 | 65 | 88 |
| 70 | 103 | 139 | 103 | 139 | 103 | 139 | 102 | 138 | 71 | 97 | 71 | 97 | 71 | 96 | 70 | 96 |
| 75 | 111 | 150 | 111 | 150 | 110 | 149 | 110 | 149 | 77 | 105 | 77 | 104 | 76 | 104 | 76 | 103 |
| 80 | 119 | 160 | 118 | 160 | 118 | 160 | 118 | 159 | 83 | 112 | 82 | 112 | 82 | 111 | 81 | 110 |
| 85 | 126 | 171 | 126 | 171 | 126 | 170 | 126 | 170 | 88 | 120 | 88 | 119 | 87 | 119 | 87 | 118 |
| 90 | 134 | 182 | 134 | 181 | 134 | 181 | 134 | 180 | 94 | 127 | 93 | 127 | 93 | 126 | 93 | 125 |
| 95 | 142 | 192 | 142 | 192 | 142 | 191 | 141 | 191 | 99 | 135 | 99 | 134 | 99 | 134 | 98 | 133 |
| 100 | 150 | 203 | 150 | 202 | 150 | 202 | 149 | 201 | 105 | 143 | 105 | 142 | 104 | 141 | 104 | 140 |

**Table A.33** Critical Values (Expressed in Percentages) of *R*-Index-50%

| One-Sided | $\alpha = 0.200$ | 0.100 | 0.050 | 0.025 | 0.010 | 0.005 | 0.001 |
|:---:|:---:|:---:|:---:|:---:|:---:|:---:|:---:|
| Two-Sided | $\alpha = 0.400$ | 0.200 | 0.100 | 0.050 | 0.020 | 0.010 | 0.002 |
| N | % | % | % | % | % | % | % |
| 5 | 15.21 | 21.67 | 26.06 | 29.22 | 32.27 | 34.01 | 36.91 |
| 6 | 13.91 | 20.01 | 24.26 | 27.40 | 30.48 | 32.26 | 35.30 |
| 7 | 12.89 | 18.68 | 22.80 | 25.89 | 28.97 | 30.78 | 33.91 |
| 8 | 12.07 | 17.59 | 21.58 | 24.61 | 27.67 | 29.49 | 32.68 |
| 9 | 11.39 | 16.67 | 20.53 | 23.51 | 26.54 | 28.36 | 31.58 |
| 10 | 10.81 | 15.88 | 19.63 | 22.54 | 25.54 | 27.36 | 30.60 |
| 11 | 10.31 | 15.19 | 18.84 | 21.69 | 24.65 | 26.46 | 29.71 |
| 12 | 9.88 | 14.59 | 18.13 | 20.93 | 23.85 | 25.64 | 28.89 |
| 13 | 9.49 | 14.05 | 17.50 | 20.24 | 23.12 | 24.90 | 28.14 |
| 14 | 9.15 | 13.57 | 16.93 | 19.62 | 22.46 | 24.22 | 27.45 |
| 15 | 8.84 | 13.13 | 16.42 | 19.05 | 21.85 | 23.59 | 26.81 |
| 16 | 8.56 | 12.74 | 15.95 | 18.53 | 21.29 | 23.01 | 26.21 |
| 17 | 8.31 | 12.38 | 15.51 | 18.05 | 20.77 | 22.48 | 25.65 |
| 18 | 8.08 | 12.04 | 15.11 | 17.61 | 20.28 | 21.98 | 25.13 |
| 19 | 7.86 | 11.74 | 14.74 | 17.19 | 19.83 | 21.51 | 24.64 |
| 20 | 7.66 | 11.45 | 14.40 | 16.80 | 19.41 | 21.07 | 24.17 |
| 21 | 7.48 | 11.19 | 14.08 | 16.44 | 19.02 | 20.65 | 23.74 |
| 22 | 7.31 | 10.94 | 13.78 | 16.11 | 18.64 | 20.26 | 23.32 |
| 23 | 7.15 | 10.71 | 13.50 | 15.79 | 18.29 | 19.89 | 22.93 |
| 24 | 7.00 | 10.49 | 13.23 | 15.49 | 17.96 | 19.55 | 22.56 |
| 25 | 6.86 | 10.29 | 12.98 | 15.21 | 17.65 | 19.21 | 22.20 |
| 26 | 6.72 | 10.09 | 12.75 | 14.94 | 17.35 | 18.90 | 21.86 |
| 27 | 6.60 | 9.91 | 12.52 | 14.68 | 17.07 | 18.60 | 21.54 |
| 28 | 6.48 | 9.74 | 12.31 | 14.44 | 16.79 | 18.32 | 21.23 |
| 29 | 6.37 | 9.57 | 12.11 | 14.21 | 16.54 | 18.04 | 20.93 |
| 30 | 6.26 | 9.42 | 11.92 | 13.99 | 16.29 | 17.78 | 20.65 |

*(continued)*

**Table A.33** Critical Values (Expressed in Percentages) of *R*-Index-50%—*cont...*

| One-Sided | $\alpha = 0.200$ | 0.100 | 0.050 | 0.025 | 0.010 | 0.005 | 0.001 |
|---|---|---|---|---|---|---|---|
| Two-Sided | $\alpha = 0.400$ | 0.200 | 0.100 | 0.050 | 0.020 | 0.010 | 0.002 |
| N | % | % | % | % | % | % | % |
| 31 | 6.16 | 9.27 | 11.73 | 13.78 | 16.06 | 17.53 | 20.38 |
| 32 | 6.06 | 9.12 | 11.56 | 13.58 | 15.83 | 17.29 | 20.11 |
| 33 | 5.97 | 8.99 | 11.39 | 13.39 | 15.61 | 17.06 | 19.86 |
| 34 | 5.88 | 8.86 | 11.23 | 13.21 | 15.41 | 16.84 | 19.62 |
| 35 | 5.80 | 8.74 | 11.08 | 13.03 | 15.21 | 16.63 | 19.38 |
| 36 | 5.72 | 8.62 | 10.93 | 12.86 | 15.02 | 16.42 | 19.16 |
| 37 | 5.64 | 8.50 | 10.79 | 12.70 | 14.83 | 16.22 | 18.94 |
| 38 | 5.57 | 8.39 | 10.65 | 12.54 | 14.65 | 16.03 | 18.73 |
| 39 | 5.49 | 8.29 | 10.52 | 12.39 | 14.48 | 15.85 | 18.52 |
| 40 | 5.43 | 8.19 | 10.39 | 12.24 | 14.32 | 15.67 | 18.33 |
| 45 | 5.12 | 7.72 | 9.82 | 11.58 | 13.57 | 14.87 | 17.43 |
| 50 | 4.85 | 7.33 | 9.33 | 11.02 | 12.92 | 14.18 | 16.65 |
| 55 | 4.63 | 7.00 | 8.91 | 10.53 | 12.36 | 13.57 | 15.96 |
| 60 | 4.43 | 6.70 | 8.55 | 10.10 | 11.87 | 13.04 | 15.36 |
| 65 | 4.26 | 6.44 | 8.22 | 9.72 | 11.43 | 12.56 | 14.82 |
| 70 | 4.10 | 6.21 | 7.93 | 9.38 | 11.04 | 12.14 | 14.33 |
| 75 | 3.96 | 6.01 | 7.66 | 9.08 | 10.68 | 11.75 | 13.89 |
| 80 | 3.84 | 5.82 | 7.42 | 8.80 | 10.36 | 11.40 | 13.49 |
| 85 | 3.72 | 5.65 | 7.21 | 8.54 | 10.06 | 11.08 | 13.12 |
| 90 | 3.62 | 5.49 | 7.01 | 8.31 | 9.79 | 10.79 | 12.78 |
| 95 | 3.52 | 5.34 | 6.82 | 8.10 | 9.54 | 10.51 | 12.46 |
| 100 | 3.43 | 5.21 | 6.65 | 7.89 | 9.31 | 10.26 | 12.17 |

*(continued)*

**Table A.33** Critical Values (Expressed in Percentages) of *R*-Index-50%—*cont...*

| One-Sided | $\alpha = 0.200$ | 0.100 | 0.050 | 0.025 | 0.010 | 0.005 | 0.001 |
|---|---|---|---|---|---|---|---|
| Two-Sided | $\alpha = 0.400$ | 0.200 | 0.100 | 0.050 | 0.020 | 0.010 | 0.002 |
| N | % | % | % | % | % | % | % |
| 110 | 3.27 | 4.97 | 6.35 | 7.53 | 8.89 | 9.80 | 11.64 |
| 120 | 3.13 | 4.76 | 6.08 | 7.22 | 8.53 | 9.40 | 11.17 |
| 130 | 3.01 | 4.57 | 5.85 | 6.94 | 8.21 | 9.05 | 10.76 |
| 140 | 2.91 | 4.41 | 5.64 | 6.70 | 7.92 | 8.73 | 10.39 |
| 150 | 2.81 | 4.26 | 5.45 | 6.47 | 7.65 | 8.45 | 10.05 |
| 160 | 2.72 | 4.12 | 5.28 | 6.27 | 7.41 | 8.19 | 9.75 |
| 170 | 2.64 | 4.00 | 5.12 | 6.09 | 7.20 | 7.95 | 9.47 |
| 180 | 2.56 | 3.89 | 4.98 | 5.92 | 7.00 | 7.73 | 9.21 |
| 190 | 2.49 | 3.79 | 4.85 | 5.76 | 6.82 | 7.53 | 8.98 |
| 200 | 2.43 | 3.69 | 4.73 | 5.62 | 6.65 | 7.34 | 8.76 |
| 210 | 2.37 | 3.60 | 4.61 | 5.49 | 6.49 | 7.17 | 8.56 |
| 220 | 2.32 | 3.52 | 4.51 | 5.36 | 6.34 | 7.01 | 8.37 |
| 230 | 2.27 | 3.44 | 4.41 | 5.24 | 6.21 | 6.86 | 8.19 |
| 240 | 2.22 | 3.37 | 4.32 | 5.14 | 6.08 | 6.72 | 8.02 |
| 250 | 2.17 | 3.30 | 4.23 | 5.03 | 5.96 | 6.58 | 7.86 |
| 260 | 2.13 | 3.24 | 4.15 | 4.94 | 5.84 | 6.46 | 7.71 |
| 270 | 2.09 | 3.18 | 4.07 | 4.84 | 5.74 | 6.34 | 7.57 |
| 280 | 2.05 | 3.12 | 4.00 | 4.76 | 5.63 | 6.23 | 7.44 |
| 290 | 2.02 | 3.07 | 3.93 | 4.68 | 5.54 | 6.12 | 7.31 |
| 300 | 1.98 | 3.02 | 3.87 | 4.60 | 5.45 | 6.02 | 7.19 |
| 310 | 1.95 | 2.97 | 3.80 | 4.52 | 5.36 | 5.92 | 7.08 |
| 320 | 1.92 | 2.92 | 3.74 | 4.45 | 5.28 | 5.83 | 6.97 |
| 330 | 1.89 | 2.88 | 3.69 | 4.39 | 5.20 | 5.74 | 6.87 |
| 340 | 1.86 | 2.84 | 3.63 | 4.32 | 5.12 | 5.66 | 6.77 |
| 350 | 1.84 | 2.79 | 3.58 | 4.26 | 5.05 | 5.58 | 6.67 |

*(continued)*

**Table A.33** Critical Values (Expressed in Percentages) of *R*-Index-50%—*cont...*

| One-Sided | $\alpha = 0.200$ | 0.100 | 0.050 | 0.025 | 0.010 | 0.005 | 0.001 |
|---|---|---|---|---|---|---|---|
| Two-Sided | $\alpha = 0.400$ | 0.200 | 0.100 | 0.050 | 0.020 | 0.010 | 0.002 |
| N | % | % | % | % | % | % | % |
| 360 | 1.81 | 2.76 | 3.53 | 4.20 | 4.98 | 5.50 | 6.58 |
| 370 | 1.79 | 2.72 | 3.48 | 4.14 | 4.91 | 5.43 | 6.49 |
| 380 | 1.76 | 2.68 | 3.44 | 4.09 | 4.85 | 5.36 | 6.41 |
| 390 | 1.74 | 2.65 | 3.39 | 4.04 | 4.78 | 5.29 | 6.33 |
| 400 | 1.72 | 2.61 | 3.35 | 3.99 | 4.72 | 5.22 | 6.25 |
| 410 | 1.70 | 2.58 | 3.31 | 3.94 | 4.67 | 5.16 | 6.17 |
| 420 | 1.68 | 2.55 | 3.27 | 3.89 | 4.61 | 5.10 | 6.10 |
| 430 | 1.66 | 2.52 | 3.23 | 3.85 | 4.56 | 5.04 | 6.03 |
| 440 | 1.64 | 2.49 | 3.19 | 3.80 | 4.51 | 4.98 | 5.96 |
| 450 | 1.62 | 2.46 | 3.16 | 3.76 | 4.46 | 4.93 | 5.90 |
| 460 | 1.60 | 2.44 | 3.12 | 3.72 | 4.41 | 4.88 | 5.83 |
| 470 | 1.58 | 2.41 | 3.09 | 3.68 | 4.36 | 4.82 | 5.77 |
| 480 | 1.57 | 2.39 | 3.06 | 3.64 | 4.32 | 4.77 | 5.71 |
| 490 | 1.55 | 2.36 | 3.03 | 3.60 | 4.27 | 4.73 | 5.66 |
| 500 | 1.54 | 2.34 | 3.00 | 3.57 | 4.23 | 4.68 | 5.60 |
| 550 | 1.46 | 2.23 | 2.86 | 3.40 | 4.03 | 4.46 | 5.34 |
| 600 | 1.40 | 2.13 | 2.74 | 3.26 | 3.86 | 4.27 | 5.12 |
| 650 | 1.35 | 2.05 | 2.63 | 3.13 | 3.71 | 4.11 | 4.92 |
| 700 | 1.30 | 1.98 | 2.54 | 3.02 | 3.58 | 3.96 | 4.74 |
| 750 | 1.25 | 1.91 | 2.45 | 2.92 | 3.46 | 3.83 | 4.58 |
| 800 | 1.21 | 1.85 | 2.37 | 2.83 | 3.35 | 3.71 | 4.44 |
| 850 | 1.18 | 1.79 | 2.30 | 2.74 | 3.25 | 3.60 | 4.31 |
| 900 | 1.15 | 1.74 | 2.24 | 2.66 | 3.16 | 3.49 | 4.19 |
| 950 | 1.11 | 1.70 | 2.18 | 2.59 | 3.08 | 3.40 | 4.08 |
| 1000 | 1.09 | 1.65 | 2.12 | 2.53 | 3.00 | 3.31 | 3.97 |

**Table A.34** Critical Values for Similarity Tests Using $R$-Index

| | $\alpha = 0.05$ | | | | | $\alpha = 0.10$ | | | |
| | $\Delta = 0.05$ | | $\Delta = 0.10$ | | | $\Delta = 0.05$ | | $\Delta = 0.10$ | |
| $n$ | $cl$ | $cu$ | $cl$ | $cu$ | $n$ | $cl$ | $cu$ | $cl$ | $cu$ |
|---|---|---|---|---|---|---|---|---|---|
| 30 | – | – | – | – | 30 | – | – | 0.4939 | 0.5061 |
| 31 | – | – | – | – | 31 | – | – | 0.4923 | 0.5077 |
| 32 | – | – | – | – | 32 | – | – | 0.4908 | 0.5092 |
| 33 | – | – | – | – | 33 | – | – | 0.4894 | 0.5106 |
| 34 | – | – | – | – | 34 | – | – | 0.4881 | 0.5119 |
| 35 | – | – | – | – | 35 | – | – | 0.4868 | 0.5132 |
| 36 | – | – | – | – | 36 | – | – | 0.4856 | 0.5144 |
| 37 | – | – | – | – | 37 | – | – | 0.4844 | 0.5156 |
| 38 | – | – | – | – | 38 | – | – | 0.4833 | 0.5167 |
| 39 | – | – | – | – | 39 | – | – | 0.4822 | 0.5178 |
| 40 | – | – | – | – | 40 | – | – | 0.4811 | 0.5189 |
| 41 | – | – | – | – | 41 | – | – | 0.4801 | 0.5199 |
| 42 | – | – | – | – | 42 | – | – | 0.4791 | 0.5209 |
| 43 | – | – | – | – | 43 | – | – | 0.4782 | 0.5218 |
| 44 | – | – | 0.4992 | 0.5008 | 44 | – | – | 0.4773 | 0.5227 |
| 45 | – | – | 0.4981 | 0.5019 | 45 | – | – | 0.4764 | 0.5236 |
| 46 | – | – | 0.4970 | 0.5030 | 46 | – | – | 0.4756 | 0.5244 |
| 47 | – | – | 0.4960 | 0.5040 | 47 | – | – | 0.4748 | 0.5252 |
| 48 | – | – | 0.4950 | 0.5050 | 48 | – | – | 0.4740 | 0.5260 |
| 49 | – | – | 0.4940 | 0.5060 | 49 | – | – | 0.4732 | 0.5268 |
| 50 | – | – | 0.4930 | 0.5070 | 50 | – | – | 0.4725 | 0.5275 |
| 51 | – | – | 0.4921 | 0.5079 | 51 | – | – | 0.4717 | 0.5283 |
| 52 | – | – | 0.4912 | 0.5088 | 52 | – | – | 0.4710 | 0.5290 |
| 53 | – | – | 0.4903 | 0.5097 | 53 | – | – | 0.4704 | 0.5296 |
| 54 | – | – | 0.4895 | 0.5105 | 54 | – | – | 0.4697 | 0.5303 |

*(continued)*

**Table A.34** Critical Values for Similarity Tests Using *R*-Index—*cont...*

| | $\alpha = 0.05$ | | | | | $\alpha = 0.10$ | | | |
|---|---|---|---|---|---|---|---|---|---|
| | $\Delta = 0.05$ | | $\Delta = 0.10$ | | | $\Delta = 0.05$ | | $\Delta = 0.10$ | |
| *n* | *cl* | *cu* | *cl* | *cu* | *n* | *cl* | *cu* | *cl* | *cu* |
| 55 | – | – | 0.4886 | 0.5114 | 55 | – | – | 0.4691 | 0.5309 |
| 56 | – | – | 0.4878 | 0.5122 | 56 | – | – | 0.4684 | 0.5316 |
| 57 | – | – | 0.4871 | 0.5129 | 57 | – | – | 0.4678 | 0.5322 |
| 58 | – | – | 0.4863 | 0.5137 | 58 | – | – | 0.4672 | 0.5328 |
| 59 | – | – | 0.4856 | 0.5144 | 59 | – | – | 0.4667 | 0.5333 |
| 60 | – | – | 0.4848 | 0.5152 | 60 | – | – | 0.4661 | 0.5339 |
| 61 | – | – | 0.4841 | 0.5159 | 61 | – | – | 0.4656 | 0.5344 |
| 62 | – | – | 0.4834 | 0.5166 | 62 | – | – | 0.4650 | 0.5350 |
| 63 | – | – | 0.4828 | 0.5172 | 63 | – | – | 0.4645 | 0.5355 |
| 64 | – | – | 0.4821 | 0.5179 | 64 | – | – | 0.4640 | 0.5360 |
| 65 | – | – | 0.4815 | 0.5185 | 65 | – | – | 0.4635 | 0.5365 |
| 66 | – | – | 0.4809 | 0.5191 | 66 | – | – | 0.4630 | 0.5370 |
| 67 | – | – | 0.4802 | 0.5198 | 67 | – | – | 0.4625 | 0.5375 |
| 68 | – | – | 0.4797 | 0.5203 | 68 | – | – | 0.4621 | 0.5379 |
| 69 | – | – | 0.4791 | 0.5209 | 69 | – | – | 0.4616 | 0.5384 |
| 70 | – | – | 0.4785 | 0.5215 | 70 | – | – | 0.4612 | 0.5388 |
| 71 | – | – | 0.4779 | 0.5221 | 71 | – | – | 0.4607 | 0.5393 |
| 72 | – | – | 0.4774 | 0.5226 | 72 | – | – | 0.4603 | 0.5397 |
| 73 | – | – | 0.4769 | 0.5231 | 73 | – | – | 0.4599 | 0.5401 |
| 74 | – | – | 0.4763 | 0.5237 | 74 | – | – | 0.4595 | 0.5405 |
| 75 | – | – | 0.4758 | 0.5242 | 75 | – | – | 0.4591 | 0.5409 |
| 76 | – | – | 0.4753 | 0.5247 | 76 | – | – | 0.4587 | 0.5413 |
| 77 | – | – | 0.4748 | 0.5252 | 77 | – | – | 0.4583 | 0.5417 |
| 78 | – | – | 0.4743 | 0.5257 | 78 | – | – | 0.4579 | 0.5421 |
| 79 | – | – | 0.4739 | 0.5261 | 79 | – | – | 0.4575 | 0.5425 |

*(continued)*

**Table A.34** Critical Values for Similarity Tests Using *R*-Index—*cont...*

| | α = 0.05 | | | | | α = 0.10 | | | |
| | Δ = 0.05 | | Δ = 0.10 | | | Δ = 0.05 | | Δ = 0.10 | |
| *n* | *cl* | *cu* | *cl* | *cu* | *n* | *cl* | *cu* | *cl* | *cu* |
|---|---|---|---|---|---|---|---|---|---|
| 80 | – | – | 0.4734 | 0.5266 | 80 | – | – | 0.4572 | 0.5428 |
| 81 | – | – | 0.4729 | 0.5271 | 81 | – | – | 0.4568 | 0.5432 |
| 82 | – | – | 0.4725 | 0.5275 | 82 | – | – | 0.4565 | 0.5435 |
| 83 | – | – | 0.4720 | 0.5280 | 83 | – | – | 0.4561 | 0.5439 |
| 84 | – | – | 0.4716 | 0.5284 | 84 | – | – | 0.4558 | 0.5442 |
| 85 | – | – | 0.4712 | 0.5288 | 85 | – | – | 0.4555 | 0.5445 |
| 86 | – | – | 0.4708 | 0.5292 | 86 | – | – | 0.4551 | 0.5449 |
| 87 | – | – | 0.4704 | 0.5296 | 87 | – | – | 0.4548 | 0.5452 |
| 88 | – | – | 0.4700 | 0.5300 | 88 | – | – | 0.4545 | 0.5455 |
| 89 | – | – | 0.4696 | 0.5304 | 89 | – | – | 0.4542 | 0.5458 |
| 90 | – | – | 0.4692 | 0.5308 | 90 | – | – | 0.4539 | 0.5461 |
| 91 | – | – | 0.4688 | 0.5312 | 91 | – | – | 0.4536 | 0.5464 |
| 92 | – | – | 0.4684 | 0.5316 | 92 | – | – | 0.4533 | 0.5467 |
| 93 | – | – | 0.4680 | 0.5320 | 93 | – | – | 0.4530 | 0.5470 |
| 94 | – | – | 0.4677 | 0.5323 | 94 | – | – | 0.4527 | 0.5473 |
| 95 | – | – | 0.4673 | 0.5327 | 95 | – | – | 0.4524 | 0.5476 |
| 96 | – | – | 0.4670 | 0.5330 | 96 | – | – | 0.4522 | 0.5478 |
| 97 | – | – | 0.4666 | 0.5334 | 97 | – | – | 0.4519 | 0.5481 |
| 98 | – | – | 0.4663 | 0.5337 | 98 | – | – | 0.4516 | 0.5484 |
| 99 | – | – | 0.4659 | 0.5341 | 99 | – | – | 0.4514 | 0.5486 |
| 100 | – | – | 0.4656 | 0.5344 | 100 | – | – | 0.4511 | 0.5489 |
| 101 | – | – | 0.4653 | 0.5347 | 101 | – | – | 0.4509 | 0.5491 |
| 102 | – | – | 0.4650 | 0.5350 | 102 | – | – | 0.4506 | 0.5494 |
| 103 | – | – | 0.4646 | 0.5354 | 103 | – | – | 0.4504 | 0.5496 |
| 104 | – | – | 0.4643 | 0.5357 | 104 | – | – | 0.4501 | 0.5499 |

*(continued)*

**Table A.34** Critical Values for Similarity Tests Using *R*-Index—*cont...*

| | $\alpha = 0.05$ | | | | | $\alpha = 0.10$ | | | |
| | $\Delta = 0.05$ | | $\Delta = 0.10$ | | | $\Delta = 0.05$ | | $\Delta = 0.10$ | |
| *n* | *cl* | *cu* | *cl* | *cu* | *n* | *cl* | *cu* | *cl* | *cu* |
|---|---|---|---|---|---|---|---|---|---|
| 105 | – | – | 0.4640 | 0.5360 | 105 | – | – | 0.4499 | 0.5501 |
| 106 | – | – | 0.4637 | 0.5363 | 106 | – | – | 0.4496 | 0.5504 |
| 107 | – | – | 0.4634 | 0.5366 | 107 | – | – | 0.4494 | 0.5506 |
| 108 | – | – | 0.4631 | 0.5369 | 108 | – | – | 0.4492 | 0.5508 |
| 109 | – | – | 0.4628 | 0.5372 | 109 | 0.4999 | 0.5001 | 0.4489 | 0.5511 |
| 110 | – | – | 0.4625 | 0.5375 | 110 | 0.4997 | 0.5003 | 0.4487 | 0.5513 |
| 111 | – | – | 0.4623 | 0.5377 | 111 | 0.4995 | 0.5005 | 0.4485 | 0.5515 |
| 112 | – | – | 0.4620 | 0.5380 | 112 | 0.4992 | 0.5008 | 0.4483 | 0.5517 |
| 113 | – | – | 0.4617 | 0.5383 | 113 | 0.4990 | 0.5010 | 0.4481 | 0.5519 |
| 114 | – | – | 0.4614 | 0.5386 | 114 | 0.4988 | 0.5012 | 0.4479 | 0.5521 |
| 115 | – | – | 0.4612 | 0.5388 | 115 | 0.4986 | 0.5014 | 0.4476 | 0.5524 |
| 116 | – | – | 0.4609 | 0.5391 | 116 | 0.4984 | 0.5016 | 0.4474 | 0.5526 |
| 117 | – | – | 0.4606 | 0.5394 | 117 | 0.4982 | 0.5018 | 0.4472 | 0.5528 |
| 118 | – | – | 0.4604 | 0.5396 | 118 | 0.4980 | 0.5020 | 0.4470 | 0.5530 |
| 119 | – | – | 0.4601 | 0.5399 | 119 | 0.4978 | 0.5022 | 0.4468 | 0.5532 |
| 120 | – | – | 0.4599 | 0.5401 | 120 | 0.4976 | 0.5024 | 0.4466 | 0.5534 |
| 121 | – | – | 0.4596 | 0.5404 | 121 | 0.4974 | 0.5026 | 0.4464 | 0.5536 |
| 122 | – | – | 0.4594 | 0.5406 | 122 | 0.4972 | 0.5028 | 0.4463 | 0.5537 |
| 123 | – | – | 0.4591 | 0.5409 | 123 | 0.4970 | 0.5030 | 0.4461 | 0.5539 |
| 124 | – | – | 0.4589 | 0.5411 | 124 | 0.4968 | 0.5032 | 0.4459 | 0.5541 |
| 125 | – | – | 0.4586 | 0.5414 | 125 | 0.4966 | 0.5034 | 0.4457 | 0.5543 |
| 126 | – | – | 0.4584 | 0.5416 | 126 | 0.4964 | 0.5036 | 0.4455 | 0.5545 |
| 127 | – | – | 0.4582 | 0.5418 | 127 | 0.4962 | 0.5038 | 0.4453 | 0.5547 |
| 128 | – | – | 0.4580 | 0.5420 | 128 | 0.4960 | 0.5040 | 0.4452 | 0.5548 |
| 129 | – | – | 0.4577 | 0.5423 | 129 | 0.4959 | 0.5041 | 0.4450 | 0.5550 |

*(continued)*

**Table A.34** Critical Values for Similarity Tests Using *R*-Index—*cont...*

| | $\alpha = 0.05$ | | | | | $\alpha = 0.10$ | | | |
| | $\Delta = 0.05$ | | $\Delta = 0.10$ | | | $\Delta = 0.05$ | | $\Delta = 0.10$ | |
| n | cl | cu | cl | cu | n | cl | cu | cl | cu |
|---|----|----|----|----|---|----|----|----|----|
| 130 | – | – | 0.4575 | 0.5425 | 130 | 0.4957 | 0.5043 | 0.4448 | 0.5552 |
| 131 | – | – | 0.4573 | 0.5427 | 131 | 0.4955 | 0.5045 | 0.4446 | 0.5554 |
| 132 | – | – | 0.4571 | 0.5429 | 132 | 0.4953 | 0.5047 | 0.4445 | 0.5555 |
| 133 | – | – | 0.4568 | 0.5432 | 133 | 0.4952 | 0.5048 | 0.4443 | 0.5557 |
| 134 | – | – | 0.4566 | 0.5434 | 134 | 0.4950 | 0.5050 | 0.4441 | 0.5559 |
| 135 | – | – | 0.4564 | 0.5436 | 135 | 0.4948 | 0.5052 | 0.4440 | 0.5560 |
| 136 | – | – | 0.4562 | 0.5438 | 136 | 0.4947 | 0.5053 | 0.4438 | 0.5562 |
| 137 | – | – | 0.4560 | 0.5440 | 137 | 0.4945 | 0.5055 | 0.4436 | 0.5564 |
| 138 | – | – | 0.4558 | 0.5442 | 138 | 0.4943 | 0.5057 | 0.4435 | 0.5565 |
| 139 | – | – | 0.4556 | 0.5444 | 139 | 0.4942 | 0.5058 | 0.4433 | 0.5567 |
| 140 | – | – | 0.4554 | 0.5446 | 140 | 0.4940 | 0.5060 | 0.4432 | 0.5568 |
| 141 | – | – | 0.4552 | 0.5448 | 141 | 0.4939 | 0.5061 | 0.4430 | 0.5570 |
| 142 | – | – | 0.4550 | 0.5450 | 142 | 0.4937 | 0.5063 | 0.4429 | 0.5571 |
| 143 | – | – | 0.4548 | 0.5452 | 143 | 0.4936 | 0.5064 | 0.4427 | 0.5573 |
| 144 | – | – | 0.4546 | 0.5454 | 144 | 0.4934 | 0.5066 | 0.4426 | 0.5574 |
| 145 | – | – | 0.4544 | 0.5456 | 145 | 0.4932 | 0.5068 | 0.4424 | 0.5576 |
| 146 | – | – | 0.4542 | 0.5458 | 146 | 0.4931 | 0.5069 | 0.4423 | 0.5577 |
| 147 | – | – | 0.4541 | 0.5459 | 147 | 0.4930 | 0.5070 | 0.4421 | 0.5579 |
| 148 | – | – | 0.4539 | 0.5461 | 148 | 0.4928 | 0.5072 | 0.4420 | 0.5580 |
| 149 | – | – | 0.4537 | 0.5463 | 149 | 0.4927 | 0.5073 | 0.4418 | 0.5582 |
| 150 | – | – | 0.4535 | 0.5465 | 150 | 0.4925 | 0.5075 | 0.4417 | 0.5583 |
| 151 | – | – | 0.4533 | 0.5467 | 151 | 0.4924 | 0.5076 | 0.4416 | 0.5584 |
| 152 | – | – | 0.4532 | 0.5468 | 152 | 0.4922 | 0.5078 | 0.4414 | 0.5586 |
| 153 | – | – | 0.4530 | 0.5470 | 153 | 0.4921 | 0.5079 | 0.4413 | 0.5587 |
| 154 | – | – | 0.4528 | 0.5472 | 154 | 0.4920 | 0.5080 | 0.4412 | 0.5588 |

*(continued)*

**Table A.34** Critical Values for Similarity Tests Using *R*-Index—*cont...*

| | α = 0.05 | | | | | α = 0.10 | | | |
| | Δ = 0.05 | | Δ = 0.10 | | | Δ = 0.05 | | Δ = 0.10 | |
| n | cl | cu | cl | cu | n | cl | cu | cl | cu |
|---|---|---|---|---|---|---|---|---|---|
| 155 | – | – | 0.4526 | 0.5474 | 155 | 0.4918 | 0.5082 | 0.4410 | 0.5590 |
| 156 | – | – | 0.4525 | 0.5475 | 156 | 0.4917 | 0.5083 | 0.4409 | 0.5591 |
| 157 | – | – | 0.4523 | 0.5477 | 157 | 0.4916 | 0.5084 | 0.4408 | 0.5592 |
| 158 | – | – | 0.4521 | 0.5479 | 158 | 0.4914 | 0.5086 | 0.4406 | 0.5594 |
| 159 | – | – | 0.4520 | 0.5480 | 159 | 0.4913 | 0.5087 | 0.4405 | 0.5595 |
| 160 | – | – | 0.4518 | 0.5482 | 160 | 0.4912 | 0.5088 | 0.4404 | 0.5596 |
| 161 | – | – | 0.4517 | 0.5483 | 161 | 0.4910 | 0.5090 | 0.4402 | 0.5598 |
| 162 | – | – | 0.4515 | 0.5485 | 162 | 0.4909 | 0.5091 | 0.4401 | 0.5599 |
| 163 | – | – | 0.4513 | 0.5487 | 163 | 0.4908 | 0.5092 | 0.4400 | 0.5600 |
| 164 | – | – | 0.4512 | 0.5488 | 164 | 0.4907 | 0.5093 | 0.4399 | 0.5601 |
| 165 | – | – | 0.4510 | 0.5490 | 165 | 0.4905 | 0.5095 | 0.4398 | 0.5602 |
| 166 | – | – | 0.4509 | 0.5491 | 166 | 0.4904 | 0.5096 | 0.4396 | 0.5604 |
| 167 | – | – | 0.4507 | 0.5493 | 167 | 0.4903 | 0.5097 | 0.4395 | 0.5605 |
| 168 | – | – | 0.4506 | 0.5494 | 168 | 0.4902 | 0.5098 | 0.4394 | 0.5606 |
| 169 | – | – | 0.4504 | 0.5496 | 169 | 0.4900 | 0.5100 | 0.4393 | 0.5607 |
| 170 | – | – | 0.4503 | 0.5497 | 170 | 0.4899 | 0.5101 | 0.4392 | 0.5608 |
| 171 | – | – | 0.4501 | 0.5499 | 171 | 0.4898 | 0.5102 | 0.4390 | 0.5610 |
| 172 | – | – | 0.4500 | 0.5500 | 172 | 0.4897 | 0.5103 | 0.4389 | 0.5611 |
| 173 | – | – | 0.4498 | 0.5502 | 173 | 0.4896 | 0.5104 | 0.4388 | 0.5612 |
| 174 | – | – | 0.4497 | 0.5503 | 174 | 0.4895 | 0.5105 | 0.4387 | 0.5613 |
| 175 | – | – | 0.4495 | 0.5505 | 175 | 0.4894 | 0.5106 | 0.4386 | 0.5614 |
| 176 | – | – | 0.4494 | 0.5506 | 176 | 0.4892 | 0.5108 | 0.4385 | 0.5615 |
| 177 | – | – | 0.4493 | 0.5507 | 177 | 0.4891 | 0.5109 | 0.4384 | 0.5616 |
| 178 | – | – | 0.4491 | 0.5509 | 178 | 0.4890 | 0.5110 | 0.4383 | 0.5617 |
| 179 | 0.4999 | 0.5001 | 0.4490 | 0.5510 | 179 | 0.4889 | 0.5111 | 0.4382 | 0.5618 |

*(continued)*

**Table A.34** Critical Values for Similarity Tests Using *R*-Index—*cont...*

| | $\alpha = 0.05$ | | | | | $\alpha = 0.10$ | | | |
| | $\Delta = 0.05$ | | $\Delta = 0.10$ | | | $\Delta = 0.05$ | | $\Delta = 0.10$ | |
| *n* | *cl* | *cu* | *cl* | *cu* | *n* | *cl* | *cu* | *cl* | *cu* |
|---|---|---|---|---|---|---|---|---|---|
| 180 | 0.4998 | 0.5002 | 0.4488 | 0.5512 | 180 | 0.4888 | 0.5112 | 0.4381 | 0.5619 |
| 181 | 0.4997 | 0.5003 | 0.4487 | 0.5513 | 181 | 0.4887 | 0.5113 | 0.4379 | 0.5621 |
| 182 | 0.4995 | 0.5005 | 0.4486 | 0.5514 | 182 | 0.4886 | 0.5114 | 0.4378 | 0.5622 |
| 183 | 0.4994 | 0.5006 | 0.4484 | 0.5516 | 183 | 0.4885 | 0.5115 | 0.4377 | 0.5623 |
| 184 | 0.4993 | 0.5007 | 0.4483 | 0.5517 | 184 | 0.4884 | 0.5116 | 0.4376 | 0.5624 |
| 185 | 0.4991 | 0.5009 | 0.4482 | 0.5518 | 185 | 0.4883 | 0.5117 | 0.4375 | 0.5625 |
| 186 | 0.4990 | 0.5010 | 0.4480 | 0.5520 | 186 | 0.4882 | 0.5118 | 0.4374 | 0.5626 |
| 187 | 0.4989 | 0.5011 | 0.4479 | 0.5521 | 187 | 0.4881 | 0.5119 | 0.4373 | 0.5627 |
| 188 | 0.4987 | 0.5013 | 0.4478 | 0.5522 | 188 | 0.4880 | 0.5120 | 0.4372 | 0.5628 |
| 189 | 0.4986 | 0.5014 | 0.4477 | 0.5523 | 189 | 0.4879 | 0.5121 | 0.4371 | 0.5629 |
| 190 | 0.4985 | 0.5015 | 0.4475 | 0.5525 | 190 | 0.4878 | 0.5122 | 0.4370 | 0.5630 |
| 191 | 0.4983 | 0.5017 | 0.4474 | 0.5526 | 191 | 0.4877 | 0.5123 | 0.4369 | 0.5631 |
| 192 | 0.4982 | 0.5018 | 0.4473 | 0.5527 | 192 | 0.4876 | 0.5124 | 0.4368 | 0.5632 |
| 193 | 0.4981 | 0.5019 | 0.4472 | 0.5528 | 193 | 0.4875 | 0.5125 | 0.4367 | 0.5633 |
| 194 | 0.4980 | 0.5020 | 0.4470 | 0.5530 | 194 | 0.4874 | 0.5126 | 0.4367 | 0.5633 |
| 195 | 0.4978 | 0.5022 | 0.4469 | 0.5531 | 195 | 0.4873 | 0.5127 | 0.4366 | 0.5634 |
| 196 | 0.4977 | 0.5023 | 0.4468 | 0.5532 | 196 | 0.4872 | 0.5128 | 0.4365 | 0.5635 |
| 197 | 0.4976 | 0.5024 | 0.4467 | 0.5533 | 197 | 0.4871 | 0.5129 | 0.4364 | 0.5636 |
| 198 | 0.4975 | 0.5025 | 0.4466 | 0.5534 | 198 | 0.4870 | 0.5130 | 0.4363 | 0.5637 |
| 199 | 0.4974 | 0.5026 | 0.4464 | 0.5536 | 199 | 0.4869 | 0.5131 | 0.4362 | 0.5638 |
| 200 | 0.4972 | 0.5028 | 0.4463 | 0.5537 | 200 | 0.4868 | 0.5132 | 0.4361 | 0.5639 |

*Note: "−" means "similarity" cannot be concluded for any observation in the situation.*

# References

Abbott, J. A. (1973). Sensory assessment of textural attributes of foods. In *Texture Measurements of Foods*. (A. Kramer and A. Szczesniak, eds.) D. Reidel, Washington, D. C.

Acton, F. S. (1959). *Analysis of Straight-line Data*. Dover, New York.

Afifi, A. A., and Azen, S. P. (1979). *Statistical Analysis: A Computer Oriented Approach*, 2nd ed. Academic Press, New York.

Agresti, A. (1990, 2002). *Categorical Data Analysis*, 2nd ed. Wiley, New York.

Agresti, A. (1992). A survey of exact inference for contingency tables. *Statistical Science*, **7**, 131–177.

Agresti, A. (1996). *An Introduction to Categorical Data Analysis*. Wiley, New York.

Aitchison, J., and Brown, J. A. C. (1957). *The Lognormal Distribution*. University Press, Cambridge, England.

Altman, D. G., and Bland, J. M. (1995). Absence of evidence is not evidence of absence. *British Medical Journal*, **311**, 485.

Amerine, M. A., Pangborn, R. M., and Roessler, E. B. (1965). *Principles of Sensory Evaluation of Food*. Academic Press, New York.

Amerine, M. A., and Roessler, E. B. (1976). *Wines: Their Sensory Evaluation*. Freeman, San Francisco, CA.

Anderson, R. L. (1953). Recent advances in finding best operating conditions. *Journal of the American Statistical Association*, **48**, 789–798.

Anderson, R. L. (1959). Use of contingency tables in the analysis of consumer preference studies. *Biometrics*, **15**, 582–590.

Anderson, S., and Hauck, W. W. (1983). A new procedure for testing equivalence in comparative bioavailability and other clinic trials. *Communications in Statistics*, A **12**, 2663–2692.

ASTM. (2001). Standard test method for directional difference test. In *Annual Book of ASTM Standards*, Vol. 15.08.E 2164–01, American Society for Testing and Materials. Philadelphia, Pennsylvania.

ASTM. (2003). Standard practice for estimating Thurstonian discriminal distances. In *Annual Book of ASTM Standards*, Vol. 15.08.E 2262–03, American Society for Testing and Materials. Philadelphia, Pennsylvania.

ASTM. (2004). Standard test method for paired preference test. In *Annual Book of ASTM Standards*, Vol. 15.08.E 2263–04, American Society for Testing and Materials. Philadelphia, Pennsylvania.

ASTM. (2004). Standard test method for sensory analysis—triangle test. In *Annual Book of ASTM Standards*, Vol. 15.08.E 1885–04, American Society for Testing and Materials. Philadelphia, Pennsylvania.

ASTM. (2005). Standard test method for same-different test. In *Annual Book of ASTM Standards*, Vol. 15.08.E 2139–05, American Society for Testing and Materials. Philadelphia, Pennsylvania.

Aust, L. B., ed. (1998). *Cosmetic Claims Substantiation*. Marcel Dekker, Inc., New York.

Aust, L. B., Oddo, P., Wild, J. E., Mills, O. H., and Deupree, J. S. (1987). The descriptive analysis of skin care products by a trained panel of judges. *J Soc Cosmet Chem*, **38**, 443–448.

Baird, J. C., and Noma, E. (1978). *Fundamentals of Scaling and Psychophysics*. Wiley, New York.

Balaam, L. N. (1963). Multiple comparisons—a sampling experiment. *Aust J Stat*, **5**, 62–84.

Bamber, D. (1975). The area above the ordinal dominance graph and the area below the receiver operating characteristic graph. *Journal of Mathematical Psychology*, **12**, 387–415.

Barclay, W. D. (1969). Factorial design in a pricing experiment. *Journal of Market Research*, **6**, 427–429.

Barel, A. O. (2001). Product testing: Moisturizers. In *Bioengineering of the Skin: Skin Biomechanics*. (P. Elsner, E. Berardesca, K. P. Wilhelm, and H. I. Maibach, eds.) CRC Press, Boca Raton, FL: pp. 241–256.

Barlow, R. E., and Proschan, F. (1965). *Mathematical Theory of Reliability*. Wiley, New York.

Basson, R. P. (1959). Incomplete block designs augmented with a repeated control. M.S. Thesis, Iowa State University, Ames, Iowa.

Baten, W. D. (1946). Organoleptic tests pertaining to apples and pears. *Food Res*, **11**, 84–94.

Bendig, A. W., and Hughes, J. B., II. (1953). Effect of amount of verbal anchoring and number of rating scale categories upon transmitted information. *J Exp Psychol*, **46**, 87–90.

Bennett, B. M., and Underwood, R. E. (1970). On McNemar's test for the $2 \times 2$ table and its power function. *Biometrics*, **26**, 339–343.

Berger, R. L. (1982). Multiparameter hypothesis testing and acceptance sampling. *Technometrics*, **24**, 295–300.

Berger, R. L., and Hsu, J. C. (1996). Bioequivalence trials, intersection-union tests and equivalence confidence sets (with discussion). *Statistical Science*, **11**, 283–319.

Berndt, U., and Elsner, P. (2001). Hardware and measuring principle: The cutometer. In *Bioengineering of the Skin: Skin Biomechanics*. (P. Elsner, E. Berardesca, K. P. Wilhelm, and H. I. Maibach, eds.) CRC Press, Boca Raton, Florida, pp. 91–97.

Best, D. J. (1993). Extended analysis for ranked data. *Aust J Statist*, **35**, 257–262.

Best, D. J. (1995). Consumer data—statistical tests for differences in dispersion. *Food Quality and Preference*, **6**, 221–225.

Bi, J. (2002). Variance of *d'* from the same-different method. *Behavior Research Methods, Instruments, & Computers*, **34**, 37–45.

Bi, J. (2003). Agreement and reliability assessments for performance of sensory descriptive panel. *Journal of Sensory Studies*, **18**, 61–76.

Bi, J. (2005). Similarity testing in sensory and consumer research. *Food Quality and Preference*, **16**, 139–149.

Bi, J. (2006). *Sensory Discrimination Tests and Measurements: Statistical Principles, Procedures and Tables*. Blackwell Publishing, Ames, Iowa.

Bi, J. (2007a). Similarity testing using paired comparison method. *Food Quality and Preference*, **18**, 500–507.

Bi, J. (2007b). Replicated similarity testing using paired comparison method. *Journal of Sensory Studies*, **22**, 176–186.

Bi, J. (2007c). Bayesian analysis for proportion with an independent background effect. *British Journal of Mathematical and Statistical Psychology*, **60**, 71–83.

Bi, J., and Ennis, D. M. (2001). The power of the A–Not A method. *Journal of Sensory Studies*, **16**, 343–359.

Bi, J., Ennis D. M., and O'Mahony, M. (1997). How to estimate and use the variance of *d'* from difference tests. *Journal of Sensory Studies*, **12**, 87–104.

Bi, J., and O'Mahony, M. (2007). Updated and extensive table for testing the significance of the R-index. *Journal of Sensory Studies*, **22**, 713–720.

Bickel, P. J., and Doksum, K. A. (1977). *Mathematical Statistics: Basic Ideas and Selected Topics*. Holden-Day, Inc., San Francisco, CA.

Bicking, C. A. (1954). Some uses of statistics in the planning of experiments. *Ind Qual Control*, **10**(4), 20–24.

Birch, M. W. (1963). Maximum likelihood in three-way contingency tables. *J R Stat Soc*, **B25**, 220–233.

Birdsall, T. G., and Peterson, W. W. (1954). Probability of a correct decision in a choice among *m* alternatives. University of Michigan: Electronic Defense Group, Quarterly Progress Report No. 10.

Birnbaum, Z. W. (1956). On a use of the Mann-Whitney statistic. In *Proceedings of the Third Berkeley Symposium on Mathematical Statistics and Probability*, Vol. **1**. (J. Neyman, ed.) University of California Press, Berkeley, California, pp. 13–17.

Bishop, Y. M. M., Fienberg, S. E., and Holland, P. W. (1975). *Discrete Multivariate Analysis: Theory and Practice*. MIT Press, Cambridge, Massachusetts.

Blackwelder, W. C. (1982). "Proving the null hypothesis" in clinical trials. *Controlled Clinical Trials*, **3**, 345–353.

Blyth, C. R. (1972). On Simpson's paradox and the sure-thing principle. *Journal of the American Statistical Association*, **67**, 364–366.

Boardman, T. J., and Moffitt, D. R. (1971). Graphical Monte Carlo Type I error rates for multiple comparison procedures. *Biometrics*, **27**, 738–744.

Bock, R. D., and Jones, L. V. (1968). *The Measurement and Prediction of Judgment and Choice*. Holden-Day, San Francisco, California.

Boen, J. R. (1972). The teaching of personal interaction in statistical consulting. *Am Stat*, **26**, 30–31.

Bofinger, E. (1985). Expanded confidence intervals. *Comm Statist Theory Methods*, **14**, 1849–1864.

Bofinger, E. (1992). Expanded confidence intervals, one-sided tests, and equivalence testing. *Journal of Biopharmaceutical Statistics*, **2**, 181–188.

Bose, R. C. (1956). Paired comparison designs for testing concordance between judges. *Biometrika*, **43**, 113–121.

Bose, R. C., and Shimamoto, T. (1952). Classification and analysis of partially balanced incomplete block designs with two associate classes. *Journal of the American Statistical Association*, **47**, 151–184.

Bowden, D. C., and Graybill, F. (1966). Confidence bands of uniform and proportional width for linear models. *Journal of the American Statistical Association*, **61**, 182–198.

Box, G. E. P. (1954). The exploration and exploitation of response surfaces: Some general considerations and examples. *Biometrices*, **10**, 16–60.

Box, G. E. P., and Hunter, J. S. (1957). Multifactor experimental designs for exploring response surfaces. *Annals of Mathematical Statistics*, **28**, 195–241.

Box, G. E. P., and Wilson, K. B. (1951). On the experimental attainment of optimum conditions. *J R Stat Soc Ser B*, **13**(1), 1–45.

Box, G. E. P., and Youle, P. V. (1955). The exploration and exploitation of response surfaces: An example of the link between the fitted surface and the basic mechanism of the system. *Biometrics*, **11**, 287–323.

Bradley, J. V. (1968). *Distribution—Free Statistical Tests*. Prentice-Hall, Englewood Cliffs, New Jersey.

Bradley, R. A. (1954a). The rank analysis of incomplete block designs. II. Additional tables for the method of paired comparisons. *Biometrika*, **41**, 502–537.

Bradley, R. A. (1954b). Incomplete block rank analysis: On the appropriateness of the model for a method of paired comparisons. *Biometrics*, **10**, 375–390.

Bradley, R. A. (1955). Rank analysis of incomplete block designs. III. Some large-sample results on estimation and power for a method of paired comparisons. *Biometrika*, **42**, 450–470.

Bradley, R. A. (1957). Comparison of difference-from-control, triangle, and duo-trio tests in taste testing: I. Comparable expected performance. Memorandum prepared for the General Foods Corporation, November 12.

Bradley, R. A. (1958a). Determination of optimum operating conditions by experimental methods. Part I. Mathematics and statistics fundamental to the fitting of response surfaces. *Ind Qual Control*, **15**(1), 16–20.

Bradley, R. A. (1958b). Triangle, duo-trio, and difference from control tests in taste testing. *Biometrics*, **14**, 566 (Abstract No. 511).

Bradley, R. A. (1963). Some relationships among sensory difference tests. *Biometrics*, **19**, 385–397.

Bradley, R. A. (1964). Applications of the modified triangle test in sensory difference trials. *Journal of Food Science*, **29**, 668–672.

Bradley, R. A. (1976). Science, statistics and paired comparisons. *Biometrics*, **32**, 213–232.

Bradley, R. A., and Harmon, T. J. (1964). The modified triangle test. *Biometrics*, **20**, 608–625.

Bradley, R. A., and Terry, M. E. (1952). Rank analysis of incomplete block designs. I. The method of paired comparisons. *Biometrika*, **39**, 324–345.

Brier, S. S. (1980). Analysis of contingency tables under cluster sampling. *Biometrika*, **67**, 591–596.

Brockhoff, P. B. (2003). The statistical power of replications in difference tests. *Food Quality and Preference*, **14**, 405–417.

Brockhoff, P. B., and Schlich, P. (1998). Handling replications in discrimination tests. *Food Quality and Preference*, **9**, 303–312.

Brown, J. (1974). Recognition assessed by rating and ranking. *British Journal of Psychology*, **65**, 13–22.

Byer, A. J., and Abrams, D. (1953). A comparison of the triangular and two-sample taste-test methods. *Food Technol (Chicago)*, **7**, 185–187.

Caincross, S. E., and Sjostrom, L. B. (1950). Flavor profiles—A new approach to flavor problems. *Food Technol*, **4**, 308–311.

Calvin, L. D. (1954). Doubly balanced incomplete block designs for experiments in which the treatment effects are correlated. *Biometrics*, **10**, 61–88.

Cameron, J. M. (1969). The statistical consultant in a scientific laboratory. *Technometrics*, **11**, 247–254.

Carmer, S. G., and Swanson, M. R. (1973). An evaluation of ten pairwise multiple comparison procedures by Monte Carlo methods. *Journal of the American Statistical Association*, **68**, 66–74.

Casella, G., and Berger, R. L. (1990). *Statistical Inference*. Wadsworth and Brooks/Cole, Pacific Grove, CA.

Castura, J. C., Findlay, C. J., and Lesschaeve, I. (2006). Monitoring calibration of descriptive sensory panels using distance from target measurements. *Food Quality and Preference*, **17**, 282–289.

Centor, R. M., and Schwartz, J. S. (1985). An evaluation of methods for estimating the area under the receiver operating characteristic (ROC) curve. *Med Decis Making*, **5**, 150–156.

Chambers, IV, E. and Wolf, M. B. (1996). *Sensory Testing Methods*, 2nd ed. ASTM Manual Series: MNL 26. American Society for Testing Materials, West Conshocken, Pennsylvania.

Chan, I. S. F., Tang, N-S., Tang, M-L., and Chan, P-S. (2003). Statistical analysis of noninferiority trials with a rate ratio in small-sample matched-pair designs. *Biometrics*, **59**, 1171–1177.

Chen, T. (1960). Multiple comparisons of population means. M. S. Thesis, Iowa State University, Ames, Iowa.

Chow, S. C., and Liu, J. P. (1992). *Design and Analysis of Bioavailability and Bioequivalence Studies*. Marcel Dekker, New York.

Chow, S. C., and Liu, J. P. (1998). *Design and Analysis of Clinical Trials: Concept and Methodologies*. John Wiley & Sons, Inc., New York.

Clatworthy, W. H. (1955). Partially balanced incomplete block designs with two associate classes and two treatments per block. *J Res Natl Bur Stand*, **54**(4), 177–190.

Cloninger, M. R, Baldwin, R. E., and Krause, G. F. (1976). Analysis of sensory rating scales. *Journal of Food Science*, **41**, 1225–1228.

Cochran, W. G. (1947). Some consequences when the assumptions for the analysis of variance are not satisfied. *Biometrics*, **3**, 22–38.

Cochran, W. G. (1950). The comparison of percentages in matched samples. *Biometrika*, **37**, 256–266.

Cochran, W. G. (1951). Testing a linear relation among variances. *Biometrics*, **7**, 17–32.

Cochran, W. G. (1954). Some methods for strengthening the common $X^2$ tests. *Biometrics*, **10**, 417–451.

Cochran, W. G., and Cox, G. M. (1957). *Experimental Designs*. Wiley, New York.

Cochran, W. G., Autrey, K. M., and Cannon, C. Y. (1941). A double change-over design for dairy cattle feeding experiments. *J Dairy Sci*, **24**, 937–951.

Cohen, J. (1960). A coefficient of agreement for nominal scales. *Educ Psychol Measure*, **20**, 37–46.

Conover, W. J. (1973). On methods of handling ties in the Wilcoxon signed-rank test. *Journal of the American Statistical Association*, **68**, 985–988.

Conover, W. J. (1974). Some reasons for not using the continuity correction on $2 \times 2$ contingency tables. *Journal of the American Statistical Association*, **69**, 374–376.

Conover, W. J. (1999). *Practical Nonparametric Statistics*, 3rd ed. John Wiley & Sons, New York.

Cornell, J. A. (1974a). More on extended complete block designs. *Biometrics*, **30**, 179–186.

Cornell, J. A. (1974b). An alternative estimator for the recovery of interblock information when using incomplete block designs. *Journal of Food Science*, **39**, 1254–1256.

Cornell, J. A., and Knapp, F. W. (1972). Sensory evaluation using composite complete–incomplete block designs. *Journal of Food Science*, **37**, 876–882.

Cornell, J. A., and Knapp, F. W. (1974). Replicated composite-incomplete block designs for sensory experiments. *Journal of Food Science*, **39**, 503–507.

Cornell, J. A., and Schreckengost, J. F. (1975). Measuring panelists' consistency using composite complete–incomplete block designs. *Journal of Food Science*, **40**, 1130–1133.

Cornfield, J. (1951). A method of estimating comparative rates from clinical data. Applications to cancer of the lung, breast and cervix. *J Natl Cancer Inst*, **11**, 1269–1275.

Cox, D. R. (1970). The continuity correction. *Biometrika*, **57**, 217.

Cragle, R. G., Myers, R. M., Waugh, R. K., Hunter, J. S., and Anderson, R. L. (1955). The effects of various levels of sodium citrate, glycerol, and equilibration time on survival of bovine spermatozoa after storage at $-79°C$. *J Dairy Sci*, **38**, 508–514.

Cramer, H. (1946). *Mathematical Method of Statistics*. Princeton University Press, Princeton, New Jersey.

Craven, B. J. (1992). A table of $d'$ for $M$-alternative odd-man-out forced-choice procedures. *Perception & Psychophysics*, **51**, 379–385.

Cross, H. R., Moen, R., and Stanfield, M. S. (1978). Training and testing of judges for sensory analysis of meat quality. *Food Technol*, July issue, 48–54.

Dalhoff, E., and Jul, M. (1965). Factors affecting the keeping quality of frozen foods. *Progr Refrig Sci Technol, Proc Int Congr Refrig, 11th, 1963*, **1**, 57–66.

Daniel, C. (1969). Some general remarks on consulting in statistics. *Technometrics*, **11**, 241–246.

Darlington, R. B. (1970). Is kurtosis really peakness? *Am Stat*, **24**, 19–22.

Darroch, J. N. (1962). Interactions in multi-factor contingency tables. *J Stat Soc B*, **24**, 251–263.

David, H. A. (1963a). The structure of cyclic paired-comparison designs. *J Aust Math Soc*, **3**, 117–127.

David, H. A (1963b). *The Method of Paired Comparisons*. Charles Griffin and Co. Limited, London.

David, H. A., and Trivedi, M. C. (1962). Pair, triangle, and duo–trio tests. Technical Report No. 55, Dept. of Statistics, Virginia Polytechnic Institute, Blacksburg, Virginia.

Davidson, R. R., and Bradley, R. A. (1970). Multivariate paired comparisons: Some large-sample results on estimation and tests of equality of preference. In *Nonparametric Techniques in Statistical Inference*

(M. L. Puri, ed.), Cambridge University Press, London and New York, pp. 111–129.

Davidson, R. R., and Farquhar, P. H. (1976). A bibliography on the method of paired comparisons. *Biometrics*, **32**, 233–244.

Davies, O. L., ed. (1967). *Design and Analysis of Industrial Experiments*. Hafner, New York.

Davies, O. L., and Hay, W. A. (1950). The construction and uses of fractional factorial designs in industrial research. *Biometrics*, **6**, 233–249.

Davis, D. J. (1952). An analysis of some failure data. *Journal of the American Statistical Association*, **47**, 113–150.

Davis, J. G., and Hanson, H. L. (1954). Sensory test methods. I. The triangle intensity (T-I) and related test systems for sensory analysis. *Food Technol (Chicago)*, **8**, 335–339.

DeBaun, R. M. (1956). Block effects in the determination of optimum conditions. *Biometrics*, **12**, 20–22.

De Lury, D. B. (1946). The analysis of Latin squares when some observations are missing. *Journal of the American Statistical Association*, **41**, 370–389.

Detre, K., and White, C. (1970). The comparison of two Poisson-distributed observations. *Biometrics*, **26**, 851–854.

Dijksterhuis, G. B. (1997). *Multivariate Data Analysis in Sensory and Consumer Science*. Blackwell Publishing, Ames, Iowa.

Dixon, W. J. (1953). Power functions of the sign test and power efficiency for normal alternatives. *Annals of Mathematical Statistics*, **24**, 467–473.

Dixon, W. J., and Massey, F. J., Jr. (1969). *Introduction to Statistical Analysis*. McGraw-Hill, New York.

Dong, J. (1998). On avoiding association paradox in contingency tables. *Journal of Systems Science and Mathematical Sciences*, **11**, 272–279.

Dorfman, D. D., and Alf, E. Jr. (1969). Maximum-likelihood estimation of parameters of signal-detection theory and determination of confidence intervals-rating-method data. *Journal of Mathematical Psychology*, **6**, 487–496.

Draper, N. R., Hunter, W. G., and Tierney, D. E. (1969). Analyzing paired comparison tests. *Journal of Market Research*, **6**, 477–480.

Dubey, S. D. (1967). Some percentile estimators for Weibull parameters. *Technometrics*, **9**, 119–129.

Ducharme, G. R., and Lepage, Y. (1986). Testing collapsibility in contingency tables. *Journal of the Royal Statistical Society, Series B*, **48**, 197–205.

Duncan, D. B. (1955). Multiple range and multiple *F* tests. *Biometrics*, **11**, 1–42.

Dunn, O. J. (1968). A note on confidence bands for a regression line over a finite range. *Journal of the American Statistical Association*, **63**, 1028–1033.

Dunnett, C. W. (1955). A multiple comparison procedure for comparing several treatments with a control. *Journal of the American Statistical Association*, **50**, 1096–1121.

Dunnett, C. W. (1964). New tables for multiple comparisons with a control. *Biometrics*, **20**, 482–491.

Dunnett, C. W., and Gent, M. (1977). Significance testing to establish equivalence between treatments, with special reference to data in the form of tables. *Biometrics*, **33**, 593–602.

Dunnett, C. W., and Gent, M. (1996). An alternative to the use of two-sided tests in clinical trials. *Statistics in Medicine*, **15**, 1729–1738.

Durbin, J. (1951). Incomplete blocks in ranking experiments. *Br J Psychol (Part II)*, **4**, 85–90.

Dykstra, O., Jr. (1956). A note on rank analysis of incomplete block designs: Applications beyond the scope of existing tables. *Biometrics*, **12**, 301–306.

Dykstra, O., Jr. (1960). Rank analysis of incomplete block designs: A method of paired comparisons employing unequal repetitions on pairs. *Biometrics*, **16**, 176–188.

EC-GCP. (1993). *Biostatistical Methodology in Clinical Trials in Applications for Marketing Authorization for Medical Products*. CPMP Working Party on Efficacy of Medical Products, Commission of the European Communities, Brussels.

Edwards, A. L. (1952). The scaling of stimuli by the method of successive intervals. *J Appl Psychol*, **36**, 118–122.

Egan, J. P. (1975). *Signal Detection Theory and ROC Analysis*. Academic Press, New York.

Eisenhart, C. (1947). The assumptions underlying the analysis of variance. *Biometrics*, **3**, 1–21.

Elliott, P. B. (1964). Tables of d'. In *Signal Detection and Recognition by Human Observers* (J. A. Swets, ed.) Wiley, New York, pp. 651–684.

Ellis, B. H. (1966). *Guide Book for Sensory Testing*. Continental Can Co., Inc., Chicago, Illinois.

Elsner, P., Berardesca, E., Wilhelm, K. P., and Maibach, H. I., eds. (2001). *Bioengineering of the Skin: Skin Biomechanics.* CRC Press, Boca Raton, Florida.

Ennis, D. M. (1993). The power of sensory discrimination methods. *Journal of Sensory Studies*, 8, 353–370.

Ennis, D. M., and Bi, J. (1998). The Beta-Binomial model: Accounting for inter-trial variation in replicated difference and preference tests. *Journal of Sensory Studies*, 13, 389–412.

Ennis, D. M., and Bi, J. (1999). The Dirichlet-multinomial model: Accounting for inter-trial variation in replicated ratings. *Journal of Sensory Studies*, 14, 321–345.

Everitt, B. S. (1968). Moments of the statistics kappa and weighted kappa. *British Journal of Mathematical and Statistical Psychology*, 21, 97–103.

Everitt, B. S. (1992). *The Analysis of Contingency Tables*, 2nd ed., Chapman & Hall, London.

FDA. (1992). "Bioavailability and Bioequivalence Requirements." In *U. S. Code of Federal Regulations* (**vol. 21**, chap. 320), U. S. Government Printing Office, Washington, D.C.

Federer, W. T. (1955). *Experimental Design*. Macmillan, New York.

Ferris, G. E. (1957). A modified Latin square design for taste testing. *Food Res*, 22, 251–258.

Ferris, G. E. (1958). The *k*-visit method of consumer testing. *Biometrics*, 14, 39–49.

Filipello, F. (1956). A critical comparison of the two-sample and triangular binomial designs. *Food Res*, 21, 239–241.

Finney, D. J. (1945). The fractional replication of factorial arrangements. *Ann Eugen*, 12, 291–301.

Finney, D. J. (1946). Recent developments in the design of field experiments. III. Fractional replication. *J Agric Sci*, 36, 184–191.

Finney, D. J. (1960). *An Introduction to the Theory of Experimental Design*. University of Chicago Press, Chicago, Illinois.

Fisher, R. A. (1934). *Statistical Methods for Research Workers*, 5th ed., Oliver and Boyd, Edinburgh.

Fisher, R. A. (1935, 1966). *The Design of Experiments*. Oliver and Boyd, London.

Fisher, R. A. (1962). Confidence limits for a cross-product ratio. *Austral J Statist*, 4, 41.

Fisher, R. A., and Yates, F. (1963). *Statistical Tables for Biological, Agricultural and Medical Research*. Hafner, New York.

Fleiss, J. L., Cohen, J., and Everitt, B. S. (1969). Large sample standard errors of kappa and weighted kappa. *Psychol Bull*, 72, 323–327.

Fleiss, J. L., and Everitt, B. S. (1971). Comparing the marginal totals of square contingency tables. *British Journal of Mathematical and Statistical Psychology*, 24, 117–123.

Folks, J. L., and Antle, C. E. (1967). Straight line confidence regions for linear models. *Journal of the American Statistical Association*, 62, 1365–1374.

Friedman, M. (1937). The use of ranks to avoid the assumption of normality implicit in the analysis of variance. *Journal of the American Statistical Association*, 32, 675–701.

Friedman, M. (1940). A comparison of alternative tests of significance for the problem of *m* rankings. *Annals of Mathematical Statistics*, 11, 86–92.

Frijters, J. E. R. (1979a). The paradox of discriminatory non-discriminators resolved. *Chemical Senses & Flavour*, 4, 355–358.

Frijters, J. E. R. (1979b). Variations of the triangular method and the relationship of its unidimensional probabilistic models to three-alternative forced-choice signal detection theory models. *British Journal of Mathematical and Statistical Psychology*, 32, 229–241.

Frijters, J. E. R. (1982). Expanded tables for conversion of sensory difference ($d'$) for the triangular method and the 3–alternative forced choice procedure. *Journal of Food Science*, 47, 139–143.

Frijters, J. E. R., Kooistra, A., and Vereijken, P. F. G. (1980). Tables of $d'$ for the triangular method and the 3–AFC signal detection procedure. *Perception & Psychophysics*, 27, 176–178.

Gacula, M., Jr., Mohan, P., Faller, J., Pollack, L., and Moskowitz, H. R. (2008). Questionnaire practice: What happens when the JAR scale is placed between two "overall acceptance" scales? *Journal of Sensory Studies*, 23, 136–147.

Gacula, M., Jr., Rutenbeck, S., Pollack, L., Resurreccion, A. V. A., and Moskowitz, H. R. (2007). The just-about-right intensity scale: Functional analyses and relation to hedonics. *Journal of Sensory Studies*, 22, 194–211.

Gacula, M. C., Jr. (1975). The design of experiments for shelf life study. *Journal of Food Science*, 40, 399–403.

Gacula, M. C., Jr. (1978). Analysis of incomplete block design with reference samples in every block. *Journal of Food Science*, 43, 1461–1466.

Gacula, M. C., Jr. (1993). *Design and Analysis of Sensory Optimization*. Blackwell Publishing, Ames, Iowa.

Gacula, M. C., Jr. (2003). Applications of SAS® programming language in sensory science. In *Viewpoints and Controversies in Sensory Science and Consumer Product Testing* (H. R. Moskowitz, A. M. Munoz, and M. C. Gacula Jr, eds.) Blackwell Publishing, Ames, Iowa, pp. 433–458.

Gacula, M. C., Jr., Altan, S. S., and Kubala, J. J. (1972). Data analysis: Empirical observations on statistical tests used in paired design. *Journal of Food Science*, **37**, 62–65.

Gacula, M. C., Jr., and Kubala, J. J. (1972). Data analysis: Interblock and intrablock estimates of variance on taste panel data. *Journal of Food Science*, **37**, 832–836.

Gacula, M. C., Jr., and Kubala, J. J. (1975). Statistical models for shelf life failures. *Journal of Food Science*, **40**, 404–409.

Gacula, M. C., Jr, Moran, M. J., and Reaume, J. B. (1971). Use of the sign test in sensory testing. *Food Prod Dev*, October.

Gacula, M. C., Jr., Parker, L., Kubala, J. J., and Reaume, J. (1974). Data analysis: A variable sequential test for selection of sensory panels. *Journal of Food Science*, **39**, 61–63.

Garner, W. R., and Hake, H. W. (1951). The amount of information in absolute judgments. *Psychol Rev*, **58**, 446–459.

Gibbons, J. O. (1964). Effect of non-normality on the power function of the sign test. *Journal of the American Statistical Association*, **59**, 142–148.

Goldsmith, C. H., and Gaylor, D. W. (1970). Three stage nested designs for estimating variance components. *Technometrics*, **12**, 487–498.

Goldthwaite, L. (1961). Failure rate study for the log-normal lifetime model. *Proc 7th Natl Symp Reliabil Qual Control*, 208–213.

Goodman, L. A., and Kruskal, W. H. (1954). Measure of association for cross-classifications. *Journal of the American Statistical Association*, **49**, 732–764.

Gourevitch, V., and Galanter, E. (1967). A significance test for one parameter isosensitivity functions. *Psychometrika*, **32**, 25–33.

Graybill, F. A., and Bowden, D. C. (1967). Linear segment confidence bands for simple linear models. *Journal of the American Statistical Association*, **62**, 403–408.

Green, D. M. (1964). General prediction relating yes-no and forced-choice results. *J Acoust Soc Amer*, **36**, 1042 (Abstract).

Green, D. M., and Birdsall, T. G. (1964). The effect of vocabulary size on articulation score. In *Signal Detection and Recognition by Human Observers* (J. A. Swets, ed.) Wiley, New York.

Green, D. M., and Swets, J. A. (1966). *Signal Detection—Theory and Psychophysics*. John Wiley, New York.

Greenberg, B. G. (1951). Why randomize? *Biometrics*, **7**, 309–322.

Gregson, R. A. M. (1960). Bias in the measurement of food preferences by triangular tests. *Occup Psychol*, **34** (4), 249–257.

Grey, D. R., and Morgan, B. J. T. (1972). Some aspects of ROC curve-fitting: Normal and logistic models. *Journal of Mathematical Psychology*, **9**, 128–139.

Gridgeman, N. T. (1955a). Taste comparisons: Two samples or three. *Food Technol (Chicago)*, **9**, 148–150.

Gridgeman, N. T. (1955b). The Bradley–Terry probability model and preference tasting. *Biometrics*, **11**, 335–343.

Gridgeman, N. T. (1959). Pair comparison, with and without ties. *Biometrics*, **15**, 382–388.

Gridgeman, N. T. (1964). Sensory comparisons: The two-stage triangle test with sample variability. *Journal of Food Science*, **29**, 112–117.

Gridgeman, N. T. (1970a). A reexamination of the two-stage triangle test for the perception of sensory difference. *Journal of Food Science*, **35**, 87–91.

Gridgeman, N. T. (1970b). Designs for multiple pair-comparison sensory tests. *Can Inst Food Technol J*, **3**, 25–28.

Griffin, R., and Stauffer, L. (1991). Product optimization in central-location testing and subsequent validation and calibration in home-use testing. *Journal of Sensory Studies*, **5**, 231–240.

Griffiths, C. E., Wang, T. S., Hamilton, T. A., et al. (1992) A photonumeric scale for the assessment of cutaneous photodamage. *Arch Dermatol*, **128**, 347–351.

Grim, A. C., and Goldblith, S. A. (1965). Some observed discrepancies in application of the triangle test to evaluation of irradiated whole-egg magma. *Food Technol (Chicago)*, **19**, 1452.

Grizzle, J. E. (1967). Continuity correction in the $X^2$-test for 2 × 2 tables. *Am Stat*, **21**, 28–32.

Guilford, J. P. (1954). *Psychometric Methods*. McGraw-Hill, New York.

Gulliksen, H., and Messick, S. (1960). *Psychological Scaling: Theory and Applications*. Wiley, New York.

Hacker, M. J., and Ratcliff, R. (1979). A revised table of $d'$ for m-alternative forced choice. *Perception & Psychophysics*, **26**, 168–170.

Hahn, G. J., and Nelson, W. (1970). A problem in the statistical comparison of measuring devices. *Technometerics*, **12**, 95–101.

Hanley, J. A. and Mcneil, B. J. (1982). The meaning and use of the area under a receiver operating characteristic (ROC) curve. *Radiology*, **43**, 29–36.

Hanson, H. L., Kline, L., and Lineweaver, H. (1951). Application of balanced incomplete block design to scoring of ten dried egg samples. *Food Technol*, **5**, 9–13.

Harman, H. H. (1967). *Modern Factor Analysis*. University of Chicago Press, Chicago and London.

Harmon, T. J. (1963). The modified triangle test. M. S. Thesis, Florid State Univ., Tallahasee, Florida.

Harries, J. K., and Smith, G. L. (1982). The two-factor triangle test. *Journal of Food Technology*, **17**, 153–162.

Harris, M., Horvitz, D. G., and Mood, A. M. (1948). On the determination of sample size in designing experiments. *Journal of the American Statistical Association*, **43**, 391–402.

Harter, H. L. (1957). Error rates and sample sizes for range tests in multiple comparisons. *Biometrics*, **13**, 511–536.

Harter, H. L. (1960). *Expected Values of Normal Order Statistics*. Technical Report 60-292, Aeronautical Research Laboratories, Wright–Patterson Air Force Base, Ohio.

Hartley, H. O. (1959). Smallest composite designs for quadratic response surfaces. *Biometrics*, **15**, 611–624.

Harvey, W. R. (1960). Least squares analysis of data with unequal subclass numbers. *U.S. Dept Agric, Agric Res Serv, ARS*, 20–28.

Harville, D. A. (1975). Experimental randomization: Who needs it? *Am Stat*, **29**, 27–31.

Hauck, W. W., and Anderson, S. (1984). A new statistical procedure for testing equivalence in two-group comparative bioavailability trials. *Journal of Pharmacokinetics and Biopharmaceutics*, **12**, 72–78.

Henderson, C. R. (1959). Design and analysis of animal science experiments. In *Techniques and Procedures in Animal Science Research*. ASAS, Champaign, IL.

Hochberg, Y., and Tamhane, A. C. (1987). *Multiple Comparison Procedures*. Wiley, New York.

Hollander, M., and Wolfe, D. A. (1999). *Nonparametric Statistical Methods*, 2nd ed., John Wiley & Sons, Inc., New York.

Hopkins, J. W. (1950). A procedure for quantifying subjective appraisals of odor, flavor and texture of foodstuffs. *Biometrics*, **6**, 1–16.

Hopkins, J. W. (1954). Incomplete block rank analysis: Some taste test results. *Biometrics*, **10**, 391–399.

Hopkins, J. W., and Gridgeman, N. T. (1955). Comparative sensitivity of pair and triad flavor intensity difference tests. *Biometrics*, **11**, 63–68.

Horsnell, G. (1969). A theory of consumer behaviors derived from repeat paired preference testing. *J R Statist Soc A*, **132**, 164–193.

Hsu, J. C. (1984). Constrained two-sided simultaneous confidence intervals for multiple comparisons with the best. *Ann Statist*, **12**, 1136–1144.

Hsu, J. C., Hwang, J. T. G., Lui, H.-K., and Ruberg, S. J. (1994). Confidence intervals associated with tests for bioequivalence. *Biometrika*, **81**, 103–114.

Hunter, E. A., Piggot, J. R., and Lee, K. Y. M. (2000). Analysis of discrimination tests. In *Proceedings: 6th Conference on Food Industry and Statistics*. January 19–21, Pau, France.

Hunter, J. S. (1958a). A discussion of rotatable designs. *ASQC Conv Trans*, 531–543.

Hunter, J. S. (1958b). Determination of optimum operating conditions by experimental methods. Part II-1. Models and Methods. *Ind Qual Control*, **15**(6), 16–24.

Hunter, J. S. (1959). Determination of optimum operating conditions by experimental methods. Part II-2, 3. Models and Methods. *Ind Qual Control*, **15**(7), 7–15; **15**(8), 6–14.

Huntsberger, D. V., and Billingsley, P. (1981). *Elements of Statistical Inference*. Allyn and Bacon, Boston, MA.

Hyams, L. (1971). The practical psychology of biostatistical consultation. *Biometrics*, **27**, 201–211.

Irwin, R. J., Hautus, M. J., and Butcher, J. (1999). An area theorem for the same-different experiment. *Perception & Psychophysics*, **61**, 766–769.

ISO. (1983). Sensory analysis—Methodology—Paired comparison test. ISO 5495. International Organization for Standardization. Geneva, Switzerland.

ISO. (1983). Sensory analysis—Methodology—Triangular test. ISO 4120. International Organization for Standardization. Geneva, Switzerland.

ISO. (1987). Sensory analysis—Methodology—"A"-"not A" test. ISO 8588. International Organization for Standardization. Geneva, Switzerland.

ISO. (1991). Sensory analysis—Methodology—Duo-trio test. ISO 10399. International Organization for Standardization. Geneva, Switzerland.

Jackson, J. E., and Fleckenstein, M. (1957). An evaluation of some statistical techniques used in the analysis of paired comparison data. *Biometrics*, **13**, 51–64.

Jedlicka, G. J., Wilcox, J. C., McCall, W. A., and Gacula, M. C., Jr. (1975). Effects of varying levels of nitrate and nitrite on *Staphylococcus aureus* growth in pepperoni. *Abst Ann Meet Am Soc Microbiol, 200.*

Jellinek, G. (1964). Introduction to and critical review of modern methods of sensory analysis (odor, taste, and flavor evaluation) with special emphasis on descriptive sensory analysis (flavor profile method). *J Nutr Diet*, **1**, 219–260.

Jobson, J. D. (1992). *Applied Multivariate Data Analysis: Categorical and Multivariate Methods*. Volume II. Springer-Verlag, New York.

John, J. A., Wolock, F. W., David, H. A., Cameron, J. M., and Speckman, J. A. (1972). Cyclic designs. *Natl Bur Stand Appl Math Ser*, No. 62.

John, P. W. M. (1963). Extended complete block design. *Aust J Stat*, **5**, 147–152.

John, P. W. M. (1971). *Statistical Design and Analysis of Experiments*. Macmillan, New York.

John, P. W. M. (1980). *Incomplete Block Designs*. Dekker, New York.

Johnson, E. E. (1963). Sample size for specified confidence interval. *Ind Qual Control*, **20**, 40–41.

Jonckheere, A. R. (1954). A distribution-free $k$-sample test against ordered alternatives. *Biometrika*, **41**, 133–145.

Jones, B., and Kenward, M. G. (1989). *Design and Analysis of Cross-Over Trials*. Chapman and Hall, New York.

Jones, L. V., Peryam, D. R., and Thurstone, L. L. (1955). Development of a scale for measuring soldier's food preferences. *Food Res*, **20**, 512–520.

Jones, L. V., and Thurstone, L. L. (1955). The psychophysics of semantics: An experimental investigation. *J Appl Psychol*, **39**, 31–36.

Kaiser, H. F. (1960). The application of electronic computers to factor analysis. *Educational & Psychological Measurement*, **20**, 141–151.

Kaplan, H. L. Macmillan, N. A., and Creelman, C. D. (1978). Tables of $d'$ for variable-standard discrimination paradigms. *Behavior Research Methods & Instrumentation*, **10**, 796–813.

Kappes, U. P., and Elsner, P. (2003). Clinical and photographic scoring of skin aging. *Skin Pharmacol Appl Skin Physiol*, **16**, 100–107.

Kastenbaum, M. A., and Hoel, D. G. (1970). Sample size requirements: Randomized block designs. *Biometrika*, **57**, 573–577.

Kastenbaum, M. A., Hoel, D. G., and Bowman, K. O. (1970). Sample size requirements: One-way analysis of variance. *Biometrika*, **57**, 421–430.

Katti, S. K. (1968). Exact distribution of chi-square test in a one-way table. *Biometrics*, **24**, 1034. (Abstr.)

Kempthorne, O. (1947). A simple approach to confounding and fractional replication in factorial experiment. *Biometrika*, **34**, 255–272.

Kempthorne, O. (1952). *The Design and Analysis of Experiments*. Wiley, New York.

Kendall, M. G. (1955). Further contributions to the theory of paired comparisons. *Biometrics*, **11**, 43–62.

Kendall, M. G. (1962, 1970). *Rank Correlation Methods*, 4th ed. Hafner Press, New York.

Kendall, M. G., and Babington-Smith, B. (1939). The problem of $m$ rankings. *The Annals of Mathematical Statistics*, **10**, 275–287.

King, S. C., and Henderson, C. R. (1954). Variance components analysis in heritability studies. *Poult Sci*, **33**, 147–154.

King, S. C., Meiselman, H. L, Hottenstein, A. W., Work, T. M., and Cronk, V. 2007. The effects of contextual variables on food acceptability: A confirmatory study. *Food Qual Prefer*, **18**, 58–65.

King, S. C., Weber, A. J, and Meiselman, H. L. (2004). The effect of meal situation, social interaction, physical environment and choice on food acceptability. *Food Qual Prefer*, **15**, 645–653.

Kirk, R. E. (1968). *Experimental Design: Procedures for the Behavioral Sciences*. Brooks/Cole, Belmont, CA.

Kissell, L. T., and Marshall, B. D. (1962). Multifactor responses of cake quality to basic ingredients ratios. *Cereal Chem*, **39**, 16–30.

Kleinman, J. C. (1973). Proportions with extraneous variance: Single and independent samples. *Journal of the American Statistical Association*, **68**, 46–54.

Koehler, K. J., and Wilson, J. R. (1986). Chi-square tests for comparing vectors of proportions for several cluster samples. *Communication in Statistics-Theory and Methods*, **15**, 2977–2990.

Kramer, C. Y. (1956). Extension of multiple range tests to group means with unequal numbers of replications. *Biometrics*, **12**, 307–310.

Kruskal, W. H., and Wallis, W. A. (1952). Use of ranks in one-criterion variance analysis. *Journal of the American Statistical Association*, **47**, 583–621.

Krutchkoff, R. G. (1967). Classical and inverse regression methods of calibration. *Technometrics*, **9**, 425–439.

Kunert, J., and Meyners, M. (1999). On the triangle test with replications. *Food Quality and Preference*, **10**, 477–482.

Larmond, E. (1967). Methods for sensory evaluation of food. *Publ Can Dept Agric*, No. 1284.

Larmond, E. (1977). Laboratory methods for sensory evaluation of food. *Publ Dept Agric*, No. 1637.

Larmond, E., Petrasovits, A., and Hill, P. (1969). Application of multiple paired comparisons in studying the effect of aging and finish on beef tenderness. *Can J Anim Sci*, **49**, 51–58.

Larnier C, Ortonne JP, Venot A, et al. (1994). Evaluation of cutaneous photodamage using a photographic scale. *Br J Dermatol*, **130**, 167–173.

Larsen, T. H., and Jemec, G. B. E. (2001). Skin mechanics and hydration. In *Bioengineering of the Skin: Skin Biomechanics*. (P. Elsner, E. Berardesca, K. P. Wilhelm, and H. I. Maibach, eds.) CRC Press, Boca Raton, Florida, pp. 199–206.

Latreille, J., Mauger, E., Ambroisine, L., Tenenhaus, M., Vincent, M., Navarro, S., and Guinot, C. (2006). Measurement of the reliability of sensory panel performances. *Food Quality and Preference*, **17**, 369–375.

Lawal, B. (2003). *Categorical Data Analysis with SAS and SPSS Application*. Lawrence Erlbaum Association Publishers, Mahwah, NJ.

Lawless, H. T., and Heymann, H. (1998). *Sensory Evaluation of Food: Principles and Practices*. Chapman & Hall, New York.

LeClerg, E. L. (1957). Mean separation by the functional analysis of variance and multiple comparisons. *U.S., Agric Res Serv, ARS* 20–23.

Lehman, S. Y. (1961). Exact and approximate distributions for the Wilcoxon statistic with ties. *Journal of the American Statistical Association*, **56**, 293–298.

Lehmann, E. L. (1975). *Nonparametrics: Statistical Methods Based on Ranks*. Holden-Day, San Francisco, CA.

Lentner, M. M. (1965). Listing expected mean square components. *Biometrics*, **21**, 459–466.

Lewis, B. N. (1962). On the analysis of interaction in multidimensional contingency tables. *J R Stat Soc*, **A125**, 88–117.

Liebetrau, A. M. (1983). *Measures of Association*. Sage, Beverly Hills, CA.

Liu, H. K. (1990). Confidence intervals in bioequivalence. In *Proc Biopharmaceut Sect*. American Statistical Association, Washington, D.C., pp. 51–54.

Liu, J. P., Hsueh, H. M., Hsieh, E. H., and Chen, J. J. (2002). Tests for equivalence or non-inferiority for paired binary data. *Statistics in Medicine*, **21**, 231–245.

Lockhart, E. E. (1951). Binomial systems and organoleptic analysis. *Food Technol (Chicago)*, **5**, 428–431.

Lombardi, G. J. (1951). The sequential selection of judges for organoleptic testing. M.S. Thesis, Virginia Polytechnic Institute, Blacksburg, Virginia.

Lorenz, K., Welsh, J., Normann, R., Beetner, G., and Frey, A. (1974). Extrusion processing of *Triticale*. *Journal of Food Science*, **39**, 572–576.

Louis, T. A., Lavori, P. W., Bailar, J. C., and Polansky, M. (1984). Crossover and self-controlled designs in clinical research. *New England Journal of Medicine*, **310**, 24–31.

Lu, Y., and Bean, J. A. (1995). On the sample size for one-sided equivalence of sensitivities based upon McNemar's test. *Statistics in Medicine*, **14**, 1831–1839.

Lucas, H. L. (1951). Bias in estimation of error in change-over trials with dairy cattle. *J Agric Sci*, **41**, 146.

Luce, R. D. 1994. Thurstone and sensory scaling: Then and now. *Psychol Rev*, **101**, 271–277.

MacDonald, I. A., and Bly, D. A. (1963). *The Determination of the Optimal Levels of Several Emulsifiers in Cake Mix Shortenings*. Atlas Chemical Industries, Inc., Chemical Division, Wilmington, DE.

Macmillan, N. A., and Creelman, C. D. (1991, 2005). *Detection Theory: A User's Guide*. Cambridge University Press, New York.

Macmillan, N. A., Kaplan, N. L., and Creelman, C. D. (1977). The psychophysics of categorical perception. *Psychological Review*, **84**, 452–471.

Mann, H. B., and Whitney, D. R. (1947). On a test of whether one of two random variables is stochastically larger than the other. *The Annals of Mathematical Statistics*, **18**, 50–60.

Mann, N. R., Schafer, R. E., and Singpurwalla, N. D. (1974). *Methods for Statistical Analysis of Reliability and Life Data*. Wiley, New York.

Mantel, N. (1963). Chi-square tests with one degree of freedom: Extensions of the Mantel-Haenszel procedure. *Journal of the American Statistical Association*, **58**, 690–700.

Mantel, N. (1976). The continuity correction. *Am Stat*, **69**, 103–104.

Mantel, N., and Haenszel, W. (1959). Statistical aspects of the analysis of data from retrospective studies of disease. *J Natl Cancer Inst*, **22**, 719–748.

Marascuilo, L. A. (1966). Large sample multiple comparisons. *Psychological Bulletin*, **65**, 280–290.

Marascuilo, L. A. (1970). Extension of the significance test for one-parameter signal detection hypotheses. *Psychometrika*, **35**, 237–243.

Maxwell, A. E. (1970). Comparing the classification of subjects by two independent judges. *British Journal of Psychiatry*, **116**, 651–655.

McKeon, J. J. (1960). Some cyclical incomplete paired comparison designs. University of North Carolina, The Psychometric Laboratory Rept. No. 24, Chapel Hill, NC.

McNeil, D. R. (1967). Efficiency loss due to grouping in distribution-free tests. *Journal of the American Statistical Association*, **62**, 954–965.

McNemar, Q. (1947). Note on the sampling error of the difference between correlated proportions of percentages. *Psychometrika*, **12**, 153–157.

Meilgaard, M., Civille, G. V., and Carr, B. T. (1991, 1999). *Sensory Evaluation Techniques*, 3rd ed. CRC Press, Boca Raton, FL.

Menon, M. V. (1963). Estimation of the shape and scale parameters of the Weibull distribution. *Technometrics*, **5**, 175–182.

Merrington, M., and Thompson, C. M. (1943). Tables of percentage points of the inverted beta (*F*) distribution. *Biometrika*, **33**, 73–99.

Metzler, C. M., and Haung, D. C. (1983). Statistical methods for bioavailability and bioequivalence. *Clin Res Practices & Drug Reg Affairs*, **1**, 109–132.

Meyners, M., and Brockhoff, P. B. (2003). The design of replicated difference tests. *Journal of Sensory Studies*, **18**, 291–324.

Miller, R. G., Jr. (1981). *Simultaneous Statistical Inference*. Springer-Verlag, New York.

Milliken, G. A., and Graybill, F. A. (1970). Extensions of the general linear hypothesis model. *Journal of the American Statistical Association*, **65**, 797–807.

Millman, J. (1967). Rules of thumb for writing the ANOVA table. *J Ed Meas*, **4**, 41–51.

Morgan, B. J. T. (1992). *Analysis of Quantal Response Data*. Chapman & Hall, London.

Morrison, D. G. (1978). A probability model for forced binary choices. *The American Statistician*, **23**(1), 23–25.

Morrissey, J. H. (1955). New method for the assignment of psychometric scale values from incomplete paired comparisons. *J Opt Soc Am*, **45**, 373–378.

Morsimann, J. E. (1962). On the compound multinomial distribution, the multivariate $\beta$-distribution and correlation among proportion. *Biometrika*, **49**, 65–82.

Moskowitz, H. R. (1970). Sweetness and intensity of artificial sweeteners. *Percept Psychophys*, **8**, 40–42.

Moskowitz, H. R. (1971). Ratio scales of acid sourness. *Percept Psychophys*, **9**, 371–374.

Moskowitz, H. R. (1983). *Product Testing and Sensory Evaluation of Foods: Marketing and R&D Approaches*. Blackwell Publishing, Ames, Iowa.

Moskowitz, H. R. (1984). *Cosmetic Product Testing: A Modern Psychophysical Approach*. Marcel Dekker, Inc., New York.

Moskowitz, H. R. (1988). *Applied Sensory Analysis of Foods*. CRC Press, Boca Raton, Florida.

Moskowitz, H. R., and Arabie, P. (1970). Taste intensity as a function of stimulus concentration and solvent viscosity. *J Texture Stud*, **1**, 502–510.

Moskowitz, H. R., Munoz, A. M., and Gacula, M. C., Jr. (2003). *Viewpoints and Controversies in Sensory Science and Consumer Product Testing*. Blackwell Publishing, Ames, Iowa.

Moskowitz, H. R., and Sidel, J. L. (1971). Magnitude and hedonic scales of food acceptability. *Journal of Food Science*, **36**, 677–680.

Mosteller, F. (1951a). Remarks on the method of paired comparisons: I. The least squares solution assuming equal standard deviations and equal correlations. *Psychometrika*, **16**, 3–9.

Mosteller, F. (1951b). Remarks on the method of paired comparisons: II. The effect of an aberrant standard deviation when equal standard deviations and equal correlations are assumed. *Psychometrika, 16,* 203–206.

Mosteller, F. (1951c). Remarks on the method of paired comparisons: III. A test of significance for paired comparisons when equal standard deviations and equal correlations are assumed. *Psychometrika, 16,* 207–218.

Muller-Cohrs, J. (1991). An improvement on commonly used tests in bioequivalence assessment. *Biometrical J, 33,* 357–360.

Myers, R. H. (1971). *Response Surface Methodology.* Allyn and Bacon, Boston, Massachusetts.

Naik, V. D. (1974). On tests of main effects and interactions in higher-way layouts in the analysis of variance random effects model. *Technometrics, 16,* 17–25.

Nam, J-M. (1997). Establishing equivalence of two treatments and sample size requirements in matched-pairs design. *Biometrics, 53,* 1422–1430.

Nelson, L. S. (1970). Tables for Wilcoxon's rank sum test in randomized blocks. *J Qual Technol, 2,* 207–218.

Nelson, W. (1968). Hazard plotting for reliability analysis of incomplete failure data. GE Res and Dev Center Rept. No. 68–C-313. Schenectady, New York.

Nelson, W. (1969). Hazard plotting for incomplete failure data. *J Qual Technol, 1,* 27–52.

Nelson, W. (1970). Hazard plotting methods for analysis of life data with different failure modes. *J Qual Technol, 2,* 126–149.

Nelson, W. (1972). Theory and applications of hazard plotting for censored failure data. *Technometrics, 14,* 945–966.

Nemenyi, P. (1963). Distribution-free multiple comparisons. Unpublished doctoral dissertation. Princeton University, Princeton, New York.

Neter, J., Kutner, M. H., Nachtsheim, C. J., and Wasserman, W. 1990. *Applied Linear Statistical Models.* Irwin, Chicago, IL.

Nielsen, M. A., Coulter, S. T., Morr, C. V., and Rosenau, J. R. (1972). Four factor response surface experimental design for evaluating the role of processing variables upon protein denaturation in heated whey systems. *J Dairy Sci, 56,* 76–83.

Noreen, D. L. (1981). Optimum decision rules for common psychophysical paradigms. In *Mathematical Psychology and Psychophysiology: Seminars in Applied Mathematics.* (S. Grossberg, ed.) American

Mathematical Society, Providence, RI: **Vol. 13,** pp. 237–280.

O'Mahony, M. (1979). Short-cut signal detection measures for sensory analysis. *Journal of Food Science, 44,* 302–303.

O'Mahony, M. (1992). Understanding discrimination tests: A user-friendly treatment of response bias, rating and ranking R-index tests and their relationship to signal detection. *Journal of Sensory Studies, 7,* 1–47.

O'Mahony, M. (1986). *Sensory Evaluation of Food: Statistical Methods and Procedures.* Marcel Dekker, New York.

O'Mahony, M., Heintz, C., and Autio, J. (1978). Signal detection difference testing of colas using a modified R-index approach. *IRCS Medical Sci, 6,* 222.

Odeh, R. E. (1977). The exact distribution of Page's L-statistic in the two-way layout. *Comm Stat Simulation Comput, 6,* 49–61.

Odesky, S. H. (1967). Handling the neutral vote in paired comparison product testing. *Journal of Market Research, 4,* 199–201.

Ogilvie, J. C., and Creelman, C. D. (1968). Maximum-likelihood estimation of receiver operating characteristic curve parameters. *Journal of Mathematical Psychology, 5,* 377–391.

Olds, E. G. (1938). Distributions of sums of squares of rank differences for small numbers of individuals. *Annals Math Stat, 9,* 133–148.

Olmstead, P. S., and Tukey, J. W. (1947). A corner test for association. *Annals of Mathematical Statistics, 18,* 495–513.

Ostle, B., and Mensing, R. (1975). *Statistics in Research.* Iowa State University Press, Ames, Iowa.

Page, E. B. (1963). Ordered hypotheses for multiple treatments: A significance test for linear ranks. *Journal of the American Statistical Association, 58,* 216–230.

Palmer, A. Z., Brady, D. E., Naumann, H. D., and Tucker, L. N. (1953). Deterioration in frozen pork as related to fat composition and storage temperature. *Food Technol (Chicago), 7,* 90–95.

Patterson, H. D. (1950). The analysis of change-over trials. *J Agric Sci, 40,* 375–380.

Patterson, H. D., and Lucas, H. L. (1962). Change-over designs. *NC State Univ Tech Bull, No. 147.*

Paul, S. R., Liang, K. Y., and Self, S. G. (1989). On testing departure from the binomial and multinomial assumptions. *Biometrics*, **45**, 231–236.

Pearce, S. C. (1960). Supplemented balance. *Biometrika*, **47**, 263–271.

Pearson, K. (1904). On the theory of contingency and its relation to association and normal correlation. *Draper's Co Res Mem Biometric Ser 1*. Reprinted [1948] in *Karl Pearson's Early Papers*, Cambridge University Press.

Pearson, E. S., and Hartley, H. O. (1958, 1970). *Biometrika Tables for Statisticians*, Vol. I, Cambridge University Press, London.

Pecore, S., Stoer, H., Hooge, S., Holschuh, N., Hulting, F., and Case, F. (2006). Degree of difference testing: A new approach incorporating control lot variability. *Food Quality and Preference*, **17**, 552–555.

Peng, K. C. (1967). *The Design and Analysis of Scientific Experiments*. Addison-Wesley, Reading, MA.

Peryam, D. R. (1958). Sensory difference tests. *Food Technol*, **12**, 231–236.

Peryam, D. R., and Girardot, H. F. (1952). Advanced taste-test method. *Food Eng*, **24**, 58–61, 194.

Peryam, D. R., and Swartz, V. W. (1950). Measurement of sensory differences. *Food Technol (Chicago)*, **4**, 390–395.

Petrinovich, L. F., and Hardyck, C. D. (1969). Error rates for multiple comparison methods: Some evidence concerning the frequency of erroneous conclusions. *Psychol Bull*, **71**, 43–54.

Piggott, J. R., and Harper, R. (1975). Ratio scales and category scales of odour intensity. *Chemical Senses and Flavor*, **1**, 307–316.

Pinnagoda, J., Tupker, R. A., and Serup, J. (1990). Guidelines for transepidermal water loss (TEWL) measurement. *Contact Dermatitis*, **22**, 164–178.

Pirie, W. R., and Hamdan, M. A. (1972). Some revised continuity corrections for discrete distributions. *Biometrics*, **28**, 693–701.

Plackett, R. L., and Burman, J. P. (1946). The design of optimum multifactorial experiments. *Biometrika*, **33**, 305–325.

Pollack, I. (1952). The assimilation of sequentially encoded information. *Am Psychol*, **6**, 266–267.

Pratt, J. W. (1959). Remarks on zeros and ties in the Wilcoxon signed rank procedures. *Journal of the American Statistical Association*, **54**, 655–667.

Prentice, R. L. (1986). Binary regression using an extended beta-binomial distribution with discussion of correlation induced by covariate measurement errors. *Journal of the American Statistical Association*, **81**, 321–327.

Putter, J. (1955). The treatment of ties in some non-parametric tests. *Annals of Mathematical Statistics*, **26**, 368–386.

Radhakrishna, S. (1965). Combination of results from several contingency tables. *Biometrics*, **21**, 86–98.

Raffensperger, E. L., Peryam, D. R., and Wood, K. R. (1956). Development of a scale for grading toughness–tenderness in beef. *Food Technol (Chicago)*, **12**, 627–630.

Rao, C. R. (1947). General methods of analysis for incomplete block designs. *Journal of the American Statistical Association*, **42**, 541–562.

Rao, C. R. (1973). *Linear Statistical Inference and Its Applications*, 2nd ed. Wiley, New York.

Ratkowsky, D. A. (1989). *Handbook of Nonlinear Regression Models*. Marcel Dekker, Inc., New York.

Read, D. R. (1964). A quantitative approach to the comparative assessment of taste quality in the confectionary industry. *Biometrics*, **20**, 143–155.

Reese, T. S., and Stevens, S. S. (1960). Subjective intensity of coffee odor. *Am J Psychol*, **73**, 424–428.

Resurreccion, A. V. A. (1998). *Consumer Sensory Testing for Product Development*. Aspen Publishers Inc., Gaithersburg, Maryland.

Robson, D. S. (1961). Multiple comparisons with a control in balanced incomplete block designs. *Technometrics* **3**, 103–105.

Roessler, E. B., Pangborn, R. M., Sidel, J. L., and Stone, H. (1978). Expanded statistical tables for estimating significance in paired-preference, paired difference, duo–trio and triangle tests. *Journal of Food Science*, **43**, 940–943.

Rona, C., and Berardesca, E. (2001). Skin biomechanics: Antiaging products. In *Bioengineering of the Skin: Skin Biomechanics*. (P. Elsner, E. Berardesca, K. P. Wilhelm, and H. I. Maibach, eds.) CRC Press, Boca Raton, Florida, pp. 231–239.

Rosner, B. (1986). *Fundamentals of Biostatistics*. Duxbury Press, Boston, Massachusetts.

Samuels, M. L. (1993). Simpson's paradox and related phenomena. *Journal of the American Statistical Association*, **88**, 81–88.

SAS Institute Inc. (2003). *SAS/STAT® User's Guide*, Version 9. 1. Cary, NC.

Satterthwaite, F. E. (1946). An approximate distribution of estimates of variance components. *Biometrics*, **2**, 110–114.

Scheffé, H. (1952). An analysis of variance for paired comparisons. *Journal of the American Statistical Association*, **47**, 381–400.

Scheffé, H. (1959). *The Analysis of Variance*. Wiley, New York.

Schneider, A. M., and Stockett, A. L. (1963). An experiment to select optimum operating conditions on the basis of arbitrary preference ratings. *Chem Eng Prog, Symp Ser*, **59**(42), 34–38.

Schuirmann, D. J. (1981). On hypothesis testing to determine if the mean of a normal distribution is contained in a known interval. *Biometrics*, **37**, 617.

Schuirmann, D. J. (1987). A comparison of the two one-sided tests procedure and the power approach for assessing the equivalent of average bioavailability. *Journal of Pharmacokinetic and Biopharmaceutics*, **15**, 657–680.

Schultz, E. F., Jr. (1955). Rules of thumb for determining expectations of mean squares in analysis of variance. *Biometrics*, **55**, 123–135.

Schutz, H. G., and Bradley, J. E. (1953). Effect of bias on preference in the difference-preference test. In *Food Acceptance Testing Methodology Symposium*, October 8–9, Chicago, IL, NAC-NRC.

Schutz, H. G., and Pilgrim, F. J. (1957). Differential sensitivity in gustation. *J Exp Psychol*, **54**, 41–48.

Searle, S. R. (1971). *Linear Models*. Wiley, New York.

Seber, G. A. F. (1984). *Multivariate Observations*. John Wiley & Sons, New York.

Seber, G. A. F., and Wild, C. J. (1989). *Nonlinear Regression*. John Wiley & Sons, New York.

Seshadri, V. (1963). Combining unbiased estimators. *Biometrics*, **19**, 163–170.

Simpson, E. H. (1951). The interpretation of interaction in contingency tables. *Journal of the Royal Statistical Society*, **B**, **13**, 238–241.

Smith, D. M. (1983). Maximum likelihood estimation of the parameters of the beta binomial distribution. *Applied Statistics*, **32**, 196–204.

Smyth, R. J., Jr. (1965). Allelic relationship of genes determining extended black, wild type and brown plumage patterns in the fowl. *Poult Sci*, **44**, 89–98.

Snedecor, G. W., and Cochran, W. G. (1967). *Statistical Methods*. Iowa State University Press, Ames, Iowa.

Steel, R. G. D. (1961). Some rank sum multiple comparison tests. *Biometrics*, **17**, 539–552.

Steel, R. G. D., and Torrie, J. H. (1980). *Principles and Procedures of Statistics*. McGraw-Hill, New York.

Stefansson, G., Kim, W. C., and Hsu, J. C. (1988). On confidence sets in multiple comparisons. In *Statistical Decision Theory and Related Topics IV* (S. S. Gupta and J. O. Berger, eds.), Springer, New York: pp. 2, 89–104.

Steiner, E. H. (1966). Sequential procedures for triangular and paired comparison tasting test. *Journal of Food Technology*, **1**, 41–53.

Stevens, S. S. (1946). On the theory of scales of measurement. *Science*, **103**, 677–680.

Stevens, S. S. (1951). Mathematics, measurement, and psychophysics. In *Handbook of Experimental Psychology*. (S. S. Stevens, ed.) Wiley, New York: pp. 1–49.

Stevens, S. S. (1953). On the brightness of lights and the loudness of sounds. *Science*, **118**, 576.

Stevens, S. S. (1956). The direct estimation of sensory magnitudes—loudness. *Am J Psychol*, **69**, 1–25.

Stevens, S. S. (1958). Problems and methods of psychophysics. *Psychol Bull*, **54**, 177–196.

Stevens, S. S. (1961a). The psychophysics of sensory function. In *Sensory Communication*. (W. A. Rosenblith, ed.) MIT Press, Cambridge, Massachusetts.

Stevens, S. S. (1961b). To honor Fechner and repeal his law. *Science*, **133**, 80–86.

Stevens, S. S. (1969). Sensory scales of taste intensity. *Percept Psychophys*, **6**, 302–308.

Stevens, S. S. (1970). Neural events and the psychophysical law. *Science*, **170**, 1043–1050.

Stewart, W. M. (1941). A note on the power of the sign test. *Annals of Mathematical Statistics*, **12**, 236–239.

Stone, H., and Sidel, J. (2004). *Sensory Evaluation Practices*, 3rd ed. Academic Press, Amsterdam.

Stone, H., and Bosley, J. P. (1964). Difference–preference testing with the duo–trio test. *Psychol Rep*, **14**, 620–622.

Stuart, A. (1955). A test for homogeneity of the marginal distribution in a two-way classification. *Biometrika*, **42**, 412–416.

Sudman, S., and Wansink, B. (2002). *Consumer Panels*. American Marketing Association, Chicago, Illinois.

Tacq, J. (1997). *Multivariate Analysis Techniques in Social Science Research*. Sage Publication, London.

Taguchi, G. (1986). *Introduction to Quality Engineering*. Asian Productivity Organization, Tokyo, Japan.

Tang, M. L., Tang N. S., Chan, I. S. F., and Chan, P. S. (2002). Sample size determination for establishing equivalence/noninferiority via ratio of two proportions in matched-pair design. *Biometrics*, **58**, 957–963.

Tang, N. S., Tang, M. L., and Chan, I. S. F. (2003). On tests of equivalence via non-unity relative risk for matched-pair design. *Statistics in Medicine*, **22**, 1217–1233.

Tang, P. C. (1938). The power function of the analysis of variance tests with tables and illustrations of their use. *Stat Res Mem*, **2**, 126–149.

Tango, T. (1998). Equivalence test and confidence interval for the difference in proportions for the paired-sample design. *Statistics in Medicine*, **17**, 891–908.

Tarone, R. E. (1979). Testing the goodness of fit of the binomial distribution. *Biometrika*, **66**, 585–590.

Terpstra, T. J. (1952). The asymptotic normality and consistency of Kendall's test against trend, when ties are present in one ranking. *Indag Math*, **14**, 327–333.

Thoni, H. (1967). Transformation of variables used in the analysis of experimental and observational data. A review. Iowa State University, Statistical Laboratory Tech. Rept. 7, Ames, Iowa.

Thurstone, L. L. (1927a). A law of comparative judgment. *Psychol Rev*, **34**, 273–286.

Thurstone, L. L. (1927b). Three psychophysical laws. *Psychol Rev*, **34**, 424–432.

Thurstone, L. L. (1959). *The Measurement of Values*. University of Chicago Press, Chicago, IL.

Tidwell, P. W. (1960). Chemical process improvement by response surface methods. *Ind Eng Chem*, **52**, 510–512.

Torgerson, W. S. (1958). *Theory and Methods of Scaling*. Wiley, New York.

Trail, S. M., and Weeks, D. L. (1973). Extended complete block designs generated by BIBD. *Biometrics*, **29**, 565–578.

Tukey, J. W. (1953). The problem of multiple comparisons. Unpublished memorandum, Princeton University, Princeton, New Jersey.

Upton, G. J. G. (1978). *The Analysis of Cross-Tabulated Data*. Wiley, New York.

Ura, S. (1960). Pair, triangle, and duo–trio tests. *Rep Stat Appl Res, Jpn Union Sci Eng*, **7**, 107–119.

Van Dantzig, D. (1951). On the consistency and the power of Wilcoxon's two sample test. *Nederlandse Akad Westensch Proc Ser A*, **54**, 1–8.

Van Kleef, E., Van Trijp, H. C. M., and Luning, P. (2006). Internal versus external preference analysis: An exploratory study on end-user evaluation. *Food Qual Pref*, **17**, 387–399.

Versfeld, N. J., Dai, H., and Green, D. M. (1996). The optimum decision rules for the oddity task. *Perception & Psychophysics*, **58**, 10–21.

Wald, A. (1947). *Sequential Analysis*. Wiley, New York.

Wallis, W. A. (1939). The correlation ratio for ranked data. *Journal of the American Statistical Association*, **34**, 533–538.

Walsh, J. E. (1946). On the power function of the sign test for slippage of means. *Annals of Mathematical Statistics*, **17**, 358–362.

Wasserman, A. E., and Talley, F. (1969). A sample bias in the evaluation of smoked frankfurters by the triangle test. *Journal of Food Science*, **34**, 99–100.

Weil, C. S. (1970). Selection of the valid number of sampling units and a consideration of their combination in toxicological studies involving reproduction, teratogenesis. *Food Comet Toxicol*, **8**, 177–182.

Weil, C. S., and Gad, S. C. (1980). Applications of methods of statistical analysis to efficient repeated-dose toxicologic tests. 2. Methods for analysis of body, liver, and kidney weight data. *Toxicol Appl Pharmacol*, **52**, 214–226.

Wellek, S. (2003). *Testing Statistical Hypotheses of Equivalence*. Chapman and Hall/CRC Press, Boca Raton, FL.

Wermuth, N. (1987). Parametric collapsibility and the lack of moderating effects in contingency tables with a dichotomous response variable. *Journal of the Royal Statistical Society, Series B*, **49**, 353–364.

Westlake, W. J. (1965). Composite designs based on irregular fractions of factorials. *Biometrics*, **21**, 324–336.

Westlake, W. J. (1972). Use of confidence intervals in analysis of comparative bioavailability trails. *J Pharmaceut Sci*, **61**, 1340–1341.

Westlake, W. J. (1976). Symmetrical confidence intervals for bioequivalence trials. *Biometrics*, **32**, 741–744.

Westlake, W. J. (1979). Statistical aspects of comparative bioavailability trials. *Biometrics*, **35**, 273–280.

Westlake, W. J. (1981). Response to T. B. L. Kirkwood: Bioequivalence testing—a need to rethink. *Biometrics*, **37**, 589–594.

White, R. F. (1975). Randomization and the analysis of variance. *Biometrics*, **31**, 552–572.

Whittemore, A. S. (1978). Collapsibility of multidimensional contingency tables. *Journal of the Royal Statistical Society, Series B*, **40**, 328–340.

Wickett, R. R. (2001). Standardization of skin biomechanical measurements. In *Bioengineering of the Skin: Skin Biomechanics*. (P. Elsner, E. Berardesca, K. P. Wilhelm, and H. I. Maibach, eds.) CRC Press, Boca Raton, Florida, pp. 179–185.

Wierenga, B. (1974). Paired comparison product testing when individual preferences are stochastic. *Applied Statistics*, **23**, 384–396.

Wilcoxon, F. (1945). Individual comparisons by ranking methods. *Biometrics*, **1**, 80–83.

Wilcoxon, F. (1946). Individual comparisons by grouped data by ranking methods. *J Econ Entomol*, **39**, 269.

Wilcoxon, F. (1947). Probability tables for individual comparisons by ranking methods. *Biometrics*, **3**, 119–122.

Wilcoxon, F. (1949). *Some Rapid Approximate Statistical Procedures*. American Cyanamid Co., Stamford Research Laboratories, Pearl River, New York.

Wilcoxon, F., Katti, S., and Wilcox, R. A. (1970). Critical values and probability levels for the Wilcoxon rank sum test and the Wilcoxon signed rank test. In *Selected Tables in Mathematical Statistics* (H. L. Harter and D. B. Owen, eds.) Markham, Chicago, IL: **Vol. 1**, pp. 171–259.

Wilcoxon, F., and Wilcox, R. A. (1964). Some rapid approximate statistical procedures. Lederle Laboratories, Pearl River, NY.

Wilks, S. S. (1935). The likelihood test of independence in contingency tables. *The Annals of Mathematical Statistics*, **6**, 190–196.

Wilks, S. S. (1938). The large-sample distribution of the likelihood ratio for testing composite hypotheses. *The Annals of Mathematical Statistics*, **9**, 60–62.

William, D. A. (1975). The analysis of binary responses from toxicological experiments involving reproduction and teratogenicity. *Biometrics*, **31**, 949–952.

Williams, E. J. (1949). Experimental designs balanced for the estimation of residual effects of treatments. *Aust J Sci Res Ser A*, **2**, 149–168.

Williams, E. J. (1950). Experimental designs balanced for pairs of residual effects. *Aust J Sci Res Ser A*, **3**, 351–363.

Wilson, J. R. (1989). Chi-square tests for overdispersion with multiparameter estimates. *Applied Statistics*, **38**, 441–454.

Wooding, W. M. (1969). The computation and use of residuals in the analysis of experimental data. *J Quality Technol*, **1**, 175–188.

Wright, S. (1968). *Genetic and Biometric Foundations*. University of Chicago Press, Chicago, Illinois.

Yates, F. (1936). Incomplete randomized blocks. *Ann Eugen*, **7**, 121–140.

Yates, F. (1940). The recovery of inter-block information in balanced incomplete block designs. *Ann Eugen*, **10**, 317–325.

Youden, W. J., and Hunter, J. S. (1955). Partially replicated Latin squares. *Biometrics*, **11**, 399–405.

Yule, G. U. (1903). Notes on the theory of association of attributes in statistics. *Biometrika*, **2**, 121–134.

# Index

# Food Science and Technology
## International Series

Maynard A. Amerine, Rose Marie Pangborn, and Edward B. Roessler, *Principles of Sensory Evaluation of Food.* 1965.

Martin Glicksman, *Gum Technology in the Food Industry.* 1970.

Maynard A. Joslyn, *Methods in Food Analysis*, second edition. 1970.

C. R. Stumbo, *Thermobacteriology in Food Processing*, second edition. 1973.

Aaron M. Altschul (ed.), *New Protein Foods*: Volume 1, *Technology, Part A*—1974. Volume 2, *Technology, Part B*—1976. Volume 3, *Animal Protein Supplies, Part A*—1978. Volume 4, *Animal Protein Supplies, Part B*—1981. Volume 5, *Seed Storage Proteins*—1985.

S. A. Goldblith, L. Rey, and W. W. Rothmayr, *Freeze Drying and Advanced Food Technology.* 1975.

R. B. Duckworth (ed.), *Water Relations of Food.* 1975.

John A. Troller and J. H. B. Christian, *Water Activity and Food.* 1978.

A. E. Bender, *Food Processing and Nutrition.* 1978.

D. R. Osborne and P. Voogt, *The Analysis of Nutrients in Foods.* 1978.

Marcel Loncin and R. L. Merson, *Food Engineering: Principles and Selected Applications.* 1979.

J. G. Vaughan (ed.), Food Microscopy. 1979.

J. R. A. Pollock (ed.), *Brewing Science*, Volume 1—1979. Volume 2—1980. Volume 3—1987.

J. Christopher Bauernfeind (ed.), *Carotenoids as Colorants and Vitamin A Precursors: Technological and Nutritional Applications.* 1981.

Pericles Markakis (ed.), *Anthocyanins as Food Colors.* 1982.

George F. Stewart and Maynard A. Amerine (ed.), *Introduction to Food Science and Technology*, second edition. 1982.

Hector A. Iglesias and Jorge Chirife, *Handbook of Food Isotherms: Water Sorption Parameters for Food and Food Components.* 1982.

Colin Dennis (ed.), *Post-Harvest Pathology of Fruits and Vegetables.* 1983.

P. J. Barnes (ed.), *Lipids in Cereal Technology.* 1983.

David Pimentel and Carl W. Hall (eds), *Food and Energy Resources.* 1984.

Joe M. Regenstein and Carrie E. Regenstein, *Food Protein Chemistry: An Introduction for Food Scientists.* 1984.

Maximo C. Gacula, Jr. and Jagbir Singh, *Statistical Methods in Food and Consumer Research.* 1984.

Fergus M. Clydesdale and Kathryn L. Wiemer (eds), *Iron Fortification of Foods.* 1985.

Robert V. Decareau, *Microwaves in the Food Processing Industry.* 1985.

S. M. Herschdoerfer (ed.), *Quality Control in the Food Industry*, second edition. Volume 1—1985. Volume 2—1985. Volume 3—1986. Volume 4—1987.

F. E. Cunningham and N. A. Cox (eds), *Microbiology of Poultry Meat Products.* 1987.

Walter M. Urbain, *Food Irradiation*. 1986.

Peter J. Bechtel, *Muscle as Food*. 1986. H. W.-S. Chan, *Autoxidation of Unsaturated Lipids*. 1986.

Chester O. McCorkle, Jr., *Economics of Food Processing in the United States*. 1987.

Jethro Japtiani, Harvey T. Chan, Jr., and William S. Sakai, *Tropical Fruit Processing*. 1987.

J. Solms, D. A. Booth, R. M. Dangborn, and O. Raunhardt, *Food Acceptance and Nutrition*. 1987.

R. Macrae, *HPLC in Food Analysis*, second edition. 1988.

A. M. Pearson and R. B. Young, *Muscle and Meat Biochemistry*. 1989.

Marjorie P. Penfield and Ada Marie Campbell, *Experimental Food Science*, third edition. 1990.

Leroy C. Blankenship, *Colonization Control of Human Bacterial Enteropathogens in Poultry*. 1991.

Yeshajahu Pomeranz, *Functional Properties of Food Components*, second edition. 1991.

Reginald H. Walter, *The Chemistry and Technology of Pectin*. 1991.

Herbert Stone and Joel L. Sidel, *Sensory Evaluation Practices*, second edition. 1993.

Robert L. Shewfelt and Stanley E. Prussia, *Postharvest Handling: A Systems Approach*. 1993.

Tilak Nagodawithana and Gerald Reed, *Enzymes in Food Processing*, third edition. 1993.

Dallas G. Hoover and Larry R. Steenson, *Bacteriocins*. 1993.

Takayaki Shibamoto and Leonard Bjeldanes, *Introduction to Food Toxicology*. 1993.

John A. Troller, *Sanitation in Food Processing*, second edition. 1993.

Harold D. Hafs and Robert G. Zimbelman, *Low-fat Meats*. 1994.

Lance G. Phillips, Dana M. Whitehead, and John Kinsella, *Structure-Function Properties of Food Proteins*. 1994.

Robert G. Jensen, *Handbook of Milk Composition*. 1995.

Yrjö H. Roos, *Phase Transitions in Foods*. 1995.

Reginald H. Walter, *Polysaccharide Dispersions*. 1997.

Gustavo V. Barbosa-Cánovas, M. Marcela Góngora-Nieto, Usha R. Pothakamury, and Barry G. Swanson, *Preservation of Foods with Pulsed Electric Fields*. 1999.

Ronald S. Jackson, *Wine Tasting: A Professional Handbook*. 2002.

Malcolm C. Bourne, *Food Texture and Viscosity: Concept and Measurement*, second edition. 2002.

Benjamin Caballero and Barry M. Popkin (eds), *The Nutrition Transition: Diet and Disease in the Developing World*. 2002.

Dean O. Cliver and Hans P. Riemann (eds), *Foodborne Diseases*, second edition. 2002.

Martin Kohlmeier, *Nutrient Metabolism*, 2003.

Herbert Stone and Joel L. Sidel, *Sensory Evaluation Practices*, third edition. 2004.

Jung H. Han, *Innovations in Food Packaging*. 2005.

Da-Wen Sun, *Emerging Technologies for Food Processing*. 2005.

Hans Riemann and Dean Cliver (eds) *Foodborne Infections and Intoxications*, third edition. 2006.

Ioannis S. Arvanitoyannis, *Waste Management for the Food Industries*. 2008.

Ronald S. Jackson, *Wine Science: Principles and Applications*, third edition. 2008.

Da-Wen Sun, *Computer Vision Technology for Food Quality Evaluation*. 2008.

Kenneth David and Paul Thompson, *What Can Nanotechnology Learn From Biotechnology?* 2008.

Elke K. Arendt and Fabio Dal Bello, *Gluten-Free Cereal Products and Beverages*. 2008.

Debasis Bagchi, *Nutraceutical and Functional Food Regulations in the United States and Around the World*, 2008.

R. Paul Singh and Dennis R. Heldman, *Introduction to Food Engineering*, fourth edition. 2008.

Zeki Berk, *Food Process Engineering and Technology*. 2009.

Abby Thompson Mike Boland and Harjinder Singh, *Milk Proteins: From Expression to Food*. 2009.

Wojciech J. Florkowski, Stanley E. Prussia, Robert L. Shewfelt and Bernhard Brueckner (eds) *Postharvest Handling*, second edition. 2009.